New Perspectives in Ornithology

New Perspectives in Ornithology

21st Century Dispatches Across the World of Birds

Edited by

SCOTT V. EDWARDS

Department of Organismic and Evolutionary Biology and Museum of Comparative Zoology, Harvard University

J. MICHAEL REED

Department of Biology, Tufts University

OXFORD
UNIVERSITY PRESS

Oxford University Press is a department of the University of Oxford.
It furthers the University's objective of excellence in research, scholarship,
and education by publishing worldwide. Oxford is a registered trade mark of
Oxford University Press in the UK and in certain other countries.

Published in the United States of America by Oxford University Press
198 Madison Avenue, New York, NY 10016, United States of America.

CIP data is on file at the Library of Congress.

ISBN 9780197787670

This volume was conceived to commemorate the 150th anniversary
of the Nuttall Ornithological Club in 2023 and is Number 30 in the NOC Publications
and Monographs Series.

DOI: 10.1093/oso/9780197787670.001.0001

Printed by Integrated Books International, United States of America

Contents

SECTION III BEHAVIOR

SECTION IV DATABASES AND CITIZEN SCIENCE

SECTION V CONSERVATION AND MANAGEMENT

List of Contributors

Oludayo Michael Akinsola, Department of Theriogenology and Production, University of Jos

Bradley Allf, Department of Forestry & Environmental Resources, North Carolina State University

Rindy C. Anderson, Department of Biological Sciences, Florida Atlantic University

María del Coro Arizmendi, Laboratorio de Ecología, UBIPRO, Facultad de Estudios Superiores Iztacala, Universidad Nacional Autónoma de México

Jin Bai, Department of Forestry & Environmental Resources, North Carolina State University

Oladeji Bamidele, International Livestock Research Institute (ILRI); and Department of Biological Sciences, Kings University

Thomas M. Brooks, International Union for Conservation of Nature and Natural Resources (IUCN), Rue Mauverney 28, Gland 1196, Switzerland

Santiago Claramunt, Department of Natural History, Royal Ontario Museum; and Department of Ecology and Evolutionary Biology, University of Toronto

Caren Cooper, Department of Forestry & Environmental Resources, North Carolina State University

Elizabeth P. Derryberry, Department of Ecology and Evolutionary Biology, University of Tennessee

Adriaan M. Dokter, Cornell Lab of Ornithology, Cornell University

Scott V. Edwards, Department of Organismic and Evolutionary Biology and Museum of Comparative Zoology, Harvard University

Chris S. Elphick, Department of Ecology & Evolutionary Biology and Center of Biological Risk, University of Connecticut

Damien R. Farine, Department of Evolutionary Biology and Environmental Science, University of Zurich; Division of Ecology and Evolution, Research School of Biology, Australian National University; and Department of Collective Behaviour, Max Planck Institute of Animal Behavior

Michelle García-Arroyo, Faculty of Biological and Environmental Sciences, Ecosystems and Environment Research Programme, University of Helsinki

Autumn-Lynn Harrison, Migratory Bird Center, Smithsonian's National Zoo and Conservation Biology Institute

Barbara Helm, Bird Migration Unit, Swiss Ornithological Institute, Sempach, Switzerland

Kyle Horton, Department of Fish, Wildlife, and Conservation Biology, Colorado State University

Samuel Ivande, Global Center for Species Survival, Indianapolis Zoo; and A. P. Leventis Ornithological Research Institute (APLORI), University of Jos Biological Conservatory

Leo Joseph, Australian National Wildlife Collection, CSIRO National Research Collections Australia

Dmitry Kishkinev, School of Life Sciences, Keele University

Gunnar R. Kramer, Department of Organismic and Evolutionary Biology, Harvard University

Frank A. La Sorte, Cornell Lab of Ornithology, Cornell University

Victor Leandro-Silva, Laboratório de Ecologia e Evolução de Aves, Departmento of Zoologia, Universitdade Federal de Pernambuco; and Programa de Pós-graduação em Etnobiologia e Conservação da Natureza, Universidade Federal Rural de Pernambuco

Ian MacGregor-Fors, Faculty of Biological and Environmental Sciences, Ecosystems and Environment Research Programme, University of Helsinki

Sebastian Moreno, Department of Environmental Conservation, University of Massachusetts

Luciano N. Naka, Laboratório de Ecologia e Evolução de Aves, Departmento of Zoologia, Universitdade Federal de Pernambuco

Adolfo G. Navarro-Sigüenza, Museo de Zoología, Departamento de Biología Evolutiva, Facultad de Ciencias, Universidad Nacional Autónoma de México

Jingmai O'Connor, Negaunee Integrative Research Center, Field Museum of Natural History

Gail L. Patricelli, Department of Ecology and Evolution, University of California

David C. Pavlacky, Jr., Bird Conservancy of the Rockies; and Department of Fish, Wildlife, and Conservation Biology, Colorado State University

Deja Perkins, Center for Geospatial Analytics, North Carolina State University

David A. Prieto-Torres, Facultad de Estudios Superiores Iztacala, Universidad Nacional Autónoma de México

Sushma Reddy, Bell Museum and Department of Fisheries, Wildlife, Conservation Biology, University of Minnesota–Twin Cities

J. Michael Reed, Department of Biology, Tufts University

Christine Rega-Brodsky, Department of Biology, Pittsburg State University

Amanda D. Rodewald, Cornell Lab of Ornithology and Department of Natural Resources and the Environment, Cornell University

Octavio R. Rojas-Soto, Laboratorio de Bioclimatología, Red de Biología Evolutiva, Instituto de Ecología A. C.

Kristen C. Ruegg, Biology Department, Colorado State University

Luis A. Sánchez-González, Museo de Zoología, Departamento de Biología Evolutiva, Facultad de Ciencias, Universidad Nacional Autónoma de México

C. Jonathan Schmitt, Museum of Comparative Zoology and Department of Organismic and Evolutionary Biology, Harvard University

Elizabeth S. C. Scordato, Department of Biological Sciences, California State Polytechnic University

Mary Caswell Stoddard, Department of Ecology and Evolutionary Biology, Princeton University

Hazell Shokellu Thompson, Conservation Society of Sierra Leone

David P. L. Toews, Department of Biology, The Pennsylvania State University

Paige S. Warren, Department of Environmental Conservation, University of Massachusetts

Abdulmojeed Yakubu, Department of Animal Science, Faculty of Agriculture/Centre for Sustainable Agriculture and Rural Development, Nasarawa State University

Alberto Yanosky, Guyra Paraguay, CONACYT

1
Introduction

Scott V. Edwards and J. Michael Reed

The field of ornithology is perhaps as vibrant now as it ever has been. Birds are widely recognized today as being one of the best-known clades of organisms on Earth; although many species of birds are still poorly known, we likely have on average more information per species and more publications per year for this group as a whole than for many other groups of organisms (except, of course, humans). Their relative ease of visibility in the field and our ability to find them by their calls and songs make them easier to see, even by novice naturalists, than many organisms. Their often-elaborate displays and, for some species, habit of perching prominently on the tops of trees make them yet easier to notice and study. It is not difficult to understand why birds are so popular among scientists as objects of study. Instead, we often wonder, with some seriousness, how is it possible for budding naturalists to become interested in mammals, or lizards and snakes, when birds are on average so much easier to see? As E. H. Forbush (the first State Ornithologist for Massachusetts, from 1908 to 1928, and author of many books, including the three-volume *Birds of Massachusetts and Other New England States*) stated in 1913, "There is no subject in the field of natural science that is of greater interest than the important position that the living bird occupies in the great plan of organic nature" (Forbush 1913, p. 1).

Maybe the popularity of birds among scientists belies an inconvenient truth: that many scientists gravitate toward birds precisely because they are easy to study. Call it laziness or the scientific path of least resistance, birds are indeed more convenient to study than many groups of organisms. For others of us, their intriguing combination of traits—egg laying, a toothless beak, yet homeothermic and covered by that most extraordinary evolutionary innovation, feathers—makes them irresistible as study subjects.

As a counterpoint to their ease of study and seeming ubiquity, it is also the case that birds are widely recognized as sensitive indicators of the health of the planet (e.g., Smits and Fernie 2013, Fraixedas et al. 2020, Harwell et al. 2019). By studying birds, we can perhaps gain a window into how and how fast Earth is changing for broader swaths of organisms. Climate change has emerged in the past several decades as the preeminent existential challenge for humanity; birds offer one relatively easy window into documenting that change (e.g., Dunn and Møller 2019, Nelson et al. 2023). One might therefore conclude that there are more publications per year using birds as tools to study climate change than for other groups of vertebrates. But, here we would be wrong: A cursory survey of the literature on climate change shows that fish, rather than birds, outnumber all other vertebrate groups when intersecting with the topic of climate change (Figure 1.1). Here, birds come in second place among

Scott V. Edwards and J. Michael Reed, *Introduction*. In: *New Perspectives in Ornithology*.
Edited by: Scott V. Edwards and J. Michael Reed, Oxford University Press. © Oxford University Press (2025).
DOI: 10.1093/oso/9780197787670.003.0001

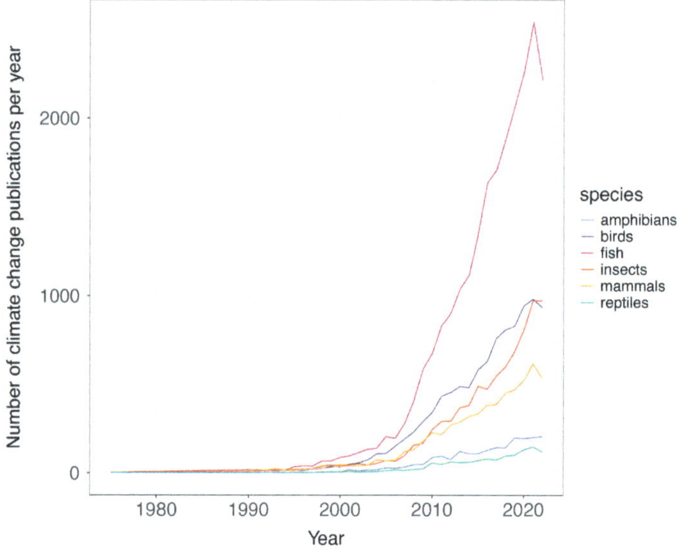

Figure 1.1 Trends in number of publications relating to climate change for different taxonomic groups of vertebrates, as retrieved from the Web of Science (https://www.webofscience.com, accessed February 20, 2023).

the five vertebrate groups surveyed, and they are closely on par with the literature on insects and climate change. These results suggest that economics might underpin some of the variation in popularity of different vertebrate groups as research subjects: Although wild birds can provide economic value through ecosystem services (Şekercioğlu et al. 2016), most bird species other than domestic chickens and turkeys have little direct economic value, at least in cultures that do not widely participate in trade or marketing in wildlife (e.g., Siriwat and Nijman 2020). Many fish species are the focus of vast fisheries industries, university departments, and associated research funding. Intriguingly, the trend of greater numbers of studies in fish versus birds was also reflected, although less strongly, in a recent survey of the number of vertebrate genomes sequenced as of 2021: Whereas research on birds had accumulated roughly 500 genomes by 2021, surprisingly, fish had accumulated more than 600 (Bravo et al. 2021). An additional factor explaining the larger number of studies of fish and climate change may be the sheer number of species to be studied: Whereas there are currently approximately 10,800 species of birds, there are more than 28,000 species of bony fish (Osteichthyes). The drivers of different numbers of studies and levels of popularity among scientists of birds versus other groups are likely complex. Nonetheless, we suspect that much of the popularity of studying birds rests in that sweet spot of accessibility, relevance to pressing issues in global change, and their sheer grace and beauty.

This book, which celebrates the 150th anniversary of the Nuttall Ornithological Club (NOC), the oldest bird club in the United States, capitalizes on and draws strength from the widespread popularity of ornithology among scientists and the broader public. Founded in 1873, the NOC survives today as an active group of birders and bird enthusiasts with a strong interest in the natural history, biodiversity, and conservation of birds. The early and sometimes conflicting impulses of the systematic cataloging of North American birds and the conservation of birds shaped the NOC and American ornithology generally (Sterling 1988). For example, the famous ornithologist Elliott Coues (1842–1899), writing in his *Field Ornithology*, proclaimed in the first sentence of Chapter 1, "The double-barrelled shot gun is your main reliance" (Coues 1874). Field ornithology in those days was focused squarely on the procurement and preparation of museum specimens for taxonomy. This statement reflects the fact that much ornithological research in the 19th century did indeed rely on museum specimens to draw generalities about size, shape, plumage, molt, geographic distributions, and evidence of occurrence—the somewhat limited purview of early ornithological studies in the United States.

Writing in 1945 in his book *Modern Bird Study*, Ludlow Griscom (1890–1959), who would serve as president of the NOC in 1952, wrote,

> As late as 1901, Robert Ridgeway, then the dean of American ornithologists . . . defined field ornithology as the collecting and stuffing of specimens. We have by no means fully recovered from this preposterously narrow definition. But when I was a boy the serious young student of birds was expected to shoot or collect his quota. . . . To be identified, the dead bird had to be in hand. . . . What people are now able to do in the way of instantly recognizing a large number of birds by song, notes, tricks of flight, shape, etc., entirely apart from their colors, seems perfectly fabulous to the uninitiated and was flatly declared to be impossible a generation ago. (Griscom 1945, p. 190)

Although the first ecological studies of birds date to the 18th century (Birkhead et al. 2014), who would have predicted that birds would become such an intense focus of ecology in the 20th and 21st centuries? The ongoing concern for the plight and conservation of birds led to the founding, primarily by women, of many important conservation groups, such as the Royal Society for the Protection of Birds in the United Kingdom in 1889 (Royal Society for the Protection of Birds 2025) and the Massachusetts Audubon Society in the United States in 1896 (Packard 1921). Such early activity on behalf of bird conservation no doubt fueled the early growth of scientific work on birds. Working mostly in the United Kingdom, from their start in the 1940s, the early ecological studies of David Lack promoted birds as useful models in the study of population regulation and life history evolution. In subsequent decades, intrepid field biologists such as Margaret Morse Nice (1883–1974) and Alexander Skutch (1904–2004) vastly improved our knowledge of the ecology and behavior of birds in both the temperate and tropical zones. The prolific ornithologist Salim Ali (1896–1987) popularized ornithology and conservation of birds in his native India. Starting in 1957, the NOC played an important role in publishing some of the leading monographs in avian biology, distributions, and ecology, including classic monographs such as Skutch's *Life Histories of Central American Highland Birds* (Skutch 1967) and Jared Diamond's *Avifauna of the Eastern Highlands of New Guinea* (Diamond 1972).

Both the NOC and ornithology generally have changed drastically since the 19th century, and even since the close of the 20th century, and we have attempted in this book to capture those changes. Perhaps the most noteworthy changes involve the increased participation in ornithology of women and researchers who were formerly underrepresented in the science. Although the need for increased diversity in ornithology has been articulated for decades by groups such as the American Ornithological Society (a merger of the former American Ornithologists Union [AOU] and Cooper Ornithological Society), it is no stretch to claim that calls for increased diversity in ornithology have multiplied exponentially in their number and urgency since the murder of the African American George Floyd by a white policeman in Minneapolis on May 25, 2020. Ornithology as a discipline was, like many scientific societies, including the NOC, heavily male-dominated for most of its history. For example, the first female president of the AOU, Francis James, assumed her post only in 1984. The first female president of the NOC was Kathleen Anderson in 1988. Scientific societies, and the discipline of ornithology as a whole, are rapidly making up for lost time by recruiting, nurturing, and celebrating contributions to the discipline by women and people of color. For example, a recent exhibit at the Concord Museum in Concord, Massachusetts, on William Brewster (1851–1919), the second president of the NOC, included an important vignette on his longtime assistant, African American Robert Gilbert (1870–1942), whose many little-known photographs and field notes are being digitized for online access by NOC in collaboration with the Massachusetts Audubon Society and the Museum of Comparative Zoology (MCZ) at Harvard University. Gilbert and his contributions to ornithology are still poorly known, but he was an elected Fellow of the AOU and was honored with an obituary in *The Auk* upon his death. The MCZ, which was founded in 1859 and has played an important role in the history of the NOC (the Club still meets monthly on Harvard campus near the MCZ), has seen fit to remove the name of MCZ founder and racist scientist Louis Agassiz (1807–1873) from one of its most prominent meeting rooms and replace it with Gilbert's name. Such small tokens of recognition fall far short of the truly transformative work that is still ongoing throughout ornithology, but they are nonetheless a significant acknowledgment that the scientific landscape and its values have changed utterly since just a few decades ago.

Another way in which ornithology has changed in recent decades is through innovations in technology, research emphases, and citizen science capacity. Ornithologists even 20 years ago could scarcely imagine the detail and precision with which we can now monitor bird populations on a global scale, whether through digital tools such as eBird, indirectly through remote sensing satellites, or at closer range via drones. In North America, decades of data from annual field surveys such as the Christmas Bird Counts and the Breeding Bird Atlases are now available online and serve as powerful databases for testing hypotheses in avian population biology and responses to climate change. Climate change itself has emerged as perhaps the dominant driver of research projects in ornithology: Whereas a prominent 1989 summary of the contributions of birds to biological understanding did not even mention climate change (Konishi et al. 1989), the behavioral, genetic, and geographical responses of birds to climate change constitute perhaps the fastest growing sectors of the field. The environmental stressors of global change, pollution, and human encroachment continue to fuel the research interests of students entering ornithology. But technological innovations within and outside of ornithology are also making the study of the basic biology of birds more exciting than it has ever been. Using scanning electron microscopes and high-energy synchrotrons, we are learning about the nanostructures of bird feathers and how they physically produce structural colors and iridescence. Genomic technologies are allowing us to sequence the entirety of avian genomes at a staggering pace and to understand the population histories of birds in unprecedented detail. We now have a trait database (AVONET; Tobias et al. 2022) that provides core information for all known species of birds, as well as a comprehensive and updatable digital phylogenetic tree for all birds that can be linked seamlessly with other data sources (Open Tree of Life; McTavish et al. 2025). Scaled-up use of museum collections and vast stores of digital environmental metadata allow detailed understanding of the connections between avian distributions, migration, and habitat variation in space and time. New fossils of dinosaurs and Mesozoic birds, many from China and known to science only in the past two decades, are transforming our understanding of the origin of birds from theropod dinosaurs (Xu et al. 2014). The transformation of ornithology by new technologies and discoveries in the past few decades is nothing short of breathtaking. These topics, and many more, are covered in detail in this volume.

Our key model when organizing this book was the landmark 1983 publication *Perspectives in Ornithology*, edited by Alan H. Brush and George A. Clark, Jr. (Brush and Clark 1983). This elegant volume was published more than 40 years ago to commemorate the centennial of the AOU; it appeared in concert with the landmark 1983 meeting of the AOU at the American Museum of Natural History. Both of us were present at that meeting, which cemented our then-budding interests in birds and transformed both our careers. We both gravitated toward the 1983 volume because of its comprehensiveness: It attempted to capture the full breadth of ornithology and did not shy away from canvassing the entire domain of the field, from the structure of the avian genome to community ecology and conservation. We also liked the "call and response" rhythm provided by the "Commentaries" presented by more senior members of the field after each chapter. These commentaries, like the ones we recruited for this volume, attempt to offer a broad overview of a major area within ornithology and to situate the chapters in that section in their broader field. In our case, these commentaries also serve to at least nod toward subdisciplines of ornithology that did not end up among the chapters in the final volume. Due to the exigencies of time, expenses, and pagination, we elected in this volume to present one commentary per section, rather than one per chapter as in the 1983 volume. The five major sections of the book—Ecology, Evolution, Behavior, Databases and Citizen Science, and Conservation and Management—no doubt reflect our particular interests as ornithologists but, it is hoped, also reflect the major swaths of modern ornithology. Several chapters embrace topics not mentioned in the 1983 volume, particularly in the areas of the social and community aspects of ornithology: citizen science, digital databases, and broadening participation in ornithology. We hope that the headings of the major sections and individual chapters stand the test of time; at the very least, they should capture our particular perception of the field of ornithology today.

We made earnest attempts to make our author list and chapter topics diverse; we know we have succeeded only partially in this. Whereas the 1983 volume had women as only 4 of 36 authors, the male:female ratio of authors of these chapters is almost right at 50:50. We attempted to capture geographic diversity of the authors, with contributions from several Latin American scientists, as well as scientists from Australia, Africa, and Europe; however, a major gap in representation is Asia. The many reviewers who generously donated their time to improving the quality of the chapters came from even more diverse regions of the world. We tried to include topics of broad and trending interest, with a strong emphasis on conservation biology, which is widely appreciated as a dominant theme in ornithology today. However, due to the vagaries of digital communication and the pressing schedules of would-be authors, several major intended topics did not make the final cut, including sexual selection, urban ecology, and more organismal topics like physiology or neuroscience. What seemed like eons of free time when we first began discussing the project in 2019 quite quickly turned into a frantic race to the finish line in 2023, and we inevitably left some wakes of wreckage and omission after what seemed like a clear, unobstructed path to the horizon at the project's inception.

Acknowledgments

We thank the many individuals who served as reviewers for each of the 19 chapters. We are grateful to Jeremy Lewis of Oxford University Press for his patience and guidance throughout the preparation of the manuscript. We thank David Larson of the NOC for his thoughts, advice, and contributions to the final product. We thank Natalia Fuentes for organizing the citations throughout the volume. Daniel Hays did the copyediting. We thank the leadership of the NOC, including Kim Peters, David Donsker, Wayne Peterson, and John Kricher for their support during this project. We thank our families for their support, patience, and tolerance of our taking many a weekend morning or afternoon to push the project forward—and of course do some birdwatching. And we thank all the authors for their contributions, patience, and understanding, especially when the editors routinely went radio-silent seemingly for months on end. To all of these individuals, and to the readers of this book, we hope it serves as an educational and representative snapshot of ornithology in 2023 and a suitable appreciation for the NOC on its 150th anniversary. Finally, we thank the following individuals for providing expert external reviews of the chapters in this book: Karlla V. C. Barbosa, Alyssa Bell, Thomas Brooks, Corey Callaghan, Victor Cazalis, Daniel Cooper, Rita Covas, Julia de Bruyn, Colleen Downs, Jan Fjeldsa, Pam Loring, Michaël Nicolaï, Jeff Podos, Amanda Rodewald, Nandadevi Cortes Rodriguez, Tom Sherry, Ben Winger, and Zhonghe Zhou.

References

Birkhead, T., J. Wimpenny, and B. Montgomerie. 2014. *Ten Thousand Birds*. Princeton University Press, Princeton, NJ.

Bravo, G. A., C. J. Schmitt, and S. V. Edwards. 2021. What have we learned from the first 500 avian genomes? *Annual Review of Ecology, Evolution, and Systematics* 52:611–663.

Brush, A. H., and G. A. Clark, Jr. 1983. *Perspectives in Ornithology: Essays Presented for the Centennial of the American Ornithologists' Union.* Cambridge University Press, Cambridge, UK.
Coues, E. 1874. *Field Ornithology: Comprising a Manual of Instruction for Procuring, Preparing and Preserving Birds and a Check List of North American Birds.* Dodd & Mead, New York.
Diamond, J. M. 1972. *Avifauna of the Eastern Highlands of New Guinea.* Nuttall Ornithological Club, Cambridge, MA.
Dunn, P. O., and A. P. Møller. 2019. *Effects of Climate Change on Birds.* 2nd ed. Oxford University Press, New York.
Forbush, E. H. 1913. *Useful Birds and Their Protection.* Massachusetts State Board of Agriculture, Boston.
Fraixedas S., A. Lindén, M. Piha, M. Cabeza, R. Gregory, and A. Lehikoinen. 2020. A state-of-the-art review on birds as indicators of biodiversity: Advances, challenges, and future directions. *Ecological Indicators* 118:106728.
Griscom, L. 1945. *Modern Bird Study.* Harvard University Press, Cambridge, MA.
Harwell, M. A., J. H. Gentile, L. D. McKinney, J. W. Tunnell, Jr., W. C. Dennison, R. H. Kelsey, K. M. Stanzel, G. W. Stunz, K. Withers, and J. Tunnell. 2019. Conceptual framework for assessing ecosystem health. *Integrated Environmental Assessment and Management* 15:544–564.
Konishi, M., S. T. Emlen, R. E. Ricklefs, and J. C. Wingfield. 1989. Contributions of bird studies to biology. *Science* 246:465–472.
McTavish E. J., J. A. Gerbracht, M. T. Holder, M. J. Iliff, D. Lepage, P. C. Rasmussen, B. D. Redelings, L. L. Sánchez Reyes, and E. T. Miller. 2025. A complete and dynamic tree of birds. *Proceedings of the National Academy of Sciences* 122:e2409658122.
Nelson, S. B. M., C. A. Ribic, N. D. Niemuth, J. Bernath-Plaisted, and B. Zuckerberg. 2023. Sensitivity of North American grassland birds to weather and climate variability. *Conservation Biology* 38: Article e14143.
Packard, W. 1921. The story of the Audubon Society. *Bulletin of the Massachusetts Audubon Society for the Protection of Birds.*
Royal Society for the Protection of Birds. 2025. Our history. Accessed May 26, 2025, from https://www.rspb.org.uk/about-us/our-history.
Şekercioğlu Ç. H., D. G. Wenny, and C. J. Whelan. 2016. *Why Birds Matter.* University of Chicago Press, Chicago.
Siriwat, P., and V. Nijman. 2020. Wildlife trade shifts from brick-and-mortar markets to virtual marketplaces: A case study of birds of prey trade in Thailand. *Journal of Asia-Pacific Biodiversity* 13:454–461.
Skutch, A. F. 1967. *Life Histories of Central American Highland Birds.* Nuttall Ornithological Club, Cambridge, MA.
Smits, J. E. G., and K. J. Fernie. 2013. Avian wildlife as sentinels of ecosystem health. *Comparative Immunology, Microbiology and Infectious Diseases* 36:333–342.
Sterling, K. B. 1988. Reviewed work: History of the Nuttall Ornithological Club, 1873–1986 by William E. Davis. *The Auk* 105:813.
Tobias, J. A., C. Sheard, A. L. Pigot, A. J. M. Devenish, J. Yang, F. Sayol, M. H. C. Neate-Clegg, N. Alioravainen, T. L. Weeks, R. A. Barber, *et al.* 2022. AVONET: morphological, ecological and geographical data for all birds. *Ecology Letters*: 25:581–597.
Xu, X., Z. Zhou, R. Dudley, S. Mackem, C.-M. Chuong, G. M. Erickson, and D. J. Varricchio. 2014. An integrative approach to understanding bird origins. *Science* 346: Article 1253293.

SECTION I
ECOLOGY

2

Commentary on Ecology in Ornithology

Amanda D. Rodewald

Natural history observations and time-intensive field or lab-based studies were once the mainstays of ornithological research, but the field of avian ecology is being rapidly transformed by innovations in technology, computation, and engagement that have during the past 10–20 years (Table 2.1). Indeed, the fresh perspectives and insights presented in the three chapters in this section reflect substantial advances in our ability to survey, monitor, and track birds at individual, population, and community levels.

Ornithologists have access to an unprecedented volume of data across broader geographies and at finer spatiotemporal resolution than previously possible (Kelling et al. 2009, La Sorte et al. 2018). Improvements in our ability to track individual animals by way of weather radar (Diehl et al. 2003), automated telemetry arrays (Motus Wildlife Tracking System arrays [MOTUS]; Taylor et al. 2017), satellites (Wikelski et al 2007), geolocators (Stutchbury et al. 2009), stable isotopes (Hobson et al. 2014), and bio-loggers (Wilmers et al. 2015) have given us sharp lenses through which we can study the movements and dynamic distributions of birds. Likewise, cutting-edge bioacoustic technology, especially related to the development of autonomous recording units and sophisticated software that detect and classify sounds, has created new possibilities to systematically or continuously monitor species that are cryptic, rare, secretive, or simply live in locations that are difficult to survey (Blumstein et al. 2011).

As bird surveys have expanded beyond the domain of professional ornithologists or highly skilled bird-watchers, our ability to collect data on the distribution, abundance, and richness of birds has changed fundamentally. Millions of people participate in citizen science projects and voluntarily submit observations and checklists from throughout the world (Bonney et al. 2009, Dickinson et al 2010). For example, eBird, a participatory science project run by the Cornell Lab of Ornithology, has amassed more than 1.4 billion observations of birds throughout the world (Sullivan et al. 2014). These data are used to estimate species abundance at fine spatial (3 km) and temporal (weekly) resolution across regional, continental, hemispheric, and global extents and estimate population trends at fine spatial resolution (27 km) across entire species ranges (Fink et al. 2024; Johnston et al. 2025, https://science.ebird.org/en/status-and-trends). Yet, generating volumes of data is one matter; managing and using those data effectively is quite another. As such, collaborations with computer scientists, data scientists, and statisticians have been instrumental in developing new ways to generate, curate, archive, share, and analyze big data (Kelling et al. 2009, Carey et al. 2019).

Amanda D. Rodewald, *Commentary on Ecology in Ornithology*. In: *New Perspectives in Ornithology*.
Edited by: Scott V. Edwards and J. Michael Reed, Oxford University Press. © Oxford University Press (2025).
DOI: 10.1093/oso/9780197787670.003.0002

Table 2.1 Examples of Innovation in Technology, Computational Methods, and Engagement/Collaborative Approaches That Are Transforming the Field of Avian Ecology

Technology	Computation	Engagement and Collaboration
Miniature transmitters	Data infrastructure and management	Participatory or community science programs
Long-range and low-power telemetry	Application programming interfaces	Crowdsourced data
Motus Wildlife Tracking System arrays (https://motus.org)	Advanced supercomputing capacity	Participant incentives
Satellite telemetry systems (e.g., ICARUS)	Machine learning	Gamification
Autonomous recording units	Deep learning/neural networks	User interfaces (e.g., Web-based platforms and apps)
Sensors/bio-loggers that monitor an animal's state and external environment (e.g., accelerometers)	Computer vision	Interdisciplinary and multi-investigator research
Radio-frequency identification devices	Sound detection and classification	Distributed experiments across regions and continents
Genomic tools	Ecological networks	Data repositories and sharing platforms (Movebank; http://www.movebank.org)
Weather radar for animal movement	Simulation models	Teaming with computer scientists and engineers
Data transfer via the Internet of Things	Optimization algorithms	
	Software programs for analysis	
	Data visualization	
	Interoperability, transparency, and reusability of data products	

Suffice to say, the study of avian ecology has been revolutionized, and the kinds of questions that can be asked are fundamentally changing (Kelling et al. 2009, Robinson et al. 2010, Hampton et al. 2013, La Sorte et al. 2018). The data revolution is unlocking the ability to describe, understand, and scale the study of ecological patterns and processes at broader geographic extents and with finer spatial and temporal resolution than ever before possible (Hampton et al. 2013). We also can detect and measure rare or unpredictable phenomena (La Sorte et al. 2018) and are starting to create global networks of animal sensors that can be used to detect and monitor environmental change (Jetz et al. 2022).

As shown in the chapters by García-Arroyo et al., Harrison and Horton, and Kramer, we can now address complex problems and questions that were difficult, if not intractable, only a decade ago (Sutherland et al. 2013). Here, I highlight several exciting areas of research at the frontier of avian ecology that are rapidly expanding due in part to recent technological and computational innovations.

Species Distributions Across the Full Annual Cycle

Unlike traditional range maps that were static and largely based on expert opinion, species ranges and distributions can now be mapped in dynamic and comparably precise ways. In Chapter 3, Kramer describes the challenge of elucidating the full annual cycle of birds, particularly when species move across hemispheres. Global participatory-science data sets such as eBird improve our ability to delineate where populations occur during breeding, nonbreeding stationary, and migratory periods, as well as the times at which birds move among them. Weather radar can be used to identify reliably important stopover locations for migrating birds (Guo et al. 2023). In combination with other methods, such as genomics, distinct population segments, ecotypes, or subspecies can be delineated across space and time (Bay et al. 2018, Mikles et al. 2020).

Shifts in Geographic Range

Not only can we now detect fine-scale heterogeneity in occurrence and abundance over space and time but also we can examine how distributions shift in response to changing climatic, environmental, or anthropogenic factors. Such approaches are particularly important for documenting and understanding the responses of species and populations to global change. An arsenal of new statistical methods makes it possible to estimate distributions of rare or recently introduced species that are not yet numerous (La Sorte et al. 2018)—something that is particularly important during the lag phase of invasions (see Chapter 5, this volume). Our improved ability to monitor the spread of nonnative birds, whether through tracking technologies or big data, facilitates study of invasion dynamics and the conditions that facilitate or discourage spread. The combination of data on abundance and population trends may signal future range expansions or contractions. In some circumstances, this information might act as an early warning system that can be used to strategically deploy control or eradication efforts for invasive species or to guide precision conservation for declining native species.

Population and Community Dynamics at Scale

With the availability of data on abundance and population trends across broad scales and at fine resolution (e.g., Johnston et al. 2025), there are new opportunities to investigate population dynamics over space and time. Recently developed integrated population models provide a unified framework to use disparate data sources to study population dynamics and estimate demographic rates (Ahrestani et al. 2017). There are a growing number of examples of big data being used to determine the consequences of species interactions to populations, communities, and landscapes. For example, Freeman et al. (2022) found that interspecific competition rather than climate best explained range shifts of species in montane regions. Martin and Bonier (2018) demonstrated globally that direct competitive interactions between closely related species reduced the persistence of subordinate species in cities and, thus, determined in part urban bird communities.

Migratory Pathways, Timing, and Strategies

As discussed by Harrison and Horton in Chapter 4, the study of migration has been radically transformed by advances in tracking technology, the availability of abundance data at weekly intervals, the use of weather radar to detect birds in flight (Robinson et al. 2010, Bauer et al. 2019), and data-sharing platforms such as Movebank (Kays et al. 2022). For the first time, we can describe the timing, pathways, and strategies used by multiple species or groups of migratory birds at the scale of continents (Dokter et al. 2011; La Sorte, Fink, et al. 2016) and understand the degree to which migration is influenced by atmospheric conditions (La Sorte et al. 2014) and ecological barriers (La Sorte and Fink 2017). Bio-loggers also can provide detailed information about a migrating bird's physiology, giving us a window into adaptations to migration and responses to climate change (Chmura et al. 2018). These lines of inquiry comprise an important dimension of the emerging discipline of aeroecology, which focuses on the manner in which airborne fauna interact with abiotic and biotic components in the atmosphere (Kunz et al. 2007).

As our understanding of migration ecology has expanded, so too has our ability to forecast when, where, and how many birds will migrate based on weather forecasts. Based on decades of past NEXRAD radar data, the Cornell Lab of Ornithology's BirdCast project employs machine-learning techniques to develop sophisticated models of migration that can then be used to make real-time predictions of migration over space and time (Farnsworth et al. 2014). Migration forecasts are used to issue alerts to individuals and municipalities so that lights can be dimmed to reduce the risk of collisions faced by migrating birds (Horton et al. 2019). These forecasts have been instrumental to establishing Lights Out programs in cities, including Houston, Dallas, Chicago, and New York City. Other applications of migration forecasts that are currently in development include reducing likelihood of air strikes and reducing risk of disease transmission to poultry (Bauer et al. 2019).

Migratory Connectivity

By integrating observational and movement data across different modalities (e.g., bioacoustics, telemetry, geolocators, isotopes, and radar), scientists are beginning to describe the migratory connectivity among population segments of migratory species during breeding and nonbreeding periods (Webster et al. 2002, Cohen et al. 2018). The Migratory Connectivity Project, led by the Smithsonian Migratory Bird Center, U.S. Geological Survey (USGS), and Georgetown University (http://migratoryconnectivityproject.org), is a consortium of scientists and science advocates working together to elucidate the full annual cycle of migratory birds and, specifically, to link breeding and nonbreeding populations. Integration with other data streams, such as mark–recapture data from the USGS Bird Banding Laboratory (Cohen et al. 2018) and stable isotopes (Fournier et al. 2017), is revealing new linkages. Establishing the migratory connectivity among population segments truly transforms our ability to identify causes of decline; the extent to which threats and other limiting factors originate in breeding, nonbreeding, or migratory periods; and where and when to direct conservation efforts (Webster et al. 2002, Webster and Marra 2005).

Upscaling the Consequences of Individual Behavior to Populations

Improvements in both the scale and resolution of behavioral data have created unparalleled opportunities to integrate different data modalities to study behavior across different contexts and scales (Smith and Pinter-Wollman 2021) and to understand the interplay between individual decisions (e.g., habitat selection) and higher levels of organization (Harrison et al. 2011). For instance, high-resolution Global Positioning System telemetry that provides locations at frequent intervals (e.g., every 5 minutes) can be used to deduce how foraging strategies and movements of individuals change as a function of the distribution and quality of habitat across landscapes (Vergara et al. 2019), which if combined with caloric expenditure data from bio-loggers could estimate the energetic consequences of alternate landscape configurations (R. Wilson et al. 2012). Radio-frequency identification devices (RFID) likewise provide extraordinary lenses for examining how ecological conditions across landscapes affect individual behavior and activity and connecting to population or community-level responses (Bonter and Bridge 2011). Methodologies also have been developed to infer social behavior and network structure from data streams collected by RFIDs (Psorakis et al. 2012), and these make possible linking sociality, movement, and the discovery of resource patches across landscapes (Aplin et al. 2012). Our ability to track individuals over long distances and time spans also facilitates detection of infrequent phenomena, such as dispersal events, and can be useful to document carryover effects (i.e., when events in one season influence individual performance or fitness in a subsequent season; Norris 2005, Norris and Marra 2007).

Eco-Evolutionary Dynamics

Evolutionary and ecological processes were once regarded as operating at vastly different timescales, but the potential for rapid evolution has cultivated strong interest in understanding reciprocal feedbacks between adaptive phenotypic change and population dynamics—a line of inquiry referred to as eco-evolutionary dynamics (Schoener 2011). We have long known that population dynamics influence selection across generations and that evolution of life history traits affects population dynamics, but empirical investigations are more tractable than ever. For example, combining traditional demographic and/or banding data with large-scale participatory science data on distribution, relative abundance, and population trends (e.g., eBird Status and Trends data products) can be used to test the extent to which climatic factors drive migratory strategies (Carbeck et al. 2022). Integrating physiological or movement data from autonomic sensors with information on demography or landscape configuration might reveal determinants of range margins of species. The ecological and evolutionary outcome of interactions between hosts and pathogens also can be studied across entire populations with lab-based research and citizen science (Hochachka et al. 2021).

Impacts of Anthropogenic Stressors

As we collect increasing volumes of data across systems, scales, and time periods, we capture more environmental heterogeneity and can, therefore, better document and predict the effects of novel anthropogenic stressors, such as artificial light at night, building collisions, energy development, and pollution/contamination, on animal physiology, individual and group behavior, population dynamics, species interactions, and communities (Tylianakis et al. 2008). From that information, we can begin to identify species traits and ecological contexts that are most strongly associated with sensitivity—and identify the thresholds above which exposure to certain stressors becomes most harmful. The availability of data also enables the study of interactive effects of anthropogenic stressors with other drivers of global change (A. Wilson et al. 2021).

Strategic Conservation

Building capacity in our ability to map the abundance and population trends of multiple species precisely and dynamically across broad geographies has never been more important from a conservation perspective. Strategically and efficiently meeting domestic and international commitments to protect 30% of the land and sea on Earth and conserve biodiversity requires sound information on where species occur. Spatial planning tools, including spatial optimization algorithms such as Marxan (Ball et al. 2009), enable prioritization of different locations based on specified targets (e.g., conserve 30% of a species' population), co-benefits (e.g., carbon storage), and costs (e.g., land area and price of land) (Sinclair et al. 2018). Recent examples include the identification of priority sites to conserve migratory birds across the full annual cycle (Schuster et al. 2019) or during nonbreeding (S. Wilson et al. 2019, 2022) or migratory periods (Lin et al. 2020). The combination of fine spatial and temporal resolution also unlocks possibilities to implement dynamic—or temporary—conservation interventions that are deployed only at specific times and locations, thereby reducing disruption of or conflict with human activities (Reynolds et al. 2017).

Conclusion

Innovations in technology and computation, coupled with the massive accumulation of data, are revolutionizing the fields of ornithology and avian ecology. Scientists can study ecological patterns and processes at broader geographic extents and with finer spatial and temporal resolution than ever before possible. Yet, as recognized by Harrison and Horton in Chapter 4, a potential pitfall for any science during times of rapid innovation is that research can be easily driven by tools rather than by questions. To avoid letting the solution become a problem, ornithologists and avian ecologists must continue to challenge themselves to be guided by questions, needs, and knowledge gaps rather than looking for the proverbial nail when holding a hammer.

Acknowledgments

Our ability to pursue many of the topics and questions highlighted here is possible because of the efforts of the hundreds of thousands of individuals who engage in participatory science projects every year, and I—along with countless others—extend my gratitude for their engagement and enthusiasm. I admire and am grateful to the many scientists and engineers whose tireless efforts and tenacity produced the innovations and advances that have revolutionized the field of avian ecology. I thank P. Arcese for his thoughtful review and feedback.

References

Ahrestani, F. S., J. F. Saracco, J. R. Sauer, K. L. Pardieck, and J. A. Royle. 2017. An integrated population model for bird monitoring in North America. *Ecological Applications* 27(3):916–924.

Aplin, L. M., D. R. Farine, J. Morand-Ferron, and B. C. Sheldon. 2012. Social networks predict patch discovery in a wild population of songbirds. *Proceedings of the Royal Society B: Biological Sciences* 279(1745):4199–4205.

Ball, I. R., H. P. Possingham, and M. Watts. 2009. Marxan and relatives: Software for spatial conservation prioritization. Pages 185–195 in A. Moilanen, K. A. Wilson, and H. Possingham, eds. *Spatial Conservation Prioritization: Quantitative Methods and Computational Tools*. Oxford University Press, New York.

Bauer, S., J. Shamoun-Baranes, C. Nilsson, A. Farnsworth, J. F. Kelly, D. R. Reynolds, A. M. Dokter, J. F. Krauel, L. B. Petterson, K. G. Horton, and J. W. Chapman. 2019. The grand challenges of migration ecology that radar aeroecology can help answer. *Ecography* 42(5):861–875.

Bay, R. A., R. J. Harrigan, V. L. Underwood, H. L. Gibbs, T. B. Smith, and K. Ruegg. 2018. Genomic signals of selection predict climate-driven population declines in a migratory bird. *Science* 359(6371):83–86.

Blumstein, D. T., D. J. Mennill, P. Clemins, L. Girod, K. Yao, G. L. Patricelli, J. L. Deppe, A. H. Krakauer, C. Clark, K. A. Cortopassi, S. F. Hanser, B. McCowan, A. M. Ali, and A. N. G. Kirschel. 2011. Acoustic monitoring in terrestrial environments using microphone arrays: Applications, technological considerations and prospectus. *Journal of Applied Ecology* 48:758–767

Bonney, R., C. B. Cooper, J. Dickinson, S. Kelling, T. Phillips, K. V. Rosenberg, and J. Shirk. 2009. Citizen science: A developing tool for expanding science knowledge and scientific literacy. *BioScience* 59(11):977–984.

Bonter, D. N., and E. S. Bridge. 2011. Applications of radio frequency identification (RFID) in ornithological research: A review. *Journal of Field Ornithology* 82(1):1–10.

Carbeck, K., T. Wang, J. M. Reid, and P. Arcese. 2022. Adaptation to climate change through seasonal migration revealed by climatic versus demographic niche models. *Global Change Biology* 28(14):4260–4275. https://doi.org/10.1111/gcb.16185

Carey, C. C., N. K. Ward, K. J. Farrell, M. E. Lofton, A. I. Krinos, R. P. McClure, K. C. Subratie, R. J. Figueiredo, J. P. Doubek, P. C. Hanson, P. Papadopoulos, and P. Arzberger. 2019.

Enhancing collaboration between ecologists and computer scientists: Lessons learned and recommendations forward. *Ecosphere* 10(5): Article e02753.

Chmura, H. E., T. W. Glass, and C. T. Williams. 2018. Biologging physiological and ecological responses to climatic variation: New tools for the climate change era. *Frontiers in Ecology and Evolution* 6: Article 92.

Cohen, E. B., J. A. Hostetler, M. T. Hallworth, C. S. Rushing, T. S. Sillett, and P. P. Marra. 2018. Quantifying the strength of migratory connectivity. *Methods in Ecology and Evolution* 9(3):513–524.

Dickinson, J. L., B. Zuckerberg, and D. N. Bonter. 2010. Citizen science as an ecological research tool: Challenges and benefits. *Annual Review of Ecology, Evolution, and Systematics* 41:149–172.

Diehl, R. H., R. P. Larkin, and J. E. Black. 2003. Radar observations of bird migration over the Great Lakes. *The Auk* 120(2):278–290.

Dokter, A. M., F. Liechti, H. Stark, L. Delobbe, P. Tabary, and I. Holleman. 2011. Bird migration flight altitudes studied by a network of operational weather radars. *Journal of the Royal Society: Interface* 8(54):30–43. http://doi.org/10.1098/rsif.2010.0116

Farnsworth, A., D. Sheldon, J. Geevarghese, J. Irvine, B. Van Doren, K. Webb, T. G. Dietterich, and S. Kelling. 2014. Reconstructing velocities of migrating birds from weather radar: A case study in computational sustainability. *AI Magazine* 35(2):31–48.

Fink, D., T. Auer, A. Johnston, M. Strimas-Mackey, S. Ligocki, O. Robinson, W. Hochachka, L. Jaromczyk, C. Crowley, K. Dunham, A. Stillman, C. Davis, M. Stokowski, P. Sharma, V. Pantoja, D. Burgin, P. Crowe, M. Bell, S. Ray, I. Davies, V. Ruiz-Gutierrez, C. Wood, and A. Rodewald. 2024. eBird status and trends, data version: 2023; Released: 2025. Cornell Lab of Ornithology, Ithaca, NY. https://doi.org/10.2173/WZTW8903.

Fournier, A. M., A. R. Sullivan, J. K. Bump, M. Perkins, M. C. Shieldcastle, and S. L. King. 2017. Combining citizen science species distribution models and stable isotopes reveals migratory connectivity in the secretive Virginia rail. *Journal of Applied Ecology* 54(2):618–627.

Freeman, B. G., M. Strimas-Mackey, and E. T. Miller. 2022. Interspecific competition limits bird species' ranges in tropical mountains. *Science* 377:416–420.

Guo, F., J. J. Buler, J. A. Smolinsky, and D. S. Wilcove. 2023. Autumn stopover hotspots and multiscale habitat associations of migratory landbirds in the eastern United States. *Proceedings of the National Academy of Sciences* 120:e2203511120. https://doi.org/10.1073/pnas.2203511120.

Hampton, S. E., C. A. Strasser, J. J. Tewksbury, W. K. Gram, A. E. Budden, A. L. Batcheller, C. S. Duke, and J. H. Porter. 2013. Big data and the future of ecology. *Frontiers in Ecology and the Environment* 11:156–162. https://doi.org/10.1890/120103

Harrison, X. A., J. D. Blount, R. Inger, D. R. Norris, and S. Bearhop. 2011. Carry-over effects as drivers of fitness differences in animals. *Journal of Animal Ecology* 80(1):4–18.

Hobson, K. A., S. L. Van Wilgenburg, J. Faaborg, J. D. Toms, C. Rengifo, A. L. Sosa, Y. Aubry, and R. Brito Aguilar. 2014. Connecting breeding and wintering grounds of Neotropical migrant songbirds using stable hydrogen isotopes: A call for an isotopic atlas of migratory connectivity. *Journal of Field Ornithology* 85(3):237–257.

Hochachka, W. M., A. P. Dobson, D. M. Hawley, and A. A. Dhondt. 2021. Host population dynamics in the face of an evolving pathogen. *Journal of Animal Ecology* 90(6):1480–1491.

Horton, K. G., C. Nilsson, B. M. Van Doren, F. A. La Sorte, A. M. Dokter, and A. Farnsworth. 2019. Bright lights in the big cities: Migratory birds' exposure to artificial light. *Frontiers in Ecology and the Environment* 17(4):209–214.

Jetz, W., G. Tertitski, R. Kays, U. Mueller, M. Wikelski, S. Åkesson, Y. Anisimov, A. Antonov, W. Arnold, F. Bairlein, and O. Baltà. 2022. Biological Earth observation with animal sensors. *Trends in Ecology & Evolution* 37(4):293–298.

Johnston, A., A. D. Rodewald, M. Strimas-Mackey, T. Auer, W. M. Hochachka, A. N. Stillman, C. L. Davis, V. Ruiz-Gutierrez, A. M. Dokter, E. T. Miller, O. Robinson, S. Ligocki, L. Oldham Jaromczyk, C. Crowley, C. L. Wood, and D. Fink. 2025. North American bird declines are greatest where species are most abundant. *Science* 388:532–537. https://doi/10.1126/science.adn4381

Kays, R., S. C. Davidson, M. Berger, G. Bohrer, W. Fiedler, A. Flack, J. Hirt, C. Hahn, D. Gauggel, B. Russell, and A. Kölzsch. 2022. The Movebank system for studying global animal movement and demography. *Methods in Ecology and Evolution* 13(2):419–431.

Kelling, S., W. M. Hochachka, D. Fink, M. Riedewald, R. Caruana, G. Ballard, and G. Hooker. 2009. Data-intensive science: A new paradigm for biodiversity studies. *BioScience* 59(7):613–620.

Kunz, T. H., S. A. Gauthreaux, N. I. Hristov, J. W. Horn, G. Jones, E. K. V. Kalko, R. P. Larkin, G. F. McCracken, S. M. Swartz, R. B. Srygley, R. Dudley, J. K. Westbrook, and M. Wikelski. 2007. Aeroecology: Probing and modeling the aerosphere. *Integrative and Comparative Biology* 48(1):1–11.

La Sorte, F. A., and D. Fink. 2017. Migration distance, ecological barriers and en-route variation in the migratory behaviour of terrestrial bird populations. Global Ecology and Biogeography 26:216–227.

La Sorte, F. A., D. Fink, W. M. Hochachka, A. Farnsworth, A. D. Rodewald, K. V. Rosenberg, B. L. Sullivan, D. W. Winkler, C. Wood, and S. Kelling. 2014. The role of atmospheric conditions in the seasonal dynamics of North American migration flyways. *Journal of Biogeography* 41(9):1685–1696.

La Sorte, F. A., D. Fink, W. M. Hochachka, and S. Kelling. 2016. Convergence of broad-scale migration strategies in terrestrial birds. *Proceedings of the Royal Society B: Biological Sciences* 283(1823): Article 20152588.

La Sorte, F. A., C. A. Lepczyk, J. L. Burnett, A. H. Hurlbert, M. W. Tingley, and B. Zuckerberg. 2018. Opportunities and challenges for big data ornithology. *The Condor: Ornithological Applications* 120(2):414–426.

Lin, H. Y., R. Schuster, S. Wilson, S. J. Cooke, A. D. Rodewald, and J. R. Bennett. 2020. Integrating season-specific needs of migratory and resident birds in conservation planning. Biological Conservation 252: Article 108826.

Martin, P. R., and F. Bonier. 2018. Species interactions limit the occurrence of urban-adapted birds in cities. *Proceedings of the National Academy of Sciences of the USA* 115(49):E11495–E11504.

Mikles, C. S., S. M. Aguillon, Y. L. Chan, P. Arcese, P. M. Benham, I. J. Lovette, and J. Walsh. 2020. Genomic differentiation and local adaptation on a microgeographic scale in a resident songbird. *Molecular Ecology* 29(22):4295–4307.

Norris, D. R. 2005. Carry-over effects and habitat quality in migratory populations. *Oikos* 109(1):178–186.

Norris, D. R., and P. P. Marra. 2007. Seasonal interactions, habitat quality, and population dynamics in migratory birds. *The Condor* 109(3):535–547.

Psorakis, I., S. J. Roberts, I. Rezek, and B. C. Sheldon. 2012. Inferring social network structure in ecological systems from spatio-temporal data streams. *Journal of the Royal Society: Interface* 9(76):3055–3066.

Reynolds, M. D., B. L. Sullivan, E. Hallstein, S. Matsumoto, S. Kelling, M. Merrifield, D. Fink, A. Johnston, W. M. Hochachka, N. E. Bruns, and M. E. Reiter. 2017. Dynamic conservation for migratory species. Science Advances 3: Article e1700707.

Robinson, W. D., M. S. Bowlin, I. Bisson, J. Shamoun-Baranes, K. Thorup, R. H. Diehl, T. H. Kunz, S. Mabey, and D. W. Winkler. 2010. Integrating concepts and technologies to advance the study of bird migration. *Frontiers in Ecology and the Environment* 8(7):354–361.

Schoener, T. W. 2011. The newest synthesis: Understanding the interplay of evolutionary and ecological dynamics. *Science* 331(6016):426–429.

Schuster, R., S. Wilson, A. D. Rodewald, P. Arcese, D. Fink, T. Auer, and J. R. Bennett. 2019. Optimizing the conservation of migratory species over their full annual cycle. Nature Communications 10: Article 1754.

Sinclair, S. P., E. J. Milner-Gulland, R. J. Smith, E. J. McIntosh, H. P. Possingham, A. Vercammen, and A. T. Knight. 2018. The use, and usefulness, of spatial conservation prioritizations. *Conservation Letters* 2018: Article e12459.

Smith, J. E., and N. Pinter-Wollman. 2021. Observing the unwatchable: Integrating automated sensing, naturalistic observations and animal social network analysis in the age of big data. *Journal of Animal Ecology* 90(1):62–75.

Stutchbury, B. J., S. A. Tarof, T. Done, E. Gow, P. M. Kramer, J. Tautin, J. W. Fox, and V. Afanasyev. 2009. Tracking long-distance songbird migration by using geolocators. *Science* 323(5916):896–896.

Sullivan, B. L., J. L. Aycrigg, J. H. Barry, R. E. Bonney, N. Bruns, C. B. Cooper, T. Damoulas, A. A. Dhondt, T. Dietterich, A. Farnsworth, and D. Fink. 2014. The eBird enterprise: An integrated approach to development and application of citizen science. *Biological Conservation* 169:31–40.

Sutherland, W. J., R. P. Freckleton, H. C. J. Godfray, S. R. Beissinger, T. Benton, D. D. Cameron, Y. Carmel, D. A. Coomes, T. Coulson, M. C. Emmerson, and R. S. Hails. 2013. Identification of 100 fundamental ecological questions. *Journal of Ecology* 101(1):58–67.

Taylor, P., T. Crewe, S. Mackenzie, D. Lepage, Y. Aubry, Z. Crysler, G. Finney, C. Francis, C. Guglielmo, D. Hamilton, and R. Holberton. 2017. The Motus Wildlife Tracking System: A collaborative research network to enhance the understanding of wildlife movement. *Avian Conservation and Ecology* 12: Article 8.

Tylianakis, J. M., R. K. Didham, J. Bascompte, and D. A. Wardle. 2008. Global change and species interactions in terrestrial ecosystems. *Ecology Letters* 11:1351–1363. doi:10.1111/j.1461-0248.2008.01250.x

Vergara, P. M., G. E. Soto, A. D. Rodewald, and M. Quiroz. 2019. Behavioral switching in Magellanic woodpeckers reveals perception of habitat quality at different spatial scales. *Landscape Ecology 34*(1):79–92.

Webster, M. S., and P. P. Marra. 2005. The importance of understanding migratory connectivity and seasonal interactions. Pages 199–208 in R. Greenberg and P. P. Marra, eds. *Birds of Two Worlds: The Ecology and Evolution of Temperate-Tropical Migration*. Johns Hopkins University Press, Baltimore, MD.

Webster, M. S., P. P. Marra, S. M. Haig, S. Bensch, and R. T. Holmes. 2002. Links between worlds: Unraveling migratory connectivity. *Trends in Ecology & Evolution*, 17(2):76–83.

Wikelski, M., R. W. Kays, N. J. Kasdin, K. Thorup, J. A. Smith, and G. W. Swenson, Jr. 2007. Going wild: What a global small-animal tracking system could do for experimental biologists. *Journal of Experimental Biology*, 210(2):181–186.

Wilmers, C. C., B. Nickel, C. M. Bryce, J. A. Smith, R. E. Wheat, and V. Yovovich. 2015. The golden age of bio-logging: How animal-borne sensors are advancing the frontiers of ecology. *Ecology* 96(7):1741–1753.

Wilson, A. A., M. A. Ditmer, J. R. Barber, N. H. Carter, E. T. Miller, L. P. Tyrrell, and C. D. Francis. (2021). Artificial night light and anthropogenic noise interact to influence bird abundance over a continental scale. *Global Change Biology* 27:3987–4004.

Wilson, R. P., F. Quintana, and V. J. Hobson. 2012. Construction of energy landscapes can clarify the movement and distribution of foraging animals. *Proceedings of the Royal Society B: Biological Sciences* 279(1730):975–980.

Wilson, S., H. Y. Lin, R. Schuster, C. Gomez, A. Gonzalez, E. Botero-Delgadillo, N. J. Bayly, J. R. Bennett, A. D. Rodewald, and V. Ruiz Gutierrez. 2022. Opportunities for the conservation of migratory birds to benefit threatened resident vertebrates in the Neotropics. *Journal of Applied Ecology* 59:653–663. doi:10.1111/1365-2664.14077

Wilson, S., R. Schuster, A. D. Rodewald, J. R. Bennett, A. C. Smith, F. A. La Sorte, P. H. Verburg, and P. Arcese. 2019. Prioritize diversity or declining species? Trade-offs and synergies in spatial planning for the conservation of migratory birds in the face of land cover change. *Biological Conservation* 239: Article 108285. https://doi.org/10.1016/j.biocon.2019.108285

3

Uncovering Factors Limiting Populations of Migratory Birds Across the Annual Cycle

Gunnar R. Kramer

Migratory birds have captivated the imagination of humans for millennia. The ebb and flow of billions of individual birds between breeding and stationary nonbreeding areas leads to the predictable, seasonal redistribution of biomass across the planet (Dokter et al. 2018). Taken together, these distinct periods (breeding, stationary nonbreeding, and periods of directional movement outside of breeding and nonbreeding areas, including dispersal) comprise the annual cycle (Ådahl et al. 2006, Marra, Cohen, et al. 2015). The annual cycle is inherent to the ecology of migratory birds because it encompasses the suite of locations and conditions experienced by individuals throughout the year. Thus, characterizing the annual cycle is fundamental to identifying the factors that regulate populations of migratory species (i.e., Leopold's "limiting factors"; 1933).

Historically, delineating the annual cycle of migratory species has been difficult due to the spatial scale across which migration often occurs (e.g., Arctic Terns [*Sterna paradisaea*] migrate up to 80,000 km annually; Egevang et al. 2010) and the lack of available tools that can safely be deployed to track birds at relevant spatial and temporal resolutions (Bridge et al. 2011). However, the miniaturization of tracking devices during recent decades now allows for tracking even the smallest migratory species across substantial portions of their annual cycles (Knight et al. 2019). This has provided opportunities to address critical knowledge gaps related to the population ecology (Hallworth et al. 2021), movement ecology (DeLuca et al. 2015), and conservation (Reynolds et al. 2017) of migratory birds. For instance, efforts to track declining populations of Swainson's Hawks (*Buteo swainsoni*) and Golden-Winged Warblers (*Vermivora chrysoptera*) led to the identification of limiting factors at stationary nonbreeding areas (Woodbridge et al. 1995, Kramer et al. 2018). For other species, such as Kirtland's Warbler (*Setophaga kirtlandii*), evidence from decades of research suggests that breeding habitat availability is the primary factor limiting population growth (Brown et al. 2017). However, recent efforts to track individuals throughout the annual cycle revealed important stationary nonbreeding areas and stopover regions that can be incorporated into full annual cycle monitoring and management strategies to improve conservation outcomes (Brown et al. 2017, Cooper et al. 2017).

The development and proliferation of modeling approaches that account for the full annual cycle have also facilitated the identification of factors limiting the

Gunnar R. Kramer, *Uncovering Factors Limiting Populations of Migratory Birds Across the Annual Cycle*.
In: *New Perspectives in Ornithology*. Edited by: Scott V. Edwards and J. Michael Reed, Oxford University Press.
 DOI: 10.1093/oso/9780197787670.003.0003

populations of migratory birds. Full-annual-cycle population models constitute a variety of modeling approaches that aim to link spatially explicit abundance or demographic data with environmental processes across seasons to identify limiting factors and predict population dynamics (Hostetler et al. 2015). For example, full-annual-cycle population models have been used to link the abundance of American Redstarts (*Setophaga ruticilla*) and Wood Thrush (*Hylocichla mustelina*) from breeding sites in eastern North America with environmental conditions experienced during the nonbreeding period (Wilson et al. 2011, Rushing et al. 2016). Similarly, full-annual-cycle models identified a trade-off associated with migratory behavior of Egyptian Vultures (*Neophron percnopterus*) wherein survival during migration was lower compared to that during stationary periods, but stationary nonbreeding period survival was higher at lower latitudes (i.e., for individuals that migrated greater distances from breeding sites; Buechley et al. 2021). Integrated population models have been used to unite demographic and abundance data (typically derived from capture–recapture and count surveys, respectively) into a single, flexible, modeling framework that can improve the accuracy and precision of parameter estimates relative to models that consider these parameters separately (Abadi et al. 2010, Schaub and Abadi 2011, Plard et al. 2019). Integrated population models have been used to quantify the relative contributions of immigration and emigration to population dynamics to identify cryptic population sinks (e.g., Weegman et al. 2016) and determine the drivers of spatial variation in population trends (e.g., Weegman et al. 2017, Saracco et al. 2022). Recent applications of integrated population models have addressed the drivers of population dynamics within the context of complex evolutionary and ecological processes (e.g., genetics and competition; Péron et al. 2012, Coulson et al. 2017, Plard et al. 2019).

For most migratory birds, the lack of basic information from across the annual cycle (e.g., population dispersion and timing of movements, reliable estimates of demographic rates, and accurate population counts) makes it difficult to determine where and when factors are acting to limit populations, thus hindering conservation efforts (Faaborg et al. 2010). Notably, many migratory species are declining (Sanderson et al. 2006). Roughly 3 billion birds, many of them migrants, have disappeared from North American skies since 1970 (Rosenberg et al. 2019). Therefore, efforts to delineate the annual cycle of migratory birds, identify limiting factors driving enigmatic declines, and reframe conservation strategies are timely and important (Wilcove and Wikelski 2008, Reynolds et al. 2017, Torstenson et al. 2024).

In this chapter, I provide a basic overview of the current state of knowledge as it pertains to the annual cycle ecology of migratory birds. I describe and define the portions of the annual cycle that are relevant to migratory birds and highlight examples of migratory species that are limited by factors during one or more periods. I also discuss the importance of carryover effects and how conditions experienced during one period of the annual cycle can affect individuals during subsequent periods. Last, I discuss the growing importance of long-term ecological studies for monitoring and describing species' specific responses to ongoing anthropogenic change and highlight the value of incorporating evolutionary perspectives into investigations of the annual cycle ecology of wild birds. Throughout this chapter, I reference but do not review the methods and technologies that have allowed ornithologists to track migratory birds throughout the annual cycle and characterize limiting factors (Bridge et al. 2011, Hostetler et al. 2015). Although I focus on migratory birds, these principles can be adapted and applied to the population ecology of nonmigratory birds and other taxa.

The Annual Cycle

Cycles are central to biological processes. The annual cycle of migratory birds consists of distinct periods that occur over a single year, including breeding, stationary nonbreeding, and periods characterized by directional movement (i.e., migration and/or dispersal; Figure 3.1; Marra, Cohen, et al. 2015). Migratory species exhibit diverse life-history strategies (e.g., delayed maturation and reproduction [Weimerskirch 1992] and facultative migration [Pagel et al. 2020]) that are important to consider within the context of annual cycle research and investigations into potentially limiting factors. Below, I describe and define the general periods of the annual cycle that are relevant to most migratory birds.

Breeding

The breeding period consists of the portion of the annual cycle during which migratory birds reproduce. The breeding period is the only portion of the annual cycle during which new individuals are produced. As a result, the notion that factors affecting reproduction are more important than factors experienced during different periods of the annual cycle of migratory birds has been pervasive during the

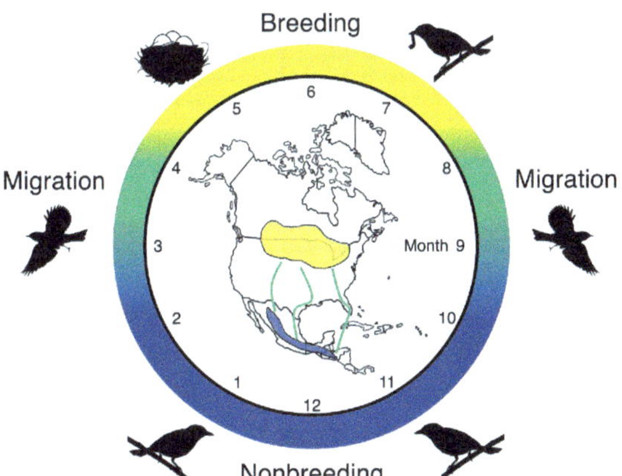

Figure 3.1 Visualization of the annual cycle for a typical North American migratory bird, including a northern breeding period (yellow), two seasonal migrations (green), and a southern nonbreeding period (blue).

Source: Silhouettes are from PhyloPic (http://www.phylopic.org; T. Michael Keesey, 2023).

past century (Pulliam and Millikan 1982), leading to a breeding-season bias in information on the annual cycle of migratory species (Marra, Cohen, et al. 2015). This bias is problematic if the factors limiting a species occur outside of the breeding period. One notable exception to this breeding-information bias is North American waterfowl (Anseriformes), which have been the focus of decades of research accounting for the full annual cycle (e.g., Lincoln 1935, Bellrose 1959, Nichols et al. 1995). In general, the bias toward breeding period research has led to detailed knowledge of the breeding ecology of many migratory species while leaving critical knowledge gaps in the basic ecology of these species outside the breeding period (Marra, Cohen, et al. 2015).

Stationary Nonbreeding

The stationary nonbreeding period is usually a prolonged sedentary period following migration away from breeding areas. Migratory birds display an array of stationary nonbreeding life-history strategies. Some species, including many waterfowl, congregate in large numbers at a few sites, whereas many songbirds (Passeriformes) occur across vast distributions in relative isolation from conspecifics (Olsen et al. 2006). Some species remain sedentary and defend a singular territory for the duration of the stationary nonbreeding period (Ritterson et al. 2021), whereas others wander (Winker et al. 1990) or move between multiple sites that can be separated by significant distances (i.e., >200 km; Heckscher et al. 2011). For many species, the stationary nonbreeding period is the longest portion of the annual cycle. Yet, it remains the least-studied period for most migratory birds (Marra, Cohen, et al. 2015). As a result, much less is known about the role that biotic and abiotic factors play in limiting populations of migratory birds during the stationary nonbreeding period (Sherry and Holmes 1996).

Migration

Migration is the period of the annual cycle during which individuals are in transit between stationary breeding and nonbreeding regions. Within the context of time, it is common for migration to comprise a relatively small proportion of the annual cycle. However, it is thought to be an especially dangerous period because it involves successfully navigating unfamiliar landscapes and barriers (Hewson et al. 2016), increased risk of exposure to novel pathogens and predators (Tian et al. 2015), and the necessity to acquire and/or strategically use resources to complete physiologically demanding flights (e.g., Gill et al. 2008). Birds migrate across various geographic scales ranging from the circumpolar migrations (e.g., Arctic Terns; Egevang et al. 2010) to local movements (e.g., ~5 km; Spruce Grouse [*Falcipennis canadensis*]; Herzog and Keppie 1980) and altitudinal movements (e.g., many species in the Neotropics; Jahn et al. 2020). Diverse migratory strategies have evolved across the avian tree of life, and variation in when (diurnal vs. nocturnal; Alerstam

2009), how (soaring vs. flapping flight; Hedenström 1993), and why (genetic vs. social inheritance mechanisms; Chernetsov et al. 2004, Liedvogel et al. 2011) birds migrate can have important implications for understanding the potential limiting factors that can impact populations during migration (Torstenson et al. 2024).

Dispersal and Molt

Dispersal and molt are important events that are often overlooked but likely warrant greater consideration in the context of describing the annual cycles of migratory species and identifying limiting factors (Luukkonen et al. 2008, Studds et al. 2008, Tonra and Reudink 2018, Taylor 2019).

There are several types of dispersal that can occur across the annual cycle of migratory birds and are relevant to identifying limiting factors. Natal dispersal involves the movement from a juvenile's natal site to their first breeding site. Natal dispersal is difficult to quantify and a frequently overlooked component of population dynamics in many migratory birds (Paradis et al. 1998). Dispersal can also occur during sedentary periods of the annual cycle (i.e., breeding and nonbreeding periods) when individuals move from a previously used site to a new breeding or stationary nonbreeding site. The fidelity that individuals exhibit to their natal sites (philopatry) or previous breeding, migratory stopover, or stationary nonbreeding sites (site fidelity) can be highly variable within and among species (Paradis et al. 1998). In cases of territorial songbirds, the strength of breeding site fidelity is frequently sex-dependent, potentially leading to biases in demographic parameter estimates between sexes (Clarke et al. 1997). Often, research occurs across a limited geographic scale such that it can be nearly impossible to differentiate dispersal outside of the study site from mortality (Koenig et al. 1996). Thus, dispersal and its effect on population dynamics remain poorly understood for many migratory species (Taylor 2019).

Molt involves the periodic replacement of feathers and is energetically costly for birds (Holmgren and Hedenström 1995). Like many other life-history traits, migratory birds exhibit diverse molting strategies that vary in their timing, extent, and frequency (Holmgren and Hedenström 1995, Tonra and Reudink 2018). Molt is important to consider within the context of the annual cycle because individuals must balance the energetic requirements associated with molting against other costly investments (e.g., migration and reproduction) potentially leading to trade-offs and carryover effects (Barshep et al. 2013, Latta et al. 2016, Møller and Nielsen 2018). Thus, like dispersal, molt is an important and likely underappreciated component of migratory birds' annual cycles (Tonra and Reudink 2018).

Limiting Factors Across the Annual Cycle

Factors limiting populations of migratory species can occur throughout the annual cycle. Sometimes, species are apparently limited by a single factor (i.e., nest site availability during the breeding period; Wiebe 2011). Perhaps more frequently, species' population dynamics are driven by complex interactions of multiple factors operating across periods of the annual cycle and differentially affecting certain segments of the population (Kramer et al. 2025). Differences in space use within and among populations throughout the annual cycle can lead to groups experiencing different conditions, complicating efforts to identify factors driving population dynamics. For example, whether breeding populations of migratory species co-occur throughout migration and the stationary nonbreeding period (i.e., exhibit strong migratory connectivity) can influence whether they experience similar conditions that could be potentially limiting (Webster et al. 2002). Populations of Golden-Winged Warblers exhibit strong migratory connectivity wherein individuals from declining eastern populations occur exclusively in northern South America, which has experienced greater rates of deforestation, during the stationary nonbreeding period, and individuals from stable populations in the Great Lakes region occur in Central America, where relatively less deforestation has occurred (Kramer et al. 2017, 2018). Golden-Winged Warblers also exhibit sexual habitat segregation during the stationary nonbreeding period such that females use lower elevation areas that are more likely to be disturbed by anthropogenic activities compared to males (Bennett et al. 2019). Similarly, migratory behavior and settlement patterns can vary by age such that limiting factors can differentially affect specific cohorts (e.g., juveniles) and shape the age-class structure of populations (e.g., Handel and Gill 2010). Elucidating drivers of population dynamics can be further complicated because conditions experienced in one period of the annual cycle can have nonlethal costs that can affect the fitness of individuals in a subsequent period (i.e., carryover effects; Harrison et al. 2011). Therefore, gathering the data required to characterize the life histories of species throughout periods of the annual cycle, although challenging to acquire, can greatly improve efforts to identify when and where limiting factors occur. Below, I describe some common abiotic, biotic, and anthropogenic limiting factors that can act to limit populations of migratory birds. I also highlight how carryover effects experienced by individuals can scale up to affect population dynamics of migratory species.

Abiotic Limiting Factors

Photoperiod and climate are two abiotic factors that can interact to affect the productivity of migratory birds (Dawson et al. 2001). In seasonal environments, photoperiod is correlated with broad temperature and precipitation patterns, and many migratory species rely on photoperiod to time their annual cycle such that they capitalize on periods of resource abundance and avoid periods of resource scarcity or inclement conditions for which they are not adapted (Dawson et al. 2001, Dawson 2008). With climate change, favorable conditions can occur earlier or later than normal, leading to increased risk that individuals mistime movements between periods of the annual cycle. These phenological mismatches can occur throughout the annual cycle and have significant impacts on the survival and productivity of migratory species (McGowan et al. 2011, Samplonius et al. 2016). Some species appear to be tracking changes in phenology, whereas others are not, leading to fitness consequences (Møller et al. 2008). However, fitness consequences can exist even for species that have effectively tracked changing environmental conditions. For example, Tree Swallows (*Tachycineta bicolor*) that advance their laying date in response to warming spring temperatures are exposed to inclement weather events that can cause mass chick mortality at twice the rate compared to historical levels (Shipley et al. 2020). The effects of phenological mismatch and level of plasticity in behaviors that might allow migratory birds to adapt to or mediate those effects are an important area of ongoing research (Kubelka et al. 2022, Torstenson et al. 2025).

Climate, independent of photoperiod, can also limit populations of migratory birds throughout the annual cycle. During the breeding period, adults may protect eggs and nestlings from extreme temperatures and precipitation (Carroll et al. 2018). However, eggs and nestlings may be susceptible to other factors, including flooding, high winds, and hail, which can reduce productivity (van de Pol et al. 2010, Hightower et al. 2018). Furthermore, extreme weather (heat, cold, wind, rain, etc.) can cause mortality in juvenile birds, reducing productivity (e.g., Bourne et al. 2020). For Nearctic–Neotropical migrants, precipitation patterns during the stationary nonbreeding period are driven by the El Niño–Southern Oscillation cycle, which affects primary production and insect abundance in Central and South America. During strong El Niño events, decreased precipitation is associated with reduced stationary nonbreeding-period survival and population declines of many Nearctic–Neotropical migrants (Sillett and Holmes 2002, Lamanna et al. 2012). Similarly, increasingly dry conditions in Spain may cause increased risk of mortality in Common Cuckoos (*Cuculus canorus*) using that region to stopover during migration to central African nonbreeding sites (Hewson et al. 2016; Figure 3.2A). Extreme weather can cause direct mortality if storms or other unfavorable weather conditions intercept migrating individuals during long-distance flights over inhospitable barriers (Newton 2007). Mass mortality associated with these events can lead to significant declines in affected populations (e.g., >50% decline in Chimney Swift [*Chaetura pelagica*] populations after encountering Hurricane Wilma during migration; Dionne et al. 2008).

Biotic Limiting Factors

Habitat availability can limit populations of migratory birds throughout the annual cycle. For example, the global population of Kirtland's Warblers grew from less than 200 pairs in the 1970s to more than 2,200 pairs following conservation efforts that increased breeding habitat availability (Brown et al. 2017; Figure 3.2B). The loss of breeding habitat was also associated with variation in population trends in Connecticut Warblers (*Oporornis agiils*; Hallworth et al. 2021), whereas population trends of *Vermivora* warblers were strongly associated with habitat loss during the stationary nonbreeding period (Kramer et al. 2018; Figure 3.2C). Species may be especially susceptible to habitat loss during migration if a significant proportion of the population congregates at stopover sites that are crucial for resting and refueling (e.g., shorebirds refueling at tidal flats in the Yellow Sea; Studds et al. 2017; Figure 3.2D). Similarly, the amount of habitat and its distribution across the landscape can affect the natal dispersal of individuals during their first breeding attempt (Harris and Reed 2002). It is worth noting that the term "habitat" often encompasses many potential

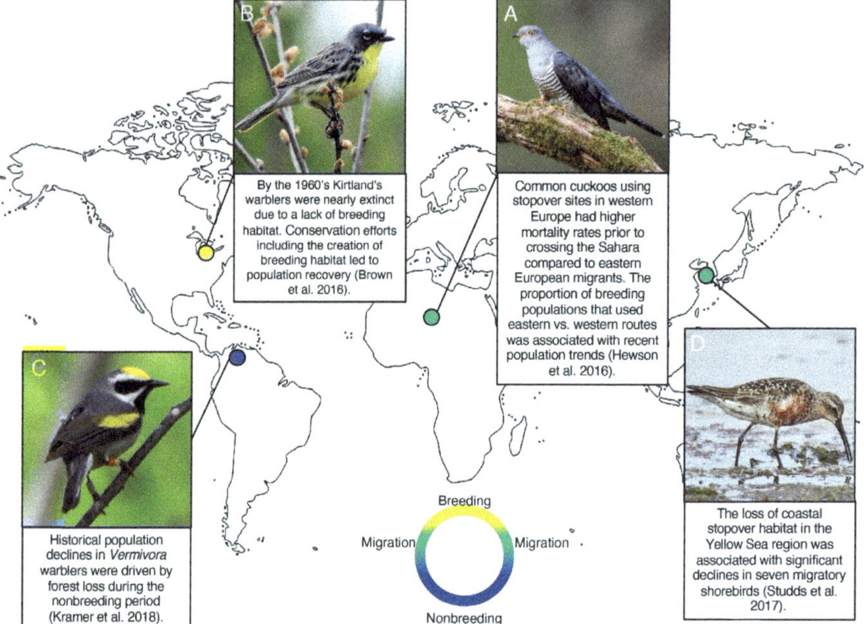

Figure 3.2 Examples of migratory birds and the limiting factors that drive their populations' dynamics throughout the world. Circles indicate the general region where limiting factors occur and the period of the annual cycle (color).

Sources: Common Cuckoo (*Cuculus canorus*) photo is courtesy of Alan Shearman via Flickr (2022, CC BY 2.0). Kirtland's Warbler (*Setophaga kirtlandii*) photo is courtesy of Joel Trick/USFWS via Flickr (2012, CC BY 2.0). Golden-Winged Warbler (*Vermivora chrysoptera*) photo by the author. Curlew Sandpiper (*Calidris ferruginea*) photo is courtesy of Andrey Gulivanov via Flickr (2022, CC BY 2.0).

limiting factors, complicating efforts to identify the specific component(s) limiting a population.

The availability of sufficient food resources to support reproduction, migration, and stationary nonbreeding-period survival can be an important limiting factor for migratory bird populations and is often connected to abiotic factors such as temperature or precipitation (Krebs et al. 2012). For example, populations of seed-eating birds breeding in the boreal forests of North America increased in abundance following years with large cone crops (Koenig and Knops 2001). In response, the productivity of raptors specialized in hunting songbirds increased due to greater availability of prey (Petty et al. 1995). Moreover, long-distance migration is energetically costly, and individuals often refuel during migration at stopover sites where the availability of sufficient food resources can be critical to their ability to complete migration. For instance, shorebirds that stopped over at Delaware Bay gained more mass and had a higher probability of survival when their stopover coincided with peak abundance of horseshoe crab eggs (*Limulus polyphemus*; McGowan et al. 2011).

Natural mortality caused by predation can limit populations of migratory birds throughout the annual cycle. Predation of nests can limit species' population growth by reducing reproductive output via direct mortality of eggs and nestlings (Martin 1993, Sherry et al. 2015). Predation is also the primary cause of juvenile mortality for songbirds after they leave the nest, and fledgling mortality can be a critical and often overlooked component of productivity (Streby and Andersen 2011). Unfamiliarity with migratory stopover sites combined with more diverse and/or abundant predator communities can increase mortality caused by predation along migratory routes (Cimprich et al. 2005, Klaassen et al. 2014). Thus, migrating birds often balance securing sufficient resources necessary to complete migration with predator avoidance (McCabe and Olsen 2015). Predation during the stationary nonbreeding period, although poorly quantified for many species, can vary across spatial scales and ecological gradients, especially if populations exhibit sexual variation in habitat selection or patterns of spatial distribution (Townsend et al. 2009, Bennett et al. 2019).

Disease and exposure to pathogens can also limit populations of migratory birds, especially during migration. Migratory species that congregate in large numbers in relatively small geographic areas (e.g., waterfowl) are especially susceptible to transmitting highly pathogenic avian influenza (Tian et al. 2015) among other diseases that are of interest within the context of human health and emerging zoonotic disease research (Reed et al. 2003).

The effects of potentially limiting factors on populations of migratory birds are often modified by population density. Density-dependent effects can act to limit populations via multiple mechanisms, including resource limitation, disease, and predation throughout the annual cycle (Hochachka and Dhondt 2000, Lok et al. 2013, Bruggeman et al. 2015). For example, high population density can increase competition for nesting sites and food resources, leading to reduced productivity during the breeding period (Gustafsson 1987, Rodenhouse et al. 2003). During migration, high population density can intensify inter- and intraspecific competition for resources (Cohen and Satterfield 2020) and increase disease transmission (Hochachka and Dhondt 2000, Tian et al. 2015). Similarly, high population densities and competition for food during the stationary nonbreeding period can reduce survival during that period (Marra, Studds, et al. 2015).

Anthropogenic Limiting Factors

Migratory birds can be killed throughout the annual cycle by (biotic and abiotic) anthropogenic factors, including introduced predators (e.g., feral cats), collisions with vehicles and structures (e.g., buildings, turbines, and antennas), and electrocutions at power lines (Loss et al. 2015). For some species, mortality caused by anthropogenic factors can reach meaningful levels (i.e., ~9% of estimated total annual populations of Yellow Rails [*Coturnicops noveboracensis*] and Swainson's Warblers [*Limnothlypis swainsonii*] are killed in collisions with communications towers during migration; Longcore et al. 2013). Evidence suggests that cats (feral and domestic) kill more birds than any other anthropogenic factor across the annual cycle, although species-specific estimates of cat-caused mortality are lacking (Loss et al. 2015). However, quantifying the effects of anthropogenic mortality on species' population dynamics remains a challenge, and the extent to which populations are limited by these factors remains unclear (Arnold and Zink 2011, Longcore et al. 2013, Loss et al. 2015, Kramer et al. 2023). Environmental pollution and exposure to contaminants across the annual cycle leading to mortality or sublethal effects have contributed to historical population declines in many species and represent a threat to future populations (e.g., Goldstein et al. 1999, Seewagen 2020, Wang et al. 2021, Mejia et al. 2024). Most notably, the use of DDT in agricultural settings throughout North America caused reproductive failure in many raptor species, leading to widespread population crashes in the mid-20th century (Bednarz et al. 1990). Legal and illegal hunting and trapping throughout migratory birds' annual cycles can also impose limits on populations (Nichols et al. 2007, Kamp et al. 2015). For example, hunting of shorebirds wintering across numerous jurisdictions in South America and the Caribbean may contribute to population declines in some species (McDuffie et al. 2022), whereas centralized management of North American waterfowl involving harvest limits based on population abundance has largely prevented population crashes (Nichols et al. 1995).

Carryover Effects: Interactions Among Seasons

Because periods of the annual cycle are inherently linked, conditions experienced in one period can affect the fitness of individuals in a subsequent period (i.e., carryover effects; Norris 2005; Figure 3.3). Documenting and quantifying carryover effects are challenging due to the difficulties involved in following individuals throughout their annual cycles. However, recent technological advances have made it easier to track individuals throughout the annual cycle and identify the impacts of carryover effects on population dynamics of migratory birds (Harrison et al. 2011).

Carryover effects can reduce the reproductive output of individuals if they cause individuals to arrive at breeding sites at the wrong time (e.g., too late) or in poor physiological condition such that they have a limited capacity to invest in reproduction (Harrison et al. 2011). During the stationary nonbreeding period, American Redstarts (*S. ruticilla*) that occupied wetter sites were in better condition and arrived at their breeding grounds earlier than individuals from drier, resource-poor sites (Marra et al. 1998). As a result, the reproductive output of males and females that occupied wetter sites during the stationary nonbreeding period was higher than that of Redstarts which occupied drier sites (Norris et al. 2004). Similarly, experiments with Greater Snow Geese (*Anser caerulescens atlanticus*) found that delaying actively migrating individuals negatively affected subsequent breeding success, but these carryover effects may be moderated during years of resource abundance on the breeding grounds (Legagneux et al. 2012). Individuals can also experience carryover effects that impact survival in a later period of the annual cycle (e.g., Latta et al. 2016). In some cases, carryover effects acquired in a previous portion of the annual cycle can be ameliorated by favorable conditions during a subsequent period (Senner et al. 2014, Briedis et al. 2018). In other instances, carryover effects can accumulate and affect individuals over multiple periods of the annual cycle (Catry et al. 2013).

Carryover effects can be sex-specific (Saino et al. 2017) and interact with other limiting factors (climate, density dependence, etc.) to affect populations of migratory

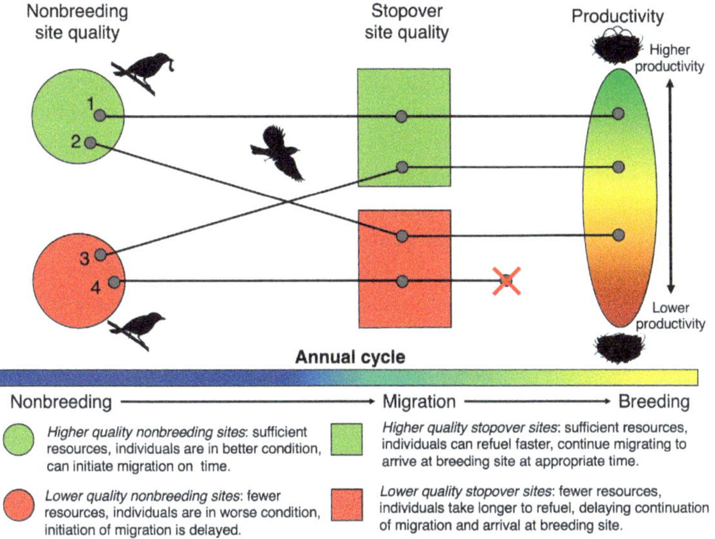

Figure 3.3 Diagram depicting a simple example of how carryover effects can impose fitness consequences on individuals across multiple periods of the annual cycle. Individuals (numbered gray circles) are shown moving through periods of the annual cycle. Nonbreeding and stopover sites used by different individuals vary in their quality. Individuals that use high-quality nonbreeding and stopover sites are more likely to have higher reproductive success during the breeding period. Conversely, individuals that occupy low-quality nonbreeding and stopover sites may be impacted by carryover effects that reduce reproductive output or experience mortality (red "x").

Source: Silhouettes are from PhyloPic (http://www.phylopic.org; T. Michael Keesey, 2023).

birds in complex ways (Morrissette et al. 2010). Exactly how important carryover effects are to the population dynamics of migratory species is still poorly understood for many species, and experimental assessments of carryover effects and their multifaceted impacts on population dynamics are warranted. Developing a clearer picture of the importance of carryover effects in migratory species' population dynamics will help address questions related to the evolution of diverse migratory strategies (O'Connor et al. 2014), the complexity of interactions between migratory birds and their environment (Norris 2005), and how best to conserve and manage migratory species in a rapidly changing world (Reynolds et al. 2017).

Conclusion and Future Challenges

Identifying the mechanisms limiting populations of migratory species is inherently challenging due to the spatial and temporal scales of migration, the myriad factors that can act to limit populations, and the capacity for those factors to interact in complex ways across periods of the annual cycle. The use of rapidly improving tracking technologies to describe the annual cycle of migratory birds has filled critical knowledge gaps related to the biology of migration at organizational scales ranging from individuals to communities (Legagneux et al. 2012, Dokter et al. 2018). Moreover, the characterization of migratory birds' annual cycles has led to major shifts in their conservation and provided opportunities to directly address limiting factors through targeted management (e.g., Reynolds et al. 2017). However, data characterizing and connecting portions of the annual cycle are lacking or nonexistent for most migratory birds, making it difficult to assess and contextualize emerging trends (Marra, Cohen, et al. 2015). For example, weak migratory connectivity between breeding and stationary nonbreeding regions (i.e., distinct breeding populations that co-occur during the stationary nonbreeding period) appears to be common among migratory species (Finch et al. 2017). However, the following are not clear: why weak migratory connectivity is common, the conditions under which strong migratory connectivity could be expected to evolve, the nature of associations between the strength of migratory connectivity and population trends, and whether anthropogenic limiting factors have shaped these patterns in the recent past (Gilroy et al. 2016, Kramer 2021, Merkle et al. 2022). Additional data from more migratory species will provide opportunities to better characterize overarching patterns at the intersections of migratory ecology, evolutionary biology, and conservation science.

Due to current costs and logistical constraints of tracking individual birds, most annual cycle studies last for several years and occur over limited spatial scales. Thus, the inferences derived from these "snapshots" into the annual cycle ecology of species likely fail to capture the full breadth of species' responses to variation in environmental conditions (Vickers et al. 2021, Nater et al. 2023). Disentangling drivers of population dynamics from the ecological noise of natural variation will require long-term monitoring of linked populations (i.e., populations with known annual-cycle distributions; Nater et al. 2023). Moreover, short-term studies may not be well-suited for characterizing how complex interactions between biotic (population density, resource availability, and interspecific competition) and abiotic factors (temperature and precipitation) shape species' population dynamics (Figure 3.4).

The development and implementation of new tools and technologies (e.g., space-based tracking and increasingly complex integrated population models) will help elucidate portions of the annual cycle that have remained enigmatic (e.g., dispersal) and improve predictions of population dynamics and species' responses to changes in habitat, climate, and other factors (Plard et al. 2019, Jetz et al. 2022). Moreover, integrating species-specific migration tracking data with other sources of data (e.g., genetic and physiological data) may aid in elucidating the mechanisms controlling migration (Toews et al. 2019, Sokolovskis et al. 2023), the plasticity of migratory behaviors (Verhoeven et al. 2022), and life-history trade-offs underlying variation in species' migratory strategies (Buechley et al. 2021). Comparative analyses across species will provide relevant insights into the shared evolutionary origins and current diversity of migratory strategies (Winger et al. 2019), their evolutionary fate in the Anthropocene (Walther et al. 2002, Both et al. 2006, Wilcove and Wikelski 2008, Weeks et al. 2020), and how best to design ecologically and evolutionarily relevant conservation strategies to prevent the loss of biodiversity (Reynolds et al. 2017, Flack et al. 2022).

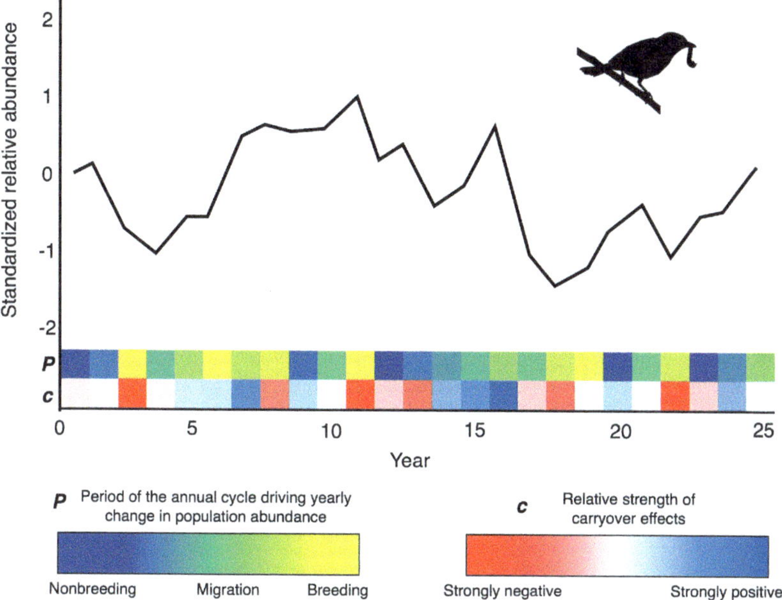

Figure 3.4 Simulated population dynamics of a migratory bird over 25 years of monitoring. The influence of different periods of the annual cycle (*P*) and the relative strength of carryover effects (*c*) for each year are presented beneath the standardized relative abundance curve (black line).

Source: Silhouette is from PhyloPic (http://www.phylopic.org; T. Michael Keesey, 2023).

Acknowledgments

I am grateful to J. M. Reed and A. D. Rodewald for comments that improved an earlier draft of this chapter. The National Science Foundation provided support during the writing of this chapter (NSF-PRFB 2109544).

References

Abadi, F., O. Gimenez, R. Arlettaz, and M. Schaub. 2010. An assessment of integrated population models: Bias, accuracy, and violation of the assumption of independence. *Ecology* 91:7–14.

Ådahl, E., P. Lundberg, and N. Jonzén. 2006. From climate change to population change: The need to consider annual life cycles. *Global Change Biology* 12:1627–1633.

Alerstam, T. 2009. Flight by night or day? Optimal daily timing of bird migration. *Journal of Theoretical Biology* 258:530–536.

Arnold, T. W., and R. M. Zink. 2011. Collision mortality has no discernible effect on population trends of North American birds. *PLoS One* 6:Article e24708.

Barshep, Y., C. D. T. Minton, L. G. Underhill, B. Erni, and P. Tomkovich. 2013. Flexibility and constraints in the molt schedule of long-distance migratory shorebirds: Causes and consequences. *Ecology and Evolution* 3:1967–1976.

Bednarz, J. C., D. Klem, Jr., L. J. Goodrich, and S. E. Senner. 1990. Migration counts of raptors at Hawk Mountain, Pennsylvania, as indicators of population trends, 1934–1986. *The Auk* 107:96–109.

Bellrose, F. C. 1959. Lead poisoning as a mortality factor in waterfowl populations. *Illinois Natural History Survey Bulletin* 27:235–288.

Bennett, R. E., A. D. Rodewald, and K. V. Rosenberg. 2019. Overlooked sexual segregation of habitats exposes female migratory landbirds to threats. *Biological Conservation* 240: Article 108266.

Both, C., S. Bouwhuis, C. M. Lessells, and M. E. Visser. 2006. Climate change and population declines in a long-distance migratory bird. *Nature* 441:81–83.

Bourne, A. R., S. J. Cunningham, C. N. Spottiswoode, and A. R. Ridley. 2020. High temperatures drive offspring mortality in a cooperatively breeding bird. *Proceedings of the Royal Society B: Biological Sciences* 287: Article 20201140.

Bridge, E. S., K. Thorup, M. S. Bowlin, P. B. Chilson, R. H. Diehl, R. W. Fléron, P. Hartl, R. Kays, J. F. Kelly, W. D. Robinson, and M. Wikelski. 2011. Technology on the move: Recent and forthcoming innovations for tracking migratory birds. *BioScience* 61:689–698.

Briedis, M., M. Krist, M. Kral, C. C. Voigt, and P. Adamik. 2018. Linking events throughout the annual cycle in a migratory bird—non-breeding period buffers accumulation of carry-over effects. *Behavioral Ecology and Sociobiology* 72:1–12.

Brown, D. J., C. A. Ribic, D. M. Donner, M. D. Nelson, C. I. Bocetti, C. M. Deloria-Sheffield, and D. Thompson. 2017. Using a full annual cycle model to evaluate long-term population viability of the conservation-reliant Kirtland's warbler after successful recovery. *Journal of Applied Ecology* 54:439–449.

Bruggeman, J. E., T. Swem, D. E. Andersen, P. L. Kennedy, and D. Nigro. 2015. Dynamics of a recovering Arctic bird population: The importance of climate, density dependence, and site quality. *Ecological Applications* 25:1932–1943.

Buechley, E. R., S. Oppel, R. Efrat, W. L. Phipps, I. Carbonell Alanís, E. Álvarez, A. Andreotti, V. Arkumarev, O. Berger-Tal, A. Bermejo Bermejo, A. Bounas, G. Ceccolini, A. Cenerini, V. Dobrev, O. Duriez, J. García, C. García-Ripollés, M. Galán, A. Gil, L. Giraud, O. Hatzofe, J. J. Iglesias-Lebrija, I. Karyakin, E. Kobierzycki, E. Kret, F. Loercher, P. López-López, Y. Miller, T. Mueller, S. C. Nikolov, J. de la Puente, N. Sapir, V. Saravia, Ç. H. Şekercioğlu, T. S. Sillett, J. Tavares, V. Urios, and P. P. Marra. 2021. Differential survival throughout the full annual cycle of a migratory bird presents a life-history trade-off. *Journal of Animal Ecology* 90:1228–1238.

Carroll, R. L., C. A. Davis, S. D. Fuhlendorf, R. D. Elmore, S. E. DuRant, and J. M. Carroll. 2018. Avian parental behavior and nest success influenced by temperature fluctuations. *Journal of Thermal Biology* 74:140–148.

Catry, P., M. P. Dias, R. A. Phillips, and J. P. Granadeiro. 2013. Carry-over effects from breeding modulate the annual cycle of a long-distance migrant: An experimental demonstration. *Ecology* 94:1230–1235.

Chernetsov, N., P. Berthold, and U. Querner. 2004. Migratory orientation of first-year white storks (*Ciconia ciconia*): Inherited information and social interactions. *Journal of Experimental Biology* 207:937–943.

Cimprich, D. A., M. S. Woodrey, and F. R. Moore. 2005. Passerine migrants respond to variation in predation risk during stopover. *Animal Behaviour* 69:1173–1179.

Clarke, A. L., B.-E. Saether, and E. Roskaft. 1997. Sex biases in avian dispersal: A reappraisal. *Oikos* 79:429–438.

Cohen, E. B., and D. A. Satterfield. 2020. "Chancing on a spectacle": Co-occurring animal migrations and interspecific interactions. *Ecography* 43:1657–1671.

Cooper, N. W., M. T. Hallworth, and P. P. Marra. 2017. Light-level geolocation reveals wintering distribution, migration routes, and primary stopover locations of an endangered long-distance migratory songbird. *Journal of Avian Biology* 48:209–219.

Coulson, T., B. E. Kendall, J. Barthold, F. Plard, S. Schindler, A. Ozgul, and J.-M. Gaillard. 2017. Modeling adaptive and nonadaptive responses of populations to environmental change. *The American Naturalist* 190:313–336.

Dawson, A. 2008. Control of the annual cycle in birds: Endocrine constraints and plasticity in response to ecological variability. *Philosophical Transactions of the Royal Society B: Biological Sciences* 363:1621–1633.

Dawson, A., V. M. King, G. E. Bentley, and G. F. Ball. 2001. Photoperiodic control of seasonality in birds. *Journal of Biological Rhythms* 16:365–380.

DeLuca, W. V., B. K. Woodworth, C. C. Rimmer, P. P. Marra, P. D. Taylor, K. P. McFarland, S. A. Mackenzie, and D. R. Norris. 2015. Transoceanic migration by a 12 g songbird. *Biology Letters* 11: Article 20141045.

Dionne, M., C. Maurice, J. Gauthier, and F. Shaffer. 2008. Impact of Hurricane Wilma on migrating birds: The case of the Chimney Swift. *Wilson Journal of Ornithology* 120:784–792.

Dokter, A. M., A. Farnsworth, D. Fink, V. Ruiz-Gutierrez, W. M. Hochachka, F. A. La Sorte, O. J. Robinson, K. V. Rosenberg, and S. Kelling. 2018. Seasonal abundance and survival of North America's migratory avifauna determined by weather radar. *Nature Ecology & Evolution* 2:1603–1609.

Egevang, C., I. J. Stenhouse, R. A. Phillips, A. Petersen, J. W. Fox, and J. R. D. Silk. 2010. Tracking of Arctic terns *Sterna paradisaea* reveals longest animal migration. *Proceedings of the National Academy of Sciences of the USA* 107:2078–2081.

Faaborg, J., R. T. Holmes, A. D. Anders, K. L. Bildstein, K. M. Dugger, S. A. Gauthreaux, P. Heglund, K. A. Hobson, A. E. Jahn, D. H. Johnson, S. C. Latta, D. J. Levey, P. P. Marra, C. L. Merkord, E. Nol, S. I. Rothstein, T. W. Sherry, T. S. Sillett, F. R. Thompson, and N. Warnock. 2010. Conserving migratory land birds in the New World: Do we know enough? *Ecological Applications* 20:398–418.

Finch, T., S. J. Butler, A. M. A. Franco, and W. Cresswell. 2017. Low migratory connectivity is common in long-distance migrant birds. *Journal of Animal Ecology* 86:662–673.

Flack, A., E. O. Aikens, A. Kölzsch, E. Nourani, K. R. S. Snell, W. Fiedler, N. Linek, H. G. Bauer, K. Thorup, J. Partecke, M. Wikelski, and H. J. Williams. 2022. New frontiers in bird migration research. *Current Biology* 32:R1187–R1199.

Gill, R. E., T. L. Tibbitts, D. C. Douglas, C. M. Handel, D. M. Mulcahy, J. C. Gottschalck, N. Warnock, B. J. McCaffery, P. F. Battley, and T. Piersma. 2008. Extreme endurance flights by landbirds crossing the Pacific Ocean: Ecological corridor rather than barrier? *Proceedings of the Royal Society B: Biological Sciences* 276:447–457.

Gilroy, J. J., J. A. Gill, S. H. M. Butchart, V. R. Jones, and A. M. A. Franco. 2016. Migratory diversity predicts population declines in birds. *Ecology Letters* 19:308–317.

Goldstein, M. I., T. E. Lacher, M. E. Zaccagnini, M. L. Parker, and M. J. Hooper. 1999. Monitoring and assessment of Swainson's hawks in Argentina following restrictions on monocrotophos use, 1996–97. *Ecotoxicology* 8:215–224.

Gustafsson, L. 1987. Interspecific competition lowers fitness in collared flycatchers *Ficedula albicollis*: An experimental demonstration. *Ecology* 68:291–296.

Hallworth, M. T., E. Bayne, E. McKinnon, O. Love, J. A. Tremblay, B. Drolet, J. Ibarzabal, S. Van Wilgenburg, and P. P. Marra. 2021. Habitat loss on the breeding grounds is a major contributor to population declines in a long-distance migratory songbird. *Proceedings of the Royal Society B: Biological Sciences* 288: Article 20203164.

Handel, C. M., and R. E. Gill. 2010. Wayward youth: Trans-Beringian movement and differential southward migration by juvenile sharp-tailed sandpipers. *Arctic* 63:273–288.

Harris, R. J., and J. M. Reed. 2002. Behavioral barriers to non-migratory movements of birds. *Annales Zoologici Fennici* 39:275–290.

Harrison, X. A., J. D. Blount, R. Inger, D. R. Norris, and S. Bearhop. 2011. Carry-over effects as drivers of fitness differences in animals. *Journal of Animal Ecology* 80:4–18.

Heckscher, C. M., S. M. Taylor, J. W. Fox, and V. Afanasyev. 2011. Veery (*Catharus fuscescens*) wintering locations, migratory connectivity, and a revision of its winter range using geolocator technology. *The Auk* 128:531–542.

Hedenström, A. 1993. Migration by soaring or flapping flight in birds: The relative importance of energy cost and speed. *Philosophical Transactions of the Royal Society of London B: Biological Sciences* 342:353–361.

Herzog, P. W., and D. M. Keppie. 1980. Migration in a local population of spruce grouse. *The Condor* 82:366–372.

Hewson, C. M., K. Thorup, J. W. Pearce-Higgins, and P. W. Atkinson. 2016. Population decline is linked to migration route in the common cuckoo. *Nature Communications* 7: Article 12296.

Hightower, J. N., J. D. Carlisle, and A. D. Chalfoun. 2018. Nest mortality of sagebrush songbirds due to a severe hailstorm. *Wilson Journal of Ornithology* 130:561–567.

Hochachka, W. M., and A. A. Dhondt. 2000. Density-dependent decline of host abundance resulting from a new infectious disease. *Proceedings of the National Academy of Sciences of the USA* 97:5303–5306.

Holmgren, N., and A. Hedenström. 1995. The scheduling of molt in migratory birds. *Evolutionary Ecology* 9:354–368.

Hostetler, J. A., T. S. Sillett, and P. P. Marra. 2015. Full-annual-cycle population models for migratory birds. *The Auk* 132:433–449.

Jahn, A. E., V. R. Cueto, C. S. Fontana, A. C. Guaraldo, D. J. Levey, P. P. Marra, and T. B. Ryder. 2020. Bird migration within the Neotropics. *The Auk* 137:1–23.

Jetz, W., G. Tertitski, R. Kays, U. Mueller, M. Wikelski, S. Åkesson, Y. Anisimov, A. Antonov, W. Arnold, F. Bairlein, O. Baltà, D. Baum, M. Beck, O. Belonovich, M. Belyaev, M. Berger, P. Berthold, S. Bittner, S. Blake, B. Block, D. Bloche, K. Boehning-Gaese, G. Bohrer, J. Bojarinova, G. Bommas, O. Bourski, A. Bragin, A. Bragin, R. Bristol, V. Brlík, V. Bulyuk, F. Cagnacci, B. Carlson, T. K. Chapple, K. F. Chefira, Y. Cheng, N. Chernetsov, G. Cierlik, S. S. Christiansen, O. Clarabuch, W. Cochran, J. M. Cornelius, I. Couzin, M. C. Crofoot, S. Cruz, A. Davydov, S. Davidson, S. Dech, D. Dechmann, E. Demidova, J. Dettmann, S. Dittmar, D. Dorofeev, D. Drenckhahn, V. Dubyanskiy, N. Egorov, S. Ehnbom, D. Ellis-Soto, R. Ewald, C. Feare, I. Fefelov, P. Fehérvári, W. Fiedler, A. Flack, M. Froböse, I. Fufachev, P. Futoran, V. Gabyshev, A. Gagliardo, S. Garthe, S. Gashkov, L. Gibson, W. Goymann, G. Gruppe, C. Guglielmo, P. Hartl, A. Hedenström, A. Hegemann, G. Heine, M. H. Ruiz, H. Hofer, F. Huber, F. Iannarilli, M. Illa, A. Isaev, B. Jakobsen, L. Jenni, S. Jenni-Eiermann, B. Jesmer, F. Jiguet, T. Karimova, N. J. Kasdin, F. Kazansky, R. Kirillin, T. Klinner, A. Knopp, A. Kölzsch, A. Kondratyev, M. Krondorf, P. Ktitorov, O. Kulikova, R. S. Kumar, C. Künzer, A. Larionov, C. Larose, F. Liechti, N. Linek, A. Lohr, A. Lushchekina, K. Mansfield, M. Matantseva, M. Markovets, P. Marra, J. F. Masello, J. Melzheimer, M. H. M. Menz, S. Menzie, S. Meshcheryagina, D. Miquelle, V. Morozov, A. Mukhin, I. Müller, T. Mueller, J. G. Navedo, R. Nathan, L. Nelson, Z. Németh, S. Newman, R. Norris, I. Okhlopkov, W. Oleś, R. Oliver, T. O'Mara, P. Palatitz, J. Partecke, R. Pavlick, A. Pedenko, J. Pham, D. Piechowski, A. Pierce, T. Piersma, W. Pitz, D. Plettemeier, I. Pokrovskaya, L. Pokrovskaya, I. Pokrovsky, M. Pot, P. Procházka, P. Quillfeldt, E. Rakhimberdiev, M. Ramenofsky, A. Ranipeta, J. Rapczyński, M. Remisiewicz, V. Rozhnov, F. Rienks, V. Rozhnov, C. Rutz, V. Sakhvon, N. Sapir, K. Safi, F. Schäuffelhut, D. Schimel, A. Schmidt, J. Shamoun-Baranes, A. Sharikov, L. Shearer, E. Shemyakin, S. Sherub, R. Shipley, Y. Sica, T. B. Smith, S. Simonov, K. Snell, A. Sokolov, V. Sokolov, O. Solomina, M. Soloviev, F. Spina, K. Spoelstra, M. Storhas, T. Sviridova, G. Swenson, P. Taylor, K. Thorup, A. Tsvey, M. Tucker, W. Turner, H. van der Jeugd, L. van Schalkwyk, M. van Toor, P. Viljoen, M. E. Visser, T. Volkmer, A. Volkov, S. Volkov, O. Volkov, J. A. C. von Rönn, B. Vorneweg, B. Wachter, J. Waldenström, M. Wegmann, A. Wehr, R. Weinzierl, J. Weppler, D. Wilcove, T. Wild, H. J. Williams, J. Wilshire, J. Wingfield, M. Wunder, A. Yachmennikova, S. Yanco, E. Yohannes, A. Zeller, C. Ziegler, A. Zięcik, and C. Zook. 2022. Biological Earth observation with animal sensors. *Trends in Ecology & Evolution* 37:293–298.

Kamp, J., S. Oppel, A. A. Ananin, Y. A. Durnev, S. N. Gashev, N. Hölzel, A. L. Mishchenko, J. Pessa, S. M. Smirenski, E. G. Strelnikov, S. Timonen, K. Wolanska, and S. Chan. 2015. Global population collapse in a superabundant migratory bird and illegal trapping in China. *Conservation Biology* 29:1684–1694.

Klaassen, R. H. G., M. Hake, R. Strandberg, B. J. Koks, C. Trierweiler, K. M. Exo, F. Bairlein, and T. Alerstam. 2014. When and where does mortality occur in migratory birds? Direct evidence from long-term satellite tracking of raptors. *Journal of Animal Ecology* 83:176–184.

Knight, S. M., G. M. Pitman, D. T. T. Flockhart, and D. R. Norris. 2019. Radio-tracking reveals how wind and temperature influence the pace of daytime insect migration. *Biology Letters* 15: Article 20190327.

Koenig, W. D., and J. M. H. Knops. 2001. Seed-crop size and eruptions of North American boreal seed-eating birds. *Journal of Animal Ecology* 70:609–620.

Koenig, W. D., D. Van Vuren, and P. N. Hooge. 1996. Detectability, philopatry, and the distribution of dispersal distances in vertebrates. *Trends in Ecology & Evolution* 11:514–517.

Kramer, G. R. 2021. Migration ecology of *Vermivora* warblers. PhD dissertation, University of Toledo.

Kramer, G. R., D. E. Andersen, D. A. Buehler, P. B. Wood, S. M. Peterson, J. A. Lehman, K. R. Aldinger, L. P. Bulluck, S. Harding, J. A. Jones, J. P. Loegering, C. Smalling, R. Vallender, and H. M. Streby. 2018. Population trends in *Vermivora* warblers are linked to strong migratory connectivity. *Proceedings of the National Academy of Sciences of the USA* 115:E3192–E3200.

Kramer, G. R., D. E. Andersen, D. A. Buehler, P. B. Wood, S. M. Peterson, J. A. Lehman, K. R. Aldinger, L. B. Bulluck, S. Harding, J. A. Jones, J. P. Loegering, C. Smalling, R. Vallender, and H. M. Streby. 2023. Exposure to risk factors experienced during migration is not associated with recent *Vermivora* warbler population trends. *Landscape Ecology* 38:2357–2380.

Kramer, G. R., H. M. Streby, S. M. Peterson, J. A. Lehman, D. A. Buehler, P. B. Wood, D. J. McNeil, J. L. Larkin, and D. E. Andersen. 2017. Nonbreeding isolation and population-specific migration patterns among three populations of golden-winged warblers. *The Condor* 119: 108–121.

Kramer, G. R., S. E. Fischer, P. J. Ruhl, E. S. Berz, R. Huffines, D. A. Aborn, and H. M. Streby. 2025. Spatial and temporal migratory connectivity of two sympatrically breeding wood-warblers with geographically discordant population trends. *Journal of Avian Biology* 2025:e03358.

Krebs, C. J., J. M. LaMontagne, A. J. Kenney, and S. Boutin. 2012. Climatic determinants of white spruce cone crops in the boreal forest of southwestern Yukon. *Botany* 90:113–119.

Kubelka, V., B. K. Sandercock, T. Székely, and R. P. Freckleton. 2022. Animal migration to northern latitudes: Environmental changes and increasing threats. *Trends in Ecology & Evolution* 37:30–41.

Lamanna, J. A., T. L. George, J. F. Saracco, M. P. Nott, and D. F. Desante. 2012. El Niño–Southern Oscillation influences annual survival of a migratory songbird at a regional scale. *The Auk* 129:734–743.

Latta, S. C., S. Cabezas, D. A. Mejia, M. M. Paulino, H. Almonte, C. M. Miller-Butterworth, and G. R. Bortolotti. 2016. Carry-over effects provide linkages across the annual cycle of a Neotropical migratory bird, the Louisiana waterthrush *Parkesia motacilla*. *Ibis* 158:395–406.

Legagneux, P., P. L. F. Fast, G. Gauthier, and J. Bêty. 2012. Manipulating individual state during migration provides evidence for carry-over effects modulated by environmental conditions. *Proceedings of the Royal Society B: Biological Sciences* 279:876–883.

Leopold, A. 1933. *Game Management*. Scribner's, New York.

Liedvogel, M., S. Åkesson, and S. Bensch. 2011. The genetics of migration on the move. *Trends in Ecology & Evolution* 26:561–569.

Lincoln, F. C. 1935. *The Waterfowl Flyways of North America*. U.S. Department of Agriculture, Washington, DC.

Lok, T., O. Overdijk, J. M. Tinbergen, and T. Piersma. 2013. Seasonal variation in density dependence in age-specific survival of a long-distance migrant. *Ecology* 94:2358–2369.

Longcore, T., C. Rich, P. Mineau, B. MacDonald, D. G. Bert, L. M. Sullivan, E. Mutrie, S. A. Gauthreaux, M. L. Avery, R. L. Crawford, A. M. Manville, E. R. Travis, and D. Drake. 2013. Avian mortality at communication towers in the United States and Canada: Which species, how many, and where? *Biological Conservation* 158:410–419.

Loss, S. R., T. Will, and P. P. Marra. 2015. Direct mortality of birds from anthropogenic causes. *Annual Review of Ecology, Evolution, and Systematics* 46:99–120.

Luukkonen, D. R., H. H. Prince, and R. C. Mykut. 2008. Movements and survival of molt migrant Canada geese from southern Michigan. *Journal of Wildlife Management* 72:449–462.

Marra, P. P., E. B. Cohen, S. R. Loss, J. E. Rutter, and C. M. Tonra. 2015. A call for full annual cycle research in animal ecology. *Biology Letters* 11: Article 20150552.

Marra, P. P., K. A. Hobson, and R. T. Holmes. 1998. Linking winter and summer events in a migratory bird by using stable-carbon isotopes. *Science* 282:1884–1886.

Marra, P. P., C. E. Studds, S. Wilson, T. S. Sillett, T. W. Sherry, and R. T. Holmes. 2015. Non-breeding season habitat quality mediates the strength of density-dependence for a migratory bird. *Proceedings of the Royal Society B: Biological Sciences* 282: Article 20150624.

Martin, T. E. 1993. Nest predation and nest sites. *BioScience* 43:523–532.

McCabe, J. D., and B. J. Olsen. 2015. Tradeoffs between predation risk and fruit resources shape habitat use of landbirds during autumn migration. *The Auk* 132:903–913.

McDuffie, L. A., K. S. Christie, A.-L. Harrison, A. R. Taylor, B. A. Andres, B. Laliberté, and J. A. Johnson. 2022. Eastern-breeding Lesser Yellowlegs are more likely than western-breeding birds to visit areas with high shorebird hunting during southward migration. *Ornithological Applications* 124: Article duab061.

McGowan, C. P., J. E. Hines, J. D. Nichols, J. E. Lyons, D. R. Smith, K. S. Kalasz, L. J. Niles, A. D. Dey, N. A. Clark, P. W. Atkinson, C. D. T. Minton, and W. Kendall. 2011. Demographic consequences of migratory stopover: Linking red knot survival to horseshoe crab spawning abundance. *Ecosphere* 2: Article art69-22.

Mejia, N., F. T. Garcia, J. Learned, J. Penniman, and S. V. Edwards. 2024. Effects of plastic ingestion on blood chemistry, gene expression and body condition in wedge-tailed shearwaters (*Ardenna pacifica*). *PeerJ* 12: Article e18566.

Merkle, J. A., B. Abrahms, J. B. Armstrong, H. Sawyer, D. P. Costa, and A. D. Chalfoun. 2022. Site fidelity as a maladaptive behavior in the Anthropocene. *Frontiers in Ecology and the Environment* 20:187–194.

Møller, A. P., and J. T. Nielsen. 2018. The trade-off between rapid feather growth and impaired feather quality increases risk of predation. *Journal of Ornithology* 159:165–171.

Møller, A. P., D. Rubolini, and E. Lehikoinen. 2008. Populations of migratory bird species that did not show a phenological response to climate change are declining. *Proceedings of the National Academy of Sciences of the USA* 105:16195–16200.

Morrissette, M., J. Bêty, G. Gauthier, A. Reed, and J. Lefebvre. 2010. Climate, trophic interactions, density dependence and carry-over effects on the population productivity of a migratory Arctic herbivorous bird. *Oikos* 119:1181–1191.

Nater, C. R., M. D. Burgess, P. Coffey, B. Harris, F. Lander, D. Price, M. Reed, and R. A. Robinson. 2023. Spatial consistency in drivers of population dynamics of a declining migratory bird. *Journal of Animal Ecology* 92:97–111.

Newton, I. 2007. Weather-related mass-mortality events in migrants. *Ibis* 149:453–467.

Nichols, J. D., F. A. Johnson, and B. K. Williams. 1995. Managing North American waterfowl in the face of uncertainty. *Annual Review of Ecology and Systematics* 26:177–199.

Nichols, J. D., M. C. Runge, F. A. Johnson, and B. K. Williams. 2007. Adaptive harvest management of North American waterfowl populations: A brief history and future prospects. *Journal für Ornithologie* 148:343–349.

Norris, R. D. 2005. Carry-over effects and habitat quality in migratory populations. *Oikos* 109:178–186.

Norris, R. D., P. P. Marra, K. T. Kyser, T. W. Sherry, and L. M. Ratcliffe. 2004. Tropical winter habitat limits reproductive success on the temperate breeding grounds in a migratory bird. *Proceedings of the Royal Society B: Biological Sciences* 271:59–64.

O'Connor, C. M., D. R. Norris, G. T. Crossin, and S. J. Cooke. 2014. Biological carryover effects: Linking common concepts and mechanisms in ecology and evolution. *Ecosphere* 5: Article art 28–11.

Olsen, B., V. J. Munster, A. Wallensten, J. Waldenström, A. D. M. E. Osterhaus, and R. A. M. Fouchier. 2006. Global patterns of influenza A virus in wild birds. *Science* 312:384–388.

Pagel, R. K., E. H. West, A. W. Jones, and H. M. Streby. 2020. Variation in individual autumn migration and winter paths of Great Lakes red-headed woodpeckers (*Melanerpes erythrocephalus*). *Animal Migration* 7:9–18.

Paradis, E., S. R. Baillie, W. J. Sutherland, and D. Gregory Richard. 1998. Patterns of natal and breeding dispersal in birds. *Journal of Animal Ecology* 67:518–536.

Péron, G., C. A. Nicolai, and D. N. Koons. 2012. Demographic response to perturbations: The role of compensatory density dependence in a North American duck under variable harvest regulations and changing habitat. *Journal of Animal Ecology* 81:960–969.

Petty, S. J., I. J. Patterson, D. I. K. Anderson, B. Little, and M. Davison. 1995. Numbers, breeding performance, and diet of the sparrowhawk *Accipiter nisus* and merlin *Falco columbarius* in relation to cone crops and seed-eating finches. *Forest Ecology and Management* 79: 133–146.

Plard, F., R. Fay, M. Kéry, A. Cohas, and M. Schaub. 2019. Integrated population models: Powerful methods to embed individual processes in population dynamics models. *Ecology* 100: Article e02715.

Pulliam, H. R., and G. C. Millikan. 1982. Social organization in the nonreproductive season. Pages 169–197 in D. S. Farner and J. R. King, eds. *Avian Biology*. Elsevier, New York.

Reed, K. D., J. K. Meece, J. S. Henkel, and S. K. Shukla. 2003. Birds, migration and emerging zoonoses: West Nile virus, Lyme disease, influenza A and enteropathogens. *Clinical Medicine & Research* 1:5–12.

Reynolds, M. D., B. L. Sullivan, E. Hallstein, S. Matsumoto, S. Kelling, M. Merrifield, D. Fink, A. Johnston, W. M. Hochachka, N. E. Bruns, M. E. Reiter, S. Veloz, C. Hickey, N. Elliott, L. Martin, J. W. Fitzpatrick, P. Spraycar, G. H. Golet, C. McColl, C. Low, and S. A. Morrison. 2017. Dynamic conservation for migratory species. *Science Advances* 3: Article e1700707.

Ritterson, J. D., D. I. King, and R. B. Chandler. 2021. Habitat-specific survival of golden-winged warblers *Vermivora chrysoptera* during the non-breeding season in an agricultural landscape. *Journal of Avian Biology* 52.

Rodenhouse, N. L., T. Scott Sillett, P. J. Doran, and R. T. Holmes. 2003. Multiple density-dependence mechanisms regulate a migratory bird population during the breeding season. *Proceedings of the Royal Society B: Biological Sciences* 270:2105–2110.

Rosenberg, K. V., A. M. Dokter, P. J. Blancher, J. R. Sauer, A. C. Smith, P. A. Smith, J. C. Stanton, A. Panjabi, L. Helft, M. Parr, and P. P. Marra. 2019. Decline of the North American avifauna. *Science* 366:120–124.

Rushing, C. S., T. B. Ryder, and P. P. Marra. 2016. Quantifying drivers of population dynamics for a migratory bird throughout the annual cycle. *Proceedings of the Royal Society B: Biological Sciences* 283: Article 20152846.

Saino, N., R. Ambrosini, M. Caprioli, A. Romano, M. Romano, D. Rubolini, C. Scandolara, F. Liechti, and J. Dunn. 2017. Sex-dependent carry-over effects on timing of reproduction and fecundity of a migratory bird. *Journal of Animal Ecology* 86:239–249.

Samplonius, J. M., E. F. Kappers, S. Brands, and C. Both. 2016. Phenological mismatch and ontogenetic diet shifts interactively affect offspring condition in a passerine. *Journal of Animal Ecology* 85:1255–1264.

Sanderson, F. J., P. F. Donald, D. J. Pain, I. J. Burfield, and F. P. J. van Bommel. 2006. Long-term population declines in Afro-Palearctic migrant birds. *Biological Conservation* 131: 93–105.

Saracco, J. F., R. L. Cormier, D. L. Humple, S. Stock, R. Taylor, and R. B. Siegel. 2022. Demographic responses to climate-driven variation in habitat quality across the annual cycle of a migratory bird species. *Ecology and Evolution* 12: Article e8934.

Schaub, M., and F. Abadi. 2011. Integrated population models: A novel analysis framework for deeper insights into population dynamics. *Journal of Ornithology* 152:S227–S237.

Seewagen, C. L. 2020. The threat of global mercury pollution to bird migration: Potential mechanisms and current evidence. *Ecotoxicology* 29:1254–1267.

Senner, N. R., W. M. Hochachka, J. W. Fox, and V. Afanasyev. 2014. An exception to the rule: Carry-over effects do not accumulate in a long-distance migratory bird. *PLoS One* 9: Article e86588.

Sherry, T. W., and R. T. Holmes. 1996. Winter habitat quality, population limitation, and conservation of Neotropical–Nearctic migrant birds. *Ecology* 77:36–48.

Sherry, T. W., S. Wilson, S. Hunter, and R. T. Holmes. 2015. Impacts of nest predators and weather on reproductive success and population limitation in a long-distance migratory songbird. *Journal of Avian Biology* 46:559–569.

Shipley, J. R., C. W. Twining, C. C. Taff, M. N. Vitousek, A. Flack, and D. W. Winkler. 2020. Birds advancing lay dates with warming springs face greater risk of chick mortality. *Proceedings of the National Academy of Sciences of the USA* 117:25590–25594.

Sillett, T. S., and R. T. Holmes. 2002. Variation in survivorship of a migratory songbird throughout its annual cycle. *Journal of Animal Ecology* 71:296–308.

Sokolovskis, K., M. Lundberg, S. Åkesson, M. Willemoes, T. Zhao, V. Caballero-Lopez, and S. Bensch. 2023. Migration direction in a songbird explained by two loci. *Nature Communications* 14:1–6.

Streby, H. M., and D. E. Andersen. 2011. Seasonal productivity in a population of migratory songbirds: Why nest data are not enough. *Ecosphere* 2: Article art78-15.

Studds, C. E., B. E. Kendall, N. J. Murray, H. B. Wilson, D. I. Rogers, R. S. Clemens, K. Gosbell, C. J. Hassell, R. Jessop, D. S. Melville, D. A. Milton, C. D. T. Minton, H. P. Possingham, A. C. Riegen, P. Straw, E. J. Woehler, and R. A. Fuller. 2017. Rapid population decline in migratory shorebirds relying on Yellow Sea tidal mudflats as stopover sites. *Nature Communications* 8: Article 14895.

Studds, C. E., T. K. Kyser, and P. P. Marra. 2008. Natal dispersal driven by environmental conditions interacting across the annual cycle of a migratory songbird. *Proceedings of the National Academy of Sciences of the USA* 105:2929–2933.

Taylor, C. M. 2019. Effects of natal dispersal and density-dependence on connectivity patterns and population dynamics in a migratory network. *Frontiers in Ecology and Evolution* 7: Article 354.

Tian, H., S. Zhou, L. Dong, T. P. Van Boeckel, Y. Cui, S. H. Newman, J. Y. Takekawa, D. J. Prosser, X. Xiao, Y. Wu, B. Cazelles, S. Huang, R. Yang, B. T. Grenfell, and B. Xu. 2015. Avian influenza H5N1 viral and bird migration networks in Asia. *Proceedings of the National Academy of Sciences of the USA* 112:172–177.

Toews, D. P. L., S. A. Taylor, H. M. Streby, G. R. Kramer, and I. J. Lovette. 2019. Selection on VPS13A linked to migration in a songbird. *Proceedings of the National Academy of Sciences of the USA* 116:18272–18274.

Tonra, C. M., and M. W. Reudink. 2018. Expanding the traditional definition of molt-migration. *The Auk* 135:1123–1132.

Torstenson, M. S., D. W. Wolfson, S. M. Safran, D. J. Walton, A. B. Hallberg, D. Kim, Y. F. Tan, G. R. Kramer, and D. E. Andersen. 2024. Conservation of North American migratory birds: insights from developments in tracking technologies. *Avian Conservation and Ecology* 19:13.

Townsend, J. M., C. C. Rimmer, J. Brocca, K. P. McFarland, and A. K. Townsend. 2009. Predation of a wintering migratory songbird by introduced rats: Can nocturnal roosting behavior serve as predator avoidance? *The Condor* 111:565–569.

van de Pol, M., B. J. Ens, D. Heg, L. Brouwer, J. Krol, M. Maier, K.-M. Exo, K. Oosterbeek, T. Lok, C. M. Eising, and K. Koffijberg. 2010. Do changes in the frequency, magnitude and timing of extreme climatic events threaten the population viability of coastal birds. *Journal of Applied Ecology* 47:720–730.

Verhoeven, M. A., A. H. J. Loonstra, A. D. McBride, W. Kaspersma, J. C. E. W. Hooijmeijer, C. Both, N. R. Senner, and T. Piersma. 2022. Age-dependent timing and routes demonstrate developmental plasticity in a long-distance migratory bird. *Journal of Animal Ecology* 91:566–579.

Vickers, S. H., A. M. A. Franco, and J. J. Gilroy. 2021. Sensitivity of migratory connectivity estimates to spatial sampling design. *Movement Ecology* 9:1–12.

Walther, G.-R., E. Post, P. Convey, A. Menzel, C. Parmesan, T. J. C. Beebee, J.-M. Fromentin, O. Hoegh-Guldberg, and F. Bairlein. 2002. Ecological responses to recent climate change. *Nature* 416:389–395.

Wang, L., G. Nabi, L. Yin, Y. Wang, S. Li, Z. Hao, and D. Li. 2021. Birds and plastic pollution: Recent advances. *Avian Research* 12:1–9.

Webster, M. S., P. P. Marra, S. M. Haig, S. Bensch, and R. T. Holmes. 2002. Links between worlds: Unraveling migratory connectivity. *Trends in Ecology & Evolution* 17:76–83.

Weegman, M. D., T. W. Arnold, R. D. Dawson, D. W. Winkler, and R. G. Clark. 2017. Integrated population models reveal local weather conditions are the key drivers of population dynamics in an aerial insectivore. *Oecologia* 185:119–130.

Weegman, M. D., S. Bearhop, A. D. Fox, G. M. Hilton, A. J. Walsh, J. L. McDonald, and D. J. Hodgson. 2016. Integrated population modelling reveals a perceived source to be a cryptic sink. *Journal of Animal Ecology* 85:467–475.

Weeks, B. C., D. E. Willard, M. Zimova, A. A. Ellis, M. L. Witynski, M. Hennen, B. M. Winger, and R. Norris. 2020. Shared morphological consequences of global warming in North American migratory birds. *Ecology Letters* 23:316–325.

Weimerskirch, H. 1992. Reproductive effort in long-lived birds: Age-specific patterns of condition, reproduction and survival in the wandering albatross. *Oikos* 64:464–473.
Wiebe, K. L. 2011. Nest sites as limiting resources for cavity-nesting birds in mature forest ecosystems: A review of the evidence. *Journal of Field Ornithology* 82:239–248.
Wilcove, D. S., and M. Wikelski. 2008. Going, going, gone: Is animal migration disappearing? *PLoS Biology* 6: Article e188.
Wilson, S., S. L. Ladeau, A. P. Toøttrup, and P. P. Marra. 2011. Range-wide effects of breeding- and nonbreeding-season climate on the abundance of a Neotropical migrant songbird. *Ecology* 92:1789–1798.
Winger, B. M., G. G. Auteri, T. M. Pegan, and B. C. Weeks. 2019. A long winter for the Red Queen: Rethinking the evolution of seasonal migration. *Biological Reviews of the Cambridge Philosophical Society* 94:737–752.
Winker, K., J. H. Rappole, and M. A. Ramos. 1990. Population dynamics of the wood thrush in southern Veracruz, Mexico. *The Condor* 92:444–460.
Woodbridge, B., K. K. Finley, and S. T. Seager. 1995. An investigation of the Swainson's hawk in Argentina. *Journal of Raptor Research* 29:202–204.

4
The Impact of Movement Ecology on Ornithology

Autumn-Lynn Harrison and Kyle Horton

Movement is fundamental to being a bird. Birds fly short and long distances: patch to patch (Hartfelder et al. 2020), country to country (Harrison et al. 2018), pole to pole (Egevang et al. 2010), the distance to the moon and back (Hoose 2012). They also walk, run, hop, skip, glide, shear, soar, paddle, dip, duck, dive, hover, flock, swarm, and swim. Movement—the change in spatial location of an object over time—is one of the class Aves' key ecological advantages. Movement determines how birds interact with their environment, with each other, and with humans.

When placing the young field of movement ecology (Nathan et al. 2008) within the old field of ornithology, there is a tendency to dismiss it as duplicative—an unnecessary new name for established fields of study, including migration, navigation, or behavioral ecology. Revealing the mysteries of avian migration is indeed a primary topic in movement ecology (Joo et al. 2022), and movement is unquestionably a behavior. However, movement ecology is an integrative field that combines organismal biology, ecology, evolution, behavior, and conservation. The ability to move determines much of what both migratory and nonmigratory birds do—how they are distributed, how their populations are maintained, and how they will survive in a changing world. It is thus unsurprising that the study of movement, facilitated by new biologging technologies and observing systems, has become a new branch of the field of ornithology. Mayr (1983) likened ornithology to a vigorously growing tree in which new branches—which may not have existed decades before—become dominant. Movement ecology fits well in this description.

In this chapter, we survey insights into how and why birds move, and we frame the discipline of movement ecology around a timeline of foundational advancements and papers as they impacted ornithology (Figure 4.1). We provide a brief summary of the initial invention of biologging devices; however, rather than focusing on tools and technologies commonly used in movement ecology studies (reviewed in Bridge et al. 2011), we emphasize the broad range of ornithological questions these tools help answer. We review topics for which movement ecology approaches have led to a deeper understanding of birds, highlighting studies that integrate multiple aspects of avian biology, ecology, and behavior. We take individual, collective, and macroecological perspectives and suggest that movement ecology enables understanding of the bird and the population, the pattern and the process (Figure 4.2). Herein, we define *macroecological* as the study of processes and patterns occurring at broad spatial scales. We discuss ethical considerations associated with studying bird movement.

Autumn-Lynn Harrison and Kyle Horton, *The Impact of Movement Ecology on Ornithology*. In: *New Perspectives in Ornithology*. Edited by: Scott V. Edwards and J. Michael Reed, Oxford University Press.
 DOI: 10.1093/oso/9780197787670.003.0004

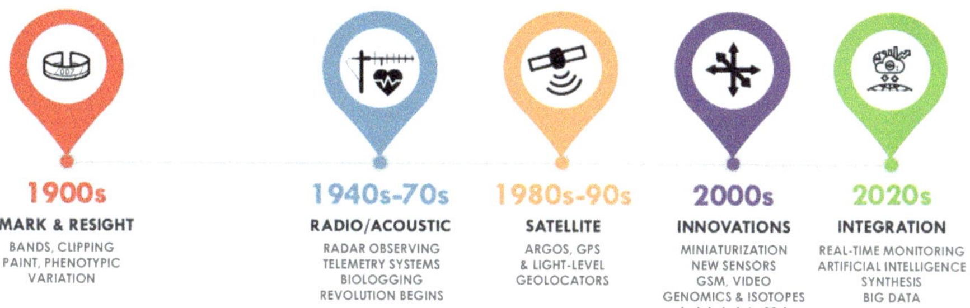

Figure 4.1 Chronology of major technological and analytical innovations in the study of avian movement ecology linked to key research reviewed in this chapter.

Finally, we provide perspectives on how the field of movement ecology could continue to impact the field of ornithology.

How Birds Move

Birds move via three main types of locomotion: in the air by powered flight; on land by walking or running; and on or in the water through paddling, swimming, and diving. The ability of birds to combine multiple forms of locomotion to exploit a variety of habitats is one of their key features and advantages.

Flight encompasses diverse movements and specializations. Most birds sustain flight through active flapping. The rapid helicopter-like movements of hovering hummingbirds are aided by high rotation in the shoulder joint and long primary feathers (Greenewalt 1960). In contrast, albatrosses have very long, narrow wings that are locked into place by a specialized tendon (Pennycuick 1982). The Wandering Albatross (*Diomedea exulans*) has the greatest wingspan—more than 100 times that of the smallest hummingbird (Bee Hummingbird, *Mellisuga helenae*). This morphology is better suited to capitalizing on winds for lift than for sustained flapping. Anatomical specializations that facilitate flight include feathers and wings, lightweight but strong skeletons, specialized tendons, and enlarged chest muscles (reviewed by Heers 2018). Flight is metabolically expensive, and compared to mammals, birds have a highly efficient respiratory system composed of lungs and air sacs that enables birds to transfer more oxygen per breath per unit body mass (Piiper and Scheid 1972). A large variety of wing morphologies, body scaling, and muscle types are demonstrated by birds and can be matched to flight styles, speeds, and foraging habitats (Rayner 1988).

Flightless birds, and other birds that fly only short distances, demonstrate a "convergent morphological adaptation association with a walking ecology" (Gaspar et al. 2020, p. 6186). Shorter wings and longer or more muscular legs tend to be traits associated with birds that walk or run as their primary form of movement. However, all birds use their legs to some extent. A new paradigm of avian locomotion is not just to focus on the trade-offs in investing in either wings or legs but also to consider the level of cooperation among the two modes of locomotion (Heers and Dial 2015).

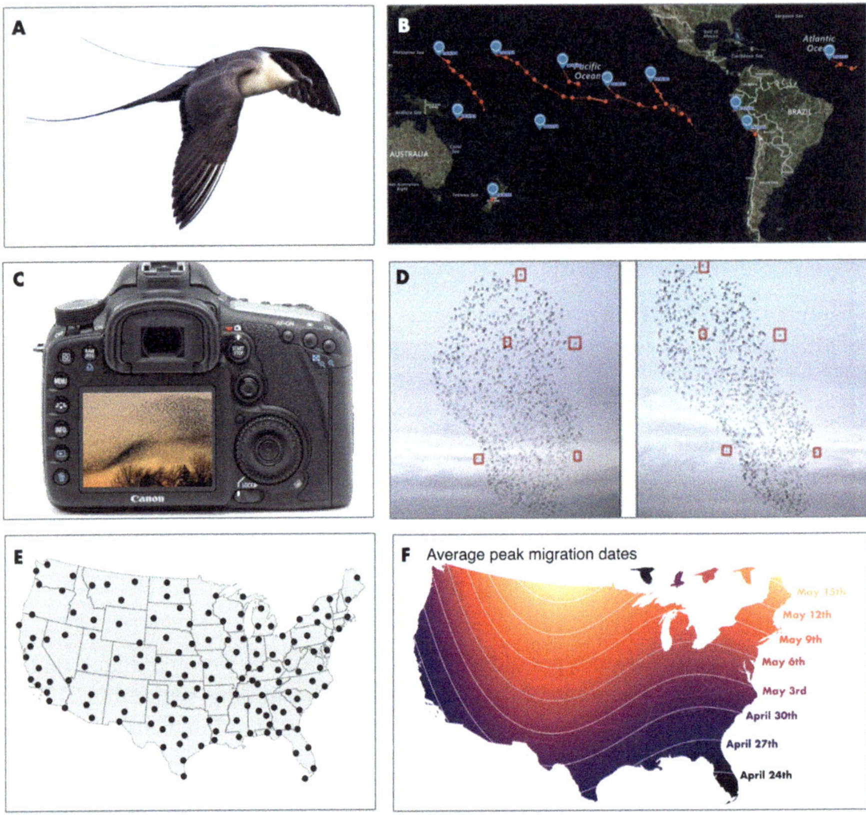

Figure 4.2 Examples of techniques for studying bird movements and data collected using these methods: (A) 5g solar satellite tag attached to a Long-tailed Jaeger to track individual movements (Photo: Kate Persons for Smithsonian Institution). (B) Ten day tails of jaeger movements during northward migration, 2020 (Created by authors). (C) high definition photography used to record collective behavior, in this case of a starling murmuration (Camera Photo: Coyau / Wikimedia Commons / CC BY-SA 3.0. Murmuration Photo: Walter Baxter / *A murmuration of starlings at Gretna* / CC BY-SA 2.0) (D) Two views of a starling flock from cameras placed 25-m apart illustrating individual's placement within it (red squares) to enable 3-D reconstruction (From Cavagna and Giardina 2008, Reproduced with permission from the Royal Statistical Society). (E) Radar network in the continental United States to record macrosystem-scale movements of bird assemblages (Created by authors). (F) Estimated migration of birds through the continental United States (Created by authors).

Sources: A, from Ballerini, Cabibbo, Candelier, Cavagna, Cisbani, Giardina, Lecomte, et al. (2008); reproduced under license from John Wiley and Sons, The Royal Statistical Society. C, based on data from Horton et al. (2020).

Fish (2016) provides an excellent review of specializations and evolution of birds that move in water. Birds that swim or dive to forage, rest, or avoid predators encompass fliers and non-fliers, but typically the best swimmers have given up the ability to either fly well or walk well. Birds may swim on the surface of the water by paddling their legs, or they may dive by using their wings or legs to propel them. Other methods of gaining speed over the surface of the water are reviewed in Heers (2018). Birds that swim tend to have webbed feet; oily feathers to aid in waterproofing; legs positioned further back on their bodies; and wings with smaller surface area relative to body size, which often is heavy. Diving birds have additional physiological specializations, which are described further below.

The ability to move in the air, on land, and underwater allows birds to use most habitats on Earth. Birds move to find suitable habitats for breeding, feeding, and rest and also to avoid predators. The interplay between the ability of a bird to move and the factors that motivate it to move can drive ecological and evolutionary processes (Dial 2003).

What Is Movement Ecology?

The study of animal movement has existed for millennia; however, the unification of tens of thousands of studies into a formal subdiscipline of ecology occurred less than 17 years ago. Ornithologist Nathan (ornithologist) et al. (2008) proposed a movement ecology paradigm that linked the internal state of an organism (its capacity to move) with external factors that affect an animal's proclivity to move (environmental and social drivers) to better understand the role movement plays in ecological and evolutionary processes. The field of movement ecology has since been broadly embraced across taxa and disciplines (Joo et al. 2022), including the formation of a scientific journal (Nathan and Giuggioli 2013). And although the study of animal movement is old, the impetus to form a new field of movement ecology likely "hinges on two 21st century advances" (Photopoulou 2017) that allowed humans to observe animals and their environments in new ways: (1) technological advances to continuously record movement paths of animals in the wild (and sometimes their internal and external states) that provided both bird-borne and remotely sensed data and (2) computing and analytical advances that facilitated the analysis of large volumes of data borne from these observing systems (for a review of statistical software packages focused on movement data, see Joo et al. 2020). The ability to integrate movement data with data sets describing the terrestrial, aquatic, and aerial habitats that birds use (e.g., remotely sensed data) has been an incredible advance.

Radar was first shown to detect flying birds in the 1940s (Lack and Varley 1945). In the 1960s, birds carrying the first radio transmitters required substantial field effort using a receiver and receiving antenna to manually relocate—night-long airplane flights and, at times, week-long forays up and down interstate highways to capture long-distance movements were not out of the question (Cochran et al. 1967, Cochran 1972). Early archival biologging devices that could record a continuous time series of bird movements revealed diving behavior of free-ranging birds for the first time (Kooyman et al. 1971). In 1984—a year after the publication of *Perspectives in Ornithology* (Brush and Clark 1983)—Trumpeter Swan (*Cygnus buccinator*), Bald Eagle (*Haliaeetus leucocephalus*), and Southern Giant Petrel (*Procellaria gigantea*) became the first birds to be tracked in the wild by satellite (Strikwerda et al. 1986).

Birds could now be followed continuously and autonomously across landscapes and seascapes. A decade later, archival light-level geolocators were invented as another means of estimating bird location (Wilson and Vandenabeele 2012). These data loggers had to be recovered to retrieve data and were less accurate than satellite tags. However, they used much smaller batteries and thus ultimately revolutionized the study of small birds throughout their full annual cycle (Stutchbury et al. 2009, McKinnon and Love 2018). Devices may integrate sensors capable of measuring the bird's state (e.g., heart rate) and the state of its associated environment (e.g., sea surface temperature).

Movement ecology studies have evolved from small to larger sample sizes and from local to hemispheric scales (Sequeira et al. 2019). Radio telemetry has experienced a recent resurgence for birds and includes local, site-specific studies and regional network arrays—for example, Motus, a coordinated and automated radio telemetry network (Taylor et al. 2017). Satellite and light-level geolocator tracking facilitate an understanding of hemispheric-scale and full-annual-cycle movements of birds. Radar surveillance networks (Gauthreaux and Belser 2003, Bauer et al. 2019) and community science initiatives (e.g., eBird; Sullivan et al. 2009) enable quantitative estimation of the flow of bird movements (and the causes and consequences of these movements) across continents and hemispheres.

Much like the way binoculars and scopes visually shed light on the behaviors of birds, these technological tools allow ornithologists to observe some aspects of birds' lives in new ways or with greater detail, and throughout their annual cycles. In turn, these new ways of "seeing" birds inspire ornithologists to develop new theories and test existing ones.

Examples of Movement Ecology's Contributions to a Broad Range of Ornithological Questions

Here, we review a sample of 14 broad questions in ornithology that have been informed by movement ecology approaches. The contributions we highlight span physiology, navigation, competition, conservation, and more. We include individual, collective, and macroecological approaches, and we favor examples that link observation of movement to other aspects of the bird's internal and external states or to long-standing ornithological questions, theories, or applications. We omit the movement of genes and we de-emphasize migratory connectivity studies because both of these topics are well-covered in Chapter 16 of this volume.

Physiological Capacity of the Individual

The physiological and energetics questions below were answered by a diverse array of scientists, including engineers, physiologists (one, Jessica Meir, who went on to become a NASA astronaut), avian ecologists, and medical doctors, demonstrating the creative and cross-disciplinary nature of movement ecology. In all cases, the ability to study physiological parameters of wild birds via biotelemetry techniques led to large advances in knowledge about the internal capacity of a bird to move. Our interpretation that avian physiological feats are remarkable may be calibrated by our own human abilities. It is increasingly apparent that much of what seems remarkable to us may be status quo in birds.

How Can Birds Fly in Hypoxic Conditions?

Birds possess many adaptations to meet the aerobic demands of flight—yet, how do birds that encounter extreme environments overcome added physiological challenges? For example, flight should become more taxing at high altitudes when low air pressure limits oxygen availability (hypoxia). The Bar-Headed Goose (*Anser indicus*) is renowned for its ability to migrate at high altitude over the Himalayan Mountains and the Tibetan Plateau (as high as 7,290 m; Hawkes et al. 2013). Physiological and morphological specializations in Bar-Headed Geese to cope with high altitude include larger lungs, high oxygen affinity in the blood via a specialized version of hemoglobin, extremely efficient breathing to improve oxygen uptake, and a greater abundance of oxidative fibers in the flight muscle (Scott et al. 2015). The first study to measure cardiorespiratory rates of Bar-Headed Geese flying in hypoxia showed that metabolic rate decreased (allowing the bird to conserve energy) but wing-beat frequency and heart rate remained unchanged from normoxia (Meir et al. 2019).

In addition, the birds may adjust their behavior to reduce their exposure to altitude. Movement ecology approaches enabled hypothesis testing in the wild about the relative importance of physiological versus behavioral adjustments. Tracking work led to the rejection of a hypothesis that geese use favorable winds to enhance their ability to climb (Hawkes et al. 2011). Geese carrying implanted loggers that measured heart rate, intra-abdominal temperature, and pressure during high-altitude migration recorded a narrow range of body temperature, suggesting the ability to thermoregulate in the cold temperatures of high altitude (Parr et al. 2019). However, Hawkes et al. (2013) showed that geese also use behavioral responses to mediate altitude's effects, including use of lower mountain passes (extreme altitudes were rare), stopovers to refuel and adjust water balance, or by flying at night when air density and oxygen availability are higher. Recent research on Great Reed Warblers (*Acrocephalus arundinaceus*; Sjöberg et al. 2021) and Great Snipes (*Gallinago media*; Lindström et al. 2021) biologgers equipped with barometric pressure sensors suggests that even small birds may make high-altitude flights more regularly than previously thought. The warblers and snipes reached maximum flight heights of 6,267 and 8,700 m above sea level, respectively.

Do Birds Sleep While Flying?

The ability of birds to sustain nonstop flight for days or even months led to competing hypotheses of the ability of birds to either sleep while flying or forgo sleep during nonstop long-distance movements (Rattenborg 2006). These hypotheses were evaluated with a "neurologger" that recorded an electroencephalogram paired with an accelerometer to measure if and when Great Frigatebirds slept while flying over ocean for 10 days without stopping (Rattenborg et al. 2016). The study confirmed that the birds did indeed sleep, albeit unihemispherically, while flying, although this comprised only approximately 3% of time spent flying. Sleep quality in flight was also inferior to sleep quality on land. Although seabirds provide many examples of sustained flight, tracking devices have also revealed the nearly 12,000-km nonstop flight of the Bar-Tailed Godwit across the Pacific Ocean (Gill et al. 2009), as well as the 314 continuous days spent aloft by swifts (Liechti et al. 2013). The growing list of impressive endurance flights awaits further confirmation of sleep in flight (Rattenborg 2006).

What Physiological Responses Occur During Diving?

The avian dive response includes multiple physiological reactions: a slowed heart rate (bradycardia), redistribution of blood toward essential organs, a reduced body temperature, and management of oxygen stores (C. Williams and Ponganis 2021). Using forced dive experiments in captivity, Scholander (1940) described multiple aspects of this response; however, research on free-ranging diving birds is relatively recent. The invention of the time depth recorder (TDR) and the creation of an ingenious experimental design in the wild (an artificial, isolated dive hole cut into the ice and covered with a heated hut) for studying physiological parameters of Weddell seals (*Leptonychotes weddellii*) (Kooyman 1965, 1966) "revolutionized marine organismal biology" (Goldbogen and Meir 2014, p. 167). Experimental studies on Emperor Penguins (*Aptenodytes forsteri*) soon followed, and TDRs were coupled with physiological equipment to record the avian dive response in the wild (Kooyman et al. 1971). A focus of many studies at the "Penguin Ranch," a field camp in Cape Washington, Antarctica, was to estimate aerobic dive limit (ADL)—the dive duration associated with the onset of post-dive blood lactate elevation (5.1 minutes for Emperor Penguins; Ponganis et al. 2010). An alternative interpretation of ADL is the time until all blood oxygen stores are depleted, estimated at a 23.1-minute dive. Emperor Penguins regularly dive between 5 and 12 minutes at the experimental dive hole and thus regularly exceed the traditional ADL estimate. A review of these decades of studies (Ponganis et al. 2010) indicates that diving Emperor Penguins regulate oxygen depletion through bradycardia as expected and by increasing swimming efficiency. In additional to physiology studies, TDRs have been regularly used since the 1990s to provide new insights on the foraging ecology of diving birds (for an early example, see T. Williams et al. 1992). Ornithologists continue to discover surprising information about bird diving. For example, a recent study of Black-Browed Albatross revealed "unexpectedly deep diving [19 m and 52 s]" by what is typically considered a surface forager (Guilford et al. 2022, p. R26). Dive shapes and the time of day suggested active propulsion and visual prey pursuit.

Social Interactions and Collective Movement

Understanding how social interactions influence movement decisions, and the consequences of these decisions, is a frontier of movement ecology (Kays et al. 2015). In the movement ecology paradigm, the external state of an animal, including a bird's social network, is a key component providing feedback to movement decision-making and ultimately to fitness (Nathan et al. 2008). Recent studies have used movement ecology approaches to answer questions about costs and benefits of sociality in birds using a range of techniques, including individual tracking, video loggers, high-speed photography, and even robotic predators.

What Is the Contribution of Innate Programming Versus Social Cues to Successful Navigation in Young Birds?

Studies comparing juveniles to adult birds have addressed questions surrounding the interplay between innate (genetically coded) and acquired (through learning and cultural transmission) knowledge on migration. In songbirds that migrate at night, there is good evidence for an inherited navigational program and additional learning over time (reviewed in Åkesson et al. 2021). In contrast, birds that migrate during the daytime over land may rely more on visual cues to navigate, such as topography or other birds. Translocation experiments in the mid-1900s (reviewed in Chernetsov et al. 2004) and a tracking study of White Storks (*Ciconia ciconia*) suggested that cultural transmission of migration information is critical for naive-to-migration soaring storks (Chernetsov et al. 2004).

A 7-year study of a reintroduced captive-bred population of Whooping Cranes (*Grus americana*) was able to disentangle the influence of genetic programming and social interactions on navigational accuracy for this species (Mueller et al. 2013). Genetic relatedness of all individuals was known, and individuals were monitored intensively during migration. Comparing successive relocations of individuals during migration, the researchers found no effect of genetic relatedness on navigational efficiency. Younger birds deviated from a straight line path more when migrating in groups without older birds and they progressively improved their migratory performance over years through social learning. Adult birds are generally known to be better navigators than young birds, largely through experience and memory. Åkesson et al. (2021) note that "the widespread use of tracking technology to study birds in different environments and geographical areas has largely contributed to our current knowledge of how migration performance improves with age" (p. 206).

How Do Birds Coordinate in Flocks?

Collective movement (Westley et al. 2018) is an integrative area of avian movement ecology involving physicists, computer scientists, engineers, and ornithologists. Flocking is a classic example of collective movement and has long been of interest to behavioral ecologists. Movement ecology studies may integrate empirical observations of bird movements with mathematical principles and information about social and environmental cues to better understand flock dynamics and their consequences. We present a few highlights from this large body of work, most of which tested long-held theories about bird flocking behavior.

Weimerskirch et al. (2001) implanted heart-rate loggers in Great White Pelicans (*Pelecanus onocrotalus*) and provided the first empirical confirmation of a model (Lissaman and Shollenberger 1970) that flying in V formation confers an energetic advantage. Using Global Positioning System (GPS) trackers and accelerometers, Portugal et al. (2014) subsequently confirmed for Northern Bald Ibis (*Geronticus eremita*) that the energy advantage is gained via trailing birds phasing the flapping of wings with the leading bird to maximize aerodynamic benefits of the updraft of air. A second ibis GPS study (Voelkl et al. 2015) evaluated the question of leadership (Lissaman and Shollenberger 1970) and Robert May's competing hypotheses about whether the lead bird would remain in position or switch places with other birds (May 1979). The authors showed that ibis cooperate in pairs by frequently switching in and out of the more arduous leading position. A fuller history of studies evaluating V formation in birds is presented by Portugal (2016).

In cluster flocks, GPS tracking coupled with statistical physics showed that Homing Pigeons (*Colomba livia*) have a well-defined hierarchy of influential leaders and followers in flight (Nagy et al. 2010), unlike the egalitarian assumptions of most simulation models. Further work suggested that multiple leaders could exist in a flock and that they tended to be exploratory birds that fly faster (Sasaki et al. 2018). Mixed-species flocks of shorebirds have been found to exhibit a cross between cluster flocking and V formation. In this "compound-V" formation (Corcoran and Hedrick 2019), the emergent property of the group appears to be a cluster; however, interactions between groups of individual members more closely resemble a V formation.

Using stereographic photography with three different points of view taken at a very high frame rate, Ballerini et al. (2008) collected the first large-scale three-dimensional data on bird flocks. For European Starlings, they revealed individual spacing rules (at least a wingspan distance from others), density rules within the flock (less dense in the center, more dense at the borders), and emergent properties of the flock as a whole (flocks were thinnest in the direction of gravity).

Is There an Optimal Group Size for Social Species?

Beyond group dynamics in migration, social dynamics are equally critical in more localized movements. For example, whereas V flocking formation confers an energetic advantage in migration, the social dynamics of more localized movements may confer an advantage of information, protection, and foraging efficiency. Vulturine Guineafowl (*Acryllium vulturinum*) are nonmigratory and highly social. They form stable, multilevel groups, and they exhibit little intra- or intergroup competition (Papageorgiou et al. 2019). This social structure is surprisingly complex for such a small-brained animal (compared to its body size). Using a combination of marked birds (97% of birds across 18 groups) and 58 GPS-tracked birds, Papageorgiou et al. (2019) described a surprisingly complex social structure for birds for the first time and showed that for Vulturine Guineafowl, it is shaped by ecological conditions (water, food, and roosting habitat availability during dry versus wet seasons). Papageorgiou and Farine (2020) further evaluated the theory that an optimal group size of Vulturine Guineafowl would maximize coordination among individuals and navigation accuracy, increase access to resources, and decrease exposure to predators. Using a GPS tracking study of 21 social groups of different sizes across five seasons, the authors predicted a nonlinear relationship between group size and benefit. The study found that groups of intermediate size (33–37 birds) were more exploratory, used a larger diversity of areas, had larger home ranges, and had more chicks than either smaller or larger groups, suggesting an optimality in collective behavior that provided these groups with a reproductive advantage. The authors indicated that "the ability to develop and test our novel hypotheses relating to collective movement of animals in the wild was only possible by conducting a large-scale multi-group tracking study."

Ecology

It is not possible to synthesize in this small section the monumental impact that movement ecology has had on the general field of avian ecology, including insights into breeding ecology; foraging ecology; and post-breeding ecology, including migration. Although studies of animal ecology often aim to describe average patterns or test the generality of overarching ecological theories, variation is considered a main currency of evolution (Darwin 1859, Lewontin 1970). Movement studies increasingly focus on variation within and across individuals and populations (Shaw 2020). Variability also underpins many topics in ornithology, including population dynamics (Sæther et al. 2004), migratory connectivity (Webster et al. 2002), and conservation (Horton et al. 2021). Below, we highlight how movement ecology approaches have been used to assess variability at multiple levels of biological organization, with a specific emphasis on migration behavior.

How Variable Are Movements Within an Individual?

Birds change their behaviors throughout their lives due to the specific constraints of each life history stage, environmental stochasticity, human impacts, or intra- and interspecific interactions. Across life history stages, much of our understanding of how, where, and why birds move comes from an understanding of the ecology of adults (for a review of this for seabirds, see Hazen et al. 2013), although this is changing. Movement ecology approaches have been used to study the ontogeny of foraging ecology, habitat selection, and migration spanning multiple avian taxa, including corvids (Loretto et al. 2016), shorebirds (Battley et al. 2020), raptors (Urios et al. 2007), and seabirds (Péron and Grémillet 2013, Weimerskirch et al. 2014). This literature shows that adult birds typically take more efficient migration routes and are more consistent in their movements across years than juveniles. However, birds can also deviate from their own established patterns, termed flexibility or plasticity. Understanding the causes and limits of intra-individual variability, and the ability of birds to respond to environmental change, are key conservation goals. Researchers have used accurate tracking methods to assess how individuals respond to rare and extreme events, including hurricanes (Wilkinson et al. 2019), a volcanic eruption (Alarcón et al. 2016), and fires (Overton et al. 2022). Perhaps the ultimate example of individual variability is demonstrated by facultative migration or migratory switching—individuals that, within a life history stage, migrate in one year but not in another. Evidence for rare switching behavior by individuals was found in a marked population of American Dippers (4 of 169 individuals; Gillis et al. 2008) and a male and female Red-Tailed Hawk tracked for 6 and 9 years, respectively (McCrary et al. 2019). A stable isotope study of Eurasian Skylarks showed that up to 45% of individuals in a Netherlands breeding population switched between migration and residency across years (Hegemann et al. 2015). This topic remains an open frontier that will benefit from further long-term studies of individuals.

How Variable Are Individuals Within Populations?

Improvements in miniaturization, price, and spatial and temporal resolution of tracking tags have led to increasingly large samples of individuals in movement ecology studies (Sequeira et al. 2019). Large sample sizes provide more statistical power to test the generality of ecological theories but have also helped illuminate the high level of individual variation and how such variation contributes to ecological processes such as migration. Many studies have explicitly evaluated the hypothesis that inter-annual differences in migration behavior (timing, speed, stops, route taken, etc.) are smaller within individuals than between them in fully migratory populations. Although individuals can adjust axes of migration behavior in response to experience and social and environmental cues (see above), intra-individual variability—even when high—is generally thought to be smaller than inter-individual variability. For some species, this hypothesis is supported from large tracking data sets—for example, Black-Tailed Godwit (Verhoeven et al. 2019), Common Tern (*Sterna hirundo*; Kürten et al. 2022), and Lesser Black-Backed Gull (*Larus fucus*; Brown et al. 2021). For other species, the relationship is either more complex—Wood Thrush (*Hylocichla mustelina*) differed in the level of individual repeatability of migration route according to season (Stanley et al. 2012)—or reversed—for Marsh Harriers (*Circus aeruginosus*), inter-annual variation in migration route was higher within individuals than across individuals, likely due to their use of a common migratory corridor (Vardanis et al. 2016). Perhaps the most striking example of variability within a population is the case of partially migratory populations, in which some individuals are resident and others migrate, which shapes population dynamics (Lundberg 2013, Acker et al. 2021).

How Variable Are Populations Within Species?

Population and species-level patterns of organization are emergent properties of individual variation. The study of avian movement ecology has expanded quickly from single-population studies to continental, hemispheric, oceanic, and even global syntheses that provide increasingly detailed understanding of population variability across space and time. Band re-encounter data showed that many species divide into population groupings that migrate together across hemispheres—Frederick Lincoln's flyway concept (Lincoln 1935). Movement data collected from multiple populations have further revealed whether separate breeding populations mix (Finch et al. 2017) or remain separated (Merkel et al. 2021) during the rest of their annual cycles (respectively weak/low versus strong/high migratory connectivity; Webster et al. 2002). Turkey vultures tracked in North America show how complex such patterns can be: Eastern birds never mixed with central or western birds, but central and western birds mixed during a portion of their annual cycles (Dodge et al. 2014). Connectivity patterns thus have important conservation implications, with some studies linking strong connectivity with land-use changes and population declines (Kramer et al. 2018, Patchett et al. 2018). Beyond description of pattern, Turbek et al. (2018) reviewed how "parallel advances in animal tracking and high-throughput genomic sequencing technology have opened previously intractable avenues of research on the evolutionary consequences of seasonal migration by allowing researchers to link migratory behavior to patterns of genetic exchange" (p. 164). They used case studies of migratory divides in three passerine species (and a salmon and eel species) to illustrate how the contribution of seasonal migration to reproductive isolation has been tested and remaining gaps in knowledge.

Macrosystems

During the past two decades, a surge of "big data" initiatives have revealed emergent properties of bird movement across regional and continental scales. Community science platforms such as eBird (Sullivan et al. 2009) and elevated access and renewed interest in remote sensing tools, such as weather surveillance radar data (Bauer et al. 2019), have led to innumerable insights into migratory phenomena. These data typically are collected at the assemblage or species level of biological organization, rather than the individual level, and are typically used to address process-based, ecological, and theoretical questions.

How Many Birds Fly Across North America and When?

Monitoring the volume and timing of assemblages of migratory birds is challenging, yet such information is critical for management and for assessing change over time. Combining eBird community science data and weather surveillance radar at the flyway scale, Horton et al. (2019) estimated that 2.1 million birds fly across the Gulf of Mexico each spring during a narrow 18-day window. No changes in numbers of birds migrating (2007–2015) or annual timing of peak migration (1995–2015) were detected, but earliest movements throughout the region advanced 1.6 days per decade. The highest advances occurred for large-bodied, shorter distance migrants. In addition to retrospective analysis, weather radar data can be useful for forecasting bird movements at a continental scale. Using 23 years of bird movement data collected by 143 weather radars, Van Doren and Horton (2018) related bird movements to atmospheric conditions to forecast bird migration across the United States. Their models explained 80% of variation in bird intensity and estimated that during peak passage, plumes of migratory birds exceed 500 million individuals per night. Accurate up to 7 days in advance, bird migration forecasts are now used in public awareness campaigns throughout the United States to switch lights off during forecasted high migration events. Using an integrated radar ecology study (Van Doren et al. 2017), artificial light at night has been shown to alter flow paths and vocalizations of birds.

What Are the Consequences of Different Migratory Strategies?

Movement ecology studies continue to test established ecological theories, including the trade-offs of long-distance migration. Migration is a risky behavior, and short-distance migrations are typically considered to be less risky (i.e., birds experience lower mortality rates during migration) than long-distance migrations (Newton

2007). However, in a study that challenged this paradigm, Dokter et al. (2018) used weather surveillance radar to compare biomass of birds passing a series of transects along the northern and southern borders of the United States in spring and autumn. They showed that North American birds migrating to Neotropics had higher return rates the following season than those that did not migrate south of the southern U.S. border (shorter distance migrants). The results of a subsequent meta-analysis of vital rates and ranges of 45 species of boreal-breeding birds (Winger and Pegan 2021) were consistent with Dokter et al.'s findings and suggested that birds making longer distance migrations to overwinter in "warmer, wetter, and greener" environments had higher survival.

How Will Migration Systems Respond to Global Change?

Community science schemes and weather surveillance radar networks have now amassed information across decades. The increasing temporal longevity of large-scale data sets is facilitating the study of how migration systems are shifting under global change. Equipped with more than 7 million community science records (eBird) encompassing 56 species across the eastern United States, Youngflesh et al. (2021) examined how changes in spring migrant arrival track annual spring plant phenology from 2002 to 2017. Migrants arrived earlier during seasons with earlier green-up, and interestingly, the correspondence was even stronger at higher latitudes, where phenology shifts tended to be strongest. Crossing over from species-specific insights to system assemblages, Horton et al. (2020) similarly examined how the timing of spring passage changed across 1995–2018 using weather surveillance radar data across the contiguous United States. Like the results of Youngflesh et al. (2021), Horton et al. found that spring migration timing incrementally advanced, and in parallel, shifts were more rapid at higher latitudes. Across these two vastly different data collection mechanisms, it is clear that migrants are cueing in to changing resources, whether indexed through changing plant phenology, temperature, or ultimately food resources, which still prove challenging to measure at macroscales. However, although shifts are clear, it remains in question whether these behavioral changes will allow migrants to effectively maintain populations and track looming changes brought on by warming global temperatures.

Conservation

Perhaps the largest promise of movement ecology is in its potential to inform conservation and management questions. Understanding how and when populations are limited, assessing the plasticity of birds, and identifying the negative impacts of humans and the positive impacts of our management interventions are all topics that can benefit from an understanding of avian movement ecology. Indeed, given that movement ecology studies often involve capture activities and equipping birds with devices (see "Critiques of Movement Ecology Approaches"), many believe that movement ecologists have an obligation to contribute movement data collected from birds to their conservation. The following examples include studies and initiatives that were motivated chiefly by the need to answer a conservation question.

When, Where, and Why Does Mortality Occur?

One of the first studies to link a movement path collected by satellite directly to an anthropogenic cause of bird mortality resulted in a chemical manufacturer committing to remove a toxic pesticide from the market in the bird's overwintering area (Line 1996). In the early 1990s, Swainson's Hawk populations were plummeting. After tracking two birds by satellite from California to Argentina in 1994 and 1995 (Woodbridge et al. 1995), researchers visited the mixed agriculture landscape in Argentina where most transmissions occurred. They counted more than 700 hawk carcasses, and in talking with landowners, it was inferred (and later confirmed; Goldstein et al. 1999) that the birds died of poisoning after consuming pesticide-laden grasshoppers considered a pest to farmers. Currently, multiple movement ecology studies have estimated bird survival rates, and sometimes anthropogenic versus natural causes of mortality, across the annual cycle (see, e.g., Klaassen et al. 2014, Senner et al. 2019, Buechley et al. 2021). Mortality rates are essential in management models and are needed to pinpoint when and where during the annual cycle populations are limited (DeSante et al. 2005; Yanco et al. 2025).

Large tracking devices may include sensors that are useful for distinguishing between mortality events and tag failure or loss. However, it can be difficult to infer mortality from many devices used on smaller species—the tags are often too small to have the battery capacity to record or transmit additional data. For archival tags, when birds fail to return to the deployment site, did the birds die or disperse? Improvements in tag miniaturization and sensor capabilities will be important to estimate this critical vital rate for tracked populations of smaller birds. Continuous estimates of migratory connectivity and risk exposure (Knight et al. 2021, McDuffie et al. 2022) can help further disentangle when and where during the annual cycle specific populations may be at highest risk.

Where Are the Important Places and What Are the Important Times to Protect or Manage?

It is challenging to protect species that spend much of their lives far away from where most humans can observe. Seabirds, for example, undertake expansive movements that cross hemispheres and numerous regulatory jurisdictions (Harrison et al. 2018, Beal et al. 2021). They were among the first birds tracked via satellite (Strikwerda et al. 1986, Jouventin and Weimerskirch 1990), and movement ecology studies revolutionized our knowledge of seabird foraging strategies, breeding season and year-round ranges, and age- and sex-based differences (Burger and Shaffer 2008). These data have also been critical for conservation (Burger and Shaffer 2008, Hays et al. 2019). BirdLife International led an early collaboration called "Tracking Ocean Wanderers" to synthesize multispecies movement data at a global level to assess potential for exposure to human impacts. A workshop in 2003 (BirdLife International 2004) assembled seabird scientists in response to an urgent conservation issue—incidental catch of procellariform seabirds in long-line fisheries (Brothers 1991). Seabirds are long-lived species with low reproductive rates; loss of adults at sea led to worrying population declines at multiple colonies and for multiple species (Weimerskirch and Jouventin 1987, Weimerskirch et al. 1997). Conservation organizations lacked the necessary data for most oceans regarding seabird movements on the high seas.

Scientists at this workshop contributed tracking data for the purpose of overlaying distributions of procellariforms on maps of fishing effort globally to identify the important places for seabirds, how they are connected, and where birds may be at highest risk of being caught in fisheries.

With movement data in hand, researchers guided changes in fisheries management. Tracking data from albatrosses and petrels in the Southern Ocean indicated that highest overlap with Patagonian toothfish (*Dissostichus eleginoides*) likely occurred during the breeding period (Croxall and Prince 1996), and this evidence led to a seasonal closure of the fishery and a 10-fold reduction in seabird bycatch rates (Croxall and Nicol 2004). From these early examples of conservation success and data sharing, the first global tracking data repository (the Seabird Tracking Database) was developed and led to many conservation successes (Hays et al. 2016). Movement data were recently used to identify a large seabird hot spot in the north Atlantic Ocean (Davies et al. 2021) and served as major evidence leading to the designation of this area as a new marine protected area in the high seas. In a turn of events, seabirds themselves are now serving as fisheries observers through GPS devices that detect ship radar and can help identify incidences of illegal fishing (Weimerskirch et al. 2020).

Critiques of Movement Ecology Approaches

The Field Is Primarily Descriptive

An indication of the field's maturation is the rapidity with which descriptive tracking studies of individual birds were first published in high-impact journals (Jouventin and Weimerskirch 1990) and now are recommended by editors more frequently for taxa-specific journals (McKinnon and Love 2018). Even amazing journeys recorded for the first time may be dismissed as "just another tracking paper/talk."

Description is important to hypothesis development and should not be undervalued. Moreover, knowledge of bird movements is still lacking for many species, is critical for conservation, and is valuable for generating public excitement about birds. However, as the examples provided in this chapter demonstrate, the true promise of movement ecology for answering major questions in ornithology is in its integrative approach. There is likely still a need to better fulfill this promise (Joo et al. 2022).

Results Are Not Generalizable or Scalable

This critique is applicable at two levels. First, in individual tracking studies, sample sizes may sometimes be too small or of too short a duration to be representative of populations, species, or inter-annual variability (Sequeira et al. 2019). Second, in macrosystem studies, either species are unknown in radar studies or population assignments and connectivity are unknown in broad-scale syntheses of community science data.

These are important considerations when seeking to make ecological or management-relevant inferences from movement studies. For biologging studies, we discuss below ways in which movement ecology can move forward to deal with the seemingly opposing demands of needing more statistical power for greater inference while at the same time reducing impact to the birds. For macrosystem studies, we hope the examples in this chapter show that big data approaches have the power to illuminate features of assemblages at the continental and hemispheric scale (Kelly et al. 2016). Approaches that integrate species-level data sets from community science initiatives such as eBird, the Audubon Christmas Bird Count, or the North American Breeding Bird Survey with assemblage-level data sets from radar surveillance networks can be especially successful at testing whether macrosystem trends scale appropriately from species to system.

The Field Has Been Driven by the Tool Rather Than the Question

This is a fair critique and a side effect of the rise of technological approaches, especially regarding the rapid increase of tracking studies of individual birds, but perhaps also for techniques such as radar ornithology or video loggers. Some researchers may enter the field simply out of the desire to track birds. Technology is enticing and fun, and birds have some of the most exciting movements on Earth. Brand new discoveries about the life histories of birds are possible in almost every new study.

In reviewing the shear breadth of questions in ornithology addressed in recent years, it is clear, however, that some questions could only have been answered by following individual birds throughout their full annual cycle, and some questions were sitting for decades unanswered only to be resolved through a biologging revolution. Important questions and hypotheses are not in limited supply. So for an ornithologist thinking introspectively, the internal battle becomes, When does knowing the answer to a scientific question outweigh the impact of a research technique to the bird, population, or species?

Techniques Used by Movement Ecologists Can Cause Impacts to Birds

Ethical considerations of capturing a bird and attaching or implanting a device were raised early in this developing field (Cochran 1972, Caccamise and Hedin 1985) and are especially important to consider when working with individuals from small or declining populations. As the use of such devices expanded, multiple studies addressed the question of nonlethal behavioral effects and lethal effects of biologging devices, often with a focus on the weight or shape of the device (Phillips et al. 2003, Barron et al. 2010, Bodey et al. 2018). Studies have also evaluated the specific effects or efficacy of attachment methods that may include glue, sutures, surgical implantation, harnesses, leg bands, patygeal clips, and collars (e.g., see the review for waterfowl by Lameris and Kleyheeg 2017).

The decision to equip a bird with a device requires careful ethical deliberation; specific permitting and approval by relevant authorities (including approvals for species listed on protection acts such as the U.S. Endangered Species Act); an assessment of the training needs and abilities of the person(s) attaching a device to a bird; and, when possible, control groups to assess affects. Additional questions that should be considered include (1) whether the information needed can only be gathered through such technology and whether the device size can be minimized to obtain that information, (2) the minimum number of individuals needed to answer the research question, (3) whether it may be duplicative of information known through non-invasive methods, and (4) whether the data collected can be shared with others for additional uses to ensure that the scientific and conservation benefits of the work are maximized. This last point includes recognizing the importance of consistent data and metadata management practices, including archiving in centralized repositories, so the data are truly available to be used and interpreted appropriately.

A promising trend in the use of biologging technology to measure bird movements is the reduction in size of attached devices and engineering improvements to small tags to increase spatial and temporal sampling resolution and extend battery life. There continues to be a need for more opportunities for training in tag attachment and for additional studies to assess and improve upon attachment methods and to evaluate tag effects.

With the increase and expanded development of avian tracking programs, some surprising trade-offs in environmental concerns may occur. For example, automated radio telemetry networks support the use of tiny tags that can help minimize effects to individuals, but they also require receiving stations to be constructed on the landscape, sometimes in pristine areas. Some researchers see opportunity in tag miniaturization and reduction in costs to increase sample size and to incorporate birds into the "Internet of Things" concept, where everyday items such as doorbells, watches, or bicycles are equipped for sensing, processing, or exchanging data through communication networks (Whitmore et al. 2015). An "Internet of Animals" has the potential to improve scientific inference from tracking studies and could benefit other areas of study, including Earth observation (Curry 2018). However, the approach often recommended by animal care committees is to use only the minimum sample needed, and the desire to equip ever-increasing numbers of birds with tracking devices may conflict with that recommendation.

How Movement Ecology Can Continue to Impact the Field of Ornithology

The movement of birds from patch to patch or hemisphere to hemisphere provides connections that permeate throughout ecosystems and ripple up and down trophic levels. As Skip Ambrose, U.S. Fish and Wildlife Service endangered species biologist, said in 1996 in *The New York Times*, "In a few years, we already have more wintering locations [of Peregrine Falcon, *Falco peregrinus*] and know more about their migration routes than in the previous 20 years of work" (Yoon 1996). The breadth of ornithological topics addressed using movement ecology approaches is astounding; the field has rapidly impacted every ornithological subdiscipline. Throughout this chapter, we have demonstrated the role of movement ecology in advancing these subdisciplines, including physiology, social interactions, ecological impacts, and conservation. By no means are our selections comprehensive in scope but, rather, work to highlight the rapid impacts of the growing field of movement ecology on ornithology. These discoveries, articulated on a backbone of technological advances (see Figure 4.1), continue to amaze our sense of biological limits and what is "normal" bird behavior.

Movement ecology research will continue to advance a deeper understanding of birds. Data sets grow and technologies advance and diversify. Free repositories, processing systems, and modeling approaches facilitate analysis (Joo et al. 2020, Kays et al. 2022). However, frontiers remain. Young life history stages and small birds remain understudied (Hazen et al. 2013, McKinnon and Love 2018). There remains a balance between the need to conduct studies with large enough sample sizes to observe an effect and the need to minimize impacts of scientific research on birds. An open frontier is to advance statistical methods that can better integrate individual-level movement data with observational data and macro-systems-level research. Finally, despite the contributions of many studies to improve conservation decision-making, most movement data sets are not yet publicly shared (Kays et al. 2022) or are not easily analyzed by downstream users.

Movement ecology moves forward in the face of rapid global change (Intergovernmental Panel on Climate Change 2023), including habitat loss, increased anthropogenic threats, and shifting climates: It remains to be fully seen how patterns of avian movement will be altered. Integrating movement ecology with deep time approaches has provided hypotheses for how birds may respond in the future (Thorup et al. 2021). Changes are already underway (Horton et al. 2020), further stretching the limits of movements and necessitating the need for integrative studies in avian movement ecology across temporal, spatial, and taxonomic scales.

References

Acker, P., F. Daunt, S. Wanless, S. J. Burthe, M. A. Newell, M. P. Harris, H. Grist, J. Sturgeon, R. L. Swann, C. Gunn, A. Payo-Payo, and J. M. Reid. 2021. Strong survival selection on seasonal migration versus residence induced by extreme climatic events. *Journal of Animal Ecology* 90:796–808.

Åkesson, S., H. Bakam, E. Martinez Hernandez, M. Ilieva, and G. Bianco. 2021. Migratory orientation in inexperienced and experienced avian migrants. *Ethology Ecology & Evolution* 33:206–229.

Alarcón, P. A. E., S. A. Lambertucci, J. A. Donázar, F. Hiraldo, J. A. Sánchez-Zapata, G. Blanco, and J. M. Morales. 2016. Movement decisions in natural catastrophes: How a flying scavenger deals with a volcanic eruption. *Behavioral Ecology* 27:75–82.

Ballerini, M., N. Cabibbo, R. Candelier, A. Cavagna, E. Cisbani, I. Giardina, V. Lecomte, A. Orlandi, G. Parisi, A. Procaccini, M. Viale, and V. Zdravkovic. 2008. Interaction ruling animal collective behavior depends on topological rather than metric distance: Evidence from a field study. *Proceedings of the National Academy of Sciences of the USA* 105:1232–1237.

Ballerini, M., N. Cabibbo, R. Candelier, A. Cavagna, E. Cisbani, I. Giardina, A. Orlandi, G. Parisi, A. Procaccini, and M. Viale. 2008. An empirical study of large, naturally occurring starling flocks: A benchmark in collective animal behaviour. *Animal Behaviour* 76:201–215.

Barron, D. G., J. D. Brawn, and P. J. Weatherhead. 2010. Meta-analysis of transmitter effects on avian behaviour and ecology. *Methods in Ecology and Evolution* 1:180–187.

Battley, P. F., J. R. Conklin, Á. M. Parody-Merino, P. A. Langlands, I. Southey, T. Burns, D. S. Melville, R. Schuckard, A. C. Riegen, and M. A. Potter. 2020. Interacting roles of breeding geography and early-life settlement in godwit migration timing. *Frontiers in Ecology and Evolution* 8: Article 52.

Bauer, S., J. Shamoun-Baranes, C. Nilsson, A. Farnsworth, J. F. Kelly, D. R. Reynolds, A. M. Dokter, J. F. Krauel, L. B. Petterson, K. G. Horton, and J. W. Chapman. 2019. The grand challenges of migration ecology that radar aeroecology can help answer. *Ecography* 42:861–875.

Beal, M., M. P. Dias, R. A. Phillips, S. Oppel, C. Hazin, E. J. Pearmain, J. Adams, D. J. Anderson, M. Antolos, J. A. Arata, J. M. Arcos, J. P. Y. Arnould, J. Awkerman, E. Bell, M. Bell, M. Carey, R. Carle, T. A. Clay, J. Cleeland, V. Colodro, M. Conners, M. Cruz-Flores, R. Cuthbert, K. Delord, L. Deppe, B. J. Dilley, H. Dinis, G. Elliott, F. De Felipe, J. Felis, M. G. Forero, A. Freeman, A. Fukuda, J. González-Solís, J. P. Granadeiro, A. Hedd, P. Hodum, J. M. Igual, A. Jaeger, T. J. Landers, M. Le Corre, A. Makhado, B. Metzger, T. Militão, W. A. Montevecchi, V. Morera-Pujol, L. Navarro-Herrero, D. Nel, D. Nicholls, D. Oro, R. Ouni, K. Ozaki, F. Quintana, R. Ramos, T. Reid, J. M. Reyes-González, C. Robertson, G. Robertson, M. S. Romdhane, P. G. Ryan, P. Sagar, F. Sato, S. Schoombie, R. P. Scofield, S. A. Shaffer, N. J. Shah, K. L. Stevens, C. Surman, R. M. Suryan, A. Takahashi, V. Tatayah, G. Taylor, D. R. Thompson, L. Torres, K. Walker, R. Wanless, S. M. Waugh, H. Weimerskirch, T. Yamamoto, Z. Zajkova, L. Zango, and P. Catry. 2021. Global political responsibility for the conservation of albatrosses and large petrels. *Science Advances* 7: Article eabd7225.

BirdLife International. 2004. *Tracking Ocean Wanderers: The Global Distribution of Albatrosses and Petrels*. BirdLife International, Cambridge, UK.

Bodey, T. W., I. R. Cleasby, F. Bell, N. Parr, A. Schultz, S. C. Votier, and S. Bearhop. 2018. A phylogenetically controlled meta-analysis of biologging device effects on birds: Deleterious effects and a call for more standardized reporting of study data. *Methods in Ecology and Evolution* 9:946–955.

Bridge, E. S., K. Thorup, M. S. Bowlin, P. B. Chilson, R. H. Diehl, R. W. Fléron, P. Hartl, R. Kays, J. F. Kelly, W. D. Robinson, and M. Wikelski. 2011. Technology on the move: Recent and forthcoming innovations for tracking migratory birds. *BioScience* 61:689–698.

Brothers, N. 1991. Albatross mortality and associated bait loss in the Japanese longline fishery in the Southern Ocean. *Biological Conservation* 55:255–268.

Brown, J. M., E. E. van Loon, W. Bouten, K. C. J. Camphuysen, L. Lens, W. Müller, C. B. Thaxter, and J. Shamoun-Baranes. 2021. Long-distance migrants vary migratory behaviour as much as short-distance migrants: An individual-level comparison from a seabird species with diverse migration strategies. *Journal of Animal Ecology* 90:1058–1070.

Brush, A. H., and G. A. Clark. 1983. *Perspectives in Ornithology: Essays Presented for the Centennial of the American Ornithologists' Union*. Cambridge University Press, Cambridge, UK.

Buechley, E. R., S. Oppel, R. Efrat, W. L. Phipps, I. Carbonell Alanís, E. Álvarez, A. Andreotti, V. Arkumarev, O. Berger-Tal, A. Bermejo Bermejo, A. Bounas, G. Ceccolini, A. Cenerini, V. Dobrev, O. Duriez, J. García, C. García-Ripollés, M. Galán, A. Gil, L. Giraud, O. Hatzofe, J. J. Iglesias-Lebrija, I. Karyakin, E. Kobierzycki, E. Kret, F. Loercher, P. López-López, Y. Miller, T. Mueller, S. C. Nikolov, J. de la Puente, N. Sapir, V. Saravia, Ç. H. Şekercioğlu, T. S. Sillett, J. Tavares, V. Urios, and P. P. Marra. 2021. Differential survival throughout the full annual cycle of a migratory bird presents a life-history trade-off. *Journal of Animal Ecology* 90:1228–1238.

Burger, A. E., and S. A. Shaffer. 2008. Application of tracking and data-logging technology in research and conservation of seabirds. *The Auk* 125:253–264.

Caccamise, D. F., and R. S. Hedin. 1985. An aerodynamic basis for selecting transmitter loads in birds. *The Wilson Bulletin* 97:306–318.

Chernetsov, N., P. Berthold, and U. Querner. 2004. Migratory orientation of first-year white storks (*Ciconia ciconia*): Inherited information and social interactions. *Journal of Experimental Biology* 207:937–943.

Cochran, W. W. 1972. Long distance tracking of birds. Pages 39–59 in SR Galler, K Schmidt-Koenig, GJ Jacobs, and RE Belleville, eds. *Animal Orientation and Navigation*. NASA SP 262. U.S. Government Printing Office, Washington, DC.

Cochran, W. W., G. G. Montgomery, and R. R. Graber. 1967. Migratory flights of *Hylocichla* thrushes in spring: A radiotelemetry study. *Living Bird* 6:25.

Corcoran, A. J., and T. L. Hedrick. 2019. Compound-V formations in shorebird flocks. *Elife* 8: Article e45071.

Croxall, J. P., and S. Nicol. 2004. Management of Southern Ocean fisheries: Global forces and future sustainability. *Antarctic Science* 16:569–584.

Croxall, J. P., and P. A. Prince. 1996. Potential interactions between wandering albatrosses and longline fisheries for Patagonian toothfish at South Georgia. *CCAMLR Science* 3: 101–110.

Curry, A. 2018. The internet of animals that could help to save vanishing wildlife. *Nature* 562:322–326.

Darwin, C. 1859. *On the Origin of Species by Means of Natural Selection, Or The Preservation of Favoured Races in the Struggle for Life*. Murray, London.

Davies, T. E., A. P. B. Carneiro, M. Tarzia, E. Wakefield, J. C. Hennicke, M. Frederiksen, E. S. Hansen, B. Campos, C. Hazin, and B. Lascelles. 2021. Multispecies tracking reveals a major seabird hotspot in the North Atlantic. *Conservation Letters* 14: Article e12824.

DeSante, D. F., M. P. Nott, and D. R. Kaschube. 2005. Monitoring, modeling, and management: Why base avian management on vital rates and how should it be done. Pages 795–804 in *Bird Conservation Implementation and Integration in the Americas: Proceedings of the Third International Partners in Flight Conference. 2002 March 20–24; Asilomar, California* (Vol. 2 Gen. Tech. Rep. PSW-GTR-191). U.S. Department of Agriculture, Forest Service, Pacific Southwest Research Station, Albany, CA.

Dial, K. P. 2003. Evolution of avian locomotion: Correlates of flight style, locomotor modules, nesting biology, body size, development, and the origin of flapping flight. *The Auk* 120:941–952.

Dodge, S., G. Bohrer, K. Bildstein, S. C. Davidson, R. Weinzierl, M. J. Bechard, D. Barber, R. Kays, D. Brandes, J. Han, and M. Wikelski. 2014. Environmental drivers of variability in the movement ecology of turkey vultures (*Cathartes aura*) in North and South America. *Philosophical Transactions of the Royal Society B: Biological Sciences* 369: Article 20130195.

Dokter, A. M., A. Farnsworth, D. Fink, V. Ruiz-Gutierrez, W. M. Hochachka, F. A. La Sorte, O. J. Robinson, K. V. Rosenberg, and S. Kelling. 2018. Seasonal abundance and survival of North America's migratory avifauna determined by weather radar. *Nature Ecology & Evolution* 2:1603–1609.

Egevang, C., I. J. Stenhouse, R. A. Phillips, A. Petersen, J. W. Fox, and J. R. D. Silk. 2010. Tracking of Arctic terns *Sterna paradisaea* reveals longest animal migration. *Proceedings of the National Academy of Sciences of the USA* 107:2078–2081.

Finch, T., S. J. Butler, A. M. A. Franco, and W. Cresswell. 2017. Low migratory connectivity is common in long-distance migrant birds. *Journal of Animal Ecology* 86:662–673.

Fish, F. E. 2016. Secondary evolution of aquatic propulsion in higher vertebrates: Validation and prospect. *Integrative and Comparative Biology* 56:1285–1297.

Gaspar, J., G. C. Gibb, and S. A. Trewick. 2020. Convergent morphological responses to loss of flight in rails (Aves: Rallidae). *Ecology and Evolution* 10:6186–6207.

Gauthreaux, S. A., Jr., and C. G. Belser. 2003. Radar ornithology and biological conservation. *The Auk* 120:266–277.

Gill, R. E., T. L. Tibbitts, D. C. Douglas, C. M. Handel, D. M. Mulcahy, J. C. Gottschalck, N. Warnock, B. J. McCaffery, P. F. Battley, and T. Piersma. 2009. Extreme endurance flights by landbirds crossing the Pacific Ocean: Ecological corridor rather than barrier. *Proceedings of the Royal Society B: Biological Sciences* 276:447–457.

Gillis, E. A., D. J. Green, H. A. Middleton, and C. A. Morrissey. 2008. Life history correlates of alternative migratory strategies in American Dippers. *Ecology* 89:1687–1695.

Goldbogen, J. A., and J. U. Meir. 2014. The device that revolutionized marine organismal biology. *Journal of Experimental Biology* 217:167–168.

Goldstein, M. I., T. E. Lacher, B. Woodbridge, M. J. Bechard, S. B. Canavelli, M. E. Zaccagnini, G. P. Cobb, E. J. Scollon, R. Tribolet, and M. J. Hopper. 1999. Monocrotophos-induced mass mortality of Swainson's Hawks in Argentina, 1995–96. *Ecotoxicology* 8:201–214.

Greenewalt, C. H. 1960. The wings of insects and birds as mechanical oscillators. *Proceedings of the American Philosophical Society* 104:605–611.

Guilford, T., O. Padget, L. Maurice, and P. Catry. 2022. Unexpectedly deep diving in an albatross. *Current Biology* 32:R26–R28.

Harrison, A.-L., D. P. Costa, A. J. Winship, S. R. Benson, S. J. Bograd, M. Antolos, A. B. Carlisle, H. Dewar, P. H. Dutton, S. J. Jorgensen, S. Kohin, B. R. Mate, P. W. Robinson, K. M. Schaefer, S. A. Shaffer, G. L. Shillinger, S. E. Simmons, K. C. Weng, K. M. Gjerde, and B. A. Block. 2018. The political biogeography of migratory marine predators. *Nature Ecology & Evolution* 2:1571–1578.

Hartfelder, J., C. Reynolds, R. A. Stanton, M. Sibiya, A. Monadjem, R. A. McCleery, and R. J. Fletcher. 2020. The allometry of movement predicts the connectivity of communities. *Proceedings of the National Academy of Sciences of the USA* 117:22274–22280.

Hawkes, L. A., S. Balachandran, N. Batbayar, P. J. Butler, B. Chua, D. C. Douglas, P. B. Frappell, Y. Hou, W. K. Milsom, S. H. Newman, D. J. Prosser, P. Sathiyaselvam, G. R. Scott, J. Y. Takekawa, T. Natsagdorj, M. Wikelski, M. J. Witt, B. Yan, and C. M. Bishop. 2013. The paradox of extreme high-altitude migration in bar-headed geese *Anser indicus*. *Proceedings of the Royal Society B: Biological Sciences* 280: Article 20122114.

Hawkes, L. A., S. Balachandran, N. Batbayar, P. J. Butler, P. B. Frappell, W. K. Milsom, N. Tseveenmyadag, S. H. Newman, G. R. Scott, P. Sathiyaselvam, J. Y. Takekawa, M. Wikelski, and C. M. Bishop. 2011. The trans-Himalayan flights of bar-headed geese (*Anser indicus*). *Proceedings of the National Academy of Sciences of the USA* 108:9516–9519.

Hays, G. C., H. Bailey, S. J. Bograd, W. D. Bowen, C. Campagna, R. H. Carmichael, P. Casale, A. Chiaradia, D. P. Costa, E. Cuevas, P. J. N. D. Bruyn, M. P. Dias, C. M. Duarte, D. C. Dunn, P. H. Dutton, N. Esteban, A. Friedlaender, K. T. Goetz, B. J. Godley, P. N. Halpin, M. Hamann, N. Hammerschlag, R. Harcourt, A.-L. Harrison, E. L. Hazen, M. R. Heupel, E. Hoyt, N. E. Humphries, C. Y. Kot, J. S. E. Lea, H. Marsh, S. M. Maxwell, C. R. McMahon, G. N. D. Sciara, D. M. Palacios, R. A. Phillips, D. Righton, G. Schofield, J. A. Seminoff, C. A. Simpfendorfer, D. W. Sims, A. Takahashi, M. J. Tetley, M. Thums, P. N. Trathan, S. Villegas-Amtmann, R. S. Wells, S. D. Whiting, N. E. Wildermann, and A. M. M. Sequeira. 2019. Translating marine animal tracking data into conservation policy and management. *Trends in Ecology & Evolution* 34:459–472.

Hays, G. C., L. C. Ferreira, A. M. M. Sequeira, M. G. Meekan, C. M. Duarte, H. Bailey, F. Bailleul, W. D. Bowen, M. J. Caley, D. P. Costa, V. M. Eguíluz, S. Fossette, A. S. Friedlaender, N. Gales, A. C. Gleiss, J. Gunn, R. Harcourt, E. L. Hazen, M. R. Heithaus, M. Heupel, K. Holland, M. Horning, I. Jonsen, G. L. Kooyman, C. G. Lowe, P. T. Madsen, H. Marsh, R. A. Phillips, D. Righton, Y. Ropert-Coudert, K. Sato, S. A. Shaffer, C. A. Simpfendorfer, D. W. Sims, G. Skomal, A. Takahashi, P. N. Trathan, M. Wikelski, J. N. Womble, and M. Thums. 2016. Key questions in marine megafauna movement ecology. *Trends in Ecology & Evolution* 31:463–475.

Hazen, E. L., S. Jorgensen, R. R. Rykaczewski, S. J. Bograd, D. G. Foley, I. D. Jonsen, S. A. Shaffer, J. P. Dunne, D. P. Costa, L. B. Crowder, and B. A. Block. 2013. Predicted habitat shifts of Pacific top predators in a changing climate. *Nature Climate Change* 3:234–238.

Hedenström, A., G. Norevik, K. Warfvinge, A. Andersson, J. Bäckman, and S. Åkesson. 2016. Annual 10-month aerial life phase in the common swift *Apus apus*. *Current Biology* 26:3066–3070.

Heers, A. M. 2018. Flight and locomotion. Pages 273–308 in M. L. Morrison, A. D. Rodewald, G. Voelker, M. R. Colón, and J. F. Prather, eds. *Ornithology: Foundation, Analysis, and Application.* Johns Hopkins University Press, Baltimore, MD.

Heers, A. M., and K. P. Dial. 2015. Wings versus legs in the avian bauplan: Development and evolution of alternative locomotor strategies. *Evolution* 69:305–320.

Hegemann, A., P. P. Marra, and B. I. Tieleman. 2015. Causes and consequences of partial migration in a passerine bird. *The American Naturalist* 186:531–546.

Hoose, P. 2012. *Moonbird: A Year on the Wind With the Great Survivor B95*. Macmillan, New York.

Horton, K. G., F. A. La Sorte, D. Sheldon, T.-Y. Lin, K. Winner, G. Bernstein, S. Maji, W. M. Hochachka, and A. Farnsworth. 2020. Phenology of nocturnal avian migration has shifted at the continental scale. *Nature Climate Change* 10:63–68.

Horton, K. G., B. M. Van Doren, H. J. Albers, A. Farnsworth, and D. Sheldon. 2021. Near-term ecological forecasting for dynamic aeroconservation of migratory birds. *Conservation Biology* 35:1777–1786.

Horton, K. G., B. M. Van Doren, F. A. La Sorte, E. B. Cohen, H. L. Clipp, J. J. Buler, D. Fink, J. F. Kelly, and A. Farnsworth. 2019. Holding steady: Little change in intensity or timing of bird migration over the Gulf of Mexico. *Global Change Biology* 25:1106–1118.

Intergovernmental Panel on Climate Change. 2023. Climate change 2023: Synthesis report. Contribution of Working Groups I, II and III to the Sixth Assessment Report of the Intergovernmental Panel on Climate Change. Intergovernmental Panel on Climate Change, Geneva, Switzerland. https://www.ipcc.ch/report/ar6/synthesis-report/

Joo, R., M. E. Boone, T. A. Clay, S. C. Patrick, S. Clusella-Trullas, and M. Basille. 2020. Navigating through the r packages for movement. *Journal of Animal Ecology* 89:248–267.

Joo, R., S. Picardi, M. E. Boone, T. A. Clay, S. C. Patrick, V. S. Romero-Romero, and M. Basille. 2022. Recent trends in movement ecology of animals and human mobility. *Movement Ecology* 10: Article 26.

Jouventin, P., and H. Weimerskirch. 1990. Satellite tracking of wandering albatrosses. *Nature* 343:746–748.

Kays, R., M. C. Crofoot, W. Jetz, and M. Wikelski. 2015. Terrestrial animal tracking as an eye on life and planet. *Science* 348: Article aaa2478.

Kays, R., S. C. Davidson, M. Berger, G. Bohrer, W. Fiedler, A. Flack, J. Hirt, C. Hahn, D. Gauggel, B. Russell, A. Kölzsch, A. Lohr, J. Partecke, M. Quetting, K. Safi, A. Scharf, G. Schneider, I. Lang, F. Schaeuffelhut, M. Landwehr, M. Storhas, L. Schalkwyk, C. Vinciguerra, R. Weinzierl, and M. Wikelski. 2022. The Movebank system for studying global animal movement and demography. *Methods in Ecology and Evolution* 13:419–431.

Kelly, J. F., K. G. Horton, and M. Sykes. 2016. Toward a predictive macrosystems framework for migration ecology. *Global Ecology and Biogeography* 25:1159–1165.

Klaassen, R. H. G., M. Hake, R. Strandberg, B. J. Koks, C. Trierweiler, K. M. Exo, F. Bairlein, and T. Alerstam. 2014. When and where does mortality occur in migratory birds? Direct evidence from long-term satellite tracking of raptors. *Journal of Animal Ecology* 83: 176–184.

Knight, E. C., A.-L. Harrison, A. L. Scarpignato, S. L. Van Wilgenburg, E. M. Bayne, J. W. Ng, E. Angell, R. Bowman, R. M. Brigham, B. Drolet, W. E. Easton, T. R. Forrester, J. T. Foster, S. Haché, K. C. Hannah, K. G. Hick, J. Ibarzabal, T. L. Imlay, S. A. Mackenzie, A. Marsh, L. P. McGuire, G. N. Newberry, D. Newstead, A. Sidler, P. H. Sinclair, J. L. Stephens, D. L. Swanson, J. A. Tremblay, and P. P. Marra. 2021. Comprehensive estimation of spatial and temporal migratory connectivity across the annual cycle to direct conservation efforts. *Ecography* 44:665–679.

Kooyman, G. L. 1965. Techniques used in measuring diving capacities of Weddell seals. *Polar Record* 12:391–394.

Kooyman, G. L. 1966. Maximum diving capacities of the Weddell seal, *Leptonychotes weddelli*. *Science* 151:1553–1554.

Kooyman, G. L., C. M. Drabek, R. Elsner, and W. B. Campbell. 1971. Diving behavior of the emperor penguin, *Aptenodytes forsteri*. *The Auk* 88:775–795.

Kramer, G. R., D. E. Andersen, D. A. Buehler, P. B. Wood, S. M. Peterson, J. A. Lehman, K. R. Aldinger, L. P. Bulluck, S. Harding, J. A. Jones, J. P. Loegering, C. Smalling, R. Vallender, and H. M. Streby. 2018. Population trends in *Vermivora* warblers are linked to strong migratory connectivity. *Proceedings of the National Academy of Sciences of the USA* 115:E3192–E3200.

Kürten, N., H. Schmaljohann, C. Bichet, B. Haest, O. Vedder, J. González-Solís, and S. Bouwhuis. 2022. High individual repeatability of the migratory behaviour of a long-distance migratory seabird. *Movement Ecology* 10: Article 5.

Lack, D., and G. C. Varley. 1945. Detection of birds by radar. *Nature* 156:446–446.

Lameris, T. K., and E. Kleyheeg. 2017. Reduction in adverse effects of tracking devices on waterfowl requires better measuring and reporting. *Animal Biotelemetry* 5:1–14.

Lewontin, R. C. 1970. The units of selection. *Annual Review of Ecology and Systematics* 1:1–18.

Liechti, F., W. Witvliet, R. Weber, and E. Bächler. 2013. First evidence of a 200-day non-stop flight in a bird. *Nature Communications* 4: Article 2554.

Lincoln, F. C. 1935. *The Waterfowl Flyways of North America.* U.S. Department of Agriculture, Washington, DC.

Lindström, Å., T. Alerstam, A. Andersson, J. Bäckman, P. Bahlenberg, R. Bom, R. Ekblom, R. H. G. Klaassen, M. Korniluk, S. Sjöberg, and J. K. M. Weber. 2021. Extreme altitude changes between night and day during marathon flights of great snipes. *Current Biology* 31:3433–3439.

Line, L. 1996. Accord is reached to recall pesticide devastating hawk. *The New York Times*: C4.

Lissaman, P. B., and C. A. Shollenberger. 1970. Formation flight of birds. *Science* 168:1003–1005.

Loretto, M.-C., S. Reimann, R. Schuster, D. M. Graulich, and T. Bugnyar. 2016. Shared space, individually used: Spatial behaviour of non-breeding ravens (*Corvus corax*) close to a permanent anthropogenic food source. *Journal of Ornithology* 157:439–450.

Lundberg, P. 2013. On the evolutionary stability of partial migration. *Journal of Theoretical Biology* 321:36–39.

May, R. M. 1979. Flight formations in geese and other birds. *Nature* 282:778–780.

Mayr, E. 1983. Introduction. Pages 1–22 in A. H. Brush and J. G. A. Clark, eds. *Perspectives in Ornithology*. Cambridge University Press, New York.

McCrary, M. D., P. H. Bloom, S. Porter, and K. J. Sernka. 2019. Facultative migration: New insight from a raptor. *Journal of Raptor Research* 53:84–90.

McDuffie, L. A., K. S. Christie, A.-L. Harrison, A. R. Taylor, B. A. Andres, B. Laliberté, and J. A. Johnson. 2022. Eastern-breeding Lesser Yellowlegs are more likely than western-breeding birds to visit areas with high shorebird hunting during southward migration. *Ornithological Applications* 124: Article duab061.

McKinnon, E. A., and O. P. Love. 2018. Ten years tracking the migrations of small landbirds: Lessons learned in the golden age of bio-logging. *The Auk* 135:834–856.

Meir, J. U., J. M. York, B. A. Chua, W. Jardine, L. A. Hawkes, and W. K. Milsom. 2019. Reduced metabolism supports hypoxic flight in the high-flying bar-headed goose (*Anser indicus*). *Elife* 8: Article e44986.

Merkel, B., S. Descamps, N. G. Yoccoz, D. Grémillet, P. Fauchald, J. Danielsen, F. Daunt, K. E. Erikstad, A. V. Ezhov, M. P. Harris, M. Gavrilo, S. H. Lorentsen, T. K. Reiertsen, G. H. Systad, T. Lindberg Thórarinsson, S. Wanless, and H. Strøm. 2021. Strong migratory connectivity across meta-populations of sympatric North Atlantic seabirds. *Marine Ecology Progress Series* 676:173–188.

Mueller, T., R. B. O'Hara, S. J. Converse, R. P. Urbanek, and W. F. Fagan. 2013. Social learning of migratory performance. *Science* 341:999–1002.

Nagy, M., Z. Akos, D. Biro, and T. Vicsek. 2010. Hierarchical group dynamics in pigeon flocks. *Nature* 464:890–893.

Nathan, R., W. M. Getz, E. Revilla, M. Holyoak, R. Kadmon, D. Saltz, and P. E. Smouse. 2008. A movement ecology paradigm for unifying organismal movement research. *Proceedings of the National Academy of Sciences of the USA* 105:19052–19059.

Nathan, R., and L. Giuggioli. 2013. A milestone for movement ecology research. *Movement Ecology* 1: Article 1.

Newton, I. 2007. Weather-related mass-mortality events in migrants. *Ibis* 149:453–467.

Overton, C. T., A. A. Lorenz, E. P. James, R. Ahmadov, J. M. Eadie, F. McDuie, M. J. Petrie, C. A. Nicolai, M. L. Weaver, D. A. Skalos, S. M. Skalos, A. L. Mott, D. A. Mackell, A. Kennedy, E. L. Matchett, and M. L. Casazza. 2022. Megafires and thick smoke portend big problems for migratory birds. *Ecology* 103: Article e03552.

Papageorgiou, D., C. Christensen, G. E. C. Gall, J. A. Klarevas-Irby, B. Nyaguthii, I. D. Couzin, and D. R. Farine. 2019. The multilevel society of a small-brained bird. *Current Biology* 29:R1120–R1121.

Papageorgiou, D., and D. R. Farine. 2020. Group size and composition influence collective movement in a highly social terrestrial bird. *Elife* 9: Article e59902.

Parr, N., C. M. Bishop, N. Batbayar, P. J. Butler, B. Chua, W. K. Milsom, G. R. Scott, and L. A. Hawkes. 2019. Tackling the Tibetan Plateau in a down suit: Insights into thermoregulation by bar-headed geese during migration. *Journal of Experimental Biology* 222: Article jeb203695.

Patchett, R., T. Finch, and W. Cresswell. 2018. Population consequences of migratory variability differ between flyways. *Current Biology* 28:R340–R341.

Pennycuick, C. J. 1982. The flight of petrels and albatrosses (Procellariiformes), observed in South Georgia and its vicinity. *Philosophical Transactions of the Royal Society B: Biological Sciences* 300:75–106.

Péron, C., and D. Grémillet. 2013. Tracking through life stages: Adult, immature and juvenile autumn migration in a long-lived seabird. *PLoS One* 8: Article e72713.

Phillips, R. A., J. C. Xavier, and J. P. Croxall. 2003. Effects of satellite transmitters on albatrosses and petrels. *The Auk* 120:1082–1090.

Photopoulou, T. 2017. Movement ecology: Stepping into the mainstream. Methods Blog, March 9. https://methodsblog.com/2017/2003/2009/movement-ecology

Piiper, J., and P. Scheid. 1972. Maximum gas transfer efficacy of models for fish gills, avian lungs and mammalian lungs. *Respiration Physiology* 14:115–124.

Ponganis, P. J., J. U. Meir, and C. L. Williams. 2010. Oxygen store depletion and the aerobic dive limit in emperor penguins. *Aquatic Biology* 8:237–245.

Portugal, S. 2016. Lissaman, Shollenberger and formation flight in birds. *Journal of Experimental Biology* 219:2778–2780.

Portugal, S. J., T. Y. Hubel, J. Fritz, S. Heese, D. Trobe, B. Voelkl, S. Hailes, A. M. Wilson, and J. R. Usherwood. 2014. Upwash exploitation and downwash avoidance by flap phasing in ibis formation flight. *Nature* 505:399–402.

Rattenborg, N. C. 2006. Do birds sleep in flight. *Naturwissenschaften* 93:413–425.

Rattenborg, N. C., B. Voirin, S. M. Cruz, R. Tisdale, G. Dell'Omo, H. P. Lipp, M. Wikelski, and A. L. Vyssotski. 2016. Evidence that birds sleep in mid-flight. *Nature Communications* 7: Article 12468.

Rayner, J. 1988. Form and function in avian flight. Pages 1–66 in R. F. Johnston, ed. *Current Ornithology*. Springer, Boston.

Sæther, B.-E., S. Engen, A. Pape Møller, H. Weimerskirch, M. E. Visser, W. Fiedler, E. Matthysen, M. M. Lambrechts, A. Badyaev, and P. H. Becker. 2004. Life-history variation predicts the effects of demographic stochasticity on avian population dynamics. *The American Naturalist* 164:793–802.

Sasaki, T., R. P. Mann, K. N. Warren, T. Herbert, T. Wilson, and D. Biro. 2018. Personality and the collective: Bold homing pigeons occupy higher leadership ranks in flocks. *Philosophical Transactions B* 373: Article 20170038.

Scholander, P. F. 1940. *Experimental Investigations on the Respiratory Function in Diving Mammals and Birds*. Hvalrådets Skrifter, Oslo, Norway.

Scott, G. R., L. A. Hawkes, P. B. Frappell, P. J. Butler, C. M. Bishop, and W. K. Milsom. 2015. How bar-headed geese fly over the Himalayas. *Physiology* 30:107–115.

Senner, N. R., M. A. Verhoeven, J. M. Abad-Gómez, J. A. Alves, J. C. E. W. Hooijmeijer, R. A. Howison, R. Kentie, A. H. J. Loonstra, J. A. Masero, A. Rocha, M. Stager, and T. Piersma. 2019. High migratory survival and highly variable migratory behavior in black-tailed godwits. *Frontiers in Ecology and Evolution* 7: Article 96.

Sequeira, A. M. M., M. R. Heupel, M. A. Lea, V. M. Eguíluz, C. M. Duarte, M. G. Meekan, M. Thums, H. J. Calich, R. H. Carmichael, D. P. Costa, L. C. Ferreira, J. Fernandéz-Gracia, R. Harcourt, A. L. Harrison, I. Jonsen, C. R. McMahon, D. W. Sims, R. P. Wilson, and G. C. Hays. 2019. The importance of sample size in marine megafauna tagging studies. *Ecological Applications* 29: Article e01947.

Shaw, A. K. 2020. Causes and consequences of individual variation in animal movement. *Movement Ecology* 8: Article 12.

Sjöberg, S., G. Malmiga, A. Nord, A. Andersson, J. Bäckman, M. Tarka, M. Willemoes, K. Thorup, B. Hansson, T. Alerstam, and D. Hasselquist. 2021. Extreme altitudes during diurnal flights in a nocturnal songbird migrant. *Science* 372:646–648.

Stanley, C. Q., M. MacPherson, K. C. Fraser, E. A. McKinnon, and B. J. Stutchbury. 2012. Repeat tracking of individual songbirds reveals consistent migration timing but flexibility in route. *PLoS One* 7: Article e40688.

Strikwerda, T. E., M. R. Fuller, W. S. Seegar, P. W. Howey, and H. D. Black. 1986. Bird-borne satellite transmitter and location program. *Johns Hopkins APL Technical Digest* 7:203–208.

Stutchbury, B. J. M., S. A. Tarof, T. Done, E. Gow, P. M. Kramer, J. Tautin, J. W. Fox, and V. Afanasyev. 2009. Tracking long-distance songbird migration by using geolocators. *Science* 323:896.

Sullivan, B. L., C. L. Wood, M. J. Iliff, R. E. Bonney, D. Fink, and S. Kelling. 2009. eBird: A citizen-based bird observation network in the biological sciences. *Biological Conservation* 142:2282–2292.

Taylor, P. D., T. L. Crewe, S. A. Mackenzie, D. Lepage, Y. Aubry, Z. Crysler, G. Finney, C. M. Francis, C. G. Guglielmo, D. J. Hamilton, R. L. Holberton, P. H. Loring, G. W. Mitchell, D. R. Norris, J. Paquet, R. A. Ronconi, J. R. Smetzer, P. A. Smith, L. J. Welch, and B. K. Woodworth. 2017. The Motus Wildlife Tracking System: A collaborative research network to enhance the understanding of wildlife movement. *Avian Conservation and Ecology* 12: Article 8.

Thorup, K., L. Pedersen, R. R. Da Fonseca, B. Naimi, D. Nogués-Bravo, M. Krapp, A. Manica, M. Willemoes, S. Sjöberg, and S. Feng. 2021. Response of an Afro-Palearctic bird migrant to glaciation cycles. *Proceedings of the National Academy of Sciences of the USA* 118: Article e2023836118.

Turbek, S. P., E. S. C. Scordato, and R. J. Safran. 2018. The role of seasonal migration in population divergence and reproductive isolation. *Trends in Ecology & Evolution* 33:164–175.

Urios, V., A. Soutullo, P. López-López, L. Cadahía, R. Limiñana, and M. Ferrer. 2007. The first case of successful breeding of a golden eagle *Aquila chrysaetos* tracked from birth by satellite telemetry. *Acta Ornithologica* 42:205–209.

Van Doren, B. M., and K. G. Horton. 2018. A continental system for forecasting bird migration. *Science* 361:1115–1118.

Van Doren, B. M., K. G. Horton, A. M. Dokter, H. Klinck, S. B. Elbin, and A. Farnsworth. 2017. High-intensity urban light installation dramatically alters nocturnal bird migration. *Proceedings of the National Academy of Sciences of the USA* 114:11175–11180.

Vardanis, Y., J.-Å. Nilsson, R. H. G. Klaassen, R. Strandberg, and T. Alerstam. 2016. Consistency in long-distance bird migration: Contrasting patterns in time and space for two raptors. *Animal Behaviour* 113:177–187.

Verhoeven, M. A., A. H. J. Loonstra, N. R. Senner, A. D. McBride, C. Both, and T. Piersma. 2019. Variation from an unknown source: Large inter-individual differences in migrating black-tailed godwits. *Frontiers in Ecology and Evolution* 7: Article 00031.

Voelkl, B., S. J. Portugal, M. Unsöld, J. R. Usherwood, A. M. Wilson, and J. Fritz. 2015. Matching times of leading and following suggest cooperation through direct reciprocity during V-formation flight in ibis. *Proceedings of the National Academy of Sciences of the USA* 112:2115–2120.

Webster, M. S., P. P. Marra, S. M. Haig, S. Bensch, and R. T. Holmes. 2002. Links between worlds: unraveling migratory connectivity. *Trends in Ecology & Evolution* 17:76–83.

Weimerskirch, H., N. Brothers, and P. Jouventin. 1997. Population dynamics of wandering albatross *Diomedea exulans* and Amsterdam albatross *D. amsterdamensis* in the Indian Ocean and their relationships with long-line fisheries: Conservation implications. *Biological Conservation* 79:257–270.

Weimerskirch, H., Y. Cherel, K. Delord, A. Jaeger, S. C. Patrick, and L. Riotte-Lambert. 2014. Lifetime foraging patterns of the wandering albatross: Life on the move. *Journal of Experimental Marine Biology and Ecology* 450:68–78.

Weimerskirch, H., J. Collet, A. Corbeau, A. Pajot, F. Hoarau, C. Marteau, D. Filippi, and S. C. Patrick. 2020. Ocean sentinel albatrosses locate illegal vessels and provide the first estimate of the extent of nondeclared fishing. *Proceedings of the National Academy of Sciences of the USA* 117:3006–3014.

Weimerskirch, H., and P. Jouventin. 1987. Population dynamics of the wandering albatross, *Diomedea exulans*, of the Crozet Islands: Causes and consequences of the population decline. *Oikos* 49:315–322.

Weimerskirch, H., J. Martin, Y. Clerquin, P. Alexandre, and S. Jiraskova. 2001. Energy saving in flight formation. *Nature* 413:697–698.

Westley, P. A. H., A. M. Berdahl, C. J. Torney, and D. Biro. 2018. Collective movement in ecology: from emerging technologies to conservation and management. *Philosophical Transactions of the Royal Society B: Biological Sciences* 373: Article 20170004.

Whitmore, A., A. Agarwal, and L. Da Xu. 2015. The Internet of Things—A survey of topics and trends. *Information Systems Frontiers* 17:261–274.

Wilkinson, B. P., Y. G. Satgé, J. S. Lamb, and P. G. R. Jodice. 2019. Tropical cyclones alter short-term activity patterns of a coastal seabird. *Movement Ecology* 7: Article 30.

Williams, C. L., and P. J. Ponganis. 2021. Diving physiology of marine mammals and birds: The development of biologging techniques. *Philosophical Transactions of the Royal Society B: Biological Sciences* 376: Article 20200211.

Williams, T. D., A. Kato, J. P. Croxall, Y. Naito, D. R. Briggs, S. Rodwell, and T. R. Barton. 1992. Diving pattern and performance in nonbreeding gentoo penguins (*Pygoscelis papua*) during winter. *The Auk* 109:223–234.

Wilson, R. P., and S. P. Vandenabeele. 2012. Technological innovation in archival tags used in seabird research. *Marine Ecology Progress Series* 451:245–262.

Winger, B. M., and T. M. Pegan. 2021. Migration distance is a fundamental axis of the slow–fast continuum of life history in boreal birds. *Ornithology* 138: Article ukab043.

Woodbridge, B., K. K. Finley, and S. T. Seager. 1995. An investigation of the Swainson's hawk in Argentina. *Journal of Raptor Research* 29:202–204.

Yanco, S. W., C. Rutz, B. Abrahms, N. W. Cooper, P. P. Marra, T. Mueller, B. C. Weeks, M. Wikelski, and R. Y. Oliver. 2025. Tracking individual animals can reveal the mechanisms of species loss. *Trends in Ecology and Evolution* 40(1):47–56.

Yoon, C. K. 1996. Tiny transmitters make it possible to track peregrines day by day. *The New York Times*: C4.

Youngflesh, C., J. Socolar, B. R. Amaral, A. Arab, R. P. Guralnick, A. H. Hurlbert, R. LaFrance, S. J. Mayor, D. A. W. Miller, and M. W. Tingley. 2021. Migratory strategy drives species-level variation in bird sensitivity to vegetation green-up. *Nature Ecology & Evolution* 5:987–994.

5

Bird Invasions in a Humanizing World

Michelle García-Arroyo, Christine Rega-Brodsky, and Ian MacGregor-Fors

Since prehistoric times, humans have used and lived alongside wild animals, domesticating those necessary to fulfill their needs and desires (Russell 2012). As humans migrated, some of such species were brought along on their journeys, and those that managed to survive under the new conditions were able to establish new populations. Even after the shift from a nomadic to a sedentary way of life, trade continued as a force that promoted the translocation of organisms (e.g., Auffray et al. 2008, Scott et al. 2021). As human communities and societies developed, the scale and rate of commerce and transportation also grew. It was with the first wave of globalization in the early 19th century that the barrier between distant regions throughout the world started to disappear, and thus the transport, introduction, and establishment of organisms of certain species appeared to grow out of control (O'Rourke and Williamson 2004). Species were transported for a wide array of reasons, ranging from involvement in the production of important goods (e.g., food, clothing, and construction) to imparting a sense of familiarity that was sought after by immigrants through the presence of known species (Seddon et al. 2012).

Notably, the translocation of species via anthropogenic means differs from the natural process through which species have shifted their geographic range distributions. Natural translocations have been an important source of evolution that has driven much of the present composition of ecosystems (Ayala 2005). Species undergoing natural translocations have usually responded to changes in the availability of resources, shifts in the local climate, and geographic dynamics. However, these geographic movements can also be a consequence of fortuitous events, such as storms and sudden changes in the landscape, or they could even be linked to the accidental transport of organisms that are parasites or otherwise are carried by other species (Nathan et al. 2008). This last mechanism of movement is the one for which humans are most frequently responsible (Seddon et al. 2012, Heinsohn 2014).

If species' natural geographic relocation has been an important source of evolution, why is it relevant that we have moved species across regions for centuries? Shifting the composition of species because of human-induced relocations (and the further establishment and population growth in many cases) has often caused important systemic alterations that were recognized as a global threat only 25 years ago (Schmitz and Simberloff 1997). After important research efforts invested in understanding this matter, biological invasions are currently considered one of the most important and main drivers of species extinctions at accelerated rates, altering species richness and composition of native communities at different spatiotemporal scales (Bellard et al. 2016).

Michelle García-Arroyo, Christine Rega-Brodsky, and Ian MacGregor-Fors, *Bird Invasions in a Humanizing World*.
In: *New Perspectives in Ornithology*. Edited by: Scott V. Edwards and J. Michael Reed, Oxford University Press.
 DOI: 10.1093/oso/9780197787670.003.0005

Numerous invasive taxa are implicated in the extinction process mentioned above, ranging from arthropods to reptiles, birds, and especially a large number of mammal species (International Union for Conservation of Nature and Natural Resources [IUCN] 2021). Although birds have not been identified among the most impactful invasive groups throughout the world (Blackburn, Lockwood, et al. 2009), they have broad geographic representation, and the magnitude and variability of their impact on native assemblages generally remain poorly understood (Kumschick and Nentwig 2010). In 2009, the most comprehensive review of avian invasions was collated by three of the leading researchers in the field (Blackburn, Lockwood, et al. 2009). They elegantly outlined the history, rationale, and ecology behind each of the invasion stages, as well as developed their own terminology. With a genetic and evolutionary approach, this synthesis paved the way for future research to advance the understanding of invasion ecology, representing a key point in the discipline regarding invasive birds.

In this chapter, we delve into the existing knowledge and supplement it with the most recent findings in the literature related to bird invasions. First, we synthesize the process that bird species invasions undergo, including its most common stages. In doing so, we aim to clarify the meaning behind many jargon terms used in the invasion biology literature. Next, we provide information on the general traits of invasive birds and focus on the available knowledge of their ecology and behavior. Finally, we revisit some of the management strategies and practices that have been proposed to control some invading bird populations, as well as some of the widely accepted and controversial ethical views related to this complex topic.

Invasion Process

In this section, our main goal is to summarize the main scenarios that can occur from the translocation of an avian species from its original distribution (also referred to as natural or prehuman) to its successful invasion (graphically summarized in Figure 5.1). In addition, we also aim to provide a clear explanation of many of the jargon terms. For the invasion process to occur, a series of events, which may differ from case to case, needs to occur; thus, different terms have been proposed and used for clarity (Blackburn et al. 2011). Notably, our definitions may vary from those of colleagues across the discipline.

The invasion process starts with *translocation*—the transportation of organisms of any given species that was not present previously in a determined area or region. The scale is an important element to consider when addressing translocation because focusing on a regional, national, or local level could result in differing interpretations. The transported individuals are often called nonnative, non-indigenous, alien, or exotic (for clarity purposes, we use the "exotic" term for this stage hereafter). If these individuals remain captive or are used for noncaptive experimental or ornamental purposes, they would remain considered as exotic. However, if such individuals or groups of them are intentionally or unintentionally released from captivity or experimental/ornamental confinement and manage to survive in the receiving ecosystem, they are considered *introduced*; obviously, all introduced species are

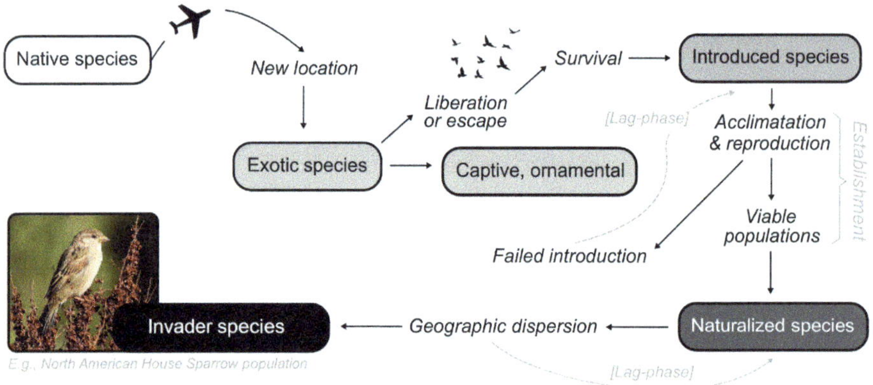

Figure 5.1 Potential stages of the process of invasion of an exotic species. The *translocation* of a species to a new location where it was not present beforehand makes it an "exotic species." From this point, two pathways are possible: If individuals of this exotic population remain in captivity, their exotic status will not change; however, if *intentional or unintentional release* takes place and the individuals *survive* in the new environment, they are now considered as "introduced species." Depending on the reproduction outcome of the latter individuals, it would be considered as a *failed introduction* if no viable populations are established or a "naturalized species" if they managed to establish a self-sustaining population. It is between these particular stages that another scenario is also a possibility—the *lag-phase*, described as a delay of population growth after *establishment*. Last, "naturalized species" are considered as such until *geographic dispersion* from those populations occurs (see "Invasion Process" for further details on scale), thus changing into "invader species" from that point on.

exotic by definition. The scale can be complicated at this point; hence, even some populations of species considered to be native to a region can be considered as invasive (we provide an in-depth discussion about this later in this section).

Several scenarios can occur after a species' introduction (see Figure 5.1). The introduced individuals could be capable of surviving, but may not be able to breed or produce enough offspring, even if numerous; thus, the introduced species may eventually disappear from that location. Accordingly, if no further individual input occurs, the species will remain absent from the area in the future. In this case, the process would end in a failed introduction. Yet, if additional individuals continue to be released and/or there is successful breeding, a process known as the "lag-phase" could be initiated. This phase corresponds to a delay in population growth after establishing local reproductive and successful self-sustaining populations (often referred to as colonization). In some cases, the released individuals lack the required traits to become successful (e.g., genetic diversity and behavioral adaptations). Based on this knowledge, even if the population of an introduced species is small, it may experience a demographic explosion in the future (Crooks 2005). Interestingly, the lag-phase can extend for decades. Although most estimations of lag-phase duration have been calculated for invasive plants, Aagaard and Lockwood (2014) found that the average duration of the lag-phase for 17 populations of invading birds in Hawaii was approximately 15 ± 11.6 years. In a recent study, the lag-phase of invasion of the House Sparrow (*Passer domesticus*) in North America was inferred in Florida using historical data (Peña-Peniche et al. 2021), taking the species close to 40 years to colonize the state fully since its documented introduction (Howell 1932).

Nevertheless, some species have swiftly acclimated and adjusted, and even adapted in some cases, to local conditions with a brief lag-phase, following an establishment process and producing viable populations. In such cases, species are considered *naturalized*, when a viable population remains in the area of introduction (again, scale matters in this stage too). The major component to becoming an invader (and therefore one that has gone through the process and identity of being exotic, introduced, and naturalized) is geographic dispersion. If individuals of naturalized populations manage to disperse and establish self-sustaining populations in locations beyond the original area of introduction, the species would now be considered an *invader* (or invading species; note that *invasive* is a term often used to refer to species that have the potential of becoming invaders; for further details, see Pereyra 2016). If the species is expanding its invasion range distribution, many failed attempts could occur, with mixed outcomes in regions with differing conditions to those it is invading from (as in the aforementioned case of the House Sparrow in Florida). In such cases, the barriers can be geographic, ecological, demographic, and/or physiological in nature, and they can represent the limit of an invader's distribution or obstacles that could be overcome by acclimation, physiological and/or behavioral plasticity, and even lag-phase processes (Blackburn et al. 2015). Propagule pressure may be a key factor increasing the probability of a population to establish successfully (Blackburn, Cassey, et al. 2009) and could be a determinant for the duration of the lag-phase (Sakai et al. 2001), as small populations are particularly affected by stochastic events that can cause the invading populations to disappear (Pimm 1989). However, even small populations may overcome stochastic events through several species' traits, related with their diet, behavior, and/or physiological plasticity, as well as the given state of resource availability at the introduction site (Vall-llosera and Sol 2009).

Some intriguing phenomena can also occur in avian invasions. As noted previously, some native species have been suggested to be able to become invaders (e.g., MacGregor-Fors et al. 2009, Valéry et al. 2009). This occurs in two main scenarios: (1) The species is considered as native to an administration boundary (often at the country level) and individuals can spread beyond its original range distribution—thus, the implied species is considered a native invader; and (2) the species is original to a type of habitat and manages to disperse and persist in anthropogenically modified landscapes (e.g., agricultural fields and urban settings), often exploiting specific resources, resulting in population increases. The latter is the case of the Great-Tailed Grackle (*Quiscalus mexicanus*), which is presumed to be native to the wetlands of the Gulf of Mexico and currently occurs from the central-southern United States to coastal Colombia (Johnson and Peer 2020). In this case, the expansion of the geographic range of the species (generally referred to as "range expansion," and species are referred to as "range-expanders") is closely related to indirect human assistance. The Great-Tailed Grackle, as well as many other species throughout the world, benefits from human-modified landscape matrices, using agricultural fields and urban mosaic settings to rapidly expand its invader distribution (Wehtje 2003). Other bird species showing similar expansion across Europe include the House Sparrow and

European Starling (*Sturnus vulgaris*). There are even hypotheses that some current "Eurasian" or "Eurasioafrican" species could have expanded their distribution along with the first human modification of landscapes. For instance, among the groups within the Eurasioafrican House Sparrow complex, signatures of recent selection in the genome of the sparrow, mainly related to adaptation to "anthropogenic niches," are absent in the Bactrianus populations (from the Middle East and Central Asia). The latter noncommensal populations migrate, are less bold, and are not associated with human settlements (Ravinet et al. 2018).

What Characterizes Invasive Birds?

A wide array of bird species with contrasting traits and from distant phylogenetic origins have been identified as successful invaders. Notably, intentional introductions—that often resulted in successful invasions—were initially (1500–1903 AD) primarily driven by European (mainly British) colonialism. However, recent introductions (1983–2000 AD) are a broader phenomenon, involving more species, countries, and processes (with range-expanding gaining terrain); thus, more recent events are generally driven by land-use change and increasing globalized economic activity (Dyer et al. 2017). Interestingly, the geographic sources of introduced bird species have shifted significantly since the mid-19th century away from Europe to the Indian subcontinent, Indochina, and sub-Saharan Africa (Dyer et al. 2017). Such dynamics could change again, because the presence of more colorful Neotropical birds in different areas of the world is also becoming more common (Su et al. 2014).

Studies have attempted to identify overarching traits related to invasive bird species. Although no clear patterns exist (Blackburn, Lockwood, et al. 2009; Sol et al. 2012), some characteristics have been identified as typical of invasive species, or at least some of them. Factors related to successfully invading species are their wide geographic distribution and high native abundance, which have often correlated with the likelihood of transport and introduction, as well as their range-expanding potential (Blackburn and Duncan 2001, Dyer et al. 2016, Evans et al. 2018). Also, propagule pressure has been proposed as a potential driver of invasion success, as has a combination of a species traits and the place of the initial introduction, where establishment occurs (Cassey et al. 2004, Duncan et al. 2014).

The species traits with the strongest evidence for contributing to invader establishment success are those related to the ability to cope with new environments (Blackburn, Cassey, et al. 2009) and/or to reproduce rapidly in environments similar to their native range (Sol et al. 2012). In many of these cases, generalist species are included (Blackburn, Lockwood, et al. 2009). Other traits, such as strategies prioritizing survival over reproduction, may also promote establishment (Sol et al. 2012). Although not a collective feature of invasive species, several studies have found that relative brain size is associated with higher rates of invasion success and establishment (Sol and Lefebvre 2000, Sol et al. 2005). In addition, increasing evidence has been published on the genetic component of species coping with new environments (e.g., Allee effects; Blackburn, Cassey, et al. 2009).

Although reviews have attempted to generalize trait patterns for invasive and invading avian species, particular attention has focused on a handful of species and their families. Three of the four bird families with the most introduced species include popular cage birds (i.e., Estrildidae, Sturnidae, and Psittacidae; Anatidae is the fourth), which reflects the importance of a shift toward unplanned introductions of species from the cage and pet trade (Cassey et al. 2016, Dyer et al. 2017). Well-documented invader populations include those of the House Sparrow, European Starling, Common Myna (*Acridotheres tristis*), and two psittacines—Rose-Ringed Parakeet (*Psittacula krameri*) and Monk Parakeet (*Myiopsitta monachus*). Here, we highlight three of these globally invading species to describe their traits, invasion processes, as well as the consequences of their introductions and invasions in more detail: The House Sparrow (Box 5.1), Common Myna (Box 5.2), and Monk Parakeet (Box 5.3).

Box 5.1 House Sparrow (*Passer domesticus*)

The House Sparrow (Figure 5.B1) is a small-sized passerine bird anthropogenically introduced to many regions throughout the world. This sparrow has been long considered to be native to Eurasia (from the British Isles, northern Scandinavia, Russia, and Siberia to north Africa, Arabia, and Burma; Lowther and Cink 2020). Although declining in some areas of its "native" distribution range (Shaw et al. 2008, Hayhow et al. 2015), the House Sparrow has established invader populations in many regions of the world. Based on the evidence that the sparrow expanded associations with human-modified landscape matrices, moving across agricultural fields and human settlements, this sparrow seems to be a range-expander in most of its current Eurasian distribution. The recent development of a human commensalism signature in the sparrow's genome shows how the noncommensal Bactrianus populations, which occur in the Middle East and Central Asian steppe, are less bold and not associated with human settlements compared to their European counterparts (Ravinet et al. 2018). Thus, this species is an interesting example of a species with different facets as an invader.

Figure 5.B1 Male House Sparrow (*Passer domesticus*) in Xalapa, Mexico.
Photograph by Ian MacGregor-Fors.

Continued

Continued

The House Sparrow has been intentionally introduced and become highly invasive in other regions of the world, with major invaded areas being eastern Australia, southern Africa, South America, and North America. Among the best recorded and studied invader populations, the North American one heads the list; thus, we focus on these populations here. The sparrow was introduced in North America in at least 16 independent events between 1850 and 1881 (Brown and Wilson 1975). The first case of House Sparrow introduction in North America, with individuals from England, took place in Brooklyn, New York (1851–1852). Further major introductions, with stock from Germany and England, took place in San Francisco, California (1871–1872) and Salt Lake City, Utah (1873–1874) (Lowther and Cink 2020). Peña-Peniche et al. (2021) geographically summarized the sparrow's historical invasions, synthesizing how the invasions occurred and how different climatic factors have played a crucial role in their expansion. To date, the House Sparrow is highly dominant and successful at exploiting urban systems and repelling potential competitor native avifauna in North America (White et al. 2005, MacGregor-Fors et al. 2010, García-Arroyo et al. 2020; but see González-Oreja et al. 2018). Interestingly, based on feeder records across the United States and Canada, populations seem to have declined in urban areas, but not in non-urban areas (Berigan et al. 2021).

It is important to note that the North and South American invasions are independent processes. However, it seems that the invasion process could be a more complex scenario, as the southernmost range of the North American invasion is already in Panama, and the northernmost range of the South American invasion is in northern, mainly coastal, Colombia. Only time will tell if the Darién Gap will keep these two complex invasions apart or if they will interact in the future.

Box 5.2 Common Myna (*Acridotheres tristis*)

The Common Myna (Figure 5.B2) is a mid-sized passerine bird native to Asia, specifically from Central Asia and South Asia to the Malay Peninsula. This member of the starling family (Sturnidae) has established populations in many areas of the world outside its native range, through either accidental or deliberate releases. Common Mynas are currently present in large portions of Australia, Madagascar, South Africa, and Israel, among other many regions. As has been recorded for other invading bird species, this myna is range-expanding at the margins of its native range distribution by exploiting human-modified landscapes (Peacock et al. 2007). The effects of this invasive bird, together with the increasing amount of knowledge generated for the species, prompted the myna's listing as one of the IUCN's world's 100 most invasive species, together with only two other bird species (i.e., Red-Vented Bulbul [*Pycnonotus cafer*] and European Starling; Lowe et al. 2000).

Figure 5.B2 Common Myna (*Acridotheres tristis*) in Puthucode, India.

Photograph by Aboodi Vesakaran (https://unsplash.com/photos/black-and-brown-bird-on-green-leaf-during-daytime-RKI_KNPc_DQ).

The success of the Common Myna has been linked with some of its life history and physiological and ecological traits, of which the following head the list: It has a broad generalist diet, often roosts and feeds in groups, has a wide climatic tolerance, has a relatively high average clutch size, and is associated with and can disperse across human-modified landscapes (Yap et al. 2002, Peacock et al. 2007). With its increasing distribution range and lack of proper control in the wildlife trade business, this myna has been recorded to pose a threat to native biodiversity not only to birds but also to cavity-related mammals (Grarock et al. 2012) and to humans (Nijman et al. 2021). Regarding their effects on native biodiversity, several aspects have been recorded. For instance, as a cavity nester, it not only competes with other species for the natural nesting and roosting space, but also preys on eggs and chicks. It has shown nesting site plasticity in urban areas because it may also nest in buildings. The myna has also been recorded to prey on adult birds and aggressively compete for food resources (Bomford and Sinclair 2002, Peacock et al. 2007). Regarding their ecological effect, the mynas have been shown to interact synergistically with some invading plants by dispersing them, while using the resources offered by these plant species (e.g., shrub verbena [*Lantana camara*]; DeGraaf and Rudis 1986).

Regarding the effect of mynas on humans, there are several ramifications. Among the most important ones are agricultural and potential health impacts. Regarding the former, this species has such an effect on crops—such as overfeeding on blueberry crops (*Vaccinium* spp.) in Australia (Bomford and Sinclair 2002)—that it is considered a pest in many regions (e.g., South Africa, Australia, and New Zealand). Although less well documented than the biodiversity and crop effects, the myna has been hypothesized to represent a threat to human health through the spread of infectious diseases. This myna has been shown to carry avian malaria, while also spreading other parasites that could cause dermatitis to people in dense human settlements, where it often roosts in large, noisy groups (Peacock et al. 2007).

Box 5.3 Monk Parakeet (*Myiopsitta monachus*)

The Monk Parakeet (Figure 5.B3) is a medium-sized psittacid natively distributed in southern South America, including Argentina, Bolivia, Brazil, Paraguay, and Uruguay. Along its native distribution, the parakeet dwells within forests and savannas, but it has also managed to thrive in plantations, orchards, and urban settlements (Davis 1974, Aramburú and Corbalán 2000, Narosky and Yzurieta 2005). One trait that makes this parakeet unique among other Psittacidae species is that it is not a cavity nester and instead constructs large domed nests with internal subdivisions (Avery et al. 2002). Among the traits that have allowed this parakeet to successfully exploit human-modified ecosystems across its native distribution are its gregarious and aggressive behaviors (Davis 1974, De Grazio 1978).

Figure 5.B3 Monk Parakeet (*Myiopsitta monachus*) in Barcelona, Spain.
Photograph by Ian MacGregor-Fors.

This parakeet has been used as an ornate bird in South America and also in many other regions throughout the world (Fitzwater 1988, Narosky and Yzurieta 2005). It has established reproductive populations in the United States, Puerto Rico, Canada, Great Britain, France, Germany, Switzerland, Austria, the former Yugoslavia, Spain, Portugal, Italy, Israel, Kenya, Japan, and Mexico (Davis 1974, Spano and Truffi 1986, Campbell 2001, Allen 2006, Roll et al. 2008, Pablo López 2009). Yet, the Monk Parakeet's major invading populations are located in North America (i.e., Mexico, the United States, and southern Canada) and Western Europe (GBIF Secretariat 2021).

The Monk Parakeet has shown a spectrum of natural history traits related to its invasiveness outside South America. Regarding its reproduction, it is socially monogamous with high extra-pair paternity rates, which, combined with its year-round nesting and gregarious behavior, often allows it to establish reproductive populations rapidly. After they are established, Monk Parakeet populations have been shown to have excellent dispersal capabilities and the ability to adjust and eventually exploit a wide variety of human disturbances and habitats (e.g., urban and agricultural). Thus, its nesting and behavioral traits, among other aspects, have allowed the species' populations to grow rapidly across their invaded range (Martín and Bucher 1993, Hyman and Pruett-Jones 1995, South and Pruett-Jones 2000, Muñoz and Real 2006, Gonçalves da Silva et al. 2010).

The parakeet has caused several problems across its invasion distribution. In South America, where its range has expanded across human-dominated systems, it has caused dramatic economic losses to agriculture (De Grazio 1978, Long 1981). Although this has also happened in countries such as Spain and Israel (Roll et al. 2008, Conroy and Senar 2009), problems have been different in the United States. There, agricultural damage is limited to some tropical fruit crops in South Florida (Tillman et al. 2000). Still, its detrimental effects diversified when it started nesting on electric utility structures, causing power outages (Avery et al. 2002, Tillman et al. 2004, Pruett-Jones et al. 2007). It is not only the economy that this parakeet has impacted; there are also ecological implications. For instance, Monk Parakeets have great potential for the dissemination of viruses (e.g., Newcastle disease virus), which could have devastating effects on wild bird communities, resulting in more potential economic losses in the poultry industry and among pet traders (Fitzwater 1988).

Ecology and Behavior of Invasive Birds

As reviewed in the previous section, when individuals of exotic species establish and colonize new locations, successful biological invasions occur (Blackburn et al. 2011). These may alter local environmental processes and the structure of native communities (e.g., nutrient cycles, trophic networks, and fire and erosion regimes; Pyšek et al. 2012, Ricciardi et al. 2013, Simberloff et al. 2013). Furthermore, invading populations can restructure local wildlife communities and even alter ecosystem processes, resulting in economic losses and/or health problems (Pimentel et al. 2005, Simberloff et al. 2013, Booy et al. 2017). In this section, we focus on the evidence regarding the ecology and behavior of exotic-to-invader birds, highlighting some intriguing aspects that have been the focus of much of the current research.

Ecology

Research on the ecology of invasive bird species and invader populations has not only spiked in the past 20 years, but also provided new insights on their plasticity and spatiotemporal population dynamics, their effects on native wildlife, synergies with other invasive species, as well as a diverse list of more specific topics. Here, we summarize the essence of research published from 2016–2021 and provide interesting examples.

Although research has shown that trait plasticity is key for the invasion potential of species, recent research has added elegantly to our growing understanding, shedding additional light on invader population densities. Evidence is growing regarding the role of tolerance to extreme weather conditions and invasion potential (Lei and Liu 2021), as well as the importance of dietary breadth in the process and establishment of successful invading species (Wang et al. 2018). With their high multifactor plasticity, some invasive species can occupy systems that few other native species can use. Such is the case of some cosmopolitan urban-related bird species (e.g., House Sparrow, Rock Dove [*Columba livia*], and European Starling; Aronson et al. 2014). These species not only occupy, but also exploit unvegetated and noisy systems with poor-quality food resources (González-Oreja et al. 2018). Most interestingly, evidence of other post-invading plasticities has been shown in the recent literature. One intriguing example is that of the Common Waxbill (*Estrilda astrild*), which has changed its breeding phenology in areas with more stable climates and food resource availability. In this case, the waxbills appear to have switched their reproduction times with resource fluctuations as a sub-Saharan invader (Beltrão et al. 2021).

The negative ecological impacts of invader bird populations were assumed in the past for many species. Some empirical information started to populate the literature approximately a decade ago, but more evidence is needed on the magnitude and role of some invading birds (which can be boosted through invader–invader synergies, reviewed below). Some species have received significant attention, such as the Rose-Ringed Parakeet in its European invasion distribution range; they dominate sites and systems, including bird feeders and nest-like cavities, and vertically segregate other potential competitors (Le Louarn et al. 2016, Mori et al. 2017, Grandi et al. 2018). Among other widely studied invader species with ecological impacts are the House Sparrow (see Box 5.1), Common Myna (see Box 5.2), Monk Parakeet (see Box 5.3), and European Starling. For example, South American studies have reported that European Starlings often cause native nesting species to abandon their nests, among other indirect effects given their immense population sizes (Ibañez et al. 2017, Jauregui et al. 2021).

As stated previously, important synergies between invader birds and invader plants have helped researchers further understand some of the scattered evidence of such published in the past. Evidence suggests that some invader birds prefer the fruits of invader plants, resulting in the wide dispersal of the implied plant and abundance rate growth for the bird (Lafleur et al. 2007, MacFarlane et al. 2016, Martin-Albarracin et al. 2018, Thibault et al. 2018, Wang et al. 2018). This results in some native birds expanding their distribution range with the presence and distribution of invader plants. For example, the Anna's Hummingbird (*Calypte anna*) has expanded its distribution along western North America mainly because of the presence of food resources offered by a mix of invading plants (and also direct human facilitation through feeders; Rowe 2018).

Recent findings suggest that some introduced-to-invader bird populations can threaten some native plants because they can drive the plants' growth rates due to overconsumption (e.g., frugivory; Bialic-Murphy et al. 2017). Not only invader bird–plant synergies but also invader bird–bird ones have been documented. For example, the Alexandrine Parakeet (*Psittacula eupatria*) has expanded its climatic niche, incorporating colder conditions compared with those in the species' native range. Ancillotto et al. (2016) suggest that this may be related to an interspecific facilitation process provided by its congeneric (*P. krameri*). The latter has invader populations throughout Europe, including colder latitudes than those of their native range distributions.

It is widely known that urban mosaic landscapes are important systems for invasives, and birds are no exception to the rule. As stated previously, the ornamental cage-bird trade market—both legal and illegal—is currently the major driver for new bird invasions. Thus, although invader birds do not dominate urban-built bird species richness (Aronson et al. 2014), they can dominate urban bird biomass (Sol et al. 2014). These species generally have a relatively broad feeding breadth; are often highly granivorous in the wild; and rely on intentionally and unintentionally provided human waste refuse, and thus are associated with heavily urbanized systems (Fava and Acosta 2016, Belinsky et al. 2019, MacGregor-Fors et al. 2020, Soh et al. 2021). Urban areas have also served as "stepping stones" in the range distribution expansion of both native and exotic invasive bird species. Recent studies have provided evidence that species' invasion potential, such as that of the Common Myna, greatly depends on urbanization (Magory Cohen and Dor 2019).

Finally, recent studies have devoted efforts to testing some general hypotheses, such as the pre-introduction selection, the propagule pressure, and the enemy release hypotheses. Research shows that for the studied species, reintroduction selection can act on a wide array of characteristics of the introduced group of individuals that can make them better candidates to become invasive, ranging from physiological to demographic traits (Baños-Villalba et al. 2021). Regarding the propagule pressure hypothesis, Moulton and Cropper (2019) showed that the hypothesis does not consistently predict the outcome of exotic bird introductions. Concerning the enemy release hypothesis, Gandoy et al. (2019) showed that for native and recently distribution range-expanded Barn Swallows (*Hirundo rustica*) in South America, the hypothesis is not supported. This suggests that additional factors determine the population growth and range expansion of this and other species.

Behavior

Birds invading new habitats often find success because of their behaviors. They generally outcompete native species due to their aggression, feeding ecology, breeding behaviors, vocalizations, and human associations. Aggression, whether against conspecifics or heterospecifics, gives invading birds competitive advantage over others, particularly for food and nesting resources, yielding species-specific differences in their breeding success and resulting distribution range expansions. For example, the aggressive Long-Legged Buzzard (*Buteo rufinus*), a newly invading species in Israel, has successfully dominated and shifted the spatial and demographic distribution of the native Short-Toed Eagle (*Circaetus gallicus*) because of similar nesting needs (Friedemann et al. 2017). Outcompeting natives in foraging also promotes the success of new invading species compared with their native counterparts, such as the range-expanding Narrow-Billed Woodcreeper (*Lepidocolaptes angustirostris*) in South America. The invading woodcreeper species has been able to exclude the native woodcreeper (*L. squamatus*) from foraging flocks because of the invader species' aggressive interactions at overlapping foraging heights of native species (Maldonado-Coelho et al. 2017). Similarly, invader populations of Common Mynas (Machovsky-Capuska et al. 2015) and Red-Billed Leiothrix (*Leiothrix lutea*) (Pereira et al. 2018) initiate aggressive behaviors toward other species at feeding locations, and both are more efficient foragers (Pereira et al. 2017), resulting in increased feeding rates for these two species.

Feeding ecology and corresponding feeding behaviors of invader birds generally differ compared with native species because of their generalist and/or opportunistic feeding patterns (Blackburn, Lockwood, et al. 2009), consumption of novel food items, and their aggression around food sources and resulting efficiency. Relatively broad diet breadth is one consistent metric positively associated with invasive species establishment, allowing for explorations of novel habitats for the invading species (Blackburn, Lockwood, et al. 2009, Wang et al. 2018). Several studies have focused on the food consumption of Common Mynas because of their impacts on the local agriculture industry. Generally, myna's foraging ecology findings include their propensity for foods high in glucose (rather than fructose and sucrose food items; Gumede and Downs 2019) and protein (Machovsky-Capuska et al. 2015, Peneaux et al. 2017). When consumed, this preferential diet has also been found to shift the myna's behavior to increase exploratory behaviors (Peneaux et al. 2017), which may significantly impact their distribution range expansion and corresponding crop destruction. This behavioral and diet flexibility allows invading species to exploit new food sources, whether from plants native to the newly invaded ecosystem, anthropogenic sources, or truly novel food sources.

In addition to their success in foraging, plasticity in breeding behaviors can result in increased breeding rates, novel nesting behaviors, nest location acquisitions, and aggression against nesting competitors; it may also allow invader birds to outcompete natives regarding their fecundity. Increased clutch size is a consistent reproductive trait of invader bird species (as reviewed by Blackburn, Lockwood, et al. 2009). However, recent studies have highlighted the importance of species-specific behavioral features that promote exotic's reproductive success in particular environments and against various heterospecifics. Increased reproductive capacities include increased brood sizes in nesting chambers (Viana et al. 2016), increased rates of re-nesting and fledging successes (Senar et al. 2019), cooperative breeding in areas of high territory density (Zeng et al. 2016), and aggression against native nesting competitors. In particular, nest site aggression against native species is a behavior well documented for mynas (Charter et al. 2016, Le Roux et al. 2016, Lermite et al. 2021). In addition, while introduced-to-invader birds may outcompete natives for nesting locations, they may also prey upon native bird eggs and nestlings. This behavior is typical of invasive Common Mynas (Hughes et al. 2017), Chimango Caracaras (*Milvago chimango*; Flores et al. 2017), and European Starlings (Massaro et al. 2013), among others.

Bird invaders may also disrupt the breeding efforts of native species through vocal competition. A variety of exotic birds have been documented to increase their calling rates (Farina et al. 2013) and consequently deter native species from vocalizing (ZoBell and Furnas 2017, Pereira et al. 2020, Hopkins et al. 2021), a behavior particularly well-documented for the Red-Billed Leiothrix in its introduced distribution range. Other species, such as the Common Myna, may also overlap native species' song frequencies (Hopkins et al. 2021), which may obscure heterospecific calls. This frequency-overlapping behavior may be particularly detrimental in urban areas. There, less acoustically plastic birds are already constrained in their song and call frequency ranges within a soundscape dominated by anthropogenic noise (Slabbekoorn 2013). Any of these vocal niche-overlapping behaviors constitute a form of breeding aggression and competition against native species, promoting successful communication only among introduced-to-invader conspecifics and their domination of the soundscape.

Management of Invasive Birds

As reviewed previously in this chapter, the impacts of bird invaders have wide-ranging and often considerable effects on the local bird communities, ecosystems, and even human health. The management of such invader populations is a massive undertaking because of the species- and location-specific challenges, the scale of invasion, the time since colonization, and varying public support with proposed or implemented actions. Before any actions take place to manage invader bird populations, even the choice of acting on such invasion can be controversial.

Public support of the management of invader populations can be described as a gradient ranging from adversarial, skeptical, neutral, and concerned to supportive of exotic species management (Courchamp et al. 2017). On one side of this gradient, those who believe that the introductions resulted due to the fault of others, and now view the invader birds as part of the local ecosystems, may challenge eradication or management efforts, mainly of charismatic species (e.g., Rose-Ringed Parakeets; Shivambu et al. 2022). Especially because some bird species have been purposely introduced by humans, the social actors with this "anti-eradication viewpoint" may believe that it is not fair or ethical to eradicate them. These social actors may believe that those who promoted the invasion are responsible for these species success in their invaded range. There are also people who may have religious or cultural beliefs for which the invading species should be retained (Courchamp et al. 2017). Conversely, an opposing group of people has taken the viewpoint that invasions were caused by humans and thus should be corrected by humans as well, especially because invading species are one of the major sources of biodiversity decline and have wide-ranging social and economic impacts (Blackburn et al. 2010, J. Russell and Blackburn 2017). For those who believe that humans are responsible for maintaining biodiversity and ensuring ecosystem function, taking action against invaders through lethal or nonlethal means is generally supported (Larson et al. 2016, Shivambu et al. 2022). Importantly, if these populations must be eradicated, it should be done following all animal welfare guidelines.

Although these ranging "schools of thought" may result in conflict, management actions can be put in place before bird invasions to avert these ethical or moral dilemmas. If prevention fails and invasions cause empirically evident harm, choices may need to be assessed regarding the introduced-to-invader species' containment, active removal, or eradication. Particularly for non-avian species, invasive species prevention is often the least costly and most proactive of the management actions (Leung et al. 2002, Burnett et al. 2006), via a series of legislative and ecological efforts to restrict exotic birds from penetrating new, "natural" systems. Before invasions occur, informing the public about the ecological consequences of invasions is imperative for not only increasing acceptance of management actions and corresponding legislation, but also to prevent the public from exacerbating the release of exotics. This is particularly important because some may view releasing captive animals into the wild as a righteous or religious action (Courchamp et al. 2017); thus, preventing these releases is one step in thwarting species invasions.

Enacting trade restrictions is another effective means of preventing invasive species movement throughout the world. The trade of wild animals, particularly the international pet trade, is one of the most important sources of invasive species worldwide (Cardador et al. 2019). Although trade restrictions may be difficult to enact in contiguous countries compared with islands, legislative actions in Europe, for instance, have been shown to be successful. For example, the European Wild Bird Trade Ban, adopted in 2005, was enacted to counter the spread of avian flu, but it was also effective at reducing invasion risks of dispersing new exotic birds (Reino et al. 2017, Cardador et al. 2019). This trade ban reduced wild bird trade volumes to approximately 10% of former levels (Reino et al. 2017), consequently shifting trade from commercialization of wild-caught to captive-bred birds, which have generally lower survival rates upon their escape (Cardador et al. 2019). The European Wild Bird Trade Ban was particularly effective at reducing the trade of passerines, while the parrot trade has recovered to 60% of pre-ban levels (Reino et al. 2017, Cardador et al. 2019). Parrots within the pet trade are particularly consequential because escaped pets are often a source of local invasions. This is especially evident as pet parrots are released in Australia (Vall-llosera and Cassey 2017a) and as they continue to be traded beyond Australia's borders (Vall-llosera and Cassey 2017b). Thus, although preventing exotic species introductions is a hopeful goal, challenges still beset these efforts because broad acceptance and implementation are needed to be effective. In addition, trade bans may also consequently shift trade to new, alternative markets (Cardador et al. 2019), such as the Monk Parakeet in Mexico (MacGregor-Fors et al. 2011), returning to the issue of to how to manage exotic species with invasive potential upon their arrival.

Containing newly introduced exotic species before viable population establishment and invasion is challenging as these species start to grow and spread. With swift action, containment can be effective for new invasions, such as the Rainbow Lorikeets (*Trichoglossus moluccanus*) in Tasmania (Robinson et al. 2020). However, containment is often challenging, especially when the invading species becomes abundant across large spatial scales, which may occur in a short period of time.

Eradication often remains the last method to counteract species invasions when solid evidence exists that a species is causing harm. Although this action is the most expensive, time-intensive, and contentious with society, it may be the only option remaining. Eradication can be approached by nonlethal or lethal means, and its efficacy is specific to the species and the situation at hand. For example, eradication of the hyper-aggressive Noisy Miner (*Manorina melanocephala*) in Australia was only successful when removal occurred before the breeding season, allowing heterospecifics time to move in and colonize the miner's once occupied breeding grounds (Crates et al. 2018). To successfully satisfy management goals, eradication efforts must be spatially and temporally targeted against invaders while informing and garnering the support of society.

Nonlethal actions such as nesting deterrents, hormonal sterilization, and live capture and translocation may be options that draw public support over lethal means; however, they are often not widely effective. Creative solutions exist, such as inhibiting reproductive hormones via chemosterilants for Monk Parakeets (Pruett-Jones et al. 2007, Avery et al. 2008) or creating artificial moving snake nest predators to deter European Starlings from nesting in nest boxes (Blackwell et al. 2021). However, these efforts may not be attainable on a large scale or only result in minimal population reductions. Although controversial, lethal means of eradication, either targeting individual species or the prevention of their reproduction, may be necessary to reduce invader populations. When necessary and properly justified, managers have used trapping and euthanasia as a common approach for many invader birds. However, the success of this method may be affected by challenges associated with their capture (especially for solitary species or those that do not nest in nest boxes), and success rates differ across species for the same trapping method. For instance, the trapping method, although successful for exotic Wild Turkey (*Meleagris gallopavo*) on Santa Cruz Island, California (Morrison et al. 2016), was not effective for managing Shiny Cowbird (*Molothrus bonariensis*) populations in Puerto Rico (Miller et al. 2016). Trapping success is also location-specific for some species. Funnel traps were an effective eradication method for Common Mynas on Frégate Island in the Seychelles (Canning 2011), whereas decoy trapping was more effective for capturing this species on other islands in the Seychelles. This indicates possible behavioral differences in this species per their location (Feare et al. 2017). Other lethal means for eradicating nuisance introduced-to-invader bird populations have been borrowed from exotic mammal management, such as toxic baiting that resulted in only moderate effectiveness for European Starlings (Werner et al. 2021); however, minimizing risk to nontarget species remains an issue.

Eradicating invaders by thwarting their reproduction has been an effective and efficient method, especially for cavity-nesting species and those that nest colonially. Nest and egg destruction, removal, and oiling are common and cost-effective. Removal of Shiny Cowbird (invasive brood parasite) eggs from endangered Yellow-Shouldered Blackbird (*Agelaius xanthomus*) in Puerto Rico was effective at reducing fledged cowbirds, which decreased the number of adult cowbirds that could parasitize further nests (Miller et al. 2016). Egg oiling, a method to limit oxygen diffusion to incubating eggs, is also effective because it keeps the adults engaged with incubation, thus reducing the opportunity to produce more clutches later in the same season. Egg oiling with widely available and inexpensive vegetable oil has successfully limited House Sparrow hatching, while also keeping adults incubating eggs for twice as long (Fernandez-Duque et al. 2019).

Whether nonlethal or lethal means of eradicating introduced-to-invader birds is appropriate for the species and situation, management should always consider and acknowledge societal perceptions, properly integrating them in management plans and strategies, which need to be revised periodically. Engaging people in a dialogue will increase their understanding of these types of birds, alert them to their impacts; support effective communication among themselves, and allow for the implementation of action plans. This is especially important because the impacts of invasions are not typically immediate, making awareness of the issue a challenge (Courchamp et al. 2017). Engaging all relevant social actors in these discussions and sharing opinions and reasons as to the species' management will reduce conflict between them and the management agency (Luna et al. 2019) while also enhancing communication and collaboration among people who differ in their ethics and their perceptions of invading birds.

Support or opposition to exotic species management depends on many factors. Generally, management is supported when the species is a nuisance or aggressive toward humans or native species (Ribeiro et al. 2021), visibly harms the ecosystem, and/or when social actors are knowledgeable of the species' invasive status (Luna et al. 2019). Opposition to management often occurs when the species is charismatic (Crowley et al. 2017, Jarić et al. 2020, Ribeiro et al. 2021, Steele and Pienaar 2021), perceived as friendly (Ribeiro et al. 2021), novel to people (Luna et al. 2019), or when people distrust expert opinion regarding their management (Jager et al. 2016). This is especially challenging because the public's support of introduced-to-invader species management efforts decreases when the target animals are birds (Bremner and Park 2007, Crowley et al. 2019, Steele and Pienaar 2021), possibly because of birds' charismatic features and their low perceived risk. In particular, species popular in the pet trade (e.g., parakeets and parrots) foster intense emotional connections to the public, which grow even further when many of these birds have a presence in popular culture. For instance, increased unregulated trade and increased availability of owls in online markets resulted in an increase in their popularity as pets in Thailand (Siriwat et al. 2020), whereas the normalization of keeping owls as pets was attributed to the "Harry Potter effect" in Indonesia (Nijman and Nekaris 2017). Anthropomorphization of exotic species may also increase their preference (Bjerke and Østdahl 2004, Borgi and Cirulli 2015). This is reflected in school-age children's heightened knowledge of exotic, large-bodied, charismatic species rather than species native to their region (De Melo et al. 2021).

Although people may be generally aware of the risks and consequences of biological invasions (Steele and Pienaar 2021), the evidence may be insufficient to influence values, attitudes, and behaviors regarding the management of bird invaders (Clayton and Myers 2009), or their impacts may not be acknowledged (Ribeiro et al. 2021). Abundant birds are the ones that urban residents interact with most often. Most park visitors could recognize the two invader species (i.e., Rock Dove and Monk Parakeet) in Santiago, Chile's public parks, with 66% of park visitors able to identify their names, compared with only 34% who could identify native species (Celis-Diez et al. 2017). Educational efforts should focus on increasing knowledge of local native species so that their value is increased compared to charismatic exotic species (Shapiro et al. 2017, De Melo et al. 2021). This can be done in formal educational settings or through environmental activities. For instance, people who participated in the NestWatch citizen science program in the United States had less favorable views of introduced-to-invader species compared to nonparticipants (Phillips et al. 2021). Similarly, park visitors in Seville, Spain, had a less favorable response to less charismatic exotic species (i.e., Rock Dove and Red Avadavat [*Amandava amandava*]) when they were provided information about the species exotic status (Luna et al. 2019).

Overall, managing exotic, introduced-to-invader, birds is no easy feat. Although the challenges regarding exotic species management are not unique to birds, due to the multifaceted choice of which management action to take and the varying levels of society support, when species are being harmful in some way, action should be taken only after extensive thought, discussion, and data-driven analyses.

References

Aagaard, K., and J. Lockwood. 2014. Exotic birds show lags in population growth. *Diversity and Distributions* 20:547–554.

Allen, C. R. 2006. Predictors of introduction success in the South Florida avifauna. *Biological Invasions* 8:491–500.

Ancillotto, L., D. Strubbe, M. Menchetti, and E. Mori. 2016. An overlooked invader? Ecological niche, invasion success and range dynamics of the Alexandrine parakeet in the invaded range. *Biological Invasions* 18:583–595.

Aramburú, R. M., and V. Corbalán. 2000. Dieta de pichones de cotorra *Myiopsitta monachus* (Aves, Psittacidae) en una población silvestre. *Ornitologia Neotropical* 11:241–246.

Aronson, M. F. J., F. A. La Sorte, C. H. Nilon, M. Katti, M. A. Goddard, C. A. Lepczyk, P. S. Warren, N. S. G. Williams, S. Cilliers, B. Clarkson, C. Dobbs, R. Dolan, M. Hedblom, S. Klotz, J. L. Kooijmans, I. Kühn, I. MacGregor-Fors, M. McDonnell, U. Mörtberg, P. Pyšek, S. Siebert, J. Sushinsky, P. Werner, and M. Winter. 2014. A global analysis of the impacts of urbanization on bird and plant diversity reveals key anthropogenic drivers. *Proceedings of the Royal Society Series B: Biological Sciences* 281: Article 20133330.

Auffray, J.-C., F. Vanlerberghe, and J. Britton-Davidian. 2008. The house mouse progression in Eurasia: A palaeontological and archaeozoological approach. *Biological Journal of the Linnean Society* 41:13–25.

Avery, M. L., E. C. Greiner, J. R. Lindsay, J. R. Newman, and S. Pruett-Jones. 2002. Monk parakeet management at electric utility facilities in south Florida. *Proceedings of the Vertebrate Pest Conference* 20:140–145.

Avery, M. L., C. A. Yoder, and E. A. Tillman. 2008. Diazacon inhibits reproduction in invasive Monk Parakeet populations. *Journal of Wildlife Management* 72:1449–1452.

Ayala, F. J. 2005. The evolution of organisms: A synopsis. Pages 1–26 in F. M. Wuketits and F. J. Ayala, eds. *Handbook of Evolution*. Wiley, New York.

Baños-Villalba, A., M. Carrete, J. L. Tella, J. Blas, J. Potti, C. Camacho, M. S. Diop, T. A. Marchant, S. Cabezas, and P. Edelaar. 2021. Selection on individuals of introduced species starts before the actual introduction. *Evolutionary Applications* 14:781–793.

Belinsky, K. L., T. C. Ellick, and S. L. LaDeau. 2019. Using a birdfeeder network to explore the effects of suburban design on invasive and native birds. *Avian Conservation and Ecology* 14: Article 2.

Bellard, C., P. Cassey, and T. M. Blackburn. 2016. Alien species as a driver of recent extinctions. *Biology Letters* 12: Article 20150623.

Beltrão, P., A. C. R. Gomes, C. I. Marques, S. Guerra, H. R. Batalha, and G. C. Cardoso. 2021. European breeding phenology of the invasive common waxbill, a sub-Saharan opportunistic breeder. *Acta Ethologica* 24:197–203.

Berigan, L. A., E. I. Greig, and D. N. Bonter. 2021. Urban House Sparrow (*Passer domesticus*) populations decline in North America. *Wilson Journal of Ornithology* 132:248–258.

Bialic-Murphy, L., O. G. Gaoue, and K. Kawelo. 2017. Microhabitat heterogeneity and a non-native avian frugivore drive the population dynamics of an island endemic shrub, *Cyrtandra dentata*. *Journal of Applied Ecology* 54:1469–1477.

Bjerke, T., and T. Østdahl. 2004. Animal-related attitudes and activities in an urban population. *Anthrozoös* 17:109–129.

Blackburn, T. M., P. Cassey, and J. L. Lockwood. 2009. The role of species traits in the establishment success of exotic birds. *Global Change Biology* 15:2852–2860.

Blackburn, T. M., and R. P. Duncan. 2001. Establishment patterns of exotic birds are constrained by non-random patterns in introduction. *Journal of Biogeography* 28:927–939.

Blackburn, T. M., J. L. Lockwood, and P. Cassey. 2009. *Avian Invasions: The Ecology and Evolution of Exotic Birds.* Oxford University Press, New York.

Blackburn, T. M., J. L. Lockwood, and P. Cassey. 2015. The influence of numbers on invasion success. *Molecular Ecology* 24:1942–1953.

Blackburn, T., N. Pettorelli, T. Katzner, M. Gompper, K. Mock, T. Garner, R. Altwegg, S. Redpath, and I. Gordon. 2010. Dying for conservation: Eradicating invasive alien species in the face of opposition. *Animal Conservation* 13:227–228.

Blackburn, T. M., P. Pyšek, S. Bacher, J. T. Carlton, R. P. Duncan, V. Jarošík, J. R. U. Wilson, and D. M. Richardson. 2011. A proposed unified framework for biological invasions. *Trends in Ecology & Evolution* 26:333–339.

Blackwell, B. F., T. W. Seamans, M. B. Pfeiffer, and B. N. Buckingham. 2021. European Starling nest-site selection given enhanced direct nest predation risk. *Wildlife Society Bulletin* 45:62–70.

Bomford, M., and R. Sinclair. 2002. Australian research on bird pests: Impact, management and future directions. *Emu – Austral Ornithology* 102:29–45.

Booy, O., A. C. Mill, H. E. Roy, A. Hiley, N. Moore, P. Robertson, S. Baker, M. Brazier, M. Bue, R. Bullock, S. Campbell, D. Eyre, J. Foster, M. Hatton-Ellis, J. Long, C. Macadam, C. Morrison-Bell, J. Mumford, J. Newman, D. Parrott, R. Payne, T. Renals, E. Rodgers, M. Spencer, P. Stebbing, M. Sutton-Croft, K. J. Walker, A. Ward, S. Whittaker, and G. Wyn. 2017. Risk management to prioritise the eradication of new and emerging invasive non-native species. *Biological Invasions* 19:2401–2417.

Borgi, M., and F. Cirulli. 2015. Attitudes toward animals among kindergarten children: Species preferences. *Anthrozoös* 28:45–59.

Bremner, A., and K. Park. 2007. Public attitudes to the management of invasive non-native species in Scotland. *Biological Conservation* 139:306–314.

Brown, N. S., and G. I. Wilson. 1975. A comparison of the ectoparasites of the House Sparrow (*Passer domesticus*) from North America and Europe. *The American Midland Naturalist* 94:154–165.

Burnett, K., B. Kaiser, B. A. Pitafi, and J. Roumasset. 2006. Prevention, eradication, and containment of invasive species: Illustrations from Hawaii. *Agricultural and Resource Economics Review* 35:63–77.

Campbell, T. S. 2001. The Monk Parakeet, *Myiopsitta monachus:* Invader of the month. Institute for Biological Invasions, University of Tennessee, Knoxville, TN.

Canning, G. 2011. Eradication of the invasive common myna, *Acridotheres tristis*, from Fregate Island, Seychelles. *Phelsuma* 19:43–53.

Cardador, L., J. L. Tella, J. D. Anadón, P. Abellán, and M. Carrete. 2019. The European trade ban on wild birds reduced invasion risks. *Conservation Letters* 12: Article e12631.

Cassey, P., T. M. Blackburn, D. Sol, R. P. Duncan, and J. L. Lockwood. 2004. Global patterns of introduction effort and establishment success in birds. *Proceedings of the Royal Society Series B: Biological Sciences* 271:S405–S408.

Cassey, P., M. Vall-Llosera, E. Dyer, and T. M. Blackburn. 2016. The biogeography of avian invasions: History, accident and market trade. Pages 37–54 in C.-C. João, ed. *Biological Invasions in Changing Ecosystems: Vectors, Ecological Impacts, Management and Predictions.* De Gruyter Open, Berlin.

Celis-Diez, J. L., C. E. Muñoz, S. Abades, P. A. Marquet, and J. J. Armesto. 2017. Biocultural homogenization in urban settings: Public knowledge of birds in city parks of Santiago, Chile. *Sustainability* 9: Article 485.

Charter, M., I. Izhaki, Y. Ben Mocha, and S. Kark. 2016. Nest-site competition between invasive and native cavity nesting birds and its implication for conservation. *Journal of Environmental Management* 181:129–134.

Clayton, S., and G. Myers. 2009. *Conservation Psychology: Understanding and Promoting Humancare for Nature.* Wiley-Blackwell, West Sussex, UK.

Conroy, M. J., and J. C. Senar. 2009. Integration of demographic analyses and decision modeling in support of management of invasive Monk Parakeets, an urban and agricultural pest. Pages 491–510 in D. L. Thomson, E. G. Cooch, and M. J. Conroy, eds. *Modeling Demographic Processes in Marked Populations*. Springer, Boston.

Courchamp, F., A. Fournier, C. Bellard, C. Bertelsmeier, E. Bonnaud, J. M. Jeschke, and J. C. Russell. 2017. Invasion biology: Specific problems and possible solutions. *Trends in Ecology & Evolution* 32:13–22.

Crates, R., A. Terauds, L. Rayner, D. Stojanovic, R. Heinsohn, C. Wilkie, and M. Webb. 2018. Spatially and temporally targeted suppression of despotic noisy miners has conservation benefits for highly mobile and threatened woodland birds. *Biological Conservation* 227:343–351.

Crooks, J. A. 2005. Lag times and exotic species: The ecology and management of biological invasions in slow-motion. *Écoscience* 12:316–329.

Crowley, S. L., S. Hinchliffe, and R. A. McDonald. 2017. Conflict in invasive species management. *Frontiers in Ecology and the Environment* 15:133–141.

Crowley, S. L., S. Hinchliffe, and R. A. McDonald. 2019. The parakeet protectors: Understanding opposition to introduced species management. *Journal of Environmental Management* 229:120–132.

Davis, L. R. 1974. The monk parakeet: A potential threat to agriculture. *Proceedings of the Vertebrate Pest Conference* 6:253–256.

DeGraaf, R. M., and D. D. Rudis. 1986. New England wildlife: Habitat, natural history, and distribution. General Technical Report NE-108. U.S. Department of Agriculture, Forest Service, Northeastern Forest Experiment Station, Broomall, PA.

De Grazio, J. W. 1978. World bird damage problems. *Proceedings of the Vertebrate Pest Conference* 8:9–24.

De Melo, E. P. C., J. Simiao-Ferreira, H. P. C. De Melo, B. S. Godoy, R. D. Daud, R. P. Bastos, and D. P. Silva. 2021. Exotic species are perceived more than native ones in a megadiverse country as Brazil. *Annals of the Brazilian Academy of Sciences* 93: Article e20191462.

Duncan, R. P., T. M. Blackburn, S. Rossinelli, and S. Bacher. 2014. Quantifying invasion risk: The relationship between establishment probability and founding population size. *Methods in Ecology and Evolution* 5:1255–1263.

Dyer, E. E., P. Cassey, D. W. Redding, B. Collen, V. Franks, K. J. Gaston, K. E. Jones, S. Kark, C. D. L. Orme, and T. M. Blackburn. 2017. The global distribution and drivers of alien bird species richness. *PLoS Biology* 15: Article e2000942.

Dyer, E. E., V. Franks, P. Cassey, B. Collen, R. C. Cope, K. E. Jones, Ç. H. Şekercioğlu, and T. M. Blackburn. 2016. A global analysis of the determinants of alien geographical range size in birds. *Global Ecology and Biogeography* 25:1346–1355.

Evans, T., S. Kumschick, Ç. H. Şekercioğlu, and T. M. Blackburn. 2018. Identifying the factors that determine the severity and type of alien bird impacts. *Diversity and Distributions* 24:800–810.

Farina, A., N. Pieretti, and N. Morganti. 2013. Acoustic patterns of an invasive species: The Red-Billed Leiothrix (*Leiothrix lutea* Scopoli 1786) in a Mediterranean shrubland. *Bioacoustics* 22:175–194.

Fava, G. A., and J. C. Acosta. 2016. Abundancia y diversidad de aves en ambientes con diferente grado de perturbacion en el monte de Argentina. *Ornitologia Neotropical* 27: Article 10.

Feare, C. J., J. van der Woude, P. Greenwell, H. A. Edwards, J. A. Taylor, C. S. Larose, P.-A. Ahlen, J. West, W. Chadwick, S. Pandey, K. Raines, F. Garcia, J. Komdeur, and A. de Groene. 2017. Eradication of common mynas *Acridotheres tristis* from Denis Island, Seychelles. *Pest Management Science* 73:295–304.

Fernandez-Duque, F., R. L. Bailey, and D. N. Bonter. 2019. Egg oiling as an effective management technique for limiting reproduction in an invasive passerine. *Avian Conservation and Ecology* 14: Article 20.

Fitzwater, W. D. 1988. Solutions to urban bird problems. *Proceedings of the Vertebrate Pest Conference* 13:254–259.

Flores, M., P. Lazo, G. Campbell, and A. Simeone. 2017. Breeding status of the Red-Tailed Tropicbird (*Phaethon rubricauda*) and threats to its conservation on Easter Island (Rapa Nui). *Pacific Science* 71:149–160.

Friedemann, G., Y. Leshem, L. Kerem, A. Bar-Massada, and I. Izhaki. 2017. Nest-site characteristics, breeding success and competitive interactions between two recently sympatric apex predators. *Ibis* 159:812–827.

Gandoy, F. A., K. Delhey, D. W. Winkler, G. Mangini, and J. I. Areta. 2019. Lower breeding success in a new range: No evidence for the enemy release hypothesis in South American Barn Swallows. *The Auk* 136: Article ukz050.

García-Arroyo, M., D. Santiago-Alarcon, J. Quesada, and I. MacGregor-Fors. 2020. Are invasive House Sparrows a nuisance for native avifauna when scarce? *Urban Ecosystems* 23:793–802.

GBIF Secretariat. 2021. *Myiopsitta monachus* (Boddaert, 1783). GBIF Backbone Taxonomy.

Gonçalves da Silva, A., J. Eberhard, T. Wright, M. Avery, and M. Russello. 2010. Data from: Genetic evidence for high propagule pressure and long-distance dispersal in monk parakeet (*Myiopsitta monachus*) invasive populations. Scholars Portal Dataverse.

González-Oreja, J. A., I. Zuria, P. Carbó-Ramírez, and G. M. Charre. 2018. Using variation partitioning techniques to quantify the effects of invasive alien species on native urban bird assemblages. *Biological Invasions* 20:2861–2874.

Grandi, G., M. Menchetti, and E. Mori. 2018. Vertical segregation by breeding ring-necked parakeets *Psittacula krameri* in northern Italy. *Urban Ecosystems* 21:1011–1017.

Grarock, K., C. R. Tidemann, J. Wood, and D. B. Lindenmayer. 2012. Is it benign or is it a pariah? Empirical evidence for the impact of the Common Myna (*Acridotheres tristis*) on Australian birds. *PLoS One* 7: Article e40622.

Gumede, S. T., and C. T. Downs. 2019. Sugar preference of invasive Common Mynas (*Sturnus tristis*). *Journal of Ornithology* 160:71–78.

Hayhow, D. B., G. Conway, M. A. Eaton, P. V. Grice, C. Hall, C. A. Holt, A. Kuepfer, D. G. Noble, S. Oppel, K. Risely, C. Stringer, D. A. Stroud, N. Wilkinson, and S. Wotton. 2015. The state of the UK's birds. RSPB, BTO,WWT, JNCC, NE, NIEA, NRW, and SNH, Sandy, Bedfordshire, UK.

Heinsohn, T. 2014. Animal translocation: Long-term human influences on the vertebrate zoogeography of Australasia (natural dispersal versus ethnophoresy). *Australian Zoologist* 32:351–376.

Hopkins, J. M., W. Edwards, J. M. Laguna, and L. Schwarzkopf. 2021. An endangered bird calls less when invasive birds are calling. *Journal of Avian Biology* 52: Article e02642.

Howell, A. H. 1932. Florida bird life. U.S. Department of Game and Fresh Water Fish.

Hughes, B. J., G. R. Martin, and S. J. Reynolds. 2017. Estimating the extent of seabird egg depredation by introduced Common Mynas on Ascension Island in the South Atlantic. *Biological Invasions* 19:843–857.

Hyman, J., and S. Pruett-Jones. 1995. Natural history of the Monk Parakeet in Hyde Park, Chicago. *Wilson Bulletin* 107:510–517.

Ibañez, L. M., J. M. Girini, F. X. Palacio, V. D. Fiorini, and D. Montalti. 2017. Interactions between the European Starling, *Sturnus vulgaris*, and native birds of Argentina for cavity use. *Revista Mexicana de Biodiversidad* 88:477–479.

International Union for Conservation of Nature and Natural Resources. 2021. *The IUCN Red List of Threatened Species*. International Union for Conservation of Nature and Natural Resources, Gland, Switzerland.

Jager, C., M. P. Nelson, L. Goralnik, and M. L. Gore. 2016. Michigan Mute Swan management: A case study to understand contentious natural resource management issues. *Human Dimensions of Wildlife* 21:189–202.

Jarić, I., F. Courchamp, R. A. Correia, S. L. Crowley, F. Essl, A. Fischer, P. González-Moreno, G. Kalinkat, X. Lambin, B. Lenzner, Y. Meinard, A. Mill, C. Musseau, A. Novoa, J. Pergl, P. Pyšek, K. Pyšková, P. Robertson, M. von Schmalensee, R. T. Shackleton, R. A. Stefansson, K. Štajerová, D. Veríssimo, and J. M. Jeschke. 2020. The role of species charisma in biological invasions. *Frontiers in Ecology and the Environment* 18:345–353.

Jauregui, A., E. Gonzalez, and L. N. Segura. 2021. Impacts of the invasive European Starling on two neotropical woodpecker species: Agonistic responses and reproductive interactions. *Emu – Austral Ornithology* 121:223–230.

Johnson, K., and B. D. Peer, eds. 2020. *Great-Tailed Grackle (Quiscalus mexicanus)*. Cornell Lab of Ornithology, Ithaca, NY.

Kumschick, S., and W. Nentwig. 2010. Some alien birds have as severe an impact as the most effectual alien mammals in Europe. *Biological Conservation* 143:2757–2762.

Lafleur, N. E., M. A. Rubega, and C. S. Elphick. 2007. Invasive fruits, novel foods, and choice: An investigation of European starling and American robin frugivory. *Wilson Journal of Ornithology* 119:429–438.

Larson, L. R., C. B. Cooper, and M. E. Hauber. 2016. Emotions as drivers of wildlife stewardship behavior: Examining citizen science nest monitors' responses to invasive House Sparrows. *Human Dimensions of Wildlife* 21:18–33.

Lei, Y., and Q. Liu. 2021. Tolerance niche expansion and potential distribution prediction during Asian openbill bird range expansion. *Ecology and Evolution* 11:5562–5574.

Le Louarn, M., B. Couillens, M. Deschamps-Cottin, and P. Clergeau. 2016. Interference competition between an invasive parakeet and native bird species at feeding sites. *Journal of Ethology* 34:291–298.

Lermite, F., S. Kark, C. Peneaux, and A. S. Griffin. 2021. Breeding success and its correlates in native versus invasive secondary cavity-nesting birds. *Emu – Austral Ornithology* 121:261–266.

Le Roux, D. S., K. Ikin, D. B. Lindenmayer, G. Bistricer, A. D. Manning, and P. Gibbons. 2016. Effects of entrance size, tree size and landscape context on nest box occupancy: Considerations for management and biodiversity offsets. *Forest Ecology and Management* 366:135–142.

Leung, B., D. M. Lodge, D. Finnoff, J. F. Shogren, M. A. Lewis, and G. Lamberti. 2002. An ounce of prevention or a pound of cure: Bioeconomic risk analysis of invasive species. *Proceedings of the Royal Society Series B: Biological Sciences* 269:2407–2413.

Long, J. L. 1981. *Introduced Birds of the World: The Worldwide History, Distribution, and Influence of Birds Introduced to New Environments.* Universe Books, New York.

Lowe, S., M. Browne, S. Boudjelas, and M. De Poorter. 2000. 100 of the world's worst invasive alien species: A selection from the Global Invasive Species Database. Paged 715–716 in D. Simberloff and M. Rejmanek, eds. *Encyclopedia of Biological Invasions.* University of California Press, Berkeley.

Lowther, P. E., and C. L. Cink. 2020. House Sparrow (*Passer domesticus*), version 1.0. *In* S. M. Billerman, ed. *Birds of the World.* Cornell Lab of Ornithology, Ithaca, NY.

Luna, Á., P. Edelaar, and A. Shwartz. 2019. Assessment of social perception of an invasive parakeet using a novel visual survey method. *NeoBiota* 46:71–89.

MacFarlane, A. E. T., D. Kelly, and J. V. Briskie. 2016. Introduced blackbirds and song thrushes useful substitutes for lost mid-sized native frugivores, or weed vectors? *New Zealand Journal of Ecology* 40:80–87.

MacGregor-Fors, I., R. Calderón–Parra, A. Meléndez–Herrada, S. López–López, and J. E. Schondube. 2011. Pretty, but dangerous! Records of non-native Monk Parakeets (*Myiopsitta monachus*) in Mexico. *Revista Mexicana de Biodiversidad* 82:1053–1056.

MacGregor-Fors, I., M. García-Arroyo, O. H. Marín-Gómez, and J. Quesada. 2020. On the meat scavenging behavior of House Sparrows (*Passer domesticus*). *Wilson Journal of Ornithology* 132:188–191.

MacGregor-Fors, I., L. Morales-Pérez, J. Quesada, and J. E. Schondube. 2010. Relationship between the presence of House Sparrows (*Passer domesticus*) and Neotropical bird community structure and diversity. *Biological Invasions* 12:87–96.

MacGregor-Fors, I., L. Vázquez, J. H. Vega-Rivera, and J. E. Schondube. 2009. Non-exotic invasion of Great-Tailed Grackles *Quiscalus mexicanus* in a tropical dry forest reserve. *Ardea* 97:367–369.

Machovsky-Capuska, G. E., A. M. Senior, S. P. Zantis, K. Barna, A. J. Cowieson, S. Pandya, C. Pavard, M. Shiels, and D. Raubenheimer. 2015. Dietary protein selection in a free-ranging urban population of common myna birds. *Behavioral Ecology* 27:219–227.

Magory Cohen, T., and R. Dor. 2019. The effect of local species composition on the distribution of an avian invader. *Scientific Reports* 9: Article 15861.

Maldonado-Coelho, M., M. Â. Marini, F. R. Do Amaral, and R. Ribon. 2017. The invasive species rules: Competitive exclusion in forest avian mixed-species flocks in a fragmented landscape. *Revista Brasileira de Ornitologia* 25:54–59.

Martín, L. F., and E. H. Bucher. 1993. Natal dispersal and first breeding age in monk parakeets. *The Auk* 110:930–933.

Martin-Albarracin, V. L., M. A. Nuñez, and G. C. Amico. 2018. Non-redundancy in seed dispersal and germination by native and introduced frugivorous birds: Implications of invasive bird impact on native plant communities. *Biodiversity and Conservation* 27:3793–3806.

Massaro, M., M. Stanbury, and J. V. Briskie. 2013. Nest site selection by the endangered black robin increases vulnerability to predation by an invasive bird. *Animal Conservation* 16:404–411.

Miller, P. S., R. C. Lacy, R. Medina-Miranda, R. López-Ortiz, and H. Díaz-Soltero. 2016. Confronting the invasive species crisis with metamodel analysis: An explicit, two-species demographic assessment of an endangered bird and its brood parasite in Puerto Rico. *Biological Conservation* 196:124–132.

Mori, E., L. Ancillotto, M. Menchetti, and D. Strubbe. 2017. "The early bird catches the nest": Possible competition between scops owls and ring-necked parakeets. *Animal Conservation* 20:463–470.

Morrison, S. A., A. J. DeNicola, K. Walker, D. Dewey, L. Laughrin, R. Wolstenholme, and N. Macdonald. 2016. An irruption interrupted: Eradication of wild turkeys *Meleagris gallopavo* from Santa Cruz Island, California. *Oryx* 50:121–127.

Moulton, M. P., and W. P. Cropper, Jr. 2019. Propagule pressure does not consistently predict the outcomes of exotic bird introductions. *PeerJ* 7: Article e7637.

Muñoz, A.-R., and R. Real. 2006. Assessing the potential range expansion of the exotic monk parakeet in Spain. *Diversity and Distributions* 12:656–665.

Narosky, S., and D. Yzurieta. 2005. *Guía para la identificación de las aves de Argentina y Uruguay*. Asociacion Ornitológica del Plata, Buenos Aires, Argentina.

Nathan, R., W. M. Getz, E. Revilla, M. Holyoak, R. Kadmon, D. Saltz, and P. E. Smouse. 2008. A movement ecology paradigm for unifying organismal movement research. *Proceedings of the National Academy of Sciences of the USA* 105:19052–19059.

Nijman, V., M. Campera, M. A. Imron, A. Ardiansyah, A. Langgeng, T. Dewi, K. Hedger, R. Hendrik, and K. A.-I. Nekaris. 2021. The role of the songbird trade as an anthropogenic vector in the spread of invasive non-native mynas in Indonesia. *Life* 11: Article 814.

Nijman, V., and K. A.-I. Nekaris. 2017. The Harry Potter effect: The rise in trade of owls as pets in Java and Bali, Indonesia. *Global Ecology and Conservation* 11:84–94.

O'Rourke, K. H., and J. G. Williamson. 2004. Once more: When did globalisation begin? *European Review of Economic History* 8:109–117.

Pablo López, R. E. 2009. Primer registro del perico argentino (*Myiopsitta monachus*) en Oaxaca, México. *Huitzil* 10:48–51.

Peacock, D. S., B. J. van Rensburg, and M. P. Robertson. 2007. The distribution and spread of the invasive alien common myna, *Acridotheres tristis* L. (Aves: Sturnidae), in southern Africa. *South African Journal of Science* 103:465–473.

Peña-Peniche, A., C. Mota-Vargas, M. García-Arroyo, and I. MacGregor-Fors. 2021. On the North American invasion of the House Sparrow and its absence in the Yucatan Peninsula. *Avian Conservation and Ecology* 16: Article 18.

Peneaux, C., G. E. Machovsky-Capuska, D. Raubenheimer, F. Lermite, C. Rousseau, T. Ruhan, J. C. Rodger, and A. S. Griffin. 2017. Tasting novel foods and selecting nutrient content in a highly successful ecological invader, the common myna. *Journal of Avian Biology* 48:1432–1440.

Pereira, P. F., C. Godinho, M. J. Vila-Viçosa, P. G. Mota, and R. Lourenço. 2017. Competitive advantages of the red-billed leiothrix (*Leiothrix lutea*) invading a passerine community in Europe. *Biological Invasions* 19:1421–1430.

Pereira, P. F., R. Lourenço, and P. G. Mota. 2018. Behavioural dominance of the invasive red-billed leiothrix (*Leiothrix lutea*) over European native passerine-birds in a feeding context. *Behaviour* 155:55–67.

Pereira, P. F., R. Lourenço, and P. G. Mota. 2020. Two songbird species show subordinate responses to simulated territorial intrusions of an exotic competitor. *Acta Ethologica* 23:143–154.

Pereyra, P. J. 2016. Revisiting the use of the invasive species concept: An empirical approach. *Austral Ecology* 41:519–528.

Phillips, T. B., R. L. Bailey, V. Martin, H. Faulkner-Grant, and D. N. Bonter. 2021. The role of citizen science in management of invasive avian species: What people think, know, and do. *Journal of Environmental Management* 280: Article 111709.

Pimentel, D., R. Zuniga, and D. Morrison. 2005. Update on the environmental and economic costs associated with alien-invasive species in the United States. *Ecological Economics* 52:273–288.

Pimm, S. L., ed. 1989. *Theories of Predicting Success and Impact of Introduced Species.* Wiley, New York.

Pruett-Jones, S., J. R. Newman, C. M. Newman, M. L. Avery, and J. R. Lindsay. 2007. Population viability analysis of monk parakeets in the United States and examination of alternative management strategies. *Human–Wildlife Conflicts* 1:35–44.

Pyšek, P., V. Jarošík, P. E. Hulme, J. Pergl, M. Hejda, U. Schaffner, and M. Vilà. 2012. A global assessment of invasive plant impacts on resident species, communities and ecosystems: The interaction of impact measures, invading species' traits and environment. *Global Change Biology* 18:1725–1737.

Ravinet, M., T. O. Elgvin, C. Trier, M. Aliabadian, A. Gavrilov, and G.-P. Sætre. 2018. Signatures of human-commensalism in the house sparrow genome. *Proceedings of the Royal Society Series B: Biological Sciences* 285: Article 20181246.

Reino, L., R. Figueira, P. Beja, M. B. Araújo, C. Capinha, and D. Strubbe. 2017. Networks of global bird invasion altered by regional trade ban. *Science Advances* 3: Article e1700783.

Ribeiro, J., I. Carneiro, A. Nuno, M. Porto, P. Edelaar, Á. Luna, and L. Reino. 2021. Investigating people's perceptions of alien parakeets in urban environments. *European Journal of Wildlife Research* 67: Article 45.

Ricciardi, A., M. F. Hoopes, M. P. Marchetti, and J. L. Lockwood. 2013. Progress toward understanding the ecological impacts of nonnative species. *Ecological Monographs* 83:263–282.

Robinson, S. A., G. B. Baker, and C. Barclay. 2020. Controlling the rainbow lorikeet in Tasmania: Is it too late? *Emu – Austral Ornithology* 120:286–294.

Roll, U., T. Dayan, and D. Simberloff. 2008. Non-indigenous terrestrial vertebrates in Israel and adjacent areas. *Biological Invasions* 10:659–672.

Rowe, L. N. 2018. *Rufous (Selasphorus rufus) and Anna's Hummingbirds (Calypte anna) Population Changes in Western Washington.* University of Washington, Seattle.

Russell, J. C., and T. M. Blackburn. 2017. Invasive alien species: Denialism, disagreement, definitions, and dialogue. *Trends in Ecology & Evolution* 32:312–314.

Russell, N. 2012. *Social Zooarchaeology: Humans and Animals in Prehistory.* Cornell University Press, Ithaca, NY.

Sakai, A. K., F. W. Allendorf, J. S. Holt, D. M. Lodge, J. Molofsky, K. A. With, S. Baughman, R. J. Cabin, J. E. Cohen, N. C. Ellstrand, D. E. McCauley, P. O'Neil, I. M. Parker, J. N. Thompson, and S. G. Weller. 2001. The population biology of invasive species. *Annual Review of Ecology and Systematics* 32:305–332.

Schmitz, D. C., and D. Simberloff. 1997. Biological invasions: A growing threat. *Issues in Science and Technology* 13:33–40.

Scott, A., R. C. Power, V. Altmann-Wendling, M. Artzy, M. A. S. Martin, S. Eisenmann, R. Hagan, D. C. Salazar-García, Y. Salmon, D. Yegorov, I. Milevski, I. Finkelstein, P. W. Stockhammer, and C. Warinner. 2021. Exotic foods reveal contact between South Asia and the Near East during the second millennium BCE. *Proceedings of the National Academy of Sciences of the USA* 118: Article e2014956117.

Seddon, P. J., W. M. Strauss, and J. Innes. 2012. Animal translocations: What are they and why do we do them? Pages 1–32 in J. G. Ewen, D. P. Armstrong, K. A. Parker, and P. J. Seddon, eds. *Reintroduction Biology: Integrating Science and Management.* Wiley, Hoboken, NJ.

Senar, J. C., J. G. Carrillo-Ortiz, A. Ortega-Segalerva, F. S. E. Dawson Pell, J. Pascual, L. Arroyo, D. Mazzoni, T. Montalvo, and B. J. Hatchwell. 2019. The reproductive capacity of Monk Parakeets *Myiopsitta monachus* is higher in their invasive range. *Bird Study* 66:136–140.

Shapiro, H. G., M. N. Peterson, K. T. Stevenson, K. N. Frew, and R. B. Langerhans. 2017. Wildlife species preferences differ among children in continental and island locations. *Environmental Conservation* 44:389–396.

Shaw, L. M., D. Chamberlain, and M. Evans. 2008. The House Sparrow *Passer domesticus* in urban areas: Reviewing a possible link between post-decline distribution and human socioeconomic status. *Journal of Ornithology* 149:293–299.

Shivambu, T. C., N. Shivambu, and C. T. Downs. 2022. Citizen science survey of non-native Rose-ringed Parakeets *Psittacula krameri* in the Durban metropole, KwaZulu-Natal, South Africa. *African Zoology* 57:90–97.

Simberloff, D., J.-L. Martin, P. Genovesi, V. Maris, D. A. Wardle, J. Aronson, F. Courchamp, B. Galil, E. García-Berthou, M. Pascal, P. Pyšek, R. Sousa, E. Tabacchi, and M. Vilà. 2013. Impacts of biological invasions: What's what and the way forward. *Trends in Ecology & Evolution* 28:58–66.

Siriwat, P., K. A. I. Nekaris, and V. Nijman. 2020. Digital media and the modern-day pet trade: A test of the "Harry Potter effect" and the owl trade in Thailand. *Endangered Species Research* 41:7–16.

Slabbekoorn, H. 2013. Songs of the city: Noise-dependent spectral plasticity in the acoustic phenotype of urban birds. *Animal Behaviour* 85:1089–1099.

Soh, M. C. K., R. Y. T. Pang, B. X. K. Ng, B. P. Y. H. Lee, A. H. B. Loo, and K. B. H. Er. 2021. Restricted human activities shift the foraging strategies of feral pigeons (*Columba livia*) and three other commensal bird species. *Biological Conservation* 253: Article 108927.

Sol, D., R. P. Duncan, T. M. Blackburn, P. Cassey, and L. Lefebvre. 2005. Big brains, enhanced cognition, and response of birds to novel environments. *Proceedings of the National Academy of Sciences of the USA* 102:5460–5465.

Sol, D., C. González-Lagos, D. Moreira, J. Maspons, and O. Lapiedra. 2014. Urbanisation tolerance and the loss of avian diversity. *Ecology Letters* 17:942–950.

Sol, D., and L. Lefebvre. 2000. Behavioural flexibility predicts invasion success in birds introduced to New Zealand. *Oikos* 90:599–605.

Sol, D., J. Maspons, M. Vall-llosera, I. Bartomeus, G. E. García-Peña, J. Piñol, and R. P. Freckleton. 2012. Unraveling the life history of successful invaders. *Science* 337:580–583.

South, J. M., and S. Pruett-Jones. 2000. Patterns of flock size, diet, and vigilance of naturalized Monk Parakeets in Hyde Park, Chicago. *The Condor* 102:848–854.

Spano, S., and G. Truffi. 1986. Il Parrocchetto dal collare, *Psittacula krameri*, allo stato libero in Europa, con particolare riferimento alle presenze in Italia, e primi dati sul Pappagallo monaco, *Myiopsitta monachus*. *Rivista Italiana di Ornitologia* 56:231–239.

Steele, Z. T., and E. F. Pienaar. 2021. Knowledge, reason and emotion: Using behavioral theories to understand people's support for invasive animal management. *Biological Invasions* 23:3513–3527.

Su, S., P. Cassey, and T. M. Blackburn. 2014. Patterns of non-randomness in the composition and characteristics of the Taiwanese bird trade. *Biological Invasions* 16:2563–2575.

Thibault, M., F. Masse, A. Pujapujane, G. Lannuzel, L. Bordez, M. A. Potter, B. Fogliani, É. Vidal, and F. Brescia. 2018. "Liaisons dangereuses": The invasive red-vented bulbul (*Pycnonotus cafer*), a disperser of exotic plant species in New Caledonia. *Ecology and Evolution* 8:9259–9269.

Tillman, E. A., A. C. Genchi, J. R. Lindsay, J. R. Newman, and M. L. Avery. 2004. Evaluation of trapping to reduce Monk Parakeet populations at electric utility facilities. Pages 126–129 in *21th Vertebrate Pest Conference* University of California, Davis.

Tillman, E. A., A. Van Doom, and M. L. Avery. 2000. Bird damage to tropical fruit in south Florida. Pages 47–59 in *Wildlife Damage Management Conferences—Proceedings*. https://digitalcommons.unl.edu/icwdm_wdmconfproc/13

Valéry, L., H. Fritz, J.-C. Lefeuvre, and D. Simberloff. 2009. Invasive species can also be native. *Trends in Ecology & Evolution* 24:585–586.

Vall-llosera, M., and P. Cassey. 2017a. Leaky doors: Private captivity as a prominent source of bird introductions in Australia. *PLoS One* 12: Article e0172851.

Vall-llosera, M., and P. Cassey. 2017b. "Do you come from a land down under?" Characteristics of the international trade in Australian endemic parrots. *Biological Conservation* 207:38–46.

Vall-llosera, M., and D. Sol. 2009. A global risk assessment for the success of bird introductions. *Journal of Applied Ecology* 46:787–795.

Viana, I. R., D. Strubbe, and J. J. Zocche. 2016. Monk parakeet invasion success: A role for nest thermoregulation and bactericidal potential of plant nest material? *Biological Invasions* 18:1305–1315.

Wang, S., Y. Ho, and L. M. Chu. 2018. Diet and feeding behavior of the critically endangered Yellow-Crested Cockatoo (*Cacatua sulphurea*) in a nonnative urban environment. *Wilson Journal of Ornithology* 130:746–754.

Wehtje, W. 2003. The range expansion of the Great-Tailed Grackle (*Quiscalus mexicanus* Gmelin) in North America since 1880. *Journal of Biogeography* 30:1593–1607.

Werner, S. J., S. T. DeLiberto, H. E. McLean, K. E. Horak, and K. C. VerCauteren. 2021. Toxicity of sodium nitrite-based vertebrate pesticides for European starlings (*Sturnus vulgaris*). *PLoS One* 16: Article e0246277.

White, J. G., M. J. Antos, J. A. Fitzsimons, and G. C. Palmer. 2005. Non-uniform bird assemblages in urban environments: The influence of streetscape vegetation. *Landscape and Urban Planning* 71:123–135.

Yap, C. A. M., N. S. Sodhi, and B. W. Brook. 2002. Roost characteristics of invasive mynas in Singapore. *Journal of Wildlife Management* 66:1118–1127.

Zeng, L., J. T. Rotenberry, M. Zuk, T. K. Pratt, and Z. Zhang. 2016. Social behavior and cooperative breeding in a precocial species: The Kalij Pheasant (*Lophura leucomelanos*) in Hawaii. *The Auk* 133:747–760.

ZoBell, V. M., and B. J. Furnas. 2017. Impacts of land use and invasive species on native avifauna of Mo'orea, French Polynesia. *PeerJ* 5: Article e3761.

SECTION II
EVOLUTION

6
Commentary on Evolution in Ornithology

Leo Joseph

> Systematics has had a remarkable renaissance during the last generation.
> —E. Mayr (1969, p. vii)

So wrote Ernst Mayr more than 50 years ago in the Preface to his 1969 book *Principles of Systematic Zoology*. How true these words remain! In 1983's *Perspectives in Ornithology* (Brush and Clark 1983), George Barrowclough's chapter and John Avise's commentary necessarily focused on what we had learned from protein evolution (Avise 1983, Barrowclough 1983). Routine DNA sequencing was still some years away, as foreshadowed in the final paragraph of Gerald Shields's chapter on the avian genome (Shields 1983). The chapters in the present book on avian evolution show the extent of the renaissance and indeed revolution that has occurred since 1983. What remains as true as ever is Barrowclough's insistence that it is process we are interested in understanding through pattern. I hope my commentary will aid in that regard and that it will help students entering the field find their way through a burgeoning, if not already bewildering, literature.

In the same opening paragraph, Mayr (1969) stated that systematics is a major integral branch of biology and that it has been important in initiating the entire field of population biology including population genetics. Heady stuff! Although always a believer in this, I have long felt that the many ornithologists who worked in ecology, physiology, behavior, and so on were probably irked by claims such as Mayr's as fanciful and bombastic. This book can show that this claim has really come into its own. I hope my commentary will show just how widely improved phylogenetic frameworks have revolutionized so much of our biological understanding of birds and of course any other group of organisms.

Despite the less than universally admired roguishness that has been well documented elsewhere in the work of Charles Sibley and Jon Ahlquist (e.g., Lewin 1988a, 1988b), it nonetheless drove a revolution in the study of higher level avian relationships that has now inspired several generations of researchers and shows no signs of stopping. In Chapter 7 of this book, Sushma Reddy's sweeping overview of where our knowledge of avian phylogeny now stands neatly illustrates where that initial revolution has taken us. She shows that despite tremendous progress such as in resolving the Magnificent Seven of robustly supported major clades, there are still some

Leo Joseph, *Commentary on Evolution in Ornithology*. In: *New Perspectives in Ornithology*.
Edited by: Scott V. Edwards and J. Michael Reed, Oxford University Press.
DOI: 10.1093/oso/9780197787670.003.0006

frustratingly intransigent problems in avian phylogeny, such as the relationships of the hoatzin *Opisthocomus hoazin*. Yet, and I stress in large part due to her own innovatively analytical work, she also has been able to provide methodological guidance in how to gather more genomic data in a targeted manner, not simply adopting a "throw more data at the problem" approach. The thought of being able to do that back when Sibley and Ahlquist were working with what is now seen as the blunt tool of DNA–DNA hybridization was more or less unimaginable. Conversely, a particular revolution in knowledge of one group of birds that Sibley and Ahlquist fueled was that of our understanding of the origin and evolution of passerines, especially the oscines, or songbirds. This now brilliant corpus of work has been reviewed by Reddy. It is a foundation with which to better understand aspects of passerine biology such as the evolution of nectarivory (Toda et al. 2021) and female song (Hall and Langmore 2017) and ornamentation (Enbody et al. 2022) while still fueling some debates about precise biogeographical history of passerines (Oliveros et al. 2019). I take the opportunity to pay tribute to one non-systematist, Eleanor Russell, whose vision saw her become an "early adopter" in her brief but powerful review of what the new knowledge of passerine systematics meant for the evolution of cooperative breeding (Russell 1989). And I cannot omit mention of how phylogeny of ratites has shown how flightlessness has evolved on more than one occasion and that these birds have not just been sitting around and rafting on Gondwanan fragments. These all deliver on that promise of systematics underpinning the rest of biology.

The roll call of papers that refine our understanding of relationships and systematic position of many, many species and genera and which arise from improved phylogenies is now vast and runs the taxonomic range from order down to species. Essentially, many of these papers make a case for why a species or genus thought to have been a member of one group of birds simply belongs in another group altogether. Too numerous to collate individually, some spectacular and still inspirational examples corroborated by non-molecular data warrant mention, however. Arguably, the showstopper of any such example is that flamingos and grebes are each other's closest living relatives. Surely unexpected to many, and no doubt seeming preposterous at first to many non-ornithologists, it is probably the most strongly corroborated of findings in recent avian phylogenetics. Non-molecular data have also provided a long list of traits supporting the alignment of these two groups, and the list extends even to their cestode parasites (G. Mayr 2011). Another finding, although now decades old, is one I never tire of—that is, the vastly improved understanding of the position of Australo-Papuan oscines in the passerine tree and that they are not derived from Northern Hemisphere ecological counterparts and namesakes (Sibley and Ahlquist 1985).

At the family level, some of these findings helped pave the way for others, such as Beresford et al. (2005) and Alström et al. (2014), to establish new families Stenostiridae and Elachuridae, respectively, within the Passerida. The former accommodates several genera, some of which have since been transferred to this family from as far away as the corvoid radiation—for example, Yellow-Bellied Fairy-Fantail *Chelidorhynx hypoxanthus* (Fuchs et al. 2009). Also, the predominantly yellow-plumaged goldeneye *Pachycare flavogriseum* of New Guinea, long thought to be a pachycephalid, is now considered an acanthizid (Norman et al. 2009). The examples are many, and they will surely continue appearing and sharpening our knowledge of avian evolution.

At the ordinal level, new work is bringing phylogenetic clarity and resolution to major groups of birds within the Magnificent Seven. Parrots are an example for which a review of multilocus studies and new ultraconserved element–based genomic work (Provost et al. 2018, Smith et al. 2023) have brought stability to how we understand relationships within and among parrots. These have enabled downstream biogeographic analyses of the entire order (Selvatti et al. 2022).

Reddy's concluding observations will inspire and encourage future students of avian phylogeny. It is true that a polytomy at the base of Neoaves might not be what we all hoped for 20 years ago, but Reddy wisely notes that this does not mean we have reached the limits of what we can know. She stresses that tomorrow's phylogeneticists and bioinformaticians are who we need to achieve more innovative ways of evaluating the data. Similarly, she advocates for improvements and new models of genomic evolution. Increases in the computation for handling the complexities of how DNA and genomes evolve will help match our capacity to produce vast amounts of data. Finally, in her concluding remarks, Reddy refers to the integration of morphology, fossils, and paleoenvironmental data in phylogenomics. This relates to how important museum collections (Chapter 11) and the ever-growing array of methods in fields such as geometric morphometrics will be (see also Bravo et al. 2019, 2021).

Another revolution in avian evolution has been in the study of fossil birds and indeed in the number of avian paleontologists working on those fossils. In Chapter 8, Jingmai O'Connor details the flow of discoveries emerging from the Mesozoic especially of northeastern China and how they relate to the early evolution of birds. Her chapter intentionally focuses on that critical time when the radiation of the enantiornithines came to an end and the explosive radiation of the Neornithes, as we know them today, began. Reddy's chapter picks up that thread from the signal in molecular data. Revolutions have been continuing in paleontology of the Neornithes from the late Cenozoic and Quaternary, however. Gerald Mayr has been a leading exponent in synthesizing and interpreting these findings (e.g., Mayr 2016). Rivaling the older Mesozoic and early Cenozoic deposits from China are the Quaternary deposits illuminating Southern Hemisphere bird evolution. In particular, deposits in Australia and New Zealand, such as at Riversleigh and St. Bathans, respectively, have seen prolific growth in the literature. Timely reviews of these fossil assemblages are provided in Worthy et al. (2007, 2017), De Pietri et al. (2016), Nguyen et al. (2016), and Worthy and Nguyen (2020).

Studies Bridging Phylogeny and Speciation

I note a trend in integrating phylogeny of entire avifaunas with landscape evolution to reveal the major drivers of the evolution of the avifauna. Some examples concern speciation in the honeyeaters (Meliphagidae) of Australia (Marki et al. 2019, Hay et al. 2022), the long-term assembly of the New Guinean avifauna (Kennedy et al. 2022), and some impressive examples from Neotropical birds (suboscines: Derryberry et al. 2011; Harvey, Aleixo, et al. 2017; Harvey, Seeholzer, et al. 2017b; Harvey et al. 2020; Singhal et al. 2021; grassland and savanna birds: van Els et al. 2021, Norambuena and van Els 2021). For example, Singhal et al. (2021) posited that understanding how and why hybridization varies across geographic regions or lineages may reveal why speciation is more frequent or occurs differently in particular situations. They found that introgression is highest where species occur in close geographic proximity and in regions with more dynamic climates since the Pleistocene. Winger (2017) and Peñalba et al. (2019) address similar issues in other taxonomic and geographic contexts. Conversely, Provost et al. (2022) have explored how landscape evolution has shaped genomic landscapes through varying levels of population structuring and admixture between North American deserts.

Speciation and Adaptation

Since Poelstra et al.'s (2014) demonstration of most of the genome moving freely across taxonomically recognized species boundaries while just a few key loci may be involved in the phenotypic differentiation, whether by neutral or adaptive processes, the genomic era has been making clear just how porous species boundaries are. Introgression and how long it can occur after speciation (Joseph, Drew, et al. 2019; Nwankwo et al. 2019) and its changing rate after speciation (Pulido-Santacruz et al. 2020) are undergoing a revitalized period of study and especially notable is the number of studies relevant to hummingbirds (e.g., Rodríguez-Gómez and Ornelas 2015, Jiménez and Ornelas 2016, Palacios et al. 2019). Simultaneously, some genes are being repeatedly identified as important in phenotypic differentiation, whereas other studies suggest more idiosyncratic appearance of other genes (review of plumage genes in Peñalba et al. 2022; review of bill morphology genes in Lamichhaney et al. 2015, 2016). Mitonuclear discordance and mitochondrial captures are being uncovered (Irwin et al. 2009, Shipham et al. 2016, Ferreira et al. 2018); the relative roles of selection, drift, and genomic architecture as their drivers surely comprise rich research areas for researchers (Bonnet et al. 2017, Zhang et al. 2019, McElroy et al. 2020). Far from automatically indicating a need to revise taxonomy (cf. Gutiérrez-Zuluaga et al. 2021, Joseph et al. 2021), introgression and mitochondrial captures are windows to fascinating molecular biology and a springboard for reconciling the complexity that different lenses of molecular variation bring to understanding an organism's single evolutionary history.

Distilling more key points from the burgeoning literature of speciation genomics has challenged me. Foremost appears to be the intersection of genomic architecture, diversity statistics and what they do or do not mean (see Cruickshank and Hahn 2014), hybrid zones, and the tension between hybridization and gene flow impeding differentiation versus adaptation and differentiation promoting it. Comparisons of radiations of closely related species (e.g., *Sporophila* seedeaters: Campagna et al. 2012, 2017; Hejase et al. 2020; *Lonchura* munias: Faust-Stryjewski and Sorenson 2017) and careful, systematic comparisons of pairs and trios of species (Delmore et al. 2018, Irwin 2018, Wang et al. 2020, Natola et al. 2022, Peñalba et al. 2022) are paving the way for more insights from similar studies. For example, islands of differentiation in the genome and areas of low recombination are frequent foci of study. How do they relate either to the possibility of being close to chromosomal features such as centromeres and telomeres (Ellegren et al. 2012) or to "speciation genes" that may restrict gene flow? Is there selection against gene flow or selection, positive or purifying, that affects linked neutral sites at specific genomic regions? Selective sweeps early in speciation may also be involved (Delmore et al. 2018, Hejase et al. 2020). Delmore et al. (2018) suggest that idiosyncratic patterns may especially apply early in speciation but that commonality and repeatability increase further along the speciation continuum. Irwin et al. (2016) proposed the "sweep-before-differentiation" model in which some genomic regions undergo selective sweeps over a broad geographic area followed by within-population, selection-induced reductions in variation. Genomic regions of high relative differentiation may have moved among populations more recently than other genomic regions. These are just some of the areas in which we are sure to see more revolutionary findings in coming years.

Phylogeography and Speciation

Phylogeography or its derivatives were not indexed in Trevor Price's superb review of speciation in birds (Price 2008), nor are they mentioned in David Toews and Elizabeth Scordato's fine review in Chapter 9 of this volume. Phylogeography's original definition as a bridge between population genetics and systematics (Avise et al. 1987) should encourage anyone interested in speciation to delve deep into how phylogeography reveals what happens to standing genetic variation before, during, and after speciation. Table 6.1 is an entrée to literature on avian phylogeographic studies worldwide and includes some newer methodological approaches and studies that were designed under the banner of comparative phylogeography (see Edwards et al. 2022, McGaughran et al. 2022, Ribas et al. 2022). Phylogeography surely is a guide to where species limits need to be reassessed and the processes driving recognition of two or more species where previously fewer were recognized. Phylogeography detects introgression that should prompt reassessment but not necessarily changes in species limits, and thus how speciation has proceeded. Edwards et al. (2016) explore in detail the incorporation of introgression and reticulate patterns in phylogeny and how they relate to phylogeography. That in turn informs understanding of the processes of gene flow in terms of different categories of markers and their role in maintaining reproductive isolation and driving speciation. Burley et al.'s (2023) analysis of the Blue-Faced Honeyeater *Entomyzon cyanotis* across a suite of present-day sea barriers in northern Australia and southern New Guinea illustrates all of those points and even expands out to involve sex chromosome evolution, which is itself an emerging field (Kawakami et al. 2014, Irwin 2018, Lopez et al. 2021). Toews and Scordato's chapter illustrates these endpoints, but I stress that they are not the only points that

Table 6.1 Phylogeographic Studies in Birds Emphasizing Newer Methodological Approaches and Comparative Phylogeographic Studies of Multiple Species Across Single Biogeographic Zones

Region, Location	References
Australo-Papua	
New Guinea–Australia	Peñalba et al. (2019), Burley et al. (2023)
Other eastern Australian systems	Joseph et al. (1995), Peñalba et al. (2017)
Eyrean Barrier	Dolman and Joseph (2012)
Carpentarian Barrier	Lee and Edwards (2008), Baldassarre et al. (2014)
Ord Barrier	Dorrington et al. (2020), Lopez et al. (2021)
South America	
Marañon Valley	Winger and Bates (2015)
Amazonia	Weir et al. (2015)
Chicamocha valley	Arbeláez-Cortés and Trujillo-Arias (2021)
Mata Atlântica	Batalha-Filho and Miyaki (2016), Brown et al. (2020), Bocalini et al. (2021), Thom et al. (2021)
Mata Atlântica vs. Amazonia	Batalha-Filho et al. (2013)
North America	
Baja California	Riddle et al. (2000), Zink et al. (2001), Zink (2002)
Trans Beringia	Zink et al. (1995), Humphries and Winker (2011), McLaughlin et al. (2020)
Overview	Zink (1996)
Africa	
Dahomey Gap	Beresford and Cracraft (1999), Salzmann and Hoelzmann (2005), Fuchs and Bowie (2015)
Eastern Arc Mountains	Bowie et al. (2009), Voelker et al. (2010)
Other	Voelker et al. (2013), (2021)
India	
Western Ghats	Robin et al. (2010), (2017)

emerge from the intellectually simple task of assessing how genetic diversity is or is not structured across species' distributions and landscapes.

Looking Ahead

Further to Toews and Scordato's challenge to study adaptation in urban environments (e.g., Luther and Baptista 2010, Caizergues et al. 2022), I venture to suggest some other "frontiers" in the study of avian evolution, particularly speciation and adaptation. My aim is to highlight where novel questions and approaches can be brought to not necessarily new areas.

Island Radiations

Eliason et al. (2021, 2022) have opened entirely new approaches to advance the increasingly well-documented patterns of diversity in island radiations of birds worldwide (Table 6.2). They have studied genomic controls of key phenotypic characters that may well enable these radiations, such as rate of brain shape evolution. This will also bring new dimensions to existing collections of anatomical skeletal specimens (Webster 2018), and it sets a call for growth in these collections (see Chapter 11).

Table 6.2 Island Radiations of Birds Emphasizing Molecular Studies of Speciation

Ocean	Island Group/Region	Taxa, Topic	References
Pacific Ocean	Oceania	*Petroica*	Kearns et al. (2016, 2020); Kearns, Joseph, et al. (2019); Kearns, Malloy, et al. (2019))
		Zosterops	Clegg et al. (2002), Manthey et al. (2020)
		Erythropitta	Irestedt et al. (2013)
		Pachycephala	Brady et al. (2022)
		Todiramphus	Eliason et al. (2021, 2022)
		Acrocephalus	Cibois et al. (2011), Kearns et al. (2022)
	Northern Melanesia	General synthesis	Mayr and Diamond (2001)
	Louisiades	*Symposiachrus*	McCullough et al. (2021)
	Greater Buka (Solomons/Bougainville)		Uy et al. (2019), Manthey et al. (2020)
	Chatham Islands	Penguins	Blokland et al. (2019)
	Galapagos	*Geospiza*	Chaves et al. (2016), Cadena et al. (2018)
Indo-West Pacific	Indonesian Archipelago/Wallacea	*Monarcha cinerascens*	Ó Marcaigh et al. (2022)
		General	Cros et al. (2020), Rheindt et al. (2020), Irestedt et al. (2025)
	Philippines	*Sarcophanops*	Campillo et al. (2020)
		Eight species	Hosner et al. (2014)
Indian Ocean	Comoros/Seychelles	*Foudia*	Warren et al. (2012)
	Madagascar	Vangidae	Jønsson et al. (2012)
		Cryptic diversity	Younger et al. (2018)
Atlantic Ocean	Caribbean Sea	Parulidae	Klein et al. (2004), Catanach et al. (2021)

Continued

Table 6.2 *Continued*

Ocean	Island Group/Region	Taxa, Topic	References
	Sao Tome and Principe	*Zosterops*	Melo et al. (2011)
		Motacillidae	Alström et al. (2015)
	Tristan da Cunha	*Nesospiza*	Ryan et al. (2013)
		Atlantisia/ Laterallus	Stervander et al. (2019)
	Macaronesia (Cape Verde, Canary, Madeira)	*Sylvia*	Illera et al. (2014)
		Parus	Dietzen et al. (2008), Hansson et al. (2014)
		Lanius	Gonzalez et al. (2008), Hernández et al. (2010)
		Columba	Gonzalez et al. (2009)
		Tyto	Cumer et al. (2022)

Desert and Savanna Birds

In Chapter 10 of this book, Luciano Naka, Victor Leandro-Silva, and Santiago Claramunt's call for more work on the arid zone diagonal of South America well extends to birds in the world's other arid zone and savanna regions. Of course, these avifaunas have been studied in North America (Zink et al. 2001, Zink 2002, Provost et al. 2022, Vázquez-Miranda et al. 2022), South America (Silva and Bates 2002, Bates et al. 2003), Africa (Ribeiro et al. 2011, 2012, 2019; Voelker et al. 2021), and Australia (Dolman and Joseph 2012, Toon et al. 2012, McElroy et al. 2020), but much remains to investigate in depth whether by extending the inventory of species studied or these avifaunas' evolutionary connections with more mesic habitats. Notable here are savannas of New Guinea and northern Australia (Joseph, Bishop, et al. 2019; Peñalba et al. 2019) and evolutionary connections among rainforests and deserts in Africa, Australia, and South America and across Palearctic deserts of northern Africa and Asian deserts generally (see Korrida and Schweizer 2014, Kakhki et al. 2016).

Mimicry

Mimicry in birds has had a "century of careful rumination" (Miller et al. 2019, p. 5). Mimicry by *Cuculus* and *Heteroscenes* hawk-cuckoos of the body form of raptors is well known (Go et al. 2021, York 2021) as is mimicry of sounds in reproductive biology (Dalziell et al. 2015). Egg mimicry by brood parasites of their hosts' eggs can be obvious in our visual spectrum or not (Starling et al. 2006, Stoddard and Stevens 2011) and includes chemistry of pigments used (Igic et al. 2012). Plumage of bird nestlings can mimic distasteful insect larvae (Londoño et al. 2022). Mimicry of adult plumage of another species is well documented in passerines (Diamond 1982, Jønsson et al. 2016, Prum and Samuelson 2016) and non-passerines (Prum and Samuelson 2012,

2016) but not without debates in woodpeckers (Miller et al. 2019, 2020; Grether 2020). Prum (2014) is a rich source of other cases of mimicry warranting attention, to which I boldly add the unsurprisingly named "mimic" honeyeaters *Microptilotis* species mostly of New Guinea. All of these examples of mimicry pose challenges to speciation research in the era of genomics. What are the ecological drivers and genomic mechanisms of all of these forms of mimicry? Are there common genes and regions of the genome involved within and among groups of birds where mimicry has evolved, or does every case have some unique genomic aspect?

Olfaction and Odor

Olfaction in birds is among the most infrequently published yet most tantalizing facets of avian biology and evolution. Nonetheless, it has been reviewed (Weldon and Rappole 1997, Steiger et al. 2008, Avilés and Amo 2018) and is well understood as important in the biology of birds as diverse as Turkey Vultures *Cathartes aura* and Greater Yellow-Headed Vultures *C. melambrotus*, and in some parrots and auklets (Hagelin et al. 2003; Hagelin 2004, 2007b, 2007a; Hagelin and Jones 2007). Given that adaptive evolution in olfactory receptors of birds has been argued (Steiger et al. 2010), we may ask questions such as the following. Nectarivory has been accompanied by evolution of morphologies and indeed social systems that help birds find and share knowledge about food sources—flowering plants—that are patchy and irregular in space and time. Is this coupling of morphology and vision all that nectarivorous birds use in locating these food sources? Do they also upregulate olfactory genes in finding flowering trees in a "sea" of forest?

Conversely, are there adaptive ramifications to some birds being noticeably odiferous? The blue-cheeked rosellas *Platycercus* species have a sweet musky smell about them detectable under field conditions and in 150-year old museum specimens (personal observation). Mihailova et al. (2014) showed that female Crimson Rosellas of the subspecies *Platycercus e. elegans* can discriminate among subspecies by odor but later showed (Mihailova et al. 2018) that this odor has a cost in increasing awareness of predators to nesting rosellas on which they prey (olfactory eavesdropping).

Brood Parasitism

Colonization of new hosts by brood parasites (e.g., in urban environments; Abernathy and Langmore 2017, Abernathy et al. 2017) has been well documented in the Neotropical cowbirds (de la Colina et al. 2016). This offers opportunities to study the earliest stages of coevolution as host and parasite evolutionarily square off against each other using defenses such as egg mimicry, egg rejection, and the use of "passwords" to distinguish their own offspring from cuckoo nestlings (Colombelli-Négrel et al. 2012). What are the relative roles of phenotypic plasticity and genomic mechanisms or constraints on these systems? Are there common loci or genomic regions such as inversions involved in them?

Brood parasitism also is where sympatric speciation in nonmarine birds has been considered most plausible (DaCosta and Sorenson 2016, 2021). Evolution of host races (gentes) through females parasitizing the nests of the host species in which they were raised is well known, but sympatric speciation requires males and females raised by the same host species to form pairs generation after generation. To date, the African indigobirds seem the most likely candidates for sympatric speciation in terrestrial birds (Sorenson et al. 2003). Further genomic study of brood parasitic systems is surely a frontier of speciation and adaptation research in birds.

Migration and Speciation

The nexus between speciation and migration has been well-served in the body of research around Swainson's Thrush *Catharus ustulatus* (Ruegg et al. 2014, Delmore et al. 2015), *Catharus* thrushes generally (Winker 2010, Everson et al. 2019), and other species (Battey and Klicka 2017). The scope is great for digging deeper into this foundation for genomic studies of migration: expanded taxonomic range of migratory birds, evolution of partial migration (Sadanandan et al. 2015), and the diversity or otherwise of genomic and environmental triggers and their interactions from one year to the next.

Seabirds and Antarctic and Arctic Birds

Study of speciation of marine birds is surely starting to reach its rich potential. Sympatric speciation, albeit largely driven by temporal isolation, warrants closer study as a mode of speciation. The nature of selection in marine environments on radiations of seabirds, especially that of penguins, is gaining attention (Friesen et al. 2007, Friesen 2015, Cole et al. 2019, Lombal et al. 2020, Vianna et al. 2020, Noll et al. 2022). In a further "surprise from the sea," Masello et al. (2019) examined the prions *Pachyptila* species, which surely must be one of the world's most evolutionarily challenging groups of birds, and suggested that Salvin's prion (*P. salvini*) could be a hybrid between the narrow-billed Antarctic prion (*P. desolata*) and broad-billed prion (*P. vittata*)—that is, a case of homoploid hybrid speciation (hybridization between two species forming a new, reproductively isolated species with no change in chromosome number). Other examples of hybrid speciation in birds have been raised in Musher et al. (2022), Dai and Feng (2023), Barrera-Guzmán et al. (2022), Brelsford et al. (2011), Mikkelsen and Weir (2023), and Irestedt et al. (2025).

Climate and Adaptation

The ever-increasing relevance of climate change means that we cannot overestimate the value of knowing how birds have evolved adaptations to climate. Avian studies have been developing a number of model systems for delving into the themes and idiosyncrasies of genomic basis to adaptation versus phenotypic plasticity. Table 6.3 is an entrée to literature.

Table 6.3 Genome-Level Studies of Climatic Adaptation in Birds

Species	Region, Location	References
Eopsaltria australis	Eastern Australia—lowland	Morales et al. (2018), Kvistad et al. (2022)
Amblyornis papuensis	New Guinea—montane	Ericson et al. (2022)
Zonotrichia capensis	South America—Andes	Cheviron and Brumfield (2012), Cheviron et al. (2014)
Cercotrichas coryphaeus	Africa—Karoo	Ribeiro et al. (2011, 2012, 2019)
Montifringilla adamsi, Pyrgilauda ruficollis, Onychostruthus taczanowskii	Eurasia	Qu et al. (2021), She et al. (2022)
Parus major	Qinghai–Tibetan Plateau	Qu et al. (2015)
Parus humilis		Qu et al. (2013)
Passer montanus		Qu et al. (2020)
Cyanoderma ruficeps		Lu et al. (2023)

Conclusion

So many other facets of avian evolution warrant discussion here—for example, ring species and circular overlaps, evolution of neo-sex chromosomes and the Z chromosome and chromosome 1A in speciation, chromosomal inversions in speciation, speciation on individual continents and major islands such as Borneo and New Guinea, the growing number of model species of birds, and the rich insights the well-established models such as *Ficedula* have given us. Space, a desire to hold on to my sanity, and exhausting the editors' patience prevent me from delving into these. I note that all of these topics and the many more one can suggest will all require one thing: adequate source material from museum collections. In Chapter 11, C. Jonathan Schmitt sets out how collections are responding to these challenges as well as the operational challenges they themselves face in continuing to do so. I hope that students of the coming decades will find something to inspire them and their work here and that they too can enjoy preparing a *Perspectives in Ornithology* volume a few decades from now.

References

Abernathy, V. E., and N. E. Langmore. 2017. The first stages of coevolution between a brood parasite and its new host: Are naïve hosts defenceless? *Emu* 117:114–129.

Abernathy, V. E., J. Troscianko, and N. E. Langmore. 2017. Egg mimicry by the Pacific koel: Mimicry of one host facilitates exploitation of other hosts with similar egg types. *Journal of Avian Biology* 48:1414–1424.

Alström, P., D. M. Hooper, Y. Liu, U. Olsson, D. Mohan, M. Gelang, H. Le Manh, J. Zhao, F. Lei, and T. D. Price. 2014. Discovery of a relict lineage and monotypic family of passerine birds. *Biology Letters* 10: Article 20131067.

Alström, P., K. A. Jønsson, J. Fjeldså, A. Ödeen, P. G. P. Ericson, and M. Irestedt. 2015. Dramatic niche shifts and morphological change in two insular bird species. *Royal Society Open Science* 2: Article 140364.

Arbeláez-Cortés, E., and N. Trujillo-Arias. 2021. Role of the Chicamocha River Canyon on the phylogeography of humid montane forest birds in Colombia. *Journal of Avian Biology* 52.

Avilés, J. M., and L. Amo. 2018. The evolution of olfactory capabilities in wild birds: A comparative study. *Evolutionary Biology* 45:27–36.

Avise, J. C. 1983. Commentary. Pages 262–270 in A. H. Brush and G. A. Clark, eds. *Perspectives in Ornithology*. Cambridge University Press, Cambridge, UK.

Avise, J. C., J. Arnold, R. Martin Ball, E. Bermingham, T. Lamb, J. E. Neigel, C. A. Reeb, and N. C. Saunders. 1987. Intraspecific phylogeography: The mitochondrial DNA bridge between population genetics and systematics. *Annual Review of Ecology and Systematics* 18:489–522.

Baldassarre, D. T., T. A. White, J. Karubian, and M. S. Webster. 2014. Genomic and morphological analysis of a semipermeable avian hybrid zone suggests asymmetrical introgression of a sexual signal. *Evolution* 68:2644–2657.

Barrera-Guzmán, A. O., A. Aleixo, M. Faccio, S. D. M. Dantas, and J. T. Weir. 2022. Gene flow, genomic homogenization and the timeline to speciation in Amazonian manakins. *Molecular Ecology* 31:4050–4066.

Barrowclough, G. 1983. Biochemical studies of microevolutionary processes. Pages 223–261 in A. H. Brush and G. A. Clark, eds. *Perspectives in Ornithology*. Cambridge University Press, Cambridge, UK.

Batalha-Filho, H., J. Fjeldså, P.-H. Fabre, and C. Y. Miyaki. 2013. Connections between the Atlantic and the Amazonian forest avifaunas represent distinct historical events. *Journal of Ornithology* 154:41–50.

Batalha-Filho, H., and C. Y. Miyaki. 2016. Late Pleistocene divergence and postglacial expansion in the Brazilian Atlantic Forest: Multilocus phylogeography of *Rhopias gularis* (Aves: Passeriformes). *Journal of Zoological Systematics and Evolutionary Research* 54:137–147.

Bates, J. M., J. G. Tello, and J. M. C. Da Silva. 2003. Initial assessment of genetic diversity in ten bird species of South American Cerrado. *Studies on Neotropical Fauna and Environment* 38:87–94.

Battey, C. J., and J. Klicka. 2017. Cryptic speciation and gene flow in a migratory songbird Species Complex: Insights from the Red-Eyed Vireo (*Vireo olivaceus*). *Molecular Phylogenetics and Evolution* 113:67–75.

Beresford, P., F. K. Barker, P. G. Ryan, and T. M. Crowe. 2005. African endemics span the tree of songbirds (Passeri): Molecular systematics of several evolutionary "enigmas." *Proceedings of the Royal Society B: Biological Sciences* 272:849–858.

Beresford, P., and J. Cracraft. 1999. Speciation in African forest robins (*Stiphrornis*): Species limits, phylogenetic relationships, and molecular biogeography. *American Museum Novitates* 3270:1–22.

Blokland, J. C., C. M. Reid, T. H. Worthy, A. J. D. Tennyson, J. A. Clarke, and R. P. Scofield. 2019. Chatham Island paleocene fossils provide insight into the palaeobiology, evolution, and diversity of early penguins (Aves, sphenisciformes). *Palaeontologia Electronica* **22**:1–92.

Bocalini, F., S. D. Bolívar-Leguizamón, L. F. Silveira, and G. A. Bravo. 2021. Comparative phylogeographic and demographic analyses reveal a congruent pattern of sister relationships between bird populations of the northern and south-central Atlantic Forest. *Molecular Phylogenetics and Evolution* 154: Article 106973.

Bonnet, T., R. Leblois, F. Rousset, and P. A. Crochet. 2017. A reassessment of explanations for discordant introgressions of mitochondrial and nuclear genomes. *Evolution* 71:2140–2158.

Bowie, R. C. K., J. Fjeldså, and J. Kiure. 2009. Multilocus molecular DNA variation in Winifred's Warbler *Scepomycter winifredae* suggests cryptic speciation and the existence of a threatened species in the Rubeho-Ukaguru Mountains of Tanzania. *Ibis* 151:709–719.

Brady, S. S., R. G. Moyle, L. Joseph, and M. J. Andersen. 2022. Systematics and biogeography of the whistlers (Aves: Pachycephalidae) inferred from ultraconserved elements and ancestral area reconstruction. *Molecular Phylogenetics and Evolution* 168: Article 107379.

Bravo, G. A., A. Antonelli, C. D. Bacon, K. Bartoszek, M. P. K. Blom, S. Huynh, G. Jones, L. Lacey Knowles, S. Lamichhaney, T. Marcussen, H. Morlon, L. K. Nakhleh, B. Oxelman, B. Pfeil, A. Schliep, N. Wahlberg, F. P. Werneck, J. Wiedenhoeft, S. Willows-Munro, and S. V. Edwards. 2019. Embracing heterogeneity: Coalescing the tree of life and the future of phylogenomics. *PeerJ* 7: Article e6399.

Bravo, G. A., C. J. Schmitt, and S. V. Edwards. 2021. What have we learned from the first 500 avian genomes? *Annual Review of Ecology, Evolution, and Systematics* 52:611–639.

Brelsford, A., B. Mila, and D. E. Irwin. 2011. Hybrid origin of Audubon's Warbler. *Molecular Ecology* 20:2380–2389.

Brown, J. L., A. Paz, M. Reginato, C. A. Renata, C. Assis, M. Lyra, M. K. Caddah, J. Aguirre-Santoro, F. d'Horta, F. Raposo Do Amaral, R. Goldenberg, K. Lucas Silva-Brandão, A. V. L. Freitas, M. T. Rodrigues, F. A. Michelangeli, C. Y. Miyaki, and A. C. Carnaval. 2020. Seeing the forest through many trees: Multi-taxon patterns of phylogenetic diversity in the Atlantic Forest hotspot. *Diversity and Distributions* 26:1160–1176.

Brush, A. H., and G. A. Clark, eds. 1983. *Perspectives in Ornithology: Essays Presented for the Centennial of the American Ornithologists' Union.* Cambridge University Press, Cambridge, UK.

Burley, J. T., S. C. M. Orzechowski, S. Y. W. Sin, and S. V. Edwards. 2023. Whole-genome phylogeography of the blue-faced honeyeater (*Entomyzon cyanotis*) and discovery and characterization of a neo-Z chromosome. *Molecular Ecology* 32:1248–1270.

Cadena, C. D., F. Zapata, and I. Jiménez. 2018. Issues and perspectives in species delimitation using phenotypic data: Atlantean evolution in Darwin's finches. *Systematic Biology* 67:181–194.

Caizergues, A. E., J. Le Luyer, A. Grégoire, M. Szulkin, J. C. Senar, A. Charmantier, and C. Perrier. 2022. Epigenetics and the city: Non-parallel DNA methylation modifications across pairs of urban–forest Great tit populations. *Evolutionary Applications* 15:149–165.

Campagna, L., P. Benites, S. C. Lougheed, D. A. Lijtmaer, A. S. Di Giacomo, M. D. Eaton, and P. L. Tubaro. 2012. Rapid phenotypic evolution during incipient speciation in a continental avian radiation. *Proceedings of the Royal Society B: Biological Sciences* 279:1847–1856.

Campagna, L., M. Repenning, L. F. Silveira, C. S. Fontana, P. L. Tubaro, and I. J. Lovette. 2017. Repeated divergent selection on pigmentation genes in a rapid finch radiation. *Science Advances* 3: Article e1602404.

Campillo, L. C., J. D. Manthey, R. C. Thomson, P. A. Hosner, and R. G. Moyle. 2020. Genomic differentiation in an endemic Philippine genus (Aves: Sarcophanops) owing to geographical isolation on recently disassociated islands. Biological *Journal* of the Linnean Society of London 131:814–821.

Catanach, T. A., M. R. Halley, J. M. Allen, J. A. Johnson, R. Thorstrom, S. Palhano, C. P. Thunder, J. C. Gallardo, and J. D. Weckstein. 2021. Systematics and conservation of an endemic radiation of Accipiter hawks in the Caribbean islands. *Ornithology* 138: Article ukab041.

Chaves, J. A., E. A. Cooper, A. P. Hendry, J. Podos, L. F. De León, J. A. M. Raeymaekers, W. O. MacMillan, and J. A. C. Uy. 2016. Genomic variation at the tips of the adaptive radiation of Darwin's finches. *Molecular Ecology* 25:5282–5295.

Cheviron, Z. A., and R. T. Brumfield. 2012. Genomic insights into adaptation to high-altitude environments. *Heredity* 108:354–361.

Cheviron, Z. A., C. Natarajan, J. Projecto-Garcia, D. K. Eddy, J. Jones, M. D. Carling, C. C. Witt, H. Moriyama, R. E. Weber, A. Fago, and J. F. Storz. 2014. Integrating evolutionary and functional tests of adaptive hypotheses: A case study of altitudinal differentiation in hemoglobin function in an Andean sparrow, *Zonotrichia capensis. Molecular Biology and Evolution* 31:2948–2962.

Cibois, A., J. S. Beadell, G. R. Graves, E. Pasquet, B. Slikas, S. A. Sonsthagen, J. C. Thibault, and R. C. Fleischer. 2011. Charting the course of reed-warblers across the Pacific islands. *Journal of Biogeography* 38:1963–1975.

Clegg, S. M., S. M. Degnan, J. Kikkawa, C. Moritz, A. Estoup, and I. P. F. Owens. 2002. Genetic consequences of sequential founder events by an island-colonizing bird. *Proceedings of the National Academy of Sciences of the USA* 99:8127–8132.
Cole, T. L., L. Dutoit, N. Dussex, T. Hart, A. Alexander, J. L. Younger, G. V. Clucas, M. J. Frugone, Y. Cherel, R. Cuthbert, U. Ellenberg, S. R. Fiddaman, J. Hiscock, D. Houston, P. Jouventin, T. Mattern, G. Miller, C. Miskelly, P. Nolan, M. J. Polito, P. Quillfeldt, P. G. Ryan, A. Smith, A. J. D. Tennyson, D. Thompson, B. Wienecke, J. A. Vianna, J. M. Waters, and J. M. W. Performed. 2019. Receding ice drove parallel expansions in Southern Ocean penguins. *Proceedings of the National Academy of Sciences of the USA* 116:26690–26696.
Colombelli-Négrel, D., M. E. Hauber, J. Robertson, F. J. Sulloway, H. Hoi, M. Griggio, and S. Kleindorfer. 2012. Embryonic learning of vocal passwords in superb fairy-wrens reveals intruder cuckoo nestlings. *Current Biology* 22:2155–2160.
Cros, E., B. Chattopadhyay, K. M. Garg, N. S. R. Ng, S. Tomassi, S. Benedick, D. P. Edwards, and F. E. Rheindt. 2020. Quaternary land bridges have not been universal conduits of gene flow. *Molecular Ecology* 29:2692–2706.
Cruickshank, T. E., and M. W. Hahn. 2014. Reanalysis suggests that genomic islands of speciation are due to reduced diversity, not reduced gene flow. *Molecular Ecology* 23:3133–3157.
Cumer, T., A. P. Machado, F. Siverio, S. I. Cherkaoui, I. Roque, R. Lourenço, M. Charter, A. Roulin, and J. Goudet. 2022. Genomic basis of insularity and ecological divergence in barn owls (*Tyto alba*) of the Canary Islands. *Heredity* 129:281–294.
DaCosta, J. M., and M. D. Sorenson. 2016. DdRAD-seq phylogenetics based on nucleotide, indel, and presence–absence polymorphisms: Analyses of two avian genera with contrasting histories. *Molecular Phylogenetics and Evolution* 94:122–135.
DaCosta, J. M., and M. D. Sorenson. 2021. Variation in the non-mimetic vocalizations of brood-parasitic indigobirds and their potential role in speciation. *Frontiers in Ecology and Evolution* 9: Article 725979.
Dai, C., and P. Feng. 2023. Multiple concordant cytonuclear divergences and potential hybrid speciation within a species complex in Asia. *Molecular Phylogenetics and Evolution* 180: Article 107709.
Dalziell, A. H., J. A. Welbergen, B. Igic, and R. D. Magrath. 2015. Avian vocal mimicry: A unified conceptual framework. *Biological Reviews* 90:643–668.
de la Colina, M. A., M. E. Hauber, B. M. Strausberger, J. C. Reboreda, and B. Mahler. 2016. Molecular tracking of individual host use in the Shiny Cowbird—a generalist brood parasite. *Ecology and Evolution* 6:4684–4696.
Delmore, K. E., S. Hübner, N. C. Kane, R. Schuster, R. L. Andrew, F. Câmara, R. Guigõ, and D. E. Irwin. 2015. Genomic analysis of a migratory divide reveals candidate genes for migration and implicates selective sweeps in generating islands of differentiation. *Molecular Ecology* 24:1873–1888.
Delmore, K. E., J. S. Lugo Ramos, B. M. Van Doren, M. Lundberg, S. Bensch, D. E. Irwin, and M. Liedvogel. 2018. Comparative analysis examining patterns of genomic differentiation across multiple episodes of population divergence in birds. *Evolution Letters* 2:76–87.
De Pietri, V. L., R. P. Scofield, A. J. D. Tennyson, S. J. Hand, and T. H. Worthy. 2016. Wading a lost southern connection: Miocene fossils from New Zealand reveal a new lineage of shorebirds (Charadriiformes) linking *Gondwanan avifaunas. Journal of Systematic Palaeontology* 14:603–616.
Derryberry, E. P., S. Claramunt, G. Derryberry, R. T. Chesser, J. Cracraft, A. Aleixo, J. Pérez-Emán, J. V. Remsen, and R. T. Brumfield. 2011. Lineage diversification and morphological evolution in a large-scale continental radiation: The Neotropical ovenbirds and woodcreepers (Aves: Furnariidae). *Evolution* 65:2973–2986.
Diamond, J. M. 1982. Mimicry of friarbirds by orioles. The Auk 99:187–196.
Dietzen, C., E. Garcia-del-Rey, G. D. Castro, and M. Wink. 2008. Phylogeography of the blue tit (*Parus teneriffae*-group) on the Canary Islands based on mitochondrial DNA sequence data and morphometrics. *Journal of Ornithology* 149:1–12.
Dolman, G., and L. Joseph. 2012. A species assemblage approach to comparative phylogeography of birds in southern Australia. *Ecology and Evolution* 2:354–369.

Dorrington, A., L. Joseph, W. Hallgren, I. Mason, A. Drew, J. M. Hughes, and D. J. Schmidt. 2020. Phylogeography of the blue-winged kookaburra *Dacelo leachii* across tropical northern Australia and New Guinea. *Emu* 120:33–45.

Edwards, S. V., S. Potter, C. J. Schmitt, J. G. Bragg, and C. Moritz. 2016. Reticulation, divergence, and the phylogeography-phylogenetics continuum. *Proceedings of the National Academy of Sciences of the USA* 113:8025–8032.

Edwards, S. V., V. V. Robin, N. Ferrand, and C. Moritz. 2022. The evolution of comparative phylogeography: Putting the geography (and more) into comparative population genomics. *Genome Biology and Evolution* 14: Article evab176.

Eliason, C. M., T. Hains, J. McCullough, M. J. Andersen, and S. J. Hackett. 2022. Genomic novelty within a "great speciator" revealed by a high-quality reference genome of the collared kingfisher (*Todiramphus chloris collaris*). *G3* 12: Article jkac260.

Eliason, C. M., J. M. McCullough, M. J. Andersen, and S. J. Hackett. 2021. Accelerated brain shape evolution is associated with rapid diversification in an avian radiation. *American Naturalist* 197:576–591.

Ellegren, H., L. Smeds, R. Burri, P. I. Olason, N. Backström, T. Kawakami, A. Künstner, H. Mäkinen, K. Nadachowska-Brzyska, A. Qvarnström, S. Uebbing, and J. B. W. Wolf. 2012. The genomic landscape of species divergence in *Ficedula* flycatchers. *Nature* 491:756–760.

Enbody, E. D., S. Y. W. Sin, J. Boersma, S. V. Edwards, S. Ketaloya, H. Schwabl, M. S. Webster, and J. Karubian. 2022. The evolutionary history and mechanistic basis of female ornamentation in a tropical songbird. *Evolution* 76:1720–1736.

Ericson, P. G. P., M. Irestedt, and Y. Qu. 2022. Demographic history, local adaptation and vulnerability to climate change in a tropical mountain bird in New Guinea. *Diversity and Distributions* 28:2565–2578.

Everson, K. M., J. F. McLaughlin, I. A. Cato, M. M. Evans, A. R. Gastaldi, K. K. Mills, K. G. Shink, S. M. Wilbur, and K. Winker. 2019. Speciation, gene flow, and seasonal migration in *Catharus* thrushes (Aves: Turdidae). *Molecular Phylogenetics and Evolution* 139: Article 106564.

Faust-Stryjewski, K., and M. D. Sorenson. 2017. Mosaic genome evolution in a recent and rapid avian radiation. *Nature Ecology and Evolution* 1:1912–1922.

Ferreira, M., A. M. Fernandes, A. Aleixo, A. Antonelli, U. Olsson, J. M. Bates, J. Cracraft, and C. C. Ribas. 2018. Evidence for mtDNA capture in the jacamar *Galbula leucogastra/chalcothorax* species-complex and insights on the evolution of white-sand ecosystems in the Amazon basin. *Molecular Phylogenetics and Evolution* 129:149–157.

Friesen, V. L. 2015. Speciation in seabirds: Why are there so many species . . . and why aren't there more? *Journal of Ornithology* 156:27–39.

Friesen, V. L., A. L. Smith, E. Gó Mez-Díaz, M. Bolton, R. W. Furness, J. Gonzá Lez-Solís, and L. R. Monteiro. 2007. Sympatric speciation by allochrony in a seabird. *Proceedings of the National Academy of Sciences of the USA* 104:18589–18594.

Fuchs, J., and R. C. K. Bowie. 2015. Concordant genetic structure in two species of woodpecker distributed across the primary West African biogeographic barriers. *Molecular Phylogenetics and Evolution* 88:64–74.

Fuchs, J., E. Pasquet, A. Couloux, J. Fjeldså, and R. C. K. Bowie. 2009. A new Indo-Malayan member of the Stenostiridae (Aves: Passeriformes) revealed by multilocus sequence data: Biogeographical implications for a morphologically diverse clade of flycatchers. *Molecular Phylogenetics and Evolution* 53:384–393.

Go, J. S., J. W. Lee, and J. C. Yoo. 2021. Variations of hawk mimicry traits in the four sympatric *Cuculus* cuckoos. *Frontiers in Ecology and Evolution* 9: Article 702263.

Gonzalez, J., G. Delgado Castro, E. Garcia-del-Rey, C. Berger, and M. Wink. 2009. Use of mitochondrial and nuclear genes to infer the origin of two endemic pigeons from the Canary Islands. *Journal of Ornithology* 150:357–367.

Gonzalez, J., M. Wink, E. Garcia-del-Rey, and G. Delgado Castro. 2008. Evidence from DNA nucleotide sequences and ISSR profiles indicates paraphyly in subspecies of the Southern Grey Shrike (*Lanius meridionalis*). *Journal of Ornithology* 149:495–506.

Grether, G. F. 2020. Convergent and divergent selection drive plumage evolution in woodpeckers. *Nature Communications* 11: Article 144.

Gutiérrez-Zuluaga, A. M., C. González-Quevedo, J. A. Oswald, R. S. Terrill, J. L. Pérez-Emán, and J. L. Parra. 2021. Genetic data and niche differences suggest that disjunct populations of *Diglossa brunneiventris* are not sister lineages. *Ornithology* 138: Article ukab015.

Hagelin, J. C. 2004. Observations on the olfactory ability of the Kakapo *Strigops habroptilus*, the critically endangered parrot of New Zealand. *Ibis* 146:161–164.

Hagelin, J. C. 2007a. Odors and chemical signalling. Pages 75–119 in B. G. Jamieson, ed. *Reproductive Biology and Phylogeny of Birds, Part B: Sexual Selection, Behavior, Conservation, Embryology, and Genetics*. CRC Press, Boca Raton, FL.

Hagelin, J. C. 2007b. The citrus-like scent of crested auklets: Reviewing the evidence for an avian olfactory ornament. *Journal of Ornithology* 148(Suppl 2):195–201.

Hagelin, J. C., and I. L. Jones. 2007. Bird odors and other chemical Substances: A defense mechanism or overlooked mode of intraspecific communication? *The Auk* 124:741–761.

Hagelin, J. C., I. L. Jones, and L. E. L. Rasmussen. 2003. A tangerine-scented social odour in a monogamous seabird. *Proceedings of the Royal Society B: Biological Sciences* 270:1323–1329.

Hall, M. L., and N. E. Langmore. 2017. Fitness costs and benefits of female song. *Frontiers in Ecology and Evolution* 5.

Hansson, B., M. Ljungqvist, J. C. Illera, and L. Kvist. 2014. Pronounced fixation, strong population differentiation and complex population history in the Canary Islands blue tit subspecies complex. *PLoS One* 9: Article e90186.

Harvey, M. G., A. Aleixo, C. C. Ribas, and R. T. Brumfield. 2017. Habitat association predicts genetic diversity and population divergence in amazonian birds. *American Naturalist* 190:631–648.

Harvey, M. G., G. A. Bravo, S. Claramunt, A. M. Cuervo, G. E. Derryberry, J. Battilana, G. F. Seeholzer, J. S. McKay, B. C. O, B. C. Faircloth, S. V. Edwards, J. Pérez-Emán, R. G. Moyle, F. H. Sheldon, A. Aleixo, B. T. Smith, R. T. Chesser, L. F. Silveira, J. Cracraft, R. T. Brumfield, and E. P. Derryberry. 2020. The evolution of a tropical biodiversity hotspot. *Science* 370:1343–1348.

Harvey, M. G., G. F. Seeholzer, B. T. Smith, D. L. Rabosky, A. M. Cuervo, and R. T. Brumfield. 2017. Positive association between population genetic differentiation and speciation rates in New World birds. *Proceedings of the National Academy of Sciences of the USA* 114:6328–6333.

Hay, E. M., M. D. McGee, and S. L. Chown. 2022. Geographic range size and speciation in honeyeaters. *BMC Ecology and Evolution* 22: Article 86.

Hejase, H. A., A. Salman-Minkov, L. Campagna, M. J. Hubisz, I. J. Lovette, I. Gronau, and A. Siepel. 2020. Genomic islands of differentiation in a rapid avian radiation have been driven by recent selective sweeps. *Proceedings of the National Academy of Sciences of the USA* 117:30554–30565.

Hernández, Á. M., F. Campos, and D. P. Padilla. 2010. Tandem repeats in the mtDNA control region of the Southern Grey Shrike endemic to the Canary Islands. *Ardeola* 57:437–441.

Hosner, P. A., L. A. Sánchez-González, A. Townsend Peterson, and R. G. Moyle. 2014. Climate-driven diversification and Pleistocene refugia in Philippine birds: Evidence from phylogeographic structure and paleoenvironmental niche modeling. *Evolution* 68:2658–2674.

Humphries, E. M., and K. Winker. 2011. Discord reigns among nuclear, mitochondrial and phenotypic estimates of divergence in nine lineages of trans-Beringian birds. *Molecular Ecology* 20:573–583.

Igic, B., P. Cassey, T. Grim, D. R. Greenwood, C. Moskát, J. Rutila, and M. E. Hauber. 2012. A shared chemical basis of avian host–parasite egg colour mimicry. *Proceedings of the Royal Society B: Biological Sciences* 279:1068–1076.

Illera, J. C., A. M. Palmero, P. Laiolo, F. Rodríguez, Á. C. Moreno, and M. Navascués. 2014. Genetic, morphological, and acoustic evidence reveals lack of diversification in the colonization process in an island bird. *Evolution* 68:2259–2274.

Irestedt, M., P. H. Fabre, H. Batalha-Filho, K. A. Jønsson, C. S. Roselaar, G. Sangster, and P. G. P. Ericson. 2013. The spatio-temporal colonization and diversification across the Indo-Pacific by a "great speciator" (Aves, *Erythropitta erythrogaster*). *Proceedings of the Royal Society B: Biological Sciences* 280: Article 20130309.

Irestedt, M., I. A. Müller, F. Thörn, L. Joseph, J. A. Nylander, B. Guinet, T. van der Valk, and K. A. Jønsson. 2025. Reticulate and hybrid speciation is promoted by environmental instability in an Indo-Pacific species complex of whistlers (Aves: Passeriformes: *Pachycephala*). *Molecular Ecology* 0:e70018.

Irwin, D. E. 2018. Sex chromosomes and speciation in birds and other ZW systems. *Molecular Ecology* 27:3831–3851.

Irwin, D. E., M. Alcaide, K. E. Delmore, J. H. Irwin, and G. L. Owens. 2016. Recurrent selection explains parallel evolution of genomic regions of high relative but low absolute differentiation in a ring species. *Molecular Ecology* 25:4488–4507.

Irwin, D. E., A. S. Rubtsov, and E. N. Panov. 2009. Mitochondrial introgression and replacement between yellowhammers (*Emberiza citrinella*) and pine buntings (*Emberiza leucocephalos*) (Aves: Passeriformes). *Biological Journal of the Linnean Society* 98:422–438.

Jiménez, R. A., and J. F. Ornelas. 2016. Historical and current introgression in a Mesoamerican hummingbird species complex: A biogeographic perspective. *PeerJ* 4: Article e1556.

Jønsson, K. A., K. Delhey, G. Sangster, P. G. P. Ericson, and M. Irestedt. 2016. The evolution of mimicry of friarbirds by orioles (Aves: Passeriformes) in Australo-Pacific archipelagos. *Proceedings of the Royal Society B: Biological Sciences* 283: Article 20160409.

Jønsson, K. A., P.-H. Fabre, S. A. Fritz, R. S. Etienne, R. E. Ricklefs, T. B. Jørgensen, J. Fjeldså, C. Rahbek, G. P. Ericson, F. Woog, E. Pasquet, and M. Irestedt. 2012. Ecological and evolutionary determinants for the adaptive radiation of the Madagascan vangas. *Proceedings of the National Academy of Sciences of the USA* 109:6620–6625.

Joseph, L., K. D. Bishop, C. A. Wilson, S. V. Edwards, B. Iova, C. D. Campbell, I. Mason, and A. Drew. 2019. A review of evolutionary research on birds of the New Guinean savannas and closely associated habitats of riparian rainforests, mangroves and grasslands. Emu 119: 317–330.

Joseph, L., C. D. Campbell, A. Drew, S. S. Brady, Á. Nyári, and M. J. Andersen. 2021. How far east can a Western Whistler go? Genomic data reveal large eastward range extension, taxonomic and nomenclatural change, and reassessment of conservation needs. *Emu* 121:90–101.

Joseph, L., A. Drew, I. J. Mason, and J. L. Peters. 2019. Introgression between non-sister species of honeyeaters (Aves: Meliphagidae) several million years after speciation. *Biological Journal of the Linnean Society* 128:583–591.

Joseph, L., C. Moritz, and A. Hugall. 1995. Molecular support for vicariance as a source of diversity in rainforest. *Proceedings of the Royal Society B: Biological Sciences* 260:177–182.

Kakhki, N. A., M. Aliabadian, and M. Schweizer. 2016. Out of Africa: Biogeographic history of the open-habitat chats (Aves, Muscicapidae: Saxicolinae) across arid areas of the Old World. *Zoologica Scripta* 45:237–251.

Kawakami, T., L. Smeds, N. Backström, A. Husby, A. Qvarnström, C. F. Mugal, P. Olason, and H. Ellegren. 2014. A high-density linkage map enables a second-generation collared flycatcher genome assembly and reveals the patterns of avian recombination rate variation and chromosomal evolution. *Molecular Ecology* 23:4035–4058.

Kearns, A. M., M. G. Campana, B. Slikas, L. Berry, T. Saitoh, A. Cibois, and R. C. Fleischer. 2022. Conservation genomics and systematics of a near-extinct island radiation. *Molecular Ecology* 31:1995–2012.

Kearns, A. M., L. Joseph, J. J. Austin, A. C. Driskell, and K. E. Omland. 2020. Complex mosaic of sexual dichromatism and monochromatism in Pacific robins results from both gains and losses of elaborate coloration. *Journal of Avian Biology* 51: Article e02404.

Kearns, A. M., L. Joseph, A. Thierry, J. F. Malloy, M. N. Cortes-Rodriguez, and K. E. Omland. 2019. Diversification of *Petroica* robins across the Australo-Pacific region: First insights into the phylogenetic affinities of New Guinea's highland robin species. *Emu* 119:205–217.

Kearns, A. M., L. Joseph, L. C. White, J. J. Austin, C. Baker, A. C. Driskell, J. F. Malloy, and K. E. Omland. 2016. Norfolk Island Robins are a distinct endangered species: Ancient DNA unlocks surprising relationships and phenotypic discordance within the Australo-Pacific Robins. *Conservation Genetics* 17:321–335.

Kearns, A. M., J. F. Malloy, M. K. Gobbert, A. Thierry, L. Joseph, A. C. Driskell, and K. E. Omland. 2019. Nuclear introns help unravel the diversification history of the Australo-Pacific *Petroica* robins. *Molecular Phylogenetics and Evolution* 131:48–54.

Kennedy, J. D., P. Z. Marki, A. H. Reeve, M. P. K. Blom, D. M. Prawiradilaga, T. Haryoko, B. Koane, P. Kamminga, M. Irestedt, and K. A. Jønsson. 2022. Diversification and community assembly of the world's largest tropical island. *Global Ecology and Biogeography* 31:1078–1089.

Klein, N. K., K. J. Burns, S. J. Hackett, and C. S. Griffiths. 2004. Molecular phylogenetic relationships among the wood warblers (Parulidae) and historical biogeography in the Caribbean Basin. *Journal of Caribbean Ornithology* 17:3–17.

Korrida, A., and M. Schweizer. 2014. Diversification across the Palaearctic desert belt throughout the Pleistocene: Phylogeographic history of the Houbara–Macqueen's bustard complex (Otididae: Chlamydotis) as revealed by mitochondrial DNA. *Journal of Zoological Systematics and Evolutionary Research* 52:65–74.

Kvistad, L., S. Falk, and L. Austin. 2022. Widespread genomic signatures of reproductive isolation and sex-specific selection in the Eastern Yellow Robin, *Eopsaltria australis*. *G3* 12: Article jkac145.

Lamichhaney, S., J. Berglund, M. S. Almén, K. Maqbool, M. Grabherr, A. Martinez-Barrio, M. Promerová, C.-J. Rubin, C. Wang, and N. Zamani. 2015. Evolution of Darwin's finches and their beaks revealed by genome sequencing. *Nature* 518:371–375.

Lamichhaney, S., F. Han, J. Berglund, C. Wang, M. S. Almén, M. T. Webster, B. R. Grant, P. R. Grant, and L. Andersson. 2016. A beak size locus in Darwin's finches facilitated character displacement during a drought. *Science* 352:470–474.

Lee, J. Y., and S. V. Edwards. 2008. Divergence across Australia's Carpentarian barrier: Statistical phylogeography of the red-backed fairy wren (*Malurus melanocephalus*). *Evolution* 62:3117–3134.

Lewin, R. 1988a. Conflict over DNA clock results. *Science* 241:1598–1600.

Lewin, R. 1988b. DNA clock conflict continues: Two researchers who apparently revolutionized bird and human/ape systematics using the technique of DNA hybridization are faced with having to reanalyze all their data. *Science* 241:1756–1759.

Lombal, A. J., J. E. O'Dwyer, V. Friesen, E. J. Woehler, and C. P. Burridge. 2020. Identifying mechanisms of genetic differentiation among populations in vagile species: Historical factors dominate genetic differentiation in seabirds. *Biological Reviews* 95:625–651.

Londoño, G. A., J. Sandoval-H, M. F. Sallam, and J. M. Allen. 2022. On the evolution of mimicry in avian nestlings. *Ecology and Evolution* 12: Article e8842.

Lopez, K. A., C. S. McDiarmid, S. C. Griffith, I. J. Lovette, and D. M. Hooper. 2021. Evaluating evidence of mitonuclear incompatibilities with the sex chromosomes in an avian hybrid zone. *Evolution* 75:1395–1414.

Lu, C. W., S. T. Huang, S. J. Cheng, C. T. Lin, Y. C. Hsu, C. T. Yao, F. Dong, C. M. Hung, and H. C. Kuo. 2023. Genomic architecture underlying morphological and physiological adaptation to high elevation in a songbird. *Molecular Ecology* 32:2234–2251.

Luther, D., and L. Baptista. 2010. Urban noise and the cultural evolution of bird songs. *Proceedings of the Royal Society B: Biological Sciences* 277:469–473.

Manthey, J. D., C. H. Oliveros, M. J. Andersen, C. E. Filardi, and R. G. Moyle. 2020. Gene flow and rapid differentiation characterize a rapid insular radiation in the southwest Pacific (Aves: Zosterops). *Evolution* 74:1788–1803.

Marki, P. Z., J. D. Kennedy, C. R. Cooney, C. Rahbek, and J. Fjeldså. 2019. Adaptive radiation and the evolution of nectarivory in a large songbird clade. *Evolution* 73:1226–1240.

Masello, J. F., P. Quillfeldt, E. Sandoval-Castellanos, R. Alderman, L. Calderón, Y. Cherel, T. L. Cole, R. J. Cuthbert, M. Marin, M. Massaro, J. Navarro, R. A. Phillips, P. G. Ryan, L. D. Shepherd, C. G. Suazo, H. Weimerskirch, Y. Moodley, and C. Russo. 2019. Additive traits lead to feeding advantage and reproductive isolation, promoting homoploid hybrid speciation. *Molecular Biology and Evolution* 36:1671–1685.

Mayr, E. 1969. *Principles of Systematic Zoology*. McGraw-Hill, New York.

Mayr, E., and J. Diamond. 2001. *The Birds of Northern Melanesia*. Oxford University Press, New York.

Mayr, G. 2011. Metaves, Mirandornithes, Strisores and other novelties: A critical review of the higher-level phylogeny of neornithine birds. *Journal of Zoological Systematics and Evolutionary Research* 49:58–76.

Mayr, G. 2016. *Avian Evolution: The Fossil Record of Birds and Its Paleobiological Significance*. Wiley.

McCullough, J. M., E. F. Gyllenhaal, X. M. Mapel, M. J. Andersen, and L. Joseph. 2021. Taxonomic implications of recent molecular analyses of Spectacled (*Symposiachrus trivirgatus*) and Spot-Winged (*S. guttula*) monarchs (Passeriformes: Monarchidae). *Emu* 121:365–371.

McElroy, K., A. Black, G. Dolman, P. Horton, L. Pedler, C. D. Campbell, A. Drew, and L. Joseph. 2020. Robbery in progress: Historical museum collections bring to light a mitochondrial capture within a bird species widespread across southern Australia, the Copperback Quail-Thrush *Cinclosoma clarum*. *Ecology and Evolution* 10:6785–6793.
McGaughran, A., L. Liggins, K. A. Marske, M. N. Dawson, L. M. Schiebelhut, S. D. Lavery, L. L. Knowles, C. Moritz, and C. Riginos. 2022. Comparative phylogeography in the genomic age: Opportunities and challenges. *Journal of Biogeography* 49:2130–2144.
McLaughlin, J. F., B. C. Faircloth, T. C. Glenn, and K. Winker. 2020. Divergence, gene flow, and speciation in eight lineages of trans-Beringian birds. *Molecular Ecology* 29:3526–3542.
Melo, M., B. H. Warren, and P. J. Jones. 2011. Rapid parallel evolution of aberrant traits in the diversification of the Gulf of Guinea white-eyes (Aves, Zosteropidae). *Molecular Ecology* 20:4953–4967.
Mihailova, M., M. L. Berg, K. L. Buchanan, and A. T. D. Bennett. 2014. Odour-based discrimination of subspecies, species and sexes in an avian species complex, the crimson rosella. *Animal Behaviour* 95:155–164.
Mihailova, M., M. L. Berg, K. L. Buchanan, and A. T. D. Bennett. 2018. Olfactory eavesdropping: The odor of feathers is detectable to mammalian predators and competitors. *Ethology* 124:14–24.
Mikkelsen, E. K., and J. T. Weir. 2023. Phylogenomics reveals that mitochondrial capture and nuclear introgression characterize skua species proposed to be of hybrid origin. *Systematic Biology* 72:78–91.
Miller, E. T., G. M. Leighton, B. G. Freeman, A. C. Lees, and R. A. Ligon. 2019. Ecological and geographical overlap drive plumage evolution and mimicry in woodpeckers. *Nature Communications* 10: Article 1602.
Miller, E. T., G. M. Leighton, B. G. Freeman, A. C. Lees, and R. A. Ligon. 2020. Reply to "Convergent and Divergent Selection in Sympatry Drive Plumage Evolution in Woodpeckers." *Nature Communications* 11: Article 145.
Morales, H. E., A. Pavlova, N. Amos, R. Major, A. Kilian, C. Greening, and P. Sunnucks. 2018. Concordant divergence of mitogenomes and a mitonuclear gene cluster in bird lineages inhabiting different climates. *Nature Ecology and Evolution* 2:1258–1267.
Musher, L. J., M. Giakoumis, J. Albert, G. Del-Rio, M. Rego, G. Thom, A. Aleixo, C. C. Ribas, R. T. Brumfield, B. T. Smith, and J. Cracraft. 2022. River network rearrangements promote speciation in lowland Amazonian birds. *Science Advances* 8: Article eabn1099.
Natola, L., S. S. Seneviratne, and D. Irwin. 2022. Population genomics of an emergent tri-species hybrid zone. *Molecular Ecology* 31:5356–5367.
Nguyen, J. M. T., S. J. Hand, and M. H. Archer. 2016. The late Cenozoic passerine avifauna from Rackham's roost site, Riversleigh, Australia. *Records of the Australian Museum* 68:201–230.
Noll, D., F. Leon, D. Brandt, P. Pistorius, C. Le Bohec, F. Bonadonna, P. N. Trathan, A. Barbosa, A. R. Rey, G. P. M. Dantas, R. C. K. Bowie, E. Poulin, and J. A. Vianna. 2022. Positive selection over the mitochondrial genome and its role in the diversification of gentoo penguins in response to adaptation in isolation. *Scientific Reports* 12: Article 3767.
Norambuena, H. V., and P. van Els. 2021. A general scenario to evaluate evolution of grassland birds in the Neotropics. *Ibis* 163:722–727.
Norman, J. A., W. E. Boles, and L. Christidis. 2009. Relationships of the New Guinean songbird genera *Amalocichla* and *Pachycare* based on mitochondrial and nuclear DNA sequences. *Journal of Avian Biology* 40:640–645.
Nwankwo, E. C., K. G. Mortega, A. Karageorgos, B. O. Ogolowa, G. Papagregoriou, G. F. Grether, A. Monadjem, and A. N. G. Kirschel. 2019. Rampant introgressive hybridization in *Pogoniulus* tinkerbirds (Piciformes: Lybiidae) despite millions of years of divergence. Biological Journal of the Linnean Society 127:125–142.
Oliveros, C. H., D. J. Field, D. T. Ksepka, F. Keith Barker, A. Aleixo, M. J. Andersen, P. Alström, B. W. Benz, E. L. Braun, M. J. Braun, G. A. Bravo, R. T. Brumfield, R. T. Chesser, S. Claramunt, J. Cracraft, A. M. Cuervo, E. P. Derryberry Aa, T. C. Glenn, M. G. Harvey Aa, P. A. Hosner, L. Joseph, R. T. Kimball, A. L. Mack Ee, C. M. Miskelly, A. T. Peterson, M. B. Robbins, F. H. Sheldon, F. Silveira, B. T. Smith, N. D. White, R. G. Moyle, B. C. Faircloth, and M. E. Alfaro. 2019. Earth

history and the passerine superradiation. *Proceedings of the National Academy of Sciences of the USA* 116:7916–7925.

Ó Marcaigh, F., D. P. O'Connell, K. Analuddin, A. Karya, N. Lawless, C. M. McKeon, N. Doyle, N. M. Marples, and D. J. Kelly. 2022. Tramps in transition: Genetic differentiation between populations of an iconic "supertramp" taxon in the Central Indo-Pacific. *Frontiers of Biogeography* 14.

Palacios, C., S. García-R, J. L. Parra, A. M. Cuervo, F. G. Stiles, J. E. McCormack, and C. D. Cadena. 2019. Shallow genetic divergence and distinct phenotypic differences between two Andean hummingbirds: Speciation with gene flow? *Auk* 136: Article ukz046.

Peñalba, J. V., L. Joseph, and C. Moritz. 2019. Current geography masks dynamic history of gene flow during speciation in northern Australian birds. *Molecular Ecology* 28:630–643.

Peñalba, J. V., I. J. Mason, R. Schodde, C. Moritz, and L. Joseph. 2017. Characterizing divergence through three adjacent Australian avian transition zones. *Journal of Biogeography* 44:2247–2258.

Peñalba, J. V., J. L. Peters, and L. Joseph. 2022. Sustained plumage divergence despite weak genomic differentiation and broad sympatry in sister species of Australian woodswallows (*Artamus* spp.). *Molecular Ecology* 31:5060–5073.

Poelstra, J. W., N. Vijay, C. M. Bossu, H. Lantz, B. Ryll, I. Müller, V. Baglione, P. Unneberg, M. Wikelski, M. G. Grabherr, and J. B. W. Wolf. 2014. The genomic landscape underlying phenotypic integrity in the face of gene flow in crows. *Science* 344:1410–1414.

Price, T. 2008. *Speciation in Birds.* Roberts, Greenwood Village, CO.

Provost, K., S. Y. Shue, M. Forcellati, and B. T. Smith. 2022. The genomic landscapes of desert birds form over multiple time scales. *Molecular Biology and Evolution* 39: Article msac200.

Provost, K. L., L. Joseph, and B. T. Smith. 2018. Resolving a phylogenetic hypothesis for parrots: Implications from systematics to conservation. *Emu-Austral Ornithology* 118:7–21.

Prum, R., and L. Samuelson. 2012. The Hairy-Downy game: A model of interspecific social dominance mimicry ∗. *Journal of Theoretical Biology* 313:42–60.

Prum, R. O. 2014. Interspecific social dominance mimicry in birds. *Zoological Journal of the Linnean Society* 172:910–941.

Prum, R. O., and L. Samuelson. 2016. Mimicry cycles, traps, and chains: The coevolution of toucan and kiskadee mimicry. *American Naturalist* 187:753–764.

Pulido-Santacruz, P., A. Aleixo, and J. T. Weir. 2020. Genomic data reveal a protracted window of introgression during the diversification of a neotropical woodcreeper radiation. *Evolution* 74:842–858.

Qu, Y., C. Chen, Y. Xiong, H. She, Y. E. Zhang, Y. Cheng, S. Dubay, D. Li, P. G. P. Ericson, Y. Hao, H. Wang, H. Zhao, G. Song, H. Zhang, T. Yang, C. Zhang, L. Liang, T. Wu, J. Zhao, Q. Gao, W. Zhai, and F. Lei. 2020. Rapid phenotypic evolution with shallow genomic differentiation during early stages of high elevation adaptation in Eurasian Tree Sparrows. *National Science Review* 7:113–127.

Qu, Y., X. C. Chunhai Chen, C. J. Chunyan Chen, and F. Chen. 2021. The evolution of ancestral and species-specific adaptations in snowfinches at the Qinghai-Tibet Plateau. *Proceedings of the National Academy of Sciences of the USA* 118: Article e2012398118.

Qu, Y., S. Tian, N. Han, H. Zhao, B. Gao, J. Fu, Y. Cheng, G. Song, P. G. P. Ericson, Y. E. Zhang, D. Wang, Q. Quan, Z. Jiang, R. Li, and F. Lei. 2015. Genetic responses to seasonal variation in altitudinal stress: Whole-genome resequencing of great tit in eastern Himalayas. *Scientific Reports* 5: Article 14256.

Qu, Y., H. Zhao, N. Han, G. Zhou, G. Song, B. Gao, S. Tian, J. Zhang, R. Zhang, X. Meng, Y. Zhang, Y. Zhang, X. Zhu, W. Wang, D. Lambert, P. G. P. Ericson, S. Subramanian, C. Yeung, H. Zhu, Z. Jiang, R. Li, and F. Lei. 2013. Ground tit genome reveals avian adaptation to living at high altitudes in the Tibetan plateau. *Nature Communications* 4: Article 2071.

Rheindt, F. E., D. M. Prawiradilaga, H. Ashari, C. Yin Gwee, G. W. X Lee, M. Yue Wu, and N. S. R Ng. 2020. A lost world in Wallacea: Description of a montane archipelagic avifauna. *Science* 367:167–170.

Ribas, C. C., S. C. Fritz, and P. A. Baker. 2022. The challenges and potential of geogenomics for biogeography and conservation in Amazonia. *Journal of Biogeography* 49:1839–1847.

Ribeiro, Â. M., P. Lloyd, and R. C. K. Bowie. 2011. A tight balance between natural selection and gene flow in a southern African arid-zone endemic bird. *Evolution* 65:3499–3514.

Ribeiro, Â. M., R. J. Lopes, and R. C. K. Bowie. 2012. Historical demographic dynamics underlying local adaptation in the presence of gene flow. *Ecology and Evolution* 2:2710–2721.
Ribeiro, Â. M., L. Puetz, N. B. Pattinson, L. Dalén, Y. Deng, G. Zhang, R. R. da Fonseca, B. Smit, and M. T. P. Gilbert. 2019. 31° South: The physiology of adaptation to arid conditions in a passerine bird. *Molecular Ecology* 28:3709–3721.
Riddle, B. R., D. J. Hafner, L. F. Alexander, and J. R. Jaeger. 2000. Cryptic vicariance in the historical assembly of a Baja California Peninsular Desert biota. *Proceedings of the National Academy of Sciences of the USA* 97:14438–14443.
Robin, V. V., A. Sinha, and U. Ramakrishnan. 2010. Ancient geographical gaps and paleo-climate shape the phylogeography of an endemic bird in the sky islands of southern India. *PLoS One* 5: Article e13321.
Robin, V. V., C. K. Vishnudas, P. Gupta, F. E. Rheindt, D. M. Hooper, U. Ramakrishnan, and S. Reddy. 2017. Two new genera of songbirds represent endemic radiations from the Shola Sky Islands of the Western Ghats, India. *BMC Evolutionary Biology* 17:1–14.
Rodríguez-Gómez, F., and J. F. Ornelas. 2015. At the passing gate: Past introgression in the process of species formation between *Amazilia violiceps* and *A. viridifrons* hummingbirds along the Mexican Transition Zone. *Journal of Biogeography* 42:1305–1318.
Ruegg, K., E. C. Anderson, J. Boone, J. Pouls, and T. B. Smith. 2014. A role for migration-linked genes and genomic islands in divergence of a songbird. *Molecular Ecology* 23:4757–4769.
Russell, E. M. 1989. Co-operative breeding: A Gondwanan perspective. *Emu – Austral Ornithology* 89:61–62.
Ryan, P. G., L. B. Klicka, K. F. Barker, and K. J. Burns. 2013. The origin of finches on Tristan da Cunha and Gough Island, central South Atlantic Ocean. *Molecular Phylogenetics and Evolution* 69:299–305.
Sadanandan, K. R., D. J. X. Tan, K. Schjølberg, P. D. Round, and F. E. Rheindt. 2015. DNA reveals long-distance partial migratory behavior in a cryptic owl lineage. *Avian Research* 6: Article 25.
Salzmann, U., and P. Hoelzmann. 2005. The Dahomey Gap: An abrupt climatically induced rain forest fragmentation in West Africa during the late Holocene. The Holocene 15:190–199.
Selvatti, A. P., A. Galvão, G. Mayr, C. Y. Miyaki, and C. A. d. M. Russo. 2022. Southern hemisphere tectonics in the Cenozoic shaped the pantropical distribution of parrots and passerines. *Journal of Biogeography* 49:1753–1766.
She, H., Z. Jiang, G. Song, P. G. P. Ericson, X. Luo, S. Shao, F. Lei, and Y. Qu. 2022. Quantifying adaptive divergence of the snowfinches in a common landscape. *Diversity and Distributions* 28:2579–2592.
Shields, G. F. 1983. Organization of the avian genome. Pages 271–290 in A. H. Brush and G. A. Clark, eds. *Perspectives in Ornithology*. Cambridge University Press, Cambridge, UK.
Shipham, A., D. J. Schmidt, L. Joseph, and J. M. Hughes. 2016. A genomic approach reinforces a hypothesis of mitochondrial capture in eastern Australian rosellas. *Auk* 134:181–192.
Sibley, C. G., and J. E. Ahlquist. 1985. The phylogeny and classification of the Australo-Papuan passerine birds. *Emu – Austral Ornithology* 85:1–14.
Silva, J. M. C., and J. M. Bates. 2002. Biogeographic patterns and conservation in the South American cerrado: A tropical savanna hotspot. *BioScience* 52:225–234.
Singhal, S., G. E. Derryberry, G. A. Bravo, E. P. Derryberry, R. T. Brumfield, and M. G. Harvey. 2021. The dynamics of introgression across an avian radiation. *Evolution Letters* 5: 568–581.
Smith, B. T., J. Merwin, K. L. Provost, G. Thom, R. T. Brumfield, M. Ferreira, W. M. Mauck, R. G. Moyle, T. F. Wright, and L. Joseph. 2023. Phylogenomic analysis of the parrots of the world distinguishes artifactual from biological sources of gene tree discordance. *Systematic Biology* 72:228–241.
Sorenson, M. D., K. M. Sefc, and R. B. Payne. 2003. Speciation by host switch in brood parasitic indigobirds. *Nature* 424:928–931.
Starling, M., R. Heinsohn, A. Cockburn, and N. E. Langmore. 2006. Cryptic gentes revealed in pallid cuckoos *Cuculus pallidus* using reflectance spectrophotometry. *Proceedings of the Royal Society B: Biological Sciences* 273:1929–1934.
Steiger, S. S., A. E. Fidler, J. C. Mueller, and B. Kempenaers. 2010. Evidence for adaptive evolution of olfactory receptor genes in 9 bird species. *Journal of Heredity* 101:325–333.

Steiger, S. S., A. E. Fidler, M. Valcu, and B. Kempenaers. 2008. Avian olfactory receptor gene repertoires: Evidence for a well-developed sense of smell in birds? *Proceedings of the Royal Society B: Biological Sciences* 275:2309–2317.

Stervander, M., P. G. Ryan, M. Melo, and B. Hansson. 2019. The origin of the world's smallest flightless bird, the Inaccessible Island Rail *Atlantisia rogersi* (Aves: Rallidae). *Molecular Phylogenetics and Evolution* 130:92–98.

Stoddard, M. C., and M. Stevens. 2011. Avian vision and the evolution of egg color mimicry in the common cuckoo. *Evolution* 65:2004–2013.

Thom, G., M. Gehara, B. T. Smith, C. Y. Miyaki, and F. R. Do Amaral. 2021. Microevolutionary dynamics show tropical valleys are deeper for montane birds of the Atlantic Forest. *Nature Communications* 12: Article 6269.

Toda, Y., M.-C. Ko, Q. Liang, E. T. Miller, A. Rico-Guevara, T. Nakagita, A. Sakakibara, K. Uemura, T. Sackton, T. Hayakawa, S. Yung, W. Sin, Y. Ishimaru, T. Misaka, P. Oteiza, J. Crall, S. V. Edwards, W. Buttemer, S. Matsumura, and M. W. Baldwin. 2021. Early origin of sweet perception in the songbird radiation. *Science* 373:226–231.

Toon, A., J. J. Austin, G. Dolman, L. Pedler, and L. Joseph. 2012. Evolution of arid zone birds in Australia: Leapfrog distribution patterns and mesic-arid connections in quail-thrush (Cinclosoma, Cinclosomatidae). *Molecular Phylogenetics and Evolution* 62:286–295.

Uy, J. A. C., E. A. Cooper, and J. A. Chaves. 2019. Convergent melanism in populations of a Solomon Island flycatcher is mediated by unique genetic mechanisms. *Emu* 119:242–250.

van Els, P., E. Zarza, L. Rocha Moreira, V. Gómez-Bahamón, A. Santana, A. Aleixo, C. C. Ribas, P. S. Do Rêgo, M. P. D. Santos, K. Zyskowski, R. O. Prum, and J. Berv. 2021. Recent divergence and lack of shared phylogeographic history characterize the diversification of neotropical savanna birds. *Journal of Biogeography* 48:1124–1137.

Vázquez-Miranda, H., R. M. Zink, and B. J. Pinto. 2022. Comparative phylogenomic patterns in the Baja California avifauna, their conservation implications, and the stages in lineage divergence. *Molecular Phylogenetics and Evolution* 171: Article 107466.

Vianna, J. A., F. A. N. Fernandes, M. J. Frugone, H. V. Figueiró, L. R. Pertierra, D. Noll, K. Bi, C. Y. Wang-Claypool, A. Lowther, and P. Parker. 2020. Genome-wide analyses reveal drivers of penguin diversification. *Proceedings of the National Academy of Sciences of the USA* 117:22303–22310.

Voelker, G., B. D. Marks, C. Kahindo, U. A'Genonga, F. Bapeamoni, L. E. Duffie, J. W. Huntley, E. Mulotwa, S. A. Rosenbaum, and J. E. Light. 2013. River barriers and cryptic biodiversity in an evolutionary museum. *Ecology and Evolution* 3:536–545.

Voelker, G., R. K. Outlaw, and R. C. K. Bowie. 2010. Pliocene forest dynamics as a primary driver of African bird speciation. *Global Ecology and Biogeography* 19:111–121.

Voelker, G., G. O. U. Wogan, J. W. Huntley, and R. C. K. Bowie. 2021. Comparative phylogeography of Southern African bird species suggests an ephemeral speciation model. *Diversity* 13: Article 434.

Wang, S., S. Rohwer, D. R. de Zwaan, D. P. L. Toews, I. J. Lovette, J. Mackenzie, and D. Irwin. 2020. Selection on a small genomic region underpins differentiation in multiple color traits between two warbler species. *Evolution Letters* 4:502–515.

Warren, B. H., E. Bermingham, Y. Bourgeois, L. K. Estep, R. P. Prys-Jones, D. Strasberg, and C. Thébaud. 2012. Hybridization and barriers to gene flow in an island bird radiation. *Evolution* 66:1490–1505.

Webster, M. S. 2018. The extended specimen: Emerging frontiers in collections-based ornithological research. *Studies in Avian Biology* 50:1–240.

Weir, J. T., M. S. Faccio, P. Pulido-Santacruz, A. O. Barrera-Guzmán, and A. Aleixo. 2015. Hybridization in headwater regions, and the role of rivers as drivers of speciation in Amazonian birds. *Evolution* 69:1823–1834.

Weldon, P. J., and J. H. Rappole. 1997. A survey of birds odorous or unpalatable to humans: Possible indications of chemical defense. *Journal of Chemical Ecology* 23:2609–2633.

Winger, B. M. 2017. Consequences of divergence and introgression for speciation in Andean cloud forest birds. *Evolution* 71:1815–1831.

Winger, B. M., and J. M. Bates. 2015. The tempo of trait divergence in geographic isolation: Avian speciation across the Marañon Valley of Peru. *Evolution* 69:772–787.

Winker, K. 2010. On the origin of species through heteropatric differentiation: A review and a model of speciation in migratory animals. *Ornithological Monographs* 69:1–30.
Worthy, T. H., V. L. De Pietri, and R. P. Scofield. 2017. Recent advances in avian palaeobiology in New Zealand with implications for understanding New Zealand's geological, climatic and evolutionary histories. *New Zealand Journal of Zoology* 44:177–211.
Worthy, T. H., and J. M. T. Nguyen. 2020. An annotated checklist of the fossil birds of Australia. *Transactions of the Royal Society of South Australia* 144:66–108.
Worthy, T. H., A. J. D. Tennyson, C. Jones, J. A. McNamara, and B. J. Douglas. 2007. Miocene waterfowl and other birds from central Otago, New Zealand. *Journal of Systematic Palaeontology* 5:1–39.
York, J. E. 2021. The evolution of predator resemblance in avian brood parasites. *Frontiers in Ecology and Evolution* 9: Article 725842.
Younger, J. L., L. Strozier, J. D. Maddox, Á. S. Nyári, M. T. Bonfitto, M. J. Raherilalao, S. M. Goodman, and S. Reddy. 2018. Hidden diversity of forest birds in Madagascar revealed using integrative taxonomy. *Molecular Phylogenetics and Evolution* 124:16–26.
Zhang, D., L. Tang, Y. Cheng, Y. Hao, Y. Xiong, G. Song, Y. Qu, F. E. Rheindt, P. Alström, C. Jia, and F. Lei. 2019. "Ghost introgression" as a cause of deep mitochondrial divergence in a bird species complex. *Molecular Biology and Evolution* 36:2375–2386.
Zink, R. M. 1996. Comparative phylogeography in North American birds. *Evolution* 50:308–317.
Zink, R. M. 2002. Methods in comparative phylogeography, and their application to studying evolution in the North American aridlands. *Integrative and Comparative Biology* 42:953–959.
Zink, R. M., A. E. Kessen, T. V. Line, and R. C. Blackwell-Rago. 2001. Comparative phylogeography of some aridland bird species. *The Condor* 103:1–10.
Zink, R. M., S. Rohwer, A. V. Andreev, and D. L. Dittmann. 1995. Trans-Beringia comparisons of mitochondrial DNA differentiation in birds. *The Condor* 97:639–649.

7
Advances in the Avian Tree of Life and Biogeography in the 21st Century

Sushma Reddy

It is not an exaggeration to say that tree of life studies are the most important contributions to organismal biology in the past two decades. This is exemplified by the numerous phylogenetic studies that have collectively revolutionized our knowledge of evolutionary patterns while leading to a more accurate view of diversity. In birds, the goal of deciphering deep phylogenetic relationships has proved to be more difficult than in other vertebrate groups and has driven much innovative research (e.g., Hackett et al. 2008, Jarvis et al. 2014, Prum et al. 2015, Suh et al. 2015, Sackton et al. 2019, Feng et al. 2020, Kuhl et al. 2020, Stiller et al. 2024, Wu et al. 2024). With concerted effort, we have made tremendous advancements in our knowledge of the deep evolutionary relationships of modern birds in the past several decades. We now have a fairly robust understanding of the bird tree of life, which has fostered a subsequent renaissance of comparative biology to study avian traits in a phylogenetic context. Yet, several pieces of this puzzle are still missing or seemingly untenable.

In this chapter, I discuss why the deep relationships in the bird tree of life were and continue to be difficult to resolve, what we know so far with confidence, and what the future potentially holds for new discoveries. This chapter focuses on higher level phylogeny of birds and on the advances made in the past several decades. Although there have been tremendous efforts to untangle the tips of the avian tree, with numerous phylogenetic studies showing that traditional groupings of families, genera, and species were incorrect or finding hidden diversity leading to the description of new taxa, covering all these changes would be quite a challenge and beyond what is possible in a single book chapter. Moreover, new discoveries are being made constantly, and novel supertrees to summarize these relationships are in development.

What may be surprising to the modern student is that much of the currently established bird tree of life knowledge is relatively new—built during just the past 20 years. Much of what was hypothesized/suspected of bird ordinal relationships before the 21st century turned out to be incorrect, leading to dramatic changes in avian classification (Sibley and Ahlquist 1990, Hackett et al. 2008, Sangster et al. 2022). Sibley and Ahlquist's (1990) study was revolutionary for producing the first large-scale evolutionary tree and fueling interest in phylogenetic and comparative analyses for decades. Sibley and Ahlquist's tree led to a major alteration of avian classification (Sibley and Monroe 1991), which prompted many studies that set out to test their proposed relationships. The impacts of Sibley and Ahlquist's work were both progressive and problematic, with many of their findings later shown to be incorrect.

Sushma Reddy, *Advances in the Avian Tree of Life and Biogeography in the 21st Century*. In: *New Perspectives in Ornithology*. Edited by: Scott V. Edwards and J. Michael Reed, Oxford University Press.
 DOI: 10.1093/oso/9780197787670.003.0007

For example, a generation ago, many ornithology students wrongly learned that New World Vultures were more closely related to storks than other raptors because of Sibley and Ahlquist (1990). Although subsequent analyses returned Cathartidae to the Accipitriformes (Hackett et al. 2008, Jarvis et al. 2014, Prum et al. 2015), this relationship remained contentious for decades. Furthermore, it is now also clear that raptors are not monophyletic because falcons and hawks are consistently recovered as separate, distantly related clades (Hackett et al. 2008, Jarvis et al. 2014, Prum et al. 2015), whereas for almost a century before they were grouped into the same order. And perhaps most dramatically, today we take for granted the odd relationship of flamingos and grebes as closest living relatives, yet when it was first proposed, it prompted wide skepticism of molecular phylogenetics (Tuinen et al. 2001, Mayr 2004a, Livezey 2011).

In this chapter, I focus largely on recent evidence from phylogenomic studies, analyses of substantial portions of the genome made possible by tremendous advances in DNA sequencing technologies. These types of studies have allowed us to have confidence in some seemingly strange relationships that contradict many long-standing traditional groupings based on phenotypic traits. Although there have been several morphological phylogenetic analyses of birds, many of their results conflict with molecular studies. Phenotypic characters can be meaningful to indicate relationships, yet given the high degree of convergent evolution in birds, many of these features are homoplastic and result in large morphological analyses having low signal-to-noise ratios (Mayr 2008). As molecular phylogenies have increased in size and depth, both in numbers of taxa and amount of sequence data, we now have consensus on several deep evolutionary relationships within modern birds. These relationships have dramatically changed our interpretations of the early evolution of avian traits and biogeography.

Why Is the Bird Tree of Life So Complicated?

Tree of life studies of many groups have uncovered challenging phylogenetic issues, yet birds have proved to be notoriously difficult, with little agreement between studies for most basal relationships (Suh 2016, Brown et al. 2017, Reddy et al. 2017, Franz et al. 2019, Sangster et al. 2022). There are several reasons why the phylogenetic relationships of birds are difficult to resolve. First, the diversity of modern birds is such that the living groups include several morphologically distinctive and cohesive clades with few intermediaries between them, resulting in a dearth of extant taxa that are available to break up long branches. Second, the shape and timing of the avian radiation, based on ample evidence, indicate that bird lineages arose extremely rapidly, resulting in an explosive diversification of all the modern orders originating near the Cretaceous–Paleogene (K–Pg) boundary approximately 65 million years ago (Jarvis et al. 2014, Brusatte et al. 2015, Claramunt and Cracraft 2015, Prum et al. 2015, Field et al. 2018). To exacerbate the situation, there are limited fossil taxa of birds from this period that can conclusively help date lineages or trace character changes (Mayr 2014, Claramunt and Cracraft 2015, Cracraft and Claramunt 2017, Mayr 2017, Field et al. 2020). Controversy surrounding some fossils as well as molecular dating models has also led to a steamy debate about the age of the avian tree (Ericson et al. 2006, 2007, Mayr 2014, Claramunt and Cracraft 2015, Cracraft and Claramunt 2017, Mayr 2017, Crouch et al. 2019, Field et al. 2020). Finally, the problems in resolving the bird tree of life are certainly not due to lack of effort. Many studies have tackled this issue using a variety of different data types, yet in almost every new study there are some elements of corroboration and usually much conflict in comparison to previous studies, making it difficult to evaluate what is real signal versus artifactual.

Finding signal or shared characters, either phenotypic or genomic, to unite major groups (e.g., orders) has been a tremendous challenge. There are few phenotypic traits in birds that are informative across orders, and several of these have later been shown to be convergent (Sibley and Ahlquist 1990, Cracraft 2002, Livezey and Zusi 2007, Hackett et al. 2008). Each genomic study has resulted in tremendously different relationships at the very base of the avian tree. A typical feature of rapid radiations is extremely short branch lengths indicating a narrow time frame of divergence or conflict in reconstructing evolutionary history. In birds, incomplete lineage sorting or ancestral polymorphisms that fail to reflect species divergences has been recognized as the cause of extreme gene tree discordance (Suh et al. 2015). Furthermore, it is sometimes unclear if this conflict is due to uninformative data or lack of signal in some types of genetic data or if systematic biases such as vastly different rates of evolution across gene regions and between clades might be leading to this issue (Suh 2016, Brown et al. 2017, Reddy et al. 2017, Braun et al. 2019). Other issues, such as long-branch attraction due to oversaturation of mutational changes and base-composition nonstationarity, have also been noted to cause systematic biases due to model misspecifications (Jeffroy et al. 2006). Some authors have even suggested that some parts of the avian tree, such as the base of Neoaves, are hard polytomies (i.e., divergences too rapid to resolve) and will remain untenable (Suh 2016).

Choice of gene regions, which changed over time given new developments in DNA sequencing technologies, dictated the amount of resolution possible for phylogenetic analyses and in some cases uncovered biases that led to the recovery of certain anomalous relationships. Early nucleotide analyses focused on mitochondrial genes due to their ease of sequencing. However, it was quickly clear that these genes suffered from substitution saturation for higher level phylogenetic signal and other model violations that resulted in aberrant relationships, which have not been corroborated with other genic data sets (Mindell et al. 1999, Pratt et al. 2009). Exons or coding regions were promising but also proved to be too slowly evolving to be informative for deep divergences within birds (Cracraft et al. 2003). Introns and other noncoding regions, seemingly neutral markers, proved to be useful across both shallow and deep divergences and quickly became markers of choice (Fain and Houde 2004, Ericson et al. 2006, Hackett et al. 2008, Jarvis et al. 2014). Concerns about homoplasy in nucleotide substitutions led researchers to explore other types of genomic changes, such as indels, inversions, or transposable elements (Braun et al. 2011, Han et al. 2011, Yuri et al. 2013, Suh et al. 2015, Houde et al. 2019). These so-called rare genomic changes, nevertheless, proved to still suffer from many of the same issues that plague analyses of nucleotides (Han et al. 2011). The advent of next-generation sequencing technologies made massive genomic data sets possible. Target capture methods enabled harnessing information from hundreds to thousands of loci across the genome, such as ultraconserved elements (McCormack et al. 2013) and anchored enrichment

(Prum et al. 2015). Ornithologists were leaders in being able to harness whole genomes earlier than most groups. The initial analyses of 48 species and subsequent expansions of the Bird 10,000 Genomes (B10K) project have produced a wealth of data for comparative genomics (Jarvis et al. 2014, Feng et al. 2020). The Jarvis et al. (2014) study corroborated many previously described relationships and recovered some novel ones with strong support. However, although this study (and its many associated publications) produced whole genomes, the amount of homologous, alignable data used for phylogeny reconstruction was roughly 3.5% of the average avian genome (Jarvis et al. 2014). The B10K group and many other ornithological studies have generated genomes for thousands of bird species, yet adding more taxa does not necessarily improve our understanding of their deep evolutionary history, and more data makes analyses computationally untenable under the current algorithms (Feng et al. 2020).

What Do We Know (So Far)?

Modern birds, defined as the crown radiation of Neornithes, comprise the only lineage of dinosaurs to survive the K–Pg mass extinction (Brusatte et al. 2015). There are a few aspects of the bird tree of life that we "know" with confidence because they have been consistently recovered during the past several decades of analyses. Living birds (Neornithes; clade 1 in Figure 7.1) share a single common ancestor and are split into two groups—Paleognathes and Neognathes (clades 2 and 3, respectively, in Figure 7.1)—based mainly on the morphology of their palates (Cracraft 1974). The paleognaths consist of the ratites, which are secondarily flightless birds, and tinamous, flighted birds restricted to South America. All other living birds are classified as neognaths, within which there is an early split between Galloanserae and Neoaves (clades 4 and 5, respectively, in Figure 7.1). These divisions have been long established and are consistently recovered. However, ordinal relationships within Neoaves have been hotly debated. This section highlights many of the changes within and between orders that are consistently found across recent phylogenomic analyses. Moreover, as these phylogenomic analyses have dictated nomenclatural changes, the definitions of various ordinal names have changed substantially in the past few decades, with some variation in how these names are used in current classifications (see Figure 7.1). Therefore, it is critical for ornithologists to keep in mind the historical context as well as the date and source to interpret the meaning of clade names when reading the past literature.

Palaeognathae

Genetic data challenged the classic model of paleognath evolution that was based on morphological and biogeographic inferences (Hackett et al. 2008, Harshman et al. 2008, Phillips et al. 2009, Smith et al. 2012, Baker et al. 2014, Mitchell et al. 2014, Grealy et al. 2017, Yonezawa et al. 2017, Cloutier et al. 2019, Sackton et al. 2019). Previously, a model of the flightless ratites (Struthioniformes, Casuariiformes,

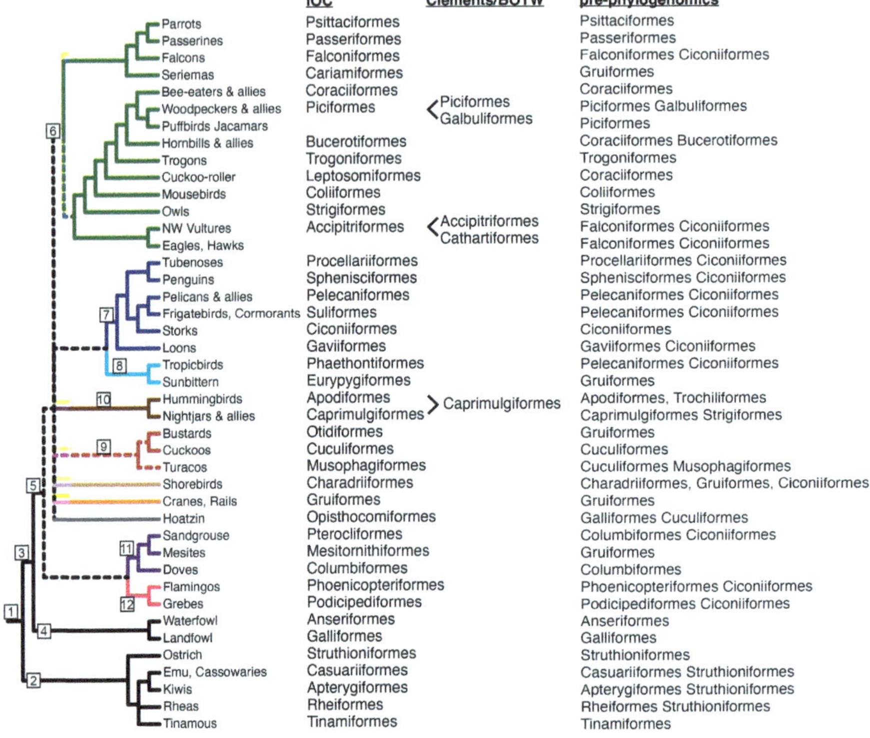

Figure 7.1 Consensus phylogeny of major phylogenomic studies (from Reddy et al. 2017) showing relationships between all extant avian orders with current ordinal names according to the International Ornithological Congress World Bird List (Gill et al. 2022), Clements Checklist of Birds of the World (Clements et al. 2021) where different, and traditional ordinal names from pre-phylogenomic studies (i.e., Peters n.d., Sibley and Monroe 1991). Solid lines indicate corroboration across Jarvis et al. (2014), Prum et al. (2015), and Reddy et al. (2017), and while dotted lines are relationships that are recovered in two but not all these studies. Numbers at nodes indicate higher level clades mentioned in the text: 1, Neornithes; 2, Paleognathes; 3, Neognathes; 4, Galloanserae; 5, Neoaves; 6, Telluraves; 7, Aequornithes; 8, Phaethontimorphae; 9, Otidimorphae; 10, Strisores; 11, Columbimorphae; 12, Mirandornithes.

Apterygiformes, and Rheiformes) diverging due to vicariance caused by the breakup of Gondwana was proposed to explain the presence of an exemplar on each of the southern continents (Cracraft 1974). However, molecular studies recovered strong evidence for flighted tinamous (Tinamiformes), a group restricted to South America, to phylogenetically nest within, rather than sister to, ratites as was traditionally assumed (Haddrath and Baker 2001, Hackett et al. 2008, Harshman et al. 2008, Phillips et al. 2009, Smith et al. 2012, Sackton et al. 2019). Although highly controversial upon initial publication (Hackett et al. 2008, Harshman et al. 2008), this revised view of paleognaths quickly received support from several studies using evidence from across the genome (McCormack et al. 2013, Jarvis et al. 2014, Prum et al. 2015, Suh 2016, Kimball et al. 2019, Kuhl et al. 2020, Sangster et al. 2022). Furthermore, ancient genomic studies brought more surprises as DNA from extinct ratites changed the story again. Genetic evidence recovered unexpected sister relationships between kiwis, restricted to New Zealand, and the elephant birds of Madagascar as well as the moas of New Zealand and tinamous (Phillips et al. 2009, Baker et al. 2014, Mitchell et al. 2014, Grealy et al. 2017, Yonezawa et al. 2017).

New insights on paleognath divergences have led to reinterpretations of their biogeography and biology. The classic Gondwanan vicariance model is no longer supported, and the current scenario includes several cross-oceanic dispersals (Mitchell et al. 2014, Yonezawa et al. 2017, Sackton et al. 2019). Dating estimates from recent studies place the timing of paleognath divergence well after the breakup of the supercontinent (Mitchell et al. 2014, Prum et al. 2015, Yonezawa et al. 2017). In addition, these changes also led to a reinterpretation of the loss of flight in ratites, with the current view that they independently lost flight either three or five times, depending on the placement of rheas, which is highly unstable despite evidence from whole genomes (Cloutier et al. 2019, Sackton et al. 2019). The more parsimonious interpretation of one loss of flight in the common ancestor of all paleognaths and then a gain of flight in tinamous, however, is considered unlikely given that most evidence points to loss of flight being an irreversible character state.

Neognathae: Galloanserae and Neoaves

Galloanserae is the name of the clade uniting Galliformes (landfowl) and Anseriformes (waterfowl). Although these two orders were long hypothesized to be closely related, this early branching of Neognathae into a split of Galloanserae and Neoaves is relatively recent and has been consistently recovered since Sibley & Ahlquist (1990). Higher level relationships within Galloanserae are relatively stable and show intriguing patterns of some families being restricted to southern continents, despite both orders being generally cosmopolitan (Eo et al. 2009, Kimball and Braun 2014, Sun et al. 2017, Wang et al. 2017). Within Galliformes, the earliest divergence is Megapodiidae (mound-builders), which are restricted to Australasia, followed by Cracidae (chachalacas, guans, and curassows) that are found only in Central and South America and the Numididae (guineafowl) that are endemic to Africa. The other two families, Odontophoridae (mostly New World quails) and Phasianidae (pheasants, peafowl, turkeys, and quails that are found almost worldwide except in Antarctica and South America), have yielded some surprises in phylogenetic relationships and their biogeographic interpretations given rapid diversification along several lineages (Kimball and Braun 2014, Hosner et al. 2016, 2017, Wang et al. 2017). Anseriformes is divided into three families: Anhimidae (screamers), consisting of three species restricted to South America; Anseranatidae (Magpie Goose), which includes a single extant species found only in Australasia; and the Anatidae (ducks, geese, and swans), found almost worldwide. Despite their economic and cultural importance to humans, there have been only a few studies examining phylogenomic relationships within most Galloanserae families, and there is still much to be learned about these diverse lineages that are found throughout the world.

Neoaves comprises 95% of modern avian diversity. The early branching events within Neoaves remain a mystery, with little consensus across many studies since cladistic analyses were first conducted for birds. Nevertheless, a closer look at recent large-scale phylogenies indicates that although relationships between major branches vary tremendously across studies, there is an emerging consensus of consistent groupings within Neoaves, namely several superordinal groups called the "Magnificent Seven" plus three "orphan orders" by Reddy et al. (2017). Below is a summary of each of these clades and a brief history of major changes in the circumscription of orders (see Figure 7.1).

Telluraves or "core landbirds" (clade 6) includes 12 orders: Passeriformes, Psittaciformes, Falconiformes, Cariamiformes, Piciformes, Coraciiformes, Bucerotiformes, Trogoniformes, Leptosomiformes, Coliformes, Strigiformes, and Accipitriformes. The general interpretation of this clade is that these groups are broadly adapted to terrestrial habitats. First uncovered in Hackett et al. (2008), Telluraves is strongly supported and recovered in all subsequent bird tree of life studies (Jarvis et al. 2014, Prum et al. 2015, Reddy et al. 2017). A long-standing question in avian phylogenetics was how a single order, the Passeriformes (passerines), was able to radiate into the most species-rich radiation comprising more than half the diversity of birds. These studies have consistently pointed to a close relationship between passerines and parrots (Hackett et al. 2008, Jarvis et al. 2014, Prum et al. 2015, Reddy et al. 2017, Braun et al. 2019), providing a comparative framework for subsequent examinations of shared features or key changes that might lead to the evolution of this incredible diversity. Other notable and consistent revelations have been the non-monophyly of several traditional orders, including the previous definition of Coraciiformes that was rendered non-monophyletic with the inclusion of Piciformes (Hackett et al. 2008, Jarvis et al. 2014, Prum et al. 2015, Reddy et al. 2017, Braun et al. 2019). This led to the redefinition of Coraciiformes to be limited to rollers, kingfishers, bee-eaters, and their allies and the recognition of three new orders—Bucerotiformes (hornbills and hoopoes), Trogoniformes (trogons), and Leptosomiformes (cuckoo-roller, a single species restricted to Madagascar)—given their distinction and monophyly (see Figure 7.1). A substantial departure from traditional understanding was the discovery that raptors (birds of prey) comprised multiple unrelated or distantly related lineages and are not a monophyletic lineage. At one point in history, diurnal raptors (falcons and hawks) were included in a single order (Falconiformes); however, phylogenomic studies quickly showed that falcons are more closely related to a clade including passerines + parrots and seriemas (Cariamiformes) than other diurnal raptors now placed in the Accipitriformes. Finally, a source of contention caused by Sibley and Ahlquist's (1990) finding that New World vultures (Cathartidae) were closely related to storks is also being laid to rest. Cathartidae is now clearly placed, with strong confidence and little doubt, with other hawks, eagles, and Old World Vultures.

The second clade is Aequornithes or core waterbirds (clade 7 in Figure 7.1), which includes six orders of birds that are primarily adapted to living in aquatic habitats: Pelecaniformes, Suliformes, Ciconiformes, Procellariformes, Sphenisciformes, and Gaviiformes. The only consistent grouping that was hypothesized in the pre-phylogenomics era and which has remained a strong association across recent studies is the one between tube-nosed seabirds and penguins. The composition of the Pelecaniformes and Ciconiiformes has been heavily debated in terms of which species belong within these orders. For instance, the traditional order Pelecaniformes united all birds with totipalmate feet, ones with webbing connecting all four toes, yet close associations between other lineages without webbing, such as pelicans and shoebill, were hotly contested (Cracraft 1985, Sibley and Ahlquist 1990). Prior to the phylogenomics era, Sibley and Ahlquist's study led to a major reshuffle of the traditional order of Ciconiiformes to include many waterbirds, such as storks, herons, pelicans, shoebill, frigatebirds, penguins, loons, tropicbirds, boobies, and cormorants, as well as New World Vultures, hawks, falcons, and shorebirds. However, all phylogenomic studies are consistent in finding that some of the major changes proposed by Sibley and Ahlquist (as in the example above) are not corroborated. Aequornithes is a clade that finds agreement across both morphological and phylogenomic data and is one of few superordinal relationships within Neoaves that has withstood the test of time as evidence from diverse data sets comes together to support the close relationships between these birds that had been historically hypothesized to be allied due to their aquatic adaptations. A surprising result in Hackett et al. (2008) and subsequent phylogenomic analyses was the inclusion of storks (Ciconiiformes) within the core waterbirds and the exclusion of grebes and tropicbirds, which were traditionally hypothesized to be part of this group (Hackett et al. 2008, Jarvis et al. 2014, Prum et al. 2015).

Phaethontimorphae (clade 8 in Figure 7.1) comprises three lineages—tropicbirds, kagu, and sunbittern—that were historically placed in other orders and have since been drastically redefined. Tropicbirds have historically been thought of as waterbirds belonging within the Pelecaniformes (Cracraft 1985). However, from Sibley and Ahlquist (1990) onward, studies have found that tropicbirds do not group with the core waterbirds. Instead, with copious phylogenomic data, it is now becoming increasingly well supported that tropicbirds are a long branch and are most closely related to a clade containing kagu and sunbittern, formerly of Gruiformes and now placed in the order Eurypygiformes. Kagu and sunbittern, both monotypic families found in New Caledonia and the Neotropics, respectively, have emerged as unlikely but very strongly supported relatives (Reddy et al. 2017). Strong support for these three phenotypically disparate lineages of tropicbird, kagu, and sunbittern is found consistently across all phylogenomic studies with both coding and noncoding data (Reddy et al. 2017, Kuhl et al. 2020). However, there is still disagreement in terms of the placement of Phaethontimorphae within Neoaves.

The fourth superordinal clade is the Otidimorphae (clade 9 in Figure 7.1), a tentative grouping of the cuckoos (Cuculiformes), bustards (Otidiformes), and turacos (Musophagiformes). Historically, in 19th- and 20th-century classifications, cuckoos and turacos had been grouped together, often in the same order. From Sibley and Ahlquist (1990) onward, it was clear that the relationship between these two superficially similar looking avian orders is complicated. Furthermore, the finding of bustards (traditionally included in the Gruiformes) being part of these complex evolutionary dynamics is a recent development only uncovered with phylogenomic analyses. Nevertheless, there is unclear support for this grouping across different studies. In Prum et al. (2015) and Jarvis et al. (2014), there is strong support for the trio to be closely related but with different relationships within: Cuckoos and bustards are more closely related than turacos in Prum et al. (2015), whereas in Jarvis et al. (2014), turacos and bustards are more closely related. However, in other analyses (Hackett et al. 2008, Reddy et al. 2017, Kuhl et al. 2020), these three orders do not consistently group together. Clearly, we need more analyses to work out whether this superordinal clade will stand the test of time and data.

The Strisores (clade 10 in Figure 7.1) is a superordinal name for a group that was variably considered one, two, or five orders. Phylogenomic studies resulted in strongly supported placement of hummingbirds and swifts (Apodiformes) within the Caprimulgiformes, rendering the traditional definition of this order paraphyletic (Reddy et al. 2017). The Caprimulgiformes formerly consisted of nocturnal, cryptic birds such as nightjars, owlet-nightjars, frogmouths, potoos, and oilbirds. The placement of the hummingbirds, classically diurnal and colorful birds, within the Caprimulgiformes is a great example of the incredible power of evolutionary processes to produce widely different forms from a shared recent common ancestor. Swifts are cryptically colored but have foot, syrinx, and other morphological similarities with hummingbirds and nightjars. Most modern classifications based on phylogenomic analyses place all members of the Strisores in a single order Caprimulgiformes (Dickinson and Christidis 2014, Clements et al. 2021); however, in some cases, Apodiformes is retained for hummingbirds, swifts, and owlet-nightjars (Gill et al. 2022), which results in a problematic and non-monophyletic definition of Caprimulgiformes in those classifications. In addition to classification confusion, there is still some uncertainty in the phylogenetic relationships within the Strisores. The topologies of Prum et al. (2015) and Reddy et al. (2017) disagree in the placement of the root of this clade, which seems to be driven by the kinds of data used. Furthermore, the placement of Strisores within Neoaves appears to also be driven by data-type effects. With exon data, Strisores is the earliest divergence and sister to a clade of all other Neoavians (Prum et al. 2015, Jarvis et al. 2014—coding tree); however, this result is not supported with noncoding loci (Hackett et al. 2008, McCormack et al. 2013, Jarvis et al. 2014, Prum et al. 2015, Suh et al. 2015, Reddy et al. 2017, Kuhl et al. 2020).

The Columbimorphae (clade 11 in Figure 7.1) consists of three orders: Columbiformes (pigeons and doves), Pterocliformes (sandgrouse), and Mesitornithiformes (mesites). Early phylogenetic studies suggested a relationship between doves and sandgrouse (Cracraft 1981), but this was discounted by Sibley and Ahlquist (1990) until later supported by several studies (Mayr and Clarke 2003, Ericson et al. 2006, Livezey and Zusi 2007, Hackett et al. 2008). Phylogenomic studies have consistently recovered these three orders in a clade and seem to have stabilized on sandgrouse and mesites being more closely related to each other than to doves (Reddy et al. 2017). Mesites, previously placed in the Gruiformes, are three species restricted to Madagascar and are an excellent example of a relictual lineage that has persisted across the Cenozoic with little evidence for substantial diversification. Sandgrouse are a group of open, dry habitat species that are found only in Asia and Africa. Pigeons and doves, on the other hand, are cosmopolitan and highly familiar birds across the world. This order includes lineages that have repeatedly dispersed from continents to islands and back, which is generally thought to be rare given potential competition due to high existing diversity on continents (Lapiedra et al. 2021). One explanation is that colonization of islands may have promoted adaptations to new environments and led to novel niche shifts that were advantageous for subsequent diversification to other regions, including continents where they could occupy new habitats (Lapiedra et al. 2021).

The final superordinal clade, named Phoenicopterimorphae or Mirandornithes (clade 12 in Figure 7.1), consists of the grebes (Podicipediformes) and flamingos (Phoenicopteriformes). The odd pairing of these very divergent groups was met with much skepticism when first proposed (as noted above) but was quickly discovered to be one of the most strongly supported relationships within Neoaves and is recovered with strong confidence across a majority of gene trees and some morphological characters (Mayr 2011, Reddy et al. 2017). In the Prum tree, this clade is part of the extended waterbird clade; however, this relationship is not recovered in studies using noncoding data (Suh 2016, Brown et al. 2017, Reddy et al. 2017). In contrast, studies using noncoding data (e.g., Jarvis et al. 2014, Reddy et al. 2017) place the Mirandornithes as sister to Columbimorphae, and together this clade is sister to all other Neoaves.

Three "orphan orders"—Opisthocomiformes (hoatzin), Charadriiformes (shorebirds), and Gruiformes (cranes and rails)—do not group consistently with any others. The relationships of the enigmatic hoatzin have long been a mystery that many studies have attempted to resolve. However, none of the previously proposed groupings (variously with Galliformes, Cuculiformes, and Musophagiformes) have garnered any support in subsequent phylogenomic studies; therefore, there is little agreement across studies on the placement of this strange bird (Sibley et al. 1988, Hackett et al. 2008, Mayr 2011, Reddy et al. 2017). The shorebird order, Charadriiformes, has seen little change since Sibley and Ahlquist (1990), which extended the ecological dimensions of this clade to include terrestrial lineages that are not necessarily along water edges. Subsequent phylogenomic studies strongly support the inclusion of two genera that were previously placed in Gruiformes—Plains-Wanderer (*Pedionomus*) and Buttonquail (*Turnix*) (Hackett et al. 2008, Prum et al. 2015, Reddy et al. 2017). The Plains-Wanderer is a highly endangered species found only in Australia and appears to a have close relationship with Thinocoridae (seedsnipes) found only in South America. The Buttonquail lineage exhibits an extreme, accelerated rate of genomic evolution compared to close relatives that has caused some confusion with genetic models (Prum et al. 2015, Reddy et al. 2017, Braun et al. 2019, Sangster et al. 2022). Last, the current definition of Gruiformes is substantially different and much reduced from what it included three decades ago. Many of the lineages placed into this grab-bag order have since been shown to have strong alliances elsewhere on the bird tree and therefore placed in other orders. Examples of these lineages that were previously placed in Gruiformes include kagu and sunbittern (now Eurypygiformes), mesites (now Mesitornithiformes), bustards (now Otidiformes), and seriemas (now Cariamiformes). What remains in Gruiformes are the cranes, rails, and their allies (Jarvis et al. 2014, Claramunt and Cracraft 2015, Prum et al. 2015). Still, there seems to be no conclusive data to place any of these three orders close to other neoavian lineages.

Passeriformes

The Passeriformes, perching birds or passerines, includes more than 6,000 species and comprises more than 60% of avian diversity. Knowledge of this part of the avian tree has probably changed the most during the past three decades of molecular phylogenetics. Hundreds of studies have shown that many classic assumptions of passerine evolution using ecology, morphology, and biogeography were inaccurate. During the past 30 years, these studies have helped reorganize the relationships of passerine lineages across all levels of classification below order. The number of recognized families has increased more than threefold, from 45 in Sibley and Monroe (1991) to 141–145 (Gill et al. 2022). Similarly, the number of recognized species has increased from 5,739 (Sibley and Monroe 1991) to 6,533 (Gill et al. 2022). These increases reflect new discoveries made with careful field surveys, but most are the result of more systematic analyses to uncover diversity previously hidden in taxonomic designations. The importance of phylogenetic analyses to examine classic assumptions of diversity, taxonomic classifications, and species limits cannot be overemphasized.

Intraordinal relationships within Passeriformes were long debated given that these lineages exhibit a high degree of convergence and are so numerous. Sibley and Ahlquist (1990) pioneered a well-supported classification that formed the basis of many subsequent studies. However, their study was shown to be problematic and biased by some assumptions in their analyses, and subsequent studies have clarified key aspects of passerine relationships, leading to reinterpretation of their evolutionary history. Sibley and Ahlquist (1990) proposed an initial split of passerines into suboscines and oscines (songbirds). Later studies showed, with high confidence, that Acanthisittidae (New Zealand Wrens) are the earliest divergence within passerines, followed by the split into suboscines and oscines (Barker et al. 2002, Cracraft et al. 2003, Hackett et al. 2008, Kimball et al. 2019, Oliveros et al. 2019). The Acanthisittidae comprises two extant, diminutive species that are endemic to New Zealand. This pattern mimics the earliest divergence in their sister group, the parrots, and led to the hypothesis that perhaps the splitting of New Zealand from the rest of Gondwana resulted in vicariant divergence in the early evolution of these two orders (Barker et al. 2004, Wright et al. 2008). Using this biogeographic calibration resulted in a hypothesis that the origin of modern birds and Neoaves occurred well into the Cretaceous, a hotly debated topic (see below). Subsequent papers have proposed later divergence times, generally at either the K–Pg boundary or even later into the Palaeogene/Eocene (Moyle et al. 2016, Crouch et al. 2019, Oliveros et al. 2019, Field et al. 2020). The suboscines (Tyranni) are two sister clades that are generally split into Old World (Eurylaimides) and New World (Tyrannides) divisions except for one aberrant species, *Sapoyoa aenigma*, which is endemic to the Neotropics but is phylogenetically nested within the Eurylaimides (Fjeldsa et al. 2003, Moyle et al. 2006, Hackett et al. 2008, Oliveros et al. 2019). The oscines or songbirds (Passeri) include the vast majority of passerines that have radiated throughout the world and are found in almost every habitat and continent except Antarctica. Within oscines, there are a number of early diverging lineages that are found exclusively in Australasia, indicating that songbirds likely originated in this region (Oliveros et al. 2019), followed by the divergence of two globally distributed clades—the Corvides and Passerides. These two clades of roughly "crow-like" and "sparrow-like" birds are each highly morphologically, ecologically, and behaviorally diverse as well as species-rich. Their dispersal out of Australasia and into Asia, Africa, and the rest of the world drove the diversification of these two clades, with colonization of each new region leading to heightened speciation rates as lineages exploited new environments (Moyle et al. 2016, Oliveros et al. 2019).

A question that has long puzzled ornithologists is why passerines are so diverse. Answers to this question have come from many directions, such as identifying unique features, synapomorphies, or key innovations that unite this clade and may explain why they are so successful. Although some morphological (Raikow 1986), genomic (Yuri et al. 2008, Feng et al. 2020), and behavioral (Fitzpatrick 1988) features have been identified, other hypotheses suggest that passerines are remarkable because of their high adaptability, as evidenced by increased speciation rates upon dispersal to new regions and high degrees of bill variation across the order (Fitzpatrick 1988, Ksepka et al. 2019, Oliveros et al. 2019). Most likely, a combination of several features contributes to the success of passerines.

Timing and Biogeography

Dramatic changes in our understanding of avian phylogeny have led to equally remarkable revisions of chronological and biogeographic interpretations. Several early molecular dating studies concluded that the divergences of many modern orders occurred deep into the Cretaceous (Cooper and Penny 1997, Cracraft 2001, Brown et al. 2008), pushing back the origins of modern birds many tens of millions of years before the appearance of fossils that were conclusively attributed to Neornithes. More recent studies have taken a more conservative approach by using a multitude of fossils along with trees generated from phylogenomic studies to estimate divergence times within modern birds (Jarvis et al. 2014, Claramunt and Cracraft 2015, Prum et al. 2015). These analyses generally agree that at least two splitting events—the divergence of Paleognaths and Neognaths and the divergence of Galloanserae and Neoaves—occurred prior to the K–Pg mass extinction. These analyses also generally agree that most Neoavian orders originated soon after the mass extinction event.

Given the ample number of unknowns—poor fossil record, uncertainty in phylogenies, and difficulties in interpreting current distributions for paleoreconstructions—it is challenging to be able to extrapolate much about the early biogeographic history of birds. There are few conclusive fossils of modern birds that predate the K–Pg boundary (Clarke et al. 2005, Field et al. 2020, Field et al. 2020). Interestingly, there are several examples of fossils that have been attributed to modern orders that are found in regions where these lineages no longer exist—for example, hummingbird fossils found in Europe (Mayr 2004b, 2007) and mousebirds in North America (Ksepka et al. 2017). Clearly, climatic and tectonic transformations throughout the Cenozoic led to dramatic shifts in distributions of many groups of birds (Saupe et al. 2019). The influence of fossil-rich sites (lagerstatten) located in northern regions and many diverse avian orders restricted to tropical distributions have led to interpretations of massive dispersal events throughout the world over many periods in the Cenozoic (Claramunt and Cracraft 2015, Cracraft and Claramunt 2017, Mayr 2017, McCullough et al. 2019, Oliveros et al. 2019). Analyses using newly developed paleoclimatic models might help shed some light on these scenarios (see Field et al. 2018, Saupe et al. 2019), yet it is difficult to assess the strengths of these reconstructions without more evidence or corroboration from other studies.

What Are Future Challenges?

If we use the criteria of consilience across different phylogenomic studies, the base of Neoaves is best represented as a polytomy (see Figure 7.1). This may not seem like much advancement from two decades ago when the phylogenomics era began; however, we have made substantial headway into the circumscription of various orders and superorders, as outlined above. Given that many of these previously controversial relationships have now stabilized and find support across multiple types of data is a tremendous leap in our confidence in some branches of the avian tree of life. Nevertheless, several ornithologists have suggested that perhaps there is a limit to how much we can decipher and further the idea that perhaps the bird phylogeny is shaped like a bush rather than tree. Others, like this author, believe that it is premature to consider that we have reached the limits of what we can know. There is much evidence that the early divergence history of modern birds was explosive, yet calling this a "bush" does not provide any more power for subsequent analyses and may discourage future studies from aspiring to more innovative ways of evaluating the data.

We have learned some critical aspects of the nature of the problem that might be useful in developing future directions for addressing the challenges of uncovering bird evolutionary history. First, the simple solution of adding more taxa and more genomic data is inadequate to resolve the remaining intractable issues. Currently, we have copious amounts of whole genome data from more than 1,000 species of birds, yet we lack the computational power to conduct thorough phylogenetic analyses of these data. Second, as mentioned previously, there are no extant species available to break up many long branches, so adding more species may be informative for tips but will likely not add much clarity to deep evolutionary history. Another issue that is increasingly becoming clear is that evaluating model fit to the data is an often overlooked yet critical issue (Reddy et al. 2017). Perhaps most substantially, many of the conflicts across avian phylogenomic studies are due to differences in the type of data used, such as coding versus noncoding. There is considerable evidence that coding genes deviate from expectations of even the most complex evolutionary models available for nucleotide data (e.g., GTR+I+G). Coding genes violate assumptions of base-composition stationarity to a much greater extent that noncoding data. Although noncoding data are better fit for currently available models, we should also be pushing for more improvements and new developments in models of genomic evolution, something that far lags in the genomic era that has tremendously facilitated generating data but has not produced a corresponding leap in computational resources to handling the complexities of how DNA and genomes evolve.

Other avenues that may be fruitful for discovering phylogenetic information and perhaps developing innovative analytical tools are comparative genomics and phenomics. The power of more in-depth comparative studies is critical for understanding avian evolution and especially their correlation of genotype–phenotype variation (Feng et al. 2020, Bravo et al. 2021). Recent analyses have identified major genomic changes, such as gene duplications and inversions, that appear to be associated with functions and traits. These types of mutations are relatively underexplored and may also be informative for phylogenetic relationships.

In addition, prospects to integrate morphology, fossils, and paleoenvironmental data with phylogenomics will no doubt yield fruitful insights about avian evolution. The recent surge of morphometrics studies shows that large-scale efforts to characterize morphology are feasible and valuable toward understanding avian adaptations over time. Efforts to construct morphological data sets for phylogenetic analyses are undoubtedly difficult yet highly valuable for interpreting anatomical features of both extant and fossil representatives. Finally, there have been many recent advances in reconstructing past climatic and environmental changes, and their implications for avian evolution are just beginning to be explored.

The power of robust phylogenies for comparative studies has never been more apparent than in the 21st century. Phylogenomic studies have inspired and aided a tremendous number of analyses to study various aspects of avian ecology, evolution, morphology, behavior, and biogeography. In the past two decades, ornithologists have made impressive strides toward understanding avian phylogenetics, and building on these efforts with novel and innovative approaches will continue to drive our quest to learn more about the incredible diversity of birds.

References

Baker, A. J., O. Haddrath, J. D. McPherson, and A. Cloutier. 2014. Genomic support for a Moa–Tinamou clade and adaptive morphological convergence in flightless ratites. *Molecular Biology and Evolution* 31:1686–1696.

Barker, F. K., G. F. Barrowclough, and J. G. Groth. 2002. A phylogenetic hypothesis for passerine birds: Taxonomic and biogeographic implications of an analysis of nuclear DNA sequence data. *Proceedings of the Royal Society B: Biological Sciences* 269:295–308.

Barker, F. K., A. Cibois, P. Schikler, J. Feinstein, and J. Cracraft. 2004. Phylogeny and diversification of the largest avian radiation. *Proceedings of the National Academy of Sciences of the USA* 101:11040–11045.

Braun, E. L., J. Cracraft, and P. Houde. 2019. Avian genomics in ecology and evolution, from the lab into the wild. Pages 151–210 in R. Kraus, ed. *Avian Genomics in Ecology and Evolution*. Springer, Cham, Switzerland.

Braun, E. L., R. T. Kimball, K.-L. Han, N. R. Iuhasz-Velez, A. J. Bonilla, J. L. Chojnowski, J. V. Smith, R. C. K. Bowie, M. J. Braun, S. J. Hackett, J. Harshman, C. J. Huddleston, B. D. Marks, K. J. Miglia, W. S. Moore, S. Reddy, F. H. Sheldon, C. C. Witt, and T. Yuri. 2011. Homoplastic microinversions and the avian tree of life. *BMC Evolutionary Biology* 11: Article 141.

Bravo, G. A., C. J. Schmitt, and S. V. Edwards. 2021. What have we learned from the first 500 avian genomes? *Annual Review of Ecology, Evolution, and Systematics* 52:1–29.

Brown, J. W., J. S. Rest, J. García-Moreno, M. D. Sorenson, and D. P. Mindell. 2008. Strong mitochondrial DNA support for a Cretaceous origin of modern avian lineages. *BMC Biology* 6:6–18.

Brown, J. W., N. Wang, and S. A. Smith. 2017. The development of scientific consensus: Analyzing conflict and concordance among avian phylogenies. *Molecular Phylogenetics and Evolution* 116:69–77.
Brusatte, S. L., J. K. O'Connor, and E. D. Jarvis. 2015. The origin and diversification of Birds. *Current Biology* 25:R888–R898.
Claramunt, S., and J. Cracraft. 2015. A new time tree reveals Earth history's imprint on the evolution of modern birds. *Science Advances* 1: Article e1501005.
Clarke, J. A., C. P. Tambussi, J. I. Noriega, G. M. Erickson, and R. A. Ketcham. 2005. Definitive fossil evidence for the extant avian radiation in the Cretaceous. *Nature* 433:305–308.
Clements, J. F., T. S. Schulenberg, M. J. Iliff, S. M. Billerman, T. A. Fredericks, J. A. Gerbracht, D. Lepage, B. L. Sullivan, and C. L. Wood. 2021. The eBird/Clements checklist of birds of the world: v2021. https://www.birds.cornell.edu/clementschecklist/download
Cloutier, A., T. B. Sackton, P. Grayson, M. Clamp, A. J. Baker, and S. V. Edwards. 2019. Whole-genome analyses resolve the phylogeny of flightless birds (Palaeognathae) in the presence of an empirical anomaly zone. *Systematic Biology* 68:937–955.
Cooper, A., and D. Penny. 1997. Mass survival of birds across the Cretaceous–Tertiary boundary: Molecular evidence. *Science* 275:1109–1113.
Cracraft, J. 1974. Phylogeny and evolution of the ratite birds. *Ibis* 116:494–521.
Cracraft, J. 1981. Toward a phylogenetic classification of the recent birds of the world (Class Aves). *The Auk* 98:681–714.
Cracraft, J. 1985. Monophyly and phylogenetic relationships of the Pelecaniformes: A numerical cladistic analysis. *The Auk* 102:834–853.
Cracraft, J. 2001. Avian evolution, Gondwana biogeography and the Cretaceous–Tertiary mass extinction event. *Proceedings of the Royal Society B: Biological Sciences* 268:459–469.
Cracraft, J. 2002. Early evolution of birds. *Nature* 331:389–390.
Cracraft, J., F. K. Barker, and A. Cibois. 2003. Avian higher-level phylogenetics and the Howard and Moore Checklist of Birds. Pages 16–21 in E. C. Dickinson, ed. *The Howard & Moore Complete Checklist of the Birds of the World*. 3rd ed. Helm, London.
Cracraft, J., and S. Claramunt. 2017. Conceptual and analytical worldviews shape differences about global avian biogeography. *Journal of Biogeography* 44:958–960.
Crouch, N. M. A., K. Ramanauskas, and B. Igić. 2019. Tip-dating and the origin of Telluraves. *Molecular Phylogenetics and Evolution* 131:55–63.
Dickinson, E. C., and L. Christidis. 2014. *The Howard and Moore Complete Checklist of the Birds of the World*. 4th ed., Vol. 2: Passerines Page (E. C. Dickinson and L. Christidis, eds.). Aves Press, Eastbourne, UK.
Eo, S. H., O. R. P. Bininda-Emonds, and J. P. Carroll. 2009. A phylogenetic supertree of the fowls (Galloanserae, Aves). *Zoologica Scripta* 38:465–481.
Ericson, P. G. P., C. L. Anderson, T. Britton, A. Elzanowski, U. S. Johansson, M. Källersjö, J. I. Ohlson, T. J. Parsons, D. Zuccon, and G. Mayr. 2006. Diversification of Neoaves: Integration of molecular sequence data and fossils. *Biology Letters* 2:543–547.
Ericson, P. G. P., C. Anderson, and G. Mayr. 2007. Hangin' on to our rocks 'n clocks: A reply to Brown et al. *Biology Letters* 3:260–261.
Fain, M. G., and P. Houde. 2004. Parallel radiations in the primary clades of birds. *Evolution* 58:2558–2573.
Feng, S., J. Stiller, Y. Deng, J. Armstrong, Q. Fang, A. H. Reeve, D. Xie, G. Chen, C. Guo, B. C. Faircloth, B. Petersen, Z. Wang, Q. Zhou, M. Diekhans, W. Chen, S. Andreu-Sánchez, A. Margaryan, J. T. Howard, C. Parent, G. Pacheco, M.-H. S. Sinding, L. Puetz, E. Cavill, Â. M. Ribeiro, L. Eckhart, J. Fjeldså, P. A. Hosner, R. T. Brumfield, L. Christidis, M. F. Bertelsen, T. Sicheritz-Ponten, D. T. Tietze, B. C. Robertson, G. Song, G. Borgia, S. Claramunt, I. J. Lovette, S. J. Cowen, P. Njoroge, J. P. Dumbacher, O. A. Ryder, J. Fuchs, M. Bunce, D. W. Burt, J. Cracraft, G. Meng, S. J. Hackett, P. G. Ryan, K. A. Jønsson, I. G. Jamieson, R. R. da Fonseca, E. L. Braun, P. Houde, S. Mirarab, A. Suh, B. Hansson, S. Ponnikas, H. Sigeman, M. Stervander, P. B. Frandsen, H. van der Zwan, R. van der Sluis, C. Visser, C. N. Balakrishnan, A. G. Clark, J. W. Fitzpatrick, R. Bowman, N. Chen, A. Cloutier, T. B. Sackton, S. V. Edwards, D. J. Foote, S. B. Shakya, F. H. Sheldon, A. Vignal, A. E. R. Soares, B. Shapiro, J. González-Solís, J. Ferrer-Obiol, J. Rozas, M. Riutort, A. Tigano, V. Friesen,

L. Dalén, A. O. Urrutia, T. Székely, Y. Liu, M. G. Campana, A. Corvelo, R. C. Fleischer, K. M. Rutherford, N. J. Gemmell, N. Dussex, H. Mouritsen, N. Thiele, K. Delmore, M. Liedvogel, A. Franke, M. P. Hoeppner, O. Krone, A. M. Fudickar, B. Milá, E. D. Ketterson, A. E. Fidler, G. Friis, Á. M. Parody-Merino, P. F. Battley, M. P. Cox, N. C. B. Lima, F. Prosdocimi, T. L. Parchman, B. A. Schlinger, B. A. Loiselle, J. G. Blake, H. C. Lim, L. B. Day, M. J. Fuxjager, M. W. Baldwin, M. J. Braun, M. Wirthlin, R. B. Dikow, T. B. Ryder, G. Camenisch, L. F. Keller, J. M. DaCosta, M. E. Hauber, M. I. M. Louder, C. C. Witt, J. A. McGuire, J. Mudge, L. C. Megna, M. D. Carling, B. Wang, S. A. Taylor, G. Del-Rio, A. Aleixo, A. T. R. Vasconcelos, C. V. Mello, J. T. Weir, D. Haussler, Q. Li, H. Yang, J. Wang, F. Lei, C. Rahbek, M. T. P. Gilbert, G. R. Graves, E. D. Jarvis, B. Paten, and G. Zhang. 2020. Dense sampling of bird diversity increases power of comparative genomics. *Nature* 587:252–257.

Field, D. J., J. Benito, A. Chen, J. W. M. Jagt, and D. T. Ksepka. 2020. Late Cretaceous neornithine from Europe illuminates the origins of crown birds. *Nature* 579:397–401.

Field, D. J., A. Bercovici, J. S. Berv, R. Dunn, D. E. Fastovsky, T. R. Lyson, V. Vajda, and J. A. Gauthier. 2018. Early evolution of modern birds structured by global forest collapse at the end-Cretaceous mass extinction. *Current Biology* 28:1825–1831.

Field, D. J., J. S. Berv, A. Y. Hsiang, R. Lanfear, M. J. Landis, and A. Dornburg. 2020. Timing the extant avian radiation: The rise of modern birds, and the importance of modeling molecular rate variation. *Bulletin of the American Museum of Natural History* 440:159–181.

Fitzpatrick, J. W. 1988. Why so many Passerine birds? A response to Raikow. *Systematic Zoology* 37:71–76.

Fjelds, J., D. Zuccon, M. Irestedt, U. S. Johansson, and P. G. P. Ericson. 2003. *Sapayoa aenigma*: A New World representative of "Old World suboscines." *Proceedings of the Royal Society B: Biological Sciences* 270:S238–S241.

Franz, N. M., L. J. Musher, J. W. Brown, S. Yu, and B. Ludäscher. 2019. Verbalizing phylogenomic conflict: Representation of node congruence across competing reconstructions of the neoavian explosion. *PLoS Computational Biology* 15: Article e1006493.

Gill, F., D. Donsker, and P. Rasmussen. 2022. IOC World Bird List (v12.1). doi:10.14344/IOC.ML.12.1.

Grealy, A., M. Phillips, G. Miller, M. T. P. Gilbert, J.-M. Rouillard, D. Lambert, M. Bunce, and J. Haile. 2017. Eggshell palaeogenomics: Palaeognath evolutionary history revealed through ancient nuclear and mitochondrial DNA from Madagascan elephant bird (*Aepyornis* sp.) eggshell. *Molecular Phylogenetics and Evolution* 109:151–163.

Hackett, S. J., R. T. Kimball, S. Reddy, R. C. K. Bowie, E. L. Braun, M. J. Braun, J. L. Chojnowski, W. A. Cox, K.-L. Han, J. Harshman, C. J. Huddleston, B. D. Marks, K. J. Miglia, W. S. Moore, F. H. Sheldon, D. W. Steadman, C. C. Witt, and T. Yuri. 2008. A phylogenomic study of birds reveals their evolutionary history. *Science* 320:1763–1768.

Haddrath, O., and A. J. Baker. 2001. Complete mitochondrial DNA genome sequences of extinct birds: Ratite phylogenetics and the vicariance biogeography hypothesis. *Proceedings of the Royal Society B: Biological Sciences* 268:939–945.

Han, K.-L., E. L. Braun, R. T. Kimball, S. Reddy, R. C. K. Bowie, M. J. Braun, J. L. Chojnowski, S. J. Hackett, J. Harshman, C. J. Huddleston, B. D. Marks, K. J. Miglia, W. S. Moore, F. H. Sheldon, D. W. Steadman, C. C. Witt, and T. Yuri. 2011. Are transposable element insertions homoplasy free? An examination using the avian tree of life. *Systematic Biology* 60:375–386.

Harshman, J., E. L. Braun, M. J. Braun, C. J. Huddleston, R. C. K. Bowie, J. L. Chojnowski, S. J. Hackett, K.-L. Han, R. T. Kimball, B. D. Marks, K. J. Miglia, W. S. Moore, S. Reddy, F. H. Sheldon, D. W. Steadman, S. J. Steppan, C. C. Witt, and T. Yuri. 2008. Phylogenomic evidence for multiple losses of flight in ratite birds. *Proceedings of the National Academy of Sciences of the USA* 105:13462–13467.

Hosner, P. A., E. L. Braun, and R. T. Kimball. 2016. Rapid and recent diversification of curassows, guans, and chachalacas (Galliformes: Cracidae) out of Mesoamerica: Phylogeny inferred from mitochondrial, intron, and ultraconserved element sequences. *Molecular Phylogenetics and Evolution* 102:320–330.

Hosner, P. A., J. A. Tobias, E. L. Braun, and R. T. Kimball. 2017. How do seemingly non-vagile clades accomplish trans-marine dispersal? Trait and dispersal evolution in the landfowl (Aves: Galliformes). *Proceedings of the Royal Society B: Biological Sciences* 284: Article 20170210.

Houde, P., E. L. Braun, N. Narula, U. Minjares, and S. Mirarab. 2019. Phylogenetic signal of indels and the Neoavian radiation. *Diversity* 11: Article 108.

Jarvis, E. D., S. Mirarab, A. J. Aberer, B. Li, P. Houde, and C. Li. 2014. Whole-genome analyses resolve early branches in the tree of life of modern birds. *Science* 346:1320–1331.

Jeffroy, O., H. Brinkmann, F. Delsuc, and H. Philippe. 2006. Phylogenomics: The beginning of incongruence? *Trends in Genetics* 22:225–231.

Kimball, R. T., and E. L. Braun. 2014. Does more sequence data improve estimates of galliform phylogeny? Analyses of a rapid radiation using a complete data matrix. *PeerJ* 2: Article e361.

Kimball, R. T., C. H. Oliveros, N. Wang, N. D. White, F. K. Barker, D. J. Field, D. T. Ksepka, R. T. Chesser, R. G. Moyle, M. J. Braun, R. T. Brumfield, B. C. Faircloth, B. T. Smith, and E. L. Braun. 2019. A phylogenomic supertree of birds. *Diversity* 11: Article 109.

Ksepka, D. T., L. Grande, and G. Mayr. 2019. Oldest finch-beaked birds reveal parallel ecological radiations in the earliest evolution of passerines. *Current Biology* 29:657–663.

Ksepka, D. T., T. A. Stidham, and T. E. Williamson. 2017. Early Paleocene landbird supports rapid phylogenetic and morphological diversification of crown birds after the K–Pg mass extinction. *Proceedings of the National Academy of Sciences of the USA* 114:8047–8052.

Kuhl, H., C. Frankl-Vilches, A. Bakker, G. Mayr, G. Nikolaus, S. T. Boerno, S. Klages, B. Timmermann, and M. Gahr. 2020. An unbiased molecular approach using 3′-UTRs resolves the avian family-level tree of life. *Molecular Biology and Evolution* 38:108–127.

Lapiedra, O., F. Sayol, J. Garcia-Porta, and D. Sol. 2021. Niche shifts after island colonization spurred adaptive diversification and speciation in a cosmopolitan bird clade. *Proceedings of the Royal Society B: Biological Sciences* 288: Article 20211022.

Livezey, B. 2011. Grebes and flamingos: Standards of evidence, adjudication of disputes, and societal politics in avian systematics. *Cladistics* 27:391–401.

Livezey, B. C., and R. L. Zusi. 2007. Higher-order phylogeny of modern birds (Theropoda, Aves: Neornithes) based on comparative anatomy: II. Analysis and discussion. *Zoological Journal of the Linnean Society* 149:1–95.

Mayr, G. 2004a. Morphological evidence for sister group relationship between flamingos (Aves: Phoenicopteridae) and grebes (Podicipedidae). *Zoological Journal of the Linnean Society* 140:157–169.

Mayr, G. 2004b. Old World fossil record of modern-type hummingbirds. *Science* 304:861–864.

Mayr, G. 2007. New specimens of the early Oligocene Old World hummingbird *Eurotrochilus inexpectatus*. *Journal of Ornithology* 148:105–111.

Mayr, G. 2008. Avian higher-level phylogeny: Well-supported clades and what we can learn from a phylogenetic analysis of 2954 morphological characters. *Journal of Zoological Systematics and Evolutionary Research* 46:63–72.

Mayr, G. 2011. Metaves, Mirandornithes, Strisores and other novelties: A critical review of the higher-level phylogeny of neornithine birds. *Journal of Zoological Systematics and Evolutionary Research* 49:58–76.

Mayr, G. 2014. The origins of crown group birds: Molecules and fossils. *Palaeontology* 57:231–242.

Mayr, G. 2017. Avian higher level biogeography: Southern Hemispheric origins or Southern Hemispheric relicts? *Journal of Biogeography* 44:956–958.

Mayr, G., and J. Clarke. 2003. The deep divergences of neornithine birds: A phylogenetic analysis of morphological characters. *Cladistics* 19:527–553.

McCormack, J. E., M. G. Harvey, B. C. Faircloth, N. G. Crawford, T. C. Glenn, and R. T. Brumfield. 2013. A phylogeny of birds based on over 1,500 loci collected by target enrichment and high-throughput sequencing. *PLoS One* 8: Article e54848.

McCullough, J. M., R. G. Moyle, B. T. Smith, and M. J. Andersen. 2019. A Laurasian origin for a pantropical bird radiation is supported by genomic and fossil data (Aves: Coraciiformes). *Proceedings of the Royal Society B: Biological Sciences/* 286: Article 20190122.

Mindell, D. P., M. D. Sorenson, D. E. Dimcheff, M. Hasegawa, J. C. Ast, and T. Yuri. 1999. Interordinal relationships of birds and other reptiles based on whole mitochondrial genomes. *Systematic Biology* 48:138–152.

Mitchell, K. J., B. Llamas, J. Soubrier, N. J. Rawlence, T. H. Worthy, J. Wood, M. S. Y. Lee, and A. Cooper. 2014. Ancient DNA reveals elephant birds and kiwi are sister taxa and clarifies ratite bird evolution. *Science* 344:898–900.

Moyle, R. G., R. T. Chesser, R. Prum, P. Schikler, and J. Cracraft. 2006. Phylogeny and evolutionary history of Old World suboscine birds (Aves, Eurylaimides). *American Museum Novitates* 3544:1–22.

Moyle, R. G., C. H. Oliveros, M. J. Andersen, P. A. Hosner, B. W. Benz, J. D. Manthey, S. L. Travers, R. M. Brown, and B. C. Faircloth. 2016. Tectonic collision and uplift of Wallacea triggered the global songbird radiation. *Nature Communications* 7: Article 12709.

Oliveros, C. H., M. J. Andersen, P. A. Hosner, W. M. M. III, F. H. Sheldon, J. Cracraft, and R. G. Moyle. 2019. Rapid Laurasian diversification of a pantropical bird family during the Oligocene–Miocene transition. *Ibis* 162:137–152.

Oliveros, C. H., D. J. Field, D. T. Ksepka, F. K. Barker, A. Aleixo, M. J. Andersen, P. Alström, B. W. Benz, E. L. Braun, M. J. Braun, G. A. Bravo, R. T. Brumfield, R. T. Chesser, S. Claramunt, J. Cracraft, A. M. Cuervo, E. P. Derryberry, T. C. Glenn, M. G. Harvey, P. A. Hosner, L. Joseph, R. T. Kimball, A. L. Mack, C. M. Miskelly, A. T. Peterson, M. B. Robbins, F. H. Sheldon, L. F. Silveira, B. T. Smith, N. D. White, R. G. Moyle, and B. C. Faircloth. 2019. Earth history and the passerine superradiation. *Proceedings of the National Academy of Sciences of the USA* 116: Article 201813206.

Peters, J. L. (n.d.). *Check-List of the Birds of the World.* Vols. 1–16. Harvard University Press, Cambridge, MA.

Phillips, M. J., G. C. Gibb, E. A. Crimp, and D. Penny. 2009. Tinamous and Moa flock together: Mitochondrial genome sequence analysis reveals independent losses of flight among ratites. *Systematic Biology* 59:90–107.

Pratt, R. C., G. C. Gibb, M. Morgan-Richards, M. J. Phillips, M. D. Hendy, and D. Penny. 2009. Toward resolving deep neoaves phylogeny: Data, signal enhancement, and priors. *Molecular Biology and Evolution* 26:313–326.

Prum, R. O., J. S. Berv, A. Dornburg, D. J. Field, J. P. Townsend, E. M. Lemmon, and A. R. Lemmon. 2015. A comprehensive phylogeny of birds (Aves) using targeted next-generation DNA sequencing. *Nature* 526:569–573.

Raikow, R. J. 1986. Why are there so many kinds of passerine birds? *Systematic Biology* 35:255–259.

Reddy, S., R. T. Kimball, A. Pandey, P. A. Hosner, M. J. Braun, S. J. Hackett, K.-L. Han, J. Harshman, C. J. Huddleston, S. Kingston, B. D. Marks, K. J. Miglia, W. S. Moore, F. H. Sheldon, C. C. Witt, T. Yuri, and E. L. Braun. 2017. Why do phylogenomic data sets yield conflicting trees? Data type influences the avian tree of life more than taxon sampling. *Systematic Biology* 66:857–879.

Sackton, T. B., P. Grayson, A. Cloutier, Z. Hu, J. S. Liu, N. E. Wheeler, P. P. Gardner, J. A. Clarke, A. J. Baker, M. Clamp, and S. V. Edwards. 2019. Convergent regulatory evolution and loss of flight in paleognathous birds. *Science* 364:74–78.

Sangster, G., E. L. Braun, U. S. Johansson, R. T. Kimball, G. Mayr, and A. Suh. 2022. Phylogenetic definitions for 25 higher-level clade names of birds. *Avian Research* 13: Article 100027.

Saupe, E. E., A. Farnsworth, D. J. Lunt, N. Sagoo, K. V. Pham, and D. J. Field. 2019. Climatic shifts drove major contractions in avian latitudinal distributions throughout the Cenozoic. *Proceedings of the National Academy of Sciences of the USA* 116:12895–12900.

Sibley, C. G., and J. Ahlquist. 1990. *Phylogeny and Classification of Birds.* Yale University Press, New Haven, CT.

Sibley, C. G., J. E. Ahlquist, and B. L. Monroe. 1988. A classification of the living birds of the world based on DNA–DNA hybridization studies. *The Auk* 105:409–423.

Sibley, C. G., and B. J. Monroe. 1991. *Distribution and Taxonomy of Birds of the World.* Yale University Press, New Haven, CT.

Smith, J. V., E. L. Braun, and R. T. Kimball. 2012. Ratite nonmonophyly: Independent evidence from 40 novel loci. *Systematic Biology* 62:35–49.

Stiller, J., S. Feng, A.-A. Chowdhury, I. Rivas-González, D. A. Duchêne, Q. Fang, Y. Deng, A. Kozlov, A. Stamatakis, S. Claramunt, J. M. T. Nguyen, S. Y. W. Ho, B. C. Faircloth, J. Haag, P. Houde, J. Cracraft, M. Balaban, U. Mai, G. Chen, R. Gao, C. Zhou, Y. Xie, Z. Huang, Z. Cao, Z. Yan, H. A. Ogilvie, L. Nakhleh, B. Lindow, B. Morel, J. Fjeldså, P. A. Hosner, R. R. da Fonseca, B. Petersen, J. A. Tobias, T. Székely, J. D. Kennedy, A. H. Reeve, A. Liker, M. Stervander, A. Antunes, D. T. Tietze,

M. F. Bertelsen, F. Lei, C. Rahbek, G. R. Graves, M. H. Schierup, T. Warnow, E. L. Braun, M. T. P. Gilbert, E. D. Jarvis, S. Mirarab, and G. Zhang. 2024. Complexity of avian evolution revealed by family-level genomes. *Nature* 629:851–860.

Suh, A. 2016. The phylogenomic forest of bird trees contains a hard polytomy at the root of Neoaves. *Zoologica Scripta* 45:50–62.

Suh, A., L. Smeds, and H. Ellegren. 2015. The dynamics of incomplete lineage sorting across the ancient adaptive radiation of Neoavian birds. *PLoS Biology* 13: Article e1002224.

Sun, Z., T. Pan, C. Hu, L. Sun, H. Ding, H. Wang, C. Zhang, H. Jin, Q. Chang, X. Kan, and B. Zhang. 2017. Rapid and recent diversification patterns in Anseriformes birds: Inferred from molecular phylogeny and diversification analyses. *PLoS One* 12: Article e0184529.

Tuinen, M. V., D. B. Butvill, J. A. W. Kirsch, and S. B. Hedges. 2001. Convergence and divergence in the evolution of aquatic birds. *Proceedings of the Royal Society B: Biological Sciences* 268:1345–1350.

Wang, N., R. T. Kimball, E. L. Braun, B. Liang, and Z. Zhang. 2017. Ancestral range reconstruction of Galliformes: The effects of topology and taxon sampling. *Journal of Biogeography* 44:122–135.

Wright, T. F., E. E. Schirtzinger, T. Matsumoto, J. R. Eberhard, G. R. Graves, J. J. Sanchez, S. Capelli, H. Muller, J. Scharpegge, G. K. Chambers, and R. C. Fleischer. 2008. A multilocus molecular phylogeny of the parrots (Psittaciformes): Support for a Gondwanan origin during the Cretaceous. *Molecular Biology and Evolution* 25:2141–2156.

Wu, S., F. E. Rheindt, J. Zhang, J. Wang, L. Zhang, C. Quan, Z. Li, M. Wang, F. Wu, Y. Qu, S. V. Edwards, Z. Zhou, and L. Liu. 2024. Genomes, fossils, and the concurrent rise of modern birds and flowering plants in the Late Cretaceous. *Proceedings of the National Academy of Sciences of the USA* 121:e2319696121.

Yonezawa, T., T. Segawa, H. Mori, P. F. Campos, Y. Hongoh, H. Endo, A. Akiyoshi, N. Kohno, S. Nishida, J. Wu, H. Jin, J. Adachi, H. Kishino, K. Kurokawa, Y. Nogi, H. Tanabe, H. Mukoyama, K. Yoshida, A. Rasoamiaramanana, S. Yamagishi, Y. Hayashi, A. Yoshida, H. Koike, F. Akishinonomiya, E. Willerslev, and M. Hasegawa. 2017. Phylogenomics and morphology of extinct paleognaths reveal the origin and evolution of the ratites. *Current Biology* 27:68–77.

Yuri, T., R. T. Kimball, E. L. Braun, and M. J. Braun. 2008. Duplication of accelerated evolution and growth hormone gene in passerine birds. *Molecular Biology and Evolution* 25:352–361.

Yuri, T., R. T. Kimball, J. Harshman, R. C. K. Bowie, M. J. Braun, J. L. Chojnowski, K.-L. Han, S. J. Hackett, C. J. Huddleston, W. S. Moore, S. Reddy, F. H. Sheldon, D. W. Steadman, C. C. Witt, and E. L. Braun. 2013. Parsimony and model-based analyses of indels in avian nuclear genes reveal congruent and incongruent phylogenetic signals. Biology 2:419–444.

8
The Origin and Early Evolution of Birds

Jingmai O'Connor

During the past 30 years, the Mesozoic fossil record of birds (Aves) has grown enormously. The astonishing discovery of thousands of well-preserved specimens from the Early Cretaceous of northeastern China has afforded an unexpectedly vivid glimpse of an early stage in avian evolution. These fossils preserve soft tissues and other rarely fossilized traces, revealing the Cretaceous diversification of birds in unprecedented detail. Together with important discoveries of non-avian dinosaurs closely related to birds from the same and older deposits, paleontologists are now able to construct a cohesive and comprehensive picture of the stepwise and homoplastic evolution of the modern bird. What has become clear is that nearly all features we think of today as "avian" were either inherited from non-avian dinosaurs or absent in the earliest birds. The multiple evolutionary origins of flight in maniraptoran dinosaurs means that some traditionally "avian" features are better referred to as "aerodynamic" features, and these evolved repeatedly, creating homoplasy that hinders interpretations regarding the immediate relationships between birds and their closest non-avian relatives. Furthermore, the rapid radiation of avian lineages in the latest Jurassic–earliest Cretaceous resulted in high homoplasy within Aves as each lineage independently refined its flight apparatus. Despite enormous gains, there remain critical gaps to be filled, and with most fossil data pertaining to the Early Cretaceous, broad evolutionary trajectories spanning the Cretaceous remain poorly understood.

Archaeopteryx and the Dinosaurian Ancestry of Birds

Just 2 years after Darwin first published *On the Origin of Species By Means of Natural Selection*, an isolated feather was described from lithographic limestones in southern Germany (von Meyer 1861). Named *Archaeopteryx lithographica*, this discovery provided the first evidence of a Late Jurassic bird that lived in what is today Bavaria but 150 million years ago (Mya) formed part of an island archipelago on the edge of the Tethys Sea (Wellnhofer 2008). Shortly after, the London specimen of *Archaeopteryx* was described (Owen 1863); with its teeth and elongate, reptilian tail yet large remiges forming a nearly modern wing morphology, *Archaeopteryx* appears to be the perfect "transitional" form, predicted by Darwin in the *Origin of Species* to exist in the fossil record. However, due to previously reported Triassic-aged theropod footprints thought to belong to giant birds, Darwin underestimated the importance of this discovery (Darwin 1866).

Jingmai O'Connor, *The Origin and Early Evolution of Birds*. In: *New Perspectives in Ornithology*.
Edited by: Scott V. Edwards and J. Michael Reed, Oxford University Press.
DOI: 10.1093/oso/9780197787670.003.0008

More than 160 years later, paleontologists recognize that the modern study of Mesozoic avian evolution began with these discoveries. *Archaeopteryx* is the first discovered Mesozoic bird, making it the oldest and most basal bird known at the time of discovery, titles this iconic taxon still holds today. The morphology of *Archaeopteryx* is such that Thomas Huxley's observations of this taxon and the theropod dinosaur *Compsognathus* led him to suggest that dinosaurs represented an intermediary between birds and reptiles and that the bird-ancestor was probably morphologically similar to small, carnivorous, bipedal dinosaurs such as *Compsognathus* (Switek 2010).

In the century that followed, a relationship between birds and dinosaurs was largely rejected primarily based on the absence of clavicles in all dinosaurs known at the time (Heilmann 1926). Instead, scientists suggested birds had descended from basal archosauromorphs such as the now defunct clade "Thecodontia" (Heilmann 1926), and some even hypothesized that birds were more closely related to the crocodilian archosaur lineage (A. Walker 1972).

In the early 1970s, John Ostrom resurrected the hypothesis that birds descended from theropod dinosaurs based on his observations of *Archaeopteryx* (Ostrom 1972, 1973, 1974, 1975, 1976) and the newly discovered dromaeosaurid theropod *Deinonychus* (Ostrom 1969). His observations included identification of a new *Archaeopteryx* (the Haarlem specimen), the first specimen ever discovered (ca. 1857) but originally identified as a pterosaur (Ostrom 1970). Despite a wealth of morphological evidence in favor of this hypothesis, including the eventual discovery of clavicles and a furcula in several groups of dinosaurs (Camp 1936, Barsbold 1983, Yates and Vasconcelos 2005), Ostrom's arguments were met with resistance by many within the scientific community. Even today, despite overwhelming evidence uncovered within the past 30 years, this hypothesis is rejected by a very small contingent of researchers (Feduccia 2002, 2013, Prum 2003).

The first feathered dinosaurs were discovered in the Lower Cretaceous Jehol deposits in northeastern China in the 1990s. The first specimen uncovered, the compsognathid *Sinosauropteryx*, preserved fuzzy, filamentous integumentary structures (proto-feathers) very unlike the complex, branching feathers present in birds today (Ji and Ji 1996). However, just 2 years later, the discovery of the basal oviraptorosaur *Caudipteryx* revealed the presence of modern-appearing pennaceous feathers on the distal forelimb preserved in a wing-like arrangement (Ji et al. 1998), thus indicating that pennaceous feathers and the wing itself are both features birds inherited from non-avian dinosaurs. Since these discoveries, feathered non-avian dinosaurs have also been recovered in Mongolia (Schweitzer et al. 1999), Germany (Rauhut et al. 2012), Canada (Zelenitsky et al. 2012), Russia (Godefroit et al. 2014), and Brazil, together elucidating the distribution of integumentary structures in the Dinosauria and revealing the homoplastic evolution of complex feather morphotypes.

These discoveries indicate that the pennaceous feather was only present in the Oviraptorosauria, Dromaeosauridae, Troodontidae, and Aves (Zhang et al. 2006, Foth et al. 2014). These clades together form a subset of Maniraptora referred to as the Pennaraptora (Figure 8.1). Pennaceous feathers form structures on the distal forelimbs and tail of oviraptorosaurs (Ji et al. 1998) and also on the hindlimbs in troodontids and dromaeosaurids (X. Xu et al. 2003, D. Hu et al. 2009). Behavioral evidence

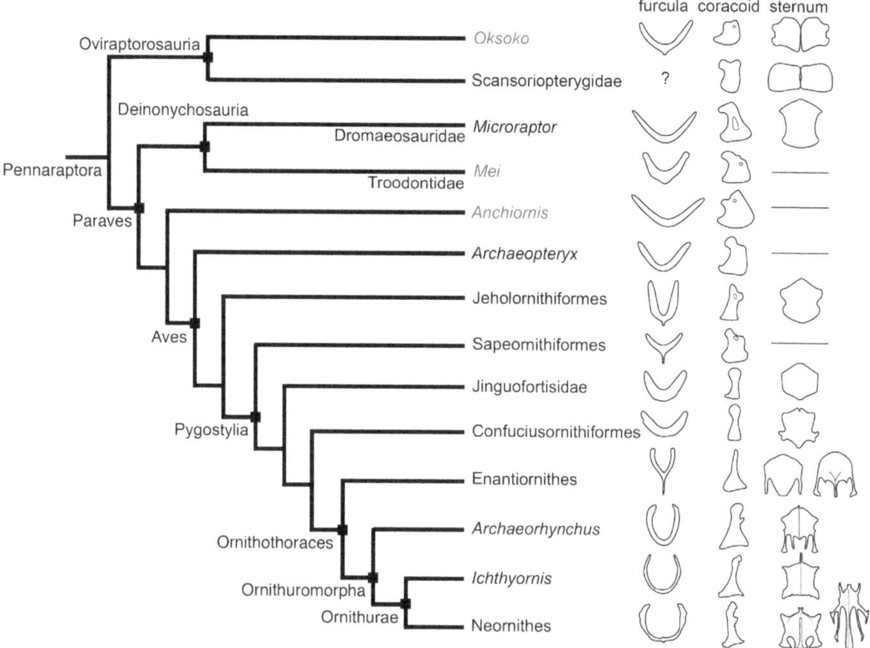

Figure 8.1 (Left) Simplified cladogram depicting the most commonly resolved relationships between major lineages of Mesozoic birds and their closest non-avian relatives. Gray indicates non-volant taxa (note that *Microraptor* and possibly *Rahonavis*, not figured, are the only probable volant members of the Dromaeosauridae). (Right) Morphology of important elements of the pectoral girdle (furcula, coracoid, and sternum) for each taxon.

also supports a close relationship among pennaraptorans. For example, these are the only dinosaurs to build open nests (Norell et al. 1995, Varricchio and Jackson 2016) and lay colored eggs (Wiemann et al. 2018), and the troodontid *Mei* is preserved sleeping like a bird with its head tucked under its forelimb (X. Xu and Norell 2004). The evolution of volant behavior in the Scansoriopterygidae (X. Xu et al. 2015), a probable basal oviraptorosaur lineage from the Late Jurassic Yanliao Biota (Gianechini et al. 2018), and in the dromaeosaurids *Microraptor* (Early Cretaceous, Jehol Biota) (X. Xu et al. 2003) and *Rahonavis* (Late Cretaceous, Maevarano Formation) (Forster et al. 1998) highlight the potential for flight in these derived maniraptorans, with their elongated forelimbs, pennaceous feathers, and relatively small body sizes (Pei et al. 2020). Evidence suggests *Archaeopteryx* may be closely aligned with the Anchiornithinae (Pei et al. 2020, Turner et al. 2021), a group that is sometimes placed within Avilae (Aves + Anchiornithinae) or alternatively interpreted as belonging to the Troodontidae (Gianechini et al. 2018). Compared to *Archaeopteryx*, the proportionately small size of the surface formed by the pennaceous forelimb feathers and their lack of asymmetry suggest *Anchiornis* was not volant (Kiat & O'Connor 2024).

As evidence from deposits in northeastern China and elsewhere continues to mount in favor of the "birds are maniraptoran theropod dinosaurs" hypothesis, five new specimens of *Archaeopteryx* have been described. The Thermopolis (10th) (Mayr et al. 2005), 11th (Foth et al. 2014), and 12th specimens (Rauhut et al. 2018) belong to private collections, although the 10th is widely available for study. The Chicago specimen is the best preserved, at least partially due to the modern preparation techniques used to remove the surrounding matrix (O'Connor et al. 2025). These specimens have revealed important new information regarding the skeletal morphology and plumage of this iconic taxon. The skull retains plesiomorphic morphologies such as the presence of an elongate postorbital forming a complete postorbital bar and an antorbital fossa perforated by maxillary and promaxillary foramina (Mayr et al. 2007, Rauhut et al. 2018). An alula was absent, tertials were present, and hindlimb feathers extended onto the tarsometatarsus. The application of new technologies has also revealed new information from previously known specimens (Kundrát et al. 2019). Computed tomography (CT) scans of the London specimen have revealed that the brain of *Archaeopteryx* was structurally capable of supporting flight with vision dominating sensory functions and an inner ear morphology indicating increased auditory and spatial capabilities relative to most non-avian maniraptorans (Alonso et al. 2004).

The postcranial skeleton of *Archaeopteryx* is characterized by a plesiomorphic axe-like scapulocoracoid that fused late in ontogeny; a wide, boomerang-shaped furcula; the absence of an ossified sternum (as in troodontids and *Anchiornis*) (Figure 8.2); an elongate forelimb with three manual claws; and a long boney tail formed by 24 free caudal vertebrae (Elzanowski 2002, O'Connor et al. 2025). Except for its large wings formed by asymmetrical, pennaceous feathers, *Archaeopteryx* essentially lacks features that would indicate it is a bird and not a non-avian dinosaur, indicating that soft tissue modifications for avian flight, such as remige asymmetry, preceded major skeletal transformations. The recent discovery of *Baminornis*, a Jurassic bird from 149–148 Ma deposits in China, with its surprisingly advanced features of the pectoral girdle, suggests avian flight evolved well before 150 Ma (Chen et al. 2025).

A Growing Diversity of Cretaceous Avians

In the century following the description of the Late Cretaceous "Odontornithes" (*Hesperornis* and *Ichthyornis*) from North America (Marsh 1880), the Mesozoic avian fossil record experienced little growth. In the 1970s, the first *Gobipteryx* specimens were described from Mongolia (Elzanowski 1974) and *Alexornis* was described from Mexico (Brodkorb 1976). Both were identified as crown group birds, although they were later reinterpreted as belonging to the Enantiornithes. This new clade was erected in 1981 from a collection of disarticulated bones found while preparing the sauropod *Saltasaurus*, collected in the Upper Cretaceous Lecho Formation in Argentina (C. Walker 1981). In the following decade, articulated partial skeletons of Cretaceous birds were uncovered in Spain (Sanz et al. 1988), China (Z. Zhou et al. 1992), and Argentina (Chiappe and Calvo 1994), and they were eventually recognized as belonging to this new group of birds. At approximately the same time,

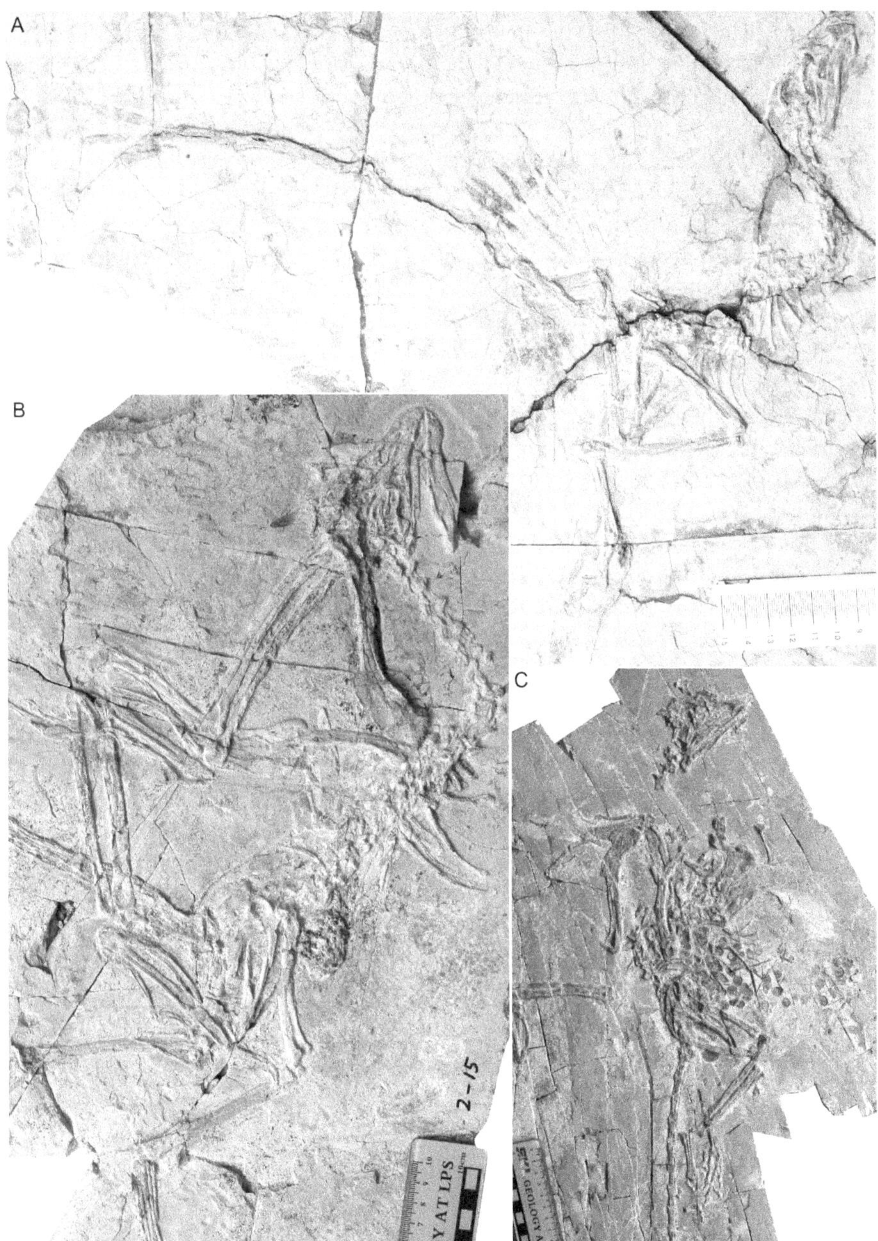

Figure 8.2 Specimens of the long-bony-tailed bird *Jeholornis* sp. (A) STM (Shandong Tianyu Museum of Nature) 3–4 preserving the two caudal plumage tracts. (B) STM2-31 preserving gastroliths interpreted as gizzard stones. (C) STM2-41 preserving plant remains ruptured from the ventriculus.

Source: X.-T. Zheng.

the first specimens of non-ornithurine ornithuromorph birds were described from Mongolia (Kurochkin 1982), China (Hou and Liu 1984, Hou and Zhang 1993), and Argentina (Chiappe 1990), and cladistic approaches demonstrated that these two clades, Ornithuromorpha and Enantiornithes, are sister taxa together forming the Ornithothoraces (Chiappe 1995) (see Figure 8.1).

The first Cretaceous non-ornithothoracine bird, *Confuciusornis sanctus* (Hou et al. 1995), was found just 1 year before the feathered *Sinosauropteryx*, the discovery of which increased the scale of excavations in the Jehol deposits enormously. During the next two decades, thousands of nearly complete specimens were uncovered, revealing a spectacular diversity of early birds (Chang et al. 2003). *Confuciusornis* is now recognized to be part of a diverse clade that first appeared 130 Ma and persisted for at least 10 My (M. Wang, O'Connor, and Zhou 2019). Two other clades of basal pygostylians have been identified, the Sapeornithiformes (Z. Zhou and Zhang 2002a) and the Jinguofortisidae (M. Wang, X. Wang, et al. 2016), as well as a lineage of long-bony-tailed birds, the Jeholornithiformes (Z. Zhou and Zhang 2002b). Dozens of species of enantiornithines and ornithuromorphs have been named such that the diversity of the Jehol avifauna accounts for half of all known Mesozoic bird species (Z. Zhou and Zhang 2006). The Jehol avifauna occurs in three successive geologic units. The 130 Ma *Protopteryx*-horizon of the Huajiying Formation (Pan et al. 2013) represents the third oldest avian bearing deposits currently known. Avian diversity increases in the 125 Ma Yixian Formation and peaks in the 120 Ma Jiufotang Formation, which records the most diverse Mesozoic avifauna currently known to science (Z. Zhou and Zhang 2006).

Deposits in Spain (Navalón et al. 2015, Knoll et al. 2018), Mongolia (Norell and Clarke 2001, Chiappe et al. 2006), and Argentina (Agnolin and Martinelli 2009) have continued to yield small numbers of specimens that are critical to understanding broader aspects of avian evolution. New species have also been named from Antarctica (Bono et al. 2016), Brazil (de Souza Carvalho et al. 2015), Canada (Hou 1999), France (Buffetaut et al. 1995), Japan (Matsuoka et al. 2002), Lebanon (Vecchia and Chiappe 2003), Madagascar (P. O'Connor and Forster 2010), Myanmar (Xing et al. 2016, 2017), Russia (Zelenkov and Averianov 2015), and the United States (Atterholt et al. 2018). However, only the Jehol deposits have produced Cretaceous-aged non-ornithothoracine birds, with the possible exception of *Fukuipteryx* from Lower Cretaceous deposits in Japan (Imai et al. 2019).

Cretaceous Long-Boney-Tailed Birds

In 1996, the raven-sized *Rahonavis* was described from the Upper Cretaceous Maevarano Formation in Madagascar (Forster et al. 1998). The animal has an elongate, bony tail, similar to that in *Archaeopteryx*, as well as a reversed hallux and remige papillae on the ulna. This was considered evidence that *Rahonavis* was a bird—a conclusion supported by cladistic analysis, which placed *Rahonavis* in a clade with *Archaeopteryx* and *Unenlagia* (Forster et al. 1996). It is now widely accepted that *Rahonavis* and *Unenlagia* are dromaeosaurids belonging to a basal clade known as the Unenlaginae (Turner et al. 2007, Pei et al. 2020). The avian characteristics of *Rahonavis* are now interpreted as evidence that may suggest this dromaeosaurid evolved some form of volant behavior independent of that in birds (Pei et al. 2020).

The only definitive long-bony-tailed birds from the Cretaceous are the Jeholornithiformes from the Jiufotang Formation (Z. Zhou and Zhang 2002b). Another taxon, *Yandangornis* (Cai and Chao 1999), is most likely a non-avian pennaraptoran (Unwin and Junchang 1997). The taxonomy of the Jeholornithiformes requires revision. In addition to *Jeholornis prima*, several other species of *Jeholornis* have been named (O'Connor et al. 2012, Lefèvre et al. 2014), as well as four additional genera—*Dalianraptor* (Gao and Liu 2005), *Jixiangornis* (Ji, Ji, Zhang, et al. 2002), *Shenzhouraptor* (Ji, Ji, You, et al. 2002), and *Kompsornis* (X. Wang et al. 2020)—but the validity of these taxa is unclear apart from *Dalianraptor*, which is based on a composite specimen and considered invalid (O'Connor et al. 2012).

Jeholornis is characterized by a short, deep skull that retains most plesiomorphic morphologies, although it does exhibit dental reduction with small, simple teeth present only in the maxilla (two present) and rostral dentary (three present) (H. Hu et al. 2022) (see Figure 8.2). Like *Archaeopteryx*, it has three manual claws, although in contrast to this stemward taxon, it has a narrow furcula with a short hypocleidium; curved scapula; strut-like coracoid with procoracoid process; ossified sternal plates medially fused into a sternum; modified sternal ribs; and tail that is more elongate, composed of 27 vertebrae (Z. Zhou and Zhang 2003b) (see Figures 8.1 and 8.2). Many of these derived features evolved in parallel to similar features in ornithuromorphs (e.g., narrow furcula, procoracoid).

The elongation of the tail contrasts with the clearly more advanced morphology of the pectoral girdle in jeholornithiforms (Z. Zhou and Zhang 2002b). The discovery of specimens preserving the caudal plumage suggests this unexpected morphology is a unique aerodynamic adaptation (O'Connor, Wang, et al. 2013). *Archaeopteryx* exhibits a "frond-like" tail with rectrices extending laterally from the caudal vertebrae the full length of the tail (Gatesy and Dial 1996). This may have contributed some lift to the flight apparatus based on the interpretation that some of the tail feathers were asymmetrical (Foth et al. 2014). In contrast, rectrices in *Jeholornis* are separated into two distinct tracts: a proximal tract forming a dorsal fan of rectrices at the base of the tail and a frond of feathers restricted to the distal-most portion of the tail (Figure 8.2A). The width of the proximal fan is proportionately similar to the width of the tail frond in *Archaeopteryx* and thus would have generated the same amount of lift while incurring less drag (O'Connor, Wang, et al. 2013). The distal tail feathers only overlap to form an airfoil in their proximal portions, after which they curve and taper. By raising and depressing the tail, the small airfoil formed by the proximally overlapping portions of these feathers would generate pitching moments that could act to stabilize *Jeholornis* during flight (O'Connor, Wang, et al. 2013). Past the point of overlap, the distal feathers are shaped by sexual selection. This unique tail morphology demonstrates the complex interplay between natural and sexual selection that has shaped avian plumage since the Early Cretaceous.

Direct evidence indicates that *Jeholornis* was herbivorous, although the precise identity of preserved stomach contents is uncertain (Z. Zhou and Zhang 2002b, O'Connor, Wang, et al. 2018) (Figure 8.2C). Originally interpreted as evidence of granivory, recent interpretations suggest frugivory (Ksepka et al. 2019, H. Hu et al. 2022). Other specimens preserve small, tight clusters of gastroliths interpreted as evidence of a gastric mill, which may suggest seasonal changes in diet (O'Connor, Wang, et al. 2018, H. Hu et al. 2022) (Figure 8.2B).

Basal Pygostylia

Three clades of basal pygostylians have been recovered from Jehol deposits. The Confuciusornithiformes is known from thousands of specimens. Four genera (*Confuciusornis*, *Changchengornis*, *Eoconfuciusornis*, and *Yangavis*) and five species (e.g., *C. sanctus* and *C. dui*) are currently considered valid (M. Wang, O'Connor, and Zhou 2019). Most known specimens are assigned to *C. sanctus* and are almost entirely recovered in the ~125 Ma Yixian Formation.

This clade is characterized by an edentulous rostrum with proportionately large, medially fused premaxillae with elongate frontal processes that contact the frontals; a furcula that is robust and wide; a scapulocoracoid that fuses early in ontogeny and further differs from the plesiomorphic condition in that the coracoid is proportionately more elongate; an ossified sternum consisting of two medially fused plates with a slight midline ridge that forms with somatic maturity and may have anchored a cartilaginous keel; a robust humerus with a massive deltopectoral perforated by a fenestra; a manus with three manual claws, with the major digit ungual reduced; and a long, robust, rod-like pygostyle perforated by foramina (Chiappe et al. 1999, Wang, O'Connor, and Zhou 2019, Q. Wu et al. 2021) (Figures 8.1 and 8.3A,B).

Specimens preserving plumage reveal that the wings were long and narrow, forming a wing shape not found in extant birds (Falk et al. 2016) (see Figure 8.3A). A propatagium was present extending from the shoulder to wrist, but the alular digit was free without alula or alular patagium (Zheng et al. 2017). Crural feathers diminished distally, not reaching the ankle (O'Connor 2020). Some specimens preserve a pair of elongate rachis-dominated racket-plumes, whereas others preserve a clear absence of rectrices, instead revealing a pygostyle surrounded only by unspecialized contour feathers (Chiappe et al. 1999) (see Figure 8.3A). These racket-plumes are interpreted as ornamental and evidence of sexual dimorphism, being only present in males (Hou et al. 1996, Chiappe et al. 2008). Although far more rare, a few specimens preserve traces of the soft tissue of the keratinous rhamphotheca that covered the premaxilla and possibly also the maxilla (Zheng et al. 2020) (see Figure 8.3B). Pedal proportions suggest a generalist ecology (Hopson 2001).

The basal *Eoconfuciusornis* is known from two specimens from the Huajiying Formation (Zhang et al. 2008, 2017). The features that characterize its younger relatives are less developed in *Eoconfuciusornis*: The premaxilla is not enlarged, yet this taxon is still fully edentulous; the humeral deltopectoral crest is smaller and imperforate; and the alular and minor digit claws are proportionately smaller although still larger than that of the major digit (Zheng et al. 2017). Both specimens preserve exceptional soft tissue, revealing the presence of melanized wing tips and ornamental tail feathers (Zheng et al. 2017). Preserved melanosomes revealed through scanning electron

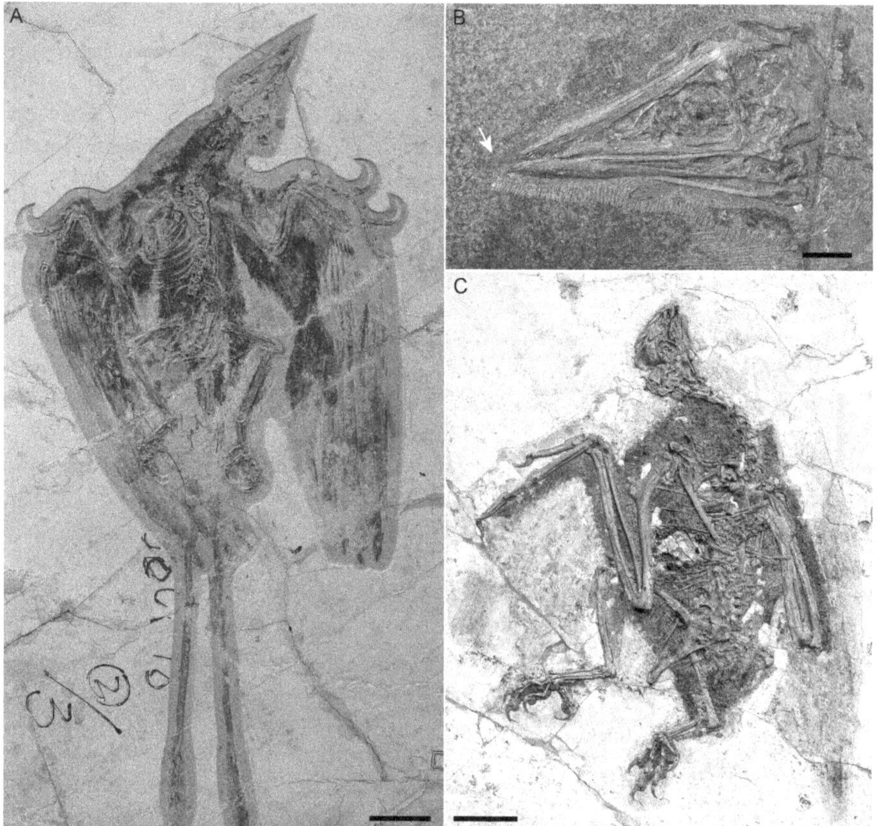

Figure 8.3 Early Cretaceous basal pygostylians. (A) *Confuciusornis sanctus* IVPP (Institute of Vertebrate Paleontology and Paleoanthropology) V13156 preserving feathers and other soft tissue (scale bar = 5 cm). (B) *Confuciusornis sanctus* IVPP V12352 preserving faint traces of the rhamphotheca, indicated by the white arrow (scale bar = 1 cm). (C) *Sapeornis chaoyangensis* 41HIII0405 (scale bar = 4 cm).
Source: L. Xu.

microscopy indicate a dark overall plumage with a rufous throat patch (Zheng et al. 2017). Transmission electron microscopy reveals hollow melanosomes that indicate the head was iridescent (Pan et al. 2021).

Whereas confuciusornithiforms dominate in the Huajiying and Yixian formations, the pygostylian *Sapeornis chaoyangensis* occurs in the Jiufotang Formation (Z. Zhou and Zhang 2002a). Although other sapeornithiforms have been described (e.g., *Didactylornis* and *Shenhiornis*), they are generally regarded as junior synonyms of *Sapeornis* (H. Hu, O'Connor, McDonald, et al. 2020). More than 100 specimens have been reported revealing *Sapeornis* to be a large bird with elongate forelimbs; a humerus with a large, perforated deltopectoral crest; and a reduced manus with only two manual claws (Z. Zhou and Zhang 2002a, 2003a) (see Figure 8.3C). The pectoral girdle remains primitive, with a wide furcula bearing a short hypocleidium, a plesiomorphically axe-like scapulocoracoid, and lacking an ossified sternum (Z. Zhou and Zhang 2002a, Zheng, O'Connor, Wang, et al. 2014) (see Figure 8.1). The skull is short and deep, reminiscent of that in scansoriopterygids and basal oviraptorosaurs (H. Hu, O'Connor, McDonald, et al. 2020). The four premaxillary teeth are robust with basal tubercles. The three maxillary teeth are similar but smaller. The two dentary teeth are very reduced and are accompanied by two empty alveoli (Y. Wang et al. 2017). It has been suggested that *Sapeornis* experienced ontogenetic tooth loss in the dentary (M. Wang et al. 2017). The pygostyle in *Sapeornis* is small, short, and ploughshare-shaped, very similar to that present in crown birds (M. Wang and O'Connor 2017). The palate of *Sapeornis* was very theropod-like, suggesting the cranium was akinetic (H. Hu et al. 2019).

Sapeornis had a long, graded, fan-shaped tail formed by 8–10 rectrices (M. Wang and O'Connor 2017). Two specimens preserve evidence of a small patch of feathers extending off the intertarsal joint (Zheng, Zhou, et al. 2013). These feathers do not overlap for most of their lengths, suggesting they are ornamental (O'Connor and Chang 2015). Direct evidence indicates *Sapeornis* was herbivorous (Zheng et al. 2011): The remains of numerous seed-like objects representing multiple morphotypes have been recovered in the crop of several specimens. Small numbers of gastroliths have also been found in the stomachs of some specimens (Z. Zhou and Zhang 2003a, Zheng et al. 2011). Analysis of its flight style suggests *Sapeornis* had thermal soaring capabilities and may be analogous to the herbivorous *Chauna* (Serrano and Chiappe 2017).

The Jinguofortisidae is known from only two specimens representing two taxa (*Chongmingia* and *Jinguofortis*). They are fully toothed with fused scapulacoracoids, robust U-shaped furculae, medially fused sternal plates, unfenestrated humeri, two manual claws, and preserve gastroliths interpreted as gizzard stones (M. Wang, Zhou, et al. 2016, M. Wang et al. 2018) (see Figure 8.1).

Enantiornithes: Dominant Landbirds of the Cretaceous

If a fossil bird is found in the Cretaceous, chances are it belongs to the Enantiornithes. This clade dominated Cretaceous terrestrial ecosystems and accounts for approximately half of the entire diversity of Mesozoic birds. More than half of all taxa are known from the Jehol Biota, where they account for half of the entire diversity of the Jehol avifauna (Z. Zhou and Zhang 2006). The clade first appears in the *Protopteryx*-horizon of the Huajiying Formation, in which a handful of taxa encapsulating significant morphological disparity have been recovered (M. Wang et al. 2014, M. Wang and O'Connor 2017, Liu et al. 2019). Several diverse (five or more species) lineages have been identified (e.g., Bohaiornithidae and Longipterygidae) (O'Connor et al. 2009, M. Wang et al. 2014), at least one of which, the Pengornithidae, spans the entirety of the deposition of the Jehol (Z. Zhou et al. 2008, X. Wang et al. 2014).

In the Early Cretaceous, enantiornithines were limited to small body sizes and arboreal habitats. Although the record of fossil birds is much poorer in the Late Cretaceous, it appears that younger enantiornithines utilized a wider range of body sizes and had achieved greater ecological diversity (Chiappe 1993, Chiappe et al. 2006, Atterholt et al. 2018). Enantiornithines show important skeletal adaptations absent in stemward clades, such as the presence of an ossified sternal keel and sternal trabeculae (Chiappe and Walker 2002). In relation to the crownward ornithuromorphs, advanced skeletal morphologies have evolved through alternative developmental pathways. For example, the enantiornithine keel is restricted to the caudal half of the sternum, and although both clades have narrow furculae, a synapomorphy of enantiornithines is that theirs is Y-shaped (and dorsolaterally excavated), whereas that of ornithuromorphs is U-shaped, similar to the condition in modern birds (O'Connor et al. 2009).

Enantiornithines are additionally characterized by a pygostyle that is cranially forked with ventrolateral processes, a poorly developed scapular cotyla that is aligned with the glenoid and acrocoracoid, an elongate minor metacarpal that extends distally further than the major metacarpal, and a reduced metatarsal IV with the distal trochlea formed by a single condyle (Chiappe and Walker 2002, M. Wang et al. 2017). The enantiornithine sternum plesiomorphically consists of paired sternal plates medially fused in mature individuals (observed in pengornithids), but in more derived taxa the sternum ossifies from four sternal anlagen (Zheng et al. 2012, O'Connor, Zheng, et al. 2015). The first to appear is a midline ossification from which the xiphoid process, caudally restricted sternal keel, and caudal half of the sternal corpus ossify. Another anlagen ossifies the rostral half of the sternal corpus, and a bilateral pair of ossifications form the lateral trabeculae (Zheng et al. 2012).

Enantiornithine skull morphology is plesiomorphic and akinetic (M. Wang et al. 2022). A complete postorbital bar is present in some taxa and was probably independently reduced numerous times (e.g., *Yuornis*) (O'Connor and Chiappe 2011, H. Hu, O'Connor, Wang, et al. 2020). The premaxillae are often unfused in Early Cretaceous taxa, and the corpus is unexpanded. The maxilla typically forms more than half the facial margin, and in basal taxa the antorbital fossa is perforated by a foramen (O'Connor and Chiappe 2011). The squamosal is unincorporated into the braincase (Sanz et al. 1997). The quadrate is bicondylar with a unique lateral crest in some taxa (Stidham and O'Connor 2021).

Toothed enantiornithines exhibit polyphyodonty (O'Connor and Chiappe 2011, Y.-H. Wu et al. 2021). A huge diversity of tooth morphologies and dental patterns are observed (O'Connor and Chiappe 2011, O'Connor 2019). Pengornithids have numerous, low-crowned teeth, whereas bohaiornithids have robust teeth that are basally expanded and apically sharply tapered and recurved with enamel grooves in at least one taxon (*Sulcavis*) (O'Connor, Zhang, et al. 2013). Longipterygids are the only Early Cretaceous taxa to exhibit some form of dental reduction, with teeth restricted to the tip of the rostrum and thus absent in the maxilla. The longipterygid rostrum is proportionately longer than in other known enantiornithines, forming between 60% and 65% the length of the skull (compared to ~50% in most other taxa) (O'Connor 2009). The Longipterygidae consists of two subclades: the Longipteryginae is characterized by overall greater body size and teeth that are large, labiolingually compressed, and recurved with crenulated distal margins (X. Wang et al. 2015); the smaller Longirostravinae have more peg-like teeth, although their small size may be obscuring morphological details (O'Connor et al. 2011). These considerable differences in tooth morphology and arrangement strongly suggest ecological resource partitioning in the Jehol (O'Connor 2019). In contrast to previous predictions, direct evidence of diet reveals Longipteryx was frugivorous (O'Connor et al. 2024). Eoalulavis from the Lower Cretaceous La Huérguina Formation at Las Hoyas in Spain preserves the remains of freshwater crustaceans in its abdominal cavity but no cranium to help elucidate the relationship between dental and rostral morphology and diet in enantiornithines (Sanz et al. 1996).

By the Late Cretaceous, enantiornithines evolved more advanced flight-related skeletal morphologies in parallel to the Ornithuromorpha, such as the rostrocaudally extensive and ventrally deep sternal keel in *Neuquenornis* (Chiappe and Calvo 1994), and increased skeletal pneumaticity, indicated by the presence of a pneumatic foramen piercing the humerus in *Elbreteornis* (Chiappe and Walker 2002). Several lineages evolved cranial modifications in parallel to neornithines. An edentulous rostrum is present in *Gobipteryx* and *Yuornis*. *Yuornis* additionally has completely reduced the lacrimal and postorbital (L. Xu et al. 2021), whereas *Gobipteryx* has somewhat reduced the maxilla (Chiappe et al. 2001). Disparity also increases with the appearance of unusual morphologies such as *Falcatakely* from the Maevarano Formation, which has a dorsoventrally deep, elongated rostrum, superficially resembling a toucan with teeth (P. O'Connor et al. 2020) (Figure 8.4D).

Understanding enantiornithine ecology is challenging not only because no living bird possesses teeth but also because the small amount of available evidence indicates that enantiornithines did things differently (Clark and O'Connor 2021). Although it is likely that these birds occupied ecological niches similar to those occupied by many extant birds, they possess morphologies not observed in living birds. This suggests that enantiornithines may have utilized familiar niches through unfamiliar strategies. For example, *Elektorornis* exhibits an elongated third pedal digit and is interpreted as a probing feeder like extant woodpeckers, only using its feet rather than its mouth (Xing, O'Connor, et al. 2019, Clark and O'Connor 2021). Enantiornithines sometimes achieve morphologies comparable to those observed in extant birds but through different elements. *Brevirostruavis* combines an elongate hyoid apparatus with a rostrum that is not elongated (Li et al. 2022), as in some woodpeckers, but the elongation of the tongue is achieved through the ceratobranchials, whereas it is typically the epibranchials that are elongated in crown birds. Similarly, rostral elongation in longiptergyids and *Falcatakely* is achieved through the maxilla, whereas crown birds elongate the premaxilla (P. O'Connor et al. 2020).

The vast majority of enantiornithines preserve characteristics associated with an arboreal lifestyle. The hallux is elongate, pedal unguals are recurved, and phalanx length increases distally within each pedal digit (Figures 8.4A,B). Ecological diversity appears to be greater in the Late Cretaceous. *Elsornis* from the Upper Cretaceous Djadochta Formation in Mongolia is interpreted as flightless based on its humeral morphology (Chiappe et al. 2006), and *Lectavis* from the Lecho Formation is interpreted as a wading taxon (Chiappe 1993) but may not be enantiornithine (O'Connor et al. 2009). *Yungavolucris* preserves an unusual morphology of the tarsometatarsus clearly indicating an ecology different from other known taxa (Chiappe 1993) (Figure 8.4C).

Figure 8.4 Cretaceous Enantiornithes. (A) *Eopengornis martini* STM24-1 (Pengornithidae) from the Lower Cretaceous ~130 Ma Huajiying Formation (scale bar = 2 cm). (B) Foot of hatchling HPG (Hupoge Amber Museum)-15-1 preserved in ~100 Ma Burmese amber (scale bar = 5 mm). (C) *Yungavolucris brevipedalis* PVL (Fundación-Instituto Miguel Lillo) 4053 from the Upper Cretaceous Lecho Formation (scale bar = 1 cm). (D) *Falcatakely forsterae* UA (University of Antananarivo) 10,015 from the Upper Cretaceous Maevarano Formation (scale bar = 1 cm).

Sources: A, X,-L. Wang; B, L.-D. Xing; C, J. O'Connor; D, P. O'Connor.

Enantiornithines already had only a single functional ovary and oviduct (Zheng, O'Connor, et al. 2013) and utilized medullary bone as a calcium reservoir for eggshell production (O'Connor, Erickson, et al. 2018). Fossilized nests indicate communal nest sites located near water (Dyke et al. 2012) and small clutches of eggs partially embedded in the sediment (Fernández et al. 2013). All available evidence indicates enantiornithines were highly precocial. Several advanced stage embryos reveal that enantiornithines hatched with their skeletons highly ossified and fully fledged (Elzanowski 1981, Z. Zhou and Zhang 2004). This developmental strategy may in turn have forced alternative adaptations because most specialized foot morphologies in crown birds are only found in altricial species (Clark and O'Connor 2021). Numerous juveniles indicate that ornamental tail feathers grew in well before skeletal or sexual maturity was achieved (Zheng et al. 2012). Sexual maturity is inferred to have preceded skeletal maturity, which took several years (O'Connor, Erickson, et al. 2018). Growth strategies varied between lineages, as evidenced by varying degrees of bone tissue vascularization, which ranges from entirely avascular parallel fibered to woven, some with intermittent periods in which fibrolamellar bone was formed (e.g., *Mirarce*). No mature specimen provides evidence that it had uninterrupted growth like that present in most living birds and first appears in Early Cretaceous ornithuromorphs (e.g., *Iteravis* and *Yanornis*) (Chinsamy et al. 1994, C. O'Connor et al. 2014, O'Connor, Wang, et al. 2015, Atterholt et al. 2018).

Only the tail plumage is well known. Although the most common tail morphology is an absence of distinct rectrices, with only wispy contour feathers surrounding the pygostyle, several ornamental tail shapes have been documented. Many taxa have a pair of elongate rachis-dominated feathers (e.g., *Bohaiornis*, *Dapingfangornis*, *Protopteryx*, and *Junornis*) (O'Connor 2020). These are fully pennaceous (streamers) in pengornithids (*Eopengornis* and *Yuanchuavis*; see Figure 8.4A) (X. Wang et al. 2014, M. Wang et al. 2021) but have distally restricted vane-forming racket-plumes in non-pengornithid enantiornithines (*Dapingfangornis*, *Orienantius*, and *Protopteryx*). Two pairs of rachis-dominated racket-plumes are present in *Paraprotopteryx* (Zheng et al. 2007). These feathers are interpreted as sexually dimorphic, a conclusion supported by their absence in specimens in which it is possible to determine sex (Zheng, Zhou, et al. 2013, Bailleul, O'Connor, et al. 2019). Burmese amber specimens have revealed that the rachis-dominated feather morphotype consists of a thin, open, C-shaped rachis lacking medullary pith (Xing et al. 2018, Carroll et al. 2019).

Chiappeavis preserves a short fan of rectrices that may have had aerodynamic function (O'Connor, X. Wang, et al. 2016), although preserved soft tissue in other specimens indicates rectricial bulbs were absent (Liu et al. 2019). *Yuanchuavis* preserves a similar fan together with a pair of fully pennaceous rachis-dominated feathers (M. Wang et al. 2021), which may alternatively suggest that the tail fan was also ornamental or that it helped offset the energetic cost of the streamers, which incur much more drag than racket-plumes (Møller and Hedenström 1999). The holotype of *Feitianius paradisi* preserves at least three different feather morphotypes together forming a complex tail morphology that suggests this taxon was polygamous (O'Connor, M. Wang, et al. 2016).

Less is known about the wing feathers, but they generally appear to form a wing that was short and broad (see Figure 8.4A). Nine or 10 primaries and 10–12 secondaries were present in specimens that permit accurate counts (O'Connor, Zheng, et al. 2020, Xing et al. 2020). Vane asymmetry is not as developed as in modern birds but greater than in confuciusornithiforms (Feo et al. 2015). Burmese amber specimens reveal patterns of spots and stripes that suggest crypsis, although this may only typify juvenal plumage, given that the enantiornithines preserved in amber are almost all immature (Xing et al. 2020).

Hindlimb feathers are quite variable. In some taxa, feathers extend fully down the pedal digits (Xing, McKellar, et al. 2019), whereas in others the feathers end just above the ankle (O'Connor 2020). One specimen has long pennaceous feathers on the tibiotarsus that probably would have formed a shaggy morphology like that present in many raptorial birds (Zhang and Zhou 2004).

Ornithuromorpha: The Rise of Modern Birds

Ornithuromorpha is the clade that includes Neornithes (crown birds) nested within. Cretaceous non-neornithine ornithuromorphs have been collected in Argentina (Agnolin and Martinelli 2009), Mongolia (Kurochkin 1982), Madagascar (Forster et al. 2002), China (Hou and Liu 1984, M. Wang et al. 2015), France (Buffetaut and Le Loeuff 1998), Antarctica (Cordes-Person et al. 2020), the United States (Marsh 1880), the United Kingdom (Seeley 1876), and Canada (Hou 1999). They appear in the fossil record 130 Ma (M. Wang et al. 2015) but form a smaller portion of the diversity relative to enantiornithines in most avifaunas until the latter clade goes extinct together with all other non-neornithine dinosaurs.

There are two Early Cretaceous exceptions: the Changma locality of the Xiagou Formation in Gansus Province, northwestern China (You et al. 2006), and the Sihedang locality of the Jiufotang Formation, both roughly 120 Ma (S. Zhou et al. 2014). These two localities are dominated by morphologically similar, semiaquatic taxa. It is unknown if the prevalence of ornithuromorphs at these two localities is due to some biological factor, such as breeding, or simply records an environment that favored these taxa (O'Connor et al. 2022).

Two important Cretaceous Lagerstätten have produced avifaunas consisting only of enantiornithines: Las Hoyas in central Spain (Sanz and Ortega 2002) and Burmese amber from the Hukawng Valley in northern Myanmar (Xing et al. 2020). Although ecology and body size could explain the absence of ornithuromorphs in Burmese amber, the absence of this clade in the Early Cretaceous wetlands recorded at Las Hoyas is puzzling. Ornithurine dominance has been suggested for some localities in the latest Cretaceous of North America that have produced a handful of fragments (Longrich 2009), but it is unclear if this will hold up with increased sampling.

Non-neornithine ornithuromorphs have superficially modern skeletons apart from a few features: the presence of teeth in some taxa, manual claws, distally contacting pubes, and the absence of a hypotarsus. Like enantiornithines, rostral elongation is achieved through elongation of the maxilla (O'Connor, M. Wang, et al. 2016). No ornithuromorph preserves a postorbital, indicating this element was very reduced, if not absent, in even the earliest members of this clade. Compared to enantiornithines, ornithuromorphs have an enlarged sternum with a well-developed keel; curved and tapered scapula; coracoid with a procoracoid process; expanded first phalanx of the major digit; reduced, weakly curved manual claws; and a small, tapered pygostyle (S. Zhou et al. 2012, 2013, M. Wang, Zhou, et al. 2016).

In contrast to enantiornithines, which exhibit limited ecological diversity, several lines of evidence indicate that Early Cretaceous ornithuromorphs utilized a variety of habitats. The hongshanornithids, a diverse clade that persisted for the entirety of the Jehol, have elongate hindlimbs interpreted as adapted for wading (O'Connor et al. 2010). The basal ornithuromorph *Archaeorhynchus* has a reduced the hallux and short toes, suggesting cursorial habits (S. Zhou et al. 2013) (Figure 8.5B). *Gansus* from the Jiufotang equivalent Xiagou Formation in northwestern China is interpreted as a volant, foot-propelled diver (Nudds et al. 2012). By the Late Cretaceous, even more extreme ecological adaptations are observed, exemplified by the flightless Hesperornithiformes, a group of globally dispersed (Europe, Mongolia, and North America), large-bodied, foot-propelled divers (Bell and Chiappe 2015).

This ecological diversity is also reflected in the structure of the digestive system. Similar to extant taxa, the piscivorous *Yanornis* has a simple crop that can hold several whole fish, which it macerates without the aid of gizzard stones, egesting bones and scales in the form of a pellet (Zheng, O'Connor, Huchzermeyer, et al. 2014) (Figure 8.5A). In contrast, the herbivorous *Eogranivora* preserves a crop full of probable seeds that forms a distinct pouch positioned low in the esophagus and a ventriculus containing a large aggregate of gizzard stones (Zheng et al. 2018).

In ornithuromorphs, the complete loss of teeth appears to be correlated with herbivory (e.g., *Eogranivora*; inferred also for *Archaeorhynchus* based on gastral mass proportions) (O'Connor 2019) (see Figure 8.5B). Tooth reduction evolved independently in many lineages. In some, teeth are absent only from the premaxilla (e.g., *Iteravis* and *Ichthyornis*) (S. Zhou et al. 2014, Field, Hanson, et al. 2018), whereas in other taxa, teeth are absent in the maxilla (*Mengciusornis*), exemplifying the numerous evolutionary pathways through which edentulism was achieved (Wang, O'Connor, Zhou, et al. 2019). Increases in dentition are tentatively correlated with piscivory: *Yanornis* has more teeth than other Early Cretaceous birds, and the Late Cretaceous ornithurine *Ichthyornis* has a more extensive tooth row than any other bird with maxillary teeth extending to the rostral margin of the orbit (O'Connor 2019). The teeth in *Hesperornis* and *Ichthyornis* preserve distal carinae (Dumont et al. 2016). In *Hesperornis*, the teeth are positioned in a communal groove without interdental bone, a morphology that has evolved multiple times both in Cretaceous birds and in non-avian dinosaurs (O'Connor et al. 2022).

An unusual feature of toothed non-neornithine ornithuromorphs is a small ossification rostromedial to the dentaries termed the predentary. This element was first recognized in *Hesperornis* and presumed present in *Ichthyornis* based on the similar morphology of the tip of the dentaries (Martin 1987). It was later identified in

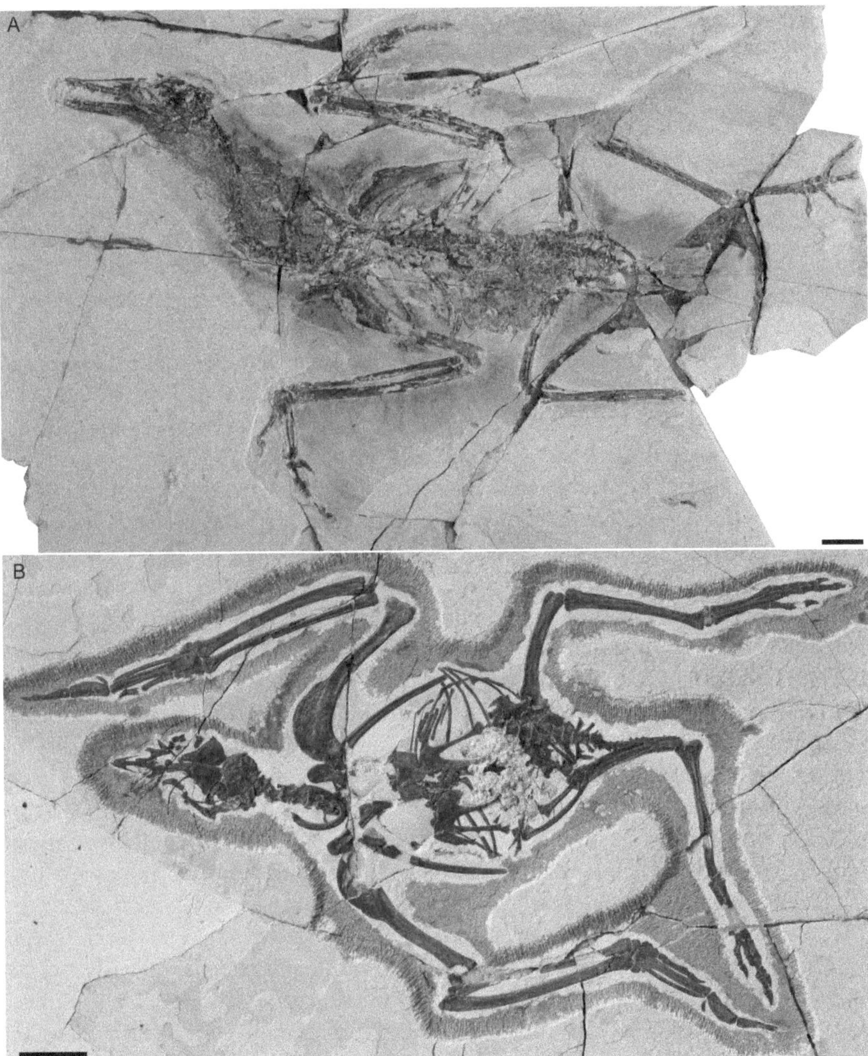

Figure 8.5 Non-neornithine ornithuromorphs from the Early Cretaceous Jehol Group. (A) *Yanornis martini* STM9-15 preserving whole fish in the esophagus and macerated fish remains in the ventriculus. (B) Edentulous *Archaeorhynchus spathula* IVPP V17075 preserving a mass of gastroliths interpreted as gizzard stones. Scale bars = 2 cm.

Sources: A, X.-T. Zheng; B, G. Wei.

ornithuromorphs from the Jehol avifauna (Z. Zhou and Martin 2011). The predentary co-occurs with an edentulous region at the distal tip of the premaxilla. Both the tip of the premaxilla and the predentary are inferred to be covered in rhamphotheca (Bailleul, Li, et al. 2019). The predentary is interpreted as a sesamoid that was proprioceptive and mobile, creating a form of mandibular kinesis unique to toothed, probably animalivorous or omnivorous, ornithuromorphs (Bailleul, Li, et al. 2019).

The pterygoid of the ichthyornithiform *Janavis* indicates the cranium was also kinetic (Benito et al. 2022).

The Radiation of Crown Birds

The diversification of Neornithes begins in the Cretaceous, although the record of this clade's humble beginnings is extremely poor and limited to the latest Cretaceous, just 2 million years before the meteor impact. The holotype of *Asteriornis maastrichtensis*, recovered from the Maastricht Formation in Belgium, consists of a nearly complete skull with a partial postcranial skeleton and is interpreted as a stem galloanseran, possessing traits typical of both Galliformes and Anseriformes (Field et al. 2020). The Antarctic bird *Vegavis* has been interpreted as a crown anseriform in some but not all cladistic analyses (Clarke et al. 2005, M. Wang et al. 2018). *Vegavis* preserves a syrinx, suggesting this feature evolved within or near the crown (Clarke et al. 2016). Most other Cretaceous neornithines are very fragmentary, and their identifications are controversial (Hope 2002). Still, *Asteriornis* and *Vegavis* indicate major evolutionary divergences had occurred in the Cretaceous between the Paleognathae and Neognathae lineages, within Neognathae between Galloanserae and Neoaves, and within Galloanserae between anseriforms and galliforms (Field et al. 2020). However, the pterygoid of *Janavis* may challenge these traditional relationships.

The Cenozoic radiation of crown birds following the meteor impact is described as explosive, with numerous modern orders thought to have appeared within the first 5–10 million years of the Paleocene (James 2005, Ksepka et al. 2017). Mousebirds (Coliiformes) are documented in 62.5 Mya deposits in New Mexico (Ksepka et al. 2017), and penguins (Sphenisciformes) have been found in 62–60 Mya deposits in New Zealand (Slack et al. 2006), confirming the rapid tempo of the crown radiation. Early Cenozoic avifaunas are best documented in Eocene Lagerstätten: the Green River Formation in North America and the Messel Shale in Germany. The 52 Ma Green River "fossil lake" avifauna records stem psittaciforms, lithornithids (paleognaths), galliforms, anseriforms, caprimulgiforms, oil birds (steatornithids), frigatebirds (fregatids), pelicaniformes, thresikornithids, ralliforms, piciforms, coraciiforms, leptosomiforms, coliiforms, and probable stem passerines (Grande 2013). The 47 Ma Messel avifauna includes stem psittaciforms, falconiforms, coraciforms, caprimulgiforms, and apodiforms, as well as gruiforms, sulids, struthionids, cariamiforms, strigiforms, alcediniforms, anseriforms, galliforms, and a possible charadriiform (Mayr 2009). These faunas suggest passerines were not a major component of the early neornithine radiation.

Selectivity of the End Cretaceous Mass Extinction

As new data emerge, factors that may explain the selectivity of the end-Cretaceous extinction begin to take shape. The same factors that facilitated the survival of Neornithes likely enabled their subsequent radiation, such that today Neornithes is the most diverse clade of amniotes. Several hypotheses have been put forth. For example, it was suggested that the global loss of forests due to an impact-induced firestorm explains the selectivity due to the arboreal habits of enantiornithines (Field, Bercovici, et al. 2018). However, this does not account for the greater ecological diversity of this clade in the Late Cretaceous nor the loss of aquatic and semiaquatic non-neornithine ornithurine birds. Differences in dietary preference and reproductive behaviors have also been suggested as possible factors explaining the extinction of non-neornithine birds (Varricchio and Jackson 2016, O'Connor 2019). However, it is most likely not a single factor that can explain the extinction selectivity but, rather, a combination of features unique to neornithines that enabled their survival and subsequent radiation.

Modern birds are the most physiologically modified amniotes on the planet. The evolution of these modifications is a story that begins in the Jurassic and takes place largely in the Cretaceous. Many traits that today are unique to living birds were in fact inherited from their close non-avian dinosaurian relatives. Others evolved in the Cretaceous as birds became increasingly adapted for powered flight, the most physically demanding form of locomotion. Ultimately, paleontologists infer the only clade to possess the entire suite of modifications was the Neornithes, facilitating their survival and subsequent success. New methods of analysis and exceptionally well-preserved specimens primarily from Asia tell this story in surprising detail.

Open nests (Norell et al. 1995), colored eggs (Wiemann et al. 2018), and contact incubation (Bi et al. 2021)—features that today are uniquely avian—are traits that appear in oviraptorosaurs and were presumably widespread in the Pennaraptora and inherited by birds. Evidence suggests that like non-avian pennaraptorans, enantiornithines laid eggs that were half buried in sediment (Schweitzer et al. 2002, Varricchio and Jackson 2016). This nest structure suggests that eggs were not manipulated after development had begun. This in turn suggests that chalazae, the membranes that hold the developing embryo in place, were absent in non-neornithine pennaraptorans, as in crocodilians (Varricchio and Jackson 2016). Aerial nests have a very low likelihood of becoming fossilized, but to date, no Cretaceous ornithuromorph nests have been found, suggesting that the chalazae may be restricted to this clade or some subset of it (i.e., Neornithes). The presence of chalazae may have allowed neornithines to rescue clutches in the wake of the impact, as has been observed by gulls following the eruption of Mt. St. Helens (Hayward et al. 1989).

The neornithine digestive system is extremely fast, lightweight, and efficient, and it is characterized by features absent in crocodilians, such as a crop, two-part stomach, ceca, and bidirectional peristalsis (Gill 2007). Numerous lineages of dinosaurs preserve tight clusters consisting of many gastroliths interpreted as evidence of a gastric mill (e.g., ceratopsian *Psittacosaurus*, ornithomimosaur *Shenzhousaurus*, and oviraptorosaur *Caudipteryx*) (Ji et al. 1998, 2003, Sereno et al. 2010). The two-part stomach structure that allows the ventriculus to become specialized into a grinding gizzard in birds was therefore inherited from non-avian dinosaurs. Evidence for a gastric mill is present in *Jeholornis*, *Sapeornis*, and numerous ornithuromorphs (O'Connor 2019).

The crop, a specialization of the esophagus that allows it to store excess food, appears to be uniquely avian, present in *Sapeornis* and several ornithuromorphs (O'Connor and Zhou 2015). In combination, these two features would allow birds to effectively feed on detritus such as seeds available during the post-impact nuclear winter (Larson et al. 2016). As noted previously, Cretaceous granivores are edentulous, which may explain why only toothless birds survived.

The digestive system of modern birds shows high phenotypic flexibility, able to drastically change proportions in response to seasonal changes in trophic resources (Gill 2007). Although based on their phylogenetic position enantiornithines should possess both a crop and the ability to evolve a gastric mill, thousands of well-preserved specimens indicate the absence of these features. The limited structure of the digestive system in enantiornithines may indicate they lacked this flexibility and were unable to utilize certain trophic resources, such as seeds, which may have been an important factor in their extinction (O'Connor, Wang, et al. 2018). Differences in body size, plumage, metabolic rates, and developmental strategy were also likely important factors in neornithine survival.

Conclusion and Future Directions

Our understanding of the Mesozoic evolution of birds has greatly expanded during the past three decades. The discovery of important new material from throughout the world and particularly northeastern China has revealed new lineages of Cretaceous birds, documented the extreme homoplasy that characterized early avian evolution, and revealed numerous aspects of plumage and other soft tissues through exceptional preservation. The application of new methods such as high-resolution CT scanning in turn has revealed previously hidden aspects of biology, such as the evolution of the brain and inner ear. Despite this enormous progress, there still exists critical gaps in the fossil record that limit current interpretations. The incredible diversity of the Huajiying avifauna indicates that filling in the 20 My gap in the fossil record between these birds and *Archaeopteryx* is necessary to understand the initial radiation of major Cretaceous lineages, exemplified by the recent discovery of *Baminornis*. The paucity of Late Cretaceous avian fossils and particularly Cretaceous neornithines limits our ability to understand broad evolutionary patterns and the origins of modern birds. As new discoveries continue unabated and new methods continue to shed light on unexpected aspects of early avian biology, in the next 30 years we can expect our knowledge of early avian evolution to expand at a similarly exciting rate.

References

Agnolin, F. L., and A. G. Martinelli. 2009. Fossil birds from the Late Cretaceous Los Alamitos Formation, Río Negro Province, Argentina. *Journal of South American Earth Sciences* 27: 42–49.

Alonso, P. D., A. C. Milner, R. A. Ketcham, M. J. Cookson, and T. B. Rowe. 2004. The avian nature of the brain and inner ear of *Archaeopteryx*. *Nature* 430:666–669.

Atterholt, J., J. H. Hutchison, and J. K. O'Connor. 2018. The most complete enantiornithine from North America and a phylogenetic analysis of the Avisauridae. *PeerJ* 6: Article e5910.

Bailleul, A. M., Z. Li, J. O'Connor, and Z. Zhou. 2019. Origin of the avian predentary and evidence of a unique form of cranial kinesis in Cretaceous ornithuromorphs. *Proceedings of the National Academy of Sciences of the USA* 116:24696–24706.

Bailleul, A. M., J. O'Connor, S. Zhang, Z. Li, Q. Wang, M. C. Lamanna, X. Zhu, and Z. Zhou. 2019. An Early Cretaceous enantiornithine (Aves) preserving an unlaid egg and probable medullary bone. *Nature Communications* 10:1275–1275.

Barsbold, R. 1983. Carnivorous dinosaurs from the Cretaceous of Mongolia. *Joint Soviet–Mongolian Palaeontological Expedition Transactions* 19:1–117.

Bell, A., and L. M. Chiappe. 2015. A species-level phylogeny of the Cretaceous Hesperornithiformes (Aves: Ornithuromorpha): Implications for body size evolution amongst the earliest diving birds. *Journal of Systematic Palaeontology* 14:239–251.

Benito, J., P.-C. Kuo, K. E. Widrig, J. W. M. Jagt, and D. J. Field. 2022. Cretaceous ornithurine supports a neognathous crown bird ancestor. *Nature* 612:100–105.

Bi, S., R. Amiot, C. Peyre de Fabrègues, M. Pittman, M. C. Lamanna, Y. Yu, C. Yu, T. Yang, S. Zhang, Q. Zhao, and X. Xu. 2021. An oviraptorid preserved atop an embryo-bearing egg clutch sheds light on the reproductive biology of non-avialan theropod dinosaurs. *Science Bulletin* 66:947–954.

Bono, R. K., J. Clarke, J. A. Tarduno, and D. Brinkman. 2016. A large Ornithurine bird (*Tingmiatornis arctica*) from the Turonian High Arctic: Climatic and evolutionary implications. *Scientific Reports* 6: Article 38876.

Brodkorb, P. 1976. *Discovery of a Cretaceous Bird, Apparently Ancestral to the Orders Coraciiformes and Piciformes (Aves: Carinatae)*. Smithsonian Institution Press, Wasington, DC.

Buffetaut, E., and J. Le Loeuff. 1998. A new giant ground bird from the Upper Cretaceous of southern France. *Journal of the Geological Society* 155:1–4.

Buffetaut, E., J. L. Loeuff, P. Mechin, and A. Mechin-Salessy. 1995. A large French Cretaceous bird. *Nature* 377:110–110.

Cai, Z., and L. Chao. 1999. A long tailed bird from the Late Cretaceous of Zhejiang. *Science in China (Series D)* 42:434–441.

Camp, C. L. 1936. A new type of small bipedal dinosaur from the Navajo Sandstone of Arizona. *University of California Publications in Geological Sciences* 24:39–53.

Carroll, N. R., L. M. Chiappe, and D. J. Bottjer. 2019. Mid-Cretaceous amber inclusions reveal morphogenesis of extinct rachis-dominated feathers. *Scientific Reports* 9: Article 18108.

Chang, M. M., P. J. Chen, Y. Q. Wang, Y. Q. Wang, Y. Wang, and D. S. Miao. 2003. *The Jehol Fossils: The Emergence of Feathered Dinosaurs, Beaked Birds and Flowering Plants*. Shanghai Scientific & Technical Publishers, Shanghai, China.

Chen, R.-S., M. Wang, L.-P. Dong, G.-W. Zhou, X. Xu, K. Deng, L.-M. Xu, C. Zhang, L.-C. Wang, H.-G. Du, et al. 2025. Earliest short-tailed bird from the Late Jurassic of China. *Nature* 638:441–448.

Chiappe, L. M. 1990. A flightless bird from the Late Cretaceous of Patagonia (Argentina). *Archosaurian Articulations* 1:73–77.

Chiappe, L. M. 1993. Enantiornithine (Aves) tarsometatarsi from the Crecaceous Lecho Formation of northwestern Argentina. *American Museum Novitates* 3083:1–27.

Chiappe, L. M. 1995. *The Phylogenetic Position of the Cretaceous Birds of Argentina: Enantiornithes and Patagopteryx deferrariisi*. Forschungsinstitut Senckenberg, Frankfurt, Germany.

Chiappe, L. M., and J. O. Calvo. 1994. *Neuquenornis volans*, a new Late Cretaceous bird (Enantiornithes: Avisauridae) from Patagonia, Argentina. *Journal of Vertebrate Paleontology* 14:230–246.

Chiappe, L. M., S. Ji, Q. Ji, and M. A. Norell. 1999. Anatomy and systematics of the Confuciusornithidae (Theropoda: Aves) from the Late Mesozoic of northeastern China. *Bulletin of the American Museum of Natural History* 242:1–89.

Chiappe, L. M., J. Marugán-Lobón, S. A. Ji, and Z. Zhou. 2008. Life history of a basal bird: Morphometrics of the Early Cretaceous Confuciusornis. *Biology Letters* 4:719–723.

Chiappe, L. M., M. Norell, and J. Clark. 2001. A new skull of *Gobipteryx minuta* (Aves: Enantiornithes) from the Cretaceous of the Gobi Desert. *American Museum Novitates* 3346:1–15.

Chiappe, L. M., S. Suzuki, G. J. Dyke, M. Watabe, K. Tsogtbaatar, and R. Barsbold. 2006. A new enantiornithine bird from the Late Cretaceous of the Gobi Desert. *Journal of Systematic Palaeontology* 5:193–208.

Chiappe, L. M., and C. A. Walker. 2002. *Skeletal Morphology and Systematics of the Cretaceous Eunantiornithes (Ornithothoraces: Enantiornithes).* University of California Press, Berkeley.

Chinsamy, A., L. M. Chiappe, and P. Dodson. 1994. Growth rings in Mesozoic birds. *Nature* 368:196–197.

Clark, A. D., and J. K. O'Connor. 2021. Exploring the ecomorphology of two Cretaceous Enantiornithines with unique pedal morphology. *Frontiers in Ecology and Evolution* 9: Article 654156.

Clarke, J. A., S. Chatterjee, Z. Li, T. Riede, F. Agnolin, F. Goller, M. P. Isasi, D. R. Martinioni, F. J. Mussel, and F. E. Novas. 2016. Fossil evidence of the avian vocal organ from the Mesozoic. *Nature* 538:502–505.

Clarke, J. A., C. P. Tambussi, J. I. Noriega, G. M. Erickson, and R. A. Ketcham. 2005. Definitive fossil evidence for the extant avian radiation in the Cretaceous. *Nature* 433:305–308.

Cordes-Person, A., C. Acosta Hospitaleche, J. Case, and J. Martin. 2020. An enigmatic bird from the lower Maastrichtian of Vega Island, Antarctica. *Cretaceous Research* 108: Article 104314.

Darwin, C. R. 1866. *On the Origin of Species by Means of Natural Selection, or the Preservation of Favoured Races in the Struggle for Life.* 4th ed. Murray, London.

de Souza Carvalho, I., F. E. Novas, F. L. Agnolín, M. P. Isasi, F. I. Freitas, and J. A. Andrade. 2015. A Mesozoic bird from Gondwana preserving feathers. *Nature Communications* 6:7141–7141.

Dumont, M., P. Tafforeau, T. Bertin, B.-A. Bhullar, D. Field, A. Schulp, B. Strilisky, B. Thivichon-Prince, L. Viriot, and A. Louchart. 2016. Synchrotron imaging of dentition provides insights into the biology of Hesperornis and Ichthyornis, the "last" toothed birds. *BMC Evolutionary Biology* 16: Article 178.

Dyke, G., M. Vremir, G. Kaiser, and D. Naish. 2012. A drowned Mesozoic bird breeding colony from the Late Cretaceous of Transylvania. *Naturwissenschaften* 99:435–442.

Elzanowski, A. 1974. Preliminary note on the palaeognthous bird from the Upper Cretaceous of Mongolia. *Palaeontologica Polonica* 30:103–109.

Elzanowski, A. 1981. Embryonic bird skeletons from the Late Cretaceous of Mongolia. *Palaeontologica Polonica* 42:147–179.

Elzanowski, A. 2002. *Archaeopterygidae (Upper Jurassic of Germany).* University of California Press, Berkeley.

Falk, A. R., T. G. Kaye, Z. Zhou, and D. A. Burnham. 2016. Laser fluorescence illuminates the soft tissue and life habits of the Early Cretaceous bird *Confuciusornis. PLoS One* 11: Article e0167284.

Feduccia, A. 2002. Birds are dinosaurs: Simple answer to a complex problem. *The Auk* 119:1187–1201.

Feduccia, A. 2013. Bird origins anew. *The Auk* 130:1–12.

Feo, T. J., D. J. Field, and R. O. Prum. 2015. Barb geometry of asymmetrical feathers reveals a transitional morphology in the evolution of avian flight. *Proceedings of the Royal Society B: Biological Sciences* 282: Article 20142864.

Fernández, M. S., R. A. García, L. Fiorelli, A. Scolaro, R. B. Salvador, C. N. Cotaro, G. W. Kaiser, and G. J. Dyke. 2013. A large accumulation of avian eggs from the Late Cretaceous of Patagonia (Argentina) reveals a novel nesting strategy in Mesozoic birds. *PLoS One* 8: Article e61030.

Field, D. J., J. Benito, A. Chen, J. W. M. Jagt, and D. T. Ksepka. 2020. Late Cretaceous neornithine from Europe illuminates the origins of crown birds. *Nature* 579:397–401.

Field, D. J., A. Bercovici, J. S. Berv, R. Dunn, D. E. Fastovsky, T. R. Lyson, V. Vajda, and J. A. Gauthier. 2018. Early evolution of modern birds structured by global forest collapse at the End-Cretaceous mass extinction. *Current Biology* 28:1825–1831.

Field, D. J., M. Hanson, D. Burnham, L. E. Wilson, K. Super, D. Ehret, J. A. Ebersole, and B.-A. S. Bhullar. 2018. Complete Ichthyornis skull illuminates mosaic assembly of the avian head. *Nature* 557:96–100.

Forster, C. A., L. M. Chiappe, D. W. Krause, and S. D. Sampson. 1996. The first Cretaceous bird from Madagascar. *Nature* 382:532–534.

Forster, C. A., L. M. Chiappe, D. W. Krause, and S. D. Sampson. 2002. *Vorona berivotrensis, a Primitive Bird From the Late Cretaceous of Madagascar*. University of California Press, Berkeley.
Forster, C. A., S. D. Sampson, L. M. Chiappe, and D. W. Krause. 1998. The theropod ancestry of birds: New evidence from the Late Cretaceous of Madagascar. *Science* 279:1915–1919.
Foth, C., H. Tischlinger, and O. W. M. Rauhut. 2014. New specimen of *Archaeopteryx* provides insights into the evolution of pennaceous feathers. *Nature* 511:79–82.
Gao, C. L., and J. Y. Liu. 2005. A new avian taxon from Lower Cretaceous Jiufotang Formation of western Liaoning. *Global Ecology* 24:313–318.
Gatesy, S. M., and K. P. Dial. 1996. From frond to fan: *Archaeopteryx* and the evolution of short-tailed birds. *Evolution* 50:2037–2048.
Gianechini, F. A., P. J. Makovicky, S. Apesteguía, and I. Cerda. 2018. Postcranial skeletal anatomy of the holotype and referred specimens of *Buitreraptor gonzalezorum* Makovicky, Apesteguía and Agnolín 2005 (Theropoda, Dromaeosauridae), from the Late Cretaceous of Patagonia. PeerJ 6: Article e4558.
Gill, F. B. 2007. *Ornithology*. 3rd ed. Freeman, New York.
Godefroit, P., S. M. Sinitsa, D. Dhouailly, Y. L. Bolotsky, A. V. Sizov, M. E. McNamara, M. J. Benton, and P. Spagna. 2014. A Jurassic ornithischian dinosaur from Siberia with both feathers and scales. *Science* 345:451–455.
Grande, L. 2013. *The Lost World of Fossil Lake*. University of Chicago Press, Chicago.
Hayward, J. L., C. J. Amlaner, and K. A. Young. 1989. Turning eggs to fossils: A natural experiment in taphonomy. *Journal of Vertebrate Paleontology* 9:196–200.
Heilmann, G. 1926. *The Origin of Birds*. H. F. & G. Witherby, Edinburgh, Scotland.
Hope, S. 2002. *The Mesozoic Radiation of Neornithes*. University of California Press, Berkeley.
Hopson, J. A. 2001. Ecomorphology of avian and nonavian theropod phalangeal proportions: Implications for the arboreal versus terrestrial origin of bird flight. Pages 211–235 in J. Gauthier and L. F. Gall, eds. *New Perspectives on the Origin and Evolution of Birds*. Peabody Museum of Natural History, New Haven, CT.
Hou, L. 1999. New hesperonithid (Aves) from the Canadian Arctic. *Vertebrata Palasiatica* 37:228–233.
Hou, L., and Z. Liu. 1984. A new fossil bird from Lower Cretaceous of Gansu and early evolution of birds. *Scientia Sinica Series B* 27:1296–1301.
Hou, L., L. D. Martin, Z. Zhou, and A. Feduccia. 1996. Early adaptive radiation of birds: Evidence from fossils from northeastern China. *Science* 274:1164–1167.
Hou, L., and J. Zhang. 1993. A new fossil bird from Lower Cretaceous of Liaoning. *Vertebrata Palasiatica* 31:217–224.
Hou, L., Z. Zhou, Y. Gu, and H. Zhang. 1995. *Confuciusornis sanctus*, a new Late Jurassic sauriurine bird from China. *Chinese Science Bulletin* 40:1545–1551.
Hu, D., L. Hou, L. Zhang, and X. Xu. 2009. A pre-*Archaeopteryx* troodontid theropod from China with long feathers on the metatarsus. *Nature* 461:640–643.
Hu, H., J. K. O'Connor, P. G. McDonald, and S. Wroe. 2020. Cranial osteology of the Early Cretaceous *Sapeornis chaoyangensis* (Aves: Pygostylia). *Cretaceous Research* 113: Article 104496.
Hu, H., J. K. O'Connor, M. Wang, S. Wroe, and P. G. McDonald. 2020. New anatomical information on the bohaiornithid *Longusunguis* and the presence of a plesiomorphic diapsid skull in Enantiornithes. *Journal of Systematic Palaeontology* 18:1481–1495.
Hu, H., G. Sansalone, S. Wroe, P. G. McDonald, J. K. O'Connor, Z. Li, X. Xu, and Z. Zhou. 2019. Evolution of the vomer and its implications for cranial kinesis in Paraves. *Proceedings of the National Academy of Sciences of the USA* 116:19571–19578.
Hu, H., Y. Wang, P. G. McDonald, S. Wroe, J. K. O'Connor, A. Bjarnason, J. J. Bevitt, X. Yin, X. Zheng, Z. Zhou, and R. B. J. Benson. 2022. Earliest evidence for fruit consumption and potential seed dispersal by birds. *eLife* 11: Article e74751.
Imai, T., Y. Azuma, S. Kawabe, M. Shibata, K. Miyata, M. Wang, and Z. Zhou. 2019. An unusual bird (Theropoda, Avialae) from the Early Cretaceous of Japan suggests complex evolutionary history of basal birds. *Communications Biology* 2: Article 399.
James, H. F. 2005. Paleogene fossils and the radiation of modern birds. *The Auk* 122:1049–1054.

Ji, Q., P. J. Currie, M. A. Norell, and J. Shu-An. 1998. Two feathered dinosaurs from northeastern China. *Nature* 393:753–761.
Ji, Q., and S. Ji. 1996. On the discovery of the earliest fossil bird in China (*Sinosauropteryx* gen. nov.) and the origin of birds. *Chinese Geology* 233:30–33.
Ji, Q., S. Ji, H. You, J. Zhang, C. Yuan, X. Ji, J. Li, and Y. Li. 2002. Discovery of an avialae bird—*Shenzhouraptor sinensis* gen. et sp. nov.—from China. *Geological Bulletin of China* 21:363–369.
Ji, Q., S. Ji, H. Zhang, H. You, J. Zhang, L. Wang, C. Yuan, and X. Ji. 2002b. A new avialian bird—*Jixiangornis orientalis* gen. et sp. nov.—from the Lower Cretaceous of western Liaoning, NE China. *Journal of Nanjing University (Natural Sciences)* 38:723–735.
Ji, Q., M. A. Norell, P. J. Makovicky, K.-Q. Gao, S. A. Ji, and C. Yuan. 2003. An early ostrich dinosaur and implications for ornithomimosaur phylogeny. *American Museum Novitates* 3420:1–19.
Kiat, Y., and J. K. O'Connor. 2024. Functional constraints on the number and shape of flight feathers. *Proceedings of the National Academy of Sciences U.S.A.* 121:1–11.
Knoll, F., L. M. Chiappe, S. Sanchez, R. J. Garwood, N. P. Edwards, R. A. Wogelius, W. I. Sellers, P. L. Manning, F. Ortega, F. J. Serrano, J. Marugán-Lobón, E. Cuesta, F. Escaso, and J. L. Sanz. 2018. A diminutive perinate European Enantiornithes reveals an asynchronous ossification pattern in early birds. *Nature Communications* 9: Article 937.
Ksepka, D. T., L. Grande, and G. Mayr. 2019. Oldest finch-beaked birds reveal parallel ecological radiations in the earliest evolution of passerines. *Current Biology* 29:657–663.
Ksepka, D. T., T. A. Stidham, and T. E. Williamson. 2017. Early Paleocene landbird supports rapid phylogenetic and morphological diversification of crown birds after the K–Pg mass extinction. *Proceedings of the National Academy of Sciences of the USA* 114:8047–8052.
Kundrát, M., J. Nudds, B. P. Kear, J. Lü, and P. Ahlberg. 2019. The first specimen of *Archaeopteryx* from the Upper Jurassic Mörnsheim Formation of Germany. *Historical Biology* 31:3–63.
Kurochkin, E. N. 1982. New order of birds from the Lower Cretaceous in Mongolia. *Doklady AN SSSR* 1982:215–218.
Larson, D. W., C. M. Brown, and D. C. Evans. 2016. Dental disparity and ecological stability in bird-like dinosaurs prior to the end-Cretaceous mass extinction. *Current Biology* 26:1325–1333.
Lefèvre, U., D. Hu, F. Escuillié, G. Dyke, and P. Godefroit. 2014. A new long-tailed basal bird from the Lower Cretaceous of north-eastern China. *Biological Journal of the Linnean Society* 113: 790–804.
Li, Z., M. Wang, T. A. Stidham, Z. Zhou, and J. Clarke. 2022. Novel evolution of a hyper-elongated tongue in a Cretaceous enantiornithine from China and the evolution of the hyolingual apparatus and feeding in birds. *Journal of Anatomy* 240:627–638.
Liu, D., L. M. Chiappe, Y. Zhang, F. J. Serrano, and Q. Meng. 2019. Soft tissue preservation in two new enantiornithine specimens (Aves) from the Lower Cretaceous Huajiying Formation of Hebei Province, China. *Cretaceous Research* 95:191–207.
Longrich, N. 2009. An ornithurine-dominated avifauna from the Belly River Group (Campanian, Upper Cretaceous) of Alberta, Canada. *Cretaceous Research* 30:161–177.
Marsh, O. C. 1880. *Odontornithes: A Monograph on the Extinct Toothed Birds of North America; With Thirty-Four Plates and Forty Woodcuts.* Government Printing Office, Washington, DC.
Martin, L. D. 1987. The beginning of the modern aviation radiation. Pages 9–19 in C. Mourer-Chauviré, ed. *L'Evolution des Oiseaux d'Apres le Temoignange des Fossiles.* Département des Sciences de la Terre, Université Claude-Bernard, Lyon, France.
Matsuoka, H., N. Kusuhashi, T. Takada, and T. Setoguchi. 2002. A clue to the Necomian vertebrate fauna: Initial results from the Kuwajima "Kaseki-kabe" (Tetori Group) in Shiramine, Ishikawa, central Japan. *Memoirs of the Faculty of Science, Kyoto University, Series of Geology and Mineralogy* 59:33–45.
Mayr, G. 2009. *Paleogene Fossil Birds.* Springer-Verlag, Berlin.
Mayr, G., B. Pohl, S. Hartman, and D. S. Peters. 2007. The tenth skeletal specimen of *Archaeopteryx.* *Zoological Journal of the Linnean Society* 149:97–116.
Mayr, G., B. Pohl, and D. S. Peters. 2005. A well-preserved *Archaeopteryx* specimen with theropod features. *Science* 310:1483–1486.
Møller, A., and A. Hedenström. 1999. Comparative evidence for costs of secondary sexual characters: Adaptive vane emargination of ornamented feathers in birds. *Journal of Evolutionary Biology* 12:296–305.

Navalón, G., J. Marugán-Lobón, L. M. Chiappe, J. Luis Sanz, and Á. D. Buscalioni. 2015. Soft-tissue and dermal arrangement in the wing of an Early Cretaceous bird: Implications for the evolution of avian flight. *Scientific Reports* 5: Article 14864.

Norell, M. A., J. M. Clark, L. M. Chiappe, and D. Dashzeveg. 1995. A nesting dinosaur. *Nature* 378:774–776.

Norell, M. A., and J. A. Clarke. 2001. Fossil that fills a critical gap in avian evolution. *Nature* 409:181–184.

Nudds, R. L., J. Atterholt, X. Wang, H. L. You, and G. J. Dyke. 2012. Locomotory abilities and habitat of the Cretaceous bird *Gansus yumenensis* inferred from limb length proportions. *Journal of Evolutionary Biology* 26:150–154.

O'Connor, C. M., D. R. Norris, G. T. Crossin, and S. J. Cooke. 2014. Biological carryover effects: Linking common concepts and mechanisms in ecology and evolution. *Ecosphere* 5: Article art28–11.

O'Connor, J. 2020. The plumage of basal birds. Pages 147–172 in C. Foth and O. W. M. Rauhut, eds. *The Evolution of Feathers*. Springer, Cham, Switzerland.

O'Connor, J. K., A. D. Clark, P.-C. Kuo, Y. Kiat, M. Fabbri, A. Shinya, C. Van Beek, J. Lu, M. Wang, and H. Han. 2025. Chicago Archaeopteryx informs on the early evolution of the avian bauplan. *Nature* 641:1201–1207.

O'Connor, J. K., A. D. Clark, F. Herrera, X. Yang, X.-L. Wang, X.-T. Zheng, H. Hu, and Z.-H. Zhou. 2024. Direct evidence of frugivory in the Mesozoic bird Longipteryx contradicts morphological proxies for diet. *Current Biology* 34:4559–4566.

O'Connor, J., G. M. Erickson, M. Norell, A. M. Bailleul, H. Hu, and Z. Zhou. 2018. Medullary bone in an Early Cretaceous enantiornithine bird and discussion regarding its identification in fossils. *Nature Communications* 9: Article 5169.

O'Connor, J., X. Wang, C. Sullivan, Y. Wang, X. Zheng, H. Hu, X. Zhang, and Z. Zhou. 2018. First report of gastroliths in the Early Cretaceous basal bird *Jeholornis*. *Cretaceous Research* 84:200–208.

O'Connor, J., X. Wang, C. Sullivan, X. Zheng, P. Tubaro, X. Zhang, and Z. Zhou. 2013. Unique caudal plumage of *Jeholornis* and complex tail evolution in early birds. *Proceedings of the National Academy of Sciences of the USA* 110:17404–17408.

O'Connor, J., Y. Zhang, L. M. Chiappe, Q. Meng, L. Quanguo, and L. Di. 2013. A new enantiornithine from the Yixian Formation with the first recognized avian enamel specialization. *Journal of Vertebrate Paleontology* 33:1–12.

O'Connor, J., and Z. Zhou. 2015. Early evolution of the biological bird: Perspectives from new fossil discoveries in China. *Journal of Ornithology* 156:333–342.

O'Connor, J. K. 2009. A systematic review of Enantiornithies (Aves: Ornithothoraces). PhD dissertation, University of Southern California, Los Angeles.

O'Connor, J. K. 2019. The trophic habits of early birds. *Palaeogeography, Palaeoclimatology, Palaeoecology* 513:178–195.

O'Connor, J. K., and H. Chang. 2015. Hindlimb feathers in paravians: Primarily "wings" or ornaments? *Biology Bulletin* 42:616–621.

O'Connor, J. K., and L. M. Chiappe. 2011. A revision of enantiornithine (Aves: Ornithothoraces) skull morphology. *Journal of Systematic Palaeontology* 9:135–157.

O'Connor, J. K., K. Q. Gao, and L. M. Chiappe. 2010. A new ornithuriomorph (Aves: Ornithothoraces) bird from the Jehol Group indicates a higher-level diversity. *Journal of Vertebrate Paleontology* 30:311–321.

O'Connor, J. K., T. A. Stidham, J. D. Harris, M. C. Lamanna, A. M. Bailleul, H. Hu, M. Wang, and H. L. You. 2022. Avian skulls represent a diverse ornithuromorph fauna from the Lower Cretaceous Xiagou Formation, Gansu Province, China. *Journal of Systematics and Evolution* 60:1172–1198.

O'Connor, J. K., C. Sun, X. Xu, X. Wang, and Z. Zhou. 2012. A new species of *Jeholornis* with complete caudal integument. *Historical Biology* 24:29–41.

O'Connor, J. K., M. Wang, and H. Hu. 2016. A new ornithuromorph (Aves) with an elongate rostrum from the Jehol Biota and the early evolution of rostralization in birds. *Journal of Systematic Palaeontology* 14:939–948.

O'Connor, J. K., M. Wang, S. Zhou, and Z. Zhou. 2015. Osteohistology of the Lower Cretaceous Yixian Formation ornithuromorph (Aves) *Iteravis huchzermeyeri*. *Palaeontologia Electronica* 18.2.35A:1–11.
O'Connor, J. K., X. Wang, L. M. Chiappe, C. Gao, Q. Meng, X. Cheng, and J. Liu. 2009. Phylogenetic support for a specialized clade of Cretaceous enantiornithine birds with information from a new species. *Journal of Vertebrate Paleontology* 29:188–204.
O'Connor, J. K., X. Wang, X. Zheng, H. Hu, X. Zhang, and Z. Zhou. 2016. An enantiornithine with a fan-shaped tail, and the evolution of the rectricial complex in early birds. *Current Biology* 26:114–119.
O'Connor, J. K., X. Zheng, Y. Pan, X. Wang, Y. Wang, X. Zhang, and Z. Zhou. 2020. New information on the plumage of *Protopteryx* (Aves: Enantiornithes) from a new specimen. *Cretaceous Research* 116: Article 104577.
O'Connor, J. K., X. T. Zheng, C. Sullivan, C. M. Chuong, X. L. Wang, A. Li, Y. Wang, X. M. Zhang, and Z. H. Zhou. 2015. Evolution and functional significance of derived sternal ossification patterns in ornithothoracine birds. *Journal of Evolutionary Biology* 28:1550–1567.
O'Connor, J. K., Z. Zhou, and F. Zhang. 2011. A reappraisal of *Boluochia zhengi* (Aves: Enantiornithes) and a discussion of intraclade diversity in the Jehol avifauna, China. Journal of Systematic Palaeontology **9**:51–63.
O'Connor, P. M., and C. A. Forster. 2010. A Late Cretaceous (Maastrichtian) avifauna from the Maevarano Formation, Madagascar. *Journal of Vertebrate Paleontology* 30:1178–1201.
O'Connor, P. M., A. H. Turner, J. R. Groenke, R. N. Felice, R. R. Rogers, D. W. Krause, and L. J. Rahantarisoa. 2020. Late Cretaceous bird from Madagascar reveals unique development of beaks. *Nature* 588:272–276.
Ostrom, J. H. 1969. Osteology of *Deinonychus antirrhopus*, an unusual theropod from the Lower Cretaceous of Montana. *Bulletin of the Peabody Museum of Natural History* 30:1–165.
Ostrom, J. H. 1970. Archaeopteryx: Notice of a "new" specimen. *Science* 170:537–538.
Ostrom, J. H. 1972. Description of the *Archaeopteryx* specimen in the Teyler Museum, Haarlem. *Proceedings of the Koninklijke Nederlandse Akademie van Wetenschappen B* 75:289–305.
Ostrom, J. H. 1973. The ancestry of birds. *Nature* 242:136–136.
Ostrom, J. H. 1974. Archaeopteryx and the origin of flight. *Quarterly Review of Biology* 49:27–47.
Ostrom, J. H. 1975. The origin of birds. *Annual Review of Earth and Planetary Sciences* 3:55–77.
Ostrom, J. H. 1976. *Archaeopteryx* and the origin of birds. *Biological Journal of the Linnean Society* 8:91–182.
Owen, R. 1863. III. On the archeopteryx of von Meyer, with a description of the fossil remains of a long-tailed species, from the lithographic stone of Solenhofen. *Philosophical Transactions of the Royal Society of London* 153:33–47.
Pan, Y., Z. Li, M. Wang, T. Zhao, X. Wang, and X. Zheng. 2021. Unambiguous evidence of brilliant iridescent feather color from hollow melanosomes in an Early Cretaceous bird. *National Science Review* **9**: Article nwab227.
Pan, Y., J. Sha, Z. Zhou, and F. T. Fürsich. 2013. The Jehol Biota: Definition and distribution of exceptionally preserved relicts of a continental Early Cretaceous ecosystem. *Cretaceous Research* 44:30–38.
Pei, R., M. Pittman, P. A. Goloboff, T. A. Dececchi, M. B. Habib, T. G. Kaye, H. C. E. Larsson, M. A. Norell, S. L. Brusatte, and X. Xu. 2020. Potential for powered flight neared by most close avialan relatives, but few crossed its thresholds. *Current Biology* 30:4033–4046.
Prum, R. O. 2003. Are current critiques of the theropod origin of birds science? Rebuttal to Feduccia (2002). *The Auk* 120:550–561.
Rauhut, O. W. M., C. Foth, and H. Tischlinger. 2018. The oldest *Archaeopteryx* (Theropoda: Avialiae): A new specimen from the Kimmeridgian/Tithonian boundary of Schamhaupten, Bavaria. *PeerJ* 6: Article e4191.
Rauhut, O. W. M., C. Foth, H. Tischlinger, and M. A. Norell. 2012. Exceptionally preserved juvenile megalosauroid theropod dinosaur with filamentous integument from the Late Jurassic of Germany. *Proceedings of the National Academy of Sciences of the USA* 109:11746–11751.

Sanz, J. L., J. F. Bonapartet, and A. Lacasa. 1988. Unusual Early Cretaceous birds from Spain. *Nature* 331:433–435.

Sanz, J. L., L. M. Chiappe, B. P. Pérez-Moreno, A. D. Buscalioni, J. J. Moratalla, F. Ortega, and F. J. Poyato-Ariza. 1996. An Early Cretaceous bird from Spain and its implications for the evolution of avian flight. *Nature* 382:442–445.

Sanz, J. L., L. M. Chiappe, B. P. Pérez-Moreno, J. J. Moratalla, F. Hernández-Carrasquilla, A. D. Buscalioni, F. Ortega, F. J. Poyato-Ariza, D. Rasskin-Gutman, and X. Martínez-Delclós. 1997. A nestling bird from the Lower Cretaceous of Spain: Implications for avian skull and neck evolution. *Science* 276:1543–1546.

Sanz, J. L., and F. Ortega. 2002. The birds from Las Hoyas. *Science Progress* 85:113–130.

Schweitzer, M. H., F. D. Jackson, L. M. Chiappe, J. G. Schmitt, J. O. Calvo, and D. E. Rubilar. 2002. Late Cretaceous avian eggs with embryos from Argentina. *Journal of Vertebrate Paleontology* 22:191–195.

Schweitzer, M. H., J. A. Watt, R. Avci, L. Knapp, L. Chiappe, M. Norell, and M. Marshall. 1999. Beta-keratin specific immunological reactivity in feather-like structures of the Cretaceous Alvarezsaurid, *Shuvuuia deserti*. *Journal of Experimental Zoology* 285:146–157.

Seeley, H. G. 1876. On the British fossil Cretaceous birds. *Quarterly Journal of the Geological Society* 32:496–512.

Sereno, P. C., Z. Xijin, and T. Lin. 2010. A new psittacosaur from Inner Mongolia and the parrot-like structure and function of the psittacosaur skull. *Proceedings of the Royal Society B: Biological Sciences* 277:199–209.

Serrano, F. J., and L. M. Chiappe. 2017. Aerodynamic modelling of a Cretaceous bird reveals thermal soaring capabilities during early avian evolution. *Journal of the Royal Society, Interface* 14: Article 20170182.

Slack, K. E., C. M. Jones, T. Ando, G. L. Harrison, R. E. Fordyce, U. Arnason, and D. Penny. 2006. Early penguin fossils, plus mitochondrial genomes, calibrate avian evolution. *Molecular Biology and Evolution* 23:1144–1155.

Stidham, T. A., and J. K. O'Connor. 2021. The evolutionary and functional implications of the unusual quadrate of *Longipteryx chaoyangensis* (Avialae: Enantiornithes) from the Cretaceous Jehol Biota of China. *Journal of Anatomy* 239:1066–1074.

Switek, B. 2010. Thomas Henry Huxley and the reptile to bird transition. *Geological Society, London, Special Publications* 343:251–263.

Turner, A. H., S. Montanari, and M. A. Norell. 2021. A new dromaeosaurid from the Late Cretaceous Khulsan Locality of Mongolia. *American Museum Novitates*: Article 3965.

Turner, A. H., D. Pol, J. A. Clarke, G. M. Erickson, and M. A. Norell. 2007. A basal dromaeosaurid and size evolution preceding avian flight. *Science* 317:1378–1381.

Unwin, D. M., and L. Junchang. 1997. On *Zhejiangopterus* and the relationships of pterodactyloid pterosaurs. *Historical Biology* 12:199–210.

Varricchio, D. J., and F. D. Jackson. 2016. Reproduction in Mesozoic birds and evolution of the modern avian reproductive mode. *The Auk* 133:654–684.

Vecchia, F. M. D., and L. M. Chiappe. 2003. First avian skeleton from the Mesozoic of northern Gondwana. *Journal of Vertebrate Paleontology* 22:856–860.

von Meyer, H. 1861. *Achaeopteryx litographica* (Vogel–Feder) und Pterodactylus von Solenhofen. *Neues Jahrbuch für Mineralogie, Geognosie, Geologie, und Petrefakten-kunde*:678–679.

Walker, A. D. 1972. New light on the origin of birds and crocodiles. *Nature* 237:257–263.

Walker, C. A. 1981. New subclass of birds from the Cretaceous of South America. *Nature* 292:51–53.

Wang, M., J. K. O'Connor, Y. Pan, and Z. Zhou. 2017. A bizarre Early Cretaceous enantiornithine bird with unique crural feathers and an ornithuromorph plough-shaped pygostyle. *Nature Communications* 8: Article 14141.

Wang, M., J. K. O'Connor, N. Z. Zelenkov, and Z. H. Zhou. 2014. A new diverse enantiornithine family (Bohaiornithidae fam. nov.) from the Lower Cretaceous of China with information from two new species. *Vertebrata Palasiatica* 52:31–76.

Wang, M., J. K. O'Connor, T. Zhao, Y. Pan, X. Zheng, X. Wang, and Z. Zhou. 2021. An Early Cretaceous enantiornithine bird with a pintail. *Current Biology* 31:4845–4852.
Wang, M., J. K. O'Connor, S. Zhou, and Z. Zhou. 2019. New toothed Early Cretaceous ornithuromorph bird reveals intraclade diversity in pattern of tooth loss. *Journal of Systematic Palaeontology* 18:631–645.
Wang, M., J. K. O'Connor, and Z. H. Zhou. 2019. A taxonomical revision of the Confuciusornithiformes (Aves: Pygostylia). *Vertebrata Palasiatica* 57:1–37.
Wang, M., T. A. Stidham, J. K. O'Connor, and Z. Zhou. 2022. Insight into the evolutionary assemblage of cranial kinesis from a Cretaceous bird. *eLife* 11: Article e81337.
Wang, M., T. A. Stidham, and Z. Zhou. 2018. A new clade of basal Early Cretaceous pygostylian birds and developmental plasticity of the avian shoulder girdle. *Proceedings of the National Academy of Sciences of the USA* 115:10708–10713.
Wang, M., X. Wang, Y. Wang, and Z. Zhou. 2016. A new basal bird from China with implications for morphological diversity in early birds. *Scientific Reports* 6: Article 19700.
Wang, M., X. Zheng, J. K. O'Connor, G. T. Lloyd, X. Wang, Y. Wang, X. Zhang, and Z. Zhou. 2015. The oldest record of ornithuromorpha from the early cretaceous of China. *Nature Communications* 6: Article 6987.
Wang, M., Z. Zhou, and S. Zhou. 2016. A new basal ornithuromorph bird (Aves: Ornithothoraces) from the Early Cretaceous of China with implication for morphology of early Ornithuromorpha. *Zoological Journal of the Linnean Society* 176:207–223.
Wang, W., and J. K. O'Connor. 2017. Morphological coevolution of the pygostyle and tail feathers in Early Cretaceous birds. *Vertebrata Palasiatica* 55:289–314.
Wang, X., J. Huang, M. Kundrát, A. Cau, X. Liu, Y. Wang, and S. Ju. 2020. A new jeholornithiform exhibits the earliest appearance of the fused sternum and pelvis in the evolution of avialan dinosaurs. *Journal of Asian Earth Sciences* 199: Article 104401.
Wang, X., J. K. O'Connor, X. Zheng, M. Wang, H. Hu, and Z. Zhou. 2014. Insights into the evolution of rachis dominated tail feathers from a new basal enantiornithine (Aves: Ornithothoraces). *Biological Journal of the Linnean Society* 113:805–819.
Wang, X., C. Shen, S. Liu, C. Gao, X. Cheng, and F. Zhang. 2015. New material of Longiptery (Aves: Enantiornithes) from the Lower Cretaceous Yixian Formation of China with the first recognized avian tooth crenulations. *Zootaxa* 3941:565–578.
Wang, Y., H. Hu, J. K. O'Connor, M. Wang, X. Xu, Z. Zhou, X. Wang, and X. Zheng. 2017. A previously undescribed specimen reveals new information on the dentition of *Sapeornis chaoyangensis*. *Cretaceous Research* 74:1–10.
Wellnhofer, P. 2008. *Archaeopteryx. Der Urvogel von Solnhofen*. Friedrich Pfeil., Munich, Germany.
Wiemann, J., T.-R. Yang, and M. A. Norell. 2018. Dinosaur egg colour had a single evolutionary origin. *Nature* 563:555–558.
Wu, Q., A. M. Bailleul, Z. Li, J. O'Connor, and Z. Zhou. 2021. Osteohistology of the scapulocoracoid of confuciusornis and preliminary analysis of the shoulder joint in aves. *Frontiers in Earth Science* 9: Article 617124.
Wu, Y.-H., L. M. Chiappe, D. J. Bottjer, W. Nava, and A. G. Martinelli. 2021. Dental replacement in Mesozoic birds: Evidence from newly discovered Brazilian enantiornithines. *Scientific Reports* 11: Article 19349.
Xing, L., P. Cockx, R. C. McKellar, and J. O'Connor. 2018. Ornamental feathers in Cretaceous Burmese amber: Resolving the enigma of rachis-dominated feather structure. *Journal of Palaeogeography* 7: Article 13.
Xing, L., R. C. McKellar, J. K. O'Connor, M. Bai, K. Tseng, and L. M. Chiappe. 2019a. A fully feathered enantiornithine foot and wing fragment preserved in mid-Cretaceous Burmese amber. *Scientific Reports* 9:927–927.
Xing, L., R. C. McKellar, M. Wang, M. Bai, J. K. O'Connor, M. J. Benton, J. Zhang, Y. Wang, K. Tseng, M. G. Lockley, G. Li, W. Zhang, and X. Xu. 2016. Mummified precocial bird wings in mid-Cretaceous Burmese amber. *Nature Communications* 7: Article 12089.
Xing, L., J. K. O'Connor, L. M. Chiappe, R. C. McKellar, N. Carroll, H. Hu, M. Bai, and F. Lei. 2019b. A new enantiornithine bird with unusual pedal proportions found in amber. *Current Biology* 29:2396–2401.

Xing, L., J. K. O'Connor, R. C. McKellar, L. M. Chiappe, K. Tseng, G. Li, and M. Bai. 2017. A mid-Cretaceous enantiornithine (Aves) hatchling preserved in Burmese amber with unusual plumage. *Gondwana Research* 49:264–277.

Xing, L., J. K. O'Connor, K. Niu, P. Cockx, H. Mai, and R. C. McKellar. 2020. A new enantiornithine (Aves) preserved in mid-Cretaceous Burmese amber contributes to growing diversity of cretaceous plumage patterns. *Frontiers in Earth Science* 8: Article 00264.

Xu, L., E. Buffetaut, J. O'Connor, X. Zhang, S. Jia, J. Zhang, H. Chang, and H. Tong. 2021. A new, remarkably preserved, enantiornithine bird from the Upper Cretaceous Qiupa Formation of Henan (central China) and convergent evolution between enantiornithines and modern birds. *Geological Magazine* 158:2087–2094.

Xu, X., and M. A. Norell. 2004. A new troodontid dinosaur from China with avian-like sleeping posture. *Nature* 431:838–841.

Xu, X., X. Zheng, C. Sullivan, X. Wang, L. Xing, Y. Wang, X. Zhang, J. K. O'Connor, F. Zhang, and Y. Pan. 2015. A bizarre Jurassic maniraptoran theropod with preserved evidence of membranous wings. *Nature* 521:70–73.

Xu, X., Z. Zhou, X. Wang, X. Kuang, F. Zhang, and X. Du. 2003. Four-winged dinosaurs from China. *Nature* 421:335–340.

Yates, A. M., and C. C. Vasconcelos. 2005. Furcula-like clavicles in the prosauropod dinosaur *Massospondylus*. *Journal of Vertebrate Paleontology* 25:466–468.

You, H.-l., M. C. Lamanna, J. D. Harris, L. M. Chiappe, J. O'Connor, S.-a. Ji, J.-c. Lü, C.-x. Yuan, D.-q. Li, X. Zhang, K. J. Lacovara, P. Dodson, and Q. Ji. 2006. A nearly modern amphibious bird from the Early Cretaceous of northwestern China. *Science* 312:1640–1643.

Zelenitsky, D. K., F. Therrien, G. M. Erickson, C. L. DeBuhr, Y. Kobayashi, D. A. Eberth, and F. Hadfield. 2012. Feathered non-avian dinosaurs from North America provide insight into wing origins. *Science* 338:510–514.

Zelenkov, N. V., and A. O. Averianov. 2015. A historical specimen of enantiornithine bird from the Early Cretaceous of Mongolia representing a new taxon with a specialized neck morphology. *Journal of Systematic Palaeontology* 14:319–338.

Zhang, F., and Z. Zhou. 2004. Leg feathers in an Early Cretaceous bird. *Nature* 431: Article 925.

Zhang, F., Z. Zhou, and M. J. Benton. 2008. A primitive confuciusornithid bird from China and its implications for early avian flight. *Science in China Series D: Earth Sciences* 51:625–639.

Zhang, F., Z. Zhou, and G. J. D'yke. 2006. Feathers and "feather-like" integumentary structures in Liaoning birds and dinosaurs. *Geological Journal* 41:395–404.

Zheng, X., L. D. Martin, Z. Zhou, D. A. Burnham, F. Zhang, and D. Miao. 2011. Fossil evidence of avian crops from the Early Cretaceous of China. *Proceedings of the National Academy of Sciences of the USA* 108:15904–15907.

Zheng, X., J. O'Connor, F. Huchzermeyer, X. Wang, Y. Wang, M. Wang, and Z. Zhou. 2013. Preservation of ovarian follicles reveals early evolution of avian reproductive behaviour. *Nature* 495:507–511.

Zheng, X., J. O'Connor, X. Wang, M. Wang, X. Zhang, and Z. Zhou. 2014. On the absence of sternal elements in Anchiornis (Paraves) and Sapeornis (Aves) and the complex early evolution of the avian sternum. *Proceedings of the National Academy of Sciences of the USA* 111:13900–13905.

Zheng, X., J. O'Connor, Y. Wang, X. Wang, Y. Xuwei, X. Zhang, and Z. Zhou. 2020. New information on the keratinous beak of Confuciusornis (Aves: Pygostylia) from two new specimens. *Frontiers in Earth Science* 8: Article 00367.

Zheng, X., J. K. O'Connor, F. Huchzermeyer, X. Wang, Y. Wang, X. Zhang, and Z. Zhou. 2014. New specimens of Yanornis indicate a piscivorous diet and modern alimentary canal. *PLoS One* 9: Article e95036.

Zheng, X., J. K. O'Connor, X. Wang, Y. Pan, Y. Wang, M. Wang, and Z. Zhou. 2017. Exceptional preservation of soft tissue in a new specimen of Eoconfuciusornis and its biological implications. *National Science Review* 4:441–452.

Zheng, X., J. K. O'Connor, X. Wang, Y. Wang, and Z. Zhou. 2018. Reinterpretation of a previously described Jehol bird clarifies early trophic evolution in the Ornithuromorpha. *Proceedings of the Royal Society B: Biological Sciences* 285: Article 20172494.

Zheng, X., X. Wang, J. O'Connor, and Z. Zhou. 2012. Insight into the early evolution of the avian sternum from juvenile enantiornithines. *Nature Communications* 3: Article 1116.
Zheng, X., Z. Zhou, X. Wang, F. Zhang, X. Zhang, Y. Wang, G. Wei, S. Wang, and X. Xu. 2013. Hind wings in basal birds and the evolution of leg feathers. *Science* 339:1309–1312.
Zheng, X., Z. Zihui, and H. O. U. Lianhai. 2007. A new enantiornitine bird with four long rectrices from the Early Cretaceous of Northern Hebei, China. *Acta Geologica Sinica* [English edition] 81:703–708.
Zhou, S., J. K. O'Connor, and M. Wang. 2014. A new species from an ornithuromorph dominated locality of the Jehol Group. *Chinese Science Bulletin* 59:5366–5378.
Zhou, S., Z. Zhou, and J. K. O'Connor. 2013. Anatomy of the basal ornithuromorph bird *Archaeorhynchus spathula* from the Early Cretaceous of Liaoning, China. *Journal of Vertebrate Paleontology* 33:141–152.
Zhou, S., Z. H. Zhou, and J. K. O'Connor. 2012. A new toothless ornithurine bird (*Schizooura lii* gen. et sp. nov) from the Lower Cretaceous of China. *Vertebrata Palasiatica* 50:9–24.
Zhou, Z., J. Clarke, and F. Zhang. 2008. Insight into diversity, body size and morphological evolution from the largest Early Cretaceous enantiornithine bird. *Journal of Anatomy* 212:565–577.
Zhou, Z., F. Jin, and J. Zhang. 1992. Preliminary report on a Mesozoic bird from Liaoning, China. *Kexue Tongbao* 5:435–437.
Zhou, Z., and L. D. Martin. 2011. Distribution of the predentary bone in Mesozoic ornithurine birds. *Journal of Systematic Palaeontology* 9:25–31.
Zhou, Z., and F. Zhang. 2002a. Largest bird from the Early Cretaceous and its implications for the earliest avian ecological diversification. *Naturwissenschaften* 89:34–38.
Zhou, Z., and F. Zhang. 2002b. A long-tailed, seed-eating bird from the Early Cretaceous of China. *Nature* 418:405–409.
Zhou, Z., and F. Zhang. 2003a. Anatomy of the primitive bird *Sapeornis chaoyangensis* from the Early Cretaceous of Liaoning, China. *Canadian Journal of Earth Sciences* 40:731–747.
Zhou, Z., and F. Zhang. 2003b. *Jeholornis* compared to *Archaeopteryx*, with a new understanding of the earliest avian evolution. *Naturwissenschaften* 90:220–225.
Zhou, Z., and F. Zhang. 2004. A precocial avian embryo from the Lower Cretaceous of China. *Science* 306:653–653.
Zhou, Z., and F. Zhang. 2006. A beaked basal ornithurine bird (Aves, Ornithurae) from the Lower Cretaceous of China. *Zoologica Scripta* 35:363–373.

9

Speciation and Adaptation in Birds

David P. L. Toews and Elizabeth S. C. Scordato

In writing the introduction to the 1987 *Perspectives in Ornithology*, Ernst Mayr was given latitude to discuss the realms of ornithology that he believed were not treated sufficiently in the subsequent chapters (Mayr 1987). One major field he noted that had not received much attention—a field, unsurprisingly, in which he was conducting his own research—was the process of avian speciation and adaptation. In the intervening years, landmark publications, particularly Trevor Price's *Speciation in Birds* (Price 2008), have signaled that the field of avian speciation has matured into a rich, diverse, and global endeavor. The goals of the work span broad scales, from the detailed study of adaptive radiation of island birds to understanding the macroevolutionary processes that generate differential diversification across continental scales.

As students of avian speciation and adaptation, we have been part of this maturing field. Most recently, we have had a front seat to the novel application of genomic tools to long outstanding questions. In this chapter, we focus on four specific areas of the extensive speciation literature (Figure 9.1), with an emphasis on studies applying molecular approaches: (1) pre- and postmating reproductive isolating barriers in birds, (2) the genomics of avian speciation and adaptation, (3) the role of contemporary and ancient gene flow in speciation, and (4) anthropogenic influences on contemporary adaptation.

As biologists focused on microevolutionary speciation research, we chose topics that we believe are on the forefront of the field—topics that have numerous unanswered questions and plenty of current debate in the literature and, most importantly, those that we believe could benefit from our perspective. We also highlight our lack of discussion of "species concepts" and avoid the intersection of speciation, adaptation, and taxonomic classification. In particular, the application of genomic data to taxonomic questions has recently reinvigorated debate about their role in "species delimitation" (e.g., Cadena and Zapata 2021).

We first focus on the most basic unit of the speciation process: reproductive isolating barriers. Unlike *Mimulus* flowers or *Drosophila* fruit flies, bird species are not particularly well-suited to experimental manipulations that tease apart specific isolating barriers. However, birds are unique among taxonomic groups in that large data sets of phylogenetic relationships (Jetz et al. 2012), morphology (Pigot et al. 2020), color (Cooney et al. 2019), and ecological traits (Wilman et al. 2014, Tobias et al. 2022) have been compiled for most of the 10,000 avian species, and enormous citizen science databases are available (Sullivan et al. 2009), enabling unprecedented comparative analyses of macroevolutionary and macroecological processes

David P. L. Toews and Elizabeth S. C. Scordato, *Speciation and Adaptation in Birds*. In: *New Perspectives in Ornithology*. Edited by: Scott V. Edwards and J. Michael Reed, Oxford University Press. © Oxford University Press (2025). DOI: 10.1093/oso/9780197787670.003.0009

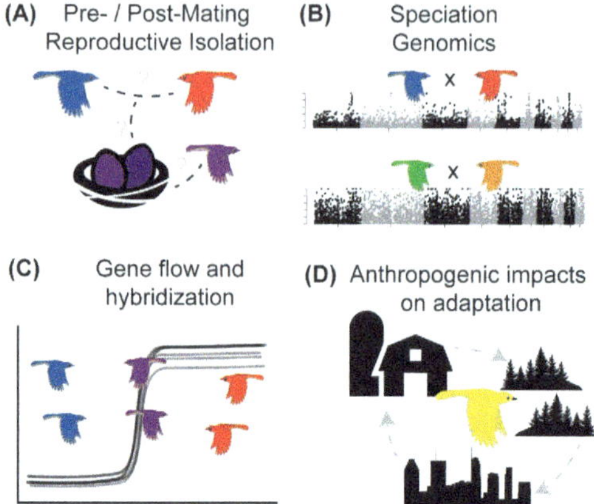

Figure 9.1 Illustration of the four main topics discussed in this chapter. The question marks in panel A represent some of the key questions regarding the outcomes of premating/postmating reproductive isolation.

and providing novel insights into the progression of the speciation process. Thus, as Ernst Mayr emphasized over his long career, deep knowledge of the natural history and geographic variation in bird species can overcome the drawbacks of avian systems—most that are not amenable to experimentation—to reveal key insights into the process of speciation.

Pre- and Postmating Reproductive Isolation in Birds

Despite vigorous debates in the literature about the prevalence of speciation with gene flow in various taxonomic groups (e.g., Smadja and Butlin 2011, Kopp et al. 2018), it is now widely recognized that most "speciation cycles" in birds begin when populations differentiate in allopatry, and they are completed when these populations maintain reproductive isolation on secondary contact (Price et al. 2014, Tobias et al. 2020). Microevolutionary speciation research in birds has therefore focused largely on uncovering the processes driving the formation, maintenance, and erosion of reproductive isolation between lineages. Reproductive barriers fall into two broad categories: premating barriers, including behavioral, ecological, and temporal isolation; and postmating barriers, including both intrinsic and extrinsic selection against hybrids.

Studies of reproductive isolation in birds have traditionally emphasized the importance of premating barriers. Extensive evidence shows that populations that differ in song, plumage, and mating display traits often exhibit preferences for their own phenotype over heterotypic phenotypes (e.g., Irwin et al. 2001, Uy et al. 2009, Turbek et al. 2021; reviewed in Uy et al. 2018). These results have led to the widespread conclusion that in birds, premating barriers, particularly those linked to mating traits, are more important to the early stages of speciation than postmating barriers (e.g., Edwards et al. 2005, Price 2008). This interpretation is bolstered by phylogenetic comparative studies showing that putatively sexually selected characters can evolve comparatively rapidly (e.g., Seddon et al. 2013, Drury et al. 2018, Cooney et al. 2019) and by studies showing that intrinsic genetic incompatibilities evolve very slowly (Price and Bouvier 2002). However, several caveats have led to recent re-examination of the relative importance of premating isolation, and particularly premating isolation due to divergent sexual selection, to speciation in birds. These are (1) limited associations between behavioral measures of premating isolation and realized assortative mating and gene flow in the wild; (2) a lack of concurrent assessment of premating isolation due to sexual versus ecological differentiation; and (3) limited empirical data on the relative importance of pre- versus postmating barriers to reproductive isolation, particularly with respect to the prevalence of weak extrinsic selection against hybrids.

Most measures of premating isolation in wild birds derive from measurement of male response to simulated male territorial intruders (e.g., through playback or mount experiments) rather than measures of female preference or tests of assortative mating and gene flow. This focus on male behavior occurs because females are often cryptic and less likely to respond to playbacks (but see, e.g., Seddon and Tobias 2007, 2010) and because it has historically been challenging to collect large-scale data on gene flow in wild populations. Because females are predicted to be the more discriminatory sex (Andersson 1994, Price 2008), it is typically assumed that if males discriminate between conspecifics and heterospecifics in the context of territory defense, females are also likely to do so in the context of mating.

Of the few studies that test discrimination of the same signals in both males and females, most do show congruence between responses, with females performing more copulation solicitations or showing lower response latencies toward conspecific versus heterospecific signals (reviewed in Uy et al. 2018). However, very few studies have directly correlated the strength of behavioral isolation measured from playback experiments or preference tests with realized measures of mate choice and gene flow in the wild (notable exceptions include the well-studied pied and collared flycatcher hybrid zones in Sweden, reviewed in Qvarnstrom et al. 2010). Furthermore, recent analyses indicate that the extent of divergence in sexually selected traits is often a poor predictor of assortative mating in hybrid zones (Hudson and Price 2014). We therefore have a limited understanding of the true strength of behavioral isolation due to mating signals in most avian systems. Part of this discrepancy may be that putative sexual signals such as olfactory cues (e.g., Mihailova et al. 2014, Huynh and Rice 2019) and female ornamentation (e.g., Lipshutz 2017, Riebel et al. 2019, Enbody et al. 2022) have historically been overlooked in studies of premating isolation. Further challenges may arise from inconsistent measurements of reproductive barriers and even varying conceptions of reproductive isolation (e.g., Westram et al. 2022). As genotyping large numbers of individuals becomes more tractable, future work that takes a more expansive view of mating traits and directly links rigorous mate choice or territorial intrusion assays to realized gene flow will provide considerable insight into the relative strength of sexually selected signals as premating reproductive barriers in wild populations.

A second challenge to our understanding of the importance of premating barriers to reproductive isolation in birds is that research has overwhelmingly focused on mating traits, and isolation due to sexual trait divergence and that due to ecological trait divergence are rarely measured in concert (Scordato et al. 2014). This may be in part because divergence in sexually selected traits such as plumage and song is relatively easy to measure, but detecting ecological divergence can be challenging, particularly if differentiation is subtle. However, exceptions occur in cases of discrete ecological differentiation, including specialization on distinct food resources and migratory divides. A classic example occurs in red crossbills (*Loxia curvirostra*), which exhibit distinct bill ecotypes that allow them to specialize on different types of pinecones (Benkman 2003). The different ecotypes flock assortatively on preferred food resources, produce divergent mating calls, and exhibit preferences for the calls of their own ecotypes (Snowberg and Benkman 2009, Smith et al. 2012). There is premating reproductive isolation due to habitat isolation, reduced immigrant fecundity, and behavioral isolation, with the relative strengths of these barriers varying depending on habitat, breeding period, and food availability (Porter and Benkman 2022).

Discrete ecological differentiation contributes to reproductive barriers early in the divergence process in another classic avian study system, the Darwin's finches. Disruptive selection in a population of medium ground finches (*Geospiza fortis*) on Santa Cruz island has led to a bimodal distribution in bill morphology (Hendry et al. 2006, 2009), assortative mating and reduced gene flow based on bill size (Huber et al. 2007, deLeón et al. 2010), and correlated divergence between bill morphology and song frequency (Huber and Podos 2006). In both Darwin's finches and crossbills, divergence in ecological traits and mating signals may thus synergistically contribute to premating isolation (Servedio et al. 2011).

Finally, migratory divides—regions where populations that exhibit different migratory behaviors breed in sympatry—are increasingly recognized as important reproductive barriers (Irwin et al. 2005, Turbek et al. 2018). Migratory divides can generate strong premating isolation if birds with different migratory strategies breed at different times (Bearhop et al. 2005, Rolshausen et al. 2009, Delmore et al. 2012, Ruegg et al. 2012, Delmore and Irwin 2014, Scordato et al. 2020), although it is important to note that no study has yet estimated the fitness costs of maladaptive hybridization within migratory divides.

The above case studies provide compelling evidence that ecological divergence can generate premating isolation via assortative mating, alone or in synergy with sexual selection. Nonetheless, these examples remain largely confined to cases of clear ecological differentiation. We still lack a more general understanding of the relative contributions of divergent ecological versus sexual traits to premating isolation in taxa that occupy less discrete ecological niches, although promising systems for future study abound, particularly in taxa distributed along ecological or elevational gradients (e.g., Zhen et al. 2017, Linck et al. 2020, Mikles et al. 2020).

Whereas microevolutionary studies have been slow to assess the relative contributions of ecological differentiation to reproductive isolation in secondary contact, recent macroevolutionary research provides insight into the roles that sexual and ecological trait divergence may play at various stages of the speciation process. For example, comparative analyses show that sexual signals can diverge more rapidly in allopatry than ecological traits (e.g., Drury et al. 2018), but transitions to sympatry occur faster in sister species pairs with greater differentiation in ecological traits (Pigot and Tobias 2013). Furthermore, a comparative analysis of more than 1,300 pairs of avian sister species showed that plumage sexual dichromatism is a good predictor of transitions from allopatry to parapatry: More dichromatic sister pairs are able to achieve range overlap more rapidly than less dichromatic pairs (Cooney et al. 2017). However, this result was confined to cases of limited parapatry (<20% range overlap), and sexual trait divergence was not associated with the large-scale transitions to sympatry necessary for the initiation of new speciation cycles. Together, these studies suggest that divergent sexual selection may be an important contributor to the maintenance of reproductive isolation at the early stages of secondary contact, but ecological differentiation is likely required to achieve extensive range overlap and kickstart subsequent speciation events (Cooney et al. 2017). Future work assessing the contributions of ecological and sexual traits to premating barriers in taxa with varying degrees of range overlap, and considering historical changes in distributions (e.g., Peñalba et al. 2019, Lombal et al. 2020), would help elucidate the microevolutionary mechanisms underlying these emergent macroevolutionary patterns. A recent example of such work occurs in Freeman et al. (2022), which found lower song discrimination in taxon pairs that interbreed in parapatry compared to pairs that do not interbreed in parapatry, implicating song discrimination in reproductive isolation. However, this study also found that faster evolution of song discrimination was not associated with higher speciation rates, suggesting that demographic, ecological, historical, or stochastic factors, rather than premating isolation, may better explain variation speciation rates.

Although the contribution of premating isolation to reproductive isolation has been more intensively studied, the role of postmating barriers in avian speciation is receiving renewed interest. It has long been suggested that because intrinsic postmating barriers evolve slowly (an average of 3–4 million years for F1 hybrid sterility to evolve in birds), speciation is often completed via the evolution of premating barriers before postmating isolation is complete (Price and Bouvier 2002). Yet, recent work on Amazonian taxa suggests that premating barriers may evolve slowly in the tropics, implicating an important role for intrinsic genetic incompatibilities in the maintenance of reproductive isolation in these groups (Pulido-Santacruz et al. 2018, Weir and Price 2019). These results raise the intriguing possibility that the relative importance of pre- versus postmating isolation may vary across different contexts in birds.

Beyond intrinsic incompatibilities, less dramatic forms of postmating isolation, such as weak extrinsic selection against hybrids, are poorly studied, potentially leading to a pervasive underestimation of the importance of postmating isolation to the maintenance of avian species boundaries. Recent modeling by Irwin (2020) underscores this point, indicating that premating isolation alone is generally insufficient to maintain species barriers in hybrid zones unless assortative mating is nearly complete. By contrast, a small (10%) reduction in hybrid fitness can maintain a stable hybrid zone. Although there are simplified assumptions about the nature of assortative mating in these models, particularly the assumption that hybrids prefer to mate with each other rather than with parentals, the results nonetheless point to a potentially critical role of moderately reduced hybrid fitness in the maintenance of avian species barriers. Future work assessing patterns of assortative mating and mate preferences in hybrid zones will yield a better understanding of the relative importance of pre- versus postmating barriers to the maintenance of avian species barriers, and may clarify currently tenuous links between the evolution of reproductive isolation and large-scale patterns of diversification (e.g., Rabosky and Matute 2013, Matute and Cooper 2021). In particular, the frequency of weak extrinsic selection against hybrids as well as hybrid mate preferences should be a major focus of future research. As discussed below, exciting advances in both genomic sequencing technology and individual-level tracking capabilities will facilitate better estimates of the extent of hybridization, mate preferences, and hybrid survival in wild populations.

Finally, it is important to remember that studies of reproductive barriers in secondary contact, although useful for identifying the factors currently maintaining species boundaries, are often not informative about the processes responsible for population divergence. The drivers of initial divergence in allopatry—the first step of the speciation cycle—remain largely understudied (Weir and Price 2019, Tobias et al. 2020; but see Scordato 2018). Debate thus continues over the relative importance of deterministic versus stochastic processes in generating the variation among populations necessary for reproductive barriers to evolve. We hope to see significant progress on this front during the coming years.

Speciation Genomics

Few areas in ornithology have made as rapid and significant progress as the application of genomic tools to questions about avian evolutionary history. In the context of speciation and adaptation, these tools have facilitated the identification of genes underlying species-defining traits, as well as genes involved in ecologically relevant phenotypes. For example, genome data have been used to directly link genetic variation at a single gene (*HGMA2*) in *Geospiza* finches to survival during a drought (Lamichhaney et al. 2015). Moreover, a host of molecular approaches have recently been used to study elevational variation in hummingbirds—both within (Lim et al. 2021) and between (Projecto-Garcia et al. 2013) species—to understand adaptation to reduced partial pressure of oxygen and temperature. Application of genomic data to questions about avian speciation has also resulted in significant debate in the community, particularly in the realms of how best to interpret patterns of divergence across the genome and the utility of genomic data in delimiting species boundaries.

Here, we focus primarily on understanding and interpreting genome scans in avian species. The first contours of an avian "landscape of divergence" were reported in a foundational paper by Ellegren et al. (2012), which quantified genomic differentiation—as measured by F_{ST}—between pied (*Ficedula hypoleuca*) and collared (*F. albicollis*) flycatchers. At the time, many in the community interpreted the pattern of heterogeneous divergence between these species—the hills and valleys punctuating the distribution of divergence—as mirroring the influence of gene flow between the two species. Given these flycatchers were known for their well-studied hybrid zone populations, it was intuitive to think that "speciation genes"—genes responsible for reproductive isolation and resistant to movement between species—were housed in the dozen or so F_{ST} "peaks" between them (i.e., regions of low gene flow). Indeed, previous studies of *Ficedula* had explicitly linked the Z chromosome to isolating barriers between the species (Sætre et al. 2003, Sæther et al. 2007). In divergence landscapes, however, although the Z chromosome had elevated differentiation (a finding since borne out in many other contrasts; Irwin 2018), there were no obvious "islands" of differentiation on the Z (Ellegren et al. 2012). More generally, there was some hope in the community that a straightforward application of genome scans might one day reveal the various "speciation genes" between numerous closely related taxa across the avian tree. However, several subsequent findings beguiled this narrative.

Although pied and collared flycatchers did hybridize (i.e., they presented the *possibility* for gene flow), subsequent studies of other highly divergent flycatchers (Burri et al. 2015), as well as other allopatric taxa with no obvious history of hybridization (Baiz et al. 2021), also showed heterogeneous patterns of divergence, and in some cases in the same regions of the genome as the pied/collared pair. We might expect "speciation genes" to be idiosyncratic among species and not so repeatable among independent groups, as presumably different traits would be under selection in different pairs. Thus, some other explanation is needed to explain this common architecture of divergence across independent taxa. Subsequently, a shared recombination landscape has emerged as the likely cause of common divergence patterns, where studies have found that chromosomal regions with reduced recombination frequency can amplify the effects of localized selection (Burri et al. 2015). The targets and causes of this selection are still unclear—and this is an active area of research—but what is clear is that in many cases divergence in F_{ST} peaks between closely related species does not automatically implicate variation in gene flow among different regions (Cruickshank and Hahn 2014).

At the same time, broad generalizations are challenging because there are several avian taxa that appear to have such low background divergence that F_{ST} peaks do likely demarcate genes involved in reproductive isolation. The best example of this is the *Sporophila* seedeaters of South America (Campagna et al. 2017, Turbek et al. 2021). With 10 predominantly sympatric species that differ from each other at a handful of genomic regions, reproductive isolation appears to be nearly complete, as mate choice experiments demonstrate (Turbek et al. 2021). Moreover, whole-genome analysis shows that those few, small chromosomal regions that do differ among species mostly contain genes involved in pigmentation, a finding that has been repeatedly borne out in other systems (Toews et al. 2016, Stryjewski and Sorenson 2017, Wang et al. 2020, Baiz et al. 2021).

What are we to make of the contrasting interpretations of divergence landscapes, such as those in flycatchers and seedeaters? What insight into the speciation process might we be able to provide in these different genomic patterns? First, we think it is helpful to qualitatively distinguish between several different patterns of divergence landscapes, which we can classify based on the chromosomal span of F_{ST} peaks: large ($\gg$1 Mb), moderate (~1 Mb), and small ($\ll$1 Mb). We have qualitatively binned a non-exhaustive list of various divergence landscapes that we have observed from several different avian systems in Table 9.1.

The group defined by the largest divergence peaks (Table 9.1, group A) consists of either fully allopatric taxa or taxa that are sympatric but are also strongly reproductively isolated. For example, the most well-studied in this group, *Ficedula* flycatchers, show a drastically reduced fitness for heterospecific pairings during hybridization

Table 9.1 Qualitatively Binned Nonexhaustive List of Various Divergence Landscapes Observed From Several Different Avian Systems

F_{ST} Peak	Reference
Group A: Large ($\gg$1 Mb)	
Acanthis spp.	Funk et al. (2021)
Catharus ustulatus ssp.	Delmore et al. (2015)
Ficedula albicollis and *F. hypoleuca*	Ellegren et al. (2012)
Hirundo rustica ssp.	Schield et al. (2021)
Poephila acuticauda	Hooper and Price (2017)
Phylloscopus trochiloides ssp.	Irwin et al. (2016)
Phylloscopus trochilus ssp.	Lundberg et al. (2017)
Selasphorus ssp.	Battey (2020)
Group B: Moderate (~1 Mb)	
Setophaga coronata ssp.	Brelsford et al. (2017)
Corvus corone ssp.	Poelstra et al. (2014)
Colaptes auratus ssp.	Aguillon et al. (2021)
Eopsaltria australis ssp.	Morales et al. (2018)
Motacilla alba ssp.	Semenov et al. (2021)
Setophaga coronata ssp.	Brelsford et al. (2017)
Group C: Few and Small ($\ll$1 Mb)	
Artamus spp.	Peñalba et al. (2022)
Lonchura spp.	Stryjewski and Sorenson (2017)
Sporophila spp.	Campagna et al. (2017)
Setophaga townsendi and *S. occidentalis*	Wang et al. (2020)
Vermivora chrysoptera and *V. cyanoptera*	Toews et al. (2016)

(i.e., very high postmating isolation; Sætre and Sæther 2010). We see pairs in group A as a type of genomic "control," illuminating the broad strokes of divergence during avian speciation, where similarities in the location of F_{ST} peaks represent a common recombination architecture, as mentioned previously.

In some cases in group A, however, such as between different migratory ecotypes of Willow Warbler, or plumage types of *Acanthis* redpolls, divergence is restricted to several exceptionally large regions, possibly indicative of chromosomal inversions (Lundberg et al. 2017, Funk et al. 2021). Inversions suppress recombination between the ancestral and inverted haplotypes, and they have been associated with species differences in non-avian clades such as *Drosophila* (Fuller et al. 2019). They have been increasingly observed in avian genomes, although whether they are a common mechanism for speciation in birds remains to be determined (Hooper and Price 2017).

The "moderate" group (Table 9.1, group B) presents with slightly more nuance. In this case, all the taxa form clearly demarcated hybrid zones and show low background genomic divergence. Admixture mapping has been used to associate species-defining phenotypes to particular F_{ST} peaks (e.g., Brelsford et al. 2017, Aguillon et al. 2021, Semenov et al. 2021), although we are not aware of any that have been directly linked to reproductive isolation per se. This is exemplified by studies of hybridizing myrtle and Audubon's warblers, where Brelsford et al. (2017) found that many F_{ST} peaks associated with color variation in several plumage phenotypes, although nearly as many did not. In these cases, the exceptionally low background divergence is likely a product of hybridization and gene flow, but divergence peaks are less predictable (i.e., fewer divergent regions are in the same chromosomal locations among the different pairs). Although many of these cases involved taxa with divergent plumage phenotypes, Morales et al. (2018) focus on a set of mitonuclear genes in the Eastern Yellow Robin in Australia—genes which could be involved in reproductive isolation. In this case, divergence is disproportionately concentrated in a region on chromosome 1A—which is enriched with many mitochondrial function genes—and appears as a "hot spot" of divergence between many avian taxa.

Finally, in Table 9.1, pairs in group C with "few and small" divergent regions are notable in two ways. First, they all occur within species "radiations," where the rate of species formation is much higher than the background tempo of diversification. Their occurrence in radiations might be a product of study bias—species radiations have long been attractive model systems—or could speak to broader, causal patterns (discussed below). Second, there is an obvious overrepresentation of genes involved in pigmentation development/deposition in the few peaks between species. Albeit within colorful radiations, the fact that these studies were not directly selecting coloration phenotypes for gene associations per se means that this overrepresentation of pigmentation genes is likely not a biased observation (which might have been a fair criticism if these studies were applying candidate gene approaches).

That the same genes have been the target of repeated divergence in independent species radiations (most notably *ASIP*, involved in melanogenesis, and *BCO2*, involved in carotenoid processing) also suggests the exciting possibility that these are not just speciation genes in the most traditional sense but also "diversification" genes: Evolution at these gene regions might be predictive of higher order patterns of species formation. Connections between genome evolution and macro-speciation patterns have been linked in cichlid fishes (Ronco et al. 2021), but correlating speciation rate and gene or genome evolution has not been attempted in any avian systems. Thus, we view this as an active area of research during the next decade.

A prescient quote from John Avise from the *Perspectives in Ornithology* volume from 1987 is a useful marker for progress in the field of avian speciation genomics (Barrowclough and Avise 1987). Avise, in describing warblers and finches that were "strikingly different in plumage coloration," (p. 266) surmised that "perhaps the ecological differences among these bird species stem from rather superficial genetic changes, involving relatively few genes" (267). Indeed, although our brief discussion of speciation genomics has featured a wide spectrum of divergence patterns—and the "rule" is likely that myriad genetic changes are often involved (i.e., Table 9.1, group A)—many avian species pairs can now be distinguished by very few genetic changes. This is, from our perspective, quite a remarkable result and—save for Avise's prediction—a rather unexpected finding.

Gene Flow and Hybridization

When the process of speciation is not complete, and two divergent taxa on semi-independent evolutionary trajectories come back into contact, hybridization and gene flow can occur (Short 1969, Grant and Grant 1992). This process has been most often studied in the context of contemporary gene flow in hybrid zones, where hybrids are formed with some frequency. Some iconic avian hybrid zones have been studied for decades, such as between the hooded and carrion crow, pied and collared flycatchers, and bullocks Icterus bullockii and Baltimore orioles I. galbula. In many cases, because reproductive isolation was *present* but not *complete*, the goals of studying these hybrid zones are to identify the barriers to gene flow that evolved early in the speciation process. What are the traits that are involved in maintaining at least some isolation between the groups? What are the genes involved in speciation?

As students of this field, we are actively asking these kinds of questions and using avian hybrid zones (as is now cliché) as "windows on the evolutionary process" (Harrison 1990). We also recognize that hybrid zones have been, and will continue to be, important natural laboratories of evolutionary biology. At the same time, we can now see the questions in our field moving beyond identifying specific barriers or genes involved in reproductive isolation to more general questions, in a "post-hybrid zone" research world. For instance, what is the timescale of gene flow? How often can we detect evidence of historical introgression between species that do not currently hybridize? How has reticulate evolution of specific genes or gene classes influenced patterns of higher level diversification, such as in adaptive radiations? In large part, we are only now positioned to address these questions because of the advancing tools associated with genomic analyses, where we are no longer reliant on contemporary hybrids and phenotypic evidence of introgression.

A notable recent application in which these kinds of data have been used to determine the timing, prevalence, and magnitude of introgression is in the Neotropical woodcreeper genus *Dendrocincla*. Pulido-Santacruz et al. (2020) generated genomic data for six species that diverged during the past 7 million years. Using a combination of phylogenetic dating and ABBA–BABA tests (a common metric to quantify ancient introgression between groups), they recovered at least five distinct introgression events. Cleverly, for those instances in which there was significant evidence of introgression, the authors compared the proportion of the genome introgressing to the age of the lineages at the estimated time of hybridization. In other words, they were able to ask, How long did gene flow last? At one extreme, they found evidence for an introgression event after approximately 2.5 million years of divergence, and in some comparisons up to 11% of the genome showed some evidence of gene flow. What is particularly notable is that the two subspecies in the study that currently form a hybrid zone (*D. fuliginosa atrirostris* and *D. f. rufoolivacea*, last sharing a common ancestor ~3.1 million years ago) did *not* show any appreciable gene movement from the introgression tests.

There has also been a new appreciation for certain classes of genes that show a disproportionate proclivity for introgressing among avian species. A key example of these "jumping genes" are the carotenoid processing genes, which are involved in the distribution and digestion of carotenoid molecules that many birds use to color their feathers with yellows, oranges, and reds. Whereas arguably one of the most celebrated examples of horizontal gene transfer involves the movement of carotenoid processing genes from fungi to herbivorous aphids (Moran and Jarvik 2010), the examples among vertebrates have been between much less divergent taxa.

First, among Darwin's finches, Enbody et al. (2021) identified a noncoding SNP in the beta carotene oxygenase 2 (*BCO2*) gene that was associated with pink versus yellow beak coloration in nestling *Geospiza* finches (i.e., a difference associated with less or more carotenoids, respectively). The polymorphism appeared to be at moderate frequencies across multiple finch species, and explicitly associated with the phenotype in several, implicating either a broadly balanced polymorphism or extensive ancient hybridization. Although gene flow at *BCO2* was not explicitly tested, there was some suggestion from a correlation between *BCO2* allele frequencies and hybridization timing that gene flow between *G. scandens* and *G. fortis* may have contributed to the gene's movement.

Second, Baiz et al. (2021) also found *BCO2* as a genomic outlier—and in this case explicitly quantified evidence of historical gene flow—across a clade of 34 New World warblers in the genus *Setophaga*. Baiz et al. (2021) showed that sequence variation at *BCO2* differed strongly from the species tree. This finding implicated gene flow among six species, all of which express extensive carotenoid-based coloration. Baiz et al. (2021) found that this introgressed region included a presumably functional amino acid substitution that is unique to these warblers, with the ancestral variant being highly conserved across other birds. Again, none of the species that show evidence of *BCO2* introgression are known to have formed contemporary hybrids or hybrid zones.

In his 1998 article titled "Hybridization in Birds," Frank Gill suggested that "we bear witness here to maturing ornithological studies of hybrids that mirror maturing theory and technology" (p. 282). At the time, that issue of *The Auk* had described a discovery of an intergeneric hybrid—dubbed a "wonder warbler" (Latta et al. 1998)—as well as a new hybrid zone between hermit and Townsend's warblers, which would eventually become a textbook example of a moving hybrid zone (Rohwer and Wood 1998). Today, we similarly see that studies of hybridization are continuing to generate insights into the process of avian speciation and adaptation. However, increasingly often, these findings are coming not from *geographic* regions of active hybridization but, rather, from *genomic* regions that show a propensity to introgress between species. This post-hybrid zone research avenue is making impressive progress. Much like the nuanced demographic history of ancient hybridization in the *Homo* lineage has revealed the tangled history of our ancestors, we believe a similarly complex story is told in the genomes of contemporary birds.

Anthropogenic Impacts on Adaptation

An exciting prospect of genetic and genomic tools is to study some of the earliest stages of divergence and adaptation. Prior studies have often relied on obvious phenotypic differences as a starting point to detect incipient speciation or adaptation (e.g., Campagna et al. 2012). However, we are now able to detect reductions in gene flow or regions of the genome putatively under selection in the absence of any clear phenotypic differentiation. Perhaps the most intriguing examples of recent divergence and adaptation come not from populations on pristine Andean peaks or far-flung islands but, rather, from cities, towns, and other heavily human-altered environments. As human activity spreads throughout the world, it fragments habitats, reorders ecological communities, and creates new environments with novel selective pressures. We would be remiss, in a chapter about speciation and adaptation in the 21st century, if we did not discuss the potential impact of human activity on the evolution, divergence, and adaptation of bird populations. We focus on recent work assessing evidence for divergence and adaptation between urban and non-urban populations, and we conclude by touching on the potential impact of long-term human activity on avian populations.

Considerable recent effort has been devoted to understanding the impact of human activity, and particularly urbanization, on evolutionary processes in wild populations (e.g., Johnson and Munshi-South 2017). Whereas many studies in the past two decades have focused on how urbanization reduces connectivity between "natural" populations, more recent research has begun to assess the genetic changes associated with colonization of urban environments. Such work offers insight into how quickly wild populations may be able to respond and adapt to ongoing (and intensifying) anthropogenic pressure. Perhaps unsurprisingly, studies of putatively neutral loci indicate that avian populations colonizing urban environments often exhibit evidence for rapid genetic bottlenecks and reductions in gene flow with non-urban populations. For example, in the early 2000s, dark-eyed juncos (*Junco hyemalis*) colonized the University of California, San Diego campus from nearby mountains and subsequently diverged at microsatellite loci (Rasner et al. 2004). Burrowing owls (*Athene cunicularia*) have independently colonized cities in Argentina and exhibit reduced gene flow with rural source populations (Mueller et al. 2018). Colonization of cities also occurred independently and is associated with genetic divergence from rural populations in European blackbirds (*Turdus merula*; Evans et al. 2009) and great tits (*Parus major*; Salmón et al. 2021).

Colonization of urban environments is often associated with phenotypic as well as genetic divergence from rural source populations, suggesting urban birds may be adapting to city life. Common phenotypic changes associated with urbanization include reduced migratory propensity (Evans et al. 2012, Greig et al. 2017), higher frequency songs (likely for enhanced transmission in areas with low-frequency road noise; Potvin et al. 2011, Derryberry et al. 2021), and changes in life history traits (Sepp et al. 2018), relative to non-urban counterparts. Perhaps the best studied "urban phenotype" in birds is boldness, wherein urban birds exhibit increased exploratory behavior and decreased risk avoidance compared to rural birds. Two candidate loci have been extensively studied in association with boldness: *DRD4*, which is linked to wariness, and *SERT*, which is associated with harm avoidance behavior. Polymorphisms at these loci are correlated with behavioral measures of boldness in a variety of species, including great tits (*P. major*; Fidler et al. 2007), blackbirds (*T. merula*; Mueller et al. 2013), black swans (*Cygnus atratus*; vanDongen et al. 2015), and dunnocks (*Prunella modularis*; Holtmann et al. 2016). These results are compelling and suggest that the boldness phenotype may be under selection in urban environments; however, variation at candidate boldness loci has not yet been linked to fitness (Lambert et al. 2021), leaving open an important area of future work.

Genome scans comparing pairs of urban and rural populations are also offering insight into candidate loci potentially associated with urban adaptations. For example, gene ontology analyses of outlier loci in urban–rural comparisons of burrowing owls show enrichment for genes associated with synapses (Mueller et al. 2020), which the authors suggest might reflect different selective pressures on behavior in urban versus rural habitats. This analysis also found a strong association between the promoter region of *SERT* and urban versus rural populations, but it had low power to detect evidence of selection on this region. Another recent study of great tits, also using multiple pairs of urban and rural populations, found evidence for selective sweeps in the same genomic regions in independent urban populations (Salmón et al. 2021). This result could not be explained by gene flow and was better attributed to recurrent selection in urban populations. Future work assessing whether these "urban outlier" loci impact fitness will shed further light on how, in what genomic regions, and over what timescale selection operates in urban populations.

To date, most work on the impact of human activity on adaptation has focused explicitly on urbanization. However, urbanization is a relatively recent phenomenon, with the first urban areas emerging during the Industrial Revolution approximately 300 years ago. By contrast, rural areas, comprising small to midsize villages and small-scale agriculture, have persisted for more than 15,000 years, representing a widespread and long-term form of human landscape modification. Indeed, some estimates indicate that most of the global biosphere has been impacted by humans for the past 12,000 years (Ellis et al. 2021). Despite this long history of anthropogenic activity, few studies have assessed how non-urban habitats have shaped patterns of adaptation and population divergence in birds. The few that have done so, however, suggest that these impacts may be profound: A study of the evolutionary history of the house sparrow (*Passer domesticus*) suggests that this species diverged and spread throughout the world concomitant with the Neolithic agricultural expansion (Ravinet et al. 2018). Likewise, demographic reconstructions of barn swallows (*Hirundo rustica*) show recent population size expansions, again concomitant with the rise and spread of human agriculture (Smith et al. 2018). Future work that looks more broadly at the myriad, long-term ways that human activity has shaped the global biosphere will help us better understand, predict, and mitigate ongoing human impacts on wild populations.

Conclusion

Perhaps the greatest advance in the study of speciation and adaptation in birds during the past 20 years has been the use of next-generation sequencing to assess patterns of differentiation, adaptation, and gene flow in wild populations. The ability to interrogate thousands of markers across the genome has liberated studies of evolution from a handful of model organisms and allowed examination of broad-scale patterns of differentiation in diverse taxa. We have begun to elucidate the contours of the genome-wide divergence landscape, identify loci underpinning key phenotypes, and track the footprints of adaptation across the genome. We are gaining insight into the windows of time over which the speciation process proceeds and discovering the regions of the genome critical to its completion. Increased adoption of whole-genome sequencing and advances in long-read sequencing technology promise to provide even greater insight into the genomic architecture of divergence and adaptation, particularly via analyses of linked selection, assessment of variation in recombination landscapes, analyses of epigenetic variation, and identification of loci underlying complex traits.

Despite these advances, one of the great gaps in our understanding of adaptation and speciation in birds remains the dearth of data linking genomic and phenotypic variation to fitness in wild populations. This gap may be filled in part as sequencing costs continue to decrease and new tracking technologies emerge. For example, we envision studies of hybrid zones in which hundreds of individuals are genotyped and tracked in each generation to assess fine-scale extrinsic selection against hybrids and identify loci associated with fitness. Such studies will fill key gaps in our understanding of the contributions of pre- and postmating isolation to species barriers.

Nonetheless, we caution that technology is not a silver bullet destined to replace the hard, muddy work of natural history. We cannot imagine a future in which boots and binoculars are entirely supplanted by molecular labs and remote tracking; rather, we are excited about the novel ways in which technology can be leveraged to provide further insight into the evolutionary phenomena uncovered by careful field studies. We need look no further than the new insights offered by old museum collections to see the continuing power of natural history, as ancient DNA analysis allows us to identify changes in population size and allele frequencies that would be otherwise impossible to detect. We thus end this genomics- and technology-heavy chapter with a call for continued, and even renewed, interest in natural history. Fieldwork can be painstaking and slow, but it is necessary to contextualize our genomic analyses and ultimately identify the variables shaping fitness in wild populations. G. E. Hutchinson wrote in 1975 that a

> modern biological education may let us down as ecologists if it does not insist . . . that a wide and quite deep understanding of organisms, past and present, is as basic a requirement as anything else. It may be best self-taught, but how often is this difficult process made harder by a misplaced emphasis on a quite specious modernity. (p. 516)

In a world under increasing threat, ornithologists with detailed understanding of their study organisms will need, more than ever, to leverage their knowledge for the preservation of the natural world.

References

Aguillon, S. M., J. Walsh, and I. J. Lovette. 2021. Extensive hybridization reveals multiple coloration genes underlying a complex plumage phenotype. *Proceedings of the Royal Society B: Biological Sciences* 288: Article 20201805.

Andersson, M. 1994. *Sexual Selection*. Princeton University Press, Princeton, NJ.

Baiz, M. D., A. W. Wood, A. Brelsford, I. J. Lovette, and D. P. L. Toews. 2021. Pigmentation genes show evidence of repeated divergence and multiple bouts of introgression in *Setophaga* warblers. *Current Biology* 31:643–649.

Barrowclough, G. F., and J. Avise. 1987. Biochemical studies of microevolutionary processes. Pages 223–270 in A. H. Brush and G. A. Clark, Jr., eds. *Perspectives in Ornithology: Essays Presented for the Centennial of the American Ornithologists' Union.* Cambridge University Press, Cambridge, UK.

Battey, C. J. 2020. Evidence of linked selection on the Z chromosome of hybridizing hummingbirds. *Evolution* 74:725–739.

Bearhop, S., W. Fiedler, R. W. Furness, S. C. Votier, S. Waldron, J. Newton, G. J. Bowen, P. Berthold, and K. Farnsworth. 2005. Assortative mating as a mechanism for rapid evolution of a migratory divide. *Science* 310:502–504.

Benkman, C. W. 2003. Divergent selection drives the adaptive radiation of crossbills. *Evolution* 57:1176–1181.

Brelsford, A., D. P. L. Toews, and D. E. Irwin. 2017. Admixture mapping in a hybrid zone reveals loci associated with avian feather coloration. *Proceedings of the Royal Society B: Biological Sciences* 284: Article 20171106.

Burri, R., A. Nater, T. Kawakami, C. F. Mugal, P. I. Olason, L. Smeds, A. Suh, L. Dutoit, S. Bures, L. Z. Garamszegi, S. Hogner, J. Moreno, A. Qvarnström, M. Ružić, S. A. Saether, G.-P. Saetre, J. Török, and H. Ellegren. 2015. Linked selection and recombination rate variation drive the evolution of the genomic landscape of differentiation across the speciation continuum of *Ficedula* flycatchers. *Genome Research* 25:1656–1665.

Cadena, C. D., and F. Zapata. 2021. The genomic revolution and species delimitation in birds (and other organisms): Why phenotypes should not be overlooked. *Ornithology* 138:1–18.

Campagna, L., P. Benites, S. C. Lougheed, D. A. Lijtmaer, A. S. D. Giacomo, M. D. Eaton, and P. L. Tubaro. 2012. Rapid phenotypic evolution during incipient speciation in a continental avian radiation. *Proceedings of the Royal Society B: Biological Sciences* 279:1847–1856.

Campagna, L., M. Repenning, L. F. Silveira, C. S. Fontana, P. L. Tubaro, and I. J. Lovette. 2017. Repeated divergent selection on pigmentation genes in a rapid finch radiation. *Science Advances* 3: Article e1602404.

Cooney, C. R., J. A. Tobias, J. T. Weir, C. A. Botero, and N. Seddon. 2017. Sexual selection, speciation and constraints on geographical range overlap in birds. *Ecology Letters* 20:863–871.

Cooney, C. R., Z. K. Varley, L. O. Nouri, C. J. A. Moody, M. D. Jardine, and G. H. Thomas. 2019. Sexual selection predicts the rate and direction of colour divergence in a large avian radiation. *Nature Communications* 10: 1773.

Cruickshank, T. E., and M. W. Hahn. 2014. Reanalysis suggests that genomic islands of speciation are due to reduced diversity, not reduced gene flow. *Molecular Ecology* 23:3133–3157.

deLeón, L. F., E. Bermingham, J. Podos, and A. P. Hendry. 2010. Divergence with gene flow as facilitated by ecological differences: Within-island variation in Darwin's finches. *Philosophical Transactions of the Royal Society B: Biological Sciences* 365:1041–1052.

Delmore, K. E., J. W. Fox, and D. E. Irwin. 2012. Dramatic intraspecific differences in migratory routes, stopover sites and wintering areas, revealed using light-level geolocators. *Proceedings of the Royal Society B: Biological Sciences* 279:4582–4589.

Delmore, K. E., S. Hübner, N. C. Kane, R. Schuster, R. L. Andrew, F. Câmara, R. Guigõ, and D. E. Irwin. 2015. Genomic analysis of a migratory divide reveals candidate genes for migration and implicates selective sweeps in generating islands of differentiation. *Molecular Ecology* 24:1873–1888.

Delmore, K. E., and D. E. Irwin. 2014. Hybrid songbirds employ intermediate routes in a migratory divide. *Ecology Letters* 17:1211–1218.

Derryberry, E. P., J. N. Phillips, G. E. Derryberry, M. J. Blum, and D. Luther. 2021. Singing in a silent spring: Birds respond to a half-century soundscape reversion during the COVID-19 shutdown. *Science* 370:575–579.

Drury, J. P., J. A. Tobias, K. J. Burns, N. A. Mason, A. J. Shultz, and H. Morlon. 2018. Contrasting impacts of competition on ecological and social trait evolution in songbirds. *PLoS Biology* 16: Article e2003563.

Edwards, S. V., S. B. Kingan, J. D. Calkins, C. N. Balakrishnan, W. B. Jennings, W. J. Swanson, and M. D. Sorenson. 2005. Speciation in birds: Genes, geography, and sexual selection. *Proceedings of the National Academy of Sciences of the USA* 102:6550–6557.

Ellegren, H., L. Smeds, R. Burri, P. I. Olason, N. Backström, T. Kawakami, A. Künstner, H. Mäkinen, K. Nadachowska-Brzyska, A. Qvarnström, S. Uebbing, and J. B. W. Wolf. 2012. The genomic landscape of species divergence in *Ficedula* flycatchers. *Nature* 491:756–760.

Ellis, E. C., N. Gauthier, K. K. Goldewijk, R. B. Bird, N. Boivin, S. Díaz, D. Q. Fuller, J. L. Gill, J. O. Kaplan, N. Kingston, H. Locke, C. N. H. McMichael, D. Ranco, T. C. Rick, M. R. Shaw, L. Stephens, J.-C. Svenning, and J. E. M. Watson. 2021. People have shaped most of terrestrial nature for at least 12,000 years. *Proceedings of the National Academy of Sciences of the USA* 118: Article e2023483118.

Enbody, E. D., S. Y. W. Sin, J. Boersma, S. V. Edwards, S. Ketaloya, H. Schwabl, M. S. Webster, and J. Karubian. 2022. The evolutionary history and mechanistic basis of female ornamentation in a tropical songbird. *Evolution* 76:1720–1736.

Enbody, E. D., C. G. Sprehn, A. Abzhanov, H. Bi, M. P. Dobreva, O. G. Osborne, C.-J. Rubin, P. R. Grant, B. R. Grant, and L. Andersson. 2021. A multispecies BCO2 beak color polymorphism in the Darwin's finch radiation. *Current Biology* 31:5597–5604.

Evans, K. L., K. J. Gaston, A. C. Frantz, M. Simeoni, S. P. Sharp, A. McGowan, D. A. Dawson, K. Walasz, J. Partecke, T. Burke, and B. J. Hatchwell. 2009. Independent colonization of multiple urban centres by a formerly forest specialist bird species. *Proceedings of the Royal Society B: Biological Sciences* 276:2403–2410.

Evans, K. L., J. Newton, K. J. Gaston, S. P. Sharp, A. McGowan, and B. J. Hatchwell. 2012. Colonisation of urban environments is associated with reduced migratory behaviour, facilitating divergence from ancestral populations. *Oikos* 121:634–640.

Fidler, A. E., K. v. Oers, P. J. Drent, S. Kuhn, J. C. Mueller, and B. Kempenaers. 2007. Drd4 gene polymorphisms are associated with personality variation in a passerine bird. *Proceedings of the Royal Society B: Biological Sciences* 274:1685–1691.

Freeman, B. G., J. Rolland, G. A. Montgomery, and D. Schluter. 2022. Faster evolution of a premating reproductive barrier is not associated with faster speciation rates in New World passerine birds. *Proceedings of the Royal Society B: Biological Sciences* 289: Article 20211514.

Fuller, Z. L., S. A. Koury, N. Phadnis, and S. W. Schaeffer. 2019. How chromosomal rearrangements shape adaptation and speciation: Case studies in *Drosophila pseudoobscura* and its sibling species *Drosophila persimilis*. *Molecular Ecology* 28:1283–1301.

Funk, E. R., N. A. Mason, S. Pálsson, T. Albrecht, J. A. Johnson, and S. A. Taylor. 2021. A supergene underlies linked variation in color and morphology in a Holarctic songbird. *Nature Communications* 12: Article 6833.

Gill, F. B. 1998. Hybridization in birds. *The Auk* 115:281–283.

Grant, P. R., and B. R. Grant. 1992. Hybridization of bird species. *Science* 256:193–197.

Greig, E. I., E. M. Wood, and D. N. Bonter. 2017. Winter range expansion of a hummingbird is associated with urbanization and supplementary feeding. *Proceedings of the Royal Society B: Biological Sciences* 284: Article 20170256.

Harrison, R. G. 1990. Hybrid zones: windows on evolutionary process. *Oxford Surveys in Evolutionary Biology* 7:69–128.

Hendry, A. P., P. R. Grant, B. R. Grant, H. A. Ford, M. J. Brewer, and J. Podos. 2006. Possible human impacts on adaptive radiation: Beak size bimodality in Darwin's finches. *Proceedings of the Royal Society B: Biological Sciences* 273:1887–1894.

Hendry, A. P., S. K. Huber, L. F. D. Len, A. Herrel, and J. Podos. 2009. Disruptive selection in a bimodal population of Darwin's finches. *Proceedings of the Royal Society B: Biological Sciences* 276:753–759.

Holtmann, B., S. Grosser, M. Lagisz, S. L. Johnson, E. S. A. Santos, C. E. Lara, B. C. Robertson, and S. Nakagawa. 2016. Population differentiation and behavioural association of the two "personality" genes DRD4 and SERT in dunnocks (*Prunella modularis*). *Molecular Ecology* 25:706–722.

Hooper, D. M., and T. D. Price. 2017. Chromosomal inversion differences correlate with range overlap in passerine birds. *Nature Ecology & Evolution* 1:1526–1534.

Huber, S. K., L. F. D. Len, A. P. Hendry, E. Bermingham, and J. Podos. 2007. Reproductive isolation of sympatric morphs in a population of Darwin's finches. *Proceedings of the Royal Society B: Biological Sciences* 274:1709–1714.

Huber, S. K., and J. Podos. 2006. Beak morphology and song features covary in a population of Darwin's finches (*Geospiza fortis*). *Biological Journal of the Linnean Society* 88:489–498.

Hudson, E. J., and T. D. Price. 2014. Pervasive reinforcement and the role of sexual selection in biological speciation. *Journal of Heredity* 105:821–833.

Hutchinson, G. E. 1975. Variations on a theme by Robert MacArthur. Pages 492–521 in M. L. Cody and J. M. Diamond, eds. Ecology and Evolution of Communities. Belknap Press–Harvard University Press, Cambridge, MA.

Huynh, A. V., and A. M. Rice. 2019. Conspecific olfactory preferences and interspecific divergence in odor cues in a chickadee hybrid zone. *Ecology and Evolution* 9:9671–9683.

Irwin, D., S. Bensch, J. Irwin, and T. Price. 2005. Speciation by distance in a ring species. *Science* 307:414–416.

Irwin, D., S. Bensch, and T. Price. 2001. Speciation in a ring. *Nature* 409:333–337.

Irwin, D. E. 2018. Sex chromosomes and speciation in birds and other ZW systems. *Molecular Ecology* 27:3831–3851.

Irwin, D. E. 2020. Assortative mating in hybrid zones is remarkably ineffective in promoting speciation. *The American Naturalist* 195:E150–E167.

Irwin, D. E., M. Alcaide, K. E. Delmore, J. H. Irwin, and G. L. Owens. 2016. Recurrent selection explains parallel evolution of genomic regions of high relative but low absolute differentiation in a ring species. *Molecular Ecology* 25:4488–4507.

Jetz, W., G. H. Thomas, J. B. Joy, K. Hartmann, and A. O. Mooers. 2012. The global diversity of birds in space and time. *Nature* 491:444–448.

Johnson, M. T. J., and J. Munshi-South. 2017. Evolution of life in urban environments. *Science* 358: Article eaam8327.

Kopp, M., M. R. Servedio, T. C. Mendelson, R. J. Safran, R. L. Rodríguez, M. E. Hauber, E. C. Scordato, L. B. Symes, C. N. Balakrishnan, D. M. Zonana, and G. S. V. Doorn. 2018. Mechanisms of assortative mating in speciation with gene flow: Connecting theory and empirical research. *The American Naturalist* 191:1–20.

Lambert, M. R., K. I. Brans, S. D. Roches, C. M. Donihue, and S. E. Diamond. 2021. Adaptive evolution in cities: Progress and misconceptions. *Trends in Ecology & Evolution* 36:239–257.

Lamichhaney, S., J. Berglund, M. S. Almén, K. Maqbool, M. Grabherr, A. Martinez-Barrio, M. Promerová, C.-J. Rubin, C. Wang, and N. Zamani. 2015. Evolution of Darwin's finches and their beaks revealed by genome sequencing. *Nature* 518:371–375.

Latta, S. C., K. C. Parkes, and J. M. Wunderle. 1998. A new intrageneric Dendroica hybrid from Hispaniola. *The Auk* 115:533–537.

Lim, M. C. W., K. Bi, C. C. Witt, C. H. Graham, and L. M. Dávalos. 2021. Pervasive genomic signatures of local adaptation to altitude across highland specialist Andean hummingbird populations. *Journal of Heredity* 112:229–240.

Linck, E., B. G. Freeman, and J. P. Dumbacher. 2020. Speciation and gene flow across an elevational gradient in New Guinea kingfishers. *Journal of Evolutionary Biology* 33:1643–1652.

Lipshutz, S. E. 2017. Divergent competitive phenotypes between females of two sex-role-reversed species. *Behavioral Ecology and Sociobiology* 71: Article 106.

Lombal, A. J., J. E. O'Dwyer, V. Friesen, E. J. Woehler, and C. P. Burridge. 2020. Identifying mechanisms of genetic differentiation among populations in vagile species: Historical factors dominate genetic differentiation in seabirds. *Biological Reviews* 95:625–651.

Lundberg, M., M. Liedvogel, K. Larson, H. Sigeman, M. Grahn, A. Wright, S. Akesson, and S. Bensch. 2017. Genetic differences between willow warbler migratory phenotypes are few and cluster in large haplotype blocks. *Evolution Letters* 1:155–168.

Matute, D. R., and B. S. Cooper. 2021. Comparative studies on speciation: 30 years since Coyne and Orr. *Evolution* 75:764–778.

Mayr, E. 1987. Introduction. Pages 1–22 in A. H. Brush and G. A. Clark, Jr., eds. *Perspectives in Ornithology: Essays Presented for the Centennial of the American Ornithologists' Union.* Cambridge University Press, Cambridge, UK.

Mihailova, M., M. L. Berg, K. L. Buchanan, and A. T. D. Bennett. 2014. Odour-based discrimination of subspecies, species and sexes in an avian species complex, the crimson rosella. *Animal Behaviour* 95:155–164.
Mikles, C. S., S. M. Aguillon, Y. L. Chan, P. Arcese, P. M. Benham, I. J. Lovette, and J. Walsh. 2020. Genomic differentiation and local adaptation on a microgeographic scale in a resident songbird. *Molecular Ecology* 29:4295–4307.
Morales, H. E., A. Pavlova, N. Amos, R. Major, A. Kilian, C. Greening, and P. Sunnucks. 2018. Concordant divergence of mitogenomes and a mitonuclear gene cluster in bird lineages inhabiting different climates. *Nature Ecology and Evolution* 2:1258–1267.
Moran, N. A., and T. Jarvik. 2010. Lateral transfer of genes from fungi underlies carotenoid production in aphids. *Science* 328:624–627.
Mueller, J. C., M. Carrete, S. Boerno, H. Kuhl, J. L. Tella, and B. Kempenaers. 2020. Genes acting in synapses and neuron projections are early targets of selection during urban colonization. *Molecular Ecology* 29:3403–3412.
Mueller, J. C., H. Kuhl, S. Boerno, J. L. Tella, M. Carrete, and B. Kempenaers. 2018. Evolution of genomic variation in the burrowing owl in response to recent colonization of urban areas. *Proceedings of the Royal Society B: Biological Sciences* 285: Article 20180206.
Mueller, J. C., J. Partecke, B. J. Hatchwell, K. J. Gaston, and K. L. Evans. 2013. Candidate gene polymorphisms for behavioural adaptations during urbanization in blackbirds. *Molecular Ecology* 22:3629–3637.
Peñalba, J. V., L. Joseph, and C. Moritz. 2019. Current geography masks dynamic history of gene flow during speciation in northern Australian birds. *Molecular Ecology* 28:630–643.
Peñalba, J. V., J. L. Peters, and L. Joseph. 2022. Sustained plumage divergence despite weak genomic differentiation and broad sympatry in sister species of Australian woodswallows (*Artamus* spp.). *Molecular Ecology* 31:5060–5073.
Pigot, A. L., C. Sheard, E. T. Miller, T. P. Bregman, B. G. Freeman, U. Roll, N. Seddon, C. H. Trisos, B. C. Weeks, and J. A. Tobias. 2020. Macroevolutionary convergence connects morphological form to ecological function in birds. *Nature Ecology & Evolution* 4:230–239.
Pigot, A. L., and J. A. Tobias. 2013. Species interactions constrain geographic range expansion over evolutionary time. *Ecology Letters* 16:330–338.
Poelstra, J. W., N. Vijay, C. M. Bossu, H. Lantz, B. Ryll, I. Müller, V. Baglione, P. Unneberg, M. Wikelski, M. G. Grabherr, and J. B. W. Wolf. 2014. The genomic landscape underlying phenotypic integrity in the face of gene flow in crows. *Science* 344:1410–1414.
Porter, C. K., and C. W. Benkman. 2022. Performance trade-offs and resource availability drive variation in reproductive isolation between sympatrically diverging crossbills. *The American Naturalist* 199:362–379.
Potvin, D. A., K. M. Parris, and R. A. Mulder. 2011. Geographically pervasive effects of urban noise on frequency and syllable rate of songs and calls in silvereyes (*Zosterops lateralis*). *Proceedings of the Royal Society B: Biological Sciences* 278:2464–2469.
Price, T. 2008. *Speciation in Birds*. Roberts, Greenwood Village, CO.
Price, T., and M. M. Bouvier. 2002. The evolution of F1 postzygotic incompatibilities in birds. *Evolution* 56:2083–2089.
Price, T. D., D. M. Hooper, C. D. Buchanan, U. S. Johansson, D. T. Tietze, P. Alström, U. Olsson, M. Ghosh-Harihar, F. Ishtiaq, S. K. Gupta, J. Martens, B. Harr, P. Singh, and D. Mohan. 2014. Niche filling slows the diversification of Himalayan songbirds. *Nature* 509:222–225.
Projecto-Garcia, J., C. Natarajan, H. Moriyama, R. E. Weber, A. Fago, Z. A. Cheviron, R. Dudley, J. A. McGuire, C. C. Witt, and J. F. Storz. 2013. Repeated elevational transitions in hemoglobin function during the evolution of Andean hummingbirds. *Proceedings of the National Academy of Sciences of the USA* 110:20669–20674.
Pulido-Santacruz, P., A. Aleixo, and J. T. Weir. 2018. Morphologically cryptic Amazonian bird species pairs exhibit strong postzygotic reproductive isolation. *Proceedings of the Royal Society B: Biological Sciences* 285: Article 20172081.
Pulido-Santacruz, P., A. Aleixo, and J. T. Weir. 2020. Genomic data reveal a protracted window of introgression during the diversification of a neotropical woodcreeper radiation. *Evolution* 74:842–858.

Qvarnström, A., A. M. Rice, and H. Ellegren. 2010. Speciation in *Ficedula* flycatchers. *Philosophical Transactions of the Royal Society B: Biological Sciences* 365:1841–1852.

Rabosky, D. L., and D. R. Matute. 2013. Macroevolutionary speciation rates are decoupled from the evolution of intrinsic reproductive isolation in *Drosophila* and birds. *Proceedings of the National Academy of Sciences of the USA* 110:15354–15359.

Rasner, C. A., P. Yeh, L. S. Eggert, K. E. Hunt, D. S. Woodruff, and T. D. Price. 2004. Genetic and morphological evolution following a founder event in the dark-eyed junco, *Junco hyemalis thurberi*. *Molecular Ecology* 13:671–681.

Ravinet, M., T. O. Elgvin, C. Trier, M. Aliabadian, A. Gavrilov, and G.-P. Sætre. 2018. Signatures of human-commensalism in the house sparrow genome. *Proceedings of the Royal Society B: Biological Sciences* 285: Article 20181246.

Riebel, K., K. J. Odom, N. E. Langmore, and M. L. Hall. 2019. New insights from female bird song: Towards an integrated approach to studying male and female communication roles. *Biology Letters* 15: Article 20190059.

Rohwer, S., and C. Wood. 1998. Three hybrid zones between Hermit and Townsend's Warblers in Washington and Oregon. *The Auk* 115:284–310.

Rolshausen, G., G. Segelbacher, K. A. Hobson, and H. M. Schaefer. 2009. Contemporary evolution of reproductive isolation and phenotypic divergence in sympatry along a migratory divide. *Current Biology* 19:2097–2101.

Ronco, F., M. Matschiner, A. Böhne, A. Boila, H. H. Büscher, A. E. Taher, A. Indermaur, M. Malinsky, V. Ricci, A. Kahmen, S. Jentoft, and W. Salzburger. 2021. Drivers and dynamics of a massive adaptive radiation in cichlid fishes. *Nature* 589:76–81.

Ruegg, K., E. C. Anderson, and H. Slabbekoorn. 2012. Differences in timing of migration and response to sexual signalling drive asymmetric hybridization across a migratory divide. *Journal of Evolutionary Biology* 25:1741–1750.

Sæther, S. A., G.-P. Sætre, T. Borge, C. Wiley, N. Svedin, G. Andersson, T. Veen, J. Haavie, M. R. Servedio, S. Bures, M. Kral, M. B. Hjernquist, L. Gustafsson, J. Traeff, and A. Qvarnströem. 2007. Sex chromosome-linked species recognition and evolution of reproductive isolation in flycatchers. *Science* 318:95–97.

Sætre, G., T. Borge, K. Lindroos, J. Haavie, B. C. Sheldon, C. Primmer, and A. Syvnen. 2003. Sex chromosome evolution and speciation in *Ficedula* flycatchers. *Proceedings of the Royal Society B: Biological Sciences* 270:53–59.

Sætre, G.-P., and S. A. Sæther. 2010. Ecology and genetics of speciation in *Ficedula* flycatchers. *Molecular Ecology* 19:1091–1106.

Salmón, P., A. Jacobs, D. Ahrén, C. Biard, N. J. Dingemanse, D. M. Dominoni, B. Helm, M. Lundberg, J. C. Senar, P. Sprau, M. E. Visser, and C. Isaksson. 2021. Continent-wide genomic signatures of adaptation to urbanisation in a songbird across Europe. *Nature Communications* 12: Article 2983.

Schield, D. R., E. S. C. Scordato, C. C. R. Smith, J. K. Carter, S. I. Cherkaoui, S. Gombobaatar, S. Hajib, S. Hanane, A. K. Hund, K. Koyama, W. Liang, Y. Liu, N. Magri, A. Rubtsov, B. Sheta, S. P. Turbek, M. R. Wilkins, L. Yu, and R. J. Safran. 2021. Sex-linked genetic diversity and differentiation in a globally distributed avian species complex. *Molecular Ecology* 30:2313–2332.

Scordato, E. S. C. 2018. Male competition drives song divergence along an ecological gradient in an avian ring species. *Evolution* 72:2360–2377.

Scordato, E. S. C., C. C. R. Smith, G. A. Semenov, Y. Liu, M. R. Wilkins, W. Liang, A. Rubtsov, G. Sundev, K. Koyama, S. P. Turbek, M. B. Wunder, C. A. Stricker, and R. J. Safran. 2020. Migratory divides coincide with reproductive barriers across replicated avian hybrid zones above the Tibetan Plateau. *Ecology Letters* 23:231–241.

Scordato, E. S. C., L. B. Symes, T. C. Mendelson, and R. J. Safran. 2014. The role of ecology in speciation by sexual selection: A systematic empirical review. *Journal of Heredity* 105:782–794.

Seddon, N., C. A. Botero, J. A. Tobias, P. O. Dunn, H. E. A. MacGregor, D. R. Rubenstein, J. A. C. Uy, J. T. Weir, L. A. Whittingham, and R. J. Safran. 2013. Sexual selection accelerates signal evolution during speciation in birds. *Proceedings of the Royal Society B: Biological Sciences* 280: Article 20131065.

Seddon, N., and J. A. Tobias. 2007. Song divergence at the edge of Amazonia: An empirical test of the peripatric speciation model. *Biological Journal of the Linnean Society* 90:173–188.
Seddon, N., and J. A. Tobias. 2010. Character displacement from the receiver's perspective: Species and mate recognition despite convergent signals in suboscine birds. *Proceedings of the Royal Society B: Biological Sciences* 277:2475–2483.
Semenov, G. A., E. Linck, E. D. Enbody, R. B. Harris, D. R. Khaydarov, P. Alström, L. Andersson, and S. A. Taylor. 2021. Asymmetric introgression reveals the genetic architecture of a plumage trait. *Nature Communications* 12: Article 1019.
Sepp, T., K. J. McGraw, A. Kaasik, and M. Giraudeau. 2018. A review of urban impacts on avian life-history evolution: Does city living lead to slower pace of life? *Global Change Biology* 24:1452–1469.
Servedio, M. R., G. S. V. Doorn, M. Kopp, A. M. Frame, and P. Nosil. 2011. Magic traits in speciation: "Magic" but not rare? *Trends in Ecology & Evolution* 26:389–397.
Short, L. L. 1969. Taxonomic aspects of avian hybridization. *The Auk* 86:84–105.
Smadja, C. M., and R. K. Butlin. 2011. A framework for comparing processes of speciation in the presence of gene flow. *Molecular Ecology* 20:5123–5140.
Smith, C. C. R., S. M. Flaxman, E. S. C. Scordato, N. C. Kane, A. K. Hund, B. M. Sheta, and R. J. Safran. 2018. Demographic inference in barn swallows using whole-genome data shows signal for bottleneck and subspecies differentiation during the Holocene. *Molecular Ecology* 27:4200–4212.
Smith, J. W., S. M. Sjoberg, M. C. Mueller, and C. W. Benkman. 2012. Assortative flocking in crossbills and implications for ecological speciation. *Proceedings of the Royal Society B: Biological Sciences* 279:4223–4229.
Snowberg, L. K., and C. W. Benkman. 2009. Mate choice based on a key ecological performance trait. *Journal of Evolutionary Biology* 22:762–769.
Stryjewski, K., and M. D. Sorenson. 2017. Mosaic genome evolution in a recent and rapid avian radiation. *Nature Ecology and Evolution* 1:1912–1922.
Sullivan, B. L., C. L. Wood, M. J. Iliff, R. E. Bonney, D. Fink, and S. Kelling. 2009. eBird: A citizen-based bird observation network in the biological sciences. *Biological Conservation* 142:2282–2292.
Tobias, J. A., J. Ottenburghs, and A. L. Pigot. 2020. Avian diversity: Speciation, macroevolution, and ecological function. *Annual Review of Ecology, Evolution, and Systematics* 51:1–28.
Tobias, J. A., C. Sheard, A. L. Pigot, A. J. M. Devenish, J. Yang, F. Sayol, M. H. C. Neate-Cleg, N. Alioravainen, T. L. Weeks, R. A. Barber. *et al.* 2022. AVONET: morphological, ecological and geographical data for all birds. *Ecology Letters*: 25:581–597.
Toews, D. P. L., S. A. Taylor, R. Vallender, A. Brelsford, B. G. Butcher, P. W. Messer, and I. J. Lovette. 2016. Plumage genes and little else distinguish the genomes of hybridizing warblers. *Current Biology* 26:2313–2318.
Turbek, S. P., M. Browne, A. S. D. Giacomo, C. Kopuchian, W. M. Hochachka, C. Estalles, D. A. Lijtmaer, P. L. Tubaro, L. F. Silveira, I. J. Lovette, R. J. Safran, S. A. Taylor, and L. Campagna. 2021. Rapid speciation via the evolution of pre-mating isolation in the Iberá seedeater. *Science* 371: Article eabc0256.
Turbek, S. P., E. S. C. Scordato, and R. J. Safran. 2018. The role of seasonal migration in population divergence and reproductive isolation. *Trends in Ecology & Evolution* 33:164–175.
Uy, J. A. C., D. E. Irwin, and M. S. Webster. 2018. Behavioral isolation and incipient speciation in birds. *Annual Review of Ecology, Evolution, and Systematics* 49:1–24.
Uy, J. A. C., R. G. Moyle, C. E. Filardi, and Z. A. Cheviron. 2009. Difference in plumage color used in species recognition between incipient species is linked to a single amino acid substitution in the melanocortin-1 receptor. *The American Naturalist* 174:244–254.
vanDongen, W. F. D., R. W. Robinson, M. A. Weston, R. A. Mulder, and P.-J. Guay. 2015. Variation at the *DRD4* locus is associated with wariness and local site selection in urban black swans. *BMC Evolutionary Biology* 15: Article 253.
Wang, S., S. Rohwer, D. R. Zwaan, D. P. L. Toews, I. J. Lovette, J. Mackenzie, and D. Irwin. 2020. Selection on a small genomic region underpins differentiation in multiple color traits between two warbler species. *Evolution Letters* 4:502–515.

Weir, J. T., and T. D. Price. 2019. Song playbacks demonstrate slower evolution of song discrimination in birds from Amazonia than from temperate North America. *PLoS Biology* 17: Article e3000478.
Westram, A. M., S. Stankowski, P. Surendranadh, and N. Barton. 2022. What is reproductive isolation? *Journal of Evolutionary Biology* 35:1143–1164.
Wilman, H., J. Belmaker, J. Simpson, C. D. L. Rosa, M. M. Rivadeneira, and W. Jetz. 2014. EltonTraits 1.0: Species-level foraging attributes of the world's birds and mammals: Ecological Archives E095-178. *Ecology* 95:2027–2027.
Zhen, Y., R. J. Harrigan, K. C. Ruegg, E. C. Anderson, T. C. Ng, S. Lao, K. E. Lohmueller, and T. B. Smith. 2017. Genomic divergence across ecological gradients in the Central African rainforest songbird (*Andropadus virens*). *Molecular Ecology* 26:4966–4977.

10
Avian Diversity and Patterns of Diversification in South America

Luciano N. Naka, Victor Leandro-Silva, and Santiago Claramunt

The Neotropics represents one of the most wonderful laboratories of biological diversification on Earth. From a global perspective, its avian diversity and endemism are unparalleled: despite representing only approximately 13% of the planet's land area, the Neotropical avifauna includes a staggering 41% (4,279 species) of the World's avian diversity, 80% of which are endemic. These include nearly half of all extant avian families and genera, with at least twice as many unique genera than any other region (Table 10.1). Nearly 150 years ago, Alfred Wallace wrote "richness combined with isolation is the predominant feature of Neotropical zoology, and no other region can approach it in the number of its peculiar family and generic types" (Wallace 1876, p. 5). Most of this diversity resides in South America, often considered "the land of birds," where avian diversity and endemism have reached their highest splendor (Ricklefs 2002). But why does South America harbor such disproportionately high avian diversity and endemism? This question fascinated Humboldt, Darwin, and Wallace in the 19th century, and it remains a key theme of research today.

South America's overwhelming diversity and endemism likely come from both historical and ecological processes. History determined how lineages formed and diversified through time, whereas ecology defines how species persist and sort throughout the landscape (Kennedy et al. 2014, Smith et al. 2014). Much of South America's endemism comes from the nearly 40 Myr of isolation that this landmass endured as an island continent, which ended at the end of the Neogene with the formation of the Isthmus of Panama (Webb 1991). This land bridge resulted in the Great American Biotic Interchange (GABI), the largest intercontinental biotic exchange ever documented (G. Simpson 1980). Major geological and climatic shifts, such as tectonic movements, mountain uplift, and hydrological processes, created ample opportunities for speciation during geological timescales. The rise of the Andes completely reconfigured South America, affecting climatic patterns throughout the continent, reorganizing river drainages, and producing novel highland habitats and ample opportunities for speciation. Finally, Pleistocene glacial cycles likely produced biome expansions and retractions, creating cycles of isolation and dispersal that may have stimulated speciation (Haffer 1969).

The great avian diversity of South America reflects the extent and variety of the continent's landscapes (Figure 1). A unique feature of South America is the juxtaposition of the longest trans-tropical mountain chain in the world (the Andes) and the largest tract of continuous tropical forest (the Amazon) (Antonelli and Sanmartín 2011).

Luciano N. Naka, Victor Leandro-Silva, and Santiago Claramunt, *Avian Diversity and Patterns of Diversification in South America*. In: *New Perspectives in Ornithology*. Edited by: Scott V. Edwards and J. Michael Reed, Oxford University Press.
 DOI: 10.1093/oso/9780197787670.003.0010

Table 10.1 Number of Bird Species, Genera, Families, and Orders in the Different Zoogeographic Realms or Regions, Including Total Number and Endemic Taxa in Parentheses

Zoogeographic	No. of Taxa (Endemic)			
Realm or Region	**Species**	**Genera**	**Families**	**Orders**
World	10,307	2,320	244	37
Neotropics	4,279 (3,395)	1,106 (699)	126 (32)	29 (2)
South America	3,401 (2,582)	931 (641)	100 (25)	28 (2)
Central America	1,448 (575)	267 (81)	101 (2)	25 (0)
Caribbean	535 (230)	262 (43)	77 (6)	24 (0)
Nearctic	1,036 (88)	448 (24)	97 (0)	25 (0)
Palearctic	1,470 (147)	491 (14)	114 (1)	28 (0)
Afrotropical	2,225 (1,672)	589 (308)	118 (18)	31 (3)
Australasian	1,948 (1,437)	594 (302)	128 (36)	25 (0)
Oceania	516 (132)	221 (12)	68 (0)	20 (0)

Sources: Species list, taxonomy, and distribution maps were obtained from BirdLife International (2020). Shapefiles of zoogeographic regions were adapted from Dinerstein et al. (2017). Regional values and levels of endemism were obtained by intersecting both data sets, using the crop function in the raster R package (Hijmans 2022) and the distinct function in the tidyverse R package (Wickham et al. 2019).

The Andes and the Amazon are bounded by two corridors of dry environments: the South American Dry and Arid Diagonals, which are flanked by two highly diverse isolated tropical and subtropical humid forests (the Choco and the Atlantic Forest). South America also harbors four mega wetlands (Amazon floodplains, Pantanal, Beni, and the Colombian–Venezuelan Llanos), four major grassland ecosystems in the lowlands (Cerrado and Pampas) and the highlands (Puna and Paramo), and several cold and alpine ecosystems (Patagonian and high-Andean steppes and forests). The fact that the Andes traverse South America latitudinally, rather than longitudinally (like the Himalayas or the Alps), results in an increased number of altitudinal gradients in a variety of climatic zones, from the equatorial tropics to the Patagonian steppes, further providing opportunities for niche specialization. Therefore, the Neotropics has more climatic variation than any other region in the world, resulting in more biomes, more ecosystems, and ultimately more habitats available for birds. Each landscape contributes to the overall species diversity of the continent, delivering a unique share of highly specialized avian endemics. Massive mountain ranges and continental-size tropical forests, however, do not necessarily mean more species, as these could be occupied by wide-ranging broad-niche species. Although biome and ecosystem diversity can explain much of South American avian diversity, geographic barriers and species interactions likely played central roles in population isolation, speciation, and species coexistence.

South American birds have been the focus of evolutionary studies since Charles Darwin concluded his 5-year round-the-world trip on the *HMS Beagle* in 1836. Evolutionary models first proposed for Andean and Amazonian avian diversification have been exported into other regions, evidencing the importance of South American biogeography for evolutionary biology (Fjeldså 1995, Anthonny et al. 2007, Jackson and Austin 2010, Harcourt and Wood 2012). However, the complexity of the continent's history and geography (Bush 1994; Ribas, Maldonado Coelho, et al. 2012) and misrepresentation of diversity by current taxonomy (Cracraft et al. 2020) have hampered the finding of general common patterns of Neotropical avian diversification. This has compelled some authors to give up altogether on the search for unifying evolutionary models (Rull and Carnaval 2020). Although simple models cannot capture all the complexity of the diversification process of the South American avifauna, the integration of genomic, geological, ecological, and organismal perspectives can shed light on our understanding of avian diversification in the continent.

Here, we discuss key aspects of avian diversity in South America, including the ecological and historical factors behind its outstanding diversity and endemism. We outline the biogeographic history of South America throughout the Cenozoic, explore patterns of species diversity evaluating the role of ecology and geography on macroecological patterns, and revisit some of the evolutionary models that attempted to explain the outstanding diversity in the "land of birds."

Origins of the South American Avifauna

The discussion about the origins of the South American avifauna is partially intertwined with the debate about the biogeographical origin of modern birds. A southern origin of modern birds is supported by the many avian groups that are distributed across former Gondwanan continents, including South America, Africa, and Australia (Cracraft 2001, Ericson 2012). However, molecular divergence time estimates indicate that major clades are too young to be the result of Gondwana's breakup, and early fossils of many of these groups, even tropical ones, have been found in North America and Europe (Feduccia 2003, G. Mayr 2011). A recent biogeographic reconstruction integrating biogeographical information of nearly all extant bird families and multiple Holarctic early fossils recovered South America as the area of origin of modern birds and proposed a biogeographic scenario to reconcile divergence times and the history of intercontinental breakup and connections (Claramunt and Cracraft 2015). However, major challenges persist related to biases imposed by the predominantly northern fossil record versus the predominantly southern extant tropical avifaunas (Cracraft and Claramunt 2017, G. Mayr 2017).

Regardless of the ultimate origin of modern birds, the South American avifauna can be viewed as an assembly of ancient autochthonous lineages and more recent immigrant groups. Ancient groups originated during the Paleocene (56–66 Mya), when South America was connected to Australia through Antarctica (Claramunt and Cracraft 2015). Given the challenges in reconstructing the geographic origin of major avian clades, it is difficult to provide a definitive list of ancient South American lineages, but Rheiformes, Tinamiformes, Anseriformes, Gruiformes, Charadriiformes, Sphenisciformes, Procellariiformes, Cariamiformes, Falconiformes, Psittaciformes, and Passeriformes are all good candidates (Claramunt and Cracraft 2015). Some reconstructions suggested an Australian origin for Psittaciformes and Passeriformes

(Oliveros et al. 2019, Selvatti et al. 2022), but these were based on analyses that considered Australia and Zealandia as a single landmass. In any case, because South America and Australia were connected, distinguishing between the two in biogeographic reconstructions may be difficult, and their ancient avifaunas were probably highly related. This connection ended with the opening of the Tasman Gateway, which separated Australia from Antarctica approximately 50 Mya (Bijl et al. 2013), and resulted in the glaciation of Antarctica in the late Eocene (Carter et al. 2017). After this connection was lost, South America began a long period of isolation that resulted in the diversification of its endemic fauna (G. Simpson 1980, Ricklefs 2002).

The isolation of South America during the middle Cenozoic, however, may not have been absolute. Paleontological evidence and time trees of mammals suggest that iconic South American groups such as caviomorph rodents and platyrrhine monkeys separated from their African closest relatives by the middle Cenozoic, suggesting transatlantic dispersal events (Poux et al. 2006, Oliveira et al. 2009, Ezcurra and Agnolin 2012). Evidence from birds is less clear. Hoatzins (Opisthocomiformes) may represent one case of middle Cenozoic colonization: Fossils from Africa and Europe suggest that hoatzins may have arrived in South America by transatlantic dispersal (G. Mayr and de Pietri 2014). Another interesting case is that of *Donacobius* (Donacobiidae), whose closest extant relatives are in Africa (Barker 2004, Alström et al. 2006) and its ancestors may have dispersed to South America in the Miocene (Oliveros et al. 2019). Transatlantic dispersal may also explain the colonization of the Old World by some South American groups. For example, Psittaciformes and suboscine passerines may have dispersed from South America to Africa sometime in the Oligocene (Selvatti et al. 2015, 2016, 2022; Smith et al. 2022). Further paleontological and biogeographic studies will shed light on these allegedly middle Cenozoic connections between South America and Africa.

The "splendid isolation" of South America ended with the emergence of the Isthmus of Panama toward the end of the Cenozoic, initiating the GABI (Marshall 1988). Although the final closure of the isthmus likely occurred approximately 3 Mya (O'Dea et al. 2016), recent evidence indicates that most of central Panama was already above sea level and very close to South America by the early Miocene (~20 Mya; Montes et al. 2012). The interchange of mammalian faunas is well documented by a rich fossil record, but the interchange of bird faunas has only recently been analyzed with some detail using molecular phylogenies and ancestral biogeographic reconstructions (Barker 2007, Weir et al. 2009, Smith and Klicka 2010, Barker et al. 2015).

Some groups seem to have colonized South America before the completion of the isthmus, likely by flying across the remaining narrow seaway. Prominently, the tanagers (Thraupidae), one of the most diverse endemic Neotropical families, likely originated in North America and arrived in South America approximately 12 Mya (Barker et al. 2015). Other likely early arrivals include clades within woodpeckers (*Melanerpes*, 5 Mya; *Veniliornis*, 6 Mya; Shakya et al. 2017), swallows (Hirundinidae: *Notiochelidon* and allies, >12 Mya; Weir et al. 2009), thrushes (Turdidae: *Entomodestes* and *Cichlopsis* solitaries, 13–8 Mya; Weir et al. 2009), mockingbirds (Mimidae, 6.5 Mya), wrens (Troglodytidae, ~8 Mya; Weir et al. 2009), and two clades of blackbirds (Icteridae: *Cacicus* and *Psaracolius*, ~8 Mya; *Pseudoleistes* and allies, 6 Mya; Weir et al. 2009). After the completion of the isthmus, dispersal rates increased dramatically (Weir et al. 2009, Smith and Klicka 2010, Bacon et al. 2015, Barker et al. 2015). Some iconic elements of the Neotropical avifauna such as Cracidae and Momotidae may not have arrived in South America until the completion of the isthmus (Hosner et al. 2016, McCullough et al. 2019).

Patterns of Species Diversity

Patterns of South American avian diversity are spatially heterogeneous. Diversity hot spots include the eastern slope of the tropical Andes; the Amazonian lowlands; the Atlantic Forest; and some isolated mountains in the eastern Andes of Colombia and Venezuela, such as Santa Marta, Perijá, and the Cordillera de la Costa (see Figure 10.1). In general, current patterns of species richness correlate with contemporary climatic factors such as temperature and precipitation; warm and wet climates sustain the highest number of species (Hawkins et al. 2003, 2012). Climate-driven models, however, are limited because they can underestimate the importance of historical factors in shaping contemporary patterns, particularly among small-ranged endemics (Rahbek 2005). Such models can also be scale dependent (Chase and Leibold 2002, Gotelli et al. 2009). At coarser scales, for example, bird species richness is significantly higher in the Andes than in the Amazon, even though the local (alpha) avian diversity at any Andean site is nowhere close to that found in the lowland humid tropics (Herzog and Kattan 2011). For instance, a 1-degree cell in the Andean foothills may accumulate more species by including several altitudinal zones, spanning different climates and habitats. Similarly, cells located along ecotones, or across both sides of Amazonian rivers, may increase species counts without real increments in alpha diversity. Therefore, patterns of regional diversity seem to be best explained by topographic complexity and ecosystem diversity, reinforcing the idea that species richness is governed by a synergism between climate and topographic heterogeneity (Rahbek and Graves 2001).

Despite these caveats, patterns of species richness seem to fit well with the species richness–energy hypothesis, which postulates that more productive environments can sustain more individuals and therefore more species (Brown 1981, Wright 1983). Although more energy can sustain more individuals, it is less clear how this may translate into supporting more species (Currie et al. 2004, Wiens and Graham 2005). Tropical forests may accommodate more species by having more narrowly partitioned ecological niches or by allowing species to carve out new niches (E. Mayr 1969). This could be achieved by increasing vertical stratification, habitat specialization, and behavioral adaptation.

Even within apparently homogeneous lowland forests, subtle topographic differences can result in considerable variation in species distribution. In Amazonian upland terra firme forest, for example, micro-topographic changes (i.e., 50-m altitudinal variation) can create large hydro-edaphic gradients that separate well-drained clayed-dominated plateaus (90% clay and 10% sand) from soaked sandy-soiled stream beds (98% sand and 2% clay) (Guillaumet 1987, Castilho et al. 2006, Gerolamo et al. 2022). These gradients impact forest structure and tree species composition,

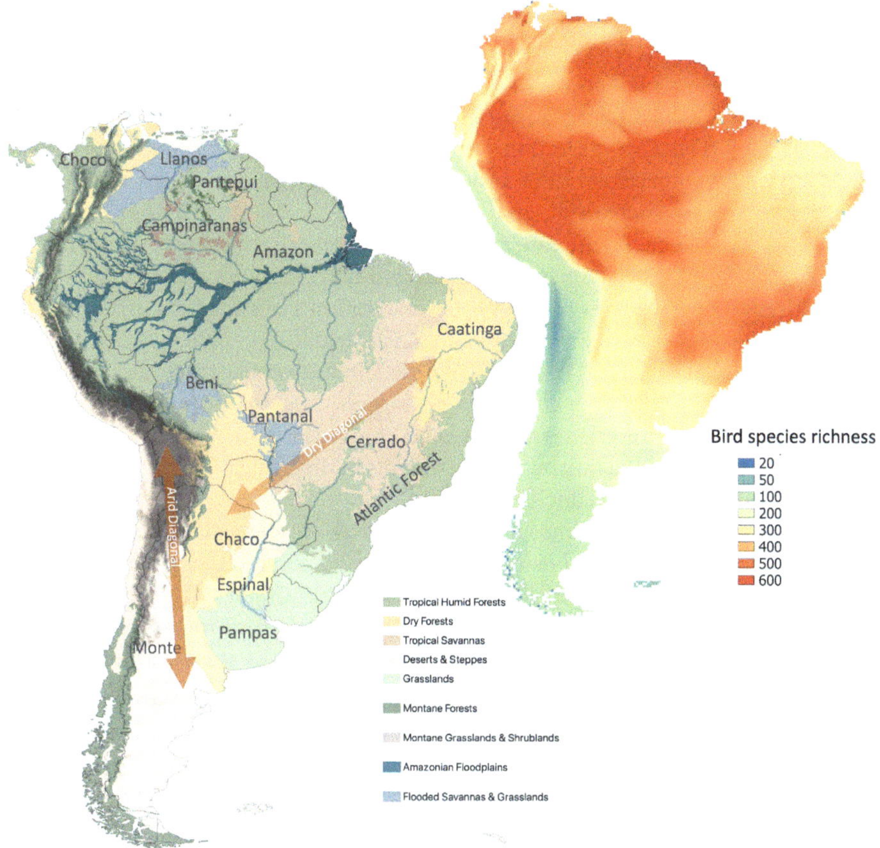

Figure 10.1 Major biomes and ecosystems of South America (left) and spatial distribution of bird species in the continent (right). Biomes and ecosystems were obtained from Olson et al. (2001). Spatial richness map is based on 3,401 individual shapefiles obtained from BirdLife International (2020). All shapefiles were overlaid into a 2.5 arc-minute grid of South America. Species richness for each cell was calculated using the sf R package (Pebesma 2018).

which directly affect avian distribution and occupancy patterns (Cintra and Naka 2012). Similarly, small altitudinal changes along Amazonian rivers may define the period of flooding along the gradient, creating a series of river-created habitats (transitional, riverine, and flooded forest, and sand-bar scrubs) in riparian environments, each with a different set of habitat specialists (Remsen and Parker 1983, Rosenberg 1990). Therefore, habitat and microhabitat ecological specialization represents a key factor in regional species diversity. Behavioral specialization, such as obligate army-ant followers, highly structured permanent mix-species flocks, or dead-leaf specialists, may also increase the number of species that coexist in a tropical forest (Klopfer and MacArthur 1961, Karr and James 1975).

General Patterns of Avian Diversification

At larger spatial and temporal scales, species richness may be more dependent on three fundamental historical processes: speciation, extinction, and dispersal (Ricklefs 1987, Wiens 2011). The interaction among these processes determines how lineages are distributed in space and time, producing the biodiversity patterns we see today (Ricklefs 1987, Goldberg et al. 2005, Wiens 2011, Weeks et al. 2016). The rate of lineage diversification, or the rate at which new lineages accumulate over time, is a key aspect of macroevolutionary dynamics and may be the cause of current diversity patterns (Mittelbach et al. 2007). Diversification rates may change over time, space, and among clades (Ricklefs 2003, 2006; Cardillo et al. 2005; Jetz et al. 2012; I. Quintero and Jetz 2018). In theory, by modeling diversification rates using fossils, clade diversity, and time-calibrated phylogenies, we can assess whether certain geographical regions and clades accumulated lineages faster than others (Helmstetter et al. 2022). However, the link between gradients of diversification rates and patterns of species richness remains unclear.

Terms such as "cradles," "museums," and "graves" of diversity have been coined to account for the spatial and temporal variation in speciation and extinction rates, which may result in differential diversification rates. *Cradles of diversity* refer to regions with exceptionally high rates of speciation that went through rapid accumulation of recent taxa, whereas *museums of diversity* refer to areas that accumulated old lineages through time, which have persisted in the landscape with relatively low extinction rates (Stebbins 1974, Gaston and Blackburn 1996). *Graves of diversity* refer to geographic regions that have endured relatively higher rates of extinctions through time (Rangel et al. 2018). The Amazon has been considered a museum of diversity by having a long history of species accumulation and relatively lower extinctions (Vargas and Dick 2020). The Andes, in contrast, are often considered both a cradle and a museum, combining frequent speciation events and old surviving lineages (Fjeldså 1995, Fjeldså et al. 2012, Rahbek et al. 2019, Perrigo et al. 2020).

Recent studies from the Andes suggest that young lineages arise more often above the tree line, whereas older species accumulate in more stable pockets of cloud forests at lower elevations (Sonne et al. 2022). At the continental scale, simulations suggested that cradles concentrate along Andean slopes, whereas graves cluster at lower elevations (Rangel et al. 2018). However, empiric data suggest that topographic complexity and climate along altitudinal gradients can both promote habitat fragmentation and provide climatic refugia, and biomes can act as both cradles and museums at the same time (Rahbek et al. 2019). It seems clear that conflicting processes could lead to similar patterns. For instance, according to Stebbins' (1974) definition, a cradle may represent an area of higher speciation or of higher species turnover, or it may even be created by differences in fossil preservation, less immigration, and different chances of origin of major clades (Vasconcelos et al. 2022). Furthermore, the terms cradle and museum seem to be related to historical and temporal climatic stability, which can be dynamic in space and time (Rangel et al. 2018). Therefore, it seems unwarranted to categorize general biomes into these terms, knowing that a given region can be a cradle during a given geological period and a museum in another.

Despite the recent interest in the role of diversification dynamics in species richness gradients, evidence from diverse groups, including birds (Jetz et al. 2012, Rabosky et al. 2015) and plants (Tietje et al. 2022), suggests that regions with higher species richness do not necessarily have higher diversification rates. Across the Americas, the highest speciation rates are found in cold, dry, and unstable areas, suggesting faster turnover and highest recent speciation and extinction rates in areas with low species diversity (Weir 2006, Weir and Schluter 2007, Harvey et al. 2020). Therefore, species seem to form faster in harsh environments but accumulate over time in more benign and stable environments, where extinction rates are lower (Pulido-Santacruz & Weir 2016). These results align with patterns of diversification through time in birds, in which speciation rates were higher during periods of global cooling (Claramunt and Cracraft 2015).

At the population level, avian phylogeographic diversity is greater in South America than in North America, which is attributed to lower population extinction rates in the tropics (Smith et al. 2017). Within South America, however, genetic diversity is higher in colder and drier environments and lower in tropical forests (Smith et al. 2017 Figure 2). These results are in agreement with those of a global meta-analysis that found higher avian subspecies richness in harsher environments (Botero et al. 2014). Finally, speciation rates have been associated with population genetic differentiation in New World birds, linking micro- and macroevolutionary patterns (Harvey et al. 2017).

Dispersal may also play an important role in the generation of diversity patterns. Even without in situ speciation, regions can accumulate species by receiving lineages from other regions. Due to niche conservatism, however, dispersal likely occurs within, rather than between, biomes. For example, phylogeographic evidence indicates historical avian exchange between the Amazon and the Atlantic Forest (Cracraft 1985, Batalha-Filho et al. 2013; Figure 10.2a), the Cerrado and northern South American savannas (da Silva 1995, van Els et al. 2021; Figure 10.2b), and the Chaco and the Caatinga (Short 1975, Prado 1992; Figure 10.2c). Data from plants suggest that lineages rarely colonize new biomes, even over long periods of macroevolutionary time (Crisp et al. 2009). Thus, niche conservatism, which is the tendency of lineages to retain their fundamental niche over time, may be key for the establishment of allopatric populations, not allowing species to venture into unsuitable regions (Wiens and Graham 2005).

Niche conservatism may vary among higher taxa (Peterson et al. 1999). For example, some groups, such as the antbirds (Thamnophilidae), have hardly ventured into cold biomes (Skutch 1996), whereas the closely related ovenbird radiation (Furnariidae) has occupied virtually every habitat in South America (Remsen 2003). Although expansions into novel environments may be rare, they may be key for ecological radiations and may have important consequences for patterns of species diversification. Therefore, characterizing the effect of dispersal across biomes and biogeographic regions may be important for understanding the generation of

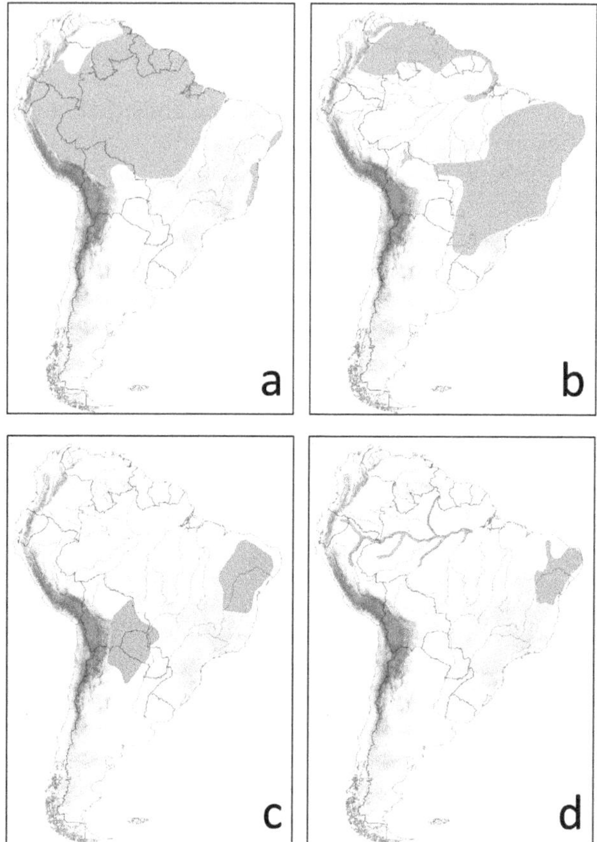

Figure 10.2 Examples of avian distributions depicting possible ancestral connections, including (a) the Screaming Piha (*Lipaugus vociferans*) occurring in the Amazon and the Atlantic Forest, (b) the Burnished-Buff Tanager (*Stilpnia cayana*) occupying the Cerrado and other open landscapes in northern South America, and (c) the Stripe-Backed Antbird (*Myrmorchilus strigilatus*) occurring in the Chaco and the Caatinga dry forests. (d) Distribution of the Lesser and Bahia Wagtail-Tyrants (*Stigmatura napensis* and *S. bahiae*) in the Amazonian floodplains and Caatinga Dry Forest, an example of niche conservatism across biomes.

diversity and endemism (Wiens and Donoghue 2004, Matos-Maraví et al. 2021, Vasconcelos et al. 2022). A recent analysis on the Caatinga dry forest avifauna from northeastern Brazil has shown that although some avian lineages likely originated in situ, most endemic lineages seemingly arrived from other open habitats (i.e., other dry forests or savannas), but a quarter of them likely colonized this dry region from adjacent humid forests (Lima et al. 2025).

Analyses of biogeographic origins and dispersal have revealed that the Amazon has been the primary source of biological diversity in South America, providing more funding lineages than any other biome across different clades (Antonelli et al. 2018). However, the colonization of a new biome may not always entail lack of niche conservatism, because lineages could potentially be searching for similar habitats or microhabitats, despite occurring in different biomes. For example, the Lesser Wagtail-Tyrant (*Stigmatura napensis*) and the Bahia Wagtail-Tyrant (*Stigmatura bahiae*), which are sometimes considered conspecific, inhabit shrubby vegetation, either early successional vegetation on river islands (*napensis*) or dry forest vegetation in the semi-arid interior of northeastern Brazil (*bahiae*) (Fitzpatrick et al. 2022; Figure 10.2d). Although the biomes and climatic envelopes of both allopatric species are very different, their local-scale habitats are quite similar, suggesting some degree of niche conservatism. Therefore, the next generation of interbiome biotic exchange analyses may benefit from accurate habitat definitions.

Biogeographic Models of Diversification

Our understanding of avian evolutionary processes in South America has made significant progress by taking a comparative approach. Although each lineage could have its own history, the aim of evolutionary biology is to find common models that apply to entire biotas, stimulating the search for universal explanations. Independent lineages with common distribution patterns abound in South America, likely due to the many hard boundaries (i.e., rivers, cordilleras, and valleys) that dissect the landscape. These common patterns were key for the delimitation of refugia, centers of dispersal, and areas of endemism (Haffer 1969, Muller 1973, Cracraft 1985; Figure 10.3a). These areas with multiple co-distributed taxa are generally interpreted as representing species that share common diversification processes (Rosen 1978, Cracraft 1985, Crother and Murray 2011). Bounded by common landscape features, species distributions were thought to be the result of a common history rather than the result of similar ecological requirements (Hovenkamp 1997). Prior to the expansion of molecular phylogenetics, biogeographers relied on area cladograms and raw species distributions to develop biogeographic models of diversification (Cracraft and Prum 1988, Brumfield and Capparella 1996, Bates et al. 1998). Once genetic data sets became available for multiple lineages, particularly from the South American tropical lowlands, it became clear that congruent distributions were masking different histories (Smith et al. 2014, Naka and Brumfield 2018). It is now believed that Neotropical biodiversity is not the product of a few historical events but, rather, the consequence of a protracted and complex history of events that interacted with evolutionary and ecological factors (Rull 2011; but also see Ribas & Aleixo 2019, Cracraft et al. 2020).

Various evolutionary models have been proposed to account for the diversification of South American birds, but two regions stand out for their global influence: the Andes and the Amazon. Given their continental extension and outstanding biodiversity, these two regions attracted the attention of early explorers, such as Alexander von Humboldt and Alfred Wallace, and their studies were key to the development of evolutionary hypotheses.

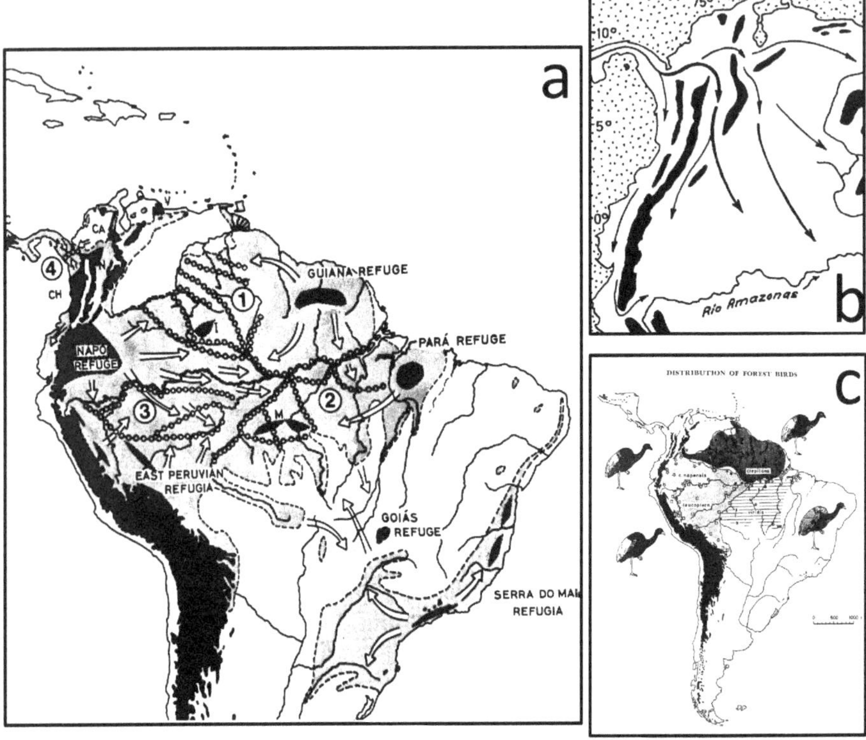

Figure 10.3 Haffer's views on speciation, dispersal, and species distributions in tropical South America. (a) Pleistocene forest refugia (dark shading) as centers of differentiation in Amazonian birds, areas of secondary contact (open circles), and forest expansion (arrows). (b) Allegedly dispersal paths of northern lowland faunas into South America during the uppermost Pliocene, prior to the final uplift of the northern Andes. (c) Distribution of the Trumpeters, *Psophia crepitans* superspecies, showing an Amazonian parapatric river-bounded distribution.

Plates obtained from Haffer (1974).

Speciation in the Andes

Building on Humboldt and Darwin's work, Frank Chapman was one of the first to recognize the role of the Andes in the diversification of birds (Chapman 1917, 1926). He argued that "a comparison of the bird-life of the Pacific coast of Colombia and northern Ecuador with that of the Tropical Zone at the eastern base of the Andes . . . induces the belief that we have here, in part, a pre-Andean fauna, the Pacific portion of which has been cut off from that of upper Amazonia by the Andean uplift" (Chapman 1917, p. 89).

Chapman provided a list of taxon replacements across the Andes and noted a differential effect of this barrier on species with different dispersal abilities (Kattan et al. 2016). An alternative scenario was offered by Haffer (1967a, 1967b, 1974), who postulated that trans- and cis-Andean disjunctions were the result of dispersal events across low passes around the northern end of the Colombian Andes prior the final rise of the Andes and during Quaternary glacial cycles (Figure 10.3b). Molecular data provided empirical evidence that the rise of the Andes likely divided ancestral populations. creating cis- and trans-Andean populations in a variety of lineages, including parrots (Ribas et al. 2005, 2007, 2012), toucans (Lutz et al. 2013), woodcreepers (Weir and Price 2011), and antbirds (Fernandes, Wink, et al. 2014). However, levels of genetic differentiation and inferred divergence times differed greatly across lineages, rejecting a single vicariant event or suggesting some post-uplift dispersal (Brumfield and Edwards 2007, Miller et al. 2008, Burney and Brumfield 2009, Smith et al. 2014).

Chapman (1917, 1926) also noted the importance of dispersal from other regions and postulated that novel Andean habitats were enriched by the influx of lineages from elsewhere tracking similar climates. Specifically, he argued that new montane environments were either occupied by lineages living immediately below the tropical/subtropical/Andean/Paramo altitudinal gradient or by birds from other regions with similar climates (Chapman 1917, p. 87). According to Chapman, whereas species from the Subtropical zone arrived from the Tropical lowlands, "nearly all those of the Paramo Zone have come from the sea-level equivalent of this zone in southern South America" (Chapman 1917, p. 88). Some of these hypotheses were corroborated, but others fared poorly under closer examination.

The hypothesis postulating that species from the Paramo derived from lineages that originated in southern South America has been at least partially supported by studies on *Muscisaxicola* ground-tyrants (Chesser 2000), *Cinclodes* furnariids (Chesser 2004, Sanín et al. 2009), and *Scytalopus* tapaculos (Cadena et al. 2020). A similar phenomenon (colonization of highlands from other areas) likely explains the presence of some Andean avian lineages in the rocky outcrops (*Campos rupestres*) of eastern Brazil in areas such as the Serra do Espinhaço or Serra Geral (Sick 1985). These include a species of *Cinclodes* (*Cinclodes pabsti*), two canasteros (*Asthenes luizae* and *A. moreirae*), and at least six species of *Scytalopus* tapaculos, all with close relatives in the Andes (Chaves et al. 2015; Figure 10.4).

On the other hand, patterns of altitudinal substitutions of closely related species along Andean altitudinal gradients remained contentious until the advent of molecular studies. Two alternative scenarios were postulated to explain these substitutions: (1) local adaptation and parapatric speciation along elevational gradients and (2) allopatric speciation and subsequent dispersal with competitive exclusion. Under the local adaptation hypothesis, parapatric speciation was the result of divergent environmental conditions along elevational gradients (Chapman 1917, Endler 1977). Another possibility is that species formed in allopatry, came into secondary contact after range expansion (E. Mayr 1969), and maintained their current ranges due to competition. The model of parapatric replacements should show a signature of sister relatedness, whereas non-monophyly or lack of nested populations along the gradient would imply that other mechanisms are at play, such as secondary contact. Molecular phylogenies showed that many montane taxa have closest relatives in the lowlands, but there is little evidence that they resulted from successive speciation due to local adaptation along altitudinal gradients (Bates and Zink 1994,

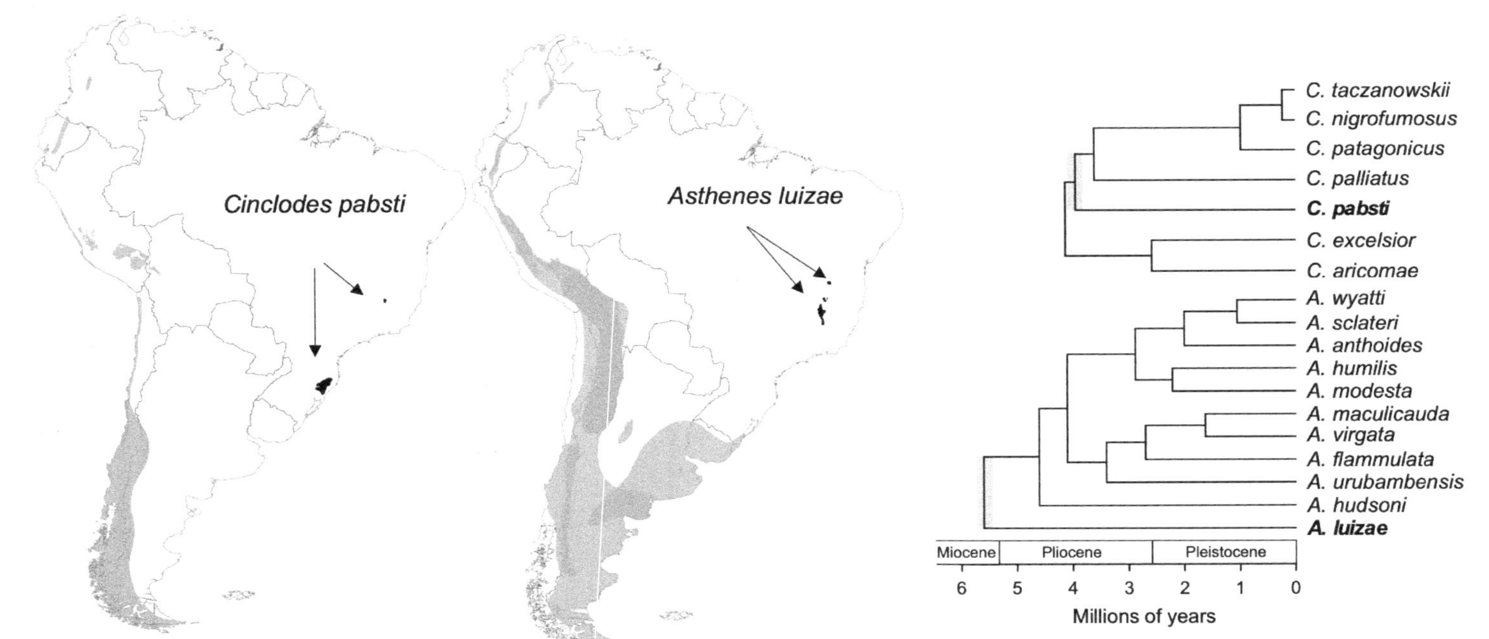

Figure 10.4 Geographic distribution, phylogenetic position, and time of divergence of the Long-Tailed Cinclodes *Cinclodes pabsti* (left) and the Cipo Canastero *Asthenes luizae* (right), which hold isolated populations in the highland rocky outrcrops of eastern Brazil, in comparison to their Andino-Patagonian relatives. The phylogenies of subclades of *Cinclodes* and *Asthenes* show early splits from Andean lineages, including Late Miocene (*C. pabsti*) and mid-Pliocene (*A. luizae*) basal splits (gray).

Phylogenetic trees extracted from Harvey et al. (2020).

García-Moreno et al. 1999, Pérez-Emán 2005, Brumfield and Edwards 2007, Cadena et al. 2007, E. Quintero et al. 2013). Therefore, the tropical Andes seem to support the second model, suggesting patterns of allopatric speciation followed by secondary contact (García-Moreno and Fjeldså 2000, Cadena et al. 2012, Caro et al. 2013, Cadena and Céspedes 2020). Ample discussions on the role of climate and interspecific competition in the maintenance of species altitudinal ranges were sparked by John Terborgh's pioneering studies in the Peruvian Andes (Terborgh 1971, Terborgh and Weske 1975). Recent analyses are in agreement that competition, not climate or ecotones, is the leading driver of narrow elevational ranges in tropical mountains (Freeman 2015, Freeman et al. 2022).

Another biogeographic pattern noted by Chapman (1926) was the presence of geographic breaks along the Andes, where closely related avian taxa living at similar altitudinal ranges replace one another latitudinally, a pattern observed in both Paramo and cloud forest avifaunas (Vuilleumier 1969, Remsen 1984, Graves 1985, Brumfield and Remsen 1996, Winger 2017). Whereas isolated Paramo highlands act as sky islands, cloud forests cover Andean humid slopes between 1,000 and 3,000 m (Webster 1995). These cloud forests are broken by several geographic barriers, including lowland gaps, dry valleys and canyons, narrow low montane passes, and high-elevation Andean ridges with inhospitable environments (Cuervo 2013, Hazzi et al. 2018). Species differentiation and taxon replacements along the Andes coincide with these breaks, of what otherwise would be a continuous belt of cloud forest (Graves 1985, Krabbe 2008, Bonaccorso 2009, Weir 2009, Cuervo 2013, Hazzi et al. 2018).

Several hypotheses have been postulated to explain latitudinal replacements along the Andes. Leading models suggest that replacements resulted from vicariant speciation—an idea supported by the multiple species that show similar geographic replacements and common boundaries (Cuervo 2013). For instance, continuous ancestral populations could have been divided by vicariance due to the formation of deep Andean valleys (Ribas et al. 2007, Chaves et al. 2011, E. Quintero et al. 2013). Alternatively, barriers could predate avian populations, and those could have gone through dispersal-mediated peripatric speciation, dispersing across already existing barriers through founder events (García-Moreno et al. 1999, Valderrama et al. 2014, Benham et al. 2015, Winger et al. 2015).

Given the monumental size of those barriers under current conditions, which include deep canyons and inhospitable drylands, biogeographers have assumed that dispersal was only possibly under past different environmental conditions, associated with climatic shifts during Pleistocene glacial cycles. According to this model, the down-slope expansion of glaciers, paramos, and cloud forests during glacial maxima facilitated connectivity across low-elevation barriers, promoting dispersal and gene flow. Conversely, up-slope shifts during interglacials promoted isolation and divergence (Vuilleumier 1969, Graves 1982, Bush et al. 2011, Ramírez-Barahona and Eguiarte 2013). Paleoecological and palynological evidence supports altitudinal shifts in the Andes, including the presence of high-altitude plants and mesic vegetation at lower altitudes during glacial periods (e.g., Hooghiemstra and Van der Hammen 2004, Hooghiemstra et al. 2006, Cárdenas et al. 2011, Groot et al. 2011, Torres et al. 2013). Andean refugia have also been suggested as a mechanism for species diversification in montane avifaunas, due to the formation of pockets of humid climates that may be stable over evolutionary timescales (Fjeldså 1995, Fjeldså et al. 1999), creating the conditions for speciation across refugia (Fjeldså et al. 2012).

It is quite possible that both processes (Andean uplift and climate change) acted in concert throughout the history of the Andean avifauna. The rise of the Andes, however, was not a continuous and steady process. Although the earliest stages of Andean uplift initiated during the early Paleocene, the rise of the Andes was particularly active at the onset of the Neogene (~25 Mya) until the late Pliocene, when it reached its current organization (Mora et al. 2010). Geological data indicate that the southern Andes are relatively older than the central and northern portions (B. Simpson 1979, Boschman 2021). Furthermore, the northern Andes are the result of more complex geological interactions, which include the Caribbean Plate and the Panama Arc (Taboada et al. 2000, Bermúdez et al. 2010, Hoorn et al. 2010). Therefore, it has been suggested that avian lineages occupied the northern Andes from the south (Vuilleumier 1986), a model that has empirical support (Doan 2003, Ribas et al. 2007, Chaves et al. 2011, E. Quintero et al. 2013, Winger et al. 2015). On the other hand, northern lineages that colonized South America after the formation of the Isthmus of Panama (e.g., many oscine passerines) imprinted an inverse northern origin and a southward colonization of the Andes, suggesting that the current Andean avifauna has a mixed origin (Cuervo 2013). Along these lines, a recent study suggests that Andean bioregions were not formed from a single biogeographical event in a certain time frame but, rather, from a combination of vicariance and dispersal events, which occurred at different time periods, both due to Andean uplift and mountain dispersal, facilitated by temperature oscillations during the Pleistocene (Hazzi et al. 2018). Understanding how these two historical processes promoted Andean speciation is key to disentangling the role of dispersal and vicariance in the speciation process (Winger et al. 2015), a theme that would resurge in other Neotropical ecosystems.

Speciation in the Amazon

The rise of the Andes had a tremendous influence on the evolution of Amazonian landscapes. During intense periods of mountain uplift, depression of foreland basins west of the Andes resulted in oceanic ingressions and the formation of mega wetlands (Pebas system) in western Amazonia (Hoorn et al. 2010, Latrubesse et al. 2010, Bicudo et al. 2019). More recently, uplifts of these foreland basins resulted in the development of the current more elevated terrains, likely achieved by the Late Miocene (Dobson et al. 2001, Harris and Mix 2002, Hoorn 2006, Hoorn et al. 2017). During the Late Miocene and Pliocene, Amazonian fluvial systems may have been poorly delimited in the lowlands, following into their current well-defined valleys during the Pleistocene (Wesselingh and Salo 2006). As mountain building progressed and reached a critical elevation of approximately 2,000 m, rainfall increased along the Andean eastern flank (Hoorn et al. 2010), providing the conditions for the growth of not only very wet montane and cloud forests but also tropical lowland forests (Poulsen et al. 2010). Therefore, the uplift of the Andes altered the regional climate in dramatic ways, changing the Amazonian landscape by reconfiguring drainage patterns, providing additional rainfall, and throwing a vast influx of sediments into the basin (Hoorn et al. 2010, Poulsen et al. 2010).

The number of evolutionary models for the Amazon increased in the 1960s when the Amazon became a hot spot for evolutionary studies, raising some of the most heated debates of tropical evolutionary biology (Haffer 1992, 2008; Colwell 2000; Knapp and Mallet 2003; Bush and Oliveira 2006; Rocha and Kaefer 2019; Baker et al. 2020). Despite the several ecosystems that comprise the Amazon, models proposed to explain current species diversity were largely restricted to upland terra firme forests. It is the distribution patterns of the avifauna of these forests that have raised interest among biogeographers since Wallace returned to Europe after his 4-year journey in 1852.

The uniqueness of this pattern consists of a patchwork of parapatric faunas established in major interfluvia, bounded by large rivers (Haffer 1969, 1974; Figure 10.3c). First described in the mid-19th century for primates (Wallace 1852), this pattern was later established for Amazonian birds, where "a number of well-marked species of birds capable of flight, yet with their range in certain directions accurately defined by great rivers" (Wallace 1876, p. 357). Today, we know that even smaller rivers can bound dozens of pairs of avian taxa, representing significant biogeographical barriers (Naka 2011, Naka et al. 2012, Fernandes, Cohn-Haft, et al. 2014).

Most diversification models proposed to explain Amazonian bird distribution patterns have relied on geographic speciation. These include at least four paleogeographic models, two climatic models, and one ecological-driven model (Haffer 2008). Paleogeographic models include the island (Emsley 1965, Nores 1999), arch (Patton et al. 2000), lagoon (Marroig and Cerqueira 1997), and riverine (Sick 1967, Capparella 1987, 1988, 1991) hypotheses, which rely on major tectonic, geomorphologic, and hydrological reconfigurations of the landscape. Climatic-driven models, such as the refuge (Haffer 1969) and disturbance vicariance (Colinvaux 1993, Bush 1994) models, are based on the long-term changes of the planet's climate and its effect on biome distributions. The ecological model proposed (the gradient hypothesis) postulates parapatric speciation across ecotones (Endler 1982).

Given the coincidence of Amazonian rivers and species distributional limits, it is not surprising that the riverine barrier hypothesis was among the first models postulated to account for Amazonian avian diversification. Although this pattern was first described by Wallace for primates and birds (1852, 1876), it was the German–Brazilian ornithologist Helmut Sick who first explicitly suggested that the genesis of Amazonian rivers could have divided ancestral ranges of widespread bird species (Sick 1967). This hypothesis was further developed two decades later by Angelo Capparella (1987, 1988, 1991). Using protein electrophoretic allozyme analyses, Capparella found genetic differentiation among river-separated avian populations, including species with and without plumage differentiation, suggesting a pervasive role of rivers in reducing gene flow. These studies encouraged a legion of Neotropical biogeographers to sample different lineages and evaluate alternative models of diversification (e.g., Marks et al. 2002; Aleixo 2004; Ribas, Aleixo, et al. 2005, 2012). The role of Amazonian rivers as primary barriers (those responsible for the diversification events) was highly contested by Haffer (1969, 1992), who insisted on teasing apart the primary and secondary role of rivers in the speciation process. Rather than dividing ancestral populations, Haffer believed that rivers were important as secondary barriers. This conclusion was later supported by comparative studies that found disparate times of divergence among co-distributed pairs of taxa on opposite sides of some Amazonian rivers, which in some cases spanned several million years, from the Miocene to the Late Pleistocene (Naka and Brumfield 2018, Silva et al. 2019). Simulations suggested that (non-adaptive) neutral processes and reduced dispersal could maintain allopatry despite repeated river crossings, allowing rivers to act as effective secondary barriers (Santorelli et al. 2022).

The most influential diversification model postulated for the Amazon was the Pleistocene refugia hypothesis (Haffer 1969). According to this model, borrowed from Australia (Keast 1961) and Africa (Moreau 1963), glaciation cycles during the Pleistocene affected tropical landscapes by modifying their rainfall patterns, creating cycles of expansion and retraction of humid forests (see Figure 10.3a). These cyclical processes resulted in the isolation and subsequent differentiation of avian populations, producing the current patchwork of parapatric distribution seen in dozens of avian lineages (see Figure 10.3c). The refuge hypothesis reigned virtually undisputed in the Neotropical speciation literature for two decades, becoming the standard explanation to account for Amazonian speciation patterns (E. Mayr and O'Hara 1986).

Palynological studies, however, failed to support some of the most critical predictions of Haffer's hypothesis, such as the idea that most Amazonian humid forests were replaced by savannas during glacial maxima (for historical reviews of this model, see Bush 1994, Colinvaux et al. 1996, Bush and Oliveira 2006, Rocha and Kaefer 2019). Distribution models of Amazonian birds projected to the Last Glacial Maximum (LGM) do not show severe retraction into multiple small refugia but, rather, a split into two large forests (Bonaccorso et al. 2006). Furthermore, comparative phylogeographic data indicate that the diversification of many Amazonian species complexes predates the Pleistocene (Silva et al. 2019). Although Haffer tried to accommodate these inconsistencies into an expanded (older) version of the refuge hypothesis (Haffer 2008), it remains clear that the diversification of Amazonian birds requires more complex scenarios than any single explanation could offer (Bush 1994, Rull 2011, Cracraft et al. 2020). The current understanding is that Quaternary glacial cycles likely influenced the distribution of forests and savannas along ecotones, but not at the core of the Amazon.

In northeastern Amazonia, the table-top mountains known as tepuis, situated in the Guiana Shield, raised early interest among biogeographers (Chapman 1931, E. Mayr and Phelps 1967, Cook 1974). These isolated mountains contain many endemics with different ages and putative origins, which include taxa likely derived from (1) the tropical lowlands and (2) long-distance colonists, possibly from the Andes or the coastal mountains from Venezuela (Chapman 1931, E. Mayr and Phelps 1967). In the "habitat shift" theory, avian forms derived from the lowlands were called "indigenous forms of tropical origin" (E. Mayr and Phelps 1967). This model involved a process quite like that proposed by Chapman in the Andes, where "at least part of the fauna of Pantepui owes its origin to a shift in ecological tolerance or preference among species that elsewhere are essentially tropical" (E. Mayr and Phelps 1967, p. 297). Alternatively, nearly half of the Pantepui avian endemics were considered long-distance colonists. Two alternative hypotheses were postulated by Chapman—what E. Mayr and Phelps (1967) called the "cool climate" theory and the "distance dispersal" theory. Chapman dismissed altogether the possibility of avian immigration on the wing and supported the "cool climate" hypothesis, which states that "the subtropical fauna of Pantepui was able to reach these mountains when the lowlands had a sub-tropical climate during the cool periods of the Pleistocene" (E. Mayr and Phelps 1967, p. 294), an idea that received considerable support (Haffer 1970, Cook 1974, Rull 2005). Alternatively, E. Mayr and Phelps (1967) supported the "distance dispersal" hypothesis, postulating that active dispersal was possible as island hoppers cruised the landscape. Recent findings in Brazilian peripheral tepuis suggest the potential of dispersal to fund new populations (Laranjeiras, Melinski, et al. 2019). A third, less supported hypothesis is the "plateau theory," which postulates that the Pantepui fauna represents a relict of a formerly widespread fauna that lived in a once continuous plateau, now dissected by erosion into separate tepuis (E. Mayr and Phelps 1967). According to this model, given that Pantepui mountains predate the Andes and the coastal Venezuelan mountains, the Pantepui would have been a source of migrants (Chapman 1931, Graves 1982).

Unfortunately, molecular data are surprisingly scarce, and those that are available offer poor resolution to test most of these hypotheses. Few molecular data support the habitat shift theory, which includes alternative hypotheses of parapatric speciation due to ecological specialization or peripatric speciation due to funder effects. Some tepui endemics, such as *Myioborus* redstarts, have their closest relatives in coastal Venezuela (Pérez-Emán 2005), whereas others, such as the Roraiman Antbird (*Myrmelastes saturatus*) or the Tepui Anshrike (*Thamnophilus insignis*), are likely related to lowland forms (Braun et al. 2005, Brumfield and Edwards 2007). Although the origin of the Tepui avifauna remains unsettled, there is considerable evidence against the plateau theory, arguing against a Tepui origin in several endemic clades (Mauck and Burns 2009, Sedano and Burns 2010, Bonaccorso and Guayasamin 2013).

In contrast to the many evolutionary models proposed for terra firme forest birds, other ecosystems, such as the Amazonian floodplains, have received little attention. Consistent with the high habitat connectivity provided by water courses, riparian birds tend to show low levels of population differentiation in the Amazon (Aleixo 2006, Cohn-Haft et al. 2007, Cadena et al. 2011, Harvey et al. 2017). Recent molecular studies, however, are starting to shed light on the patterns and processes of diversification in flooded forest birds. For example, three flooded forest specialist antbirds in the genera *Thamnophilus*, *Myrmoborus*, and *Myrmotherula* show reduced levels of gene flow in areas where floodplain forest was likely interrupted during the Middle and Late Pleistocene due to climate-driven changes in fluvial dynamics (Thom et al. 2020). River sedimentation may also have pervasive influence in species distributions. Amazonian rivers are characterized by their distinct colors, including black-, clear-, and white-water rivers, according to the amount of sediments and tannines they carry (Junk et al. 2011). Habitats flooded by sediment-rich white-water rivers tend to have taller and highly productive *várzea* forest, whereas nutrient-poor black-water rivers have lower, less productive *igapó* forest. Therefore, river sediments influence species composition in avian communities (Laranjeiras, Naka, et al. 2019). Recent studies have shown that tributaries of different colors can act as biogeographical barriers for some bird species, isolating populations and inhibiting gene flow (Naka et al. 2007, 2020; Luna et al. 2022). Riparian birds can respond to interruptions in the continuity of riparian habitats. For example, the separation of the Amazon basin from the Paraná basin in the late Miocene or early Pliocene correlates with speciation events in several river-adapted lineages in the genera *Furnarius*, *Stigmatura*, and *Serpophaga* (Claramunt 2014). River network rearrangements may also result in genetic differentiation and speciation (Ribas, Aleixo, et al. 2012; Luna et al. 2022; Musher et al. 2022). We strongly recommend visiting the thorough analyses on the role of Amazonian rives in the diversification process of birds, provided by Ribas, Sawakuchi, et al. (2025).

Lowland Open Vegetation Formations

The South American dry diagonal, the vast corridor of open formations that extends from northeastern Brazil to central Argentina, includes three key Neotropical ecosystems: the Caatinga, the Cerrado, and the Chaco. The Caatinga and the Chaco are the two largest nuclei of seasonally dry forests and have ecological and biogeographic affinities with other Neotropical dry forests (Pennington et al. 2009), whereas the Cerrado is the dominant savanna woodlands of central Brazil (Silva and Bates 2002). The allegedly peripheral retractions of humid forests during glacial periods likely affected the dynamics of open formations, resulting in a continental expansion of dry habitats, connecting nonforest ecosystems across South America (Haffer 1987, Silva and Bates 2002). Influenced by Haffer's studies, Prado and Gibbs (1993) postulated the "Pleistocene Arch" hypothesis, which proposes that current isolated patches of Neotropical dry forests were connected during glacial maxima. Like Haffer, these authors provide a very recent temporal framework, suggesting that these connections occurred during the LGM, only 21,000 years ago.

Biogeographical evidence of this ancestral connection is offered by the multiple bird species shared between open habitats north and south of Amazonia and across currently isolated patches of dry forest. The scarce molecular studies on the subject in birds seem to partially support the idea of ancestral connections. For example, molecular data are consistent with a clockwise route of colonization for the Burnished-Buff Tanager (*Stilpnia cayana*), suggesting a Cerrado origin and a westward expansion into Bolivia and across seasonally dry forests at the base of the Andes and into Guyana, from where they expanded into coastal northeastern Brazil (Savit and Bates 2015). Similarly, currently disjunct subspecies of the Rufous-Fronted Thornbird (*Phacellodomus rufifrons*) were connected in the recent past, probably during the Middle and Late Pleistocene (Corbett et al. 2020). Paleoclimate models indicate greater connectivity during the LGM, decreasing through the mid-Holocene toward the present (Savit and Bates 2015, Corbett et al. 2020). On the other hand, an ancestral Caatinga/Chaco connection seems to be significantly older for some lineages, such as the Stripe-Backed Antbird (*Myrmorchilus strigilatus*) or the Greater Wagtail-Tyrant (*Stigmatura budytoides*) (L. N. Naka, unpublished data). However, it seems likely that Pleistocene glacial cycles largely affected avian populations along the dry diagonal. The Narrow-Billed Woodcreeper (*Lepidocolaptes angustirostris*) shows a strong genetic structure associated with climatic stable areas along the diagonal, suggesting that the genetic structure of this species was driven by Pleistocene climatic oscillations (Rocha et al. 2020). Despite the potential of common diversification, comparative population genetic analyses of widely distributed savanna and dry forest bird species found little congruence in temporal and geographic patterns of genetic diversity, suggesting idiosyncratic histories of connection and dispersal (Ritter et al. 2021, van Els et al. 2021, Lima-Rezende et al. 2022). For a complete review of avian evolutionary models in South American grasslands, see Norambuena and Van Els (2021).

The idea of humid forest retractions and expansions during glacial cycles has also been central to several avian diversification models in the Atlantic Forest and subtropical South America (Nores 1992; Cabanne et al. 2008, 2016). Forest expansion during humid periods likely allowed biotic exchanges between Andean montane forests and Atlantic forests, explaining the distribution of dozens of sister taxa currently separated by seasonal forests and open formations. Population genetic analyses and niche paleomodels are consistent with Quaternary or pre-Quaternary connections or dispersal events (Cabanne et al. 2019, Trujillo-Arias et al. 2020). Greater connectivity likely occurred during glacial maxima, presumably due to an increase in humidity along the intervening lowlands. Under this perspective, Andean and Atlantic forests may be currently acting as "interglacial refugia" (Trujillo-Arias et al. 2020). Finally, glaciations likely played a major role in speciation among Patagonian birds (Vuilleumier 1991), but very little information has been generated for this corner of the continent.

The Role of Dispersal

Dispersal is a key process in the generation of diversity patterns (Ricklefs 1987, Wiens 2011). The movement of organisms through the landscape, or the lack thereof, is deeply involved in the generation of geographic patterns of distribution and endemism. Moreover, dispersal plays a complicated role in speciation. In classic vicariant models, the lack of dispersal across a barrier can trigger population differentiation and speciation, but dispersalist models suggest that dispersal may also trigger speciation if it increases the chances of colonization of new areas, especially across barriers (Claramunt et al. 2012).

The role of dispersal in shaping avian diversity patterns in South America remains unclear. One of the main explanations for the lack of congruence in diversification patterns predicted by vicariant models (refugia, rivers, etc.) is the idiosyncratic movement of species across the landscape (Brumfield 2012, Smith et al. 2014, Cabanne et al. 2016). On the other hand, early population genetic studies revealed high genetic structure and low levels of gene flow among Amazonian understory birds, suggesting low levels of dispersal (Capparella 1988, Bates 2000). Moreover, recent dispersal-challenge experiments revealed dramatically different capacities of tropical birds to cross water bodies (Moore et al. 2008, Naka et al. 2022). Surprisingly, approximately 30% of the species tested were unable to sustain flight for 100 m, confirming that

Amazonian rivers likely represent insurmountable barriers (Naka et al. 2022). Even narrow gaps over land may inhibit the movement of tropical forest birds (Laurance et al. 2004, Ibarra-Macias et al. 2011).

Recent studies have explored the effects of dispersal on South American avian species richness patterns, using ecological characteristics as surrogates for species' dispersal ability. Habitat specialization and microhabitat (forest understory) correlated with higher species and subspecies richness in a wide sample of Amazonian birds (Salisbury et al. 2012, Smith et al. 2014). Low dispersal tendencies among tropical birds have been attributed to behavioral factors such as a fear of venturing into open spaces (Diamond, 1981; Laurance et al. 2004). However, recent comparative analyses revealed that crossing success is better predicted by proxies for flight aerodynamic efficiency derived from wing morphology such as the hand–wing index and the aspect ratio, suggesting that physical capabilities influence avian dispersal ability (Claramunt et al. 2012, 2022; Naka et al. 2022). Other important ecological and evolutionary phenomena, such as natal dispersal distances, migration distances, and geographic range sizes, have also been linked to flight efficiency and wing morphology (Claramunt 2021, Capurucho et al. 2020, Arango et al. 2022, Weeks et al. 2022, Chu and Claramunt 2023).

That a species' dispersal ability can be assessed from its wing morphology opens the possibility for analyzing the role of dispersal on diversification in a more rigorous way and for testing contrasting predictions of different evolutionary models. For example, alternative modes of speciation make distinctive predictions regarding the relationship between dispersal ability and speciation rates. Under vicariant models, speciation rates should be greater in lineages with low dispersal ability, whereas under dispersal-induced speciation, speciation rates may be higher in lineages with high dispersal ability (Claramunt et al. 2012). Because of the potential combination of these two effects, in addition to the effect of range expansions and extinctions, the relationship between dispersal ability and diversification may be hump-shaped, with speciation rates being maximized at intermediate levels of dispersal (Claramunt et al. 2012). Macroevolutionary models revealed an overall negative relationship between dispersal and speciation rates in Furnariidae, suggesting the predominant effect of vicariant speciation in this family, but with a hint of lower speciation rates for species with extremely low dispersal capabilities (Claramunt et al. 2012). At the population genetics level, low dispersal should correlate with lower gene flow and greater population structuring, but the empirical evidence remains ambiguous (Harvey et al. 2017, Smith et al. 2017). The combination of flight efficiency proxies and genomic-scale data sets should improve estimates of population genetic parameters and time-calibrated phylogenies, increasing our ability to analyze the role of dispersal in diversification and species richness patterns.

Conclusion

The megadiverse avifauna of South America has inspired generations of researchers. The initial documentation of major distributional patterns gave rise to a rich body of theory aimed at explaining them. The first generation of hypotheses, such as those provided by Chapman, Mayr, Haffer, and Sick, was mostly based on avian distribution patterns and traditional taxonomies. Throughout this chapter, we have shown that some common themes repeat across different biomes, including the presence of physical barriers (mountains, rivers, and valleys), climatic shifts, and species ecologies and interactions. How much each factor contributes to the final assemblage of current biotas remains a central research topic.

With the advent of DNA sequencing and phylogenetic and population genetics analysis, researchers started to test historical predictions of the different classic models. New geological data and paleoreconstruction methods also increased our ability to visualize the historical geographic scenarios in which diversification took place. However, progress toward testing different models has been slow. Initial genetic data based on few genetic markers (mostly mitochondrial DNA) were enough for describing major breaks in the genetic structure of populations across space but insufficient for revealing complex population genetic histories. Next-generation sequencing technologies are solving this issue, and currently thousands of genomic regions, if not entire genomes, are being used for evaluating population structure, patterns of gene flow, histories of population sizes, and genome landscape adaptation in greater detail.

The use of genomic-scale data has revealed some strong patterns but also considerable complexity, and our ability to infer processes from data is still limited. This may be due in part to limitations in our analytical approaches, as popular methods used to analyze genomic data still make relatively simple assumptions about population structure, such as ignoring isolation by distance effects or assuming that individuals come from a discrete set of ancestral populations instead of a species tree (Bradburd and Ralph 2019, Jiao et al. 2021, Hibbins and Hahn 2022). Also, establishing a temporal framework for population subdivision and speciation is challenging, and it is an area in which further improvements are needed to attain higher accuracy and precision.

On the other hand, empirical findings may not match the predictions because the history and the processes shaping biogeographic patterns may be much more complex than our simplistic models propose. More recent studies are departing from traditional models and considering the simultaneous effects of geomorphological and climatic changes, and their interactions, to explain observed patterns of differentiation and genetic variation, but other factors may still be neglected. The processes governing species and population divergence are complex and often multifactorial, likely affected by population size, generation time, mutation rates, and dispersal ability.

Embracing differences in ecology and dispersal ability among species may add new insights to the investigation of biogeographic diversification. For example, some models assume that birds move together with their habitats, whereas others conceive birds as independently moving organisms. The "true history" we try to model may lie somewhere in the middle of such extremes; accepting biodiversity as a gradient will help us explore more possibilities, with the humble eye of Humboldt or Wallace. We believe that the confluence of genomic data sets, historical landscape models, and data on ecological requirements and dispersal capabilities of species has great potential to put to the test more complex models of biographic diversification. South America, the land of birds, is the perfect natural laboratory to conduct such tests.

References

Aleixo, A. 2004. Historical diversification of a terra-firme forest bird superspecies: A phylogeographic perspective on the role of different hypotheses of Amazonian diversification. *Evolution* 58(6):1303–1317.

Aleixo, A. 2006. Historical diversification of floodplain forest specialist species in the Amazon: A case study with two species of the avian genus *Xiphorhynchus* (Aves: Dendrocolaptidae). *Biological Journal of the Linnean Society* 89(2):383–395.

Alström, P., P. G. Ericson, U. Olsson, and P. Sundberg. 2006. Phylogeny and classification of the avian superfamily Sylvioidea. *Molecular Phylogenetics and Evolution* 38(2):381–397.

Anthony, N. M., M. Johnson-Bawe, K. Jeffery, S. L. Clifford, K. A. Abernethy, C. E. Tutin, . . . M. W. Bruford. 2007. The role of Pleistocene refugia and rivers in shaping gorilla genetic diversity in central Africa. *Proceedings of the National Academy of Sciences of the USA* 104(51):20432–20436.

Antonelli, A., and I. Sanmartín. 2011. Why are there so many plant species in the Neotropics? *Taxon* 60(2):403–414.

Antonelli, A., A. Zizka, F. A. Carvalho, R. Scharn, C. D. Bacon, D. Silvestro, and F. L. Condamine. 2018. Amazonia is the primary source of Neotropical biodiversity. *Proceedings of the National Academy of Sciences of the USA* 115(23):6034–6039.

Arango, A., J. Pinto-Ledezma, O. Rojas-Soto, A. M. Lindsay, C. D. Mendenhall, and F. Villalobos. 2022. Hand–Wing Index as a surrogate for dispersal ability: The case of the Emberizoidea (Aves: Passeriformes) radiation. *Biological Journal of the Linnean Society* 137(1):137–144.

Bacon, C. D., D. Silvestro, C. Jaramillo, B. T. Smith, P. Chakrabarty, and A. Antonelli. 2015. Biological evidence supports an early and complex emergence of the Isthmus of Panama. *Proceedings of the National Academy of Sciences of the USA* 112(19):6110–6115.

Baker, P. A., S. C. Fritz, D. S. Battisti, C. W. Dick, O. M. Vargas, G. P. Asner, . . . I. Prates. 2020. Beyond refugia: New insights on Quaternary climate variation and the evolution of biotic diversity in tropical South America. Pages 51–70 in V. Rull and A. Carnaval, eds. *Neotropical Diversification: Patterns and Processes*. Springer, Cham, Switzerland.

Barker, F. K. 2004. Monophyly and relationships of wrens (Aves: Troglodytidae): A congruence analysis of heterogeneous mitochondrial and nuclear DNA sequence data. *Molecular Phylogenetics and Evolution* 32(2):486–504.

Barker, F. K. 2007. Avifaunal interchange across the Panamanian isthmus: Insights from *Campylorhynchus* wrens. *Biological Journal of the Linnean Society* 90(4):687–702.

Barker, F. K., K. J. Burns, J. Klicka, S. M. Lanyon, and I. J. Lovette. 2015. New insights into New World biogeography: An integrated view from the phylogeny of blackbirds, cardinals, sparrows, tanagers, warblers, and allies. *The Auk* 132(2):333–348.

Batalha-Filho, H., J. Fjeldså, P. H. Fabre, and C. Y. Miyaki. 2013. Connections between the Atlantic and the Amazonian forest avifaunas represent distinct historical events. *Journal of Ornithology* 154(1):41–50.

Bates, J. M. 2000. Allozymic genetic structure and natural habitat fragmentation: Data for five species of Amazonian forest birds. *The Condor* 102(4):770–783.

Bates, J. M., S. J. Hackett, and J. Cracraft. 1998. Area-relationships in the Neotropical lowlands: An hypothesis based on raw distributions of Passerine birds. *Journal of Biogeography* 25(4):783–793.

Bates, J. M., and R. M. Zink. 1994. Evolution into the Andes: Molecular evidence for species relationships in the genus *Leptopogon*. *The Auk* 111(3):507–515.

Benham, P. M., A. M. Cuervo, J. A. McGuire, and C. C. Witt. 2015. Biogeography of the Andean metaltail hummingbirds: Contrasting evolutionary histories of tree line and habitat-generalist clades. *Journal of Biogeography* 42(4):763–777.

Bermúdez, M. A., B. P. Kohn, P. A. van der Beek, M. Bernet, P. B. O'Sullivan, and R. Shagam. 2010. Spatial and temporal patterns of exhumation across the Venezuelan Andes: Implications for Cenozoic Caribbean geodynamics. *Tectonics* 29(5): Article TC5009.

Bicudo, T. C., V. Sacek, R. P. de Almeida, J. M. Bates, and C. C. Ribas. 2019. Andean tectonics and mantle dynamics as a pervasive influence on Amazonian ecosystem. *Scientific Reports* 9(1):1–11.

Bijl, P. K., J. A. Bendle, S. M. Bohaty, J. Pross, S. Schouten, L. Tauxe, . . . M. Yamane. 2013. Eocene cooling linked to early flow across the Tasmanian Gateway. *Proceedings of the National Academy of Sciences of the USA* 110(24):9645–9650.

BirdLife International. 2020. *Handbook of the Birds of the World and BirdLife International Digital Checklist of the Birds of the World*. Version 5.0

Bonaccorso, E. 2009. Historical biogeography and speciation in the Neotropical highlands: Molecular phylogenetics of the jay genus *Cyanolyca*. *Molecular Phylogenetics and Evolution* 50(3):618–632.

Bonaccorso, E., and J. M. Guayasamin. 2013. On the origin of Pantepui montane biotas: A perspective based on the phylogeny of *Aulacorhynchus* toucanets. *PLoS One* 8(6): Article e67321.

Bonaccorso, E., I. Koch, and A. T. Peterson. 2006. Pleistocene fragmentation of Amazon species' ranges. *Diversity and Distributions* 12(2):157–164.

Boschman, L. M. 2021. Andean mountain building since the Late Cretaceous: A paleoelevation reconstruction. *Earth-Science Reviews* 220: Article 103640.

Botero, C. A., R. Dor, C. M. McCain, and R. J. Safran. 2014. Environmental harshness is positively correlated with intraspecific divergence in mammals and birds. *Molecular Ecology* 23(2): 259–268.

Bradburd, G. S., and P. L. Ralph. 2019. Spatial population genetics: It's about time. *Annual Review of Ecology, Evolution, and Systematics* 50:427–449.

Braun, M. J., M. L. Isler, P. R. Isler, J. M. Bates, and M. B. Robbins. 2005. Avian speciation in the Pantepui: The case of the Roraiman Antbird (*Percnostola* [*Schistocichla*] "*leucostigma*" saturata). *The Condor* 107(2):327–341.

Brown, J. H. 1981. Two decades of homage to Santa Rosalia: Toward a general theory of diversity. *American Zoologist* 21(4):877–888.

Brumfield, R. T. 2012. Inferring the origins of lowland Neotropical birds. *The Auk* 129(3):367–376.

Brumfield, R. T., and A. P. Capparella. 1996. Historical diversification of birds in northwestern South America: A molecular perspective on the role of vicariant events. *Evolution* 50(4): 1607–1624.

Brumfield, R. T., and S. V. Edwards. 2007. Evolution into and out of the Andes: A Bayesian analysis of historical diversification in *Thamnophilus* antshrikes. *Evolution* 61(2):346–367.

Brumfield, R. T., and J. V. Remsen Jr. 1996. Geographic variation and species limits in *Cinnycerthia* wrens of the Andes. *The Wilson Bulletin* 108(2):205–227.

Burney, C. W., and R. T. Brumfield. 2009. Ecology predicts levels of genetic differentiation in Neotropical birds. *The American Naturalist* 174(3):358–368.

Bush, M. B. 1994. Amazonian speciation: A necessarily complex model. *Journal of Biogeography* 21:5–17.

Bush, M. B., W. D. Gosling, and P. A. Colinvaux. 2011. Climate and vegetation change in the lowlands of the Amazon Basin. Pages 61–84 in M. Bush, J. Flenley, and W. Gosling, eds. *Tropical Rainforest Responses to Climatic Change*. Springer, Berlin.

Bush, M. B., and P. E. D. Oliveira. 2006. The rise and fall of the refugial hypothesis of Amazonian speciation: A paleoecological perspective. *Biota Neotropica* 6(1).

Cabanne, G. S., L. Calderón, N. Trujillo Arias, P. Flores, R. Pessoa, F. M. d'Horta, and C. Y. Miyaki. 2016. Effects of Pleistocene climate changes on species ranges and evolutionary processes in the Neotropical Atlantic Forest. *Biological Journal of the Linnean Society* 119(4):856–872.

Cabanne, G. S., L. Campagna, N. Trujillo-Arias, K. Naoki, I. Gómez, C. Y. Miyaki, . . . P. L. Tubaro. 2019. Phylogeographic variation within the Buff-browed Foliage-gleaner (Aves: Furnariidae: *Syndactyla rufosuperciliata*) supports an Andean–Atlantic forests connection via the Cerrado. *Molecular Phylogenetics and Evolution* 133:198–213.

Cabanne, G. S., F. M. d'Horta, E. H. Sari, F. R. Santos, and C. Y. Miyaki. 2008. Nuclear and mitochondrial phylogeography of the Atlantic forest endemic *Xiphorhynchus fuscus* (Aves: Dendrocolaptidae): Biogeography and systematics implications. *Molecular Phylogenetics and Evolution* 49(3):760–773.

Cadena, C. D., and L. N. Céspedes. 2020. Origin of elevational replacements in a clade of nearly flightless birds: Most diversity in tropical mountains accumulates via secondary contact following allopatric speciation. Pages 635–659 in V. Rull and A. Carnaval, eds. *Neotropical Diversification: Patterns and Processes*. Springer, Cham, Switzerland.

Cadena, C. D., N. Gutiérrez-Pinto, N. Dávila, and R. T. Chesser. 2011. No population genetic structure in a widespread aquatic songbird from the Neotropics. *Molecular Phylogenetics and Evolution* 58:540–545.

Cadena, C. D., J. Klicka, and R. E. Ricklefs. 2007. Evolutionary differentiation in the Neotropical montane region: Molecular phylogenetics and phylogeography of *Buarremon* brush-finches (Aves, Emberizidae). *Molecular Phylogenetics and Evolution* 44(3):993–1016.

Cadena, C. D., K. H. Kozak, J. P. Gomez, J. L. Parra, C. M. McCain, R. C. Bowie, . . . C. H. Graham. 2012. Latitude, elevational climatic zonation and speciation in New World vertebrates. *Proceedings of the Royal Society B: Biological Sciences* 279(1726):194–201.

Capparella, A. P. 1987. Effects of riverine barriers on genetic differentiation of Amazonian forest undergrowth birds (Peru). Doctoral dissertation, Louisiana State University and Agricultural & Mechanical College.

Capparella, A. P. 1988. Genetic variation in Neotropical birds: Implications for the speciation process. *Acta Congressus Internationalis Ornithologici* 19:1658–1664.

Capparella, A. P. 1991. Neotropical avian diversity and riverine barriers. *Acta Congressus Internationalis Ornithologici* 20:307–316.

Capurucho, J. M., M. V. Ashley, B. R. Tsuru, J. C. Cooper, and J. M. Bates. 2020. Dispersal ability correlates with range size in Amazonian habitat-restricted birds. *Proceedings of the Royal Society B: Biological Sciences* 287: Article 20201450.

Cárdenas, M. L., W. D. Gosling, S. C. Sherlock, I. Poole, R. T. Pennington, and P. Mothes. 2011. The response of vegetation on the Andean flank in western Amazonia to Pleistocene climate change. *Science* 331(6020):1055–1058.

Cardillo, M., C. D. L. Orme, and I. P. Owens. 2005. Testing for latitudinal bias in diversification rates: An example using New World birds. *Ecology* 86(9):2278–2287.

Caro, L. M., P. C. Caycedo-Rosales, R. C. K. Bowie, H. Slabbekoorn, and C. D. Cadena. 2013. Ecological speciation along an elevational gradient in a tropical passerine bird? *Journal of Evolutionary Biology* 26(2):357–374.

Carter, A., T. R. Riley, C. D. Hillenbrand, and M. Rittner. 2017. Widespread Antarctic glaciation during the late Eocene. *Earth and Planetary Science Letters* 458:49–57.

Castilho, C. V., W. E. Magnusson, R. N. O. de Araújo, R. C. Luizao, F. J. Luizao, A. P. Lima, and N. Higuchi. 2006. Variation in aboveground tree live biomass in a central Amazonian forest: Effects of soil and topography. *Forest Ecology and Management* 234(1–3):85–96.

Chapman, F. M. 1917. The distribution of bird-life in Colombia . *Bulletin of the American Museum of Natural History* 36:1–728.

Chapman, F. M. 1926. The distribution of bird-life in Ecuador. *Bulletin of the American Museum of Natural History* 55:1–784.

Chapman, F. M. 1931. The upper zonal bird-life of Mts. Roraima and Duida. *Bulletin of the American Museum of Natural History* 63:1–135.

Chase, J. M., and M. A. Leibold. 2002. Spatial scale dictates the productivity–biodiversity relationship. *Nature* 416(6879):427–430.

Chaves, A. V., G. H. Freitas, M. F. Vasconcelos, and F. R. Santos. 2015. Biogeographic patterns, origin and speciation of the endemic birds from eastern Brazilian mountaintops: A review. *Systematics and Biodiversity* 13(1):1–16.

Chesser, R. T. 2000. Evolution in the high Andes: The phylogenetics of *Muscisaxicola* ground-tyrants. *Molecular Phylogenetics and Evolution* 15(3):369–380.

Chesser, R. T. 2004. Systematics, evolution, and biogeography of the South American ovenbird genus *Cinclodes*. *The Auk* 121(3):752–766.

Chu, J. J., and S. Claramunt. 2023. Determinants of natal dispersal distances in North American birds. *Ecology and Evolution* 13(2): Article e9789.

Cintra, R., and L. N. Naka. 2012. Spatial variation in bird community composition in relation to topographic gradient and forest heterogeneity in a central Amazonian rainforest. *International Journal of Ecology* 1: Article 435671.

Claramunt, S. 2014. Phylogenetic relationships among *Synallaxini* spinetails (Aves: Furnariidae) reveal a new biogeographic pattern across the Amazon and Paraná river basins. *Molecular Phylogenetics and Evolution* 78:223–231.

Claramunt, S. 2021. Flight efficiency explains differences in natal dispersal distances in birds. *Ecology* 102(9): Article e03442.

Claramunt, S., and J. Cracraft. 2015. A new time tree reveals Earth history's imprint on the evolution of modern birds. *Science Advances* 1(11): Article e1501005.
Claramunt, S., E. P. Derryberry, J. V. Remsen, Jr., and R. T. Brumfield. 2012. High dispersal ability inhibits speciation in a continental radiation of passerine birds. *Proceedings of the Royal Society B: Biological Sciences* 279(1733):1567–1574.
Claramunt, S., M. Hong, and A. Bravo. 2022. The effect of flight efficiency on gap-crossing ability in Amazonian forest birds. *Biotropica* 54(4):860–868.
Cohn-Haft, M., L. N. Naka, and A. M. Fernandes. 2007. Padrões de distribuição da avifauna da várzea dos rios Solimões e Amazonas. Conservação da várzea: identificação e caracterização de regiões biogeográficas. *Manaus, Ibama, ProVárzea* 356:287–323.
Colinvaux, P. A. 1993. Pleistocene biogeography and diversity in tropical forests of South America. Pages 473–499 in P. Goldblatt, ed. *Biological Relationships Between Africa and South America*. Yale University Press, New Haven, CT.
Colinvaux, P. A., P. E. De Oliveira, J. E. Moreno, M. C. Miller, and M. B. Bush. 1996. A long pollen record from lowland Amazonia: Forest and cooling in glacial times. *Science* 274(5284):85–88.
Colwell, R. K. 2000. A barrier runs through it . . . or maybe just a river. *Proceedings of the National Academy of Sciences of the USA* 97(25):13470–13472.
Cook, R. E. 1974. Origin of the highland avifauna of southern Venezuela. *Systematic Biology* 23(2):257–264.
Corbett, E. C., G. A. Bravo, F. Schunck, L. N. Naka, L. F. Silveira, and S. V. Edwards. 2020. Evidence for the Pleistocene arc hypothesis from genome-wide SNPs in a Neotropical dry forest specialist, the Rufous-fronted Thornbird (Furnariidae: *Phacellodomus rufifrons*). *Molecular Ecology* 29(22):4457–4472.
Cracraft, J. 1985. Historical biogeography and patterns of differentiation within the South American avifauna: Areas of endemism. *Ornithological Monographs* 36:49–84.
Cracraft, J. 2001. Avian evolution, Gondwana biogeography and the Cretaceous–Tertiary mass extinction event. *Proceedings of the Royal Society B: Biological Sciences* 268(1466):459–469.
Cracraft, J., and S. Claramunt. 2017. Conceptual and analytical worldviews shape differences about global avian biogeography. *Journal of Biogeography* 44(4):958–960.
Cracraft, J., and R. O. Prum. 1988. Patterns and processes of diversification: Speciation and historical congruence in some Neotropical birds. *Evolution* 42(3):603–620.
Cracraft, J., C. C. Ribas, F. M. d'Horta, J. Bates, R. P. Almeida, A. Aleixo, . . . P. Baker. 2020. The origin and evolution of Amazonian species diversity. Pages 225–244 in V. Rull and A. Carnaval, eds. *Neotropical Diversification: Patterns and Processes*. Springer, Cham, Switzerland.
Crisp, M. D., M. T. Arroyo, L. G. Cook, M. A. Gandolfo, G. J. Jordan, M. S. McGlone, . . . H. P. Linder. 2009. Phylogenetic biome conservatism on a global scale. *Nature* 458(7239):754–756.
Crother, B. I., and C. M. Murray. 2011. Ontology of areas of endemism. *Journal of Biogeography* 38(6):1009–1015.
Cuervo, A. M. 2013. Evolutionary assembly of the Neotropical montane avifauna. Doctoral dissertation, Louisiana State University, Baton Rouge, LA.
Currie, D. J., G. G. Mittelbach, H. V. Cornell, R. Field, J. F. Guégan, B. A. Hawkins, . . . J. R. G. Turner. 2004. Predictions and tests of climate-based hypotheses of broad-scale variation in taxonomic richness. *Ecology Letters* 7(12):1121–1134.
da Silva, J. M. C. 1995. Biogeographic analysis of the South American Cerrado avifauna. *Steenstrupia* 21:49–67.
Diamond, J. M. 1981. Flightlessness and fear of flying in island species. *Nature* 293:507–508.
Dinerstein, E., D. Olson, A. Joshi, C. Vynne, N. D. Burgess, E. Wikramanayake, . . . M. Saleem. 2017. An ecoregion-based approach to protecting half the terrestrial realm. *BioScience* 67(6):534–545.
Doan, T. M. 2003. A south-to-north biogeographic hypothesis for Andean speciation: Evidence from the lizard genus *Proctoporus* (Reptilia, Gymnophthalmidae). *Journal of Biogeography* 30(3):361–374.
Dobson, D. M., G. R. Dickens, and D. K. Rea. 2001. Terrigenous sediment on Ceara Rise: A Cenozoic record of South American orogeny and erosion. *Palaeogeography, Palaeoclimatology, Palaeoecology* 165(3–4):215–229.

Emsley, M. G. 1965. Speciation in *Heliconius* (Lep., Nymphalidae): Morphology and geographic distribution. *Zoologica* 50:191–254.
Endler, J. A. 1977. *Geographic Variation, Speciation, and Clines.* Princeton University Press, Princeton, NJ.
Endler, J. A. 1982. Problems in distinguishing historical from ecological factors in biogeography. *American Zoologist* 22(2):441–452.
Ericson, P. G. 2012. Evolution of terrestrial birds in three continents: Biogeography and parallel radiations. *Journal of Biogeography* 39(5):813–824.
Ezcurra, M. D., and F. L. Agnolin. 2012. A new global palaeobiogeographical model for the late Mesozoic and early Tertiary. *Systematic Biology* 61:553–566.
Feduccia, A. 2003. "Big bang" for tertiary birds? *Trends in Ecology & Evolution* 18(4):172–176.
Fernandes, A. M., M. Cohn-Haft, T. Hrbek, and I. P. Farias, 2014. Rivers acting as barriers for bird dispersal in the Amazon. *Revista Brasileira de Ornitologia* 22(4):363–373.
Fernandes, A. M., M. Wink, C. H. Sardelli, and A. Aleixo. 2014. Multiple speciation across the Andes and throughout Amazonia: The case of the spot-backed antbird species complex (*Hylophylax naevius/Hylophylax naevioides*). *Journal of Biogeography* 41(6):1094–1104.
Fitzpatrick, J. W., P. F. D. Boesman, and H. F. Greeney. 2022. Lesser Wagtail-Tyrant (*Stigmatura napensis*), version 1.2. In S. M. Billerman, ed. *Birds of the World.* Cornell Lab of Ornithology, Ithaca, NY. https://doi.org/10.2173/bow.lewtyr1.01.2
Fjeldså, J. 1995. Geographical patterns of neoendemic and older relict species of Andean forest birds: The significance of ecologically stable areas. Pages 89–102 in S. P. Churchill, H. Basley, E. Forero, and J. L. Luteyn, eds. *Biodiversity and Conservation of Neotropical Montane Forests.* New York Botanical Garden, New York.
Fjeldså, J., R. C. Bowie, and C. Rahbek. 2012. The role of mountain ranges in the diversification of birds. *Annual Review of Ecology, Evolution and Systematics* 43:249–265.
Fjeldså, J., E. Lambin, and B. Mertens. 1999. Correlation between endemism and local ecoclimatic stability documented by comparing Andean bird distributions and remotely sensed land surface data. *Ecography* 22(1):63–78.
Freeman, B. G. 2015. Competitive interactions upon secondary contact drive elevational divergence in tropical birds. *The American Naturalist* 186(4):470–479.
Freeman, B. G., M. Strimas-Mackey, and E. T. Miller. 2022. Interspecific competition limits bird species' ranges in tropical mountains. *Science* 377(6604):416–420.
García-Moreno, J., P. Arctander, and J. Fjeldså. 1999. Strong diversification at the treeline among *Metallura* hummingbirds. *The Auk* 116(3):702–711.
García-Moreno, J., and J. Fjeldså. 2000. Chronology and mode of speciation in the Andean avifauna. *Bonner Zoological Monographs* 46:25–46.
Gaston, K. J., and T. M. Blackburn. 1996. The tropics as a museum of biological diversity: An analysis of the New World avifauna. *Proceedings of the Royal Society B: Biological Sciences* 263(1366):63–68.
Gerolamo, C. S., F. R. Costa, A. R. Zuntini, A. Vicentini, L. G. Lohmann, J. Schietti, . . . A. Nogueira. 2022. Hydro-edaphic gradient and phylogenetic history explain the landscape distribution of a highly diverse clade of lianas in the Brazilian Amazon. *Frontiers in Forests and Global Change* 5: Article 809904.
Goldberg, E. E., K. Roy, R. Lande, and D. Jablonski. 2005. Diversity, endemism, and age distributions in macroevolutionary sources and sinks. *The American Naturalist* 165(6):623–633.
Gotelli, N. J., M. J. Anderson, H. T. Arita, A. Chao, R. K. Colwell, S. R. Connolly, . . . M. R. Willig. 2009. Patterns and causes of species richness: A general simulation model for macroecology. *Ecology Letters* 12(9):873–886.
Graves, G. R. 1982. Speciation in the carbonated flower-piercer (*Diglossa carbonaria*) complex of the Andes. *The Condor* 84(1):1–14.
Graves, G. R. 1985. Elevational correlates of speciation and intraspecific geographic variation in plumage in Andean forest birds. *The Auk* 102(3):556–579.
Groot, M. H. M., L. J. Lourens, H. Hooghiemstra, M. Vriend, J. C. Berrio, E. Tuenter, . . . W. Westerhoff. 2011. Ultra-high resolution pollen record from the northern Andes reveals rapid shifts in montane climates within the last two glacial cycles. *Climate of the Past* 7(1):299–316.

Guillaumet, J. L. 1987. Some structural and floristic aspects of the forest. *Experientia* 43(3):241–251.
Haffer, J. 1967a. Some allopatric species pairs of birds in north-western Colombia. *The Auk* 84(3):343–365.
Haffer, J. 1967b. *Speciation in Colombian Forest Birds West of the Andes.* American Museum of Natural History, New York.
Haffer, J. 1969. Speciation in Amazonian forest birds: Most species probably originated in forest refuges during dry climatic periods. *Science* 165(3889):131–137.
Haffer, J. 1970. Geologic–climatic history and zoogeographic significance of the Uraba region in northwestern Colombia. *Caldasia* 10:603–636.
Haffer, J. 1974. *Avian Speciation in Tropical South America, With a Systematic Survey of the Toucans (Ramphastidae) and jacamars (Galbulidae).* Nuttall Ornithological Club, Cambridge, MA.
Haffer, J. 1987. Quaternary history of tropical America. Pages 1–18 in T. C. Whitmore and G. T. Prance, eds. *Biogeography and Quaternary History in Tropical America.* Oxford University Press, New York.
Haffer, J. 1992. On the "river effect" in some forest birds of southern Amazonia. *Boletim do Museu Paraense Emílio Goeldi. Nova Série. Zoologia, 8*(1), 217–245.
Haffer, J. 2008. Hypotheses to explain the origin of species in Amazonia. *Brazilian Journal of Biology* 68:917–947.
Harcourt, A. H., and M. A. Wood. 2012. Rivers as barriers to primate distributions in Africa. *International Journal of Primatology* 33(1):168–183.
Harris, S. E., and A. C. Mix. 2002. Climate and tectonic influences on continental erosion of tropical South America, 0–13 Ma. *Geology* 30(5):447–450.
Harvey, M. G., G. A. Bravo, S. Claramunt, A. M. Cuervo, G. E. Derryberry, J. Battilana, . . . E. P. Derryberry. 2020. The evolution of a tropical biodiversity hotspot. *Science* 370(6522): 1343–1348.
Harvey, M. G., G. F. Seeholzer, B. T. Smith, D. L. Rabosky, A. M. Cuervo, and R. T. Brumfield. 2017. Positive association between population genetic differentiation and speciation rates in New World birds. *Proceedings of the National Academy of Sciences of the USA* 114(24):6328–6333.
Hawkins, B. A., C. M. McCain, T. J. Davies, L. B. Buckley, B. L. Anacker, H. V. Cornell, . . . P. R. Stephens. 2012. Different evolutionary histories underlie congruent species richness gradients of birds and mammals. *Journal of Biogeography* 39(5):825–841.
Hawkins, B. A., E. E. Porter, and J. A. Felizola Diniz-Filho. 2003. Productivity and history as predictors of the latitudinal diversity gradient of terrestrial birds. *Ecology* 84(6):1608–1623.
Hazzi, N. A., J. S. Moreno, C. Ortiz-Movliav, and R. D. Palacio. 2018. Biogeographic regions and events of isolation and diversification of the endemic biota of the tropical Andes. *Proceedings of the National Academy of Sciences of the USA* 115(31):7985–7990.
Helmstetter, A. J., S. Glemin, J. Käfer, R. Zenil-Ferguson, H. Sauquet, H de Boer, . . . F. L. Condamine. 2022. Pulled diversification rates, lineages-through-time plots, and modern macroevolutionary modeling. *Systematic Biology* 71(3):758–773.
Herzog, S. K., and G. H. Kattan. 2011. Patterns of diversity and endemism in the birds of the tropical Andes. Pages 245–259 in S. K. Herzog, R. Martinez, P. M. Jorgensen, and H. Tiessen, eds. *Climate Change and Biodiversity in the Tropical Andes.* McArthur Foundation, Paris.
Hibbins, M. S., and M. W. Hahn. 2022. Phylogenomic approaches to detecting and characterizing introgression. *Genetics* 220(2): Article iyab173.
Hijmans, R. J. 2022. raster: Geographic data analysis and modeling. R package version 3.5-15. https://CRAN.R-project.org/package=raster.
Hooghiemstra, H., and T. Van der Hammen. 2004. Quaternary Ice-Age dynamics in the Colombian Andes: Developing an understanding of our legacy. *Philosophical Transactions of the Royal Society B: Biological Sciences* 359(1442):173–181.
Hooghiemstra, H., V. M. Wijninga, and A. M. Cleef. 2006. The paleobotanical record of Colombia: Implications for biogeography and biodiversity. *Annals of the Missouri Botanical Garden* 93:297–324.
Hoorn, C. 2006. The birth of the mighty Amazon. *Scientific American* 294(5):52–59.
Hoorn, C., G. R. Bogotá-A, M. Romero-Baez, E. I. Lammertsma, S. G. Flantua, E. L. Dantas, . . . F. Chemale, Jr. 2017. The Amazon at sea: Onset and stages of the Amazon River from a marine

record, with special reference to Neogene plant turnover in the drainage basin. *Global and Planetary Change* 153:51–65.

Hoorn, C., F. P. Wesselingh, H. Ter Steege, M. A. Bermudez, A. Mora, J. Sevink, . . . A. Antonelli. 2010. Amazonia through time: Andean uplift, climate change, landscape evolution, and biodiversity. *Science* 330(6006):927–931.

Hosner, P. A., E. L. Braun, and R. T. Kimball. 2016. Rapid and recent diversification of curassows, guans, and chachalacas (Galliformes: Cracidae) out of Mesoamerica: Phylogeny inferred from mitochondrial, intron, and ultraconserved element sequences. *Molecular Phylogenetics and Evolution* 102:320–330.

Hovenkamp, P. 1997. Vicariance events, not areas, should be used in biogeographical analysis. *Cladistics* 13(1–2):67–79.

Ibarra-Macias, A., W. D. Robinson, and M. S. Gaines. 2011. Experimental evaluation of bird movements in a fragmented neotropical landscape. *Biological Conservation* 144:703–712.

Jackson, N. D., and C. C. Austin. 2010. The combined effects of rivers and refugia generate extreme cryptic fragmentation within the common ground skink (*Scincella lateralis*). *Evolution* 64(2):409–428.

Jetz, W., G. H. Thomas, J. B. Joy, K. Hartmann, and A. O. Mooers. 2012. The global diversity of birds in space and time. *Nature* 491(7424):444–448.

Jiao, X., T. Flouri, and Z. Yang. 2021. Multispecies coalescent and its applications to infer species phylogenies and cross-species gene flow. *National Science Review* 8(12): Article nwab127.

Junk, W. J., M. T. F. Piedade, J. Schöngart, M. Cohn-Haft, J. M. Adeney, and F. Wittmann. 2011. A classification of major naturally-occurring Amazonian lowland wetlands. *Wetlands* 31(4):623–640.

Karr, J. R., and F. C. James. 1975 Eco-morphological configuration and convergent evolution in species communities. Pages 258–291 in M. L. Cody and J. M. Diamond, eds. *Ecology and Evolution of Communities*. Harvard University Press, Cambridge, MA.

Kattan, G. H., S. A. Tello, M. Giraldo, and C. D. Cadena. 2016. Neotropical bird evolution and 100 years of the enduring ideas of Frank M. Chapman. *Biological Journal of the Linnean Society* 117(3):407–413.

Keast, A. 1961. Bird speciation on the Australian continent. *Bulletin of the Museum of Comparative Zoology at Harvard College* 123:403–495.

Kennedy, J. D., Z. Wang, J. T. Weir, C. Rahbek, J. Fjeldså, and T. D. Price. 2014. Into and out of the tropics: The generation of the latitudinal gradient among New World passerine birds. *Journal of Biogeography* 41(9):1746–1757.

Klopfer, P. H., and R. H. MacArthur. 1961. On the causes of tropical species diversity: Niche overlap. *The American Naturalist* 95(883):223–226.

Knapp, S., and J. Mallet. 2003. Refuting refugia? *Science* 300(5616):71–72.

Krabbe, N. 2008. Arid valleys as dispersal barriers to high-Andean forest birds in Ecuador. *Cotinga* 29:28–30.

Laranjeiras, T. O., R. D. Melinski, L. N. Naka, G. A. Leite, G. R. Lima, J. A. d'Affonseca-Neto, and M. Cohn-Haft. 2019. Three bird species new to Brazil from the Serra da Mocidade, a remote mountain in Roraima. *Revista Brasileira de Ornitologia* 27(4):275–283.

Laranjeiras, T. O., L. N. Naka, and M. Cohn-Haft. 2019. Using river color to predict Amazonian floodplain forest avifaunas in the world's largest blackwater river basin. *Biotropica* 51(3):330–341.

Latrubesse, E. M., M. Cozzuol, S. A. da Silva-Caminha, C. A. Rigsby, M. L. Absy, and C. Jaramillo. 2010. The Late Miocene paleogeography of the Amazon Basin and the evolution of the Amazon River system. *Earth-Science Reviews* 99(3-4):99–124.

Laurance, S. G., P. C. Stouffer, and W. F. Laurance. 2004. Effects of road clearings on movement patterns of understory rainforest birds in Central Amazonia. *Conservation Biology* 18:1099–1109.

Lima, H. S., G. A. Bravo, D. Astúa, D. Mariz, S. V. Edwards, and L. N. Naka. 2025. Origins and Diversification of the Caatinga Dry Forest Endemic Avifauna. *Journal of Biogeography*: e70003.

Lima-Rezende, C. A., G. S. Cabanne, A. V. Rocha, M. Carboni, R. M. Zink, and R. Caparroz. 2022. A comparative phylogenomic analysis of birds reveals heterogeneous differentiation processes among Neotropical savannas. *Molecular Ecology* 31(12):3451–3467.

Luna, L. W., C. C. Ribas, and A. Aleixo. 2022. Genomic differentiation with gene flow in a widespread Amazonian floodplain-specialist bird species. *Journal of Biogeography* 49(9):1670–1682.

Lutz, H. L., J. D. Weckstein, J. S. Patané, J. M. Bates, and A. Aleixo. 2013. Biogeography and spatio-temporal diversification of *Selenidera* and *Andigena* Toucans (Aves: Ramphastidae). *Molecular Phylogenetics and Evolution* 69(3):873–883.

Marks, B. D., S. J. Hackett, and A. P. Capparella. 2002. Historical relationships among Neotropical lowland forest areas of endemism as determined by mitochondrial DNA sequence variation within the Wedge-billed Woodcreeper (Aves: Dendrocolaptidae: *Glyphorynchus spirurus*). *Molecular Phylogenetics and Evolution* 24(1):153–167.

Marroig, G., and R. Cerqueira. 1997. Plio-Pleistocene South American history and the Amazon Lagoon hypothesis: A piece in the puzzle of Amazonian diversification. *Journal of Comparative Biology* 2(2):103–119.

Marshall, L. G. 1988. Land mammals and the great American interchange. *American Scientist* 76(4):380–388.

Matos-Maraví, P., N. Wahlberg, A. V. Freitas, P. Devries, A. Antonelli, and C. M. Penz. 2021. Mesoamerica is a cradle and the Atlantic Forest is a museum of Neotropical butterfly diversity: Insights from the evolution and biogeography of Brassolini (Lepidoptera: Nymphalidae). *Biological Journal of the Linnean Society* 133(3):704–724.

Mauck, W. M., III, and K. J. Burns. 2009. Phylogeny, biogeography, and recurrent evolution of divergent bill types in the nectar-stealing flowerpiercers (Thraupini: *Diglossa* and *Diglossopis*). *Biological Journal of the Linnean Society* 98(1):14–28.

Mayr, E. 1969. Bird speciation in the tropics. *Biological Journal of the Linnean Society* 1(1–2):1–17.

Mayr, E., and R. J. O'Hara. 1986. The biogeographic evidence supporting the Pleistocene forest refuge hypothesis. *Evolution* 40(1):55–67.

Mayr, E., and W. H. Phelps. 1967. The origin of the bird fauna of the south Venezuelan highlands. *Bulletin of the American Museum of Natural History* 136:309–327.

Mayr, G. 2011. Two-phase extinction of "Southern Hemispheric" birds in the Cenozoic of Europe and the origin of the Neotropic avifauna. *Palaeobiodiversity and Palaeoenvironments* 91(4):325–333.

Mayr, G. 2017. Avian higher level biogeography: Southern Hemispheric origins or Southern Hemispheric relicts? *Journal of Biogeography* 44(4):956–958.

Mayr, G., and V. L. De Pietri. 2014. Earliest and first Northern Hemispheric hoatzin fossils substantiate Old World origin of a "Neotropic endemic." *Naturwissenschaften* 101(2):143–148.

McCullough, J. M., R. G. Moyle, B. T. Smith, and M. J. Andersen. 2019. A Laurasian origin for a pantropical bird radiation is supported by genomic and fossil data (Aves: Coraciiformes). *Proceedings of the Royal Society B: Biological Sciences* 286(1910): Article 20190122.

Miller, M. J., E. Bermingham, J. Klicka, P. Escalante, F. S. R. Do Amaral, J. T. Weir, and K. Winker. 2008. Out of Amazonia again and again: Episodic crossing of the Andes promotes diversification in a lowland forest flycatcher. *Proceedings of the Royal Society B: Biological Sciences* 275(1639):1133–1142.

Mittelbach, G. G., D. W. Schemske, H. V. Cornell, A. P. Allen, J. M. Brown, M. B. Bush, . . . M. Turelli. 2007. Evolution and the latitudinal diversity gradient: Speciation, extinction and biogeography. *Ecology Letters* 10(4):315–331.

Montes, C., A. Cardona, R. McFadden, S. E. Morón, C. A. Silva, S. Restrepo-Moreno, . . . J. A. Flores. 2012. Evidence for middle Eocene and younger land emergence in central Panama: Implications for isthmus closure. *GSA* Bulletin 124(5–6):780–799.

Moore, R. P., W. D. Robinson, I. J. Lovette, and T. R. Robinson. 2008. Experimental evidence for extreme dispersal limitation in tropical forest birds. *Ecology Letters* 11:960–968.

Mora, A., P. Baby, M. Roddaz, M. Parra, S. Brusset, W. Hermoza, and N. Espurt. 2010. Tectonic history of the Andes and sub-Andean zones: Implications for the development of the Amazon drainage basin. Pages 38–60 in C. Hoorn and F. P. Wesselingh, eds. *Amazonia, Landscape and Species Evolution: A Look Into the Past.* Blackwell, Hoboken, NJ.

Moreau, R. E. 1963. Vicissitudes of the African biomes in the Late Pleistocene. *Proceedings of the Zoological Society of London* 141(2):395–421.

Muller, P. 1973. The dispersal centers of terrestrial vertebrates in the Neotropical realm: A study in the evolution of the Neotropical biota and its native landscape. *Biogeographica* 2:1–250.
Musher, L. J., M. Giakoumis, J. Albert, G. Del-Rio, M. Rego, G. Thom, . . . J. Cracraft. 2022. River network rearrangements promote speciation in lowland Amazonian birds. *Science Advances* 8(14): Article eabn1099.
Naka, L. N. 2011. Avian distribution patterns in the Guiana Shield: Implications for the delimitation of Amazonian areas of endemism. *Journal of Biogeography* 38(4):681–696.
Naka, L. N., C. L. Bechtoldt, L. M. P. Henriques, and R. T. Brumfield. 2012. The role of physical barriers in the location of avian suture zones in the Guiana Shield, northern Amazonia. *The American Naturalist* 179(4):E115–E132.
Naka, L. N., and R. T. Brumfield. 2018. The dual role of Amazonian rivers in the generation and maintenance of avian diversity. *Science Advances* 4(8): Article eaar8575.
Naka, L. N., M. Cohn-Haft, A. Whittaker, J. M. Barnett, and M. D. F. Torres. 2007. Avian biogeography of Amazonian flooded forests in the Rio Branco Basin, Brazil. *Wilson Journal of Ornithology* 119(3):439–449.
Naka, L. N., B. M. da S. Costa, G. Rodrigues Lima, and S. Claramunt. 2022. Riverine barriers as obstacles to dispersal in Amazonian birds. *Frontiers in Ecology and Evolution* 10: Article 846975.
Naka, L. N., T. O. Laranjeiras, G. R. Lima, A. C. Plaskievicz, D. Mariz, B. M. Da Costa, . . . M. Cohn-Haft. 2020. The Avifauna of the Rio Branco, an Amazonian evolutionary and ecological hotspot in peril. *Bird Conservation International* 30(1):21–39.
Norambuena, H. V., and P. Van Els. 2021. A general scenario to evaluate evolution of grassland birds in the Neotropics. *Ibis* 163:722–727.
Nores, M. 1992. Bird speciation in subtropical South America in relation to forest expansion and retraction. *The Auk* 109(2):346–357.
Nores, M. 1999. An alternative hypothesis for the origin of Amazonian bird diversity. *Journal of Biogeography* 26(3):475–485.
O'Dea, A., H. A. Lessios, A. G. Coates, R. I. Eytan, S. A. Restrepo-Moreno, A. L. Cione, . . . J. B. Jackson. 2016. Formation of the Isthmus of Panama. *Science Advances* 2(8): Article e1600883.
Oliveira, F. B. D., E. C. Molina, and G. Marroig. 2009. Paleogeography of the South Atlantic: A route for primates and rodents into the New World? Pages 55–68 in P. A. Garber, A. Estrada, J. C. Bicca-Marques, E. W. Heymann, and K. B. Strier, eds. *South American Primates.* Springer, New York.
Oliveros, C. H., D. J. Field, D. T. Ksepka, F. K. Barker, A. Aleixo, M. J. Andersen, . . . B. C. Faircloth. 2019. Earth history and the passerine superradiation. *Proceedings of the National Academy of Sciences of the USA* 116(16):7916–7925.
Olson, D. M., E. Dinerstein, E. D. Wikramanayake, N. D. Burgess, G. V. Powell, E. C. Underwood, . . . C. J. Loucks. 2001. Terrestrial ecoregions of the world: A new map of life on Earth: A new global map of terrestrial ecoregions provides an innovative tool for conserving biodiversity. *BioScience* 51(11):933–938.
Patton, J. L., M. N. F. Da Silva, and J. R. Malcolm. 2000. Mammals of the Rio Juruá and the evolutionary and ecological diversification of Amazonia. *Bulletin of the American Museum of Natural History* 244:1–306.
Pebesma, E. 2018. Simple features for R: Standardized support for spatial vector data. *The R Journal* 10(1):439–446.
Pennington, R.T., Lavin, M. and Oliveira-Filho, A., 2009. Woody plant diversity, evolution, and ecology in the tropics: perspectives from seasonally dry tropical forests. *Annual Review of Ecology, Evolution, and Systematics* 40(1): 437–457.
Pérez-Emán, J. L. 2005. Molecular phylogenetics and biogeography of the Neotropical redstarts (*Myioborus*; Aves, Parulinae). *Molecular Phylogenetics and Evolution* 37(2):511–528.
Perrigo, A., C. Hoorn, and A. Antonelli. 2020. Why mountains matter for biodiversity. *Journal of Biogeography* 47(2):315–325.
Peterson, A. T., J. Soberón, and V. Sánchez-Cordero. 1999. Conservatism of ecological niches in evolutionary time. *Science* 285(5431):1265–1267.
Poulsen, C. J., T. A. Ehlers, and N. Insel. 2010. Onset of convective rainfall during gradual late Miocene rise of the central Andes. *Science* 328(5977):490–493.

Poux, C., P. Chevret, D. Huchon, W. W. De Jong, and E. J. Douzery. 2006. Arrival and diversification of caviomorph rodents and platyrrhine primates in South America. *Systematic Biology* 55(2):228–244.
Prado, D. E. 1992. A Critical Evaluation of the Floristic Links Between Chaco and Caatingas Vegetation in South America. Doctoral dissertation, University of St. Andrews, Fife, UK.
Prado, D. E., and P. E. Gibbs. 1993. Patterns of species distributions in the dry seasonal forests of South America. *Annals of the Missouri Botanical Garden* 328:902–927.
Pulido-Santacruz, P., and J. T. Weir. 2016. Extinction as a driver of avian latitudinal diversity gradients. *Evolution* 70(4):860–872.
Quintero, E., C. C. Ribas, and J. Cracraft. 2013. The Andean *Hapalopsittaca* parrots (Psittacidae, Aves): An example of montane–tropical lowland vicariance. *Zoologica Scripta* 42(1):28–43.
Quintero, I., and W. Jetz. 2018. Global elevational diversity and diversification of birds. *Nature* 555(7695):246–250.
Rabosky, D. L., P. O. Title, and H. Huang. 2015. Minimal effects of latitude on present-day speciation rates in New World birds. *Proceedings of the Royal Society B: Biological Sciences* 282(1809): Article 20142889.
Rahbek, C. 2005. The role of spatial scale and the perception of large-scale species-richness patterns. *Ecology Letters* 8(2):224–239.
Rahbek, C., M. K. Borregaard, A. Antonelli, R. K. Colwell, B. G. Holt, D. Nogues-Bravo, . . . J. Fjeldså. 2019. Building mountain biodiversity: Geological and evolutionary processes. *Science* 365(6458):1114–1119.
Rahbek, C., and G. R. Graves. 2001. Multiscale assessment of patterns of avian species richness. *Proceedings of the National Academy of Sciences of the USA* 98(8):4534–4539.
Ramírez-Barahona, S., and L. E. Eguiarte. 2013. The role of glacial cycles in promoting genetic diversity in the Neotropics: The case of cloud forests during the Last Glacial Maximum. *Ecology and Evolution* 3(3):725–738.
Rangel, T. F., N. R. Edwards, P. B. Holden, J. A. F. Diniz-Filho, W. D. Gosling, M. T. P. Coelho, . . . R. K. Colwell. 2018. Modeling the ecology and evolution of biodiversity: Biogeographical cradles, museums, and graves. *Science* 361(6399): Article eaar5452.
Remsen, J. V., Jr. 1984. High incidence of "leapfrog" pattern of geographic variation in Andean birds: Implications for the speciation process. *Science* 224(4645):171–173.
Remsen, J. V., Jr. 2003. Family Furnariidae (ovenbirds). Pages 162–357 in J. Del Hoyo, A. Elliott, and D. A. Christie, eds. *Handbook of the Birds of the World, Vol. 8: Broadbills to Tapaculos*. Lynx Edicions, Barcelona, Spain.
Remsen, J. V., Jr., & T. A., Parker III. 1983. Contribution of river-created habitats to bird species richness in Amazonia. *Biotropica* 15(3):223–231.
Ribas, C. C., and A. Aleixo. 2019. Diversity and evolution of Amazonian birds: Implications for conservation and biogeography. *Anais da Academia Brasileira de Ciências* 91: Article e20190218.
Ribas, C. C., A. Aleixo, A. C. Nogueira, C. Y. Miyaki, and J. Cracraft. 2012. A palaeobiogeographic model for biotic diversification within Amazonia over the past three million years. *Proceedings of the Royal Society B: Biological Sciences* 279(1729):681–689.
Ribas, C. C., R. Gaban-Lima, C. Y. Miyaki, and J. Cracraft. 2005. Historical biogeography and diversification within the Neotropical parrot genus *Pionopsitta* (Aves: Psittacidae). *Journal of Biogeography* 32(8):1409–1427.
Ribas, C. C., M. Maldonado Coelho, B. T. Smith, G. S. Cabanne, F. M. d'Horta, and L. N. Naka. 2012. Towards an integrated historical biogeography of the Neotropical lowland avifauna: Combining diversification analysis and landscape evolution. *Ornitología Neotropical* 23:187–206.
Ribas, C. C., R. G. Moyle, C. Y. Miyaki, and J. Cracraft. 2007. The assembly of montane biotas: Linking Andean tectonics and climatic oscillations to independent regimes of diversification in *Pionus* parrots. *Proceedings of the Royal Society B: Biological Sciences* 274(1624):2399–2408.
Ribas, C. C., A. O. Sawakuchi, R. P. Almeida, F. N. Pupim, M. A. Rego, R. Batista, and L. L, Knowles. 2025. The role of rivers in the origin and future of Amazonian biodiversity. *Nature Reviews Biodiversity* 1(1): 14–31.
Ricklefs, R. E. 1987. Community diversity: Relative roles of local and regional processes. *Science* 235(4785):167–171.

Ricklefs, R. E. 2002. Splendid isolation: Historical ecology of the South American passerine fauna. *Journal of Avian Biology* 33(3):207–211.
Ricklefs, R. E. 2003. Global diversification rates of passerine birds. *Proceedings of the Royal Society B: Biological Sciences* 270(1530):2285–2291.
Ricklefs, R. E. 2006. Global variation in the diversification rate of passerine birds. *Ecology* 87(10):2468–2478.
Ritter, C. D., L. A. Coelho, J. M. Capurucho, S. H. Borges, C. Cornelius, and C. C. Ribas. 2021. Sister species, different histories: Comparative phylogeography of two bird species associated with Amazonian open vegetation. *Biological Journal of the Linnean Society* 132(1):161–173.
Rocha, A. V., Cabanne, G. S. Cabanne, A. Aleixo, L. F. Silveira, P. Tubaro, and R. Caparroz. 2020. Pleistocene climatic oscillations associated with landscape heterogeneity of the South American dry diagonal explains the phylogeographic structure of the narrow-billed woodcreeper (*Lepidocolaptes angustirostris*, Dendrocolaptidae). *Journal of Avian Biology* 51(9): Article 02537.
Rocha, D. G. D., and I. L. Kaefer. 2019. What has become of the refugia hypothesis to explain biological diversity in Amazonia? *Ecology and Evolution* 9(7):4302–4309.
Rosen, D. E. 1978. Vicariant patterns and historical explanation in biogeography. *Systematic Zoology* 27(2):159–188.
Rosenberg, G. H. 1990. Habitat specialization and foraging behavior by birds of Amazonian river islands in northeastern Peru. *The Condor* 92(2):427–443.
Rull, V. 2005. Biotic diversification in the Guayana Highlands: A proposal. *Journal of Biogeography* 32(6):921–927.
Rull, V. 2011. Neotropical biodiversity: timing and potential drivers. *Trends in Ecology & Evolution* 26(10):508–513.
Rull, V., and A. C. Carnaval, Eds. 2020. *Neotropical Diversification: Patterns and Processes.* Springer, Berlin.
Salisbury, C. L., N. Seddon, C. R. Cooney, and J. A. Tobias. 2012. The latitudinal gradient in dispersal constraints: Ecological specialisation drives diversification in tropical birds. *Ecology Letters* 15(8):847–855.
Sanín, C., C. D. Cadena, J. M. Maley, D. A. Lijtmaer, P. L. Tubaro, and R. T. Chesser. 2009. Paraphyly of *Cinclodes fuscus* (Aves: Passeriformes: Furnariidae): Implications for taxonomy and biogeography. *Molecular Phylogenetics and Evolution* 53(2):547–555.
Santorelli, S., Jr., W. E. Magnusson, C. P. de Deus, and T. H. Keitt. 2022. Neutral processes and reduced dispersal across Amazonian rivers may explain how rivers maintain species diversity after secondary contact. *Perspectives in Ecology and Conservation* 20:151–158.
Savit, A. Z., and J. M. Bates. 2015. Right around the Amazon: The origin of the circum-Amazonian distribution in *Tangara cayana*. *Folia Zoologica* 64(3): 273–283.
Sedano, R. E., and K. J. Burns. 2010. Are the Northern Andes a species pump for Neotropical birds? Phylogenetics and biogeography of a clade of Neotropical tanagers (Aves: Thraupini). *Journal of Biogeography* 37(2):325–343.
Selvatti, A. P., A. Galvão, G. Mayr, C. Y. Miyaki, and C. A. D. M. Russo. 2022. Southern Hemisphere tectonics in the Cenozoic shaped the pantropical distribution of parrots and passerines. *Journal of Biogeography* 49(10):1753–1766.
Selvatti, A. P., A. Galvao, A. G. Pereira, L. Pedreira Gonzaga, and C. A. D. M. Russo. 2016. An African origin of the Eurylaimides (Passeriformes) and the successful diversification of the ground-foraging pittas (Pittidae). *Molecular Biology and Evolution* 34(2):483–499.
Selvatti, A. P., L. P. Gonzaga, and C. A. de Moraes Russo. 2015. A Paleogene origin for crown passerines and the diversification of the Oscines in the New World. *Molecular Phylogenetics and Evolution* 88, 1–15.
Shakya, S. B., J. Fuchs, J. M. Pons, and F. H. Sheldon. 2017. Tapping the woodpecker tree for evolutionary insight. *Molecular Phylogenetics and Evolution* 116;182–191.
Short, L. L. 1975. A zoogeographic analysis of the South American Chaco avifauna. *Bulletin of the American Museum of Natural History* 154:163–352.
Sick, H. 1967. Rios e enchentes na Amazônia como obstáculo para a avifauna. *Atas do Simpósio Sobre a Biota Amazônica* 5:495–520.

Sick, H. 1985. Observations on the Andean–Patagonian component of southeastern Brazil's avifauna. *Ornithological Monographs* 36:233–237.
Silva, J. M. C., and J. M. Bates. 2002. Biogeographic patterns and conservation in the South American Cerrado: A tropical savanna hotspot. *BioScience* 52(3):225–234.
Silva, S. M., A. T. Peterson, L. Carneiro, T. C. T. Burlamaqui, C. C. Ribas, T. Sousa-Neves, . . . A. Aleixo. 2019. A dynamic continental moisture gradient drove Amazonian bird diversification. *Science Advances* 5(7): Article eaat5752.
Simpson, B. B. 1979. Quaternary biogeography of the high montane regions of South America. Pages 157–188 in W. E. Duellman, ed. *The South American Herpetofauna: Its Origin, Evolution, and Dispersal*. Monographs of the Museum of Natural History University of Kansas, Lawrence, KS.
Simpson, G. G. 1980. *Splendid Isolation. The Curious History of South American Mammals*. Yale University Press, New Haven, CT.
Skutch, A. F. 1996. *Antbirds and Ovenbirds: Their Lives and Homes* (Vol. 31). University of Texas Press, Austin, TX.
Smith, B. T., and J. Klicka. 2010. The profound influence of the Late Pliocene Panamanian uplift on the exchange, diversification, and distribution of New World birds. *Ecography* 33(2):333–342.
Smith, B. T., J. E. McCormack, A. M. Cuervo, M. Hickerson, A. Aleixo, C. D. Cadena, . . . R. T. Brumfield. 2014. The drivers of tropical speciation. *Nature* 515(7527):406–409.
Smith, B. T., J. Merwin, K. L. Provost, G. Thom, R. T. Brumfield, M. Ferreira, . . . L. Joseph. 2022. Phylogenomic analysis of the parrots of the world distinguishes artifactual from biological sources of gene tree discordance. *Systematic Biology* 72(1):228–241.
Smith, B. T., G. F. Seeholzer, M. G. Harvey, A. M. Cuervo, and R. T. Brumfield. 2017. A latitudinal phylogeographic diversity gradient in birds. *PLoS Biology* 15(4): Article e2001073.
Sonne, J., B. Dalsgaard, M. K. Borregaard, J. Kennedy, J. Fjeldså, and C. Rahbek. 2022. Biodiversity cradles and museums segregating within hotspots of endemism. *Proceedings of the Royal Society B: Biological Sciences* 289(1981): Article 20221102.
Stebbins, G. L. 1974. *Flowering Plants: Evolution Above the Species Level*. Harvard University Press, Cambridge, MA.
Taboada, A., L. A. Rivera, A. Fuenzalida, A. Cisternas, H. Philip, H. Bijwaard, . . . C. Rivera. 2000. Geodynamics of the northern Andes: Subductions and intracontinental deformation (Colombia). *Tectonics* 19(5):787–813.
Terborgh, J. 1971. Distribution on environmental gradients: Theory and a preliminary interpretation of distributional patterns in the avifauna of the Cordillera Vilcabamba, Peru. *Ecology* 52(1):23–40.
Terborgh, J., and J. S. Weske. 1975. The role of competition in the distribution of Andean birds. *Ecology* 56(3):562–576.
Thom, G., A. T. Xue, A. O. Sawakuchi, C. C. Ribas, M. J. Hickerson, A. Aleixo, and C. Miyaki. 2020. Quaternary climate changes as speciation drivers in the Amazon floodplains. *Science Advances* 6(11): Article eaax4718.
Tietje, M., A. Antonelli, W. J. Baker, R. Govaerts, S. A. Smith, and W. L. Eiserhardt. 2022. Global variation in diversification rate and species richness are unlinked in plants. *Proceedings of the National Academy of Sciences of the USA* 119(27): Article e2120662119.
Torres, V., H. Hooghiemstra, L. Lourens, and P. C. Tzedakis. 2013. Astronomical tuning of long pollen records reveals the dynamic history of montane biomes and lake levels in the tropical high Andes during the Quaternary. *Quaternary Science Reviews* 63:59–72.
Trujillo-Arias, N., M. J. Rodríguez-Cajarville, E. Sari, C. Y. Miyaki, F. R. Santos, C. C. Witt, . . . G. S. Cabanne. 2020. Evolution between forest macrorefugia is linked to discordance between genetic and morphological variation in Neotropical passerines. *Molecular Phylogenetics and Evolution* 149: Article 106849.
Valderrama, E., J. L. Pérez-Emán, R. T. Brumfield, A. M. Cuervo, and C. D. Cadena. 2014. The influence of the complex topography and dynamic history of the montane Neotropics on the evolutionary differentiation of a cloud forest bird (*Premnoplex brunnescens*, Furnariidae). *Journal of Biogeography* 41(8):1533–1546.
van Els, P., E. Zarza, L. Rocha Moreira, V. Gómez-Bahamón, A. Santana, A. Aleixo, . . . J. Berv. 2021. Recent divergence and lack of shared phylogeographic history characterize the diversification of neotropical savanna birds. *Journal of Biogeography* 48(5):1124–1137.

Vargas, O. M., and C. W. Dick. 2020. Diversification history of Neotropical Lecythidaceae, an ecologically dominant tree family of Amazon rain forest. Pages 791–809 in V. Rull and A. Carnaval, eds. *Neotropical Diversification: Patterns and Processes*. Springer, Cham, Switzerland.
Vasconcelos, T., B. C. O'Meara, and J. M. Beaulieu. 2022. Retiring "cradles" and "museums" of biodiversity. *The American Naturalist* 199(2):194–205.
Vuilleumier, F. 1969. Pleistocene speciation in birds living in the high Andes. *Nature* 223(5211):1179–1180.
Vuilleumier, F. 1986. Origins of the tropical avifaunas of the high Andes. Pages 586–622 in F. Vuilleumier and M. Moasterio, eds. *High Altitude Tropical Biogeography*. Oxford University Press, New York.
Vuilleumier, F. 1991. A quantitative survey of speciation phenomena in Patagonian birds. *Ornitología Neotropical* 2:5–28.
Wallace, A. R. 1852. On the monkeys of the Amazon. *Proceedings of the Zoological Society of London* 20:107–110.
Wallace, A. R. 1876. *The Geographical Distribution of Animals*. Vol. 1. Harper and Brothers, New York.
Webb, S. D. 1991. Ecogeography and the great American interchange. *Paleobiology* 17(3):266–280.
Webster, G. L. 1995. The panorama of Neotropical cloud forests. Pages 53–77 in S. P. Churchill, H. Balslev, E. Forero, and J. L. Luteyn, eds. *Biodiversity and Conservation of Neotropical Montane Forests. Neotropical Montane Forest Biodiversity and Conservation Symposium 1*. New York Botanical Garden, New York.
Weeks, B. C., S. Claramunt, and J. Cracraft. 2016. Integrating systematics and biogeography to disentangle the roles of history and ecology in biotic assembly. *Journal of Biogeography* 43(8):1546–1559.
Weeks, B. C., B. K. O'Brien, J. J. Chu, S. Claramunt, C. Sheard, and J. A. Tobias. 2022. Morphological adaptations linked to flight efficiency and aerial lifestyle determine natal dispersal distance in birds. *Functional Ecology* 36(7):1681–1689.
Weir, J. T. 2006. Divergent timing and patterns of species accumulation in lowland and highland neotropical birds. *Evolution* 60(4):842–855.
Weir, J. T. 2009. Implications of genetic differentiation in neotropical montane forest birds. *Annals of the Missouri Botanical Garden* 96(3):410–433.
Weir, J. T., E. Bermingham, and D. Schluter. 2009. The Great American Biotic Interchange in birds. *Proceedings of the National Academy of Sciences of the USA* 106(51):21737–21742.
Weir, J. T., and M. Price. 2011. Andean uplift promotes lowland speciation through vicariance and dispersal in *Dendrocincla* woodcreepers. *Molecular Ecology* 20(21):4550–4563.
Weir, J. T., and D. Schluter. 2007. The latitudinal gradient in recent speciation and extinction rates of birds and mammals. *Science* 315(5818):1574–1576.
Wesselingh, F., and J. A. Salo. 2006. A Miocene perspective on the evolution of the Amazonian biota. *Scripta Geologica* 133:439–458.
Wickham, H., M. Averick, J. Bryan, W. Chang, L. D. A. McGowan, R. François, . . . H. Yutani. 2019. Welcome to the Tidyverse. *Journal of Open Source Software* 4(43): Article 1686.
Wiens, J. J. 2011. The causes of species richness patterns across space, time, and clades and the role of "ecological limits." *Quarterly Review of Biology* 86(2):75–96.
Wiens, J. J., and M. J. Donoghue. 2004. Historical biogeography, ecology and species richness. *Trends in Ecology & Evolution* 19(12):639–644.
Wiens, J. J., and C. H. Graham. 2005. Niche conservatism: Integrating evolution, ecology, and conservation biology. *Annual Review of Ecology, Evolution, and Systematics* 36:519–539.
Winger, B. M. 2017. Consequences of divergence and introgression for speciation in Andean cloud forest birds. *Evolution* 71(7):1815–1831.
Winger, B. M., P. A. Hosner, G. A. Bravo, A. M. Cuervo, N. Aristizábal, L. E. Cueto, and J. M. Bates. 2015. Inferring speciation history in the Andes with reduced-representation sequence data: An example in the bay-backed antpittas (Aves; Grallariidae; *Grallaria hypoleuca* sl). *Molecular Ecology* 24:6256–6277.
Wright, D. H. 1983. Species-energy theory: an extension of species-area theory. *Oikos* 41:496–506.

11
Natural History Collections for 21st-Century Ornithology

C. Jonathan Schmitt

Natural history collections of bird specimens have always been foundational to the field of ornithology. Although rooted in ancient skills such as hunting and mummification, the scientific institutions we know as natural history collections first appeared as recently as the 16th and 17th centuries in Europe. Although these early collections were small, poorly curated, and lost to history, they quickly became the inspiration that fueled ornithology's growth as a scientific discipline. Fast forwarding to the present, dogged dedication to preserving specimens and technological advances have expanded the influence of natural history collections so much that bird specimens are now studied far outside the traditional domains of ornithology. Nevertheless, natural history collections are not free from controversy and bureaucracy, and faced with a biodiversity crisis, they are under increasing pressure to meet the traditional goals of documenting avian diversity, the evolving needs of cutting-edge research, and the necessity to educate future generations. In this chapter, I briefly review the history of natural history collections and highlight the ways they are driving current ornithological research. I conclude with comments on the challenges, such as declines in specimen acquisition (Figure 11.1), that threaten to undermine museums' ability to meet the goals of research in the 21st century.

A Brief History

The acquisition and preservation of dead birds, what could be called bird collecting, are ancient methodologies that predate natural history collections or ornithology itself. In antiquity, bird collecting was perhaps most proficiently and prolifically executed by the Egyptians, who mummified millions of birds, many of which remain well preserved nearly 4,500 years later (Ikram 2012). Other human cultures throughout the world, such as the Pueblo people of the American Southwest and pre-Columbian societies of western South America, also preserved dead birds so well that their remains are being studied hundreds of years later (Watson et al. 2015, Capriles et al. 2021). The considerable effort that went into acquiring and preserving birds through millennia highlights the depth of human curiosity in birds; however, according to Stresemann (1975), no one had successfully preserved dead birds in Europe until the early decades of the 16th century. In fact, during the 15th to 17th centuries, it was

C. Jonathan Schmitt, *Natural History Collections for 21st-Century Ornithology*. In: *New Perspectives in Ornithology*.
Edited by: Scott V. Edwards and J. Michael Reed, Oxford University Press. © Oxford University Press (2025).
DOI: 10.1093/oso/9780197787670.003.0011

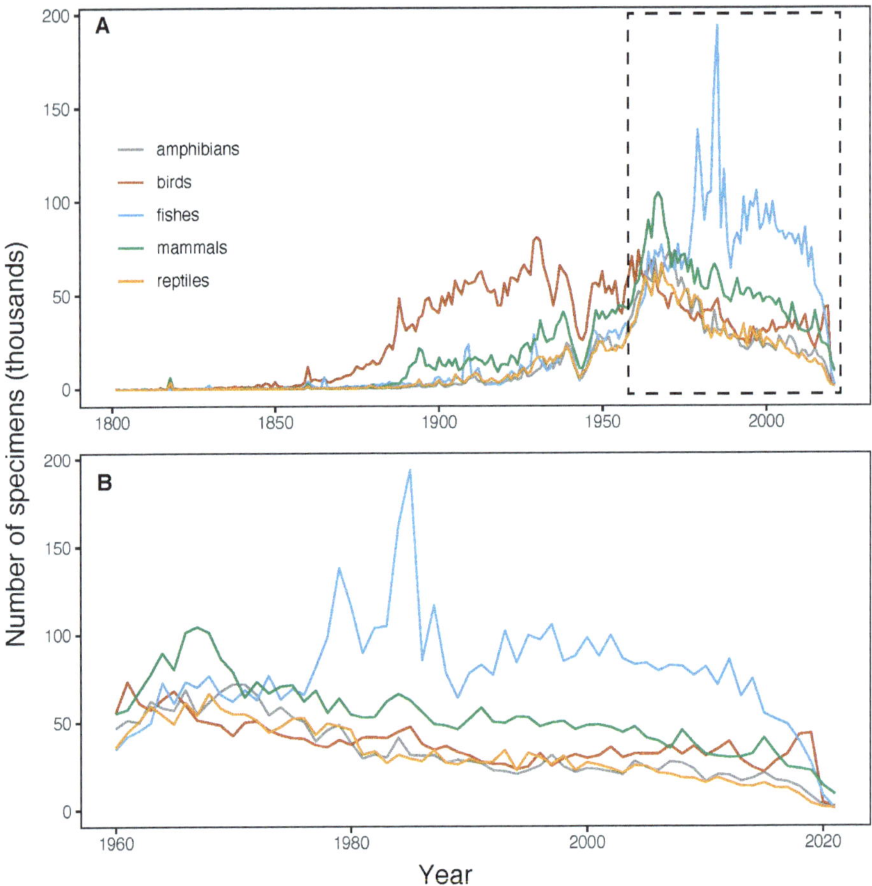

Figure 11.1 Specimen acquisition of vertebrates by natural history collections through time. These data represent a subset of the world's collections available through the Global Biodiversity Information Facility (GBIF), highlighting not only the limitations of current specimen portals but also an alarming decline in specimen acquisition of birds and other vertebrates during the past six decades. Panel A shows the number of vertebrate specimens added to natural history museums since 1840, and the dashed rectangle denotes the past six decades depicted in panel B.

Reproduced from data representing more than 245 natural history collections worldwide that were acquired by Rohwer et al. (2022) through the GBIF.

widely considered easier and cleaner to paint dead birds than preserve them, and avian diversity was thus inadequately documented by collections of bird art. Stresemann (1975) attributed the oldest written protocol for preserving dead birds in Europe to Belon (1555), who described how a bird can be preserved by carefully removing and salting its skin. Belon (1555) also commented how resourceful sailors used similar methods to preserve exotic birds on their travels so that they could sell them as curiosities back in Europe. This was the start of what Stresemann (1975) called "exotic ornithology."

In many ways, the exotic phase of the 16th and 17th centuries casts a dark shadow on the history of ornithology and natural history collections. Technological advances in sailing and navigation ushered in an age of global ornithological exploration, albeit in concert with imperialism and colonialism. Therefore, to discuss the history of museums and ornithology, we must acknowledge the fact that natural history museums were founded and grew at the expense of great human suffering and injustice. Furthermore, people of the era viewed bird specimens as symbols of status and wealth rather than scientific infrastructure. For example, possession of exotic birds, whether live or dead, became popular among the European elite in the decades following Columbus' 1493 return to Spain, where he wowed the crowds in Barcelona with live parrots (Stresemann 1975). Many dead and live birds were brought back to Europe during the next 100 years; however, most disappeared into private aviaries or the curios and cabinets of the rich, where few were ever examined by contemporary experts. Indeed, it was not until the end of the 16th century when wealthy amateurs had the idea of making a museum out of their bird specimens. Conceptually, these early private collections, often maintained by apothecaries, marked the auspicious beginning of what would be natural history museums; however, most, if not all, specimens from the 16th and 17th centuries eventually degraded due to poor preservation and maintenance.

Even as late as the 17th and 18th centuries, there were few places in Europe that kept collections of preserved birds; however, preserving dead birds was becoming more common, and much needed attention was given to making specimens last longer. Of particular importance was the realization that moisture and humidity quickly caused specimens to decay. For example, Stresemann (1975) notes cabinets and bottles that could be shut tightly or sealed as well as the preservation of specimens in alcohol as important advances of the time. Such progress in techniques of preservation made the prospects of collecting specimens and maintaining them in museums practical and even enjoyable. As collections of bird specimens became more common, a desire as well as a necessity to organize them also arose. For nearly 2,000 years, attempts to classify birds had progressed little from Aristotle's philosophically based classification that emphasized function and behavior over form. Notably, the comparison of avian form was aided by examining bird specimens side-by-side and marked a paradigm shift toward Linnean classifications that were the antecedents of current avian taxonomy. Collections of bird specimens therefore played an important role in the development of taxonomy, a cornerstone of modern biology.

Encouraged by the catalogs of bird species compiled by Linneaus, Buffon, and Brisson, a zeal for discovering new species of bird characterized 18th- and 19th-century Europe. Expensively equipped expeditions of the time routinely included naturalists to collect and document biodiversity, and even Charles Darwin and Alfred Russell Wallace collected bird specimens while employed as naturalists on their famed journeys. As a result of these expeditions, natural history collections grew rapidly in both Europe and the United States. By the end of the 19th century, contemporary museums such as the British Museum, Smithsonian National Museum of Natural History, Museum of Comparative Zoology, and American Museum of Natural History had large collections of birds and were early hubs of ornithological research. The first half of 20th century saw the continued growth of natural history collections, especially university-based collections; however, specimen acquisition has declined worldwide since the 1960s (Rohwer et al. 2022) (see Figure 11.1). Meanwhile, technological advances and discoveries during the past 100 years have significantly broadened the study and importance of bird specimens outside the traditional domains of taxonomy and anatomy.

Natural History Collections and Modern Research

The history of natural history collections and ornithology are intertwined, both being shaped by technological advances but also supporting the growth of the other. However, now more than ever, scientific specimens of birds housed in natural history collections also represent critical scientific infrastructure for studies far outside of ornithology. The value of scientific specimens of birds in diverse domains of science is a direct result of the myriad sample types and data that can be derived from them (Figure 11.2). Below, I summarize some of the diverse areas of research that rely on specimens of birds and comment on how researchers at natural history collections must continue collecting birds if they are to meet the needs of current and future research.

Taxonomy

One of the most traditional uses of scientific specimens in the biological sciences has been in the taxonomic classification and documentation of avian diversity. Just as in the days of Linnaeus, bird species are classified and named based on scientific specimens (Krell and Wheeler 2014, Ceriaco et al. 2016). Although it is true that birds are the most taxonomically well-documented class of vertebrates, our understanding of avian diversity is by no means complete—a fact evidenced by many recently discovered new genera (Pacheco et al. 1996, Lane et al. 2021), species (Isler et al. 2020, Krabbe et al. 2020, Rheindt et al. 2020, Tyler et al. 2020), and subspecies (Rheindt et al. 2020) of birds. Although the rate of discovery of new avian taxa is highest in the tropics, new research continually identifies evolutionary distinct, often cryptic populations, even in well-studied North America taxa (Young et al. 2000, Benkman et al. 2009, Klicka et al. 2016). The importance of these discoveries is further underscored by the biodiversity crisis that will force some species to extinction, perhaps before they are described and studied. Collectively, none of these efforts to document avian diversity would be possible without scientific specimens of birds and their continued collection.

Systematics and Phylogenetics

Although related to taxonomy, systematics and phylogenetics deal with not only describing diversity but also understanding the relationship between populations, species, and lineages through time. Avian systematists have historically relied on morphological characteristics to infer the phylogenetic relationships between species and higher level groups such as families and orders (reviewed by Livezey and

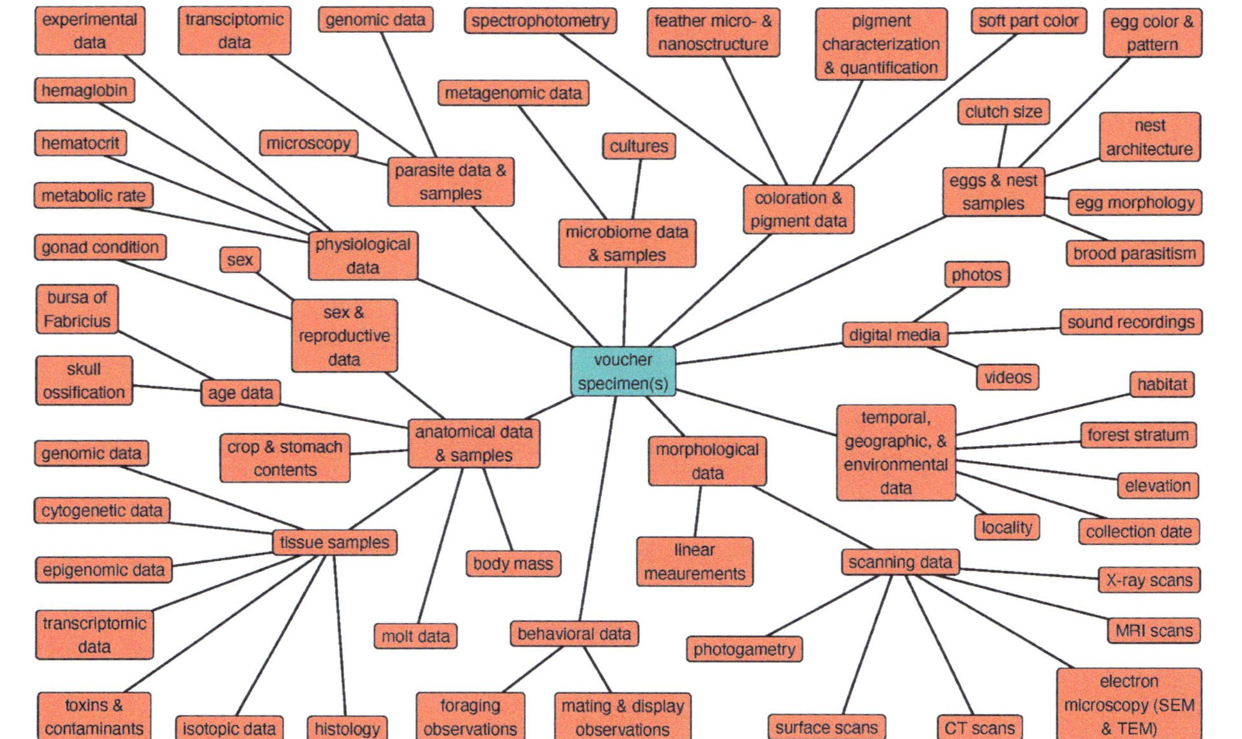

Figure 11.2 Network showing many of the myriad sample types and data that can be derived from museum voucher specimens of birds. The centrality of the voucher specimen in the network highlights the centrality of museum specimens of birds to diverse areas of research.

Zusi 2007). During the DNA sequencing revolution of the past four decades, avian systematists have increasingly turned to DNA sequence information to reconstruct the phylogenetic relationships of the avian tree of life (Sibley and Ahlquist 1990, Hackett et al. 2008, Faircloth et al. 2012, McCormack et al. 2013, Jarvis et al. 2014, Prum et al. 2015, Oliveros et al. 2019, Harvey et al. 2020). In either case, scientific specimens are the gold standard for systematic and phylogenetic studies of birds—either morphological characters from physical specimens or DNA sequence information derived from the tissue samples now standardly preserved in association with physical specimens (Peterson et al. 2007). Unfortunately, specimen-vouchered tissue samples of all currently recognized bird species still do not exist (Stoeckle and Winker 2009). The situation is even worse when considering the availability of tissue from avian subspecies, further hindering our understanding of avian diversity and evolution. These limitations can occasionally be overcome due to new technologies that allow for the extraction and sequencing of DNA from toepad or skin samples taken from historical specimens (McCormack et al. 2017, Tsai et al. 2020). Although the DNA recovered from historical specimens is usually highly fragmented, many genomic markers and even whole genomes now can be sequenced routinely and assembled.

Phylogeography

As a "bridge between population genetics and systematics," phylogeography focuses explicitly on the genetic, geographical, and ecological processes that generate the contemporary geographic distribution of populations and species (Avise et al. 1987). Avian phylogeographers sequence DNA from scientific specimens and their associated tissue samples to understand the mechanisms that generated the incredible diversity of modern birds and explain the observed spatial patterns of avian diversity. For example, through ever advancing DNA sequencing technology and population genetic methods, researchers are able to infer past events in the history of a species or population, such as divergence, gene flow, or population size fluctuations, and connect them to geographical or environmental events, including mountain or island formation and sea-level changes (Harvey and Brumfield 2015, Klicka et al. 2016, Oswald et al. 2017, Battey et al. 2018). Collectively, these inferences have profound implications for our understanding of how avian diversity evolves and is distributed, providing a framework for understanding how birds might respond to pressing contemporary concerns such as habitat destruction and climate change. At its core, phylogeography is based on the geographical distributions of species and populations, highlighting the need for scientific specimens across spatial scales. Unfortunately, the sampling of bird specimens and tissues remains geographically sparse even for common avian species or subspecies (Stoeckle and Winker 2009), a deficiency that only the continued collection of bird specimens and tissues can address.

Global Change Biology

Although historic specimens provide an important baseline or snapshot of diversity at a given time in the past, many studies of birds and environmental change require temporal sampling of specimens. A salient example of how bird specimens document change in the environment is demonstrated by DuBay and Fuldner (2017), who measured bird specimens collected during the past century and found that the environmentally acquired soot on feathers was a sensitive indicator of air pollution levels when the bird was collected. Measurements of soot preserved on the feathers of bird specimens documented a decrease in black carbon air pollution through time—an important change in the environment that is relevant not only to birds but also to human health. Temporal series of bird specimens are already yielding important results; however, future research documenting change through time will be dependent on the continued acquisition of new specimens. This is particularly true in the context of climate change and other human impacts on the environment we are witnessing in the Anthropocene (Schmitt et al. 2019), and newly acquired scientific specimens will provide a resource for documenting and understanding human-mediated changes going forward.

Conservation

An abundant literature demonstrates how scientific specimens, and the continued collection of them, inform bird conservation (reviewed by Remsen 1995). Scientific specimens are a rich source of information on distribution, (micro)habitat association, diet, as well as annual cycles of breeding, molt, and migration—data that are frequently accessed from natural history museums by conservation biologists (Collar and Rudyanto 2003). For example, bird specimen data from more than 60 museums were used to assess the conservation status of threatened birds in the Americas (Clinton-Eitniear 1994), and the International Union for Conservation of Nature regularly incorporates data derived from specimens in its assessments of avian populations (Collar and Rudyanto 2003). More generally, specimens and specimen collection are critical for conservation because scientific specimens form the basis for evaluating and establishing the taxonomic units of conservation—species and subspecies. As briefly mentioned above, molecular studies based on tissue samples of bird specimens regularly identify cryptic species, subspecies, and genetically distinct populations that are of conservation concern (Moritz 1994). A recent example is provided by a study of Bell's Vireo (*Vireo bellii*) that found genomic evidence for the existence of two distinct species, each with two subspecies, totaling four evolutionarily distinct conservation units (Klicka et al. 2016).

Host–Pathogen Biology and Infectious Disease

Scientific specimens and associated tissue samples of birds connect the fields of parasitology, virology, and infectious disease research with ornithology by providing important insight into birds as hosts of diverse parasites and viruses (Lutz et al. 2017). One of the basic yet most important uses of scientific specimens in the study of parasitology is the discovery and description of parasite diversity. New species of parasites, particularly endoparasites, cannot be described or studied without collecting, preserving, and dissecting bird specimens (Lutz et al. 2017). A recent study on helminths (parasitic worms) infecting vertebrates estimated that 85–95% of species are still undescribed, with the majority of undescribed diversity existing in birds and fish (Carlson et al. 2020). Furthermore, Carlson et al. (2020) estimate that it will take centuries to completely describe the diversity of helminths at current rates of discovery. The status of species discoveries of other parasite groups is similarly bleak, highlighting the need for increased collection of birds and their parasites. The conservation implications of parasite infections are well known, but host–parasite interactions also have broader implications in ecology, evolution, and even human health. For example, museum specimens of birds are also an increasingly important resource in studies of avian malaria (Fecchio et al. 2019), and studies of avian malaria based on museum tissue samples are making important discoveries about how abiotic and biotic factors shape the distribution and evolution of birds and their parasites (Barrow et al. 2019, McNew et al. 2021). In light of the worldwide COVID-19 pandemic, it is imperative to note that museum specimens provide critical infrastructure for studies of human health. A geographically relevant example is the essential role that specimens and tissues of small mammals played in identifying (Nichol et al. 1993, Yates et al. 2002) and monitoring (Glass et al. 2006) the "Sin Nombre" virus (Hantaviridae: Order Bunyavirales) that emerged in the four corners of the American Southwest. Although this example relates to mammal specimens, numerous avian diseases, such as influenza and West Nile and Newcastle viruses, pose equally serious human health risks and are regularly studied using scientific specimens of birds (reviewed by Suarez and Tsutsui 2004). Going forward, continued collection of bird specimens will be important for efforts to identify, track, and potentially prevent novel zoonotic diseases.

Genomics, Transcriptomics, and Epigenomics

Modern sequencing technologies have revolutionized biology by allowing researchers to sequence thousands of genes or even whole genomes. In 2012, the genomes of only 2 bird species had been sequenced; as of January 2021, this number has grown to 514 (Bravo et al. 2021). Nearly all these genomes were sequenced from tissue samples associated with scientific specimens in natural history collections. Furthermore, the Bird 10K (B10K) project (Zhang et al. 2014), an offshoot of the Vertebrate 10K (G10K) project (Genome 10K Consortium 2009), relies almost entirely on museum-persevered tissues to accomplish its bold goal of sequencing the genome of all species of birds. A list of the species whose genomes are already sequenced or currently being sequenced is available from the B10K website (https://b10k.genomics.cn/species.html). The scientific studies resulting from the B10K project (Jarvis et al. 2014, Zhang et al. 2014, Feng et al. 2020) and avian genomes sequenced by other researchers (Sackton et al. 2019, Vianna et al. 2020) are far-reaching and impactful, a fact evidenced by their publication in some of the top scientific journals, such as *Science*, *Nature*, and *Proceedings of the National Academy of Sciences of the United States of America*. Unfortunately, most currently existing tissue samples were collected before the importance of tissue quality was widely appreciated and as a result are insufficient for many modern genomic techniques (Edwards et al. 2005), thus placing a ceiling on the quality of genome assemblies that can be generated (Card et al. 2021). Furthermore, the presence of a germline restricted chromosome in some passerines (Pigozzi and Solari 1998, Kinsella et al. 2019, Torgasheva et al. 2019) is a fascinating area of research that depends directly on high-quality gonad samples which are not available for most species.

Related to genomics, and equally impacted by modern sequencing technologies, are the fields of transcriptomics and epigenomics. Epigenomics is the study of genome modifications, such as histones, that affect gene expression, whereas transcriptomics is the study of gene expression. Transcriptomic and epigenomic studies of birds usually assay tissue-specific levels of gene expression and genome modification; however, developments in single-cell sequencing technologies give the opportunity to conduct cutting-edge research at the level of individual cells. Currently, the quality of tissue samples and sampling schemes required for transcriptomic and epigenomic studies is a limitation to their application in wild birds. With these limitations in mind, natural history collections are now preserving high-quality samples of multiple tissue types per specimen collected (Card et al. 2021); however, the sampling of such tissues across avian diversity, geographic space, and time is still extremely sparse. To meet the demands of future genomic studies of birds, researchers at natural history collections will need to prioritize the preservation of diverse tissue types, with an emphasis on tissue quality.

Isotopes, Anatomy, and Bioengineering

Arguably the most valuable quality of a specimen is the unrealized or inaccessible data and insight they might hold in the future. Advancements in technology continually expand the diversity and quality of data that can be obtained from even the oldest specimens. Isotopic analysis is one example of how technological advances have extended the use of scientific specimens. The ability to measure isotopes from feathers, claws, and even frozen tissue samples has enriched studies of migratory connectivity, diets, and ecological studies (reviewed by Rocque and Winker 2005). The domain of specimen-based research has been further expanded by modern scanning technologies. For example, contrast tomography (CT) scanning of fluid-preserved specimens now allows the detailed visualization and measurement of skeletal, muscular, and cardiovascular morphologies in three dimensions, something that was impossible only a few years ago. A National Science Foundation (NSF)-funded project called oVert aims to produce freely available three-dimensional (3D) CT scans of every vertebrate genus (https://www.floridamuseum.ufl.edu/overt). For birds, a major constraint facing the oVert project is the dearth of fluid-preserved specimens, a limitation that will have to be overcome by continued collecting. oBird, a partner project also funded by NSF, is using a 3D modeling technique called photogrammetry to produce 3D models of bird specimens that allow researchers to accurately measure the external morphology as well as quantify the color and pattern of bird plumage (Medina et al. 2020). Finally, museum specimens are increasingly appreciated as an invaluable resource in the emerging field of biomimetic design, in which naturally occurring architectures and systems inspire technological innovation (Gibson 2005, Gad-el-Hak 2019).

Prospects for the 21st Century

Many of the challenges facing natural history collections in the 21st century (National Academies of Sciences, Engineering, and Medicine 2020) are those identified by Barlow and Flood (1983): permitting, training, and funding. In addition, natural history collections face the challenge of digitizing specimen data, an important step toward open science, as well as education and outreach. Now well into the 21st century, and with a biodiversity crisis looming, natural history collections are also faced with a catch-22: Specimen-based research is increasing but specimen acquisition is decreasing (Rohwer et al. 2022) (see Figure 11.1).

Although bird collections and bird collecting are supported and justified by a large body of peer-reviewed literature (Winker et al. 1991; Goodman and Lanyon 1994; Remsen 1995; Suarez and Tsutsui 2004; Winker 2004, 2005; Winker et al. 2010; Joseph 2011; Rocha et al. 2014; Gropp 2018; Bakker et al. 2020), the continued collection of birds and other animals has been criticized (Bekoff and Elzanowski 1997, Minteer et al. 2014). Anti-collecting sentiments are well intentioned but ignore the fact that collecting of bird specimens has minimal impacts even on local populations. In fact, mortality from bird collecting represents only a small fraction of total human-caused bird mortality, yet ironically it is the only source of bird mortality that contributes toward advancing our knowledge of avian biology. Nevertheless, increasing restrictions on bird collecting and specimen transportation at the state, national, and international levels stymie scientific discovery, especially in the poorly sampled tropics. Unless the need for continued collecting of birds is broadly accepted by the ornithological community and permitting agencies, a major challenge facing natural history collections in the 21st century will be collecting enough new specimens to meet the demands of current and future research.

Traditionally, all biologists, or at least zoologists and botanists, were trained in the identification of organisms through the use of scientific specimens. Although scientific specimens still lie at the core of biology curricula throughout the United States, their use in undergraduate and graduate education is declining. Therefore, a challenge increasingly facing museum scientists and educators is how best to use specimens in the classroom. This challenge is already being met by undergraduate programs structured around teaching biological concepts and principles with specimen-based pedagogical exercises (Cook et al. 2014, 2016; Powers et al. 2014; Lacey et al. 2017; Bakker et al. 2020). Although these efforts span many taxonomic groups, birds and bird specimens are featured prominently. Decade-long engagement of undergraduate students in the Museum of Vertebrate Zoology collections at the University of California, Berkeley also provides a promising model for extending the educational value of specimens (Hiller et al. 2017). Finally, involving undergraduate students in ornithological expeditions, including the collection of specimens, is an undervalued opportunity to educate and train future generations of ornithologists (Winkler et al. 2017). Continued efforts to use bird specimens in education and public outreach will be important for natural history collections going forward.

In order to keep up with the digital era and the open science movement, natural history collections must digitize to survive in the 21st century. Although a centralized database of bird specimen data from all museums in the world is the idealistic end goal, several challenges face natural history collections. Data from modern specimens get digitized in real time at many collections, yet others struggle with digitizing a large backlog of historic specimens. Furthermore, differences in data quality and format between collections can cause inconsistencies when databases are merged, and the merging process used by online portals is inefficient and duplication prone. A final challenge associated with digitization is the development of better methods for linking specimen data to their derivate data types such as DNA sequence data and attributing specimens to collectors, preparators, and research products (National Academies of Sciences, Engineering, and Medicine 2020).

Funding has always been and continues to be a challenge facing natural history collections. To serve society's educational and research needs, modern natural history collections struggle with both short- and long-term financial stability. In the short term, collections have a minimal operating budget for staff and supplies required for curation, specimen collection and preparation, digitization of data, processing of loans, and facilitating physical access to collection space and specimens. On the other hand, maintaining and updating both physical and digital infrastructure, hiring additional personnel, and navigating ever complex requirements for permits are some of the many long-term challenges facing museums. Despite advances in the use of specimens in research and education already mentioned, natural history collections are still generally undervalued, and some institutions have even cut financial support for their collections. Many collections in the United States receive grants from NSF's Collections in Support of Biological Research program; however, other external funding is scarce. Recommendations proposed to overcome these financial challenges include business training for administrators, implementing pay-for-use models, and exploring nontraditional funding sources (National Academies of Sciences, Engineering, and Medicine 2020). Minimally, natural history collections need to better communicate their mission as research and educational institutions and more effectively demonstrate the value of their mission to stakeholders and society in general.

In light of the challenges already mentioned, natural history collections can rise to meet the needs of current and future research. One route to success will be to emphasize data quantity and quality over specimen quantity. Traditionally, ornithologists have preserved bird specimens as skins, a preservation type in which the bird's skin is removed and stuffed with cotton. Although skins allow for many of the studies described above, the preservation of skeletal material or fluid-preserved specimens (submerged in ethanol or formalin) is also increasingly needed for other studies (Livezey 2003, Olson 2003, James 2017). Unfortunately, skeletal specimens and fluid-preserved specimens of birds are underrepresented in museum collections across taxonomic, spatial, and temporal scales (Livezey 2003, Olson 2003, Causey and Trimble 2005, James 2017), highlighting a need for continued collecting of these specimen types. More broadly, preserving holistic, multipart bird specimens with diverse tissue samples, associated parasites, and other nontraditional parts (spread wings, syrinxes, stomach contents, etc.) will help museums maximize the research value of each specimen (Webster 2017, Schindel and Cook 2018).

Conclusion

Natural history collections, the specimens, the archive, and the continued collection of scientific specimens of birds are critical to scientific research, a fact no less true today than it was in the era of Linnaeus, Darwin, and other early biologists. Indeed, scientific specimens, as well as the natural history collections curating them, are used more extensively and across more domains of science than ever, a trend that will only accelerate as technological innovations enable future generations of scientists to unlock novel data and inferences from specimens. Bird specimens also represent important educational materials for enriching the training of future biologists as well as stimulating the general public's interest in biodiversity. Nevertheless, challenges lie ahead of natural history collections going forward—the societal importance and use of specimens are increasing, yet acquisition of new specimens is in decline. It is difficult to see how these diverging trajectories are sustainable, and serious efforts to reverse the latter trend will ultimately determine whether museums can continue to meet the goals of current and future research.

References

Avise, J. C., J. Arnold, R. M. Ball, E. Bermingham, T. Lamb, J. E. Neigel, C. A. Reeb, and N. C. Saunders. 1987. Intraspecific phylogeography: The mitochondrial DNA bridge between population genetics and systematics. *Annual Review of Ecology and Systematics* 18:489–522.

Bakker, F. T., A. Antonelli, J. A. Clarke, J. A. Cook, S. V. Edwards, P. G. P. Ericson, S. Faurby, N. Ferrand, M. Gelang, R. G. Gillespie, M. Irestedt, K. Lundin, E. Larsson, P. Matos-Maraví, J. Müller, T. von Proschwitz, G. K. Roderick, A. Schliep, N. Wahlberg, J. Wiedenhoeft, and M. Källersjö. 2020. The Global Museum: Natural history collections and the future of evolutionary science and public education. *PeerJ* 8: Article e8225.

Barlow, J. C., and N. J. Flood. 1983. Research collections in ornithology—A reaffirmation. Pages 37–54 in A. H. Brush and G. A. Clark, Jr., eds. *Perspectives in Ornithology*. Cambridge University Press, Cambridge, UK.

Barrow, L. N., S. M. McNew, N. Mitchell, S. C. Galen, H. L. Lutz, H. Skeen, T. Valqui, J. D. Weckstein, and C. C. Witt. 2019. Deeply conserved susceptibility in a multi-host, multi-parasite system. *Ecology Letters* 22:987–998.

Battey, C. J., E. B. Linck, K. L. Epperly, C. French, D. L. Slager, P. W. Sykes, Jr., and J. Klicka. 2018. A migratory divide in the Painted Bunting (*Passerina ciris*). *The American Naturalist* 191:259–268.

Bekoff, M., and A. Elzanowski. 1997. Collecting birds: The importance of moral debate. *Bird Conservation International* 7:357–361.

Belon, P. 1555. *L'histoire de la nature des oyseaux [microform]: avec leur descriptions & naifs portraicts retirez du naturel: Escrite en sept livres/par Pierre Belon du Mans*. Cauellat, Paris.

Benkman, C. W., J. W. Smith, P. C. Keenan, T. L. Parchman, and L. Santisteban. 2009. A new species of the Red Crossbill (Fringillidae: *Loxia*) from Idaho. *The Condor* 111:169–176.

Bravo, G. A., C. J. Schmitt, and S. V. Edwards. 2021. What have we learned from the first 500 avian genomes? *Annual Review of Ecology, Evolution, and Systematics* 52:611–639.

Capriles, J. M., C. M. Santoro, R. J. George, E. Flores Bedregal, D. J. Kennett, L. Kistler, and F. Rothhammer. 2021. Pre-Columbian transregional captive rearing of Amazonian parrots in the Atacama Desert. *Proceedings of the National Academy of Sciences of the USA* 118: Article e2020020118.

Card, D. C., B. Shapiro, G. Giribet, C. Moritz, and S. V. Edwards. 2021. Museum genomics. *Annual Review of Genetics* 55:633–659.

Carlson, C. J., T. A. Dallas, L. W. Alexander, A. L. Phelan, and A. J. Phillips. 2020. What would it take to describe the global diversity of parasites? *Proceedings of the Royal Society B: Biological Sciences* 287: Article 20201841.

Causey, D., and J. Trimble. 2005. Old bones in new boxes: Osteology collections in the new millennium. *The Auk* 122:971–979.

Ceriaco, L. M. P., E. E. Gutierrez, A. Dubois, C. S. Abdala, A. S. Alqarni, K. Adler, E. A. Adriano, E. Aescht, I. Agarwal, S. Agatha, D. Agosti, A. J. C. Aguiar, J. J. M. Aguiar, D. Ahrens, A. Aleixo, M. J. Alves, F. R. Do Amara, N. Ananjeva, M. C. Andrade, M. B. de Andrade, F. Andreone, P. P. U. Aquino, P. B. Araujo, H. Arnaud, J. Arroyave, W. Arthofer, T. J. Artois, D. Astua, C. Azevedo, J. C. Bagley, D. Baldo, H. M. Barber-James, E. V. Barmann, C. Bastos-Silveira, M. F. Bates, A. M. Bauer, F. Bauer, R. C. Benine, D. J. Bennett, B. Bentlage, B. Berning, D. Bharti, C. Biondo, J. Birindelli, T. Blick, G. Boano, F. A. Bockmann, W. Bogdanowicz, W. Bohme, E. Borgo, L. Borkin, M. Ricardo, R. Bour, W. R. Branch, C. A. Brasileiro, J. K. Braun, G. A. Bravo, L. Brendonck, G. R. R. Brito, M. R. Britto, P. A. Buckup, D. Burckhardt, U. Burkhardt, S. D. Busack, L. A. Campos, A. Canard, E. M. Cancello, U. Caramaschi, J. M. Carpenter, M. Carr, R. Carrenho, A. Cartaxana, M. A. Carvajal, G. S. Carvalho, M. R. de Carvalho, A. Chaabane, C. Chagas, P. Chakrabarty, K. Chandra, S. Chatzimanolis, S. W. Chordas, A. U. Christoff, F. Cianferoni, S. Claramunt, D. Cogalniceanu, B. B. Collette, G. R. Colli, T. J. Colston, W. Conradie, J. Constant, R. Constantino, J. A. Cook, D. Cordeiro, A. M. Correia, F. P. D. Cotterill, B. Coyner, M. A. Cozzuol, J. Cracraft, A. Crottini, G. Cuccodoro, F. F. Curcio, C. D. d'Acoz, G. D'Elia, C. D'Haese, I. Das, A. Datovo, A. Datta-Roy, P. David, J. G. Day, J. D. Daza, L. J. de Bisthoven, I. J. D. de la Vina, C. de Muizon, M. de Pinna, V. D. Piacentini, R. O. de Sa, M. de Vivo, J. Decher, W. Dekoninck, J. H. C. Delabie, M. Delfino, G. B. Delmastro, T. Delsinne, C. Denys, W. Denzer, L. Desutter-Grandcolas, K. Deuti, T. D. de Resbecq, F. Di Dario, V. Dinets, C. DoNascimiento, D. A. Donoso, G. Doria, R. C. Drewes, E. Drouet, M. Duarte, M. C. Durette-Desset, F. Dusoulier, S. K. Dutta, M. S. Engel, M. Epstein, M. Escalona, J. A. Esselstyn, K. Eto, J. Faivovich, R. L. Falaschi, Z. H. Falin, E. I. Faundez, A. Feijo, R. M. Feitosa, D. S. Fernandes, M. Fikacek, B. L. Fisher, M. J. FitzPatrick, D. Forero, I. Franz, H. Freitag, T. Fretey, U. Fritz, C. Gallut, S. Gao, G. S. T. Garbino, B. R. Garcete-Barrett, L. Garcia-Prieto, F. J. Garcia, P. C. A. Garcia, A. L. Gardner, S. L. Gardner, R. Garrouste, M. F. Geiger, T. C. Giarla, V. Giri, M. Glaubrecht, R. C. Glotzhober, F. S. P. Godoi, S. Gofas, P. R. Goncalves, J. Gong, V. H. Gonzalez, J. A. Gonzalez-Oreja, E. Gonzalez-Santillan, E. Gonzalez-Soriano, S. M. Goodman, P. Grandcolas, L. Grande, E. Greenbaum, R. Gregorin, H. Grillitsch, L. L. Grismer, P. Grootaert, S. Grosjean, F. M. Guarino, J. M. Guayasamin, B. Guenard, L. Guevara, M. Guidoti, D. Gupta, V. Gvozdik, C. F. B. Haddad, J. Hallermann, A. Hassanin, A. Hausmann, L. R. Heaney, M. P. Heinicke, K. M. Helgen, K. Henle, A. Hirschmann, M. W. Holmes, M. Holynska, R. Holynski, G. Hormiga, B. A. Huber, J. P. Hugot, R. Hutterer, D. Iskandar, J. B. Iverson, P. Jager, R. Janssen, F. Jerep, R. Jocque, K. H. Jungfer, J. L. Justine, R. G. Kamei, M. J. Kaminski, M. Karner, T. Kearney, R. Khot, M. Kieckbusch, J. Kohler, K. P. Koepfli, E. Kondorosy, L. Krogmann, T. K. Krolow, M. Kruger, C. Kucharzewski, S. O. Kullander, S. Kumar, A. Kupfer, M. Kuramoto, O. Kurina, A. Kury, S. Kvist, E. La Marca, A. La Terza, R. Laval, T. E. Lacher, C. J. E. Lamas, M. R. Lambert, B. Landry, F. Langeani, J. A. Langone, J. E. Lattke, E. O. Lavilla, T. Leenders, D. C. Lees, Y. L. R. Leite, T. Lehmann, M. G. Lhano, B. K. Lim, X. F. Lin, I. Lobl, C. A. S. de Lucenade Lucena, M. S. de Lucena, P. Lucinda, N. K. Lujan, P. Luporini, D. R. Luz, J. D. Lynch, L. F. Machado, S. Mahony, L. R. Malabarba, M. Manuel-Santos, J. Marinho, M. A. Marini, A. C. Marques, M. P. Marques, O. Mateus, M. Matsui, T. Mazuch, J. McCranie, R. C. McKellar, C. D. McMahan, S. Mecke, K. Meissner, M. A. Mendoza-Becerril, C. A. Mendoza-Palmero, S. Merker, M. Mezzasalma, J. M. Midgley, J. Miller, M. J. Miller, M. M. Mincarone, J. Minet, A. Miralles, T. P. Miranda, A. D. Missoup, D. Modry, J. Molinari, A. Monadjem, O. Montreuil, R. Moratelli, C. R. Moreira, F. F. F. Moreira, C. Mourer-Chauvire, P. R. Mulieri, T. A. Munroe, S. I. Naomi, F. Nascimento, W. A. Nassig, L. Neifar, A. L. Netto-Ferreira,

A. Niamir, S. V. Nielsen, S. S. Nihei, A. Nistri, A. Oceguera-Figueroa, G. Odierna, A. Ohler, A. A. Ojanguren-Affilastro, F. F. de Oliveirade Oliveirade Oliveira, M. L. de Oliveira, O. M. P. de Oliveira, S. S. Oliveira, L. E. Olson, G. O. Ong'Ondo, N. Orlov, C. P. Ornelas-Garcia, H. Ortega, M. Ortega-Andrade, H. Ota, A. Pariselle, P. Passos, M. N. L. Pastana, B. D. Patterson, L. D. Patitucci, J. L. Patton, A. C. Pavan, S. E. Pavan, M. Pavia, P. L. V. Peloso, A. Pelzer, M. O. Pereyra, A. Perez-Gonzalez, B. Perez-Luz, C. H. F. Perez, J. K. Peterhans, A. T. Peterson, J. Petillon, T. K. Philips, O. Picariello, M. R. Pie, T. G. Pikart, R. H. Pine, U. Pinheiro, L. C. Pinho, A. P. Pinto, L. P. Costa, R. Poggi, J. P. Pombal, M. Prabhu, E. Prendini, L. Prendini, J. Purushothaman, R. A. Pyron, P. Quintela-Alonso, A. S. Quinteros, M. Quiroga-Carmona, W. Rabitsch, J. Raffaelli, J. C. Rage, H. Rajaei, M. J. Ramirez, M. A. Raposo, L. H. R. Py-Daniel, J. Y. Rasplus, B. C. Ratcliffe, S. Reddy, R. E. Reis, J. V. Remsen, L. R. Richards, I. Richling, T. Robillard, M. S. Rocha, R. M. Rocha, D. Rodder, M. O. Rodel, F. P. Rodrigues, E. Rodriguez, D. S. Rogers, F. J. M. Rojas-Runjaic, B. Roll, A. L. Rosenberger, J. Rowley, A. S. Roza, M. Ruedi, J. Salazar-Bravo, N. J. Salcedo, Y. Samyn, S. E. Santana, L. Santoferrara, B. F. Santos, C. M. D. Santos, J. C. Santos, M. P. D. Santos, E. J. Sargis, W. E. Schargel, B. Schatti, M. D. Scherz, B. C. Schlick-Steiner, R. C. Schmidt, T. Schmitt, R. Schodde, C. S. Schoeman, S. Schweiger, C. F. Schwertner, E. C. J. Seamark, T. B. F. Semedo, M. K. Shin, C. D. Siler, L. F. Silveira, W. B. Simison, M. Simoes, J. W. Sites, B. T. Smith, K. T. Smith, W. B. Song, A. Soulier-Perkins, L. M. Sousa, J. S. Sparks, S. N. Stampar, F. M. Steiner, J. S. Steyer, M. L. J. Stiassny, T. Stoeck, R. Stopiglia, J. W. Streicher, M. J. Sturaro, P. Stys, L. Swierk, A. Taeger, D. M. Takiya, D. C. Taphorn, M. Tavares, V. D. Tavares, P. J. Taylor, J. G. Tello, P. Teta, F. Tillack, R. M. Timm, T. Tokaryk, A. Tominaga, J. F. R. Tonini, L. Tornabene, O. Torres-Carvajal, J. Townsend, J. F. Trape, M. T. Rodrigues, R. Trusch, E. Tschopp, L. M. Vieira, P. F. Victoriano, L. J. Vitt, P. Wagner, G. J. Watkins-Colwell, T. Weisse, F. P. Werneck, W. C. Wheeler, D. E. Wilson, K. C. W. Valero, P. L. Wood, N. Woodman, H. D. Y. Quetzalli, N. Yoshikawa, H. Zaher, T. Ziegler, J. Zima, R. M. Zink, and G. Zug. 2016. Photography-based taxonomy is inadequate, unnecessary, and potentially harmful for biological sciences. *Zootaxa* 4196:435–445.

Clinton-Eitniear, J. 1994. Threatened birds of the Americas (the ICBP/IUCN Red Data Book). *AFA Watchbird* 21:52–53.

Collar, N. J., and Rudyanto. 2003. The archive and the ark: Bird specimen data in conservation status assessment. *Bulletin of the British Ornithologists' Club* 123A:95–113.

Cook, J. A., S. V. Edwards, E. A. Lacey, R. P. Guralnick, P. S. Soltis, D. E. Soltis, C. K. Welch, K. C. Bell, K. E. Galbreath, C. Himes, J. M. Allen, T. A. Heath, A. C. Carnaval, K. L. Cooper, M. Liu, J. Hanken, and S. Ickert-Bond. 2014. Natural history collections as emerging resources for innovative education. *BioScience* 64:725–734.

Cook, J. A., E. A. Lacey, S. M. Ickert-Bond, E. P. Hoberg, K. E. Galbreath, K. C. Bell, S. E. Greiman, B. S. McLean, and S. V. Edwards. 2016. From museum cases to the classroom: Emerging opportunities for specimen-based education. *Archives of Zoological Museum of Lomonosov, Moscow State University* 54:787–799.

DuBay, S. G., and C. C. Fuldner. 2017. Bird specimens track 135 years of atmospheric black carbon and environmental policy. *Proceedings of the National Academy of Sciences of the USA* 114:11321–11326.

Edwards, S. V., S. Birks, R. T. Brumfield, and R. Hanner. 2005. Future of avian genetic resources collections: Archives of evolutionary and environmental history. *The Auk* 122:979–984.

Faircloth, B. C., J. E. McCormack, N. G. Crawford, M. G. Harvey, R. T. Brumfield, and T. C. Glenn. 2012. Ultraconserved elements anchor thousands of genetic markers spanning multiple evolutionary timescales. *Systematic Biology* 61:717–726.

Fecchio, A., M. D. Collins, J. A. Bell, E. A. Garcia-Trejo, L. A. Sanchez-Gonzalez, J. H. Dispoto, N. H. Rice, and J. D. Weckstein. 2019. Bird tissues from museum collections are reliable for assessing avian haemosporidian diversity. *Journal of Parasitology* 105:446–453.

Feng, S., J. Stiller, Y. Deng, J. Armstrong, Q. Fang, A. H. Reeve, D. Xie, G. Chen, C. Guo, B. C. Faircloth, B. Petersen, Z. Wang, Q. Zhou, M. Diekhans, W. Chen, S. Andreu-Sánchez, A. Margaryan, J. T. Howard, C. Parent, G. Pacheco, M.-H. S. Sinding, L. Puetz, E. Cavill, Â. M. Ribeiro, L. Eckhart, J. Fjeldså, P. A. Hosner, R. T. Brumfield, L. Christidis, M. F. Bertelsen, T. Sicheritz-Ponten, D. T. Tietze, B. C. Robertson, G. Song, G. Borgia, S. Claramunt, I. J. Lovette, S. J. Cowen, P. Njoroge, J. P. Dumbacher, O. A. Ryder, J. Fuchs, M. Bunce, D. W. Burt, J. Cracraft, G. Meng, S. J. Hackett,

P. G. Ryan, K. A. Jønsson, I. G. Jamieson, R. R. da Fonseca, E. L. Braun, P. Houde, S. Mirarab, A. Suh, B. Hansson, S. Ponnikas, H. Sigeman, M. Stervander, P. B. Frandsen, H. van der Zwan, R. van der Sluis, C. Visser, C. N. Balakrishnan, A. G. Clark, J. W. Fitzpatrick, R. Bowman, N. Chen, A. Cloutier, T. B. Sackton, S. V. Edwards, D. J. Foote, S. B. Shakya, F. H. Sheldon, A. Vignal, A. E. R. Soares, B. Shapiro, J. González-Solís, J. Ferrer-Obiol, J. Rozas, M. Riutort, A. Tigano, V. Friesen, L. Dalén, A. O. Urrutia, T. Székely, Y. Liu, M. G. Campana, A. Corvelo, R. C. Fleischer, K. M. Rutherford, N. J. Gemmell, N. Dussex, H. Mouritsen, N. Thiele, K. Delmore, M. Liedvogel, A. Franke, M. P. Hoeppner, O. Krone, A. M. Fudickar, B. Milá, E. D. Ketterson, A. E. Fidler, G. Friis, Á. M. Parody-Merino, P. F. Battley, M. P. Cox, N. C. B. Lima, F. Prosdocimi, T. L. Parchman, B. A. Schlinger, B. A. Loiselle, J. G. Blake, H. C. Lim, L. B. Day, M. J. Fuxjager, M. W. Baldwin, M. J. Braun, M. Wirthlin, R. B. Dikow, T. B. Ryder, G. Camenisch, L. F. Keller, J. M. DaCosta, M. E. Hauber, M. I. M. Louder, C. C. Witt, J. A. McGuire, J. Mudge, L. C. Megna, M. D. Carling, B. Wang, S. A. Taylor, G. Del-Rio, A. Aleixo, A. T. R. Vasconcelos, C. V. Mello, J. T. Weir, D. Haussler, Q. Li, H. Yang, J. Wang, F. Lei, C. Rahbek, M. T. P. Gilbert, G. R. Graves, E. D. Jarvis, B. Paten, and G. Zhang. 2020. Dense sampling of bird diversity increases power of comparative genomics. *Nature* 587:252–257.

Gad-el-Hak, I. 2019. Fluid–structure interaction for biomimetic design of an innovative lightweight turboexpander. *Biomimetics* 4: Article 27.

Genome 10K Consortium. 2009. Genome 10K: A proposal to obtain whole-genome sequence for 10,000 vertebrate species. *Journal of Heredity* 100:659–674.

Gibson, L. J. 2005. Biomechanics of cellular solids. *Journal of Biomechanics* 38:377–399.

Glass, G. E., T. M. Shields, C. A. Parmenter, D. Goade, J. N. Mills, J. A. Cook, and T. L. Yates. 2006. Predicted hantavirus risk in 2006 for the southwestern U.S. *Occasional Papers, Museum of Texas Tech University* 225:1–16.

Goodman, S. M., and S. M. Lanyon. 1994. Scientific collecting. *Conservation Biology* 8:314–315.

Gropp, R. E. 2018. Specimens, collections, and tools for future biodiversity-related research. *BioScience* 68:3–4.

Hackett, S. J., R. T. Kimball, S. Reddy, R. C. K. Bowie, E. L. Braun, M. J. Braun, J. L. Chojnowski, W. A. Cox, K. L. Han, J. Harshman, C. J. Huddleston, B. D. Marks, K. J. Miglia, W. S. Moore, F. H. Sheldon, D. W. Steadman, C. C. Witt, and T. Yuri. 2008. A phylogenomic study of birds reveals their evolutionary history. *Science* 320:1763–1768.

Harvey, M. G., G. A. Bravo, S. Claramunt, A. M. Cuervo, G. E. Derryberry, J. Battilana, G. F. Seeholzer, J. S. McKay, B. C. O'Meara, B. C. Faircloth, S. V. Edwards, J. Pérez-Emán, R. G. Moyle, F. H. Sheldon, A. Aleixo, B. T. Smith, R. T. Chesser, L. F. Silveira, J. Cracraft, R. T. Brumfield, and E. P. Derryberry. 2020. The evolution of a tropical biodiversity hotspot. *Science* 370: 1343–1348.

Harvey, M. G., and R. T. Brumfield. 2015. Genomic variation in a widespread Neotropical bird (*Xenops minutus*) reveals divergence, population expansion, and gene flow. *Molecular Phylogenetics and Evolution* 83:305–316.

Hiller, A. E., C. Cicero, M. J. Albe, T. L. W. Barclay, C. L. Spencer, M. S. Koo, R. C. K. Bowie, and E. A. Lacey. 2017. Mutualism in museums: A model for engaging undergraduates in biodiversity science. *PLoS Biology* 15: Article e2003318.

Ikram, S. 2012. An eternal aviary: Bird mummies from ancient Egypt. Pages 41–48 in R. Bailleul-Iesuer, ed. *Between Heaven and Earth: Birds in Ancient Egypt.* Oriental Institute of the University of Chicago, Chicago.

Isler, M. L., R. T. Chesser, M. B. Robbins, A. M. Cuervo, C. D. Cadena, and P. A. Hosner. 2020. Taxonomic evaluation of the *Grallaria rufula* (Rufous Antpitta) complex (Aves: Passeriformes: Grallariidae) distinguishes sixteen species. *Zootaxa* 4817: Article 4817.1.1.

James, H. F. 2017. Getting under the skin: A call for specimen-based research on the internal anatomy of birds 1. Pages 11–22 in M. S. Webster, ed. *The Extended Specimen: Emerging Frontiers in Collections-Based Ornithological Research.* CRC Press, Boca Raton, FL.

Jarvis, E. D., S. Mirarab, A. J. Aberer, B. Li, P. Houde, C. Li, S. Y. W. Ho, B. C. Faircloth, B. Nabholz, J. T. Howard, A. Suh, C. C. Weber, R. R. d. Fonseca, J. Li, F. Zhang, H. Li, L. Zhou, N. Narula, L. Liu, G. Ganapathy, B. Boussau, M. S. Bayzid, V. Zavidovych, S. Subramanian, T. Gabaldón, S. Capella-Gutiérrez, J. Huerta-Cepas, B. Rekepalli, K. Munch, M. Schierup, B. Lindow, W. C. Warren, D.

Ray, R. E. Green, M. W. Bruford, X. Zhan, A. Dixon, S. Li, N. Li, Y. Huang, E. P. Derryberry, M. F. Bertelsen, F. H. Sheldon, R. T. Brumfield, C. V. Mello, P. V. Lovell, M. Wirthlin, M. P. C. Schneider, F. Prosdocimi, J. A. Samaniego, A. M. V. Velazquez, A. Alfaro-Núñez, P. F. Campos, B. Petersen, T. Sicheritz-Ponten, A. Pas, T. Bailey, P. Scofield, M. Bunce, D. M. Lambert, Q. Zhou, P. Perelman, A. C. Driskell, B. Shapiro, Z. Xiong, Y. Zeng, S. Liu, Z. Li, B. Liu, K. Wu, J. Xiao, X. Yinqi, Q. Zheng, Y. Zhang, H. Yang, J. Wang, L. Smeds, F. E. Rheindt, M. Braun, J. Fjeldsa, L. Orlando, F. K. Barker, K. A. Jønsson, W. Johnson, K.-P. Koepfli, S. O'Brien, D. Haussler, O. A. Ryder, C. Rahbek, E. Willerslev, G. R. Graves, T. C. Glenn, J. McCormack, D. Burt, H. Ellegren, P. Alström, S. V. Edwards, A. Stamatakis, D. P. Mindell, J. Cracraft, E. L. Braun, T. Warnow, W. Jun, M. T. P. Gilbert, and G. Zhang. 2014. Whole-genome analyses resolve early branches in the tree of life of modern birds. *Science* 346:1320–1331.

Joseph, L. 2011. Museum collections in ornithology: Today's record of avian biodiversity for tomorrow's world. *Emu* 111:i–xii.

Kinsella, C. M., F. J. Ruiz-Ruano, A.-M. Dion-Côté, A. J. Charles, T. I. Gossmann, J. Cabrero, D. Kappei, N. Hemmings, M. J. P. Simons, J. P. M. Camacho, W. Forstmeier, and A. Suh. 2019. Programmed DNA elimination of germline development genes in songbirds. *Nature Communications* 10: Article 5468.

Klicka, L. B., B. E. Kus, P. O. Title, and K. J. Burns. 2016. Conservation genomics reveals multiple evolutionary units within Bell's Vireo (*Vireo bellii*). *Conservation Genetics* 17:455–471.

Krabbe, N. K., T. S. Schulenberg, P. A. Hosner, K. V. Rosenberg, T. J. Davis, G. H. Rosenberg, D. F. Lane, M. J. Andersen, M. B. Robbins, C. D. Cadena, T. Valqui, J. F. Salter, A. J. Spencer, F. Angulo, and J. Fjeldså. 2020. Untangling cryptic diversity in the high Andes: Revision of the *Scytalopus* [*magellanicus*] complex (Rhinocryptidae) in Peru reveals three new species. *The Auk* 137: Article ukaa003.

Krell, F. T., and Q. D. Wheeler. 2014. Specimen collection: Plan for the future. *Science* 344:815–816.

Lacey, E. A., T. T. Hammond, R. E. Walsh, K. C. Bell, S. V. Edwards, E. R. Ellwood, R. Guralnick, S. M. Ickert-Bond, A. R. Mast, J. E. McCormack, A. K. Monfils, P. S. Soltis, D. E. Soltis, and J. A. Cook. 2017. Climate change, collections and the classroom: Using big data to tackle big problems. *Evolution: Education and Outreach* 10: Article 2.

Lane, D. F., M. A. Aponte Justiniano, R. S. Terrill, F. E. Rheindt, L. B. Klicka, G. H. Rosenberg, C. J. Schmitt, and K. J. Burns. 2021. A new genus and species of tanager (Passeriformes, Thraupidae) from the lower Yungas of western Bolivia and southern Peru. *Ornithology* 138: Article ukab059.

Livezey, B. C. 2003. Avian spirit collections: Attitudes, importance and prospects. *Bulletin of the British Ornithologists' Club* 123A:35–51.

Livezey, B. C., and R. L. Zusi. 2007. Higher-order phylogeny of modern birds (Theropoda, Aves: Neornithes) based on comparative anatomy: II. Analysis and discussion. *Zoological Journal of the Linnean Society* 149:1–95.

Lutz, H. L., V. V. Tkach, J. D. Weckstein, and M. S. Webster. 2017. Methods for specimen-based studies of avian symbionts. In Pages 157–183 in Webster, M. S., ed. *The Extended Specimen: Emerging Frontiers in Collections-Based Ornithological Research*. CRC Press, Boca Raton, FL.

McCormack, J. E., M. G. Harvey, B. C. Faircloth, N. G. Crawford, T. C. Glenn, and R. T. Brumfield. 2013. A phylogeny of birds based on over 1,500 loci collected by target enrichment and high-throughput sequencing. *PLoS One* 8: Article e54848.

McCormack, J. E., F. Rodríguez-Gómez, W. L. E. Tsai, B. C. Faircloth, F. Rodríguez-Gómez, W. L. E. Tsai, and B. C. Faircloth. 2017. Transforming museum specimens into genomic resources. Pages 143–156 in Webster, M. S., ed. *The Extended Specimen: Emerging Frontiers in Collections-Based Ornithological Research*. CRC Press, Boca Raton, FL.

McNew, S. M., L. N. Barrow, J. L. Williamson, S. C. Galen, H. R. Skeen, S. G. DuBay, A. M. Gaffney, A. B. Johnson, E. Bautista, and P. Ordoñez. 2021. Contrasting drivers of diversity in hosts and parasites across the tropical Andes. *Proceedings of the National Academy of Sciences of the USA* 118: Article e2010714118.

Medina, J. J., J. M. Maley, S. Sannapareddy, N. N. Medina, C. M. Gilman, and J. E. McCormack. 2020. A rapid and cost-effective pipeline for digitization of museum specimens with 3D photogrammetry. *PLoS One* 15: Article e0236417.

Minteer, B. A., J. P. Collins, and R. Puschendorf. 2014. Specimen collection: Plan for the future: Response. *Science* 344:816.

Moritz, C. 1994. Defining evolutionary-significant units for conservation. *Trends in Ecology & Evolution* 9:373–375.

National Academies of Sciences, Engineering, and Medicine. 2020. *Biological Collections: Ensuring Critical Research and Education for the 21st Century*. National Academies Press, Washington, DC.

Nichol, S. T., C. F. Spiropoulou, S. Morzunov, P. E. Rollin, T. G. Ksiazek, H. Feldmann, A. Sanchez, J. Childs, S. Zaki, and C. J. Peters. 1993. Genetic identification of a hantavirus associated with an outbreak of acute respiratory illness. *Science* 262:914–917.

Oliveros, C. H., D. J. Field, D. T. Ksepka, F. K. Barker, A. Aleixo, M. J. Andersen, P. Alström, B. W. Benz, E. L. Braun, M. J. Braun, G. A. Bravo, R. T. Brumfield, R. T. Chesser, S. Claramunt, J. Cracraft, A. M. Cuervo, E. P. Derryberry, T. C. Glenn, M. G. Harvey, P. A. Hosner, L. Joseph, R. T. Kimball, A. L. Mack, C. M. Miskelly, A. T. Peterson, M. B. Robbins, F. H. Sheldon, L. F. Silveira, B. T. Smith, N. D. White, R. G. Moyle, and B. C. Faircloth. 2019. Earth history and the passerine superradiation. *Proceedings of the National Academy of Sciences of the USA* 116:7916–7925.

Olson, S. L. 2003. Development and uses of avian skeleton collections. *Bulletin of the British Ornithologists' Club*. 123A:26–34.

Oswald, J. A., I. Overcast, W. M. Mauck Iii, M. J. Andersen, and B. T. Smith. 2017. Isolation with asymmetric gene flow during the nonsynchronous divergence of dry forest birds. *Molecular Ecology* 26:1386–1400.

Pacheco, J. F., B. M. Whitney, and L. P. Gonzaga. 1996. A new genus and species of furnariid (Aves: Furnariidae) from the cocoa-growing region of southeastern Bahia, Brazil. *Wilson Bulletin* 108:397–434.

Peterson, A. T., R. G. Moyle, Á. S. Nyári, M. B. Robbins, R. T. Brumfield, and J. V. Remsen. 2007. The need for proper vouchering in phylogenetic studies of birds. *Molecular Phylogenetics and Evolution* 45:1042–1044.

Pigozzi, M. I., and A. J. Solari. 1998. Germ cell restriction and regular transmission of an accessory chromosome that mimics a sex body in the zebra finch, *Taeniopygia guttata*. *Chromosome Research* 6:105–113.

Powers, K. E., L. A. Prather, J. A. Cook, J. Woolley, H. L. Bart, A. K. Monfils, and P. Sierwald. 2014. Revolutionizing the use of natural history collections in education. *Science Education Review* 13:24–33.

Prum, R. O., J. S. Berv, A. Dornburg, D. J. Field, J. P. Townsend, E. M. Lemmon, and A. R. Lemmon. 2015. A comprehensive phylogeny of birds (Aves) using targeted next-generation DNA sequencing. *Nature* 526:569–573.

Remsen, J. V. 1995. The importance of continued collecting of bird specimens to ornithology and bird conservation. *Bird Conservation International* 5:146–180.

Rheindt, F. E., D. M. Prawiradilaga, H. Ashari, Suparno, C. Y. Gwee, G. W. X. Lee, M. Y. Wu, and N. S. R. Ng. 2020. A lost world in Wallacea: Description of a montane archipelagic avifauna. *Science* 367:167–170.

Rocha, L. A., A. Aleixo, G. Allen, F. Almeda, C. C. Baldwin, M. V. L. Barclay, J. M. Bates, A. M. Bauer, F. Benzoni, C. M. Berns, M. L. Berumen, D. C. Blackburn, S. Blum, F. Bolaños, R. C. K. Bowie, R. Britz, R. M. Brown, C. D. Cadena, K. Carpenter, L. M. Ceríaco, P. Chakrabarty, G. Chaves, J. H. Choat, K. D. Clements, B. B. Collette, A. Collins, J. Coyne, J. Cracraft, T. Daniel, M. R. de Carvalho, K. de Queiroz, F. Di Dario, R. Drewes, J. P. Dumbacher, A. Engilis, M. V. Erdmann, W. Eschmeyer, C. R. Feldman, B. L. Fisher, J. Fjeldså, P. W. Fritsch, J. Fuchs, A. Getahun, A. Gill, M. Gomon, T. Gosliner, G. R. Graves, C. E. Griswold, R. Guralnick, K. Hartel, K. M. Helgen, H. Ho, D. T. Iskandar, T. Iwamoto, Z. Jaafar, H. F. James, D. Johnson, D. Kavanaugh, N. Knowlton, E. Lacey, H. K. Larson, P. Last, J. M. Leis, H. Lessios, J. Liebherr, M. Lowman, D. L. Mahler, V. Mamonekene, K. Matsuura, G. C. Mayer, H. Mays, J. McCosker, R. W. McDiarmid, J. McGuire, M. J. Miller, R. Mooi, R. D. Mooi, C. Moritz, P. Myers, M. W. Nachman, R. A. Nussbaum, D. Ó. Foighil, L. R. Parenti, J. F. Parham, E. Paul, G. Paulay, J. Pérez-Emán, A. Pérez-Matus, S. Poe, J. Pogonoski, D. L. Rabosky, J. E. Randall, J. D. Reimer, D. R. Robertson, M. O. Rödel, M. T. Rodrigues, P. Roopnarine, L. Rüber, M. J. Ryan, F. Sheldon, G. Shinohara, A. Short, W. B. Simison, W. F. Smith-Vaniz, V. G.

Springer, M. Stiassny, J. G. Tello, C. W. Thompson, T. Trnski, P. Tucker, T. Valqui, M. Vecchione, E. Verheyen, P. C. Wainwright, T. A. Wheeler, W. T. White, K. Will, J. T. Williams, G. Williams, E. O. Wilson, K. Winker, R. Winterbottom, and C. C. Witt. 2014. Specimen collection: An essential tool. *Science* 344:814–815.

Rocque, D. A., and K. Winker. 2005. Use of bird collections in contaminant and stable-isotope studies. *The Auk* 122:990–994.

Rohwer, V. G., Y. Rohwer, and C. B. Dillman. 2022. Declining growth of natural history collections fails future generations. *PLoS Biology* 20: Article e3001613.

Sackton, T. B., P. Grayson, A. Cloutier, Z. Hu, J. S. Liu, N. E. Wheeler, P. P. Gardner, J. A. Clarke, A. J. Baker, M. Clamp, and S. V. Edwards. 2019. Convergent regulatory evolution and loss of flight in paleognathous birds. *Science* 364:74–78.

Schindel, D. E., and J. A. Cook. 2018. The next generation of natural history collections. *PLoS Biology* 16: Article e2006125.

Schmitt, C. J., J. A. Cook, K. R. Zamudio, and S. V. Edwards. 2019. Museum specimens of terrestrial vertebrates are sensitive indicators of environmental change in the Anthropocene. *Philosophical Transactions of the Royal Society B: Biological Sciences* 374: Article 20170387.

Sibley, C. G., and J. E. Ahlquist. 1990. *Phylogeny and Classification of Birds: A Study in Molecular Evolution.* Yale University Press, New Haven, CT.

Stoeckle, M., and K. Winker. 2009. A global snapshot of avian tissue collections: State of the enterprise. *The Auk* 126:684–687.

Stresemann, E. 1975. *Ornithology From Aristotle to the Present/Erwin Stresemann*; translated by H. J. Epstein and C. Epstein; edited by G. William Cottrell; with a foreword and an epilogue on American ornithology by Ernst Mayr. Harvard University Press, Cambridge, MA.

Suarez, A. V., and N. D. Tsutsui. 2004. The value of museum collections for research and society. *BioScience* 54:66–74.

Torgasheva, A. A., L. P. Malinovskaya, K. S. Zadesenets, T. V. Karamysheva, E. A. Kizilova, E. A. Akberdina, I. E. Pristyazhnyuk, E. P. Shnaider, V. A. Volodkina, A. F. Saifitdinova, S. A. Galkina, D. M. Larkin, N. B. Rubtsov, and P. M. Borodin. 2019. Germline-restricted chromosome (GRC) is widespread among songbirds. *Proceedings of the National Academy of Sciences of the USA* 116:11845–11850.

Tsai, W. L. E., M. E. Schedl, J. M. Maley, and J. E. McCormack. 2020. More than skin and bones: Comparing extraction methods and alternative sources of DNA from avian museum specimens. *Molecular Ecology Resources* 20:1220–1227.

Tyler, J., M. T. Bonfitto, G. V. Clucas, S. Reddy, and J. L. Younger. 2020. Morphometric and genetic evidence for four species of gentoo penguin. *Ecology and Evolution* 10:13836–13846.

Vianna, J. A., F. A. N. Fernandes, M. J. Frugone, H. V. Figueiró, L. R. Pertierra, D. Noll, K. Bi, C. Y. Wang-Claypool, A. Lowther, and P. Parker. 2020. Genome-wide analyses reveal drivers of penguin diversification. *Proceedings of the National Academy of Sciences of the USA* 117:22303–22310.

Watson, A. S., S. Plog, B. J. Culleton, P. A. Gilman, S. A. LeBlanc, P. M. Whiteley, S. Claramunt, and D. J. Kennett. 2015. Early procurement of scarlet macaws and the emergence of social complexity in Chaco Canyon, NM. *Proceedings of the National Academy of Sciences of the USA* 112:8238–8243.

Webster, M. S., ed. 2017. *The Extended Specimen: Emerging Frontiers in Collections-Based Ornithological Research.* CRC Press, Boca Raton, FL.

Winker, K. 2004. Natural history museums in a postbiodiversity era. *BioScience* 54:455–459.

Winker, K. 2005. Bird collections: Development and use of a scientific resource. *The Auk* 122:966–971.

Winker, K., B. A. Fall, J. T. Klicka, D. F. Parmelee, and B. T. Harrison. 1991. The importance of avian collections and the need for continued collecting. *Loon* 63:238–246.

Winker, K., J. M. Reed, P. Escalante, R. A. Askins, C. Cicero, G. E. Hough, and J. Bates. 2010. The importance, effects, and ethics of bird collecting. *The Auk* 127:690–695.

Winkler, D. W., T. M. Pegan, E. R. Gulson-Castillo, J. I. Byington, J. P. Hruska, S. C. Orzechowski, B. M. Van Doren, E. I. Greig, and E. M. Wood. 2017. Student-led expeditions as an educational and collections-building enterprise 1. Pages 185–200 in M. S. Webster, ed. *The Extended Specimen: Emerging Frontiers in Collections-Based Ornithological Research.* CRC Press, Boca Raton, FL.

Yates, T. L., J. N. Mills, C. A. Parmenter, T. G. Ksiazek, R. R. Parmenter, J. R. Vande Castle, C. H. Calisher, S. T. Nichol, K. D. Abbott, J. C. Young, M. L. Morrison, B. J. Beaty, J. L. Dunnum, R. J. Baker, J. Salazar-Bravo, and C. J. Peters. 2002. The ecology and evolutionary history of an emergent disease: Hantavirus pulmonary syndrome: Evidence from two El Niño episodes in the American Southwest suggests that El Niño–driven precipitation, the initial catalyst of a trophic cascade that results in a delayed density-dependent rodent response, is sufficient to predict heightened risk for human contraction of hantavirus pulmonary syndrome. *BioScience* 52:989–998.

Young, J. R., C. E. Braun, S. J. Oyler-McCance, J. W. Hupp, and T. W. Quinn. 2000. A new species of sage-grouse (Phasianidae: *Centrocercus*) from southwestern Colorado. *Wilson Bulletin* 112:445–453.

Zhang, G., E. D. Jarvis, and M. T. P. Gilbert. 2014. A flock of genomes. *Science* 346:1308–1309.

SECTION III
BEHAVIOR

12
Commentary on Behavior in Ornithology

Gail L. Patricelli

Behavior is not simply one among many traits of an organism. It is the crucial interface where the inside of the organism meets and interacts with the external environment—where the rubber meets the road. Within individual organisms, behavior is the culmination of interacting genetic, physiological, morphological, cognitive, and developmental processes, and behavior will impact how those processes play out during the organism's lifetime. Behavior is also the crux where these proximate processes respond to—and shape—the biotic and physical environment experienced by the organism. Behavior therefore creates direct links among individuals, species, and communities of organisms forming an ecosystem. For all of these reasons, behavior is a critical driver and target of evolutionary change.

To the consternation of some invertebrate biologists, our knowledge about animal behavior comes disproportionately from studies of birds (M. Rosenthal, et al. 2017). There is a reason for this: Not only are birds engaging to watch but also they provide excellent model systems to study physiological, evolutionary, and ecological processes relevant to all organisms. Many species of birds, although certainly not all, are relatively observable in the wild, and some are amenable to captivity in the lab. Many birds are active in the daytime, like most scientists. The dominant modes of communication in birds are visual and acoustic, also like most scientists. This makes bird communication easier and more intuitive to observe, measure, and study than species that rely heavily on olfactory communication, such as many insects, fish, and mammals. For these reasons and more, birds include many of the most commonly chosen study organisms for research in animal behavior.

In the following four chapters in this section, you will learn about the cutting edge of research on key aspects of bird behavior—social behavior, acoustic communication, coloration, and migration—highlighting the breakthroughs in technology that have contributed to these advances. These breakthroughs in technology during the past few decades have been truly revolutionary. The "-omics" revolutions (genomics, transcriptomics, and metabolomics) have allowed researchers to better understand, for example, the complex genetic, hormonal, and developmental mechanisms underlying behaviors and behavior-related traits such as plumage coloration. At the organismal level, the technology needed to track, observe, and measure behavior is improving by leaps and bounds, as are the computational methods and statistical models used to extract and analyze data.

As an animal behaviorist and ornithologist, when I reflect on the state of research in my field, advances in technology emerge front and center in my mind. This is not

Gail L. Patricelli, *Commentary on Behavior in Ornithology*. In: *New Perspectives in Ornithology*.
Edited by: Scott V. Edwards and J. Michael Reed, Oxford University Press. © Oxford University Press (2025).
DOI: 10.1093/oso/9780197787670.003.0012

just because gadgets are fun (they definitely are) but because technology is critical to helping us overcome the biases and limitations of our own senses and experiences and to broadening the questions we ask. For example, the generalization I made a few paragraphs ago—that olfactory communication is relatively unimportant to most birds—has been challenged by recent work demonstrating a role of scent in the social communication of seabirds, songbirds, and others (Balthazart and Taziaux 2009, Whittaker and Hagelin 2020, Jennings et al. 2022). Indeed, the more we look for chemical communication in birds, the more we find, making this an exciting frontier of research in bird behavior. We are only now reaching this frontier in part due to scientists' reliance on our own sensory systems and experiences to guide the questions we ask. Overcoming these limitations with technology is a recurring theme in the chapters in this section.

Advances in technology are also pushing ornithology, like other fields of science, into the era of "big data," allowing us to scale up from our detailed studies of individual components (e.g., genes or individual animals) to thinking about systems, with complex networks of interacting components that function as a whole. Such systems-level analyses require lots of data and the computing power to process it, as well as the creativity of researchers who can synthesize and make sense out of it all. For all these reasons, it is an exciting time to be an ornithologist and an animal behaviorist.

New Technology to Study Movements Large and Small

Some of the most dramatic of the recent technological advances have involved tracking animal movements—an essential tool in the study of behavior. Indeed, definitions of *behavior* often center on movements of whole organisms or parts of the body. Therefore, an understanding of how animals move in the wild is critical to understanding all aspects of behavior, from physiological and genetic mechanisms of behavior to social grouping, breeding, foraging, and migration.

Most common methods of tracking use animal-affixed telemetry tags, small devices that collect or transmit data about movements and other measures. Since the 1960s, wildlife biologists have relied on radio tags, which transmit signature VHF frequencies that can be detected with portable directional antennas. These systems are relatively inexpensive, but they typically require enormous effort to yield coarse data over a small spatial scale. After four decades of incremental advances, the past few decades have seen exponential improvement in the hardware, software, and affordability of wildlife telemetry, causing a sea change in the study of animal movement (see Chapter 13, Box 13.2). Telemetry tags have been miniaturized to less than 1 g, small enough to fit even hummingbirds and small songbirds. Some telemetry tags now yield precise locations with the onboard Global Positioning System (GPS), satellite Doppler, or light-level monitoring, or they collect complex data from onboard accelerometers, physiological monitors, or microphones. Tags may store or transmit data to other tags, to land-based receiver arrays (e.g., Motus; Taylor et al. 2017), or to satellites (e.g., ARGOS; Hofman et al. 2019). Alternatively, tiny passive integrated transponders (PIT tags) can be placed on color bands and detected by receivers placed strategically, for example, to detect visits to nests or feeders.

Technological improvements in computing power and artificial intelligence (AI) have also allowed us to use older methods of data collection in new ways. For example, audio and video recordings—critical tools in the study of animal behavior since the days of wax cylinders and stop-motion cameras—can now be used to track the movements of birds in captivity or in the wild using freely available software (Panadeiro et al. 2021). For example, machine learning (trainable AI) algorithms have been developed for pose estimation, tracking the relative position of multiple body parts during biomechanical studies of flight or dance displays (e.g. DEEPLabCut [Mathis et al. 2018] and SLEAP [Pereira et al. 2022]). On a larger scale, algorithms can efficiently track position and orientation of multiple interacting individuals throughout the day on videos (e.g., UMATrack [Yamanaka and Takeuchi 2018] and TRex [Walter and Couzin 2021]). Arrays of GPS-synchronized audio recorders can be used for acoustic localization, triangulating bird positions from vocalizations for behavioral studies; automatic detection of vocalizations by machine learning is making this data-heavy method more feasible on a larger scale (Blumstein et al. 2011, Rhinehart et al. 2020). Similar processes have made acoustic monitoring a more powerful tool for conservation, allowing detection of bird vocalizations at night, in remote locations, and over long periods of time (Salamon et al. 2016, Kahl et al. 2021, Lostanlen et al. 2022).

Somewhat ironically, improvements in computing power during the past few decades have brought us back to the beginning—making observations by eye and ear. As discussed in Chapter 16, community science approaches, such as eBird, harness the power of the masses to collect rich databases of bird observations, including behaviors (Sullivan et al. 2009, La Sorte et al. 2018). These data help researchers monitor changes in abundance, habitat choice, migrations, singing behaviors, and nesting locations. These tools also engage the public in the process of data collection, building investment in the outcomes of research and conservation efforts, benefiting both humans and birds (Dickinson and Bonney 2012, McKinley et al. 2017).

With these advances in technology, we now find ourselves awash in data about animal movements, which has necessitated the development of sophisticated new methods to analyze and interpret these data. This study of *movement ecology* is another exciting frontier of ornithology (Nathan et al. 2022).

Unsurprisingly, these advances in technology and analysis have revolutionized the study of animal migration (Chapter 16), providing detailed information about routes and timing of migration and helping identify critical locations for conservation. But the impact reaches far beyond studies of migration, opening new avenues for research in many fields, including those discussed in the following chapters. In the study of social behavior (Chapter 13), for example, technology allows researchers to examine social networks; flocking behaviors; and how these behaviors relate to differences in physiology, genes, and experience. In the study of bird coloration (Chapter 15), for example, movement data allow researchers to examine how the expression of colorful courtship displays relates to foraging and diet, territory quality, social experiences, display tactics, and environmental conditions such as ambient light. In the study of acoustic communication (Chapter 14), for example, movement data can be linked to vocalizations recorded on microphone arrays, to track where and when song learning and practice occur.

In the following sections, I summarize key points made by the authors of each chapter in this section on bird behavior, discussing connections between the topics and adding a few particularly exciting examples from my perspective as a researcher of bird communication and sexual selection.

Social Behavior

All sexually reproducing species are social at least long enough to breed; many birds go far beyond this, having complex social relationships throughout their lives. Until fairly recently, much of that complexity was hidden from us. In Chapter 13, Damien Farine makes the case that practical limitations on observing birds have biased research toward the study of breeding in altricial birds, where social behavior happens in the nest and can be observed by eye or with fixed cameras. Simply stated, parents and chicks that walk away shortly after hatching are more challenging to watch than those that stay put. And nonbreeding behaviors, which often involve rapid movements and migrations, have been difficult to track in the field until recently.

Developments in remote telemetry and automatic tracking from videos, discussed above and in Chapter 13, have given us a new view into the lives of birds and have challenged long-standing ideas about bird social behavior—that avian social structure is simpler than that of social mammals, and simpler in precocial than in altricial species. Recent tracking of large groups of precocial and altricial birds has revealed complex, multilevel social structures, from families to cohesive groups of unrelated individuals, fission–fusion dynamics among groups, and flocks with multiple interacting species (Hobson et al. 2014, Martínez et al. 2018, Papageorgiou and Farine 2021). Another exciting dimension of avian social complexity is the study of collective behavior—examining how individual behaviors scale up into complex emergent properties of groups, such as murmurations of European Starlings (*Sturnus vulgaris*; Ballerini et al. 2008). Farine argues that these new findings suggest that the lack of observed social complexity in birds may tell us more about the difficulty of studying social behavior in birds than about bird social complexity.

One factor long thought to limit the complexity of bird social lives was the limits of the "bird brain"—a term not typically used as a compliment. Ample research shows that bird brains, although small, can be quite powerful, with brain-to-body-size ratio and numbers of neurons in some species comparable to those of large-brained mammals (Iwaniuk and Nelson 2003, Emery 2006, Olkowicz et al. 2016). But, even in bird species with more modestly powered brains, recent research reveals complex, multilevel societies, challenging our understanding of the relationship between social complexity and cognitive capacity (Emery et al. 2007, Papageorgiou and Farine 2021).

Coloration and Color Vision

The topic of bird coloration may seem misplaced in a section focused on behavior, but bird coloration and color vision are linked to nearly every facet of animal behavior—whether a species is camouflaged, like tree creepers, or has gaudy colors and plumage, like birds of paradise, and whether a species detects colors from red to ultraviolet, like hummingbirds, or is tuned for low-light vision, like owls. The behavior of each species will, in turn, impact the evolution of coloration and color vision. Changes in coloration may also have the power to drive the formation of new species, as color diverges in response to mate preferences or ecological conditions.

In Chapter 15, Mary Caswell Stoddard discusses many exciting frontiers of research in avian coloration and color vision. Once again, this is an area in which new technology is opening new doors, allowing us to ask and answer new questions. This knowledge has important implications for conservation and engineering, such as the development of glass more visible to birds to reduce window strikes, or the engineering of materials that mimic the structural coloration of birds (Håstad and Ödeen 2014, Dumanli and Savin 2016, Xiao et al. 2020, Brown et al. 2021).

Many of the recent advances in the study of coloration have focused on the genetics, physiology, and physics of feather and skin colors; as we learn more about these mechanisms, new possibilities open in the study of bird evolution. One of the most striking examples of this is the study of paleocoloration. Our understanding of dinosaur appearance has changed a great deal since the green-brown beasts many of us grew up seeing in books, movies, and museum dioramas. We now know that many avian and non-avian dinosaurs sported various types of filamentous, plumaceous, or pennaceous feathers. Scientists are using what we know about the mechanisms of coloration in living birds to infer the color of these feathers, as well as scales, skin, and even eggshells, using increasingly sophisticated methods to image the fossilized microstructure of melanosomes and keratin (Wiemann et al. 2017, Roy et al. 2019, Vinther 2020). Stoddard discusses how these studies challenge the notion that all early birds and their eggs were drab.

During mate choice, potential partners often assess the yellow, orange, and red plumage produced by carotenoid pigments; these pigments have received a great deal of attention by researchers interested in bird behavior because carotenoids are sequestered from the diet and their expression in plumage is related to the bearer's health, parasite infection, and parental care behaviors (Hill 1991). This is a textbook example of an honest indicator of mate "quality," but the mechanism that links carotenoid color to quality remains controversial. I have been excited to follow recent developments in our understanding of this link, which was often assumed to be a resource trade-off, with poor foragers unable to gather enough carotenoid from their diet for both feather coloration and immune system function. Recent research has challenged this simplistic notion, finding little support for this trade-off (Koch et al. 2018, 2019). Rather, there is increasing support for a shared-pathway mechanism, where the conversion of dietary carotenoids into feather pigments is linked to the efficiency of mitochondrial function through shared biochemical pathways (Koch et al. 2017, Hill et al. 2019, Powers and Hill 2021). Although controversy remains, these results suggest that the expression of carotenoid-based signals may serve as an index of metabolic health. Interestingly, metabolic health may also affect the visual assessment of these signals via the carotenoid pigments in the retina (Caves et al. 2020, Price-Waldman and Stoddard 2021, Toomey and Ronald 2021). This opens a new line of research into the source of variation in mitochondrial function and the possibility that mitochondrial compatibility among divergent populations may play a role in speciation (Koch et al. 2021).

Stoddard also highlights multiple paradigm shifts underway in our thinking about bird color vision and perception. Recent research has revealed far more variability than expected in the color vision capabilities among bird species, due to differences in genes and gene expression (Kelber 2019, Price-Waldman and Stoddard 2021). Furthermore, there is variation among species in how the brain perceives and interprets visual information, including colors and patterns. For example, Zebra Finches (*Taeniopygia guttata*), like humans, may perceive gradual changes along an orange-red and blue-green continuum not as gradual but, rather, as changes in category, which may affect mate choice decisions based on color (Caves et al. 2018, Zipple et al. 2019). We still know little about how such cognitive biases may shape perception and behavioral responses, although this aspect of *receiver psychology* has long been identified as a critical gap in our knowledge (Guilford and Dawkins 1991, Rowe 2013, Ryan and Cummings 2013, Mendelson et al. 2016). This is one area in which human biases can inform rather than hinder research—for example, studies of human perception to improve marketing and advertisements have inspired animal behaviorists to ask whether similar perceptual biases are reflected in the courtship advertisements of nonhuman animals (Bateson and Healy 2005). For example, biases in the perception of color and the relative size of objects may explain the structure of the bower used during courtship by Great Bowerbirds (*Chlamydera nuchalis*), as well as the placement of gathered decorations in front of the bower (Endler et al. 2010, Kelley and Endler 2012). Another carryover from the study of human perception is the use of eye tracking, a technique long used to study visual attention for marketing (Wedel and Pieters 2008). Yorzinski et al. (2013, 2017) developed eye trackers to examine how the famously elaborate train of a displaying peacock is assessed by peahens and rival peacocks. Tools such as this have promise to reveal how complex displays such as the peacock's train can be favored by sexual selection, as well as similarities and differences in how humans and birds perceive the world.

Acoustic Communication

Birds are famously vocal, with *calls* used to mediate social behaviors such as parental care, flock movement, foraging, and predator avoidance, and *songs* used for courtship and territorial interactions during breeding. In Chapter 14, Elizabeth Derryberry and Rindy Anderson tackle the long-held misconception that male songbirds sing loudly and female songbirds listen quietly, showing us a much more complicated reality. The authors review the growing evidence of variation among species in which sex sings—in some species only males sing, but in other species both males and females sing as solos or in duets (Langmore 1998, Hall 2004). Comparative studies show that singing by both sexes was the ancestral state for songbirds, with male-only singing evolving more recently and more commonly in temperate species (Riebel et al. 2005, Odom et al. 2014, Price 2015). Male-only singing was long viewed as the norm for a variety of reasons, including the fact that research was done mostly by men in temperate regions. But even in some temperate species, female song was present but largely ignored (Garamszegi et al. 2007).

In addition to overlooking female song, researchers have largely overlooked "soft songs," which are used for quiet, close-range communication (Dabelsteen et al. 1998, Reichard and Welklin 2015). A growing body of research shows that soft songs play important roles, both as threats between rival males and during courtship between pairmates (Anderson et al. 2008, Reichard and Anderson 2015). The focus on loud male song is easy to understand: It is easier to detect and record in the field. Measuring amplitude (i.e., loudness) of vocalizations in the field is extremely challenging because measurements are strongly affected by distance and angle of the bird relative to the microphone, distortion due to propagation through the environment, and the equipment used for measurement. As a result of these and other factors, there is still surprisingly little known about the role of amplitude in bird communication (Dabelsteen et al. 1993, Patricelli et al. 2007, Brumm 2009).

Derryberry and Anderson make a strong case that for both soft songs and female songs, a great deal of observational study is needed to fill in the gaps in our natural history knowledge of familiar species, combined with experimental studies examining function and comparative studies to further examine the evolutionary roots of these phenomena (Odom and Benedict 2018, Riebel et al. 2019). The research reviewed in Chapter 14 on overlooked aspects of acoustic communication is both a cautionary and a hopeful tale, demonstrating the myopia of "common sense," when based on what scientists see and hear every day, and the power of new technology to open minds to overcome these biases.

Information about acoustic communication in birds becomes increasingly important as the human-built world gets increasingly loud (Barber et al. 2009, Senzaki et al. 2020). One of the most active areas of research on bird communication investigates the impact of noise pollution on the structure and efficacy of bird song. This research began after Slabbekoorn and Peet (2003) found that Great Tits (*Parus major*) in urban areas sing at higher frequency (i.e., pitch) in noise. Studies have since found examples of birds increasing repetition and amplitude of vocalizations in noise and avoiding noise by singing before rush hour and during quiet gaps (Brumm and Slabbekoorn 2005, Patricelli and Blickley 2006). The COVID-19 pandemic provided a fascinating opportunity, as quick-thinking researchers recorded urban White-Crowned Sparrows (*Zonotrichia leucophrys*) singing during the 2020 lockdown, which reduced human activity and thus noise; these songs were similar to songs from quieter areas, suggesting a surprising level of flexibility in singing behavior (Derryberry et al. 2020). Whether these changes are evidence of birds avoiding the impacts of noise or symptoms of our impact is still largely unknown. A related frontier of research in urban ecology investigates the impact of artificial light at night on singing behaviors, sleep, health, and development of birds living in human-impacted areas (Miller 2006, Kempenaers et al. 2010, Dominoni et al. 2013). Researchers are just starting to investigate how noise and light pollution, which often go together, may interact to affect birds (Swaddle et al. 2015, Willems et al. 2022).

Migration

The central challenge of studying bird migration is obvious: Birds fly, they are fast, and they cover large distances. We cannot simply follow individual birds to determine where they go. Since the 1800s, ornithologists have been learning about the vast scale of movements of some migrating species from observations or recoveries of banded birds. But using observations alone, we cannot learn the detailed pathways, timing, stopover locations, altitude, or the methods of navigation. With current technology, we can gain insights that early ornithologists could only dream of achieving. In Chapter 16, Kristen Ruegg, Dmitry Kishkinev, and Barbara Helm highlight novel insights in the study of bird migration and the advances in technology that have enabled them.

The tracking and telemetry technology discussed above and more (e.g., stable isotopes, landscape genomics, and radar) have revealed the vast scale and complexity of bird migrations. Many of these revelations strike wonder, such as the more than 10,000-km twice-yearly migration across the Pacific Ocean of the bar-tailed godwit (Gill et al. 2009). Technology has also been critical to learning about nighttime migration. For example, Gayk and Mennill (2023) used microphone arrays to localize the source of nighttime flight calls in three-dimensional space, revealing which species migrate together in mixed-species flocks, and convergence in the flight calls among these species. Acoustic monitoring promises to be a powerful tool in population estimation for species that migrate by night (Salamon et al. 2016, Lostanlen et al. 2022). Understanding night migration is critical in efforts to measure the impacts of climate change on migration timing and to reduce mortality from window strikes and other human-made hazards (Horton et al. 2020).

Bird migrations are among the most astounding feats of navigation in the natural world, but many fundamental genetic, cognitive, and sensory processes involved with migration remain unexplained. How do birds do it? Advances in tracking technology have provided insights into the relative importance of landmarks, olfactory cues, the sun, the stars, and the Earth's magnetic field in guiding navigation, as well as the physiological mechanisms that allow birds to sense these cues (Chernetsov et al. 2011, Wikelski et al. 2015, Kishkinev et al. 2016). The combination of tracking technology and new methods in genomic analysis has also yielded insights into the role of genes and experience in determining the timing and routes of migration, whether the birds migrate or remain resident, and the consequences of these differences for reproductive isolation and population divergence (Franchini et al. 2017, Turbek et al. 2022).

Bird migrations have captured the imaginations of scientists and nonscientists alike and helped build an appreciation for the wonders of nature. Learning about bird migrations is also an excellent reminder of the interconnectedness of our world—the importance of conserving habitat throughout the world to keep our local ecosystems intact.

The Whole Bird in Context

As any animal behaviorist will state, behavior is complicated. It results from interactions between complicated mechanisms, inputs, and feedbacks, including sensory physiology, perception, and cognition. This is especially true in the study of animal communication, which involves feedbacks between multiple individuals often in a complex social and physical environment. Many of the exciting frontiers in the study of communication involve grappling with the evolutionary implications of this complexity and variability over time, over space, and among interacting parts of a system (Hebets et al. 2016, Patricelli and Hebets 2016, M. Rosenthal et al. 2018).

Arguably the most complex communication in birds occurs during courtship, where one or both sexes produce displays that can integrate sounds, scents, colors, and dances to tap into the aesthetic sense of their potential mate to elicit copulation. To understand the evolution of these courtship displays, researchers typically measure each signal component—for example, quantifying the amplitude and pitch of the vocalizations, the speed and consistency of the dance moves, reflectance and size of color patches, and length of the feathers. These data can be used to examine how variation among individuals in each component relates to signaler condition or mating success or to variation among species in a phylogenetic comparative analysis. Such work has been very informative, but it risks overlooking the dynamic and interactive nature of courtship and also the evolution of courtship as a complex, multisensory whole.

How do we proceed from studying components to studying a whole? Ligon et al. (2018) studied one of the most spectacular and complex families of birds, the birds of paradise (Paradisaeidae), using a novel approach to examine how complexity per se evolves among species and among courtship components. To do so, they collected painstaking measurements for 40 species from museum skins and also from video and audio recordings from the Macaulay Library at Cornell University. To make measurements comparable among species, they adapted methods from information theory to calculate complexity for different courtship components (acoustic, visual, and behavioral). They found that instead of trade-offs between complexity in acoustic, color, and behavioral signaling, there were positive correlations among modalities, with integrated suites of traits evolving in tandem as a *courtship phenotype*. They argue that this functional overlap and interdependency reflects system-level *robustness* and may promote diversification (Hebets et al. 2016). Wilkins et al. (2015) used a similarly holistic approach to examine how variation in sexual displays among individuals is related to fitness in a single (less flashy) species, the North American Barn Swallow (*Hirundo rustica erythrogaster*). In this study, the authors examined the pattern of correlations among pairs of traits to calculate the *phenotype network*, identifying modules of traits important for male–male competition versus female choice, and some redundancy between them.

Analyses such as those by Ligon et al. (2018) and Wilkins et al. (2015) provide a path toward viewing courtship systems and identifying patterns in the evolution of complexity. Ongoing improvements in computing power and machine learning will allow automation of such analyses on larger numbers of species, tapping into the ever-increasing availability of online databases, such as libraries of audio and video recordings and three-dimensional color scans of museum skins (Thomas et al. 2016, Mathis et al. 2018, Yamanaka and Takeuchi 2018, Janisch et al. 2021). Exciting days are ahead for research into how sexual selection has led to the astonishing variation in avian sexual signals.

When considering in a holistic way the courtship displays of birds of paradise as well as Barn Swallows, the word *beauty* comes to mind. We cannot always pinpoint which combination of traits evoke our sense of beauty, because it is a subjective aesthetic experience that combines sensory, emotional, and intellectual response. As scientists, we are trained to resist the temptation to anthropomorphize—assuming that nonhuman animals share our subjective experience of the world—in order to minimize the impacts of our own biases on our research. The chapters in this volume give ample support to the notion that our human biases can mislead us and narrow our view. However, there is a danger that we may overcorrect, ignoring similarities in behavior that can provide new insights and ideas for research. Darwin (1871) first proposed that female birds possess an "almost human degree of taste" and are "most excited or attracted by the most beautiful, or melodious, or gallant males" (p. 123). Some scientists interested in sexual selection have criticized this aesthetic aspect of Darwin's theory (Jones and Ratterman 2009). More recently, there has been a push to embrace the concept of beauty in thinking about the evolution of courtship displays (Endler 2012; Prum 2012, 2017; G. Rosenthal 2017; Ryan 2018; Rodríguez 2020).

Prum (2017) makes the case that female preference for beauty is Darwin's forgotten theory, ignored by most modern research on sexual selection. In particular, Prum argues that the "Sexy Son" process (also called Fisherian or Runaway sexual selection; R. Fisher 1930, Lande 1981) leads to the evolution of beauty for beauty's sake; he rejects as incompatible with beauty the alternative hypothesis that sexual displays evolve as indicators of mate quality, arguing (correctly, in my opinion) that this is too often assumed without evidence.

Others have viewed a focus on beauty as an extension of the "umwelt" concept in animal behavior—the foundational idea that to understand an organism's behavior (in this case, mate choice), we must understand its sensory and perceptual world (von Uexküll 1934). In this view, the sense of beauty is part of the mechanism of mate preference, compatible with sexual displays that evolve by any process, whether they are indicators of quality or are simply sexy (Patricelli et al. 2019).

Conceptualizing sexual signals through the lens of beauty moves us away from thinking about sexual displays only as a form of manipulation by displayers (usually but not always the male) or information extraction by receivers (usually but not always females)—concepts that are important in understanding the ultimate drivers of sexual signaling but that are incomplete. Regardless of what (if anything) the displays indicate about the displayer, the concept of beauty forces us to reckon with the courtship experience as a whole, shaped over evolutionary time by choosy receivers to suit their evolving aesthetic tastes. Beauty also calls attention to the need to study the genetic and physiological mechanisms by which sensory stimuli and cognitive processes are linked to the reward system of the brain, which evokes a pleasurable desire for more, and whether these links evolved in the context of mate choice or were co-opted into it. This is an exciting frontier in studies of sexual selection (H. Fisher et al. 2002, Riters 2011, Earp and Maney 2012), seeking to understand what a sense of beauty actually *is* and how it might evolve in humans and birds alike to motivate behaviors.

Conclusion

The study of behavior encompasses much of what draws us to birds: their beautiful songs, colors, and dances; their astonishing feats of endurance; their social engagements from co-parenting to mixed-species flocking; and more. As humans alter the planet at a quickening pace—with increasing urbanization, habitat loss, and climate change—behavior will often be the first response, either allowing birds to adjust to the changes or not. These behavioral changes are often the first thing that we, as scientists and bird lovers, can see and measure as evidence of human impacts (it is the figurative and literal "canary in the coalmine"). As we work to protect birds and their habitats, we will need to draw on all of our knowledge from the research described in these chapters—the curiosity-driven basic research about sexual selection, bioacoustics, coloration, foraging, group living, navigation, and so on, and the conservation-focused research about responses to human impacts, critical habitats and flyways, methods of population monitoring, and so on. And, importantly, we will also need scientists, science communicators, and engaged bird lovers to build public support for conservation by spreading an appreciation for the beauty of birds and their critical role in the health of our ecosystems.

References

Anderson, R. C., W. A. Searcy, S. Peters, and S. Nowicki. 2008. Soft song in song sparrows: Acoustic structure and implications for signal function. *Ethology* 114:662–676.

Ballerini, M., N. Calbibbo, R. Candeleir, A. Cavagna, E. Cisbani, I. Giardina, V. Lecomte, A. Orlandi, G. Parisi, A. Procaccini, M. Viale, and V. Zdravkovic. 2008. Interaction ruling animal collective behavior depends on topological rather than metric distance: Evidence from a field study. *Proceedings of the National Academy of Sciences of the USA* 105:1232–1237.

Balthazart, J., and M. Taziaux. 2009. The underestimated role of olfaction in avian reproduction? *Behavioural Brain Research* 200:248–259.

Barber, J. R., K. R. Crooks, and K. M. Fristrup. 2009. The costs of chronic noise exposure for terrestrial organisms. *Trends in Ecology & Evolution* 25:180–189.

Bateson, M., and S. D. Healy. 2005. Comparative evaluation and its implications for mate choice. *Trends in Ecology & Evolution* 20:659–664.

Blumstein, D. T., D. J. Mennill, P. Clemins, L. Girod, K. Yao, G. Patricelli, J. L. Deppe, A. H. Krakauer, C. Clark, K. A. Cortopassi, S. F. Hanser, B. McCowan, A. M. Ali, and A. N. G. Kirschel. 2011. Acoustic monitoring in terrestrial environments using microphone arrays: Applications, technological considerations and prospectus. *Journal of Applied Ecology* 48:758–767.

Brown, B. B., S. Santos, and N. Ocampo-Penuela. 2021. Bird-window collisions: Mitigation efficacy and risk factors across two years. *PeerJ* 9: Article e11867.

Brumm, H. 2009. Song amplitude and body size in birds. *Behavioral Ecology and Sociobiology* 63:1157–1165.

Brumm, H., and H. Slabbekoorn. 2005. Acoustic communication in noise. *Advances in the Study of Behavior* 35:151–209.

Caves, E. M., P. A. Green, M. N. Zipple, S. Peters, S. Johnsen, and S. Nowicki. 2018. Categorical perception of colour signals in a songbird. *Nature* 560:365–367.

Caves, E. M., L. E. Schweikert, P. A. Green, M. N. Zipple, C. Taboada, S. Peters, S. Nowicki, and S. Johnsen. 2020. Variation in carotenoid-containing retinal oil droplets correlates with variation in perception of carotenoid coloration. *Behavioral Ecology and Sociobiology* 74: Article 93.

Chernetsov, N., D. Kishkinev, V. Kosarev, and C. V. Bolshakov. 2011. Not all songbirds calibrate their magnetic compass from twilight cues: A telemetry study. *Journal of Experimental Biology* 214:2540–2543.

Dabelsteen, T., O. N. Larsen, and S. B. Pedersen. 1993. Habitat-induced degradation of sound signals: Quantifying the effects of communication sounds and bird location on blur ratio, excess attenuation, and signal-to-noise ratio in blackbird song. *Journal of the Acoustical Society of America* 93:2206–2220.

Dabelsteen, T., P. K. McGregor, H. M. Lampe, N. E. Langmore, and J. Holland. 1998. Quiet song in song birds: An overlooked phenomenon. *Bioacoustics* 9:89–105.

Darwin, C. 1871. *The Descent of Man, and Selection in Relation to Sex*. Vol. 2. Murray, London.

Derryberry, E. P., J. N. Phillips, G. E. Derryberry, M. J. Blum, and D. Luther. 2020. Singing in a silent spring: Birds respond to a half-century soundscape reversion during the COVID-19 shutdown. *Science* 370:575–579.

Dickinson, J. L., and R. Bonney. 2012. *Citizen Science: Public Participation in Environmental Research*. Cornell University Press, Ithaca, NY.

Dominoni, D., M. Quetting, and J. Partecke. 2013. Artificial light at night advances avian reproductive physiology. *Proceedings of the Royal Society B: Biological Sciences* 280: Article 20123017.

Dumanli, A. G., and T. Savin. 2016. Recent advances in the biomimicry of structural colours. *Chemical Society Reviews* 45:6698–6724.

Earp, S. E., and D. L. Maney. 2012. Birdsong: Is it music to their ears? *Frontiers in Evolutionary Neuroscience* 4: Article 14.

Emery, N. 2006. Review. Cognitive ornithology: The evolution of avian intelligence. *Philosophical Transactions of the Royal Society B: Biological Sciences* 361:23–43.

Emery, N. J., A. M. Seed, A. M. von Bayern, and N. S. Clayton. 2007. Cognitive adaptations of social bonding in birds. *Philosophical Transactions of the Royal Society B: Biological Sciences* 362:489–505.

Endler, J. A. 2012. Bowerbirds, art and aesthetics: Are bowerbirds artists and do they have an aesthetic sense? *Communicative & Integrative Biology* 5:281–283.

Endler, J. A., L. C. Endler, and N. R. Doerr. 2010. Great bowerbirds create theaters with forced perspective when seen by their audience. *Current Biology* 20:1679–1684.

Fisher, H., A. Aron, D. Mashek, H. Li, G. Strong, and L. L. Brown. 2002. The neural mechanisms of mate choice: A hypothesis. *Neuro Endocrinology Letters* 23(Suppl 4):92–97.

Fisher, R. A. 1930. *The Genetical Theory of Natural Selection*. Clarendon, Oxford, UK.

Franchini, P., I. Irisarri, A. Fudickar, A. Schmidt, A. Meyer, M. Wikelski, and J. Partecke. 2017. Animal tracking meets migration genomics: Transcriptomic analysis of a partially migratory bird species. *Molecular Ecology* 26:3204–3216.

Garamszegi, L. Z., D. Z. Pavlova, M. Eens, and A. P. Moller. 2007. The evolution of song in female birds in Europe. *Behavioral Ecology* 18:86–96.

Gayk, Z. G., and D. J. Mennill. 2023. Acoustic similarity of flight calls corresponds with composition and structure of mixed-species flocks of migrating birds: Evidence from a 3-D microphone array. *Philosophical Transactions of the Royal Society B: Biological Sciences* 378: Article 20220114.

Gill, R. E., T. L. Tibbitts, D. C. Douglas, C. M. Handel, D. M. Mulcahy, J. C. Gottschalck, N. Warnock, B. J. McCaffery, P. F. Battley, and T. Piersma. 2009. Extreme endurance flights by landbirds crossing the Pacific Ocean: Ecological corridor rather than barrier? *Proceedings of the Royal Society B: Biological Sciences* 276:447–457.

Guilford, T., and M. S. Dawkins. 1991. Receiver psychology and the evolution of animal signals. *Animal Behaviour* 42:1–14.

Hall, M. L. 2004. A review of hypotheses for the functions of avian duetting. *Behavioral Ecology and Sociobiology* 55:415–430.

Håstad, O., and A. Ödeen. 2014. A vision physiological estimation of ultraviolet window marking visibility to birds. *PeerJ* 2: Article e621.

Hebets, E. A., A. B. Barron, C. N. Balakrishnan, M. E. Hauber, P. H. Mason, and K. L. Hoke. 2016. A systems approach to animal communication. *Proceedings of the Royal Society B: Biological Sciences* 283: Article 20152889.

Hill, G. E. 1991. Plumage coloration is a sexually selected indicator of male quality. *Nature* 350:337–339.

Hill, G. E., W. R. Hood, Z. Ge, R. Grinter, C. Greening, J. D. Johnson, N. R. Park, H. A. Taylor, V. A. Andreasen, M. J. Powers, N. M. Justyn, H. A. Parry, A. N. Kavazis, and Y. Zhang. 2019. Plumage redness signals mitochondrial function in the house finch. *Proceedings of the Royal Society B: Biological Sciences* 286: Article 20191354.

Hobson, E. A., M. L. Avery, and T. F. Wright. 2014. The socioecology of Monk Parakeets: Insights into parrot social complexity. *The Auk* 131:756–775.

Hofman, M. P. G., M. W. Hayward, M. Heim, P. Marchand, C. M. Rolandsen, J. Mattisson, F. Urbano, M. Heurich, A. Mysterud, J. Melzheimer, N. Morellet, U. Voigt, B. L. Allen, B. Gehr, C. Rouco, W. Ullmann, O. Holand, N. H. Jorgensen, G. Steinheim, F. Cagnacci, M. Kroeschel, P. Kaczensky, B. Buuveibaatar, J. C. Payne, I. Palmegiani, K. Jerina, P. Kjellander, O. Johansson, S. LaPoint, R. Bayrakcismith, J. D. C. Linnell, M. Zaccaroni, M. L. S. Jorge, J. E. F. Oshima, A. Songhurst, C. Fischer, R. T. McBride, Jr., J. J. Thompson, S. Streif, R. Sandfort, C. Bonenfant, M. Drouilly, M. Klapproth, D. Zinner, R. Yarnell, A. Stronza, L. Wilmott, E. Meisingset, M. Thaker, A. T. Vanak, S. Nicoloso, R. Graeber, S. Said, M. R. Boudreau, A. Devlin, R. Hoogesteijn, J. A. May-Junior, J. C. Nifong, J. Odden, H. B. Quigley, F. Tortato, D. M. Parker, A. Caso, J. Perrine, C. Tellaeche, F. Zieba, T. Zwijacz-Kozica, C. L. Appel, I. Axsom, W. T. Bean, B. Cristescu, S. Periquet, K. J. Teichman, S. Karpanty, A. Licoppe, V. Menges, K. Black, T. L. Scheppers, S. C. Schai-Braun, F. C. Azevedo, F. G. Lemos, A. Payne, L. H. Swanepoel, B. V. Weckworth, A. Berger, A. Bertassoni, G. McCulloch, P. Sustr, V. Athreya, D. Bockmuhl, J. Casaer, A. Ekori, D. Melovski, C. Richard-Hansen, D. van de Vyver, R. Reyna-Hurtado, E. Robardet, N. Selva, A. Sergiel, M. S. Farhadinia, P. Sunde, R. Portas, H. Ambarli, R. Berzins, P. M. Kappeler, G. K. Mann, L. Pyritz, C. Bissett, T. Grant, R. Steinmetz, L. Swedell, R. J. Welch, D. Armenteras, O. R. Bidder, T. M. Gonzalez, A. Rosenblatt, S. Kachel, and N. Balkenhol. 2019. Right on track? Performance of satellite telemetry in terrestrial wildlife research. *PLoS One* 14: Article e0216223.

Horton, K. G., F. A. La Sorte, D. Sheldon, T.-Y. Lin, K. Winner, G. Bernstein, S. Maji, W. M. Hochachka, and A. Farnsworth. 2020. Phenology of nocturnal avian migration has shifted at the continental scale. *Nature Climate Change* 10:63–68.

Iwaniuk, A. N., and J. E. Nelson. 2003. Developmental differences are correlated with relative brain size in birds: A comparative analysis. *Canadian Journal of Zoology* 81:1913–1928.

Janisch, J., C. Mitoyen, E. Perinot, G. Spezie, L. Fusani, and C. Quigley. 2021. Video recording and analysis of avian movements and behavior: Insights from courtship case studies. *Integrative and Comparative Biology* 61:1378–1393.

Jennings, S. L., B. A. Hoover, S. Y. Wa Sin, and S. E. Ebeler. 2022. Feather chemicals contain information about the major histocompatibility complex in a highly scented seabird. *Proceedings of the Royal Society B: Biological Sciences* 289: Article 20220567.

Jones, A. G., and N. L. Ratterman. 2009. Mate choice and sexual selection: What have we learned since Darwin? *Proceedings of the National Academy of Sciences of the USA* 106 (Suppl 1):10001–10008.

Kahl, S., C. M. Wood, M. Eibl, and H. Klinck. 2021. BirdNET: A deep learning solution for avian diversity monitoring. *Ecological Informatics* 61: Article 101236.

Kelber, A. 2019. Bird colour vision—from cones to perception. *Current Opinion in Behavioral Sciences* 30:34–40.

Kelley, L. A., and J. A. Endler. 2012. Illusions promote mating success in great bowerbirds. *Science* 335:335–338.

Kempenaers, B., P. Borgstrom, P. Loes, E. Schlicht, and M. Valcu. 2010. Artificial night lighting affects dawn song, extra-pair siring success, and lay date in songbirds. *Current Biology* 20:1735–1739.

Kishkinev, D., D. Heyers, B. K. Woodworth, G. W. Mitchell, K. A. Hobson, and D. R. Norris. 2016. Experienced migratory songbirds do not display goal-ward orientation after release following a cross-continental displacement: An automated telemetry study. *Scientific Reports* 6: Article 37326.

Koch, R. E., K. L. Buchanan, S. Casagrande, O. Crino, D. K. Dowling, G. E. Hill, W. R. Hood, M. McKenzie, M. M. Mariette, D. W. A. Noble, A. Pavlova, F. Seebacher, P. Sunnucks, E. Udino, C. R. White, K. Salin, and A. Stier. 2021. Integrating mitochondrial aerobic metabolism into ecology and evolution. *Trends in Ecology & Evolution* 36:321–332.

Koch, R. E., G. E. Hill, and B. Sandercock. 2018. Do carotenoid-based ornaments entail resource trade-offs? An evaluation of theory and data. *Functional Ecology* 32:1908–1920.

Koch, R. E., C. C. Josefson, and G. E. Hill. 2017. Mitochondrial function, ornamentation, and immunocompetence. *Biological Reviews of the Cambridge Philosophical Society* 92:1459–1474.

Koch, R. E., M. Staley, A. N. Kavazis, D. Hasselquist, M. B. Toomey, and G. E. Hill. 2019. Testing the resource trade-off hypothesis for carotenoid-based signal honesty using genetic variants of the domestic canary. *Journal of Experimental Biology* 222: Article jeb188102.

Lande, R. 1981. Models of speciation by sexual selection of polygenic traits. *Proceedings of the National Academy of Sciences of the USA* 78:3721–3725.

Langmore, N. E. 1998. Functions of duet and solo songs of female birds. *Trends in Ecology & Evolution* 13:136–140.

La Sorte, F. A., C. A. Lepczyk, J. L. Burnett, A. H. Hurlbert, M. W. Tingley, and B. Zuckerberg. 2018. Opportunities and challenges for big data ornithology. *The Condor* 120:414–426.

Ligon, R. A., C. D. Diaz, J. L. Morano, J. Troscianko, M. Stevens, A. Moskeland, T. G. Laman, and E. Scholes, 3rd. 2018. Evolution of correlated complexity in the radically different courtship signals of birds-of-paradise. *PLoS Biology* 16: Article e2006962.

Lostanlen, V., A. Cramer, J. Salamon, A. Farnsworth, B. M. Van Doren, S. Kelling, and J. P. Bello. 2022. BirdVox: Machine listening for bird migration monitoring. *bioRxiv*.

Martínez, A. E., H. S. Pollock, J. P. Kelley, and C. E. Tarwater. 2018. Social information cascades influence the formation of mixed-species foraging aggregations of ant-following birds in the Neotropics. *Animal Behaviour* 135:25–35.

Mathis, A., P. Mamidanna, K. M. Cury, T. Abe, V. N. Murthy, M. W. Mathis, and M. Bethge. 2018. DeepLabCut: Markerless pose estimation of user-defined body parts with deep learning. *Nature Neuroscience* **21**:1281–1289.

McKinley, D. C., A. J. Miller-Rushing, H. L. Ballard, R. Bonney, H. Brown, S. C. Cook-Patton, D. M. Evans, R. A. French, J. K. Parrish, T. B. Phillips, S. F. Ryan, L. A. Shanley, J. L. Shirk, K. F. Stepenuck, J. F. Weltzin, A. Wiggins, O. D. Boyle, R. D. Briggs, S. F. Chapin, D. A. Hewitt, P. W. Preuss, and M. A. Soukup. 2017. Citizen science can improve conservation science, natural resource management, and environmental protection. *Biological Conservation* 208:15–28.

Mendelson, T. C., C. L. Fitzpatrick, M. E. Hauber, C. H. Pence, R. L. Rodríguez, R. J. Safran, C. A. Stern, and J. R. Stevens. 2016. Cognitive phenotypes and the evolution of animal decisions. *Trends in Ecology & Evolution* 31:850–859.

Miller, M. W. 2006. Apparent effects of light pollution on singing behavior of American robins. *The Condor* 108:130–139.

Nathan, R., C. T. Monk, R. Arlinghaus, T. Adam, J. Alós, M. Assaf, H. Baktoft, C. E. Beardsworth, M. G. Bertram, A. I. Bijleveld, T. Brodin, J. L. Brooks, A. Campos-Candela, S. J. Cooke, K. Ø. Gjelland, P. R. Gupte, R. Harel, G. Hellström, F. Jeltsch, S. S. Killen, T. Klefoth, R. Langrock, R. J. Lennox, E. Lourie, J. R. Madden, Y. Orchan, I. S. Pauwels, M. Říha, M. Roeleke, U. E. Schlägel, D. Shohami, J. Signer, S. Toledo, O. Vilk, S. Westrelin, M. A. Whiteside, and I. Jarić. 2022. Big-data approaches lead to an increased understanding of the ecology of animal movement. *Science* 375: Article eabg1780.

Odom, K. J., and L. Benedict. 2018. A call to document female bird songs: Applications for diverse fields. *The Auk* 135:314–325.

Odom, K. J., M. L. Hall, K. Riebel, K. E. Omland, and N. E. Langmore. 2014. Female song is widespread and ancestral in songbirds. *Nature Communications* 5: Article 3379.

Olkowicz, S., M. Kocourek, R. K. Lucan, M. Portes, W. T. Fitch, S. Herculano-Houzel, and P. Nemec. 2016. Birds have primate-like numbers of neurons in the forebrain. *Proceedings of the National Academy of Sciences of the USA* 113:7255–7260.

Panadeiro, V., A. Rodriguez, J. Henry, D. Wlodkowic, and M. Andersson. 2021. A review of 28 free animal-tracking software applications: Current features and limitations. *Lab Animal* 50:246–254.

Papageorgiou, D., and D. R. Farine. 2021. Multilevel societies in birds. *Trends in Ecology & Evolution* 36:15–17.

Patricelli, G., and J. Blickley. 2006. Avian communication in urban noise: Causes and consequences of vocal adjustment. *The Auk* 123:639–649.

Patricelli, G. L., M. S. Dantzker, and J. W. Bradbury. 2007. Differences in acoustic directionality among vocalizations of the male red-winged blackbird (*Agelaius pheoniceus*) are related to function in communication. *Behavioral Ecology and Sociobiology* 61:1099–1110.

Patricelli, G. L., and E. A. Hebets. 2016. New dimensions in animal communication: The case for complexity. *Current Opinion in Behavioral Sciences* 12:80–89.

Patricelli, G. L., E. A. Hebets, and T. C. Mendelson. 2019. Book review of Prum, R. O. 2018. *The evolution of beauty: How Darwin's forgotten theory of mate choice shapes the animal world—and us* (2017), Doubleday, 428 pages, ISBN: 9780385537216. *Evolution* 73:115–124.

Pereira, T. D., N. Tabris, A. Matsliah, D. M. Turner, J. Li, S. Ravindranath, E. S. Papadoyannis, E. Normand, D. S. Deutsch, Z. Y. Wang, G. C. McKenzie-Smith, C. C. Mitelut, M. D. Castro, J.

D'Uva, M. Kislin, D. H. Sanes, S. D. Kocher, S. S. Wang, A. L. Falkner, J. W. Shaevitz, and M. Murthy. 2022. SLEAP: A deep learning system for multi-animal pose tracking. *Nature Methods* 19: 486–495.
Powers, M. J., and G. E. Hill. 2021. A review and assessment of the shared-pathway hypothesis for the maintenance of signal honesty in red ketocarotenoid-based coloration. *Integrative and Comparative Biology* 61:1811–1826.
Price, J. J. 2015. Rethinking our assumptions about the evolution of bird song and other sexually dimorphic signals. *Frontiers in Ecology and Evolution* 3: Article 40.
Price-Waldman, R., and M. C. Stoddard. 2021. Avian coloration genetics: Recent advances and emerging questions. *Journal of Heredity* 112:395–416.
Prum, R. O. 2012. Aesthetic evolution by mate choice: Darwin's really dangerous idea. *Philosophical Transactions of the Royal Society B: Biological Sciences* 367:2253–2265.
Prum, R. O. 2017. *The Evolution of Beauty: How Darwin's Forgotten Theory of Mate Choice Shapes the Animal World—and Us.* Doubleday, New York.
Reichard, D. G., and R. C. Anderson. 2015. Special Issue: Whispered communication. *Animal Behaviour* 105:253–265.
Reichard, D. G., and J. F. Welklin. 2015. On the existence and potential functions of low-amplitude vocalizations in North American birds. *The Auk* 132:156–166.
Rhinehart, T. A., L. M. Chronister, T. Devlin, and J. Kitzes. 2020. Acoustic localization of terrestrial wildlife: Current practices and future opportunities. *Ecology and Evolution* 10:6794–6818.
Riebel, K., M. L. Hall, and N. E. Langmore. 2005. Female songbirds still struggling to be heard. *Trends in Ecology & Evolution* 20:419–420.
Riebel, K., K. J. Odom, N. E. Langmore, and M. L. Hall. 2019. New insights from female bird song: Towards an integrated approach to studying male and female communication roles. *Biology Letters* 15: Article 20190059.
Riters, L. V. 2011. Pleasure seeking and birdsong. *Neuroscience and Biobehavioral Reviews* 35:1837–1845.
Rodríguez, R. L. 2020. Back to the basics of mate choice: The evolutionary importance of Darwin's sense of beauty. *Quarterly Review of Biology* 95:289–309.
Rosenthal, G. G. 2017. *Mate Choice: The Evolution of Sexual Decision Making From Microbes to Humans.* Princeton University Press, Princeton, NJ.
Rosenthal, M. F., M. Gertler, A. D. Hamilton, S. Prasad, and M. C. B. Andrade. 2017. Taxonomic bias in animal behaviour publications. *Animal Behaviour* 127:83–89.
Rosenthal, M. F., M. R. Wilkins, D. Shizuka, and E. A. Hebets. 2018. Dynamic changes in display architecture and function across environments revealed by a systems approach to animal communication. *Evolution* 72:1134–1145.
Rowe, C. 2013. Receiver psychology: A receiver's perspective. *Animal Behaviour* 85:517–523.
Roy, A., M. Pittman, E. T. Saitta, T. G. Kaye, and X. Xu. 2019. Recent advances in amniote palaeocolour reconstruction and a framework for future research. *Biological Reviews of the Cambridge Philosophical Society* 95:22–50.
Ryan, M. J. 2018. *A Taste for the Beautiful: The Evolution of Attraction.* Princeton University Press, Princeton, NJ.
Ryan, M. J., and M. E. Cummings. 2013. Perceptual biases and mate choice. *Annual Review of Ecology, Evolution, and Systematics* 44:437–459.
Salamon, J., J. P. Bello, A. Farnsworth, M. Robbins, S. Keen, H. Klinck, and S. Kelling. 2016. Towards the automatic classification of avian flight calls for bioacoustic monitoring. *PLoS One* 11: Article e0166866.
Senzaki, M., J. R. Barber, J. N. Phillips, N. H. Carter, C. B. Cooper, M. A. Ditmer, K. M. Fristrup, C. J. W. McClure, D. J. Mennitt, L. P. Tyrrell, J. Vukomanovic, A. A. Wilson, and C. D. Francis. 2020. Sensory pollutants alter bird phenology and fitness across a continent. *Nature* 587:605–609.
Slabbekoorn, H., and M. Peet. 2003. Ecology: Birds sing at a higher pitch in urban noise. *Nature* 424: Article 267.
Sullivan, B. L., C. L. Wood, M. J. Iliff, R. E. Bonney, D. Fink, and S. Kelling. 2009. eBird: A citizen-based bird observation network in the biological sciences. *Biological Conservation* 142:2282–2292.

Swaddle, J. P., C. D. Francis, J. R. Barber, C. B. Cooper, C. C. M. Kyba, D. D. Dominoni, G. Shannon, E. Aschehoug, S. E. Goodwin, A. Y. Kawahara, D. Luther, K. Spoelstra, M. Voss, and T. Longcore. 2015. A framework to assess evolutionary responses to anthropogenic light and sound. *Trends in Ecology & Evolution* 30:550–560.

Taylor, P. D., T. L. Crewe, S. A. Mackenzie, D. Lepage, Y. Aubry, Z. Crysler, G. Finney, C. M. Francis, C. G. Guglielmo, D. J. Hamilton, R. L. Holberton, P. H. Loring, G. W. Mitchell, D. R. Norris, J. Paquet, R. A. Ronconi, J. R. Smetzer, P. A. Smith, L. J. Welch, and B. K. Woodworth. 2017. The Motus Wildlife Tracking System: A collaborative research network to enhance the understanding of wildlife movement. *Avian Conservation and Ecology* 12: Article 8.

Thomas, G. H., J. A. Bright, and C. R. Cooney. 2016. Mark my bird [Data set]. Natural History Museum.

Toomey, M. B., and K. L. Ronald. 2021. Avian color expression and perception: Is there a carotenoid link? *Journal of Experimental Biology* 224: Article jeb203844.

Turbek, S. P., D. R. Schield, E. S. C. Scordato, A. Contina, X.-W. Da, Y. Liu, Y. Liu, E. Pagani-Núñez, Q.-M. Ren, C. C. R. Smith, C. A. Stricker, M. Wunder, D. M. Zonana, and R. J. Safran. 2022. A migratory divide spanning two continents is associated with genomic and ecological divergence. *Evolution* 76:722–736.

Vinther, J. 2020. Reconstructing vertebrate paleocolor. *Annual Review of Earth and Planetary Sciences* 48:345–375.

von Uexküll, J. 1934. *Streifzuge durch die Umwelten von Tieren und Menschen.* Ein Rowohlt, Hamburg, Germany.

Walter, T., and I. D. Couzin. 2021. TRex, a fast multi-animal tracking system with markerless identification, and 2D estimation of posture and visual fields. *eLife* 10: Article e64000.

Wedel, M., and R. Pieters. 2008. Eye tracking for visual marketing. *Foundations and Trends® in Marketing* 1:231–320.

Whittaker, D. J., and J. C. Hagelin. 2020. Female-based patterns and social function in avian chemical communication. *Journal of Chemical Ecology* 47:43–62.

Wiemann, J., T. R. Yang, P. N. Sander, M. Schneider, M. Engeser, S. Kath-Schorr, C. E. Muller, and P. M. Sander. 2017. Dinosaur origin of egg color: Oviraptors laid blue-green eggs. *PeerJ* 5: Article e3706.

Wikelski, M., E. Arriero, A. Gagliardo, R. A. Holland, M. J. Huttunen, R. Juvaste, I. Mueller, G. Tertitski, K. Thorup, M. Wild, M. Alanko, F. Bairlein, A. Cherenkov, A. Cameron, R. Flatz, J. Hannila, O. Huppop, M. Kangasniemi, B. Kranstauber, M. L. Penttinen, K. Safi, V. Semashko, H. Schmid, and R. Wistbacka. 2015. True navigation in migrating gulls requires intact olfactory nerves. *Scientific Reports* 5: Article 17061.

Wilkins, M. R., D. Shizuka, M. B. Joseph, J. K. Hubbard, and R. J. Safran. 2015. Multimodal signalling in the North American barn swallow: A phenotype network approach. *Proceedings of the Royal Society B: Biological Sciences* 282: Article 20151574.

Willems, J. S., J. N. Phillips, and C. D. Francis. 2022. Artificial light at night and anthropogenic noise alter the foraging activity and structure of vertebrate communities. *Science of the Total Environment* 805: Article 150223.

Xiao, M., M. D. Shawkey, and A. Dhinojwala. 2020. Bioinspired melanin-based optically active materials. *Advanced Optical Materials* 8: Article 2000932.

Yamanaka, O., and R. Takeuchi. 2018. UMATracker: An intuitive image-based tracking platform. *Journal of Experimental Biology* 221: Article jeb182469.

Yorzinski, J. L., G. L. Patricelli, J. S. Babcock, J. M. Pearson, and M. L. Platt. 2013. Through their eyes: Selective attention in peahens during courtship. *Journal of Experimental Biology* 216:3035–3046.

Yorzinski, J. L., G. L. Patricelli, S. Bykau, and M. L. Platt. 2017. Selective attention in peacocks during assessment of rival males. *Journal of Experimental Biology* 220:1146–1153.

Zipple, M. N., E. M. Caves, P. A. Green, S. Peters, S. Johnsen, and S. Nowicki. 2019. Categorical colour perception occurs in both signalling and non-signalling colour ranges in a songbird. *Proceedings of the Royal Society B: Biological Sciences* 286: Article 20190524.

13
Social Behavior in Birds

Damien R. Farine

The study of social behaviors has generated significant insights into how organisms can respond to—and are subsequently shaped by—environmental pressures. For example, we now know that new behaviors can be socially learned to allow for rapid and flexible responses in changing environments and that animal collectives can easily solve complex problems even when individuals have cognitive limitations. However, although social behavior is extremely widespread in birds, the study of birds' social behavior has remained predominantly focused on studying birds during breeding, overemphasizing the importance of behaviors such as cooperative breeding. In this chapter, I explain how the breeding biology of birds might have contributed to this bias, review new insights gained from recent studies of sociality in nonbreeding birds, and draw out some gaps in our knowledge of social behaviors in birds. Finally, I make the case that comparative studies both within and across species of birds provide for many opportunities to explore how the environment shapes sociality.

Sociality and the Reproductive Biology in Birds

The central focus in studies of social behavior in birds has been their reproductive behavior, especially cooperative breeding (Skutch 1935, Stacey and Koenig 1990, Koenig and Dickinson 2016). Although cooperation is a major topic in evolutionary biology (Dugatkin 1997, Koenig and Dickinson 2004, 2016, Shen et al. 2017), and kin selection is a major guiding hypothesis (Eberhard 1975, J. Brown 1987, Cockburn 1988), the reasons for the disproportionate focus on studying social behaviors during the breeding season are also largely pragmatic. The breeding biology of birds—nesting, external egg-laying, incubating, and (at least in altricial species) parental care at the nest—provides excellent conditions for conducting highly detailed observations, such as those needed to determine how the social environment drives reproductive investment (e.g., Dixit et al. 2017) and individual contributions to raising young (e.g., Boland et al. 1997). Traditional ornithological techniques, such as color banding, provide a simple means for identifying individuals, which when combined with (altricial) birds returning frequently to their nest means that data can be readily collected from direct observations or via images (e.g., video recordings). This contrasts with the challenges of having to follow small and highly mobile individuals that can move over large home ranges, which is the case of nonbreeding behavior or in species with precocial chicks. Furthermore, having chicks in a nest enables marking and sampling of individual nestlings, facilitating quantification of

Damien R. Farine, *Social Behavior in Birds*. In: *New Perspectives in Ornithology*. Edited by: Scott V. Edwards and J. Michael Reed, Oxford University Press. © Oxford University Press (2025). DOI: 10.1093/oso/9780197787670.003.0013 fitness (e.g., nestling weight and fledging success). These logistical advantages have made altricial birds an unparalleled model for studying reproductive social biology (Russell 2004) and have resulted in studies of birds being highly biased to the breeding versus nonbreeding periods (Marra et al. 2015).

The role of provisioning at the nest in shaping the focus of studies on sociality of birds is evident from the survey of non-kin-based cooperative breeding systems in birds by Riehl (2013). Among the 213 cooperatively breeding bird species for which data on group structure were available, only 4 species were reported to have precocial young (1.8%). This proportion is much smaller than the estimated 11% of bird species that have precocial young (N. Wang and Kimball 2016). One reason for this difference could be that precocial species live more solitary lifestyles. A comparative analysis suggests a strong link between cooperative breeding (a measure of sociality) and altriciality (N. Wang and Kimball 2016), with an estimated 11% of all altricial species versus just 4% of all precocial species reported as exhibiting cooperative care (Cockburn 2006, Scheiber et al. 2017). However, cooperatively breeding precocial species are widely distributed across the avian phylogenetic tree, meaning that they would have to represent several independent evolutionary transitions toward cooperative care (N. Wang and Kimball 2016). If such repeated transitions have taken place, then more species might be cooperative than currently detected. For example, Kalij Pheasants (*Lophura leucomelanos*) introduced onto Hawaii, where they live at high densities, were observed to breed cooperatively, despite there being no knowledge of cooperative breeding in their native range (Zeng et al. 2016). This begs the question: Does altriciality actually favor cooperative breeding (e.g., through extended opportunity for providing care [Langen 2000]) or are social behaviors such as cooperative breeding simply more difficult to detect—and therefore underreported—in the absence of a nest on which to focus post-hatching data collection? Although that question is not the specific focus of this chapter (for further discussion, see Nyaguthii et al. 2025), it illustrates how the convenience of studying birds at the nest may have skewed our understanding of their broader biology.

The breeding biology of birds, and corresponding ways that it has facilitated or hindered data collection, is likely to have shaped at least some of our present conclusions on social behavior of birds in general. For example, the ease by which Zebra Finches (*Taeniopygia guttata*) can be kept, bred, and studied in captive colonies is largely why they have played a disproportionate role in driving our current understanding of avian social behavior (Box 13.1). Similarly, the logistical advantages of adult birds provisioning nestling probably explains why ornithology has focused on cooperative breeding, in contrast to, for example, primatology, which has focused on broader aspects of social behavior (e.g., Box 1984, Lee 2001, Mitani et al. 2012). However, recent technological advances in individual tracking and automated data collection (Figure 13.1, Box 13.2) are opening up opportunities to investigate social behaviors beyond the nest. These advances, coupled with novel analytical tools such as social network analysis (Farine and Whitehead 2015), allow for integration of insights and concepts traditionally studied in the context of primate sociality into ornithological studies, such as social complexity (Scheiber et al. 2017, Papageorgiou and Farine 2021), the formation and maintenance of social bonds (Teunissen et al. 2018, Maldonado-Chaparro et al. 2021), social learning and culture (Whiten 2021), social cognition (Bugnyar 2013), the evolution of language (Suzuki et al. 2017, ten Cate 2018), and intergroup conflicts (Barve et al. 2020). In addition, with the acknowledgment that birds likely have cognitive capacities comparable to those of large-brained mammals (Olkowicz et al. 2016), new insights into birds' social behaviors beyond the nest are being generated that have ramifications for our understanding of the consequences of social behavior both at and away from the nest.

Box 13.1 The Zebra Finch: A Model System for the Study of Avian Sociality

The Zebra Finch (*Taeniopygia guttata*) is rapidly emerging as a model for the integrative study of social behavior. Living highly gregarious social lives—forming loose colonies that can easily be replicated in aviaries—makes Zebra Finches ideal for a wide range of studies. As a result, researchers across different fields are building up a very complete picture of what Zebra Finches do socially, and how they do it.

Like many social species, juvenile Zebra Finches make use of their social environments to learn how to act like adult Zebra Finches. This is important especially in male Zebra Finches that learn their song from their father (Bohner 1983, Zann 1990) or other closely associated males (Boogert et al. 2018), and who's song will guide their future pair bond formation. One experiment that combined cross-fostering of juveniles across populations of Zebra Finches with different song cultures with fine-scale tracking of social dynamics found very strong social assortment of male–female pairs by song culture background, resulting in strong reproductive assortment by culture (D. Wang et al. 2022). But how are songs learned? Zebra Finches have been widely used for studying the neural circuitry that underlies learning, with one study even able to use optogenetic methods to control learning of song timing (Zhao et al. 2019). Furthermore, the process of social learning itself is not completely passive, with experiments using video playback paradigms showing that nonvocal feedback from adult females during song development provides an important guide shaping the resulting learned song (Carouso-Peck and Goldstein 2019).

Studies have also revealed some of the hormonal mediators of social life in Zebra Finches. Young Zebra Finches that experienced elevated stress levels during early life are more gregarious but maintain weaker social bonds with others (Boogert et al. 2014, Brandl et al. 2019), and they avoid learning foraging skills from their parents (Farine, Spencer, et al. 2015). Such stress can have important downstream consequences, as the partners of Zebra Finches that experienced early life stress had higher rates of mortality than those with non-stressed partners (Monaghan et al. 2012). In a social context, stress can also mediate other processes, including learning and disease transmission. For example, Zebra Finches with elevated corticosterone were more likely to become spreaders of disease (e.g., West Nile virus) (Gervasi et al. 2017). Fine-scale tracking of colonies also found that social ties can form the substrate for stress to transmit to individuals that are not exposed to an environmental stressor (Brandl & Farine 2024). However, stress is not the only important mediator of social behavior in Zebra Finches: Individuals that had experimentally reduced vasotocin production were less gregarious (Kelly et al. 2011) and more aggressive (Kelly and Goodson 2014) than those with regular production. These studies highlight how, in just a single species, research has uncovered a breadth of drivers and consequences linked to the social lives of individual birds.

Figure 13.1 Technological advances are facilitating larger scale and longer term automated data collection for birds. (A) Two Vulturine Guineafowl (*Acryllium vulturinum*) are fitted with solar-powered GPS tags that collect at least one burst of 10 GPS locations every 5 minutes and can collect extensive bursts of 1 Hz (one point per second) for continuous locations. The ability to repeatedly collect proximity data across many individuals can reveal details of population-level structure and dynamics that are not accessible to human observers. (B) A nestling great tit (*Parus major*) fitted with a uniquely numbered metal ring (right leg) and a PIT tag (left leg) a few days before fledging. The ability to individually detect this individual when it comes in close proximity to an RFID reader (see Box 13.2) allows for lifetime monitoring of movements, social networks, and survival. (C) A group of sociable weavers (*Philetairus socius*) whose identities were extracted from the image using a deep learning algorithm trained for individual recognition. In the future, this technology may enable markerless tracking of individuals without ever having to trap them, providing the opportunity for truly population-level data to be collected at unprecedented scales.

Photo C courtesy of André Marques Condeço Ferreira.

Box 13.2 Technological Advances for Studying Sociality in Birds

Studies of behavior in birds have long been made possible by being able to fit individuals with unique combinations of color bands, thereby removing the hurdle of having to remember and recognize each animal individually. However, there are substantial challenges with studying social behavior in birds—many a field researcher has experienced their study animals flying off over the horizon. Thus, historical methods have limited the scope of our studies. However, due to the development of smartphones, tracking technologies are now becoming lighter and cheaper, and thus more widely applicable to birds.

Contemporary studies of wild birds have generally used one of three automated tracking approaches. The first are passive integrated transponder (PIT) tags, which are small (<0.1 g) microchips that can be read by radio frequency identification (RFID) readers (Bonter and Bridge 2011). When a tag enters the field of a reader, it produces a unique disturbance that is translated into a unique code. PIT tags can be fitted as leg rings or implanted under the skin, and RFID readers fitted to bird feeders (e.g., Aplin, Farine, et al. 2015) and nest boxes (e.g., Mariette et al. 2011) can record who visits, when, and with whom. However, this approach is limited to sampling individuals in fixed locations, generally limiting observations of social behavior to a feeding or nesting context. A second approach is to detect who is in proximity with whom. Proximity loggers, weighing approximately 1 g, both send signals and detect signals sent by nearby tags, allowing studies to infer spatial proximity (Levin et al. 2015). However, although fixed detectors can provide some information about spatial proximity, the accuracy in terms of distance and location is relatively limited, and errors are large relative to the scale at which small birds interact. A final approach is GPS tagging. Small and lightweight GPS tags are becoming increasingly available, and they provide data on the position and movement of individuals and (when multiple individuals are tagged) the proximity to others (He et al. 2023). Furthermore, when fitted onto the back of birds and equipped with a solar panel, they can allow long-term observations at continuous high resolutions.

Although tag-based tracking of individual birds remains largely the norm, future research is likely to increasingly rely on image-based approaches. Captive studies (Nagy et al. 2013, Alarcon-Nieto et al. 2018) have used machine-recognizable markers to track individual birds in videos. This approach has, for example, revealed the drivers behind alloparental care in Zebra Finches (Ogino et al. 2021). Markerless image-based methods are also likely to become increasingly feasible in the wild by using computers trained (i.e., using deep learning) to recognize individual birds based on their unique plumage characteristics (Ferreira et al. 2020). Importantly, this approach will facilitate data collection to move beyond historical approaches (e.g., color bands) through automation and to move beyond current methods (e.g., PIT tags) by detecting groups in context that extend away from feeders or nest boxes, such as communal trees or watering holes. Image-based methods therefore have the potential to revolutionize the study of social behavior in birds.

Studying Nonbreeding Social Behavior

A major challenge that comes with studying nonbreeding social behavior (often called wintering behavior; Figure 13.2) concerns scale. Many pair-breeding species that hold territories centered around their nests during the breeding season abandon these during the winter to form roaming flocks containing both conspecifics and heterospecifics. Accordingly, studying wintering behavior requires expanding the spatial scale of data collection to wintering ranges, which are often much larger than breeding territories. This challenge may be exacerbated in species that shift social structure from breeding territoriality to nonbreeding flocking, as these are likely to express disproportionate increases in their nonbreeding home range sizes—including in species that are otherwise resident. For example, a study conducted in central Monte Desert, Argentina (Sagario and Cueto 2014), found that the Rufous-Collared Sparrows (*Zonotrichia capensis*), which regularly form flocks outside of the breeding season, significantly increase their ranging area in this season. They contrast with sympatric Many-Colored Chaco Finches (*Saltatricula multicolor*) that predominantly remain in pairs throughout the year and express almost no discernible change in their ranging behaviors across seasons. Why some species express these changes in behavior while others do not remains an outstanding question. Studying this and other social behaviors during nonbreeding seasons requires overcoming the logistical challenge of following birds over large areas.

A second major methodological challenge for studying social behavior is the need to mark many individuals and record data on multiple individuals at once. Data might include recordings of who is present in a flock, the interactions among individuals, or aspects of their group movements. Birds are generally ideal for overcoming this challenge because it is common practice to fit individual birds with unique combinations of colored leg bands, allowing flock membership or interactions among individuals to be recorded by human observers. A few studies illustrate the use of this approach for studying nonbreeding social behavior. One study found that Golden-Crowned Sparrows (*Zonotrichia atricapilla*), winter migrants to California, form distinct social communities (sets of individual birds that consistently flock together despite having the opportunity to flock with others) during the winter and—strikingly—that birds maintained the same communities across years (Shizuka et al. 2014). This means that not only are these migrants returning to the same area each winter but also they are choosing to associate with the same flock mates each year. Why they do this remains unknown. Other species express surprising patterns during breeding that hint at important social processes—such as where to breed and with whom—taking place outside of reproduction. Slender-Billed Gulls (*Chroicocephalus genei*) breed on ephemeral river islands, and colonies can relocate hundreds of kilometers each year despite maintaining stable membership across years. Yet, breeding failures—for example, due to the colony becoming flooded—precipitate a turnover of membership in the following year (Francesiaz et al. 2017). Such studies highlight the significant potential for groups of birds to express important social behaviors during the nonbreeding season.

The problem of observing social behavior compounds when flocks are large and highly dynamic. In many species, nonbreeding flocks fission and fusion, meaning

Figure 13.2 Breeding (inner) and nonbreeding (outer) perspectives of social behavior in birds. Across the annual seasonal cycle, most bird species change between social systems, from maintaining strict membership in pairs or small family groups during breeding to participating in larger flocks during nonbreeding. The latter, often called the wintering flocking period, can represent a substantial part of an individual's lifetime. During this time, flock membership (and the resulting structure of the social network) can vary dynamically, often becoming more densely connected during harsher periods. However, studies across many species are finding important structure in winter flocks, including the maintenance (or early development) of pair bonds, consistent membership to the same flock across years, and consistent and preferred associations between breeding groups that are characteristic of multilevel societies, thereby linking or feeding into the social structure during the breeding seasons.

that individuals join and leave flocks continuously over time, underpinning a constant restructuring of the population. In species such as Great Tits (*Parus major*), this process of flock membership turnover can take place over as little as 10 minutes (Farine, Firth, et al. 2015), suggesting at first glance that flocks are random aggregations. Early observations that pair bonds might be maintained across years (Kluijver

1951) hinted otherwise, but conclusions on flock membership were reported to be too challenging to make from visual observations alone (Hinde 1952). One solution to observing wintering flocks came in the form of passive integrated transponder (PIT) tags, which are small and widely used microchips that generate a unique code when in contact with the magnetic field of a radio frequency identification (RFID) antenna (see Box 13.2). Fitting PIT tags on thousands of Great Tits and Blue Tits (*Cyanistes caeruleus*) and recording their flock membership at RFID loggers placed on grids of bird feeders revealed that individuals have distinct preferences in terms of their expression of social behaviors (Aplin, Firth, et al. 2015, Hillemann et al. 2019) that are linked to their personality and those of their associates (Aplin et al. 2013). The resulting social structure impacts what information individuals acquire, which can lead to long-term differences (or traditions) in behavior across populations (Aplin, Farine, et al. 2015). Individual and dyadic preferences during the nonbreeding season also have carryover effects into the breeding season, including on territory acquisition (Farine and Sheldon 2015), pair formation (Beck et al. 2021), breeding neighborhoods (Firth and Sheldon 2016, Beck et al. 2020), and extra-pair paternity (Beck et al. 2020). Such insights demonstrate that studying social behavior in different seasons in isolation will clearly miss some large pieces of the puzzle.

Although studies of songbirds have successfully used color bands or detections of birds visiting feeders together to infer flock membership, in many populations the social dynamics take place over scales that are intractable for observational methods alone. For example, Sulphur-Crested Cockatoos (*Cacatua galerita*) have learned to open bins—which are located outside every Australian house—to access food. Here, citizen science helped map how this knowledge spread to different parts of the greater Sydney area (Klump et al. 2021) and to track the resulting behavioral arms race between bin owners and the mess-generating cockatoos (Klump et al. 2022). Similarly, wintering Common Ravens (*Corvus corax*) can exploit resources in a range spanning thousands of kilometers (Loretto et al. 2016). It is also well-established that Common Ravens, which live in societies that exhibit fission–fusion dynamics, have the cognitive capacity to form a range of different social relationships (Boucherie et al. 2019), including strong associations with a small number of affiliates (Braun and Bugnyar 2012). Fitting Global Positioning System (GPS) trackers (see Box 13.2) to individual birds revealed that strong associations are maintained at the landscape scale, with the same individual ravens repeatedly co-occurring together at locations hundreds of kilometers apart (Loretto et al. 2017). Similar insights have also been gained from migrating birds. For example, young Whooping Cranes (*Grus americana*) rely on following experienced adults to learn migration routes (Mueller et al. 2013), whereas parents (especially fathers) head the V formation during migration in White-Fronted Goose (*Anser a. albifrons*) families (Kolzsch et al. 2020). In smaller species, geolocators successfully tracked European Bee-Eaters' (*Merops apiaster*) social migration on a global scale, revealing that groups can migrate together and even reform after long periods of spatial separation (Dhanjal-Adams et al. 2018). Maintaining social bonds during migration is likely to generate substantial consequences. For example, juvenile Caspian Terns (*Hydroprogne caspia*) that stay close to a parent are more likely to survive their first migration (Byholm et al. 2022), whereas social learning of migration routes might result in a loss of cultural knowledge if population sizes decline. Using technologies that record where animals move has highlighted how sociality is maintained in many highly mobile species, and deploying biologgers to multiple animals is beginning to reveal the consequences of social interactions during movements.

An excellent example of the progression of technology facilitating increasingly fine-scale observations of inter-individual interactions is in the study of collective behavior. Birds represent some of the most dazzling examples of collective behavior. The first studies to address questions of how birds can achieve such majestic displays used stereo photography to reconstruct structure and individual trajectories in murmurations of Common Starlings (*Sturnus vulgaris*), revealing that these were generated by simple movement rules in which individuals responded only to the movement of their neighbors (Ballerini et al. 2008). Follow-up studies used PIT tags to track the movement of individuals between feeders to reveal that flocks could benefit from containing a mixture of bolder and shyer individuals, with the former leading flocks to new resources and the latter contributing to the maintenance of group cohesion (Aplin et al. 2014). Recently, a study fitted GPS tags to each individual in several groups of Vulturine Guineafowl (*Acryllium vulturinum*) to investigate how social groups are structured (Musciotto et al. 2022). Combining whole-group GPS tracking with implanted heart rate loggers also revealed that individuals individuals trying to lead their group experience elevated heart rates—especially when there is substantial directional conflict among group members (Brandl et al. 2025). These costs are paid because democratic decision-making allows all group members to gain benefits from group-living when such conflicts of interest arise (Papageorgiou and Farine 2020b). These types of behaviors are likely to provide critical survival benefits during periods of environmental harshness, such as winter or dry seasons. Continued technological (and analytical) advances, combined with more intensive deployment of biologgers (He et al. 2023), will be critical for gaining more detailed information not just on individuals but also on *how* inter-individual interactions underpin collective outcomes.

Insights on the social lives of birds make it clear that nonbreeding birds engage in extensive social interactions, that these interactions result in a rich diversity of social systems, and that nonbreeding sociality has consequences on reproductive processes ranging from pair formation to offspring survival. We should hardly be surprised, as studies on birds have provided some of the earliest evidence for many of the key mechanisms that generate benefits when living socially. For example, (1) early experiments with herons (Krebs 1974) and observations of colonies of cliff swallows (C. Brown 1988) highlighted that birds make extensive use of, and benefit from, social information when foraging; (2) observations of flocks of Woodpigeons (*Columba palumbus*) being attacked by Northern Goshawks (*Accipiter gentilis*) demonstrated how larger groups can detect predators earlier, thereby reducing the chances that any group member is killed (Kenward 1978); and (3) passerine bird song provided what we now know to be among the earliest examples of culture outside of humans, with birds socially learning territorial song from their fathers, leading to spatial clustering of song types (Slater 2003). Furthermore, studies conducted on social interactions between species (Goodale et al. 2020), including alarm calling and response (Magrath et al. 2015), the spread of information about novel food sources (Farine, Aplin, et al. 2015), and mobbing of predators (Krams and Krama 2002), demonstrate that social benefits underpinning avian sociality can arise via direct fitness (Farine et al. 2012). However, many of these benefits would arise regardless of who individuals are being social with. Yet, by collecting finer-resolution data over large spatial scales, and combining these with analytical tools that can explicitly capture the social structure of populations, it is now very clear that individuals are choosing who to associate and interact with, even when forming large flocks, when re-forming flocks after breeding, and when forming flocks with heterospecifics. This raises new questions about the cognitive abilities of birds.

"Bird Brains"

It has generally been assumed that the structure, and resulting complexity, of avian societies is largely limited to either family groups or predominately anonymous aggregations (ranging in temporal stability from winter flocks to breeding colonies). In mammals, and especially primates, many species form stable social groups, within which multiple members can raise their own offspring (plural breeding[1]) and different types of relationships (e.g., affiliative and dominance) are expressed and known among group members (Kappeler 2019, Prox and Farine 2020). The formation of similar plural breeding groups is considered to be relatively rare in birds. For example, although nearly half of cooperatively breeding species can have non-kin group members, it is likely that fewer than 5% breed plurally (Riehl 2013). The greater scarcity of plural breeding in birds relative to mammals[2] is likely due to the fact that being oviparous (external egg-laying) allows female birds to restrict access to reproductive resources (e.g., a nest), which is more challenging for viviparous species to achieve (Raihani and Clutton-Brock 2010). As a consequence, birds are typically thought to have higher organizational complexity (i.e., societies with extensive alloparental care and reproductive division between group members) and lower relational complexity (i.e., societies with "well-defined dominance hierarchies, competitive alliances and other behavioural tactics used to maintain social status"; Lukas and Clutton-Brock 2018, p. 1129). A growing body of evidence is now demonstrating that birds can also express high relational complexity.

One reason why relational complexity has been overlooked in avian societies is because of their so-called bird brains. Living in a society in which individuals track a number of different types and qualities of relationships requires advanced cognitive

[1] Plural breeding groups are stable in membership when not breeding and comprise multiple reproductive units (males and/or females) that nest separately but raise offspring within the group, where they can also receive help (plural cooperative breeders). Plural breeding is distinct from colonial breeding, in which many birds nest nearby but do not otherwise form a cohesive group, and from the more common communal breeding (or joint-nesting), in which multiple males or females share the same nest.

[2] This may also be a definitional problem. Many bird species spend more than half of their adult lives in social groups that have relatively high stability in membership, but they nest away from the group. If the same social groups re-form after breeding, a case could be made for plural breeding. However, the stability of nonbreeding social group membership is rarely tracked year to year, meaning that plural breeding could be underdetected. Some distinction may also be made between species in which parents return to the group to raise offspring communally and those that return to the group only after offspring become independent.

skills (Aureli et al. 2008) and, consequently, a large brain (van Schaik et al. 2004, Dunbar 2009). Comparative work in mammals has further suggested that the need for a large brain is specifically associated with relational complexity (Lukas and Clutton-Brock 2018), whereas cooperative breeding (higher organizational complexity) does not seem to be associated with larger brain sizes or advanced cognitive abilities (Thornton and McAuliffe 2015, Lukas and Clutton-Brock 2018). Hence, the cost of transporting a large brain could feasibly be limiting birds' abilities to express high levels of relational complexity.

Recent studies provide clear counterpoints to the assumptions about birds' small brains. Detailed mapping of the avian brain has shown that birds such as corvids and parrots can have as many neurons as some primates (Olkowicz et al. 2016) and that birds have a brain region with an architecture resembling the neocortex found in mammalian brains (Stacho et al. 2020). These two findings suggest that birds should have the cognitive capacity to express social behaviors that parallel those of many mammals, including primates. For example, both Vulturine Guineafowl and Sulphur-Crested Cockatoos use strategic decisions when deciding on how much to invest in dominance interactions, investing more when individuals are close in rank (Dehnen, Papageorgiou, et al. 2022, Penndorf 2025). Experiments in ravens have further shown that they are capable of having third-party knowledge of relationships within their group (Boucherie et al. 2019). Third-party knowledge involves an individual knowing both its own relationship to others (e.g., who is more or less dominant) and the relationship among the other individuals themselves. Thus, individual birds are clearly capable of understanding different types of social relationships and their social position within a wider social network. The question is, therefore, can birds express the most complex form of societies that are found in mammals: multilevel societies? Evidence for this has been hiding in plain sight for several decades (Papageorgiou and Farine 2021), and it took recent technological advances, allowing for quantitative data on the structure of a whole population of birds, to convincingly demonstrate this.

Complex Societies in Birds

Multilevel societies represent a social structure comprising multiple different levels (or tiers), with each level having a distinct and stable social character. Outside of humans, these were first described in hamadryas baboons (*Papio hamadryas*), where individuals live in one-male, multi-female units, with these units then joining other units to form clans, then bands, and then troops (Kummer 1968). The different levels of associations are thought to exist to allow these baboons to express flexible group sizes in response to changing environmental conditions (Schreier and Swedell 2012). Lower (and more frequently expressed) levels typically have higher average relatedness among individuals, facilitating investment in more cooperative behaviors. The different levels are also considered to generate a significant cognitive challenge because they require individuals to simultaneously track multiple different types of relationships (Schreier and Swedell 2012). As a result, such societies were only ever described, and therefore thought possible, in large-brained mammals (Grueter et al. 2020).

In the savannah of East Africa, living among the baboons and elephants is a species that exhibits another surprisingly complex social life, the Vulturine Guineafowl, a large and predominantly terrestrial bird. Despite their striking plumage, having a highly gregarious lifestyle (they are always found in large groups), and the extensive research being conducted on social behavior of African fauna, this species has until recently completely escaped the attention of scientists. Since 2016, individuals of the 20 social groups living around the Mpala Research Centre in Kenya have been fitted with solar-powered GPS tags capable of collecting moment-by-moment movement tracks of each group every day of the year. Without GPS devices, an army of field assistants would have been required to collect such large-scale data, which were critical to revealing that Vulturine Guineafowl live in a multilevel society (Papageorgiou et al. 2019).

Vulturine Guineafowl society comprises three different levels (Papageorgiou et al. 2019). Data from the census observations revealed an intermediate level, "the group," that is almost completely stable during nonbreeding seasons, with group sizes ranging from 15 to 65 individuals (Papageorgiou and Farine 2020a). The lowest level becomes apparent during breeding, revealing that groups contain multiple reproductive units, with pairs split from the group temporarily to breed (Nyaguthii et al. 2025). Males return to their group when the female starts incubating, and the female returns with her precocial chicks soon after they hatch (or after nest failure). During breeding, the group identity is maintained through nonbreeders and subadult birds. The upper level becomes apparent outside of breeding. Extensive GPS tracking data revealed that groups are not territorial, use different areas under different seasonal conditions (Papageorgiou et al. 2021), and maintain nonrandom group-to-group associations both at communal roosts and during the day. Comparison of the social preferences among groups to their home range overlap suggests that social preferences among groups (i.e., the upper level) are not explained by the opportunity that groups have to encounter one another (Papageorgiou et al. 2019).

The presence of a multilevel society in guineafowl—of all bird species—may seem surprising. After all, the collective noun for members of the Numididae family is *a confusion of guineafowl*. Furthermore, guineafowl, like other species in the order Galliformes, are predominantly large-bodied due to their terrestrial lifestyles. This has, correspondingly, resulted in a particularly small brain to body size ratio (Olkowicz et al. 2016) and thus the assumption that they are unlikely to have the intelligence needed to express such complex social behaviors. As such, recent findings from Vulturine Guineafowl sociality patterns raise many questions, including whether high levels of cognition are necessary for animals to form multilevel societies or whether many more animals are indeed capable of solving the challenges of such a social life than we had so far assumed.

The insights gained through large-scale and high-resolution GPS tracking of Vulturine Guineafowl have allowed us to see previous research through a new lens. The alignment of perspective with those used for decades in primatology (Papageorgiou and Farine 2021) has already led researchers to seek out (and find) further evidence for multilevel societies. One clear example is the Superb Fairywren (*Malurus cyaneus*) (Camerlenghi et al. 2022). Here, breeding groups spend the nonbreeding season together by forming supergroups with other breeding groups, and they interact nonrandomly with other groups and supergroups to form larger communities. A complementary comparative study, comprising species in the Acanthizidae family (comprising fairywrens and Australian thornbills, among others) as well as 16 other Australian and New Zealand bird families known to include at least one cooperative breeding species, also found descriptive evidence matching features of multilevel societies in 23 species (of 74 species that had sufficient descriptions of the social system) (Camerlenghi et al. 2022). Descriptions of potential multilevel societies were also significantly linked with cooperative breeding (Camerlenghi et al. 2022), suggesting that organizational and relational complexity may be favored by similar environmental, social, or genetic drivers across the breeding and nonbreeding seasons. For example, avian multilevel societies may be favored by a high average local relatedness—or population viscosity. But is relatedness an essential driver for multilevel societies?

Mixed-Species Flocking

There are few more striking moments to an ornithologist than a forest that seemed completely empty just moments ago coming alive with birds of all sizes and colors. These "bird waves" (Kajiki et al. 2018) reveal a close spatiotemporal proximity maintained by individuals of different species—that is, mixed-species flocks—as they move together through the environment. Many bird species join such mixed-species flocks (a survey of the literature as part of my PhD dissertation suggested that more than 50% of passerines and near-passerines joined mixed-species flocks with some regularity), meaning that mixed-species groups are far more common (and contain far more species per flock) in birds than in other taxonomic groups (Goodale et al. 2017). Forming such flocks has important consequences for individuals; for example wintering male White-Breasted Nuthatches (*Sitta carolinensis*) showed significant loss of body condition after the removal of parids from their environment (Dolby and Grubb 1998). They are also particularly interesting for studies of social behavior because by definition, participants are unrelated. Yet, mixed-species societies have received surprisingly little attention relative to, for example, cooperative breeding.

Although the vast majority of mixed-species flocks have relatively unstable membership, there are some notable exceptions that are of particular interest to the study of avian sociality. In the Peruvian Amazon, Munn and Terborgh (1979) identified a core of 12 species that maintained, and communally defended, a shared territory. Each species was limited to one family group, and the pattern was repeated in neighboring territories (with some territories containing different subsets of the whole species composition). Strikingly, one complementary study found that such mixed-species territories in French Guiana were maintained over a period of 17 years (Martinez and Gomez 2013). A similar effort to characterize avian communities across a 36-year time span also found remarkable consistency across time (Martinez et al. 2023). These results suggest that mixed-species societies can have a long-term stability comparable to that of many intraspecific systems.

One exciting possibility is that mixed-species flocks could represent an additional upper level to avian multilevel societies. In Amazonian flocks, each species is usually represented by one pair or family group, with the flock potentially representing the second level—making these flocks a multilevel society with stability at both levels. In the woodlands of Australia, mixed-species flocks are also common, and they are often dominated by members of the Acanthizidae family (Bell 1980). Many of these species are already suspected of forming multilevel societies with conspecifics (Camerlenghi et al. 2022). For example, several species of small Australian thornbills, including the Buff-Rumped Thornbill (*Acanthiza reguloides*) and Striated Thornbill (*Acanthiza lineata*), breed in cooperative groups that then form clans during the nonbreeding season (Bell and Ford 1986). These then form mixed-species communities with each other and with a number of other species, including fairywrens (*Malurus* spp.), other thornbills, and pair-living Australasian robins (family Petroicidae) that have nonrandom relationships at the individual level (Farine and Milburn 2013). Such communities could represent a third or fourth level to some avian multilevel societies.

That mixed-species flocks could form the upper level social unit of a society may seem surprising. However, these species are unlikely to be randomly drawn from the local community. Across a range of mixed-species flocking systems, species appear to choose to flock with more similar species (Mammides et al. 2015) or to converge their niche toward others in their flock (Farine and Milburn 2013). But what would be the glue that drives such species to form and maintain such associations, increase competition with unrelated individuals, and drive potential investment in individual recognition systems extending to heterospecifics? Because these are heterospecific associations, we can explicitly discount indirect fitness benefits as causal drivers (Farine et al. 2012). In fact, the pattern of decreasing similarity at increasing levels of flocking (with conspecifics, and then with heterospecifics) mirrors those of relatedness in multilevel societies decreasing at increasing levels, as found in Guinea baboons (*Papio papio*) (Patzelt et al. 2014) and bottlenose dolphins (*Tursiops aduncus*) (Randic et al. 2012). Thus, across taxa, associations at upper levels are likely to be maintained predominately by direct mutualistic benefits (e.g., predator avoidance). That such associations should form across species boundaries then simply becomes a question of logistics.

One question that arises is why maintain nonrandom associations with heterospecifics? There are a number of intriguing possibilities. For example, although some benefits that individuals gain by flocking with heterospecifics are likely to be relatively simple (e.g., dilution of risk), the potential to benefit may be maximized through higher risk investments, such as alarm calling, mobbing of predators, or cooperative territorial defense. Such investments often require an element of "trust." Two studies show evidence consistent with increased cooperation among familiar individuals. The first involves the differential response by Splendid and Variegated Fairywrens (*Malurus splendens* and *M. lamberti*) to playbacks of the individual heterospecifics with whom they frequently associate on their largely overlapping territories versus heterospecifics from neighboring territories (Johnson et al. 2018). The second involves increasing initiation of risky mobbing by chaffinches (*Fringilla coelebs*) as they became more familiar with other birds in their community (Krams and Krama 2002). Thus, as with other forms of cooperation, maintaining nonrandom relationships is likely to allow individuals from all participating species to accrue greater benefits by avoiding freeloaders.

Mixed-species flocks of birds represent a remarkable opportunity for understanding key mechanisms underpinning social life, at least in vertebrates. Flocks vary in temporal stability of membership, in size and composition, and in seasonality, all of which could be evaluated to form rich comparative studies. Those at the more stable end also provide opportunities to better reveal the upper bounds of cooperation and the different forms of sociality that can be achieved via direct benefits alone (e.g., Dolby and Grubb 1998), how mixed-species flocks coordinate behaviors and navigate their landscape together (e.g., Farine et al. 2014), and the role of the environment in shaping such social structures (He et al. 2019). The broadening focus of social behavior in birds thus provides many new and exciting opportunities to expand our general understanding of social evolution and its consequences.

New Opportunities

Learning the Social Life

The race to reproduce can belie some surprising twists and turns. Many social bird species have a protracted juvenile development period. Sulphur-Crested Cockatoos, which are highly social and maintain significant social associations within a large open society (Aplin et al. 2021), only become sexually mature at age 7 years and probably start breeding much later than this. Similarly, Vulturine Guineafowl—despite being the ideal size for most predators and living in a predator-rich environment—have delayed maturity, with females usually dispersing at approximately 18–24 months (Klarevas-Irby et al. 2021) and males not getting any breeding opportunities for several years. What drives these delays?

Although it has long been hypothesized that the increasing fecundity that comes with age (Hamann and Cooke 1987) or experience (Komdeur 1996) can benefit delayed reproduction, it is likely to also take time to develop the social skills needed to maximize success in animal societies (Armansin et al. 2020). Studies on primates (Joffe 1997) and other mammals (Holekamp et al. 2007) have long considered the importance of an extended juvenile period for learning not only about the physical environment but also about the social environment. Young birds may also need to learn important social skills. For example, Golden-Crowned Sparrows and Sulphur-Crested Cockatoos tailor their social interactions depending on their level of familiarity. The former rely on plumage badges to infer their relative competitive ability with strangers but use prior experience when interacting with familiar conspecifics (Chaine et al. 2018), whereas the latter are more likely to escalate dominance interactions with strangers based on relative size and with familiar individuals based on relative dominance rank (Penndorf 2025). Furthermore, third-party knowledge of relationships in ravens only starts to be expressed some months post-fledging (Loretto et al. 2012). The need to gain experience is thought to explain delayed acquisition of adult plumage (which usually takes 7 years) by male Satin Bowerbird (*Ptilonorhynchus violaceus*), as looking like a female reduces aggression by other males, thus providing juveniles the opportunity to inspect bowers and learn this important component of courtship from adult males (Collis and Borgia 1993). The ability to conduct lifetime tracking of individual birds, and potentially the vast availability of long-term data on social species (Koenig and Dickinson 2016), opens exciting opportunities to shed light on the importance of the teenage years for acquiring the social skills—such as the strategic use of dominance interactions or decisions about who to interact with—required later in life.

Carryover and Transgenerational Effects

The breeding season marks both a successful endpoint for parents and the start of life for hatchlings, thereby highlighting a continuum across which social behaviors can influence individuals. Although I have advocated for more studies on nonbreeding social behavior in birds, there are also many promising avenues of research related to the ecology of breeding that are going to be necessary to assembling the bigger picture view of birds' social lives. For example, dominance ranks established among Barnacle Geese (*Branta leucopsis*) in the first days of life are remarkably resilient to individual changes in traits that predict dominance outcomes among strangers (Black and Owen 1987), and these early life effects in Black-Capped Chickadees (*Poecile atricapillus*) predict lifetime reproductive outcomes (Schubert et al. 2007). Given the importance of dominance across life history stages, such as dispersal, territory acquisition, access to mates, and access to resources when raising young, it sets the scene for extensive carryover effects in social systems (i.e., the rich get richer). When carryover effects from early life translate to reproductive outcomes, there is also extensive scope for parents to set up their offspring on particular pathways, resulting in a range of potential transgenerational effects (Dehnen, Arbon, et al. 2022).

Birds provide a unique opportunity for partitioning the relative contributions of social and genetic effects on lifetime reproductive outcomes. One pathway, social rank inheritance, has been widely studied in mammals (Holekamp and Smale 1991), but it has received much less attention in birds. Yet, because much of the reproductive cycle in birds occurs externally—that is, eggs are laid before they are incubated—they provide opportunities for experimental manipulations of important developmental factors, such as the timing of hatching and the natal social environment (Dehnen, Arbon, et al. 2022). The timing of hatching, which is commonly linked to better body condition in the mothers (Bety et al. 2003, Descamps et al. 2011), predicts early life social rank acquisition in Barnacle Geese (Black and Owen, 1987). This forms a clear pathway for social rank inheritance that can be manipulated by delaying incubation in some nests (e.g., temporarily using fake eggs) and by swapping eggs between nests (cross-fostering) to quantify the contribution of genetic effects, pre- and post-laying parental investments, and social effects. Such experimental approaches, conducted in avian societies with multigenerational groups could be used to test key predictions about intergenerational effects (e.g., McNamara and Houston 1996).

The Importance of Long-Term Social Bonds

There is now clear and widespread evidence from many social mammals, particularly among philopatric females, that long-term social bonds can transcend seasons and play an important role in maintaining health and successful reproduction (Snyder-Mackler et al. 2020). Is the same true for male birds? Emerging evidence from a range of species suggests so. For example, male reproductive success in lekking Wire-Tailed Manakin (*Pipra filicauda*) (Dakin and Ryder 2020) and Lance-Tailed Manakins (*Chiroxiphia lanceolate*) (DuVal 2007) has been linked to the maintenance of stable long-term social partners. Such social dynamics could be linked to high social viscosity (peaks of relatedness within an individual's local environment), driven by limited male dispersal in many birds, and underlie important social processes in avian systems. By contrast, a number of communal breeding species, such as Greater Anis (*Crotophaga major*), have low within-group relatedness (Riehl 2021), highlighting the potential importance of social benefits for maintaining stable group membership.

Expanding the temporal scope of studies of social behavior is revealing that long-term social relationships are present in many birds. For example, in the multilevel society of Superb Fairywrens, an intermediate level (supergroups) is formed between groups led by males that were previously part of the same breeding group (Camerlenghi et al. 2022). In Barnacle Geese, whereas females drop and reform prior social bonds between breeding and nonbreeding seasons, males maintain their strong social bonds throughout (Kurvers et al. 2020). In Zebra Finches, individuals from the same colony that reproduced more synchronously maintained long-term social bonds that carried over into future reproductive periods (Brandl et al. 2021). Finally, ravens maintain their social bonds despite moving over large geographical areas (Loretto et al. 2017). Why would such bonds be important? In one experimental study, Zebra Finch colonies temporarily split into subgroups, resulting in reduced collective performance when foraging as a group (Maldonado-Chaparro et al. 2018). Over longer terms, social bonds may be important in reducing conflicts among competitors—for example, through cooperative division of space. An example of this is the Seychelles Warblers (*Acrocephalus sechellensis*), where males fought less, and had lower telomere attrition, when they were more familiar or related to their neighbors (Bebbington et al. 2017). Among species with seasonal changes in social structures, long-term social bonds may, in and of themselves, represent cooperative behaviors that allow mutualistic benefits—like "dear enemy" effects—to be carried over from one season to the next.

Conclusion

The recent increase in focus on, and evidence for, birds' social behaviors outside of breeding has demonstrated that the idiom "bird brain" is rather more reflective of humans' biased assumptions than it is about the actual capacity for birds to express (social) intelligence. Across all body (and brain) sizes, birds have been shown to be capable of maintaining social relationships that extend beyond their immediate family, and sometimes even beyond their own species. There is evidence that these relationships are mediated by a range of drivers, including early life experience, and that they have consequences for their social partners and beyond. Yet, work on the social behavior of birds outside laboratory settings still has to catch up with field studies of mammals, especially primates. Although the shift in seasonal focus has resulted in diverging perspectives of social behavior emerging among fields of research, many parallels are likely to exist. Overcoming the current bias in research by addressing birds' broader sociality—throughout both reproductive and nonreproductive periods—will open up opportunities, such as larger scale comparative work (e.g., Camerlenghi et al. 2022) and experimental manipulations, underpinning more general insights. For example, by controlling where individuals could feed, experiments have been able to manipulate the social structure of bird communities (Firth and Sheldon 2015), with consequences on processes such as information transmission (Firth et al. 2016). Such approaches will demonstrate the capacity for studies on birds to produce definitive answers to long-standing questions, such as the role of ecological processes in shaping social evolution.

Acknowledgments

I thank the editors for inviting me to write this chapter. I am also grateful to the members of my research group and to Sjouke Kingma for their helpful feedback.

References

Alarcon-Nieto, G., J. M. Graving, J. A. Klarevas-Irby, A. A. Maldonado-Chaparro, I. Mueller, and D. R. Farine. 2018. An automated barcode tracking system for behavioural studies in birds. *Methods in Ecology and Evolution* 9:1536–1547.

Aplin, L. M., D. R. Farine, R. P. Mann, and B. C. Sheldon. 2014. Individual-level personality influences social foraging and collective behaviour in wild birds. *Proceedings of the Royal Society B: Biological Sciences* 281: Article 20141016.

Aplin, L. M., D. R. Farine, J. Morand-Ferron, A. Cockburn, A. Thornton, and B. C. Sheldon. 2015. Experimentally induced innovations lead to persistent culture via conformity in wild birds. *Nature* 518:538–541.

Aplin, L. M., D. R. Farine, J. Morand-Ferron, E. F. Cole, A. Cockburn, and B. C. Sheldon. 2013. Individual personalities predict social behaviour in wild networks of great tits (*Parus major*). *Ecology Letters* 16:1365–1372.

Aplin, L. M., J. A. Firth, D. R. Farine, B. Voelkl, R. A. Crates, A. Culina, C. J. Garroway, C. A. Hinde, L. R. Kidd, I. Psorakis, N. D. Milligan, R. Radersma, B. L. Verhelst, and B. C. Sheldon. 2015. Consistent individual differences in the social phenotypes of wild great tits, *Parus major*. *Animal Behaviour* 108:117–127.

Aplin, L. M., R. E. Major, A. Davis, and J. M. Martin. 2021. A citizen science approach reveals long-term social network structure in an urban parrot, *Cacatua galerita*. *Journal of Animal Ecology* 90:222–232.

Armansin, N. C., A. J. Stow, M. Cantor, S. T. Leu, J. A. Klarevas-Irby, A. A. Chariton, and D. R. Farine. 2020. Social barriers in ecological landscapes: The social resistance hypothesis. *Trends in Ecology & Evolution* 35:137–148.

Aureli, F., C. M. Schaffner, C. Boesch, S. K. Bearder, J. Call, C. A. Chapman, R. Connor, A. Di Fiore, R. I. M. Dunbar, S. P. Henzi, K. Holekamp, A. H. Korstjens, R. Layton, P. Lee, J. Lehmann, J. H. Manson, G. Ramos-Fernandez, K. B. Strier, and C. P. Van Schaik. 2008. Fission–fusion dynamics: New research frameworks. *Current Anthropology* 49:627–654.

Ballerini, M., N. Calbibbo, R. Candeleir, A. Cavagna, E. Cisbani, I. Giardina, V. Lecomte, A. Orlandi, G. Parisi, A. Procaccini, M. Viale, and V. Zdravkovic. 2008. Interaction ruling animal collective behavior depends on topological rather than metric distance: Evidence from a field study. *Proceedings of the National Academy of Sciences of the USA* 105:1232–1237.

Barve, S., A. S. Lahey, R. M. Brunner, W. D. Koenig, and E. L. Walters. 2020. Tracking the warriors and spectators of acorn woodpecker wars. *Current Biology* 30:R982–R983.

Bebbington, K., S. A. Kingma, E. A. Fairfield, H. L. Dugdale, J. Komdeur, L. G. Spurgin, and D. S. Richardson. 2017. Kinship and familiarity mitigate costs of social conflict between Seychelles warbler neighbors. *Proceedings of the National Academy of Sciences of the USA* 114:E9036–E9045.

Beck, K. B., D. R. Farine, and B. Kempenaers. 2020. Winter associations predict social and extra-pair mating patterns in a wild songbird. *Proceedings of the Royal Society B: Biological Sciences* 287: Article 20192606.

Beck, K. B., D. R. Farine, and B. Kempenaers. 2021. Social network position predicts male mating success in a small passerine. *Behavioral Ecology* 32:856–864.

Bell, H. L. 1980. Composition and seasonality of mixed-species feeding flocks of insectivorous birds in the Australian Capital Territory. *Emu* 80:227–232.

Bell, H. L., and H. A. Ford. 1986. A comparison of the social-organization of 3 syntopic species of Australian thornbill, *Acanthiza*. *Behavioral Ecology and Sociobiology* 19:381–392.

Bety, J., G. Gauthier, and J. F. Giroux. 2003. Body condition, migration, and timing of reproduction in snow geese: A test of the condition-dependent model of optimal clutch size. *American Naturalist* 162:110–121.

Black, J. M., and M. Owen. 1987. Determinants of social rank in goose flocks: Acquisition of social rank in young geese. *Behaviour* 102:129–146.

Bohner, J. 1983. Song learning in the zebra finch (*Taeniopygia guttata*): Selectivity in the choice of a tutor and accuracy of song copies. *Animal Behaviour* 31:231–237.

Boland, C. R. J., R. Heinsohn, and A. Cockburn. 1997. Deception by helpers in cooperatively breeding white-winged choughs and its experimental manipulation. *Behavioral Ecology and Sociobiology* 41:251–256.

Bonter, D. N., and E. S. Bridge. 2011. Applications of radio frequency identification (RFID) in ornithological research: A review. *Journal of Field Ornithology* 82:1–10.

Boogert, N. J., D. R. Farine, and K. A. Spencer. 2014. Developmental stress predicts social network position. *Biology Letters* 10: Article 20140561.

Boogert, N. J., R. F. Lachlan, K. A. Spencer, C. N. Templeton, and D. R. Farine. 2018. Stress hormones, social associations and song learning in zebra finches. *Philosophical Transactions of the Royal Society B: Biological Sciences* 373: Article 20170290.

Boucherie, P. H., M. C. Loretto, J. J. M. Massen, and T. Bugnyar. 2019. What constitutes "social complexity" and "social intelligence" in birds? Lessons from ravens. *Behavioral Ecology and Sociobiology* 73: Article 12.

Box, H. O. 1984. *Primate Behaviour and Social Ecology*. Chapman & Hall, London.

Brandl, H. B., D. R. Farine, C. Funghi, W. Schuett, and S. C. Griffith. 2019. Early-life social environment predicts social network position in wild zebra finches. *Proceedings of the Royal Society B: Biological Sciences* 286: Article 20182579.

Brandl, H. B., and D. R. Farine (2024). Stress in the social environment: behavioural and social consequences of stress transmission in bird flocks. *Proceedings B*, 291(2034), 20241961.

Brandl, H. B., S. C. Griffith, D. R. Farine, and W. Schuett. 2021. Wild zebra finches that nest synchronously have long-term stable social ties. *Journal of Animal Ecology* 90:76–86.

Brandl, H. B., J. A. Klarevas-Irby, D. Zuñega, C. Hansen Wheat, C. Christensen, F. Omengo, C. Nzomo, W. Cherono, B. Nyaguthii, and D. R. Farine 2025. The physiological cost of leadership in collective movements. *Current Biology*. https://doi.org/10.1016/j.cub.2025.06.065

Braun, A., and T. Bugnyar. 2012. Social bonds and rank acquisition in raven nonbreeder aggregations. *Animal Behaviour* 84:1507–1515.

Brown, C. R. 1988. Enhanced foraging efficiency through information centers: A benefit of coloniality in cliff swallows. *Ecology* 69:602–613.

Brown, J. L. 1987. *Helping and Communal Breeding in Birds*. Princeton University Press, Princeton, NJ.

Bugnyar, T. 2013. Social cognition in ravens. *Comparative Cognition & Behavior Reviews* 8:1–12.
Byholm, P., M. Beal, N. Isaksson, U. Lotberg, and S. Akesson. 2022. Paternal transmission of migration knowledge in a long-distance bird migrant. *Nature Communications* 13: Article 1566.
Camerlenghi, E., A. McQueen, K. Delhey, C. N. Cook, S. A. Kingma, D. R. Farine, A. Peters, and N. Pinter-Wollman. 2022. Cooperative breeding and the emergence of multilevel societies in birds. *Ecology Letters* 25:766–777.
Carouso-Peck, S., and M. H. Goldstein. 2019. Female social feedback reveals non-imitative mechanisms of vocal learning in zebra finches. *Current Biology* 29:631–636.
Chaine, A. S., D. Shizuka, T. A. Block, L. Zhang, and B. E. Lyon. 2018. Manipulating badges of status only fools strangers. *Ecology Letters* 21:1477–1485.
Cockburn, A. 1988. Evolution of helping behavior in cooperatively breeding birds. *Annual Review of Ecology and Systematics* 29:141–177.
Cockburn, A. 2006. Prevalence of different modes of parental care in birds. *Proceedings of the Royal Society B: Biological Sciences* 273:1375–1383.
Collis, K., and G. Borgia. 1993. The costs of male display and delayed plumage maturation in the satin bowerbird (*Ptilonorhynchus violaceus*). *Ethology* 94:59–71.
Dakin, R., and T. B. Ryder. 2020. Reciprocity and behavioral heterogeneity govern the stability of social networks. *Proceedings of the National Academy of Sciences of the USA* 117:2993–2999.
Dehnen, T., J. J. Arbon, D. R. Farine, and N. J. Boogert. 2022. How feedback and feed-forward mechanisms link determinants of social dominance. *Biological Reviews* 97:1210–1230.
Dehnen, T., D. Papageorgiou, B. Nyaguthii, W. Cherono, J. Penndorf, N. J. Boogert, and D. R. Farine. 2022. Costs dictate strategic investment in dominance interactions. *Philosophical Transactions of the Royal Society B: Biological Sciences* 377: Article 20200447.
Descamps, S., J. Bety, O. P. Love, and H. G. Gilchrist. 2011. Individual optimization of reproduction in a long-lived migratory bird: A test of the condition-dependent model of laying date and clutch size. *Functional Ecology* 25:671–681.
Dhanjal-Adams, K. L., S. Bauer, T. Emmenegger, S. Hahn, S. Lisovski, and F. Liechti. 2018. Spatiotemporal group dynamics in a long-distance migratory bird. *Current Biology* 28:2824–2830.
Dixit, T., S. English, and D. Lukas. 2017. The relationship between egg size and helper number in cooperative breeders: A meta-analysis across species. *PeerJ* 5: Article e4028.
Dolby, A. S., and T. C. Grubb. 1998. Benefits to satellite members in mixed-species foraging groups: An experimental analysis. *Animal Behaviour* 56:501–509.
Dugatkin, L. A. 1997. *Cooperation Among Animals: An Evolutionary Perspective.* Oxford University Press, Oxford, UK.
Dunbar, R. I. M. 2009. The social brain hypothesis and its implications for social evolution. *Annals of Human Biology* 36:562–572.
DuVal, E. H. 2007. Social organization and variation in cooperative alliances among male lance-tailed manakins. *Animal Behaviour* 73:391–401.
Eberhard, M. J. W. 1975. The evolution of social behavior by kin selection. *Quarterly Review of Biology* 50:1–33.
Farine, D. R., L. M. Aplin, C. J. Garroway, R. P. Mann, and B. C. Sheldon. 2014. Collective decision making and social interaction rules in mixed-species flocks of songbirds. *Animal Behaviour* 95:173–182.
Farine, D. R., L. M. Aplin, B. C. Sheldon, and W. Hoppitt. 2015. Interspecific social networks promote information transmission in wild songbirds. *Proceedings of the Royal Society B: Biological Sciences* 282: Article 20142804.
Farine, D. R., J. A. Firth, L. M. Aplin, R. A. Crates, A. Culina, C. J. Garroway, C. A. Hinde, L. R. Kidd, N. D. Milligan, I. Psorakis, R. Radersma, B. Verhelst, B. Voelkl, and B. C. Sheldon. 2015. The role of social and ecological processes in structuring animal populations: A case study from automated tracking of wild birds. *Royal Society Open Science* 2: Article 150057.
Farine, D. R., C. J. Garroway, and B. C. Sheldon. 2012. Social network analysis of mixed-species flocks: Exploring the structure and evolution of interspecific social behaviour. *Animal Behaviour* 84:1271–1277.

Farine, D. R., and P. J. Milburn. 2013. Social organisation of thornbill-dominated mixed-species flocks using social network analysis. *Behavioral Ecology and Sociobiology* 67:321–330.

Farine, D. R., and B. C. Sheldon. 2015. Selection for territory acquisition is modulated by social network structure in a wild songbird. *Journal of Evolutionary Biology* 28:547–556.

Farine, D. R., K. A. Spencer, and N. J. Boogert. 2015. Early-life stress triggers juvenile zebra finches to switch social learning strategies. *Current Biology* 25:2184–2188.

Farine, D. R., and H. Whitehead. 2015. Constructing, conducting and interpreting animal social network analysis. *Journal of Animal Ecology* 84:1144–1163.

Ferreira, A. C., L. R. Silva, F. Renna, H. B. Brandl, J. P. Renoult, D. R. Farine, R. Covas, and C. Doutrelant. 2020. Deep learning-based methods for individual recognition in small birds. *Methods in Ecology and Evolution* 11:1072–1085.

Firth, J. A., and B. C. Sheldon. 2015. Experimental manipulation of avian social structure reveals segregation is carried over across contexts. *Proceedings of the Royal Society B: Biological Sciences* 282: Article 20142350.

Firth, J. A., and B. C. Sheldon. 2016. Social carry-over effects underpin trans-seasonally linked structure in a wild bird population. *Ecology Letters* 19:1324–1332.

Firth, J. A., B. C. Sheldon, and D. R. Farine. 2016. Pathways of information transmission among wild songbirds follow experimentally imposed changes in social foraging structure. *Biology Letters* 12: Article 20160144.

Francesiaz, C., D. R. Farine, C. Laforge, A. Bechet, N. Sadoul, and A. Besnard. 2017. Familiarity drives social philopatry in an obligate colonial breeder with weak interannual breeding-site fidelity. *Animal Behaviour* 124:125–133.

Gervasi, S. S., S. C. Burgan, E. Hofmeister, T. R. Unnasch, and L. B. Martin. 2017. Stress hormones predict a host superspreader phenotype in the West Nile virus system. *Proceedings of the Royal Society B: Biological Sciences* 284: Article 20171090.

Goodale, E., G. Beauchamp, and G. Ruxton. 2017. *Mixed-Species Groups of Animals: Behavior, Community Structure, and Conservation*. Academic Press, San Diego, CA.

Goodale, E., H. Sridhar, K. E. Sieving, P. Bangal, G. J. Colorado, D. R. Farine, E. W. Heymann, H. H. Jones, I. Krams, A. E. Martinez, F. Montano-Centellas, J. Munoz, U. Srinivasan, A. Theo, and K. Shanker. 2020. Mixed company: A framework for understanding the composition and organization of mixed-species animal groups. *Biological Reviews* 95:889–910.

Grueter, C. C., X. G. Qi, D. Zinner, T. Bergman, M. Li, Z. F. Xiang, P. F. Zhu, A. B. Migliano, A. Miller, M. Krutzen, J. Fischer, D. I. Rubenstein, T. N. C. Vidya, B. G. Li, M. Cantor, and L. Swedell. 2020. Multilevel organisation of animal sociality. *Trends in Ecology & Evolution* 35:834–847.

Hamann, J., and F. Cooke. 1987. Age effects on clutch size and laying dates of individual female lesser snow geese *Anser caerulescens*. *Ibis* 129:527–532.

He, P., J. A. Klarevas-Irby, D. Papageorgiou, C. Christensen, E. D. Strauss, and D. R. Farine. 2023. A guide to sampling design for GPS-based studies of animal societies. *Methods in Ecology and Evolution* 14:1887–1905.

He, P., A. A. Maldonado-Chaparro, and D. R. Farine. 2019. The role of habitat configuration in shaping social structure: A gap in studies of animal social complexity. *Behavioral Ecology and Sociobiology* 73: Article 9.

Hillemann, F., E. F. Cole, D. R. Farine, and B. C. Sheldon. 2019. Wild songbirds exhibit consistent individual differences in inter-specific social behaviour. *bioRxiv*: Article 746545.

Hinde, R. A. 1952. The behaviour of the great tit (*Parus major*) and some other related species. *Behaviour Supplement*:III–201.

Holekamp, K. E., S. T. Sakai, and B. L. Lundrigan. 2007. Social intelligence in the spotted hyena (*Crocuta crocuta*). *Philosophical Transactions of the Royal Society B: Biological Sciences* 362:523–538.

Holekamp, K. E., and L. Smale. 1991. Dominance acquisition during mammalian social development: The inheritance of maternal rank. *American Zoologist* 31:306–317.

Joffe, T. H. 1997. Social pressures have selected for an extended juvenile period in primates. *Journal of Human Evolution* 32:593–605.

Johnson, A. E., C. Masco, and S. Pruett-Jones. 2018. Song recognition and heterospecific associations between 2 fairy-wren species (Maluridae). *Behavioral Ecology* 29:821–832.

Kajiki, L. N., F. Montano-Centellas, G. Mangini, G. J. Colorado, and M. E. Fanjul. 2018. Ecology of mixed-species flocks of birds across gradients in the Neotropics. *Revista Brasileira de Ornitologia* 26:82–89.

Kappeler, P. M. 2019. A framework for studying social complexity. *Behavioral Ecology and Sociobiology* 73: Article 13.

Kelly, A. M., and J. L. Goodson. 2014. Social functions of individual vasopressin–oxytocin cell groups in vertebrates: What do we really know? *Frontiers in Neuroendocrinology* 35:512–529.

Kelly, A. M., M. A. Kingsbury, K. Hoffbuhr, S. E. Schrock, B. Waxman, D. Kabelik, R. R. Thompson, and J. L. Goodson. 2011. Vasotocin neurons and septal V-1a-like receptors potently modulate songbird flocking and responses to novelty. *Hormones and Behavior* 60:12–21.

Kenward, R. E. 1978. Hawks and doves: Factors affecting success and selection in goshawk attacks on woodpigeons. *Journal of Animal Ecology* 47:449–460.

Klarevas-Irby, J. A., M. Wikelski, and D. R. Farine. 2021. Efficient movement strategies mitigate the energetic cost of dispersal. *Ecology Letters* 24:1432–1442.

Kluijver, H. N. 1951. The population ecology of the great tit. *Ardea* 39:1–135.

Klump, B. C., R. E. Major, D. R. Farine, J. M. Martin, and L. M. Aplin. 2022. Is bin-opening in cockatoos leading to an innovation arms race with humans? *Current Biology* 32:R910–R911.

Klump, B. C., J. M. Martin, S. Wild, J. K. Horsch, R. E. Major, and L. M. Aplin. 2021. Innovation and geographic spread of a complex foraging culture in an urban parrot. *Science* 373:456–460.

Koenig, W. D., and J. L. Dickinson. 2004. *Ecology and Evolution of Cooperative Breeding in Birds.* Cambridge University Press, Cambridge, UK.

Koenig, W. D., and J. L. Dickinson. 2016. *Cooperative Breeding in Vertebrates: Studies of Ecology, Evolution, and Behavior.* Cambridge University Press, Cambridge, UK.

Kolzsch, A., A. Flack, G. J. D. M. Muskens, H. Kruckenberg, P. Glazov, and M. Wikelski. 2020. Goose parents lead migration V. *Journal of Avian Biology* 51: Article 02392.

Komdeur, J. 1996. Influence of helping and breeding experience on reproductive performance in the Seychelles warbler: A translocation experiment. *Behavioral Ecology* 7:326–333.

Krams, I., and T. Krama. 2002. Interspecific reciprocity explains mobbing behaviour of the breeding chaffinches, *Fringilla coelebs. Proceedings of the Royal Society B: Biological Sciences* 269:2345–2350.

Krebs, J. R. 1974. Colonial nesting and social feeding as strategies for exploiting food resources in the Great Blue Heron (*Ardea herodias*). *Behaviour*51:99–134.

Kummer, H. 1968. *Social Organization of Hamadryas Baboons.* Karger, Basel, Switzerland.

Kurvers, R. H. J. M., L. Prox, D. R. Farine, C. Jongeling, and L. Snijders. 2020. Season-specific carryover of early life associations in a monogamous bird species. *Animal Behaviour* 164:25–37.

Langen, T. A. 2000. Prolonged offspring dependence and cooperative breeding in birds. *Behavioral Ecology* 11:367–377.

Lee, P. C. 2001. *Comparative Primate Socioecology.* Cambridge University Press, Cambridge, UK.

Levin, I. I., D. M. Zonana, J. M. Burt, and R. J. Safran. 2015. Performance of encounternet tags: Field tests of miniaturized proximity loggers for use on small birds. *PLoS One* 10: Article e0137242.

Loretto, M.-C., S. Reimann, R. Schuster, D. M. Graulich, and T. Bugnyar. 2016. Shared space, individually used: Spatial behaviour of non-breeding ravens (*Corvus corax*) close to a permanent anthropogenic food source. *Journal of Ornithology* 157:439–450.

Loretto, M.-C., O. N. Fraser, and T. Bugnyar. 2012. Ontogeny of social relations and coalition formation in common ravens. *International Journal of Comparative Psychology* 25:180–194.

Loretto, M.-C., R. Schuster, C. Itty, P. Marchand, F. Genero, and T. Bugnyar. 2017. Fission-fusion dynamics over large distances in raven non-breeders. Scientific Reports 7: Article 380.

Lukas, D., and T. Clutton-Brock. 2018. Social complexity and kinship in animal societies. *Ecology Letters* 21:1129–1134.

Magrath, R. D., T. M. Haff, P. M. Fallow, and A. N. Radford. 2015. Eavesdropping on heterospecific alarm calls: From mechanisms to consequences. *Biological Reviews* 90:560–586.

Maldonado-Chaparro, A. A., G. Alarcon-Nieto, J. A. Klarevas-Irby, and D. R. Farine. 2018. Experimental disturbances reveal group-level costs of social instability. *Proceedings of the Royal Society B: Biological Sciences* 285: Article 1577.
Maldonado-Chaparro, A. A., W. Forstmeier, and D. R. Farine. 2021. Relationship quality underpins pair bond formation and subsequent reproductive performance. *Animal Behaviour* 182: 43–58.
Mammides, C., J. Chen, U. M. Goodale, S. W. Kotagama, S. Sidhu, and E. Goodale. 2015. Does mixed-species flocking influence how birds respond to a gradient of land-use intensity? *Proceedings of the Royal Society B: Biological Sciences* 282: Article 20151118.
Mariette, M. M., E. C. Pariser, A. J. Gilby, M. J. L. Magrath, S. R. Pryke, and S. C. Griffith. 2011. Using an electronic monitoring system to link offspring provisioning and foraging behavior of a wild passerine. *The Auk* 128:26–35.
Marra, P. P., E. B. Cohen, S. R. Loss, J. E. Rutter, and C. M. Tonra. 2015. A call for full annual cycle research in animal ecology. *Biology Letters* 11: Article 20150552.
Martinez, A. E., and J. P. Gomez. 2013. Are mixed-species bird flocks stable through two decades? *American Naturalist* 181:E53–E59.
Martinez, A. E., J. M. Ponciano, J. P. Gomez, T. Valqui, J. Novoa, M. Antezana, G. Biscarra, E. Camerlenghi, B. H. Carnes, R. Huayanca Munarriz, and E. Parra. 2023. The structure and organisation of an Amazonian bird community remains little changed after nearly four decades in Manu National Park. *Ecology Letters* 26:335–346.
McNamara, J. M., and A. I. Houston. 1996. State-dependent life histories. *Nature* 380:215–221.
Mitani, J. C., J. Call, P. M. Kappeler, R. A. Palombit, and J. B. Silk. 2012. *The Evolution of Primate Societies*. Chicago University Press, Chicago.
Monaghan, P., B. J. Heidinger, L. D'Alba, N. P. Evans, and K. A. Spencer. 2012. For better or worse: Reduced adult lifespan following early-life stress is transmitted to breeding partners. *Proceedings of the Royal Society B: Biological Sciences* 279:709–714.
Mueller, T., R. B. O'Hara, S. J. Converse, R. P. Urbanek, and W. F. Fagan. 2013. Social learning of migratory performance. *Science* 341:999–1002.
Munn, C. A., and J. W. Terborgh. 1979. Multi-species territoriality in Neotropical foraging flocks. *Condor* 81:338–347.
Musciotto, F., D. Papageorgiou, F. Battiston, and D. R. Farine. 2022. Beyond the dyad: Uncovering higher-order structure within cohesive animal groups. *bioRxiv*: https://doi.org/10.1101/2022.05.30.494018.
Nagy, M., G. Vasarhelyi, B. Pettit, I. Roberts-Mariani, T. Vicsek, and D. Biro. 2013. Context-dependent hierarchies in pigeons. *Proceedings of the National Academy of Sciences of the USA* 110:13049–13054.
Nyaguthii, B., T. Dehnen, J. A. Klarevas-Irby, D. Papageorgiou, J. Kosgey, and D. R. Farine. (2025). Cooperative and plural breeding by the precocial Vulturine Guineafowl. *Ibis*, 167(3), 695–710.
Ogino, M., A. A. Maldonado-Chaparro, and D. R. Farine. 2021. Drivers of alloparental provisioning of fledglings in a colonially breeding bird. *Behavioral Ecology* 32:316–326.
Olkowicz, S., M. Kocourek, R. K. Lucan, M. Portes, W. T. Fitch, S. Herculano-Houzel, and P. Nemec. 2016. Birds have primate-like numbers of neurons in the forebrain. *Proceedings of the National Academy of Sciences of the USA* 113:7255–7260.
Papageorgiou, D., C. Christensen, G. E. C. Gall, J. A. Klarevas-Irby, B. Nyaguthii, I. D. Couzin, and D. R. Farine. 2019. The multilevel society of a small-brained bird. *Current Biology* 29:R1120–R1121.
Papageorgiou, D., and D. R. Farine. 2020a. Group size and composition influence collective movement in a highly social terrestrial bird. *eLife* 9: Article e59902.
Papageorgiou, D., and D. R. Farine. 2020b. Shared decision-making allows subordinates to lead when dominants monopolize resources. *Science Advances* 6: Article eaba5881.
Papageorgiou, D., and D. R. Farine. 2021. Multilevel societies in birds. *Trends in Ecology & Evolution* 36:15–17.
Papageorgiou, D., D. Rozen-Rechels, B. Nyaguthii, and D. R. Farine. 2021. Seasonality impacts collective movements in a wild group-living bird. *Movement Ecology* 9: Article 38.

Patzelt, A., G. H. Kopp, I. Ndao, U. Kalbitzer, D. Zinner, and J. Fischer. 2014. Male tolerance and male–male bonds in a multilevel primate society. *Proceedings of the National Academy of Sciences of the USA* 111:14740–14745.

Penndorf, J., D. R. Farine, J. M. Martin, and L. M. Aplin. (2025). Parrot politics: social decision-making in wild parrots relies on both individual recognition and intrinsic markers. *Royal Society Open Science*, 12(5), 241542.

Prox, L., and D. Farine. 2020. A framework for conceptualizing dimensions of social organization in mammals. *Ecology and Evolution* 10:791–807.

Raihani, N. J., and T. H. Clutton-Brock. 2010. Higher reproductive skew among birds than mammals in cooperatively breeding species. *Biology Letters* 6:630–632.

Randic, S., R. C. Connor, W. B. Sherwin, and M. Krutzen. 2012. A novel mammalian social structure in Indo-Pacific bottlenose dolphins (*Tursiops* sp.): Complex male alliances in an open social network. *Proceedings of the Royal Society B: Biological Sciences* 279:3083–3090.

Riehl, C. 2013. Evolutionary routes to non-kin cooperative breeding in birds. *Proceedings of the Royal Society B: Biological Sciences* 280: Article 20132245.

Riehl, C. 2021. Evolutionary origins of cooperative and communal breeding: Lessons from the crotophagine cuckoos. *Ethology* 127:827–836.

Russell, A. 2004. Mammals: Comparisons and contrasts. Pages 210–227 in W. D. Koenig and J. L. Dickinson, eds. *Ecology and Evolution of Cooperative Breeding in Birds.* Cambridge University Press, Cambridge, UK.

Sagario, M. C., and V. R. Cueto. 2014. Seasonal space use and territory size of resident sparrows in the central Monte Desert, Argentina. *Ardeola* 61:153–159.

Scheiber, I. B. R., B. M. Weiss, S. A. Kingma, and J. Komdeur. 2017. The importance of the altricial–precocial spectrum for social complexity in mammals and birds: A review. *Frontiers in Zoology* 14: Article 3.

Schreier, A. L., and L. Swedell. 2012. Ecology and sociality in a multilevel society: Ecological determinants of spatial cohesion in hamadryas baboons. *American Journal of Physical Anthropology* 148:580–588.

Schubert, K. A., D. J. Mennill, S. M. Ramsay, K. A. Otter, P. T. Boag, and L. M. Ratcliffe. 2007. Variation in social rank acquisition influences lifetime reproductive success in black-capped chickadees. *Biological Journal of the Linnean Society* 90:85–95.

Shen, S. F., S. T. Emlen, W. D. Koenig, and D. R. Rubenstein. 2017. The ecology of cooperative breeding behaviour. *Ecology Letters* 20:708–720.

Shizuka, D., A. S. Chaine, J. Anderson, O. Johnson, I. M. Laursen, and B. E. Lyon. 2014. Across-year social stability shapes network structure in wintering migrant sparrows. *Ecology Letters* 17:998–1007.

Skutch, A. F. 1935. Helpers at the nest. *The Auk* 52:257–273.

Slater, P. J. B. 2003. Fifty years of bird song research: A case study in animal behaviour. *Animal Behaviour* 65:633–639.

Snyder-Mackler, N., J. R. Burger, L. Gaydosh, D. W. Belsky, G. A. Noppert, F. A. Campos, A. Bartolomucci, Y. C. Yang, A. E. Aiello, A. O'Rand, K. M. Harris, C. A. Shively, S. C. Alberts, and J. Tung. 2020. Social determinants of health and survival in humans and other animals. *Science* 368: Article- eaax9553.

Stacey, P. B., and W. D. Koenig. 1990. *Cooperative Breeding in Birds: Long Term Studies of Ecology and Behaviour.* Cambridge University Press, Cambridge, UK.

Stacho, M., C. Herold, N. Rook, H. Wagner, M. Axer, K. Amunts, and O. Gunturkun. 2020. A cortex-like canonical circuit in the avian forebrain. *Science* 369: Article eabc5534.

Suzuki, T. N., D. Wheatcroft, and M. Griesser. 2017. Wild birds use an ordering rule to decode novel call sequences. *Current Biology* 27:R753–R755.

ten Cate, C. 2018. The comparative study of grammar learning mechanisms: Birds as models. *Current Opinion in Behavioral Sciences* 21:13–18.

Teunissen, N., S. A. Kingma, M. L. Hall, N. H. Aranzamendi, J. Komdeur, and A. Peters. 2018. More than kin: Subordinates foster strong bonds with relatives and potential mates in a social bird. *Behavioral Ecology* 29:1316–1324.

Thornton, A., and K. McAuliffe. 2015. Cognitive consequences of cooperative breeding? A critical appraisal. *Journal of Zoology* 295:12–22.

van Schaik, C. P., S. Preuschoft, and D. P. Watts. 2004. Great ape social systems. Pages 190–209 in A. E. Russon and D. R. Begun, eds. *The Evolution of Thought: Evolutionary Origins of Great Ape Intelligence*. Cambridge University Press, Cambridge, UK.

Wang, D. P., W. Forstmeier, D. R. Farine, A. A. Maldonado-Chaparro, K. Martin, Y. F. Pei, G. Alarcon-Nieto, J. A. Klarevas-Irby, S. W. Ma, L. M. Aplin, and B. Kempenaers. 2022. Machine learning reveals cryptic dialects that explain mate choice in a songbird. *Nature Communications* 13: Article 1630.

Wang, N., and R. T. Kimball. 2016. Re-evaluating the distribution of cooperative breeding in birds: Is it tightly linked with altriciality? *Journal of Avian Biology* 47:724–730.

Whiten, A. 2021. The burgeoning reach of animal culture. *Science* 372: Article eabe6514.

Zann, R. 1990. Song and call learning in wild zebra finches in south-east Australia. *Animal Behaviour* 40:811–828.

Zeng, L., J. T. Rotenberry, M. Zuk, T. K. Pratt, and Z. Zhang. 2016. Social behavior and cooperative breeding in a precocial species: The Kalij Pheasant (*Lophura leucomelanos*) in Hawaii. *The Auk* 133:747–760.

Zhao, W. C., F. Garcia-Oscos, D. Dinh, and T. F. Roberts. 2019. Inception of memories that guide vocal learning in the songbird. *Science* 366:83–89.

14
Birdsong Function
The Established Perspective, Advances, and an Expanded View

Elizabeth P. Derryberry and Rindy C. Anderson

It is a truth universally acknowledged that a singing bird in possession of a good territory must be in want of a wife. He belts out his song to tempt prospective mates and to warn off would-be competitors from far and wide.

This universal truth or—to use a scientifically more appropriate term—established perspective is recapitulated in texts about song across decades—from the writings of early ornithologists to contemporary scientific research articles. For instance, John James Audubon, an early ornithologist, wrote, "The notes of the Wood Thrush are doubly pleasing. . . . One would think that each individual is anxious to excel his distant rival" (Audubon 1834, pp. 373–374). And from one of our own first papers, "As the song of most songbirds has two functions—to attract females as mates and to repel rival males" (Derryberry 2007, p. 1939). Even the artist Banksy depicts birdsong as a bird in a tree with a bullhorn (Figure 14.1). In fact, the criteria historically used to define birdsong are components of this established perspective: songs are used exclusively or primarily by males; songs are broadcast vocalizations given by males when defending a territory or attracting a mate; and songs are loud, complex vocalizations (Spector 1994). The following is one of the most cited definitions: Song is primarily a male trait used in the breeding season for mate attraction and male–male contest (Catchpole and Slater 2003). Thus, this perspective appears to be universally accepted, but do ornithologists think this perspective is universal?

Yes, but not always.

Ornithologists have long known that female birds sing, song has functions other than mate attraction and competition (think, "It's your turn to watch the kids"), and that there are "soft" songs. These aspects of song are not new knowledge, so why are they so often missing in how song is commonly described? One reason is that the preponderance of studies on birdsong focus on the male bird shouting from a treetop for a mate—he's difficult to ignore. Such studies have been invaluable and are one of the reasons why birdsong is among the best studied of behaviors (Bateson and Laland 2013). Another reason is that female songs and soft songs were often considered less common or aberrant behaviors, but that viewpoint is shifting. In this chapter, we highlight studies that branch off from the established view of song (Figure 14.2) and compile recent evidence that challenges three of the criteria traditionally used to

Elizabeth P. Derryberry and Rindy C. Anderson, *Birdsong Function*. In: *New Perspectives in Ornithology*.
Edited by: Scott V. Edwards and J. Michael Reed, Oxford University Press. © Oxford University Press (2025).
DOI: 10.1093/oso/9780197787670.003.0014

Figure 14.1 Photo of Banksy's mural "Bird Singing in a Tree" in an empty parking lot at the corner of Erie and Mission St. in San Francisco, CA 94, 103.

Photo by Eva Blue 2010. Creative Common license (CC BY 2.0, https://www.flickr.com/photos/evablue/4574226834/in/photolist-7Yd7g5-7Yd7gU-7Y9RLc-7Y9RKi).

define song: song is a male trait; song is dual function (attract mates, repel rivals); song is a loud, long-distance signal. We consider evidence for song as a female trait; song as a multifunction trait; and song as a quiet, close-range signal. We discuss the merit of the established view of song and the merit of expanding it. We highlight fruitful avenues of future research that will continue to change how we think about birdsong and hope that these will inspire new and old ornithologists alike.

Is Song a Male Trait?

Yes, but not always. Birdsong has long been a textbook example of sexual dimorphism, a clear-cut difference between the sexes (Price 2015), but quite a lot of female birds sing (Riebel et al. 2005). So, why do many answer "yes" to this question?

The idea that song is a male trait stems in part from historical biases (Kroodsma and Miller 1996, Austin et al. 2021). One historical bias is the geographical location of the researchers who study song. If you hear a bird singing in a North American forest, like Audubon's friend the Wood Thrush, odds are it is a male bird. However,

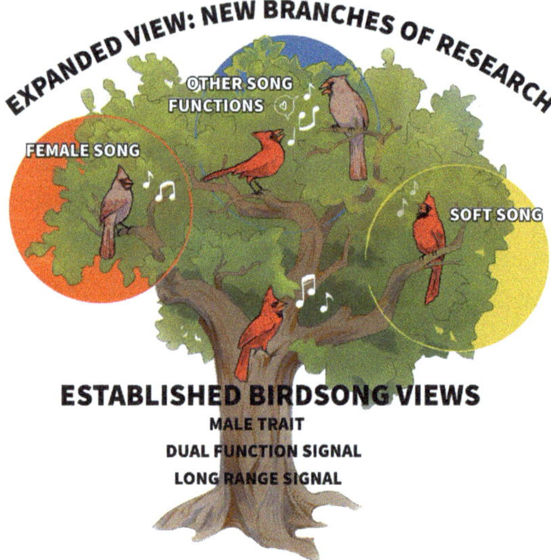

Figure 14.2 Conceptual illustration of the established and expanded view of birdsong function. The established view is on the trunk of the tree as a male trait that is dual function and a long-range signal. The expanded view branches out into studies that consider female song; other song functions, such as pair bonding; and soft songs as potent signals at close range.

Illustration by Ruth Simberloff, University of Tennessee.

if you instead walk along a forest edge in Panama, you are likely to hear a Bay Wren (*Cantorchilus nigricapillus*) female leading her male in a duet (Levin 1996a). Given that most folks who study birdsong have lived and worked in North America and Europe (thankfully this is changing), where male song is more prevalent than female song, it is perhaps not surprising that song is traditionally considered a male trait. However, even in historical hot spots for songbird research, there are many species in which females produce song (Garamszegi et al. 2007). So why have these examples of female song been effectively ignored for so long?

That brings us to our second historical bias: Most early ornithologists and behavioral biologists were men. Female songs were dismissed as rare or the outcome of individuals having too much of a "male" hormone (Thorpe 1961, Byers and King 2000, Catchpole and Slater 2003). In fact, the emergence of concerted research on female song has been led by women (Haines et al. 2020). This is not to say that contemporary male biologists dismiss female song (e.g., Price 2015) but, rather, that work on female song has been greatly limited and slowed by historical biases (Riebel et al. 2019, Haines et al. 2020, Austin et al. 2021, Rose et al. 2021). For instance, such biases contribute to a paucity of recordings of female song (Riebel et al. 2019). For at least 3,500 songbird species, we do not even know whether females sing or not.

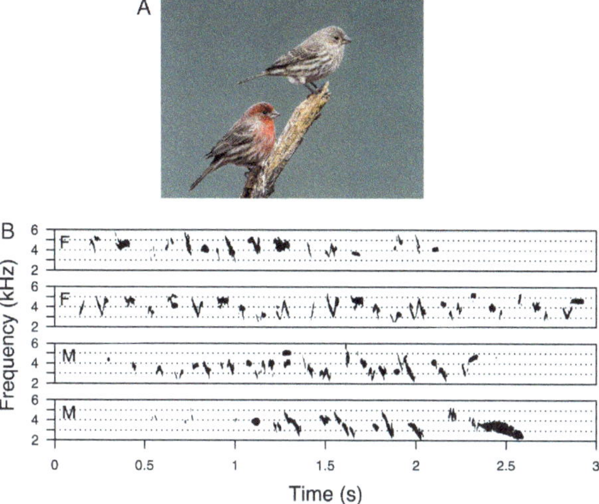

Figure 14.3 Female song can be as complex as male song. (Above) Photograph of female (top) and male (bottom) House Finch and (below) sample spectrograms from songs of female (top two rows) and male (bottom two rows) House Finches from southern New York.

Sources: (Above) Jim Merritt/Macaulay Library at the Cornell Lab of Ornithology (ML221643821). (Below) Spectrograms created from data in Kornreich et al. (2020).

This issue is exacerbated by the difficulty of identifying females in the field. If females look like males, as is often the case in species with female song, it can be difficult to determine if it is a male or a female bird singing. More detailed recording of monomorphic species can yield surprising results, however—such as the discovery that female Stripe-Headed Sparrows (*Peucaea r. ruficauda*) produce more songs, and more complex songs, than males (Illes 2015). If females do look different from males, they are often (but not always) more camouflaged. Yet, even drab females may perform complex songs. The female House Finch (*Haemorhous mexicanus*) is a drab gray-brown with a plain face, whereas her mate is orangy-red with color on crown, throat, and breast; yet, her song is just as complex (Kornreich et al. 2020) (Figure 14.3). Difficulty identifying and recording female song may also stem from females tending to sing less often (Wilkins et al. 2020), more quietly, or in different contexts from males (Odom and Benedict 2018). But, this does not mean that such songs are not biologically important (Austin et al. 2021). Combined, these issues have historically limited the study of female song, but efforts have recently ramped up (Riebel et al. 2019).

Less than 20 years ago, a number of biologists—many of them women long stating the need for such work (Riebel et al. 2005)—set out to test the radical idea that female song is widespread and ancestral in songbirds (Odom et al. 2014). They hypothesized that female song is widespread because female song is relatively common among tropical birds (Slater and Mann 2004) and tropical birds make up most of songbird biodiversity (Jetz et al. 2012, Fjeldså 2013). They also hypothesized that female song could be ancestral (i.e., females have been singing just as long as males; Riebel et al. 2005) because songbirds are thought to have originated in Australasia (Edwards and Boles 2002, Ericson et al. 2002, Barker et al. 2004), where many species have female song (Robinson 1949). In other words, if female song is common in the evolutionary home of songbirds, it stands to reason that female song could date back to the origin of songbirds.

Testing these hypotheses seems daunting—there are more than 5,000 songbird species and relatively few recordings of female song. The authors were quite clever, though, as they realized that they could just leave out the group of songbirds with the most species (the Passerida, which contains nearly 4,000 of the approximately 5,000 songbird species). That group evolved recently and is "nested" within the phylogeny (Barker et al. 2004), so it tells us very little about what happened early on in songbird diversification. Narrowing the scope of the problem let them focus on just 44 of the approximately 112 songbird families, revealing that female song occurs in more than two-thirds of those 44 families (Odom et al. 2014). Females sing in many species of Bushshrikes, Orioles, Drongos, Whistlers, and Thornbills (and the list goes on). In other words, females sing in 229 of the 323 species examined. That's a lot! And these singing females are not found solely in tropical regions. Females also sing in nearly half of European passerine species (Garamszegi et al. 2007). In other words, female song is not rare nor a hormonal aberration. Female song can be heard worldwide (Odom et al. 2014), upending the idea that song is primarily a male trait.

To test their second hypothesis, the authors then extrapolated back in time to determine whether the common ancestor of all songbirds might have had female song. Alternatively, female song may have evolved later than male song. No matter which way they ran the analysis, results suggested that female songbirds have been producing songs for just as long as male songbirds (Odom et al. 2014). This finding is revolutionary because it challenges the idea that song originally evolved to attract females.

Sexual selection on males has long been thought to be the primary driver of song evolution (Andersson 1994) and sex differences in song (Box 14.1; Darwin 1859). Males produce loud, complex songs because such songs repel rival males and because females prefer such traits (Searcy and Andersson 1986). However, female song is widespread and ancestral, so sexual selection on males alone cannot explain why and how song evolved. Furthermore, because sex differences in song appear to evolve through losses in females rather than gains in males, sexual selection on males alone cannot generate sexual dimorphism in song (Price 2019). The evolution of song and sex differences in song (see Box 14.1) are clearly the product of selective forces acting on both male and female song. Considering this, and our relative lack of knowledge of female song function or the function of song in general in many tropical species, there is a clear need to reconsider the functional roles of song as well (Riebel et al. 2019, Odom et al. 2025).

Box 14.1 Sexual Dimorphism of Songs: Functional Implications

In many Neotropical bird species, female song is the norm (Sick 1997, Stutchbury and Morton 2001, Morton 2019). For instance, female song is present in all 235 species of antbirds (family Thamnophilidae; Zimmer and Isler 2003) and has now been recorded in most species. In some antbird species, female and male songs are completely different and could be mistaken as songs from different species (Figures 14.B.1a and 14.B.1b). In contrast, in other species, female and male songs are similarly structured, with only small differences in pitch (Figures 14.B.1c and 14.B.1d). High variation in the degree of song sexual dimorphism suggests a complex evolutionary history. In species with high song sexual dimorphism, female and male songs may convey different messages. For example, differently structured songs may be intended for different receivers and may help birds defend different resource types, such as mates or territories (Morton and Derrickson 1996, Morton et al. 2000). Conversely, in species with low song sexual dimorphism, songs may have similar functions (Tobias et al. 2011, Beco et al. 2021). The story becomes even more complex when other signal modalities are considered. Ornamented females in less sexually dichromatic species respond strongly toward songs of social pairs and female and male intruders, whereas less ornamented females from more sexually dichromatic species respond only toward social pairs and female intruders (Macedo et al. 2021). In contrast, males in all species respond toward all intruder types (Macedo et al. 2021). These examples suggest that more ornamented females use songs and plumage displays in similar contexts as males (Macedo et al. 2021), indicating that both sexes may be under similar social selection pressures (*sensu* West-Eberhard 1983).

Investigating female and male songs in the tropics presents opportunities to discover new functions and associated behaviors (Morton and Derrickson 1996, Stutchbury and Morton 2001, Kroodsma et al. 2019) as well as challenges because many taxa, such as antbirds, are threatened with extinction (Zimmer and Isler 2003, Reinert et al. 2007, Mattos et al. 2009, Del-Rio et al. 2015). In addition, taxonomy is not well resolved in many tropical taxa, with new species only recently being described (Whitney and Haft 2013, Moncrieff et al. 2018) and previously recognized genera or species being split into many different ones to better reflect actual evolutionary lineages (Isler et al. 2013, Burns et al. 2014, Johnson et al. 2021). Further complicating the advancement of birdsong science in the tropics, conducting fieldwork in countries such as Brazil and Colombia can be dangerous, with researchers exposed to deadly terrain, robbery, and even murder (Conniff 2021). Despite these challenges, investigating female and male songs in the tropics will advance frontiers of our knowledge of ecology and evolution of birdsong.

Continued

Continued

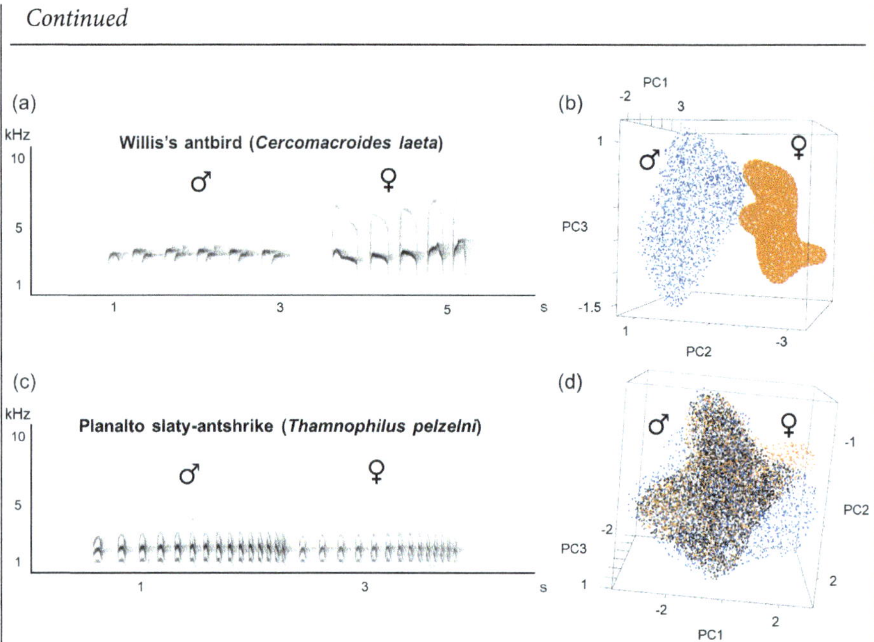

Figure 14.B.1 Representative spectrograms and three-dimensional (3D) volumes in acoustic space of duets of social pairs of antbirds. Examples with high and low song sexual dimorphism are the Willis's Antbird (*Cercomacroides laeta*) (a and b) and the Planalto Slaty-Antshrike (*Thamnophilus pelzelni*) (c and d), respectively. 3D volumes represent acoustic variation of females (yellow) and males (blue). In panel d, overlap between female and male volumes is shown in black. In panel b, there is no overlap between volumes. Volumes were obtained using the R package "hypervolume" (Blonder et al. 2015). Axes that delimit volumes are components of a principal component analysis calculated with note-by-note measurements of peak frequency (Hz), frequency bandwidth (Hz), and note duration (seconds) of female and male songs.

The box text and figure were created by Gabriel Macedo.

Is Song a Dual-Function Signal?

Yes, but not always. The best studied functions of song are those of (1) repelling rival males and (2) attracting females as mates, but if we consider female song and song in tropical species, we find that song has many more functions than the traditional two.

Of course, in many songbirds, male song does function to repel rivals and to attract mates.

The best evidence that male song functions to defend territories from competitors comes from two types of experimental designs: One silences the male's song by muting the male, and the other removes the male and replaces him with a speaker broadcasting song ("speaker occupation") (Nowicki et al. 1998). Males that have been muted experience more intrusions onto their territories and are more likely to lose their territory relative to unmuted controls (Peek 1972, McDonald 1989, Westcott 1992). In speaker occupation studies, song alone effectively keeps competitors out of the territory (Göransson et al. 1974, Krebs 1977, Yasukawa 1981, Falls 1988) and acts as a form of acoustic mate guarding (Dowling and Webster 2018). Territories with speakers broadcasting song have fewer intrusions and remain unoccupied longer than territories from which males are removed but songs are not broadcast. In other words, singing a song is akin to posting a "keep out" sign.

There is also strong evidence that male song functions to attract mates. Similar "removal" experiments provide good evidence that females are attracted to male song. Song alone can increase female visitation to nest boxes (Eriksson and Wallin 1986), and males that have had their social partner removed (i.e., are unmated) attract a new female by singing (Morton et al. 2000). Male song can even stimulate females to mate (reviewed in Searcy 1992). À la Barry Manilow, songs are sexually inviting. Taken together, these early experiments provide strong evidence that males produce song to both repel rival males and attract female mates.

But do female songs serve these same two functions? What little has been done with female song suggests that it is used more often to repel rivals than to attract mates.

Females often use song to defend breeding resources in species that compete fiercely for space or nest sites (Arcese et al. 1988) or in species that defend territories year-round, which increases the likelihood that a mate might be regularly absent or die (Morton 2019). For instance, much of what we know about female song comes from study of it in duetting species in the tropics (Box 14.1), in which males and females defend territories year-round by alternating their songs or by singing in such close coordination that their songs sound like one song (Langmore 1998). An elegant set of removal experiments in duetting Bay Wrens demonstrated that female song functions in territorial aggression between females and that females can defend their territories alone (Levin 1996a). Female song does not have to be directed only at female competitors; it can be used to repel rivals of either sex (Hall 2000, Macedo et al. 2021).

Evidence continues to accumulate that females use song in territorial defense. This is the case in temperate zone species, including the European Robin (*Erithacus rubecula*; Hoelzel 1986), the Yellow Warbler (*Dendroica petechia*; Hobson and Sealy 1990), the White-Crowned Sparrow (*Zonotrichia leucophrys*; Baptista et al. 1993), the Great Reed Warbler (*Acrocephalus arundinaceus*; Bensch and Hasselquist 1992), and the Starling (*Sturnus vulgaris*; Sandell and Smith 1997). Yet, for all these species, there are many more in-depth studies of the function of male territorial song than of female territorial song, highlighting the need for more song studies even in these well-studied species. More recent work demonstrates that female song is also used in territorial defense in tropical and south-temperate species, including the Black Coucal (*Centropus grillii*; Geberzahn et al. 2009), Happy Wren (*Pheugopedius felix*; Templeton et al. 2011), Yellow-Breasted Boubous (*Laniarius atroflavus*; Wheeldon et al. 2021), Mexican antthrushes (*Formicarius moniliger*; Kirschel et al. 2020), and the Superb Fairywren (*Malurus cyaneus*; Cooney and Cockburn 1995). Female Superb Fairywrens are thought to sing so much in part because their partners are regularly gone in search of sex with other females (Dalziell and Cockburn 2008). Talk about holding down the fort!

Like males, females can also use song to defend mates in addition to territories. Eastern Whipbird (*Psophodes olivaceus*) females are more likely to keep singing with their mate if they think an unknown visitor to their territory is another female, but not if they think that visitor is male (Rogers et al. 2006). This advertisement of the male's mated status also appears to be a function of female song in the Slate-Colored Boubou (*Laniarius funebris*; Sonnenschein and Reyer 1983) and the Crimson-Breasted Shrike (*Laniarius atrococcineus*; van den Heuvel et al. 2014). However, experimental tests in other species do not find support for this function. For example, in Eastern Bluebirds, females do not differ in their response to male and female simulated intruders, making it unlikely that they are mate guarding, even though mate guarding would be advantageous in this system with its high rate of extra-pair young (Rose et al. 2019).

Selection may also favor female song as a signal to attract mates when there is strong competition among females for mates. In the Alpine Accentor (*Prunella collaris*)—a polygynandrous species in which both males and females compete for mates—female songs elicit approaches from males but not other females, and females sing only when fertile (Langmore et al. 1996). There is even evidence for mate attraction in monogamous species: In the Dusky Antbird (*Cercomacroides tyrannina*), females sing sex-specific songs to attract a new male after their mate has been removed (Morton et al. 2000). However, removal experiments have also demonstrated that female song is not used for mate attraction in a number of species, including the Bay Wren (Levin 1996a, 1996b), Northern Cardinal (*Cardinalis cardinalis*; McElroy and Ritchison 1996), Dunnock (*Prunella modularis*; Langmore and Davies 1997), and the Superb Fairywren (Cain and Langmore 2015). It thus appears that mate attraction may not be a common function of female song. However, studies of female song continue to turn up unexpected findings. For instance, experiments with duetting Tropical Boubou (*Laniarius aethiopicus*) to test the territorial function of song found that females synchronize their songs with male solo playbacks, much to the annoyance of their mates, which immediately attempt to jam rivals' songs (Grafe and Bitz 2004). This finding suggests that mated females may be using song in a mate attraction context. Clearly, there is still much work to be done to test the attraction function of female song, including testing male preferences for female song and whether female song may stimulate male reproductive physiology (Riebel et al. 2019).

Altogether, studies to date clearly indicate that females—like males—produce songs to compete for sexual resources, both in terms of territories and in terms of mates. However, these two functions alone do not explain female song in a comprehensive way.

Females do so much more with song than just compete over sexual resources. Comparing the structure, timing, and context of female and male songs can begin to reveal their similarities and differences (Riebel et al. 2019). This approach often reveals that female song serves multiple functions, many of which are not associated specifically with repelling rivals or acquiring mates (Mennill and Vehrencamp 2008, Odom et al. 2016). These "extra" functions can be binned in different ways (Langmore 1998, Riebel et al. 2019), but they generally include functions for communication within pairs and to offspring.

Perhaps not surprising to many human mothers, female birds spend a lot of time singing to coordinate day-to-day family social and care activities. For species with long-term pair bonds, female song may be important in strengthening those bonds.

For instance, Eastern Bluebird pairs that remain together for multiple seasons have higher reproductive success, and females use song to strengthen those pair bonds (Rose et al. 2019). In fact, female Bluebirds share more song types with their mate than with nonmates, suggesting that song type sharing may be integral to within-pair communication (Sikora et al. 2021). Similarly, species of Neotropical wrens (Troglodytidae) with long breeding seasons have more coordinated duets, suggesting that duet coordination likely signals the strength of the pair bond (Keenan et al. 2020). Yet another function of female song may be to communicate with offspring, although there are only a handful of studies on this usage. The best example is that female Black-Headed Grosbeaks (*Pheucticus melanocephalus*) sing to locate and feed young fledglings, and fledglings approach playbacks of their parents singing (Ritchison 1983). Females can also use song to coordinate biparental care with their mate. For instance, Northern Cardinal (*Cardinalis cardinalis*) females sing to inform their mate when to visit the nest. Males are more likely to approach the nest with food for nestlings if the female sings, although if she matches his song type, he does not approach (Halkin 1997). A similar use of song to coordinate care is found in the Slate-Colored Boubou (Sonnenschein and Reyer 1983). Because many of these studies were early one-offs, and yet social monogamy is widespread in birds (>80% of species), song as a signal to coordinate biparental care is a ripe area for new research.

Of course, in many of these cases, the male is also singing to coordinate pair behavior, so might male song also have more than two functions? The historical bias of studying song in temperate species has limited not only our understanding of female song but also that of male song by focusing on species that sing primarily during the breeding season (Riebel et al. 2019). In many tropical systems, females and males sing year-round, even when not breeding. In these systems, song must then function in nonsexual contexts in both males and females (Tobias et al. 2011). These nonsexual functions of song do not fit neatly into the traditional framework of sexual selection. Mary Jane West-Eberhard recognized this issue many decades ago (West-Eberhard 1979, 1983). She pointed out that ornaments, such as song, can be used not only to compete for sexual resources but also for nonsexual resources and that such competition is prevalent in both sexes. Her theory of social selection extended sexual selection theory to include social competition for resources other than mates. Social selection theory postulates that traits evolve via social competition to gain access to resources, including (under sexual selection) mates. This hypothesis was revolutionary, yet still understudied, and can help us think about the evolution of song in both males and females (Tobias et al. 2011).

We conclude that although "dual function" may be a good descriptor of song, it does not encapsulate the full range of how songs are used for many species. Shifting the general view of song to multifunction will yield a wealth of new questions, as we discuss in the "Future Directions" section.

Is Song a Long-Distance Signal?

Yes, but not always. We have made much progress in understanding the very obvious, loud songs that songbirds produce, but songs can also be powerful signals at close range and at low amplitude.

Songbirds belt out their songs to reach listeners both near and far. The loudness of birdsong when experienced at close range is often shocking to bird enthusiasts and scientists alike—wow what a singer! It is no surprise, then, that birdsong is commonly referred to as a long-distance mating and advertisement signal. Singers must advertise their ownership of a territory, attract mates, and repel rivals. To accomplish these functions, song must be of sufficient amplitude to travel long distances, ideally well beyond the singer's territory, and to be detected and perceived in noise. Birdsongs do tend to be loud, and they do tend to have a large "active space" (the volume in which a signal can be detected and perceived). Yet there are important exceptions to this pattern: when songs are quiet, come-hither coos, or whispered threats of attack. Birdsong is indeed a long-distance signal, except when it's not.

Many researchers have been fascinated with the loudness of birdsong and with how far the songs propagate across the landscape (Brenowitz 1982, Wiley and Richards 1982, Dabelsteen et al. 1993, Holland et al. 1998, Naguib and Wiley 2001, Brumm 2002, Dooling and Blumenrath 2013). Brenowitz (1982) measured song amplitudes of the Red-Winged Blackbird (*Agelaius phoeniceus*) in the field at a mean of 90.8 decibels (dB; the way we quantify sound pressure level), and occasionally exceeding 93 dB. Our ears perceive this as roughly as loud as a motorcycle at a distance of 25 feet (https://www.iacacoustics.com). Brenowitz measured these amplitudes at 1-m distance from the bird and then estimated the maximum active space of redwing song, without wind, to be approximately 189 m. This estimate of active space corresponds to the maximum distance across two redwing territories, leading Brenowitz to conclude that redwing song is adapted in its acoustic structure and amplitude to communicate a male's presence across his neighbors' territories. Redwing song is, then, a long-distance signal.

A twist on this theme is found in the full song of the Eurasian Blackbird, *Turdus merula*. Blackbird song begins with the "motif" composed of loud, low-frequency, nearly pure tones, followed by the "twitter" composed of relatively quieter, complex notes that span a wide frequency range (Dabelsteen 1984). Dabelsteen, Pedersen, and colleagues showed through a series of studies in the laboratory and in the field that the motif and twitter parts of Blackbird songs are adapted for communication at long and short range and that the use of the two song parts supports this assertion (Dabelsteen and Pedersen 1990). Motif parts are emitted on their own only when the singer perceives he is alone, and twitter parts are emitted on their own in long bouts during short-range interactions, such as when courting females (Snow 1958, Dabelsteen and Pedersen 1988) or threatening rival males (Dabelsteen and Pedersen 1985, 1990). Twitter parts of Blackbird songs also have acoustic structure that appears to restrict transmission, with twitters including much higher frequencies than motifs in addition to being sung more quietly (Dabelsteen 1984). Also, measurements of transmission show that quiet twitter sounds do not transmit much beyond a single territory, whereas the louder motif sounds in full song typically travel over two to four territories (Dabelsteen et al. 1993). Taken together, studies of Blackbird song suggest that this signal is adapted to communicate to both nearby and distant receivers of both sexes.

Another recent twist on the theme of song as a long-distance mating signal comes from the spectacular scream songs of the White Bellbird (*Procnias albus*). Podos and Cohn-Haft (2019) report the loudest vocal amplitudes for any bird, with type 1 songs

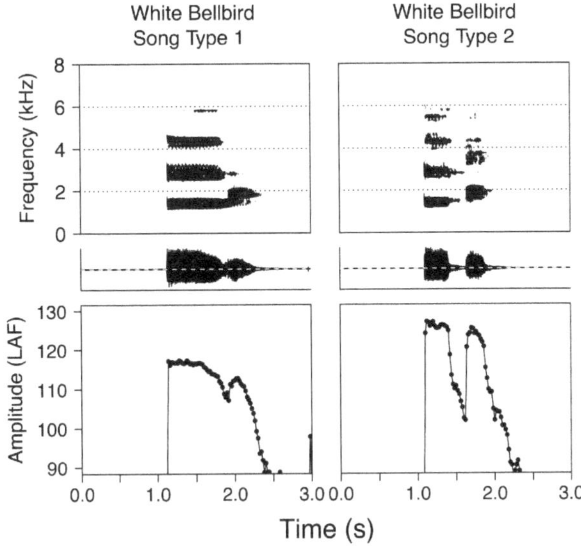

Figure 14.4 Loud song as a close-range signal. White Bellbird song type 1 and song type 2 (top) spectrograms (black indicates −26 dB from peak), (middle) waveforms, and (bottom) amplitude envelopes that show sound pressure level A-weighted, fast detector (LAF) peak (L_{peak}) plotted for 20-msec time segments, corrected for baseline noise levels and extrapolated to 1-m distance.

Source: Data from Podos and Cohn-Haft (2019) and J. Podos (personal communication).

averaging 116.7 dB(A) L_{peak}, and type 2 songs roaring in at 125.4 dB(A) L_{peak}. This exceeds the threshold for pain in humans and can cause immediate hearing damage. Clearly, these explosive songs, delivered from dead branches above the canopy, can travel long distances. And yet, these screams are a short-range courtship signal. Females are observed to join males on their display perches within a few meters, and males then deliver their loudest type 2 songs, swiveling their heads to face the female directly as they scream the second note (Figure 14.4). Although females retreat a bit as the male begins his scream, they experience peak amplitudes at the ear estimated at roughly 113 dB(A), presumably loud enough to cause hearing damage. It is unclear if the spectacular scream songs of the White Bellbird also function as long-distance signals, but they do appear to represent the unique case of an exceptionally loud song adapted to court females at close range.

We know that singers can sing loudly and can also vary the amplitude of their songs. The broadcast songs of many species have been found to vary considerably in amplitude both within and between males of a species (Brumm and Todt 2004, Brumm 2009, Brumm et al. 2025). In the Chipping Sparrow (*Spizella passerina*), for example, males adjust their songs according to social context: "Dawn song" is directed to other males at close range, whereas the louder "day song" is broadcast to distant birds (Liu and Kroodsma 2007). Not only can singers vary the loudness of their songs but also loudness matters to receivers. Within-species variation in amplitude may affect female choice and male–male competition, making amplitude the object of sexual selection. For example, females prefer louder songs in the Zebra Finch (Ritschard et al. 2010) and the Red-Winged Blackbird (Searcy 1996). In the Chaffinch (*Fringilla coelebs*) and the White-Crowned Sparrow, high-amplitude songs elicited a stronger territorial response from resident males, suggesting that song amplitude is important in assessing rival males (Brumm and Ritschard 2011, Luther et al. 2017) and that a loud intruder is likely to be a greater threat than a quieter one.

Despite the predominance of research describing the long-distance nature of birdsong, recent decades have brought a focus to singing behavior that is clearly not designed for long-distance mating and advertisement because it is produced at lower amplitudes. The Zebra Finch provides a case in point. Male Zebra Finches sing a quiet song intended for nearby listeners (Zann 1996). This song has two forms: "Directed song" is a courtship song directed at a specific female, whereas "undirected song" is sung without an obvious intended receiver (Sossinka and Böhner 1980). Recent integrative work by Loning et al. (2021) suggests that the songs of wild Zebra Finches are in fact short-range signals. These authors made calibrated song amplitude measurements and conducted song transmission experiments in the Zebra Finch native habitat, combining these data with information on the species' social behavior and hearing physiology. Integrating the results showed that the songs are a short-range signal reaching receivers an average of 9 m distance and maximally to 14 m, in stark contrast to songbird species whose songs transmit over much larger distances. There is also evidence that mated Zebra Finches perform quiet, coordinated vocal duets at the nest after a period of separation (Elie et al. 2010). These private displays appear to support pair-bond maintenance and may even allow mate recognition. Although these duets involve calls and not song, they reinforce the point that vocal signals can evolve to function at close range.

Such "private" vocal exchanges have long been recognized in songbirds (Dabelsteen et al. 1998, Titus 1998, Morton 2000), although we have made more progress in understanding the context and acoustic structure of these whispers than why the low amplitude is adaptive, especially when they are used to threaten. Dabelsteen and colleagues (1998) were among the first to draw attention to the "overlooked phenomenon" of quiet songs, describing their behavioral context and structure for several species. They offered an early hypothesis to explain quiet songs, pointing out social contexts and scenarios in which selection would favor a private exchange of information between signaler and receiver in order to avoid eavesdroppers. Experimental studies followed, showing that quiet songs are used by a number of songbird species in contexts of both courtship and competition (Box 14.2). The Dark-Eyed Junco, for example, sings "short-range songs" along with courtship displays when in close proximity to fertile females and also when engaged in close-range disputes with other males (Titus 1998, Reichard et al. 2011). Reichard and colleagues (2013) further showed that Juncos sing two categories of quiet songs, one that differs in acoustic structure from loud broadcast song only in amplitude and another with a strikingly distinct acoustic structure. A similar pattern exists in the "soft songs" of the Song Sparrow (Anderson et al. 2008), with "crystallized" soft songs being quiet versions of loud broadcast song types, and "warbled" soft songs being composed of unique notes and phrases with a seemingly rambling, disorganized structure. Unlike the Junco, however, soft songs in the Song Sparrow are strongly associated with aggressive interactions between males and not in the context of courting females (Nice 1943), and they are reliable threat signals in Song Sparrows (Searcy et al. 2006, Akçay et al. 2011, Anderson et al. 2012), in the closely related Swamp Sparrows (Ballentine et al. 2008), in Black-Throated Blue Warblers *Setophaga caerulescens* (Hof and Hazlett 2010), in Brownish-Flanked Bush Warblers *Cettia fortipes* (Xia et al. 2013), and in Bachman's Sparrows (Ali and Anderson 2018).

Box 14.2 Threats and Flirtations at Close Range: Structure and Function of Low-Amplitude Song

Beginning with Margaret Morse Nice, who made brief mention of soft song in her accounts of male–male territorial behavior in Song Sparrows (Nice 1943), birdsong researchers have been aware that songbirds sometimes whisper instead of shout. Variously termed "quiet song," "whisper song," and "soft song," among other descriptors, researchers became fascinated with the question, Why signal softly?, and they began to design studies to answer this apparent Darwinian puzzle.

Two papers published in 1998 (Dabelsteen et al. 1998, Titus 1998) were among the first to note that quiet singing is common but understudied in birds and that it provides an untapped opportunity to explore how selection shapes communication signals. Since that time, students of birdsong have turned their attention to this enigmatic mode of communication, yielding numerous examples of low-amplitude song in songbirds (Akçay et al. 2015, Reichard and Anderson 2015, Reichard and Welklin 2015).

The oddity of quiet singing is that its hushed and discrete form seems antithetical to birdsong's two primary functions: to impress potential mates or to threaten rivals. This puzzle has led to several hypotheses about why selection might favor low-amplitude signaling in birds (reviewed in Akçay et al. 2015). Studies have also identified stark differences in the acoustic structure of loud and quiet songs, suggesting that quiet songs may serve supporting or distinct functions compared to their loud counterparts (Figure 14.B.2).

A second puzzle presented by quiet song is that the songs often function as reliable whispered threats (see main text). Signal reliability can be maintained in signals that are physically difficult or impossible to cheat or are intrinsically costly to produce. The whispered threats of songbirds appear to be neither, raising the question, What keeps them honest? One answer is provided by the receiver retaliation hypothesis: Bluffing is dangerous because a dishonest threat will provoke a costly attack from a stronger opponent. The key prediction of this idea is that signal receivers will respond more aggressively to quiet songs than to loud ones. Evidence supporting this prediction has been found for several species, including Song Sparrow (Anderson et al. 2012, Templeton et al. 2012), Brownish-Flanked Bush Warblers (Xia et al. 2013), and for the quiet calls of the Corn Crake *Rex rex* (Ręk and Osiejuk 2011). Thus, a songbird that threatens a rival with quiet song had better be prepared to back up the threat with a fight.

Continued

Continued

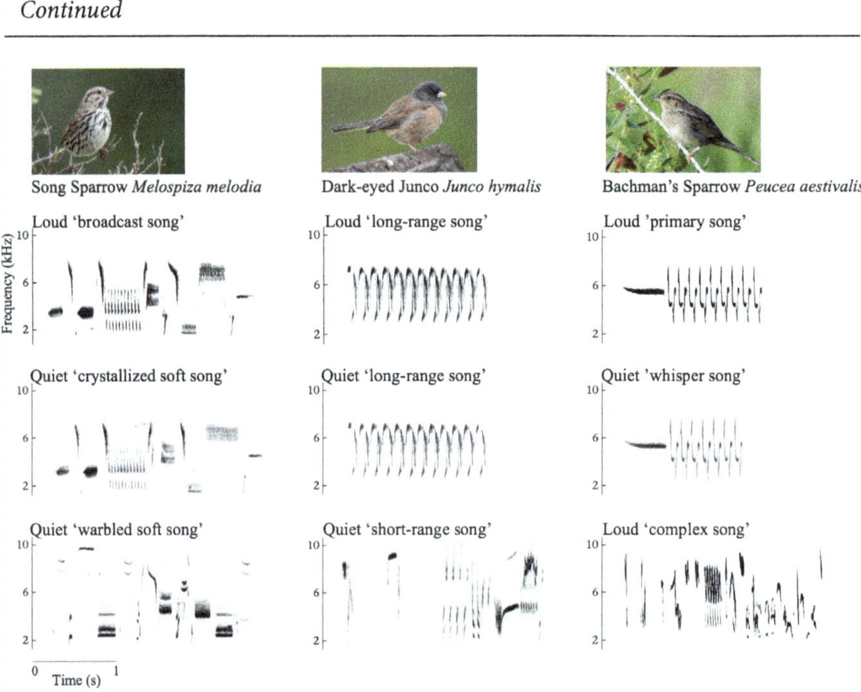

Figure 14.B.2 Sonograms showing examples of loud and quiet songs produced by three songbird species. All three sing loud songs designed to travel long distances and function to advertise territorial ownership and to attract potential mates (top). All three also sing quiet versions of these broadcast songs (middle). Song Sparrows and Bachman's Sparrows sing quiet songs almost exclusively in the context of intense male–male agonistic interactions. Male Juncos respond with equal aggressiveness to loud and quiet long-range songs, suggesting that both serve an agonistic function. All three species also sing acoustically distinct types of songs that are substantially more complex than broadcast song types in terms of the number and diversity of song elements, and they span a greater frequency range (bottom). These songs are sung exclusively quiet in Song Sparrows and Juncos, and exclusively loud in Bachman's Sparrow. These oddities are sung during male–male competitions in Song Sparrows and Bachman's Sparrow, and during close-range courtship in Juncos.

Photographs of Song Sparrow (2018) and Dark-Eyed Junco by Becky Matsubara (CC BY 2.0 https://www.flickr.com/photos/beckymatsubara), and photograph of Bachman's Sparrow by Richard Cassell.

It turns out that quiet song is much more common and widespread in songbirds than we ever thought (Reichard and Anderson 2015, Reichard and Welklin 2015). Why should it be advantageous to whisper courtship and threat signals in some situations, yet belt out these same sorts of signals in others? The eavesdropping avoidance hypothesis (Dabelsteen et al. 1998) provides an intuitively appealing result, yet studies testing this idea so far have not yielded support (Searcy and Nowicki 2006, Akçay et al. 2016). Another explanation is provided by Akçay and colleagues (2011), termed the "readiness hypothesis." The posture required for singing loudly in songbirds (head thrown back, bill open wide) reduces the singer's ability to keep eyes on his rival and puts him at a higher risk of attack and injury. It might be advantageous in such close-range encounters to sing with beak closed and eyes straight ahead, with the result that songs are necessarily quiet.

We conclude that birdsong is a long-distance mating signal, except when it's not. We now recognize that the fabulously ostentatious "broadcast" songs of so many species are not the full story and that sometimes selection favors muttered threats and whispered flirtations.

Future Directions

In branching out beyond the quintessential male bird singing in a tree, it is perhaps not surprising to find that we still have a lot to learn about birdsong. In fact, we both learned that even experts on birdsong, such as ourselves, still have a lot to learn.

We mention a few of the many additional gaps in knowledge about female song and song function (Riebel et al. 2019) just to get you curious. For instance, much is known about how male song varies in rate, structure, and repertoire size, and yet next to nothing is known about this for females in most species (e.g., Odom et al. 2016). We also know very little about how age and status might affect female song: Do females improve or decline in song performance as they age (e.g., Pavlova et al. 2010)? We know a lot about how males use song to identify and interact with other males (reviewed in Searcy and Nowicki 2010), and relatively recent studies demonstrate that females can do the same (e.g., Templeton et al. 2013, Hall et al. 2015, Sikora et al. 2021), although not always with the same outcome (Brunton et al. 2008). And these are questions generated just from thinking about adult females. If we turn our attention to juveniles, we again know very little about how females acquire song (Riebel 2016). Who do females learn from and when do they learn? Do they need to interact with an adult to develop a normal song or to learn when and where to sing it (Rivera-Cáceres et al. 2018)? Answering these types of questions could help uncover the evolutionary origins of song learning in birds (Rivera-Cáceres and Templeton 2019) because female song is ancestral (Odom et al. 2014).

To answer this multitude of questions, there is an urgent need for more recordings of female song (Benedict and Odom 2017, Odom and Benedict 2018), as well as methods to compare song structures between the sexes (Odom et al. 2021). When such recordings reach a critical mass within a group of birds (see an example in Box 14.1), it becomes possible to really explore the function and development of female song. An excellent approach, as suggested by Riebel and colleagues (2019), is to pair phylogenetic comparative studies with experimental studies. The former can reveal correlations between female song, as well as sexual song dimorphism, with important aspects of ecology and social behavior (Keenan et al. 2020, Beco et al. 2021).

These correlations can then be tested using experimental approaches to ask questions about how song develops as well as how it functions in both sexual and nonsexual contexts. Given that male song alone has fueled scientific research for more than a century, female song is sure to keep the field active for decades to come.

It is well established that birdsong can be loud—a long-distance signal shaped to attract and court females and to deter rivals. We need more studies that examine song amplitude beyond the loud versus quiet dichotomy and rather treat amplitude as a continuum (Brumm et al. 2025). We should also focus on better understanding the physical, physiological, social, and ontogenetic factors that might constrain song amplitude in the context of courtship or territory defense (Zollinger and Brumm 2015). In particular, we have much work to do to understand the adaptive value of singing quietly during escalated aggressive interactions. And why do some species sing quiet songs that are acoustically distinct from their louder songs? Does the structure of quiet song reduce its transmission distance in addition to being quiet? And, because these songs presumably function to intimidate or impress, are quiet songs physically challenging to produce? Although it is difficult to argue that singing quietly should be more costly or difficult than singing loudly based on amplitude alone, it is possible that the distinct acoustic structure of some quiet songs requires vocal acrobatics showcased at close range. Another intriguing question that needs attention is, To what degree is singing at different amplitudes learned behavior? Are these adjustments innate responses to social context, or must they be learned through interactions with adult tutors, as is song itself? Understanding how selection has acted through the environment, predation, sexual selection, and/or physical constraints to produce quiet song provides an intriguing opportunity for studying how animal communication systems evolve.

We hope this reminder of the B side of song encourages researchers as well as the thousands of bird enthusiasts throughout the world to branch out from listening to those loud males conveniently belting out their song from the treetops. We also hope that this chapter will remind all to pause, as we have begun to do, when about to describe song as a loud vocalization given by male birds in resource defense and mate attraction. Instead, remember to mention that song is sung by both males and females, is multifunctional, and sometimes is most potent when given at a whisper. Reconceptualizing song in an inclusive framework (Austin et al. 2021, Rose et al. 2021), and studying these "not always" aspects of song, will continue to change what we know about song and how we study it.

Acknowledgments

We thank Ruth Simberloff for the illustration of Figure 14.2, Jeff Podos for providing data for Figure 14.4, Graham Derryberry for coding Figures 14.3 and 14.4, and Gabriel Macedo for his contributions to writing Box 14.1 and for providing Figure 14.B.1. We also thank members of the Derryberry and Anderson labs for providing comments on earlier drafts of this chapter. We acknowledge support from the National Science Foundation (IOS-1827290 to EPD) and the Fulbright Scholars Program (to GM).

References

Akçay, Ç., R. C. Anderson, S. Nowicki, M. D. Beecher, and W. A. Searcy. 2015. Quiet threats: Soft song as an aggressive signal in birds. *Animal Behaviour* 105:267–274.

Akçay, Ç., A. Clay, S. E. Campbell, and M. D. Beecher. 2016. The sparrow and the hawk: Aggressive signaling under risk of predation. *Behavioral Ecology* 27:601–607.

Akçay, Ç., M. E. Tom, D. Holmes, S. E. Campbell, and M. D. Beecher. 2011. Sing softly and carry a big stick: Signals of aggressive intent in the song sparrow. *Animal Behaviour* 82:377–382.

Ali, S., and R. Anderson. 2018. Song and aggressive signaling in Bachman's Sparrow. *The Auk* 135:521–533.

Anderson, R. C., W. A. Searcy, M. Hughes, and S. Nowicki. 2012. The receiver-dependent cost of soft song: A signal of aggressive intent in songbirds. *Animal Behaviour* 83:1443–1448.

Anderson, R. C., W. A. Searcy, S. Peters, and S. Nowicki. 2008. Soft song in song sparrows: Acoustic structure and implications for signal function. *Ethology* **114**:662–676.

Andersson, M. 1994. *Sexual Selection*. Princeton University Press, Princeton, NJ.

Arcese, P., P. K. Stoddard, and S. M. Hiebert. 1988. The form and function of song in female song sparrows. *The Condor* 90:44–50.

Audubon, J. J. 1834. *Ornithological Biography: Or an Account of the Habits of the Birds of the United States of America; Accompanied by Descriptions of the Objects Represented in the Work Entitled The Birds of America, and Interspersed With Delineations of American Scenery and Manners.* 5 Vol. Il. Q. Vol. IA: Black, London; Vols. 2–5: A. & C. Black, London.

Austin, V. I., A. H. Dalziell, N. E. Langmore, and J. A. Welbergen. 2021. Avian vocalisations: The female perspective. *Biological Reviews* 96:1484–1503.

Ballentine, B., W. A. Searcy, and S. Nowicki. 2008. Reliable aggressive signalling in swamp sparrows. *Animal Behaviour* 75:693–703.

Baptista, L. F., P. W. Trail, B. B. DeWolfe, and M. L. Morton. 1993. Singing and its functions in female white-crowned sparrows. *Animal Behaviour* 46:511–524.

Barker, F. K., A. Cibois, P. Schikler, J. Feinstein, and J. Cracraft. 2004. Phylogeny and diversification of the largest avian radiation. *Proceedings of the National Academy of Sciences of the USA* 101:11040–11045.

Bateson, P., and K. N. Laland. 2013. Tinbergen's four questions: An appreciation and an update. *Trends in Ecology & Evolution* 28:712–718.

Beco, R., L. F. Silveira, E. P. Derryberry, and G. A. Bravo. 2021. Ecology and behavior predict an evolutionary trade-off between song complexity and elaborate plumages in antwrens (Aves, Thamnophilidae). *Evolution* 75:2388–2410.

Benedict, L., and K. Odom. 2017. Listening to nature's divas. *Birding* 49:34–43.

Bensch, S., and D. Hasselquist. 1992. Evidence for active female choice in a polygynous warbler. *Animal Behaviour* 44:301–311.

Blonder, B., C. B. Morrow, S. Brown, G. Butruille, D. Chen, A. Laini, D. J. Harris, and C. Violet. 2025. Hypervolume: High dimensional geometry, set operations, projection, and inference using kernel density estimation, support vector machines, and convex hulls. R package version 3.1.5. https://github.com/bblonder/hypervolume.

Brenowitz, E. A. 1982. The active space of red-winged blackbird song. *Journal of Comparative Physiology* 147:511–522.

Brumm, H. 2002. Sound radiation patterns in nightingale (*Luscinia megarhynchos*) songs. *Journal für Ornithologie* 143:468–471.

Brumm, H. 2009. Song amplitude and body size in birds. *Behavioral Ecology and Sociobiology* 63:1157–1165.

Brumm, H., E. de Framond, E. Kikodze, L. de Framond. 2025. A comparative analysis of song amplitude across and within bird species. *Proc. R. Soc. B* 292: 20251088.

Brumm, H., and M. Ritschard. 2011. Song amplitude affects territorial aggression of male receivers in chaffinches. *Behavioral Ecology* 22:310–316.

Brumm, H., and D. Todt. 2004. Male–male vocal interactions and the adjustment of song amplitude in a territorial bird. *Animal Behaviour* 67:281–286.

Brunton, D. H., B. Evans, T. Cope, and W. Ji. 2008. A test of the dear enemy hypothesis in female New Zealand bellbirds (*Anthornis melanura*): Female neighbors as threats. *Behavioral Ecology* 19:791–798.

Burns, K. J., A. J. Shultz, P. O. Title, N. A. Mason, F. K. Barker, J. Klicka, S. M. Lanyon, and I. J. Lovette. 2014. Phylogenetics and diversification of tanagers (Passeriformes: Thraupidae), the largest radiation of Neotropical songbirds. *Molecular Phylogenetics and Evolution* 75:41–77.

Byers, B. E., and D. I. King. 2000. Singing by female Chestnut-Sided Warblers. *Wilson Journal of Ornithology* 112:547–550.

Cain, K. E., and N. E. Langmore. 2015. Female and male song rates across breeding stage: Testing for sexual and nonsexual functions of female song. *Animal Behaviour* 109:65–71.

Catchpole, C. K., and P. J. B. Slater. 2003. *Bird Song: Biological Themes and Variations.* Cambridge University Press, Cambridge, UK.

Conniff, R. 2021. The Wall of the Dead: A memorial to fallen naturalists. https://strangebehaviors.wordpress.com/2011/01/14/the-wall-of-the-dead.

Cooney, R., and A. Cockburn. 1995. Territorial defence is the major function of female song in the superb fairy-wren, *Malurus cyaneus. Animal Behaviour* 49:1635–1647.

Dabelsteen, T. 1984. An analysis of the full song of the blackbird Turdus merula with respect to message coding and adaptations for acoustic communication. *Ornis Scandinavica* 15:227–239.

Dabelsteen, T., O. N. Larsen, and S. B. Pedersen. 1993. Habitat-induced degradation of sound signals: Quantifying the effects of communication sounds and bird location on blur ratio, excess attenuation, and signal-to-noise ratio in blackbird song. *Journal of the Acoustical Society of America* 93:2206–2220.

Dabelsteen, T., P. K. McGregor, H. M. Lampe, N. E. Langmore, and J. Holland. 1998. Quiet song in song birds: An overlooked phenomenon. *Bioacoustics* 9:89–105.

Dabelsteen, T., and S. B. Pedersen. 1985. Correspondence between messages in the full song of the blackbird *Turdus merula* and meanings to territorial males, as inferred from responses to computerized modifications of natural song. *Zeitschrift für Tierpsychologie* 69:149–165.

Dabelsteen, T., and S. B. Pedersen. 1988. Song parts adapted to function both at long and short ranges may communicate information about the species to female blackbirds *Turdus merula. Ornis Scandinavica* 19:195–198.

Dabelsteen, T., and S. B. Pedersen. 1990. Song and information about aggressive responses of blackbirds, *Turdus merula*: Evidence from interactive playback experiments with territory owners. *Animal Behaviour* 40:1158–1168.

Dalziell, A. H., and A. Cockburn. 2008. Dawn song in superb fairy-wrens: A bird that seeks extrapair copulations during the dawn chorus. *Animal Behaviour* 75:489–500.

Darwin, C. 1859. *On the Origin of Species by Means of Natural Selection, Or The Preservation of Favoured Races in the Struggle for Life.* Murray, London.

Del-Rio, G., M. A. Rêgo, and L. F. Silveira. 2015. A multiscale approach indicates a severe reduction in Atlantic Forest wetlands and highlights that São Paulo Marsh Antwren is on the brink of extinction. *PLoS One* 10: Article e0121315.

Derryberry, E. P. 2007. Evolution of bird song affects signal efficacy: An experimental test using historical and current signals. *Evolution* 61:1938–1945.

Dooling, R. J., and S. H. Blumenrath. 2013. Avian sound perception in noise. Pages 229–250 in H. Brumm, ed. *Animal Communication and Noise.* Springer, Berlin.

Dowling, J., and M. S. Webster. 2018. Acoustic and physical mate guarding have different effects on intruder behaviour in a duetting songbird. *Animal Behaviour* 135:69–75.

Edwards, S. V., and W. E. Boles. 2002. Out of Gondwana: The origin of passerine birds. *Trends in Ecology & Evolution* 17:347–349.

Elie, J. E., M. M. Mariette, H. A. Soula, S. C. Griffith, N. Mathevon, and C. Vignal. 2010. Vocal communication at the nest between mates in wild zebra finches: A private vocal duet? *Animal Behaviour* 80:597–605.

Ericson, P. G. P., L. Christidis, A. Cooper, M. Irestedt, J. Jackson, U. S. Johansson, and J. A. Norman. 2002. A Gondwanan origin of passerine birds supported by DNA sequences of the endemic New Zealand wrens. *Proceedings of the Royal Society B: Biological Sciences* 269:235–241.

Eriksson, D., and L. Wallin. 1986. Male bird song attracts females—A field experiment. *Behavioral Ecology and Sociobiology* 19:297–299.
Falls, J. B. 1988. Does song deter territorial intrusion in white-throated sparrows (*Zonotrichia albicollis*)? *Canadian Journal of Zoology* 66:206–211.
Fjeldså, J. 2013. The global diversification of songbirds (Oscines) and the build-up of the Sino-Himalayan diversity hotspot. *Chinese Birds* 4:132–143.
Garamszegi, L. Z., D. Z. Pavlova, M. Eens, and A. P. Moller. 2007. The evolution of song in female birds in Europe. *Behavioral Ecology* 18:86–96.
Geberzahn, N., W. Goymann, C. Muck, and C. Ten Cate. 2009. Females alter their song when challenged in a sex-role reversed bird species. *Behavioral Ecology and Sociobiology* 64:193–204.
Göransson, G., G. Högstedt, J. Karlsson, H. Källander, and S. Ulfstrand. 1974. Sångens roll för revirhållandet hos näktergal Luscinia luscinia–några experiment med playback–teknik. *Vår fågelvärld* 33:201–209.
Grafe, T. U., and J. H. Bitz. 2004. Functions of duetting in the tropical boubou, *Laniarius aethiopicus*: Territorial defence and mutual mate guarding. *Animal Behaviour* 68:193–201.
Haines, C. D., E. M. Rose, K. J. Odom, and K. E. Omland. 2020. The role of diversity in science: A case study of women advancing female birdsong research. *Animal Behaviour* 168:19–24.
Halkin, S. L. 1997. Nest-vicinity song exchanges may coordinate biparental care of northern cardinals. *Animal Behaviour* 54:189–198.
Hall, M. L. 2000. The function of duetting in magpie-larks: Conflict, cooperation, or commitment? *Animal Behaviour* 60:667–677.
Hall, M. L., M. R. D. Rittenbach, and S. L. Vehrencamp. 2015. Female song and vocal interactions with males in a neotropical wren. *Frontiers in Ecology and Evolution* 3: Article 12.
Hobson, K. A., and S. G. Sealy. 1990. Female song in the Yellow Warbler. *Condor* 92:259–261.
Hoelzel, A. R. 1986. Song characteristics and response to playback of male and female robins *Erithacus rubecula. Ibis* 128:115–127.
Hof, D., and N. Hazlett. 2010. Low-amplitude song predicts attack in a North American wood warbler. *Animal Behaviour* 80:821–828.
Holland, J., T. Dabelsteen, S. B. Pedersen, and O. N. Larsen. 1998. Degradation of wren *Troglodytes troglodytes* song: Implications for information transfer and ranging. *Journal of the Acoustical Society of America* 103:2154–2166.
Illes, A. E. 2015. Context of female bias in song repertoire size, singing effort, and singing independence in a cooperatively breeding songbird. *Behavioral Ecology and Sociobiology* 69:139–150.
Isler, M. L., G. A. Bravo, and R. T. Brumfield. 2013. Taxonomic revision of Myrmeciza (Aves: Passeriformes: Thamnophilidae) into 12 genera based on phylogenetic, morphological, behavioral, and ecological data. *Zootaxa* 3717:469–497.
Jetz, W., G. H. Thomas, J. B. Joy, K. Hartmann, and A. O. Mooers. 2012. The global diversity of birds in space and time. *Nature* 491:444–448.
Johnson, O., J. T. Howard, and R. T. Brumfield. 2021. Systematics of a Neotropical clade of dead-leaf-foraging antwrens (Aves: Thamnophilidae; Epinecrophylla). *Molecular Phylogenetics and Evolution* 154: Article 106962.
Keenan, E. L., K. J. Odom, M. Araya-Salas, K. G. Horton, M. Strimas-Mackey, M. A. Meatte, N. I. Mann, P. J. B. Slater, J. J. Price, and C. N. Templeton. 2020. Breeding season length predicts duet coordination and consistency in Neotropical wrens (Troglodytidae). *Proceedings of the Royal Society B: Biological Sciences* 287: Article 20202482.
Kirschel, A. N. G., Z. Zanti, Z. T. Harlow, E. E. Vallejo, M. L. Cody, and C. E. Taylor. 2020. Females don't always sing in response to male song, but when they do, they sing to males with higher-pitched songs. *Animal Behaviour* 166:129–138.
Kornreich, A., M. Youngblood, P. C. Mundinger, and D. C. Lahti. 2020. Female song can be as long and complex as male song in wild House Finches (*Haemorhous mexicanus*). *Wilson Journal of Ornithology* 132:840–849.
Krebs, J. R. 1977. Song and territory in the Great Tit *Parus major*. Pages 47–62 in B. Stonehouse and C. Perrins, eds. *Evolutionary Ecology*. Macmillan, London.

Kroodsma, D. E., and E. H. Miller. 1996. *Ecology and Evolution of Acoustic Communication in Birds.* Comstock, Ithaca, NY.

Kroodsma, D. E., J. M. E. Vielliard, and F. G. Stiles. 2019. Study of bird sounds in the Neotropics: Urgency and opportunity. Pages 269–282 in E. K. Donald and H. M. Edward, eds. *Ecology and Evolution of Acoustic Communication in Birds.* Cornell University Press, Ithaca, NY.

Langmore, N. E. 1998. Functions of duet and solo songs of *female birds. Trends in Ecology & Evolution* 13:136–140.

Langmore, N. E., and N. B. Davies. 1997. Female dunnocks use vocalizations to compete for males. *Animal Behaviour* 53:881–890.

Langmore, N. E., N. B. Davies, B. J. Hatchwell, and I. R. Hartley. 1996. Female song attracts males in the alpine accentor *Prunella collaris. Proceedings of the Royal Society B: Biological Sciences* 263:141–146.

Levin, R. N. 1996a. Song behaviour and reproductive strategies in a duetting wren, *Thryothorus nigricapillus*: I. Removal experiments. *Animal Behaviour* 52:1093–1106.

Levin, R. N. 1996b. Song behaviour and reproductive strategies in a duetting wren, *Thryothorus nigricapillus*: II. Playback experiments. *Animal Behaviour* 52:1107–1117.

Liu, W.-C., and D. E. Kroodsma. 2007. Dawn and daytime singing behavior of chipping sparrows (*Spizella passerina*). *The Auk* 124:44–52.

Loning, H., S. C. Griffith, and M. Naguib. 2021. Zebra finch song is a very short-range signal in the wild: Evidence from an integrated approach. *Behavioral Ecology* 33:37–46.

Luther, D. A., R. Danner, J. Danner, K. Gentry, and E. P. Derryberry. 2017. The relative response of songbirds to shifts in song amplitude and song minimum frequency. *Behavioral Ecology* 28:391–397.

Macedo, G., G. A. Bravo, R. S. Marcondes, E. P. Derryberry, and C. Biondo. 2021. Differences in plumage coloration predict female but not male territorial responses in three antbird sister species pairs. *Animal Behaviour* 182:107–124.

Mattos, J. C. F., M. M. Vale, M. B. Vecchi, and M. A. S. Alves. 2009. Abundance, distribution and conservation of the Restinga Antwren *Formicivora littoralis. Bird Conservation International* 19:392–400.

McDonald, M. V. 1989. Function of song in Scott's seaside sparrow, *Ammodramus maritimus peninsulae. Animal Behaviour* 38:468–485.

McElroy, D. B., and G. Ritchison. 1996. Effect of mate removal on singing behavior and movement patterns of female northern cardinals. *Wilson Bulletin* 108:550–555.

Mennill, D. J., and S. L. Vehrencamp. 2008. Context-dependent functions of avian duets revealed by microphone-array recordings and multispeaker playback. *Current Biology* 18:1314–1319.

Moncrieff, A. E., O. Johnson, D. F. Lane, J. R. Beck, F. Angulo, and J. Fagan. 2018. A new species of antbird (Passeriformes: Thamnophilidae) from the Cordillera Azul, San Martín, Peru. *The Auk* 135:114–126.

Morton, E. S. 2000. An evolutionary view of the origins and functions of avian vocal communication. *Japanese Journal of Ornithology* 49:69–78, 99.

Morton, E. S. 2019. A comparison of vocal behavior among tropical and temperate passerine birds. Pages 258–268 in E. K. Donald and H. M. Edward, eds. *Ecology and Evolution of Acoustic Communication in Birds.* Cornell University Press, Ithaca, NY.

Morton, E. S., and K. C. Derrickson. 1996. Song ranging by the dusky antbird, *Cercomacra tyrannina*: Ranging without song learning. *Behavioral Ecology and Sociobiology* 39:195–201.

Morton, E. S., K. C. Derrickson, and B. J. M. Stutchbury. 2000. Territory switching behavior in a sedentary tropical passerine, the dusky antbird (*Cercomacra tyrannina*). *Behavioral Ecology* 11:648–653.

Naguib, M., and R. H. Wiley. 2001. Estimating the distance to a source of sound: Mechanisms and adaptations for long-range communication. *Animal Behaviour* 62:825–837.

Nice, M. M. 1943. *Studies in the Life History of the Song Sparrow: II. The Behavior of the Song Sparrow and Other Passerines.* Dover, New York.

Nowicki, S., W. A. Searcy, and M. Hughes. 1998. The territory defense function of song in song sparrows: A test with the speaker occupation design. *Behaviour* 135:615–628.

Odom, K. J., M. Araya-Salas, L. Benedict, *et al.* 2025. Global incidence of female birdsong is predicted by territoriality and biparental care in songbirds. *Nat Commun* 16: 6157 (2025). https://doi.org/10.1038/s41467-025-60810-5

Odom, K. J., M. Araya-Salas, J. L. Morano, R. A. Ligon, G. M. Leighton, C. C. Taff, A. H. Dalziell, A. C. Billings, R. R. Germain, and M. Pardo. 2021. Comparative bioacoustics: A roadmap for quantifying and comparing animal sounds across diverse taxa. *Biological Reviews* 96:1135–1159.

Odom, K. J., and L. Benedict. 2018. A call to document female bird songs: Applications for diverse fields. *The Auk* 135:314–325.

Odom, K. J., M. L. Hall, K. Riebel, K. E. Omland, and N. E. Langmore. 2014. Female song is widespread and ancestral in songbirds. *Nature Communications* 5: Article 3379.

Odom, K. J., K. E. Omland, D. R. McCaffrey, M. K. Monroe, J. L. Christhilf, N. S. Roberts, and D. M. Logue. 2016. Typical males and unconventional females: Songs and singing behaviors of a tropical, duetting oriole in the breeding and non-breeding season. *Frontiers in Ecology and Evolution* 4: Article 14.

Pavlova, D., R. Pinxten, and M. Eens. 2010. Age-related changes of song traits in female European Starlings (*Sturnus vulgaris*). *Animal Biology* 60:43–59.

Peek, F. W. 1972. An experimental study of the territorial function of vocal and visual display in the male red-winged blackbird (*Agelaius phoeniceus*). *Animal Behaviour* 20:112–118.

Podos, J., and M. Cohn-Haft. 2019. Extremely loud mating songs at close range in white bellbirds. *Current Biology* 29:R1068–R1069.

Price, J. J. 2015. Rethinking our assumptions about the evolution of bird song and other sexually dimorphic signals. *Frontiers in Ecology and Evolution* 3: Article 40.

Price, J. J. 2019. Sex differences in song and plumage color do not evolve through sexual selection alone: New insights from recent research. *Journal of Ornithology* 160:1213–1219.

Reichard, D. G., and R. C. Anderson. 2015. Why signal softly? The structure, function and evolutionary significance of low-amplitude signals. *Animal Behaviour* 105:253–265.

Reichard, D. G., R. J. Rice, E. M. Schultz, and S. E. Schrock. 2013. Low-amplitude songs produced by male dark-eyed juncos (*Junco hyemalis*) differ when sung during intra-and inter-sexual interactions. *Behaviour* 150:1183–1202.

Reichard, D. G., R. J. Rice, C. C. Vanderbilt, and E. D. Ketterson. 2011. Deciphering information encoded in birdsong: Male songbirds with fertile mates respond most strongly to complex, low-amplitude songs used in courtship. *The American Naturalist* 178:478–487.

Reichard, D. G., and J. F. Welklin. 2015. On the existence and potential functions of low-amplitude vocalizations in North American birds. *The Auk* 132:156–166.

Reinert, B. L., M. R. Bornschein, and C. Firkowski. 2007. Distribuição, tamanho populacional, hábitat e conservação do bicudinho-do-brejo *Stymphalornis acutirostris* Bornschein, Reinert e Teixeira, 1995 (Thamnophilidae). *Revista Brasileira de Ornitologia* 15:493–519.

Ręk, P., and T. S. Osiejuk. 2011. Nonpasserine bird produces soft calls and pays retaliation cost. *Behavioral Ecology* 22:657–662.

Riebel, K. 2016. Understanding sex differences in form and function of bird song: The importance of studying song learning processes. *Frontiers in Ecology and Evolution* 4: Article 62.

Riebel, K., M. L. Hall, and N. E. Langmore. 2005. Female songbirds still struggling to be heard. *Trends in Ecology & Evolution* 20:419–420.

Riebel, K., K. J. Odom, N. E. Langmore, and M. L. Hall. 2019. New insights from female bird song: Towards an integrated approach to studying male and female communication roles. *Biology Letters* 15: Article 20190059.

Ritchison, G. 1983. The function of singing in female black-headed grosbeaks (*Pheucticus melanocephalus*): Family-group maintenance. *The Auk* 100:105–116.

Ritschard, M., K. Riebel, and H. Brumm. 2010. Female zebra finches prefer high-amplitude song. *Animal Behaviour* 79:877–883.

Rivera-Cáceres, K. D., E. Quirós-Guerrero, M. Araya-Salas, C. N. Templeton, and W. A. Searcy. 2018. Early development of vocal interaction rules in a duetting songbird. *Royal Society Open Science* 5: Article 171791.

Rivera-Cáceres, K. D., and C. N. Templeton. 2019. A duetting perspective on avian song learning. *Behavioural Processes* 163:71–80.

Robinson, A. 1949. The biological significance of bird song in Australia. *Emu-Austral Ornithology* 48:291–315.
Rogers, A. C., N. E. Langmore, and R. A. Mulder. 2006. Function of pair duets in the eastern whipbird: Cooperative defense or sexual conflict? *Behavioral Ecology* 18:182–188.
Rose, E. M., D. A. Coss, C. D. Haines, S. A. Danquah, C. E. Studds, and K. E. Omland. 2019. Why do females sing? Pair communication and other song functions in eastern bluebirds. *Behavioral Ecology* 30:1653–1661.
Rose, E. M., N. H. Prior, and G. F. Ball. 2021. The singing question: Re-conceptualizing birdsong. *Biological Reviews* 97:326–342.
Sandell, M. I., and H. G. Smith. 1997. Female aggression in the European starling during the breeding season. *Animal Behaviour* 53:13–23.
Searcy, W. A. 1992. Measuring responses of female birds to male song. Pages 175–189 in P. McGregor, ed. *Playback and Studies of Animal Communication*. Springer, New York.
Searcy, W. A. 1996. Sound-pressure levels and song preferences in female red-winged blackbirds (*Agelaius phoeniceus*) (Aves, Emberizidae). *Ethology* 102:187–196.
Searcy, W. A., R. C. Anderson, and S. Nowicki. 2006. Bird song as a signal of aggressive intent. *Behavioral Ecology and Sociobiology* 60:234–241.
Searcy, W. A., and M. Andersson. 1986. Sexual selection and the evolution of song. *Annual Review of Ecology and Systematics* 17:507–533.
Searcy, W. A., and S. Nowicki. 2006. Signal interception and the use of soft song in aggressive interactions. *Ethology* 112:865–872.
Searcy, W. A., and S. Nowicki. 2010. *The Evolution of Animal Communication*. Princeton University Press, Princeton, NJ.
Sick, H. 1997. *Ornitologia Brasileira* [*Brazilian ornithology*]. Editora Nova Fronteira, Rio de Janeiro, Brazil.
Sikora, J. G., M. J. Moyer, K. E. Omland, and E. M. Rose. 2021. Large female song repertoires and within-pair song type sharing in a temperate breeding songbird. *Ethology* 127:166–175.
Slater, P. J. B., and N. I. Mann. 2004. Why do the females of many bird species sing in the tropics? *Journal of Avian Biology* 35:289–294.
Snow, D. W. 1958. The breeding of the Blackbird *Turdus merula* at Oxford. *Ibis* 100:1–30.
Sonnenschein, E., and H. U. Reyer. 1983. Mate-guarding and other functions of antiphonal duets in the Slate-Coloured Boubou (*Laniarius funebris*) 1. *Zeitschrift für Tierpsychologie* **63**:112–140.
Sossinka, R., and J. Böhner. 1980. Song types in the zebra finch *Poephila guttata castanotis* 1. *Zeitschrift für Tierpsychologie* 53:123–132.
Spector, D. A. 1994. Definition in biology: The case of "Bird Song." *Journal of Theoretical Biology* 168:373–381.
Stutchbury, B. J., and E. S. Morton. 2001. *Behavioral Ecology of Tropical Birds*. Academic Press, San Diego, CA.
Templeton, C. N., Ç. Akçay, S. E. Campbell, and M. D. Beecher. 2012. Soft song is a reliable signal of aggressive intent in song sparrows. *Behavioral Ecology and Sociobiology* 66:1503–1509.
Templeton, C. N., A. A. Ríos-Chelén, E. Quirós-Guerrero, N. I. Mann, and P. J. B. Slater. 2013. Female happy wrens select songs to cooperate with their mates rather than confront intruders. *Biology Letters* 9: Article 20120863.
Templeton, C. N., K. D. Rivera-Cáceres, N. I. Mann, and P. J. B. Slater. 2011. Song duets function primarily as cooperative displays in pairs of happy wrens. *Animal Behaviour* 82:1399–1407.
Thorpe, W. H. 1961. *Bird-Song: The Biology of Vocal Communication and Expression in Birds*. Cambridge University Press, Cambridge, UK.
Titus, R. C. 1998. Short-range and long-range songs: Use of two acoustically distinct song classes by dark-eyed juncos. *The Auk* 115:386–393.
Tobias, J. A., V. Gamarra-Toledo, D. García-Olaechea, P. C. Pulgarin, and N. Seddon. 2011. Year-round resource defence and the evolution of male and female song in suboscine birds: Social armaments are mutual ornaments. *Journal of Evolutionary Biology* 24:2118–2138.
van den Heuvel, I. M., M. I. Cherry, and G. M. Klump. 2014. Land or lover? Territorial defence and mutual mate guarding in the crimson-breasted shrike. *Behavioral Ecology and Sociobiology* 68:373–381.

Westcott, D. 1992. Inter- and intra-sexual selection: The role of song in a lek mating system. *Animal Behaviour* 44:695–703.
West-Eberhard, M. J. 1979. Sexual selection, social competition, and evolution. *Proceedings of the American Philosophical Society* 123:222–234.
West-Eberhard, M. J. 1983. Sexual selection, social competition, and speciation. *Quarterly Review of Biology* 58:155–183.
Wheeldon, A., P. Szymański, A. Surmacki, and T. S. Osiejuk. 2021. Song type and song type matching are important for joint territorial defense in a duetting songbird. *Behavioral Ecology* 32:883–894.
Whitney, B. M., and M. E. C. Haft. 2013. Fifteen new species of Amazonian birds. Pages 225–239 in *Handbook of the Birds of the World: Special Volume: New Species and Global Index*. Field Guides, Austin, TX.
Wiley, R. H., and D. G. Richards. 1982. Adaptations for acoustic communication in birds: Transmission and signal detection. Pages 131–181 in D. E. Kroodsma and E. H. Miller, eds. *Acoustic Communication in Birds*, Vol. 1. Academic Press, New York.
Wilkins, M. R., K. J. Odom, L. Benedict, and R. J. Safran. 2020. Analysis of female song provides insight into the evolution of sex differences in a widely studied songbird. *Animal Behaviour* 168:69–82.
Xia, C., J. Liu, P. Alström, Q. Wu, and Y. Zhang. 2013. Is the soft song of the brownish-flanked bush warbler an aggressive signal? *Ethology* 119:653–661.
Yasukawa, K. 1981. Song and territory defense in the red-winged blackbird. *The Auk* 98:185–187.
Zann, R. A. 1996. *The Zebra Finch: A Synthesis of Field and Laboratory Studies*. Oxford University Press, New York.
Zimmer, K. J., and M. L. Isler. 2003. Family Thamnophilidae (typical antbirds). Pages 448–681 in J. Del Hoyo, A. Elliott, and D. A. Christie, eds. *Handbook of the Birds of the World: Vol. 8. Broadbills to Tapaculos*. Lynx Edicions, Barcelona, Spain.
Zollinger, S. A., and H. Brumm. 2015. Why birds sing loud songs and why they sometimes don't. *Animal Behaviour* 105:289–295.

15
New Frontiers in Avian Color Research

Mary Caswell Stoddard

Long before they piqued the interest of Darwin and his contemporaries (Bortolotti 2006), colorful birds influenced the art, literature, and ceremonies of many cultural groups throughout the world (Tidemann and Gosler 2010). The appeal is understandable. Birds are the most colorful land vertebrates, and they live on every continent and occupy a vast range of ecological niches. As research on animal coloration blossomed in the 20th century—and then skyrocketed in recent decades (Cuthill et al. 2017)—birds have played a central role. A landmark two-volume tome on bird coloration published in 2006 (Hill and McGraw 2006a, b) provided a comprehensive overview of the mechanisms, functions, and evolution of bird color. In the years since, avian color research has continued on a steep growth trajectory. Relative to 2006, the most conspicuous advances have been in genetics: Next-generation sequencing and access to high-quality bird genomes are revolutionizing our understanding of avian color. However, recent breakthroughs go well beyond genetics. Here, I outline seven new frontiers in avian color research, with a focus on recent literature (of the past 10–15 years). In each of these areas—diversity and evolution, genetics, phenomics, unusual mechanisms and functions, color vision, climate change, and conservation and innovation—researchers are propelling the field forward at a rapid pace. In doing so, they are employing integrative approaches and achieving a fully multidisciplinary vision for avian color research (Bennett and Théry 2007). For each frontier, I review recent progress and highlight opportunities for the future, as we look ahead to the next era of avian color research.

Exploring the Diversity and Evolution of the Avian Color Gamut

The diverse colors in avian feathers (Figure 15.1), bare parts (e.g., skin, bills, wattles, gapes, and eyes), and eggs (Figure 15.2) arise from pigmentary and structural mechanisms (reviewed in Hill and McGraw 2006a, Burns et al. 2017, Price-Waldman and Stoddard 2021). The major pigments in bird feathers are melanins and carotenoids (Figures 15.1A–15.1D). Birds make eumelanin and pheomelanin—which produce black/brown and rusty red colors, respectively—in their bodies (Figures 15.1A and 15.1B). By contrast, birds must acquire carotenoids—which generate red, yellow, and orange colors—from the diet (Figures 15.1C and 15.1D). After digestion, carotenoids can be incorporated directly into the feathers or further metabolized. In addition to melanins and carotenoids, some birds evolved unusual pigments to color their feathers (McGraw 2006): psittacofulvins in parrots (Figure 15.2A); (probable) pterins

Mary Caswell Stoddard, *New Frontiers in Avian Color Research*. In: *New Perspectives in Ornithology*.
Edited by: Scott V. Edwards and J. Michael Reed, Oxford University Press. © Oxford University Press (2025).
DOI: 10.1093/oso/9780197787670.003.0015

Figure 15.1 Birds produce plumage colors using diverse mechanisms. Melanin pigments are responsible for (A) the black and gray plumage of the Swallow-Tailed Gull (*Creagrus furcatus*) and (B) the brown plumage of the Burrowing Owl (*Athene cunicularia*). Carotenoid pigments, which are acquired in the diet and deposited in feathers, are found in (C) the pink plumage of the Roseate Spoonbill (*Platalea ajaja*) and (D) the orange and yellow plumage of the Western Tanager (*Piranga ludoviciana*). Structural colors, produced by the interaction of light with nanoscale structures in the feather, are responsible for (E) the blue plumage of the Blue Jay (*Cyanocitta cristata*) and (F) the magenta, iridescent gorget of the Broad-Tailed Hummingbird (*Selasphorus platycercus*). Photographs by M. C. Stoddard.

in penguins (Thomas, McGoverin, et al. 2013); porphyrins in turacos, owls, bustards, and nightjars; and even vitamin A in a tropical starling (Galván et al. 2019). Pigments also play a role in the colors of bare parts and eggs (Figures 15.2B–15.2D). Colorful bare parts often contain melanins or carotenoids (Iverson and Karubian 2017, Nicolaï et al. 2020); in some birds, the iris is colored by chromatophores packed with purine and pteridine pigments (Oliphant et al. 1992, Corbett et al. 2022). The major eggshell pigments—protoporphyrin (brown) and biliverdin (blue-green)—are porphyrins (Sparks 2011).

Structural colors offer birds another route to strikingly vibrant phenotypes (reviewed in Prum 2006, Saranathan and Finet 2021) (Figures 15.1E and 15.1F). In feather barbs, coherent scattering of light by quasi-ordered arrangements of spongy keratin and air produces non-iridescent ultraviolet, violet, blue, and green plumage (Prum 2006). In feather barbules, coherent scattering of light by arrays of air, keratin, and melanin produces iridescent plumage (Prum 2006). In bare parts, coherent scattering by dermal collagen arrays in the skin produces a range of diverse colors (Prum and Torres 2003). Finally, many bird colors stem from the interaction of pigment and structure (Shawkey and D'Alba 2017). This is the case in Blue-Footed Boobies (*Sula nebouxii*), in which carotenoids and structurally colored skin combine to color their eponymous feet (Velando et al. 2006) (Figure 15.2D).

How did these different color-producing mechanisms evolve in birds? An initial estimate of the avian plumage gamut (Stoddard and Prum 2011)—the full range of feather colors produced by birds—revealed that early birds were probably drably colored with melanins. Over time, the independent evolution of carotenoid pigments and structural colors in multiple lineages allowed birds to greatly expand the range of possible plumage colors. More recent studies—buoyed by comprehensive phylogenies of birds (Jetz et al. 2012, Prum et al. 2015)—have deepened our understanding of how avian color evolved. Broadscale comparative studies of carotenoid plumage (Thomas et al. 2014), carotenoid bare parts (Davis and Clarke 2022), carotenoid metabolic networks (Morrison and Badyaev 2018), and iridescent plumage (Nordén et al. 2021, Venable et al. 2022) have helped pinpoint the origins and distributions of color across birds. In parallel, comparative analyses across large radiations have explored plumage evolution in the contexts of sexual selection and natural selection (Dale et al. 2015, Dunn et al. 2015, Cooney et al. 2019), ecological factors (Delhey et al. 2019, E. Miller et al. 2019, Cooney et al. 2022, He et al. 2022), and speciation rate (Price-Waldman et al. 2020, Beltrán et al. 2021, Cally et al. 2021).

Alongside these developments in birds, the field of paleocolor exploded (reviewed in Roy et al. 2020, Vinther 2020). Several recent paleocolor studies are changing the way we think about bird color evolution. These include discoveries of early feather iridescence—for example, in non-avian theropod dinosaurs *Microraptor* (Li et al. 2012) and *Caihong* (Hu et al. 2018) and in stem trogons (Nordén et al. 2018)—and unappreciated structural color in the glossy feathers of fossil and extant paleognaths (Eliason and Clarke 2020). In addition, eggshell pigments—once thought to be a unique feature of bird eggs—were discovered in non-avian dinosaurs (Wiemann et al. 2017, 2018, 2019; but see Shawkey and D'Alba 2019), raising the intriguing possibility that many dinosaurs evolved egg color in response to pressure from predators or brood parasites. Finally, although carotenoid pigments have not been detected

Figure 15.2 Several color-producing mechanisms in birds are ripe for future study. These include unusual pigments, such as the psittacofulvins that give parrots—including (A) the Red-Masked Parakeet (*Psittacara erythrogenys*)—their red, yellow, and/or orange plumage colors. In addition, relatively little is known about the genetic bases of color in eggs and bare parts, examples of which are diverse across birds. Shown here are (B) the glossy turquoise eggs of the Great Tinamou (*Tinamus major*), (C) the red gular pouch of the Great Frigatebird (*Fregata minor*), and (D) the bright blue feet of the Blue-Footed Booby (*Sula nebouxii*).

Photographs by M. C. Stoddard (A, C, D) and David Ocampo Rincón (B).

in dinosaur fossils, such discoveries could be on the horizon if better methods are developed (Vinther 2020), with recent work suggesting that carotenoids may have been present in the skin—but probably not the feathers—of non-avian dinosaurs (Davis and Clarke 2022). Overall, research on the origins of color in birds and other dinosaurs has set the stage for breakthroughs in other domains, including genetics and genomics.

Uncovering the Genetic Bases of Avian Color

During the past decade, the number of bird species for which reference genomes are available has ballooned from 2 to more than 500 (Bravo et al. 2021). We are living in the golden age of avian genomics, which is radically transforming our understanding of avian color (reviewed in Price-Waldman and Stoddard [2021] and summarized in this section). Initial work on the genetic underpinnings of plumage color focused on melanin. Using candidate gene approaches, researchers uncovered several genes influencing melanin expression in birds, including melanocortin-1 receptor (*MC1R*), agouti-signaling protein (*ASIP*), and tyrosinase (*TYR*) (Mundy 2005, Hubbard et al. 2010, Roulin and Ducrest 2013). These early insights translated to robust progress in decoding the melanin pathway (McNamara et al. 2021), in part because melanin is ubiquitous in vertebrates. In contrast, much less is known about the genes controlling other colorful pigments in birds.

Encouragingly, times are changing. High-throughput sequencing—increasingly applied to non-model systems (Funk and Taylor 2019, Orteu and Jiggins 2020)—has helped unlock the genetic bases of two pigment groups: carotenoids, which are widespread in birds, and psittacofulvins, which are unique to parrots. Studies in domestic "red factor" canaries (Lopes et al. 2016)—a hybrid of Red Siskins (*Spinus cucullatus*) and Common Canaries (*Serinus canaria*)—and "yellowbeak" Zebra Finch (*Taeniopygia guttata*) mutants (Mundy et al. 2016) revealed that *CYP2J19*, which encodes a cytochrome P450 enzyme that converts dietary yellow carotenoids to red ketocarotenoids, plays a vital role in the red coloration of feathers and beaks; see also Toomey et al. (2022), which demonstrated that a second enzyme, *BDH1L*, acts together with *CYP2J19* to achieve this conversion in birds. Subsequent work on diverse wild bird species (e.g., Khalil et al. 2020, Aguillon et al. 2021) suggests that *CYP2J19* has a prominent effect on red carotenoid colors in diverse avian lineages. Another key gene controlling carotenoid color is *BCO2*, which encodes the enzyme β-carotene oxygenase 2. This enzyme breaks down colored carotenoids into colorless apocarotenoids, so its reduced activity can result in pale or yellow coloration. After the discovery that a mutation in *BCO2* produces yellow legs in chickens (*Gallus gallus domesticus*) (Eriksson et al. 2008), *BCO2* was linked to interspecific carotenoid plumage differences in *Vermivora* wood warblers (Toews et al. 2016) and *Setophaga* warblers (Baiz et al. 2021). Remarkably, *BCO2* also appears to influence sexual dichromatism in *mosaic* canaries (Gazda et al. 2020) and control a color polymorphism in nestling beak color of Darwin's Finches (Thraupidae) (Enbody et al. 2021). In parrots, bright red, orange, and yellow plumage colors stem not from diet-derived carotenoids but, rather, from psittacofulvin pigments (Figure 15.2B). The genetic mechanisms responsible for psittacofulvins were elusive until a recent study (Cooke et al. 2017) pinpointed a gene controlling yellow plumage color in budgerigars: the polyketide synthase gene (*MuPKS*). This gene is present in birds without psittacofulvin pigments, suggesting that parrots co-opted *MuPKS* (which serves an unknown function in other birds) for yellow pigmentation by modifying its expression (Cooke et al. 2017, Mundy 2018).

New insights into the genetics of plumage color are enhancing our understanding of speciation. An unexpected revelation has been that in multiple avian systems, just a handful of pigmentation genes differentiate closely related species (reviewed in Funk and Taylor 2019). For example, a genome-wide comparison of Blue-Winged (*Vermivora cyanoptera*) and Golden-Winged Warblers (*V. chrysoptera*), which hybridize in eastern North America, showed only six divergent genomic regions—four of which could be linked to genes related to feather development or pigmentation, including *ASIP* and *BCO2* (Toews et al. 2016). Differences in *ASIP* regulation correlated perfectly with differences in throat color, which is black in the Golden-Winged Warbler and yellow in the Blue-Winged Warbler (Toews et al. 2016). *ASIP* and *BCO2* are associated with melanin- and carotenoid-based plumage differences, respectively, across 36 *Setophaga* warbler species (Baiz et al. 2021)—and for two of these species, the single gene block ASIP-RALY determines the colors of multiple plumage patches on the body (S. Wang et al. 2020). An emerging theme is that gene expression, rather than changes in the gene protein-coding regions themselves, dictates variation in plumage color among closely related species (Funk and Taylor 2019).

Genetic analyses are also helping solve a puzzle about the origin of carotenoid-processing genes in birds. Carotenoid pigments are not restricted to the bare parts and feathers of many birds: They also exist in their eyes. Birds and turtles possess retinal cone oil droplets in the retina; in birds, three of the four color cone types are associated with oil droplets pigmented with carotenoids, which serve to filter certain wavelengths and refine color discrimination (reviewed in Toomey and Corbo 2017). Astaxanthin—the red carotenoid coloring the oil droplet of the long-wave-sensitive (LWS) cone type in birds—is modified by *CYP2J19*, the same ketocarotenoid gene responsible for red carotenoid-based color in the plumage and beaks of several bird species (Emerling 2018; Twyman, Andersson, et al. 2018; Twyman, Prager, et al. 2018). *CYP2J19* and its orthologs appear to be widely distributed across birds and turtles, respectively—even in species that lack red coloration (Twyman et al. 2016; Twyman, Andersson, et al. 2018). This suggests that *CYP2J19* initially evolved to facilitate oil droplet pigmentation in the context of color vision and was later co-opted independently in birds and turtles for carotenoid-based color in the integument (Lopes et al. 2016; Twyman et al. 2016; Twyman, Andersson, et al. 2018). Further work on the shared genetic mechanism for carotenoid-mediated color vision and integument coloration is likely to be fruitful, especially as we learn more about the effects of oil droplets on perception (Caves et al. 2020, Toomey and Ronald 2021).

Despite dramatic advances in our understanding of feather pigmentation genetics—and their links to speciation and color vision—many mysteries about avian color genetics remain. Three areas are particularly ripe for future study. First, the genetic mechanisms underlying structural color (Figures 15.1E and 15.1F) in birds are virtually unknown, in part because the developmental processes are still enigmatic (reviewed in Saranathan and Finet 2021). Structural colors in bird feathers are hypothesized to arise through self-assembly processes that result in nanoscale quasi-ordered arrays of keratin and air in feather barbs (non-iridescent structural colors) or layers of melanosomes and keratin in feather barbules (iridescent structural colors). Given this, genes associated with keratin and melanosome morphology may prove to play important roles in structural color development. Support for this idea comes from a new transcriptomic study in Superb Starlings (*Lamprotornis superbus*); relative to non-iridescent reddish feathers, genes related to cellular organization, keratin, and melanosome shape were upregulated in iridescent blue feathers (Rubenstein et al. 2021). Second, identifying the genes controlling plumage patterns—across the body (macropatterning) and within body regions or feathers (micropatterning)—represents a critical next step for avian color genetics research (Inaba and Chuong 2020, Mason and Bowie 2020, Price-Waldman and Stoddard 2021). Progress here may be imminent: A recent study in Estrildid finches (Hidalgo et al. 2022) revealed that diverse color patterns can emerge from differential regulation of genes associated with distinct "color domains" and their underlying embryonic skin prepatterns. Third, the genetic bases of bare part and egg coloration are largely unknown (Figures 15.2B–15.2D), but tantalizing studies—including those implicating *SLC2A11B* in the pearl iris colors of pigeons (*Columba livia*) (Andrade et al. 2021, Maclary et al. 2021, Si et al. 2021) and *SLCO1B3* in the blue eggs of chickens (Z. Wang et al. 2013, Wragg et al. 2013) but not ducks (Chen et al. 2020, Liu et al. 2020)—suggest that there is still a great deal to discover about the genes influencing these traits.

Quantifying the Phenotype in New Ways

As the genetic mechanisms underlying avian color come into focus, the need for high-quality descriptions of avian phenotypes will only intensify. A useful way of conceptualizing plumage color is in terms of the hierarchical levels of a feather (Terrill and Schultz 2023): Feathers have a characteristic color, pattern, and arrangement on a bird's body, and each individual feather has a macrostructure, microstructure, nanostructure, and molecular composition—all of which can influence color. The toolkit available to researchers for quantifying bird color has never been more robust (reviewed in Burns et al. 2017), with methods for analyzing feathers at each hierarchical level. For example, at the molecular level, high-performance liquid chromatography and Raman spectroscopy are the main technologies for identifying avian pigments. At the nano- and microscales, transmission electron microscopy and scanning electron microscopy enable detailed study of a feather's barbules and barbs. Of these techniques, Raman spectroscopy is noteworthy for its relatively recent emergence on the avian color scene. A nondestructive, laser-based technique, Raman spectroscopy has been used to analyze carotenoid (Thomas, McGraw, et al. 2013, Thomas et al. 2014), melanin (Galván and Jorge 2015), psittacofulvin (Tay et al. 2021), and eggshell pigments (Thomas et al. 2015) in diverse taxa. The method has also shed light on "color-tuning" (Burns et al. 2017), when identical pigments held in different molecular configurations may produce strikingly different plumage colors (Mendes-Pinto et al. 2012, Barnsley et al. 2018).

Turning now to more macroscale assessments of avian color, spectrophotometry has long been the gold standard for measuring the reflectance properties of feathers across bird-visible wavelengths (300–700 nm) (Andersson and Prager 2006). It transformed the study of avian color by revealing unappreciated ultraviolet reflectance in the plumage of diverse species (Bennett and Théry 2007). However, spectrophotometry suffers from a major drawback: It measures color over a small area (point source) and does not capture spatial information. Consequently, carefully calibrated digital cameras modified to capture ultraviolet light have become popular for quantifying avian color in a way that preserves two-dimensional (2D) and three-dimensional (3D) spatial information (Stevens et al. 2007). This approach—often called multispectral imaging because light is captured across multiple bands, such as red, green, blue, and ultraviolet—and the impressive software tools it spawned (Troscianko and

Stevens 2015, Berg et al. 2020) unlocked new ways of analyzing spatial patterns in animals (reviewed in Stoddard and Osorio 2019), especially in birds (reviewed in Mason and Bowie 2020). Multispectral cameras are accessible and efficient, making high-throughput studies of avian color possible at an unprecedented scale. In a recent study, multispectral cameras were used to photograph more than 24,000 bird specimens representing more than 4,500 species at the Natural History Museum at Tring, United Kingdom (He et al. 2022). After using a deep learning approach to segment the images (separating the specimen from the image background), the researchers analyzed the overall ultraviolet reflectance of passerine species and discovered that more ultraviolet-reflecting species tend to inhabit forests or locations with high incident ultraviolet radiation. The disadvantage of multispectral imaging is that it lacks spectral resolution; because reflectance is captured in a few bands, it is not possible to determine the reflectance spectrum for a given pixel in the image.

Hyperspectral imaging offers both spectral and spatial resolution—and is likely the next frontier in animal color quantification. A hyperspectral imaging system divides the light spectrum into a series of narrow bands (representing a few wavelengths) so that the final image can be visualized as a 3D stack (or data cube) in which each layer (or slice) represents reflectance in a narrow band. Each pixel in the final image corresponds to a high-resolution reflectance spectrum (Foster and Amano 2019). Because hyperspectral cameras are costly and the data sets are cumbersome, their application to animal color studies has generally been limited—but studies on cuttlefish (Chiao et al. 2011) and crab (Russell and Dierssen 2015) camouflage, jewel beetle (*Sternocera aequisignata*) iridescence (Kjernsmo et al. 2020), and fish biodiversity (Kolmann et al. 2021) illustrate the promise of a hyperspectral approach. Going one step further, mapping hyperspectral data to 3D models of specimens would enable a whole new way of interpreting color, pattern, and morphology. In a pioneering study, Kim et al. (2012) showed that this is possible: They integrated hyperspectral imaging and 3D scanning to produce a hyperspectral 3D model of a Stella's Lorikeet (*Charmosyna stellae goliathina*). Avian color researchers have been slow to adopt the highly customized and technical pipeline of Kim et al. (2012), but off-the-shelf hyperspectral cameras—coupled with photogrammetry (a cheaper option than 3D scanning; J. Medina et al. 2020)—provide a powerful alternative, as demonstrated in a recent study by Hogan and Stoddard (2024). Focusing on the extravagant plumage of a rare hybrid Bird-of-Paradise (*Cicinnurus magnificus* x *Cicinnurus regius*), they developed a novel pipeline for capturing and analyzing hyperspectral data between 325–700 nm—a range covering much of the bird-visible light spectrum—and showed how these data can be combined with detailed 3D models.

The elephant bird in the room here is that spectrophotometry, multispectral imaging, and hyperspectral imaging are methods well-suited to stationary animals—but birds move. Hummingbirds epitomize the problem: The appearance of a male's iridescent throat feathers during a courtship display will often depend on his position relative to the female and the sun. In many cases, a complete description of a bird's colorful phenotype should include signaling geometry and motion (Rosenthal 2007, Hutton et al. 2015, Echeverri et al. 2021). Recent studies of hummingbird shuttle (Simpson and McGraw 2018a, 2018b, 2019) and dive (Hogan and Stoddard 2018) displays—combining field-based approaches with multispectral imaging—represent steps in this direction. Nonetheless, quantifying iridescence remains tricky (Meadows et al. 2011; Gruson, Andraud, et al. 2019; Stuart-Fox et al. 2021). Looking ahead, virtual reality and computer animation techniques (Chouinard-Thuly et al. 2017, Stowers et al. 2017) could revolutionize some aspects of avian color research. As a first step toward creating "bird vision" virtual models of bird specimens, A. Miller et al. (2022) recently combined photogrammetry, multispectral imaging, and animation to produce animated 3D digital models containing color information relevant to birds (in the ultraviolet and human-visible spectrum). In the future, researchers could reproduce and dissect complex courtship displays in a virtual space or design sophisticated 3D stimuli for color vision experiments—with appropriate caveats. Most display screens do not emit ultraviolet light, so ensuring that screen-based stimuli appear realistic to birds is as challenging today as it was 20 years ago (Cuthill et al. 2000, Chouinard-Thuly et al. 2017).

Appreciating Unusual Mechanisms and Functions of Avian Color

The diverse mechanisms and functions of avian color have been reviewed broadly (Hill and McGraw 2006a, 2006b; Hill 2018; Roy et al. 2020; Terrill and Schultz 2023) and with specific attention to genetic mechanisms (Ng and Li 2018, Price-Waldman and Stoddard 2021), carotenoids (Toews et al. 2017, Toomey and Corbo 2017), melanins (Galván and Solano 2016), structural color (Saranathan and Finet 2021), pattern (Funk and Taylor 2019, Inaba and Chuong 2020, Mason and Bowie 2020), eggs (Kilner 2006), and bare parts (Iverson and Karubian 2017, Corbett et al. 2022). Here, I highlight recent breakthroughs related to atypical or mysterious color mechanisms and functions.

On the mechanistic front, discoveries about pigments, nanostructures, pigment–structure interactions, and microstructural modifications are updating our ideas about how birds make color. Just two eggshell pigments—blue–green biliverdin and red–brown protoporphyrin—have long been deemed responsible for all egg colors in birds (Hanley et al. 2015). However, researchers overturned this dogma when they discovered two novel pigments in tinamou eggshells: yellow–brown bilirubin from the green eggs of the Elegant Crested Tinamou (*Eudromia elegans*) and red-orange uroerythrin from the brown eggs of the Spotted Nothura (*Nothura maculosa*) (Hamchand et al. 2020). Both pigments co-occur with biliverdin and are sensitive to light exposure, a feature that could be adaptive if color-changing eggs help communally nesting females locate fresh nests. Novel carotenoid pigments were also recently discovered in the purple feathers of cotingas (Cotingidae) and in the yellow and crimson feathers of broadbills (Eurylaimidae) (reviewed in LaFountain et al. 2015).

New nanostructures are also being discovered. A 3D photonic crystal unknown in bird feathers—called a single gyroid—was recently described in the blue and green feather barbs of Blue-Winged Leafbirds (*Chloropsis cochinchinensis*) (Saranathan et al. 2021). Single gyroid networks are more highly ordered than most "typical" non-iridescent structural colors in birds, which stem from quasi-ordered arrangements of keratin and air in feather barbs. In iridescent feathers, too, detailed studies of nanostructures have been revealing. For example, melanosomes—the melanin-filled nanostructures in iridescent feather barbules—come in a variety of shapes (thin solid rods, hollow rods, solid platelets, hollow platelets) that differ from the thick ancestral rods in plain black feathers. Why? Building on key insights about melanosome diversity in select clades (Eliason et al. 2013; Maia et al. 2013; Gammie 2014; Eliason et al. 2015; Gruson, Elias, et al. 2019; Eliason et al. 2020), Nordén et al. (2021) mapped melanosome attributes onto a broad-scale phylogeny of 280 species. This, combined with optical modeling and plumage measurements, revealed that the ticket to brilliant iridescence is not a particular melanosome shape per se but, rather, melanosomes arranged in thin melanin layers. Thin melanin layers can be achieved by making melanosomes thin, hollow or platelet-shaped, such that birds have a flexible nanostructural toolkit for iridescent color production. In addition to barbs and barbules, structural mechanisms in the central shaft of the feather—the rachis—have recently been shown to produce the glossy sheen of Cassowary (*Casuarius casuarius*) feathers (Eliason and Clarke 2020) and the blue color of Great Argus (*Argusianus argus*) flight feathers (Eliason et al. 2023).

Finally, two overlooked but important influences on avian color—interactions of different mechanisms (reviewed in Shawkey et al. 2009, Shawkey and D'Alba 2017; see also Justyn and Weaver 2022) and effects of feather microstructure (reviewed in Price-Waldman and Stoddard 2021, Terrill and Schultz 2023)—are receiving increased attention. For example, the green feather colors of Budgerigars (*Melopsittacus undulatus*) arise from interactions of a structural blue-green color (from quasi-ordered spongy keratin arrays) and a yellow psittacofulvin pigment (D'Alba et al. 2012). The pigment absorbs the blue wavelengths of the structural color so that the resulting color appears green. This twist on the classic explanation—mixing of structural blue and pigmentary yellow to make green—suggests that there is still much to learn about combination pigment-structure colors, which in birds can produce colors unattainable by pigment or structure alone (Stoddard and Prum 2011). Feather microstructure—the shape of the rachis, barbs, and barbules—has recently been shown to alter the glossiness, saturation, hue, and absorbance properties of feathers (reviewed in Terrill and Schultz 2023). For example, in the male Lawes's Parotia (*Parotia lawesii*), whose shimmery breast plate resembles a disco ball, the unusual boomerang shape of the barbules amplifies the iridescent effect (Stavenga et al. 2011). In male *Ramphocelus* tanagers, enlarged, oval-shaped barbs enhance the saturation of carotenoid-based feathers (McCoy et al. 2021). We currently lack detailed comparative studies of feather microstructure across genera and families, so this area is ripe for future research. A final surprise about color-producing mechanisms in feathers comes from a recent study of tanagers (Price-Waldman et al. 2025). In these vibrantly colored birds, a "hidden" layer of white or black achromatic plumage is concealed beneath carotenoid-pigmented or structurally-colored feather tips. Remarkably, the color of the hidden layer varies systematically over the body: white plumage underlies carotenoid color and enhances its brightness, while black plumage underlies structural color and enhances its saturation. Thus, tanagers evolved an optical trick to make their feather colors really pop—a strategy that may be widespread. Hidden feather layers appear to be common in passerines and probably shaped the evolution of plumage color in ways we are just beginning to appreciate.

On the functional front, a wealth of new discoveries reminds us that bird color has evolved not just for mate choice (where research efforts are largely focused) but also for camouflage, thermoregulation, social signaling, mimicry, and more. What follows is a small sampling from a rich smorgasbord of recent studies on avian color function. In the realm of mate choice, Amat et al. (2011) discovered that Greater Flamingos (*Phoenicopterus roseus*) use carotenoid-containing preen gland secretions as "makeup" to make their feathers more colorful, a compelling case of cosmetic coloration (Delhey et al. 2007). Mayani-Parás et al. (2015) demonstrated a new mechanism of egg camouflage in Blue-Footed Boobies, which coat their blue–white eggs with soil to make them match their background substrate.

Drawing an intriguing link between color and thermoregulation, Delhey et al. (2021) showed that migratory bird species have paler plumage than nonmigratory species; presumably, migrants are under strong selection to stay cool during long-distance flights. Similarly, I. Medina et al. (2018) revealed that Australian bird species in hot and dry environments have feathers that reflect more near-infrared (NIR) light, helping to keep them cool. Although invisible to most animals, NIR reflectance (700–2,500 nm) is highly correlated with reflectance in the bird-visible wavelengths (300–700 nm), so selection for thermoregulatory efficiency in the NIR could impact signal production in the bird-visible spectrum. Finally, the iridescent feathers of sunbirds (Nectariniidae) heat up more than unstructured melanin-based feathers; this is likely due to the high overall melanin content of melanosomes in iridescent feathers (Rogalla et al. 2021). For a review of thermal effects on plumage color, see Rogalla et al. (2022).

In a comprehensive field study of color and social behavior, Falk et al. (2021) unraveled a mystery in White-Necked Jacobins (*Florisuga mellivora*), in which approximately 20% of adult females have male-like plumage. Male-like females received less aggression and enjoyed greater access to feeders, supporting the hypothesis that male-like ornamentation evolved in female Jacobins to reduce social harassment. Finally, Londoño et al. (2015) described a stunning example of possible Batesian mimicry in Cinereous Mourner (*Laniocera hypopyrra*) chicks. Covered in unusual bright orange downy feathers, the chicks closely resemble toxic, hairy caterpillars in the vicinity—which could deter predators.

Embracing New Perspectives on Avian Color Vision

In keeping with their impressive capacity for color production, birds have excellent color vision. Birds have four single color cone types, each characterized by an opsin photopigment (Hart and Hunt 2007, Bowmaker 2008). The cones include the ultraviolet- or violet-sensitive cone type (UVS or VS, containing the SWS1 opsin), the "blue" shortwave-sensitive cone type (SWS, containing the SWS2 opsin), the "green" medium-wave-sensitive cone type (MWS, containing the RH2 opsin), and the "red" LWS cone type (containing the LWS opsin). Because we humans often (erroneously) consider our sensory worlds to be the norm, we tend to think of birds as having enhanced, turbocharged color vision. In fact, birds retained the four-color cone type system of early vertebrate ancestors (reviewed in Baden and Osorio 2019). It is humans who are somewhat deviant; our comparatively meager color vision (with three color cone types) results from the loss of two color cones early in mammalian evolution followed by a gain via duplication of a third color cone in Old World primates (reviewed in Davies et al. 2012).

In birds, each color cone type contains an oil droplet (Toomey et al. 2015, Toomey and Corbo 2017). The oil droplet is transparent in the UVS/VS cone but colored by carotenoids in the SWS, MWS, and LWS cone types, where it serves as an optical filter; it narrows the spectral sensitivities of those cones, leading to more refined color discrimination. In addition to the four single color cone types, birds have rods (containing RH1 rhodopsin) for low-light vision and double cones (containing LWS opsin and a UV-blocking oil droplet) thought to mediate achromatic perception (Hart and Hunt 2007). Bird color vision has been reviewed in several classic (Goldsmith 1990, 2006; Kelber et al. 2003; Cuthill 2006; Hart and Hunt 2007) and recent papers (Bloch 2016, Toomey and Corbo 2017, Baden and Osorio 2019, Kelber 2019, Toomey and Ronald 2021). In this section, I focus on several emerging new perspectives on avian color vision.

First, molecular and genomic studies of the cone opsins are painting a more nuanced picture of interspecific variation in color perception. The prevailing view has been that the spectral sensitivities of the opsin pigments in bird color cones are largely invariant across diverse species—except with respect to SWS1, which comes in two varieties (based on a few amino acid substitutions) and confers either UVS (peak pigment sensitivity 355–373 nm) or VS (peak pigment sensitivity 402–426 nm) vision (Hart and Hunt 2007). In other words, except for the UVS/VS distinction, different bird species vary little in their color vision. However, this appears to be an oversimplification. Even in closely related species, there can be small coding differences in opsin gene sequences; for example, Bloch and colleagues identified six amino acid differences in the SWS2 genes of Old World and New World warblers (Bloch, Morrow, et al. 2015) and six amino acid differences in the RH2 genes of brightly colored *Setophaga* warblers compared to other Parulidae species (Bloch, Price, et al. 2015). When the different SWS2 and RH2 genes were expressed in vitro, the amino acid changes did not influence the absorbance spectra of the opsins—so whether these amino acid substitutions actually affect color vision is unclear. Nevertheless, opsin-based spectral tuning in birds might be more variable than previously appreciated.

Even so, to focus purely on opsin coding variation may be a mistake. Differences in opsin expression, rather than in opsin genes themselves, might influence color perception—a possibility suggested by a further study of New World warblers (Bloch 2015). Intriguingly, Bloch (2015) found that opsin expression levels of SWS2 differed between the sexes and across warbler species. In females, SWS2 expression correlated with the degree of sexual dichromatism across species—hinting that sexual selection may have shaped female color perception in this New World warbler group. Oil droplet filtering can also have a major effect on how different species perceive color. For example, Toomey et al. (2016) showed that shifts to UVS vision, which evolved from ancestral VS vision many times in birds, involve not only well-documented changes in the SWS1 gene sequence (UVS/VS cone) but also changes to the oil droplet filtering of the SWS (blue) cone. In UVS birds, modified carotenoid content in the oil droplets of SWS cones shifts SWS sensitivity toward shorter wavelengths, which enhances color discrimination by ensuring that the cones are relatively evenly spaced across the light spectrum. Even among individuals of the same species, variation in oil droplet carotenoid content might influence color vision (Caves et al. 2020), although the effects are probably subtle in most contexts (Toomey and Ronald 2021). Finally, genomic and transcriptomic studies have revealed that some opsin genes have been wholesale lost or pseudogenized in various groups, including some owls, vultures, kites, kiwis, penguins, and passerines (reviewed in Kelber 2019). As an aside, opsin genes appear to be difficult to sequence, yielding unexpected patterns of presence and absence in several comparative genomic studies (Borges et al. 2015, Feng et al. 2020). Taken together, the studies discussed here are dismantling the view that birds share the same color vision experience, and future work will undoubtedly add fuel to the fire.

A second paradigm shift underway involves understanding color processing *beyond* the color cones. For color vision to work, the outputs of the color photoreceptors have to be compared: This is a fundamental principle of color vision and an underlying assumption of animal color vision models (Kelber 2016, Renoult et al. 2017). Opponent processing, which occurs when photoreceptor outputs are compared antagonistically by neurons, is mediated by bipolar, horizontal, and retinal ganglion cells in mammal systems (mice and primates are the best studied) (Baden and Osorio 2019). However, in birds, we have no knowledge of the (probable) opponent channels (Kelber 2019, Price et al. 2019). It is difficult to overstate how vast this knowledge gap is given that birds "in many ways are the pinnacle of vertebrate retinal complexity" (Baden et al. 2020, p. 16). Electrophysiological recordings in birds are challenging, which has hindered progress, but some insights may come from turtles, which, like birds, have four single color cone types. Rocha et al. (2008) demonstrated that at least five color-opponent channels mediated by ganglion cells contribute to turtle color vision.

Uncovering the neural basis of avian color vision should be a top priority moving forward because it will help clarify whether birds are, in fact, tetrachromatic. Birds are often considered to be tetrachromatic because they have four color cone types, but inferring the dimensionality of color vision from the number of cone types is problematic (Bloch 2016, Jacobs 2018). Many behavioral and color-matching experiments suggest avian tetrachromacy—which is further supported by the recent demonstration that wild hummingbirds can discriminate nonspectral colors such as ultraviolet + green and ultraviolet + red (Stoddard et al. 2020). Nevertheless, we lack the definitive color-mixing experiments and neural studies required to confirm avian tetrachromacy—although these may be close at hand. In addition, recent work on Zebra Finches suggests that some birds may have categorical perception (Caves et al. 2018, 2021; Zipple et al. 2019), which enhances discrimination of colors belonging to different categories. Examining higher level processes—which go beyond retinal processing—will be a vital, if challenging, goal for future work.

A last emerging perspective is this: We must study avian color vision in the wild, where birds grapple with the real-world challenges of finding food and mates in complex environments. For good reason, most psychophysical experiments on bird color vision have been performed in indoor labs. However, we now have the tools to take these experiments out of the lab and into the field. Inspired by classic work by Goldsmith and colleagues (Goldsmith and Goldsmith 1979, Goldsmith 1980, Goldsmith et al. 1981), Stoddard et al. (2020) conducted color vision tests on free-flying Broad-Tailed Hummingbirds (*Selasphorus platycercus*) in the Colorado Rocky Mountains. They designed lightweight, field-portable TetraColorTubes capable of displaying a broad range of bird-visible colors, such as ultraviolet + green. By training hummingbirds to associate a colored light with a reward, they demonstrated that wild hummingbirds can discriminate a broad range of nonspectral colors that we humans can only imagine. Moving forward, "wild psychophysics" experiments will provide a powerful complement to lab-based approaches and should shed light on how birds respond to color in nature.

Understanding How Avian Color Will Be Impacted by Climate Change

How will avian color be affected by climate change? This is one of the most pressing questions about bird coloration—but there may be no simple answer. Two ecogeographic rules are relevant to color and climate (reviewed in Delhey et al. 2019, 2020; Romano et al. 2019). The first is Gloger's rule, which suggests that darker species should occur in humid, wet, and hot environments. Darker animals may be better camouflaged in these vegetation-rich habitats, although the mechanisms underlying Gloger's rule are unclear (Delhey 2019). The second is Bogert's rule, or the "thermal melanism hypothesis," which suggests that darker species should occur in cold environments, where additional pigmentation helps animals warm up quickly and cool down slowly. So as temperatures rise, will animals get darker (consistent with Gloger's rule) or lighter (consistent with Bogert's rule)? Arguments have been made for both scenarios (Delhey et al. 2020; Tian and Benton 2020a, 2020b), but the general consensus is that there are complex interactions between Gloger's and Bogert's rules, so the predicted effects of climate change on animal color are thorny—and will vary in different geographic regions. Neither rule addresses pale-colored species (or morphs) that are cryptic against snow; with reduced snow cover due to warming, these animals could darken to avoid a camouflage mismatch (Mills et al. 2018, Zimova et al. 2018).

Several studies on plumage highlight the complex relationships between color and climate change. In Finland, researchers studying polymorphic Tawny Owls (*Strix aluco*), which come in rusty brown and gray morphs, documented a sharp increase in the frequency of rusty morphs in recent decades (Karell et al. 2011). In snowy years, rusty morphs (with more pheomelanin pigment) survived less well than gray morphs. With fewer recent snowy years due to climate change, rusty morphs proliferated—evidence, the researchers argue, of a microevolutionary response to climate change. Why rusty morphs survived poorly in snowy years—and are now on the rise—is not obvious (Karell et al. 2011; Solonen, 2021a, 2021b). Rusty morphs may be poorly camouflaged in snow—an idea supported by an online citizen science study (Koskenpato et al. 2020)—but the main predators of Tawny Owls are nocturnal Eagle Owls (*Bubo bubo*) that do not rely on vision for hunting. Alternatively, rusty color may be linked to another trait (e.g., immune function, metabolism, or energetic demands) that is the actual target of selection in snowy winters. In contrast to Tawny Owls, barn owls (*Tyto* spp.)—which occur in brown and white morphs—may get lighter as temperatures rise. A recent study of three evolutionarily distinct barn owl lineages revealed that darker (brown) morphs generally occurred in wet and cold regions, consistent with Gloger's rule (for precipitation) and Bogert's rule, respectively (Romano et al. 2019). As the climate warms, the frequency of light-colored morphs is expected to increase, although dark morphs may prevail in high-humidity regions. In this sense, melanin-based pigments—like those responsible for barn owl color—could serve as a proxy for current and future adaptation to climate change.

Egg coloration may also be impacted by climate change. A recent study of egg color in 634 globally distributed species showed that birds tend to lay darker (typically brown) eggs in cold regions and lighter eggs (typically white or blue) in warm, dry regions, consistent with Bogert's rule (Wisocki et al. 2019). However, this is a twist on the traditional "thermal melanism hypothesis" because the pigments in bird eggs—brown protoporphyrin and blue biliverdin—lack melanin (Delhey 2020). Nonetheless, darker eggs heated up quickly and cooled down slowly in experiments, further indicating that temperature is a major driver of egg color variation across birds. As temperatures rise, will egg colors become lighter? Will climate-induced egg color change disrupt camouflage? Eggs are a highly tractable system for exploring these questions. Comparing distribution data of breeding birds across decades could reveal—similar to findings in dragonflies (Zeuss et al. 2014)—that climate change is already selecting for lighter egg colors.

As we have seen, climate change can alter heritable color traits. But climate change can transform bird plumage colors in nonheritable ways, too. A remarkable analysis of more than 1,300 bird specimens from museum collections showed that black carbon (or soot)—a major contributor to climate change because it traps heat in the atmosphere—darkened the feathers of five common species in the United States Rust Belt (DuBay and Fuldner 2017). The birds were sootiest in the early 1900s, when black carbon levels peaked in this region, and cleanest in the early 2000s, by which point black carbon levels (but not total coal consumption, which persisted in new forms) had declined. The fact that bird specimens provided clues about pollution over a span of 135 years is a powerful testament to museum collections—and a sobering reminder of human impacts on avian color and communication.

In addition to climate change, other human activities, such as urbanization and pollution, can influence bird color in unexpected ways (reviewed in Leveau 2021). Individual feral pigeons in more urban habitats tend to have darker, more melanin-rich plumage (Obukhova 2011), potentially as an adaptation to parasite loads (Jacquin et al. 2011). Individual male Northern Cardinals (*Cardinalis cardinalis*; Jones et al. 2010) and Eurasian Kestrel (*Falco tinnunculus*) chicks (Sumasgutner et al. 2018) are duller in more urban landscapes, perhaps because cities change access to carotenoid-based food sources. Eggs and even nests are changing color as a result of human activities. The colors of individual Herring Gull (*Larus argentatus*; Hanley and Doucet 2012) and Eurasian Sparrowhawk (*Accipiter nisus*; Jagannath et al. 2007) eggs reflect the degree of contamination with harmful compounds, including a metabolite of the insect pesticide dichlorodiphenyltrichloroethane (DDT). Depressingly, plastic litter (in a variety of unnatural colors) was prevalent in the nests of breeding seabirds—on a West African island uninhabited by humans (Tavares et al. 2019). How these changes in egg and nest color will affect camouflage, incubation behavior, and chick provisioning is largely unknown.

Applying Avian Color Research to Conservation and Innovation

The applications of animal color research are diverse and far-reaching (Caro et al. 2017), and birds provide plenty of inspiration. In two broad domains—conservation and innovation—research on bird vision and coloration is unlocking new possibilities. Collisions with glass windows and buildings is a major threat to avian life, killing hundreds of millions of birds each year in the United States (Loss et al. 2014). A deeper appreciation of bird vision—including ultraviolet sensitivity and pattern perception—spurred the development of bird-friendly window treatments (e.g., glass with embedded ultraviolet and fritted patterns) (Klem and Saenger 2013, Sheppard 2019, Swaddle et al. 2020, Brown et al. 2021), which can be highly effective. At the Javits Center in New York City, bird-friendly glass reduced bird fatalities by approximately 90%—a success story that may be repeated in light of new citywide regulations requiring "bird-safe" construction in future buildings (Piselli 2020). However, there is still much to learn about how birds perceive glass. For example, birds possess either a UVS or VS color cone type, so ultraviolet-reflecting window markings will not look the same to all species and probably appear invisible to VS birds under most lighting conditions (Håstad and Ödeen 2014).

Other promising interventions—such as light-emitting diodes, which can prevent bird collisions with communication towers, wind turbines, and power lines—are being deployed, motivated by careful studies of color vision in target species (Doppler et al. 2015, Goller et al. 2018). Sometimes the goal is to attract rather than deter birds. Painted decoys have been used to successfully recruit seabirds to desirable breeding locations, but these studies (reviewed in Friesen et al. 2017) rarely consider whether decoys look realistic to bird vision—and (spoiler alert) they probably do not (Stoddard et al. 2019). However, sensory ecology-based conservation approaches are rapidly becoming mainstream (Friesen et al. 2017, Dominoni et al. 2020, Elmer et al. 2021). A final point about conservation is that humans are drawn to colorful animals (Curtin and Papworth 2020), including colorful and intricately patterned birds (Lišková and Frynta 2013, Lišková et al. 2015). Conservationists can capitalize on this preference when selecting flagship species for fundraising campaigns and educational programs (Garnett et al. 2018).

One of the most exciting areas of applied avian color research involves the design of synthetic photonic structures, which are used in optical lenses, sensors, display screens, fiber-optic cables, solar cells, fabrics, and paints. In recent decades, the field of photonics, which explores the flow and manipulation of light, has been galvanized by examples of structural color in nature. Excellent reviews (Dumanli and Savin 2016, McDougal et al. 2019, Xiao et al. 2020) have highlighted the ways in which engineers have been influenced (or not) by structural colors in nature, particularly in birds. A key message is that although humans *can* produce impressive optical materials without mimicking nature, a bioinspired approach is often advantageous because natural structural colors are multifunctional, dynamic, adaptive, and eco-friendly. In birds, non-iridescent blue and green plumage colors arise from quasi-ordered arrangements of keratin and air in feather barbs; the sphere- and channel-like structures appear to self-assemble through liquid–liquid phase separation (Dufresne et al. 2009, Prum et al. 2009). Because producing self-assembled synthetic materials is desirable but challenging—particularly at the nano-length scales possible in feathers—deriving lessons from birds could prove to be game-changing (Fernández-Rico et al. 2022, Saranathan and Finet 2021, Sicher et al. 2021). In addition, researchers have recently discovered a new way in which birds produce weakly iridescent blues and greens. The feather barbs of blue-winged leafbirds contain networks of single gyroids (Saranathan et al. 2021), a type of 3D photonic crystal that has excellent light-capturing properties. Single gyroids are very difficult to make artificially, but leafbirds, whose single gyroids probably arise through self-assembly, could inspire new synthetic approaches.

Engineers are also borrowing design ideas from melanin-based materials (reviewed in Xiao et al. 2020), including the iridescent feathers of birds, in which melanin-packed organelles (melanosomes) are precisely arranged in feather barbules to produce iridescence. For example, Khudiyev et al. (2014) used the 2D photonic crystal structure of iridescent green Mallard (*Anas platyrhynchos*) feathers as inspiration for fabricating photonic crystal fibers with similar optical and material properties. Khudiyev et al. (2014) relied on a "top-down" synthetic approach not resembling natural systems per se, but "bottom-up," bio-inspired methods are gaining traction (Dumanli and Savin 2016). A striking example of this is a study by Xiao et al. (2015) demonstrating that synthetic melanin nanoparticles (similar in size to melanosomes in iridescent duck and peacock feathers) could be self-assembled to create structurally colored films ranging from red to green. In addition to iridescence, the discovery of super-black feathers (McCoy et al. 2018)—with modified barbules that trap light via multiple scattering and absorption—in at least 15 avian families (McCoy and Prum 2019) could motivate new classes of ultra-absorbent biomimetic materials.

A last point is that avian color research could have profound effects on biomedicine in the future. Birds are emerging as a model system for studying carotenoid processing (Toews et al. 2017, Toomey and Corbo 2017) and melanin synthesis (Galván and Solano 2016, McNamara et al. 2021), both of which have direct links to human health. For example, carotenoids are an important part of the human diet and can reduce the risk of many diseases, including cardiovascular disease, eye disease, and some cancers (Krinsky and Johnson 2005, Langi et al. 2018). Melanin is present in human skin, hair, and eyes, where it can have protective effects against ultraviolet radiation and oxidative stress. Moreover, because melanin is generally compatible with the human body and sensitive to heat, synthetic melanin-based nanoparticles have a range of applications in biosensing and photothermal therapies (d'Ischia et al. 2015, Xiao et al. 2020). Finally, birds—with their excellent tetrachromatic color perception—could inspire better ways to restore vision, through glasses (Gómez-Robledo et al. 2018) or bionic devices (Nowik et al. 2020), to visually impaired patients.

Conclusion

As I write this, a Northern Cardinal and a Blue Jay (*Cyanocitta cristata*) (Figure 15.1E) are jockeying for access to the backyard birdseed. Why is one bird red and the other blue? In some respects, we can now answer this question with exceptional precision. The cardinal is red because its (newly sequenced) genome contains *CYP2J19* and corresponding regulatory elements (Sin et al. 2020). The Blue Jay is blue because its feather barbs contain nanostructures and a basal melanin layer that together scatter and absorb light, as in the Steller's Jay (*Cyanocitta stelleri*) (Shawkey and Hill 2006). However, many mysteries remain. Did non-avian dinosaurs have carotenoid-colored feathers? What genetic mechanisms control blue structural color? How did different selection pressures and historical contingencies result in Northern Cardinals being red and Blue Jays being blue? As we move into the next phase of animal color research, answers to these questions are within reach. We must also be mindful of our historic fixation on colorful male birds. Work on drab plumage (Marcondes and Brumfield 2019) and on female plumage (Falk et al. 2021, Fargevielle et al. 2023)—long overlooked in the field of avian color—should be priorities for the future. The best research on avian color will come if we diversify our ideas about model systems and model scientists (Duffy et al. 2021). A potent example of this comes from birdsong research, in which female birdsong was largely ignored until women scientists started studying it (Haines et al. 2020). Here, I have highlighted seven new frontiers in avian color research—a dynamic, multidisciplinary, and quickly evolving field that will benefit from creative new recruits. So—to the students reading this—welcome. There has never been a better time to study avian color, and the future looks bright.

Acknowledgments

Thank you to Rosalyn Price-Waldman, Klara Nordén and Lazarena Lazarova for constructive feedback, and thank you to David Ocampo Rincón for contributing Figure 15.2B.

References

Aguillon, S. M., J. Walsh, and I. J. Lovette. 2021. Extensive hybridization reveals multiple coloration genes underlying a complex plumage phenotype. *Proceedings of the Royal Society B: Biological Sciences* 288: Article 20201805.

Amat, J. A., M. A. Rendón, J. Garrido-Fernández, A. Garrido, M. Rendón-Martos, and A. Pérez-Gálvez. 2011. Greater flamingos *Phoenicopterus roseus* use uropygial secretions as make-up. *Behavioral Ecology and Sociobiology* 65:665–673.

Andersson, S., and M. Prager. 2006. Quantifying colors. Pages 41–89 in G. E. Hill and K. J. McGraw, eds. *Bird Coloration: Volume 1. Mechanisms and Measurements.* Harvard University Press, Cambridge, MA.

Andrade, P., M. A. Gazda, P. M. Araújo, S. Afonso, J. A. Rasmussen, C. I. Marques, R. J. Lopes, M. T. P. Gilbert, and M. Carneiro. 2021. Molecular parallelisms between pigmentation in the avian iris and the integument of ectothermic vertebrates. PLoS Genetics 17: Article e1009404.

Baden, T., T. Euler, and P. Berens. 2020. Understanding the retinal basis of vision across species. *Nature Reviews Neuroscience* 21:5–20.

Baden, T., and D. Osorio. 2019. The retinal basis of vertebrate color vision. *Annual Review of Vision Science* 5: 177–200.

Baiz, M. D., A. W. Wood, A. Brelsford, I. J. Lovette, and D. P. L. Toews. 2021. Pigmentation genes show evidence of repeated divergence and multiple bouts of introgression in *Setophaga* warblers. *Current Biology* 31:643–649.

Barnsley, J. E., E. J. Tay, K. C. Gordon, and D. B. Thomas. 2018. Frequency dispersion reveals chromophore diversity and colour-tuning mechanism in parrot feathers. *Royal Society Open Science* 5: Article 172010.

Beltrán, D. F., A. J. Shultz, and J. L. Parra. 2021. Speciation rates are positively correlated with the rate of plumage color evolution in hummingbirds. Evolution 75:1665–1680.

Bennett, A. T. D., and M. Théry. 2007. Avian color vision and coloration: Multidisciplinary evolutionary biology. *The American Naturalist* 169:1–6.

Berg, C. P. v. d., J. Troscianko, J. A. Endler, N. J. Marshall, and K. L. Cheney. 2020. Quantitative colour pattern analysis (QCPA): A comprehensive framework for the analysis of colour patterns in nature. *Methods in Ecology and Evolution* 11:316–332.

Bloch, N. I. 2015. Evolution of opsin expression in birds driven by sexual selection and habitat. *Proceedings of the Royal Society B: Biological Sciences* 282: Article 20142321.

Bloch, N. I. 2016. The evolution of opsins and color vision: Connecting genotype to a complex phenotype. *Acta Biológica Colombiana* 21:481–494.

Bloch, N. I., J. M. Morrow, B. S. W. Chang, and T. D. Price. 2015. SWS2 visual pigment evolution as a test of historically contingent patterns of plumage color evolution in warblers. *Evolution* 69:341–356.

Bloch, N. I., T. D. Price, and B. S. W. Chang. 2015. Evolutionary dynamics of Rh2 opsins in birds demonstrate an episode of accelerated evolution in the New World warblers (*Setophaga*). *Molecular Ecology* 24:2449–2462.

Borges, R., I. Khan, W. E. Johnson, M. T. P. Gilbert, G. Zhang, E. D. Jarvis, S. J. O'Brien, and A. Antunes. 2015. Gene loss, adaptive evolution and the co-evolution of plumage coloration genes with opsins in birds. *BMC Genomics* 16:751.

Bortolotti, G. R. 2006. Natural selection and coloration: Protection, concealment, advertisement, or deception. Pages 3–35 in G. E. Hill and K. J. McGraw, eds. *Bird Coloration: Volume 2. Function and Evolution.* Harvard University Press, Cambridge, MA.

Bowmaker, J. K. 2008. Evolution of vertebrate visual pigments. *Vision Research* 48:2022–2041.

Bravo, G. A., C. J. Schmitt, and S. V. Edwards. 2021. What have we learned from the first 500 avian genomes? *Annual Review of Ecology, Evolution, and Systematics* 52:1–29.

Brown, B. B., S. Santos, and N. Ocampo-Penuela. 2021. Bird-window collisions: Mitigation efficacy and risk factors across two years. *PeerJ* 9: Article e11867.

Burns, K. J., K. J. McGraw, A. J. Shultz, M. C. Stoddard, D. B. Thomas, and M. W. Webster. 2017. Advanced methods for studying pigments and coloration using avian specimens. *Studies in Avian Biology* 50:23–55.

Cally, J. G., D. Stuart-Fox, L. Holman, J. Dale, and I. Medina. 2021. Male-biased sexual selection, but not sexual dichromatism, predicts speciation in birds. *Evolution* 75:931–944.

Caro, T., M. C. Stoddard, and D. Stuart-Fox. 2017. Animal coloration research: Why it matters. *Philosophical Transactions of the Royal Society B: Biological Sciences* 372: Article 20160333.

Caves, E. M., P. A. Green, M. N. Zipple, D. Bharath, S. Peters, S. Johnsen, and S. Nowicki. 2021. Comparison of categorical color perception in two estrildid finches. *The American Naturalist* 197:190–202.

Caves, E. M., P. A. Green, M. N. Zipple, S. Peters, S. Johnsen, and S. Nowicki. 2018. Categorical perception of colour signals in a songbird. *Nature* 560:365–367.

Caves, E. M., L. E. Schweikert, P. A. Green, M. N. Zipple, C. Taboada, S. Peters, S. Nowicki, and S. Johnsen. 2020. Variation in carotenoid-containing retinal oil droplets correlates with variation in perception of carotenoid coloration. *Behavioral Ecology and Sociobiology* 74: Article 93.

Chen, L., X. Gu, X. Huang, R. Liu, J. Li, Y. Hu, G. Li, T. Zeng, Y. Tian, X. Hu, L. Lu, and N. Li. 2020. Two cis-regulatory SNPs upstream of ABCG2 synergistically cause the blue eggshell phenotype in the duck. *PLoS Genetics* 16: Article e1009119.

Chiao, C. C., J. K. Wickiser, J. J. Allen, B. Genter, and R. T. Hanlon. 2011. Hyperspectral imaging of cuttlefish camouflage indicates good color match in the eyes of fish predators. *Proceedings of the National Academy of Sciences of the USA* 108:9148–9153.

Chouinard-Thuly, L., S. Gierszewski, G. G. Rosenthal, S. M. Reader, G. Rieucau, K. L. Woo, R. Gerlai, C. Tedore, S. J. Ingley, J. R. Stowers, J. G. Frommen, F. L. Dolins, and K. Witte. 2017. Technical and conceptual considerations for using animated stimuli in studies of animal behavior. *Current Zoology* 63:5–19.

Cooke, T. F., C. R. Fischer, P. Wu, T.-X. Jiang, K. T. Xie, J. Kuo, E. Doctorov, A. Zehnder, C. Khosla, C.-M. Chuong, and C. D. Bustamante. 2017. Genetic mapping and biochemical basis of yellow feather pigmentation in budgerigars. *Cell* 171:427–439.

Cooney, C. R., Y. He, Z. K. Varley, L. O. Nouri, C. J. A. Moody, M. D. Jardine, A. Liker, T. Székely, and G. H. Thomas. 2022. Latitudinal gradients in avian colourfulness. *Nature Ecology & Evolution* 6:622–629.

Cooney, C. R., Z. K. Varley, L. O. Nouri, C. J. A. Moody, M. D. Jardine, and G. H. Thomas. 2019. Sexual selection predicts the rate and direction of colour divergence in a large avian radiation. *Nature Communications* 10: Article 1773.

Corbett, E. C., R. T. Brumfield, and B. C. Faircloth. 2022. Bird eye color: A rainbow of variation, a spectrum of explanations. *EcoEvoRxiv*.

Curtin, P., and S. Papworth. 2020. Coloring and size influence preferences for imaginary animals, and can predict actual donations to species-specific conservation charities. *Conservation Letters* 13: Article e12723.

Cuthill, I. C. 2006. Color perception. Pages 3–40 in G. E. Hill and K. J. McGraw, eds. *Bird Coloration: Volume 1. Mechanisms and Measurements*. Harvard University Press, Cambridge, MA.

Cuthill, I. C., W. L. Allen, K. Arbuckle, B. Caspers, G. Chaplin, M. E. Hauber, G. E. Hill, N. G. Jablonski, C. D. Jiggins, A. Kelber, J. Mappes, J. Marshall, R. Merrill, D. Osorio, R. Prum, N. W. Roberts, A. Roulin, H. M. Rowland, T. N. Sherratt, J. Skelhorn, M. P. Speed, M. Stevens, M. C. Stoddard, D. Stuart-Fox, L. Talas, E. Tibbetts, and T. I. M. Caro. 2017. The biology of color. Science 357: Article eaan0221.

Cuthill, I. C., N. S. Hart, J. C. Partridge, A. T. D. Bennett, and S. Hunt. 2000. Avian colour vision and avian video playback experiments. *Acta Ethologica* 3:29–37.

D'Alba, L., L. Kieffer, and M. D. Shawkey. 2012. Relative contributions of pigments and biophotonic nanostructures to natural color production: A case study in budgerigar (*Melopsittacus undulatus*) feathers. *Journal of Experimental Biology* 215:1272–1277.

Dale, J., C. J. Dey, K. Delhey, B. Kempenaers, and M. Valcu. 2015. The effects of life history and sexual selection on male and female plumage colouration. *Nature* 527:367–370.

Davies, W. I. L., S. P. Collin, and D. M. Hunt. 2012. Molecular ecology and adaptation of visual photopigments in craniates. *Molecular Ecology* 21:3121–3158.

Davis, S. N., and J. A. Clarke. 2022. Estimating the distribution of carotenoid coloration in skin and integumentary structures of birds and extinct dinosaurs. *Evolution* 76:42–57.

Delhey, K. 2019. A review of Gloger's rule, an ecogeographical rule of colour: Definitions, interpretations and evidence. *Biological Reviews of the Cambridge Philosophical Society* 94: 1294–1316.

Delhey, K. 2020. Darker eggs feel the heat. *Nature Ecology & Evolution* 4:22–23.

Delhey, K., J. Dale, M. Valcu, and B. Kempenaers. 2019. Reconciling ecogeographical rules: Rainfall and temperature predict global colour variation in the largest bird radiation. *Ecology Letters* 22:726–736.

Delhey, K., J. Dale, M. Valcu, and B. Kempenaers. 2020. Why climate change should generally lead to lighter coloured animals. *Current Biology* 30:R1406–R1407.

Delhey, K., J. Dale, M. Valcu, and B. Kempenaers. 2021. Migratory birds are lighter coloured. *Current Biology* 31:R1511–R1512.

Delhey, K., A. Peters, and B. Kempenaers. 2007. Cosmetic coloration in birds: Occurrence, function, and evolution. *The American Naturalist* 169:S145–S158.

d'Ischia, M., K. Wakamatsu, F. Cicoira, E. D. Mauro, J. C. Garcia-Borron, S. Commo, I. Galván, G. Ghanem, K. Kenzo, P. Meredith, A. Pezzella, C. Santato, T. Sarna, J. D. Simon, L. Zecca, F. A. Zucca, A. Napolitano, and S. Ito. 2015. Melanins and melanogenesis: from pigment cells to human health and technological applications. *Pigment Cell & Melanoma Research* 28:520–544.

Dominoni, D. M., W. Halfwerk, E. Baird, R. T. Buxton, E. Fernández-Juricic, K. M. Fristrup, M. F. McKenna, D. J. Mennitt, E. K. Perkin, B. M. Seymoure, D. C. Stoner, J. B. Tennessen, C. A. Toth, L. P. Tyrrell, A. Wilson, C. D. Francis, N. H. Carter, and J. R. Barber. 2020. Why conservation biology can benefit from sensory ecology. *Nature Ecology & Evolution* 4:502–511.

Doppler, M. S., B. F. Blackwell, T. L. DeVault, and E. Fernndez-Juricic. 2015. Cowbird responses to aircraft with lights tuned to their eyes: Implications for birdaircraft collisions. *The Condor* 117:165–177.

DuBay, S. G., and C. C. Fuldner. 2017. Bird specimens track 135 years of atmospheric black carbon and environmental policy. *Proceedings of the National Academy of Sciences of the USA* 114:11321–11326.

Duffy, M. A., C. García-Robledo, S. P. Gordon, N. A. Grant, D. A. Green, A. Kamath, R. M. Penczykowski, M. Rebolleda-Gómez, N. Wale, and L. Zaman. 2021. Model systems in ecology, evolution, and behavior: A call for diversity in our model systems and discipline. *The American Naturalist* 198:53–68.

Dufresne, E. R., H. Noh, V. Saranathan, S. G. J. Mochrie, H. Cao, and R. O. Prum. 2009. Self-assembly of amorphous biophotonic nanostructures by phase separation. *Soft Matter* 5:1792–1795.

Dumanli, A. G., and T. Savin. 2016. Recent advances in the biomimicry of structural colours. *Chemical Society Reviews* 45:6698–6724.

Dunn, P. O., J. K. Armenta, and L. A. Whittingham. 2015. Natural and sexual selection act on different axes of variation in avian plumage color. *Science Advances* 1: Article e1400155.

Echeverri, S. A., A. E. Miller, J. Chen, E. W. McQueen, M. Plakke, M. Spicer, K. L. Hoke, M. C. Stoddard, and N. I. Morehouse. 2021. How signaling geometry shapes the efficacy and evolution of animal communication systems. *Integrative and Comparative Biology* 61:787–813.

Eliason, C. M., P.-P. Bitton, and M. D. Shawkey. 2013. How hollow melanosomes affect iridescent colour production in birds. *Proceedings of the Royal Society B: Biological Sciences* 280: Article 20131505.

Eliason, C. M., and J. A. Clarke. 2020. Cassowary gloss and a novel form of structural color in birds. *Science Advances* 6: Article eaba0187.

Eliason, C. M., J. A. Clarke, and S. A. Kane. 2023. Wrinkle nanostructures generate a novel form of blue structural color in great argus flight feathers. *iScience* 26: Article 105912.

Eliason, C. M., R. Maia, J. L. Parra, and M. D. Shawkey. 2020. Signal evolution and morphological complexity in hummingbirds (Aves: Trochilidae). *Evolution* 74:447–458.

Eliason, C. M., R. Maia, and M. D. Shawkey. 2015. Modular color evolution facilitated by a complex nanostructure in birds. *Evolution* 69:357–367.

Elmer, L. K., C. L. Madliger, D. T. Blumstein, C. K. Elvidge, E. Fernández-Juricic, A. Z. Horodysky, N. S. Johnson, L. P. McGuire, R. R. Swaisgood, and S. J. Cooke. 2021. Exploiting common senses: Sensory ecology meets wildlife conservation and management. *Conservation Physiology* 9: Article coab002.

Emerling, C. A. 2018. Independent pseudogenization of CYP2J19 in penguins, owls and kiwis implicates gene in red carotenoid synthesis. *Molecular Phylogenetics and Evolution* 118:47–53.

Enbody, E. D., C. G. Sprehn, A. Abzhanov, H. Bi, M. P. Dobreva, O. G. Osborne, C.-J. Rubin, P. R. Grant, B. R. Grant, and L. Andersson. 2021. A multispecies BCO2 beak color polymorphism in the Darwin's finch radiation. *Current Biology* 31:5597–5604.

Eriksson, J., G. Larson, U. Gunnarsson, B. Bed'hom, M. Tixier-Boichard, L. Strömstedt, D. Wright, A. Jungerius, A. Vereijken, E. Randi, P. Jensen, and L. Andersson. 2008. Identification of the yellow skin gene reveals a hybrid origin of the domestic chicken. *PLoS Genetics* 4: Article e1000010.

Falk, J. J., M. S. Webster, and D. R. Rubenstein. 2021. Male-like ornamentation in female hummingbirds results from social harassment rather than sexual selection. *Current Biology* 31:4381–4387.

Fargevielle, A., A. Grégoire, D. Gomez, and C. D'outrelant. 2023. Evolution of female colours in birds: The role of female cost of reproduction and paternal care. *Journal of Evolutionary Biology* 36:579–588.

Feng, S., J. Stiller, Y. Deng, J. Armstrong, Q. Fang, A. H. Reeve, D. Xie, G. Chen, C. Guo, B. C. Faircloth, B. Petersen, Z. Wang, Q. Zhou, M. Diekhans, W. Chen, S. Andreu-Sánchez, A. Margaryan, J. T. Howard, C. Parent, G. Pacheco, M.-H. S. Sinding, L. Puetz, E. Cavill, Â. M. Ribeiro, L. Eckhart, J. Fjeldså, P. A. Hosner, R. T. Brumfield, L. Christidis, M. F. Bertelsen, T. Sicheritz-Ponten, D. T.

Tietze, B. C. Robertson, G. Song, G. Borgia, S. Claramunt, I. J. Lovette, S. J. Cowen, P. Njoroge, J. P. Dumbacher, O. A. Ryder, J. Fuchs, M. Bunce, D. W. Burt, J. Cracraft, G. Meng, S. J. Hackett, P. G. Ryan, K. A. Jønsson, I. G. Jamieson, R. R. D. Fonseca, E. L. Braun, P. Houde, S. Mirarab, A. Suh, B. Hansson, S. Ponnikas, H. Sigeman, M. Stervander, P. B. Frandsen, H. V. D. Zwan, R. V. D. Sluis, C. Visser, C. N. Balakrishnan, A. G. Clark, J. W. Fitzpatrick, R. Bowman, N. Chen, A. Cloutier, T. B. Sackton, S. V. Edwards, D. J. Foote, S. B. Shakya, F. H. Sheldon, A. Vignal, A. E. R. Soares, B. Shapiro, J. González-Solís, J. Ferrer-Obiol, J. Rozas, M. Riutort, A. Tigano, V. Friesen, L. Dalén, A. O. Urrutia, T. Székely, Y. Liu, M. G. Campana, A. Corvelo, R. C. Fleischer, K. M. Rutherford, N. J. Gemmell, N. Dussex, H. Mouritsen, N. Thiele, K. Delmore, M. Liedvogel, A. Franke, M. P. Hoeppner, O. Krone, A. M. Fudickar, B. Milá, E. D. Ketterson, A. E. Fidler, G. Friis, Á. M. Parody-Merino, P. F. Battley, M. P. Cox, N. C. B. Lima, F. Prosdocimi, T. L. Parchman, B. A. Schlinger, B. A. Loiselle, J. G. Blake, H. C. Lim, L. B. Day, M. J. Fuxjager, M. W. Baldwin, M. J. Braun, M. Wirthlin, R. B. Dikow, T. B. Ryder, G. Camenisch, L. F. Keller, J. M. DaCosta, M. E. Hauber, M. I. M. Louder, C. C. Witt, J. A. McGuire, J. Mudge, L. C. Megna, M. D. Carling, B. Wang, S. A. Taylor, G. Del-Rio, A. Aleixo, A. T. R. Vasconcelos, C. V. Mello, J. T. Weir, D. Haussler, Q. Li, H. Yang, J. Wang, F. Lei, C. Rahbek, M. T. P. Gilbert, G. R. Graves, E. D. Jarvis, B. Paten, and G. Zhang. 2020. Dense sampling of bird diversity increases power of comparative genomics. *Nature* 587: 252–257.

Fernández-Rico, C., T. Sai, A. Sicher, R. W. Style, and E. R. Dufresne. 2022. Putting the squeeze on phase separation. *JACS Au* 2:66–73.

Foster, D. H., and K. Amano. 2019. Hyperspectral imaging in color vision research: Tutorial. *Journal of the Optical Society of America A* 36:606–627

Friesen, M. R., J. R. Beggs, and A. C. Gaskett. 2017. Sensory-based conservation of seabirds: A review of management strategies and animal behaviours that facilitate success. *Biological Reviews* 92:1769–1784.

Funk, E. R., and S. A. Taylor. 2019. High-throughput sequencing is revealing genetic associations with avian plumage color. *The Auk* 136:1–7.

Galván, I., and A. Jorge. 2015. Dispersive Raman spectroscopy allows the identification and quantification of melanin types. *Ecology and Evolution* 5:1425–1431.

Galván, I., K. Murtada, A. Jorge, Á. Ríos, and M. Zougagh. 2019. Unique evolution of vitamin A as an external pigment in tropical starlings. *Journal of Experimental Biology* 222: Article jeb205229.

Galván, I., and F. Solano. 2016. Bird integumentary melanins: Biosynthesis, forms, function and evolution. *International Journal of Molecular Sciences* 17: Article 520.

Gammie, K. 2014. The evolution of iridescent plumage in the Galliformes: Proximate mechanisms and ultimate functions [Master's thesis]. No. 4908. University of Windsor, Windsor, Ontario, Canada.

Garnett, S. T., G. B. Ainsworth, and K. K. Zander. 2018. Are we choosing the right flagships? The bird species and traits Australians find most attractive. *PLoS One* 13: Article e0199253.

Gazda, M. A., P. M. Araújo, R. J. Lopes, M. B. Toomey, P. Andrade, S. Afonso, C. Marques, L. Nunes, P. Pereira, S. Trigo, G. E. Hill, J. C. Corbo, and M. Carneiro. 2020. A genetic mechanism for sexual dichromatism in birds. *Science* 368:1270–1274.

Goldsmith, T. H. 1980. Hummingbirds see near ultraviolet light. *Science* 207:786–788.

Goldsmith, T. H. 1990. Optimization, constraint, and history in the evolution of eyes. *Quarterly Review of Biology* 65:281–322.

Goldsmith, T. H. 2006. What birds see. *Scientific American* 295:68–75.

Goldsmith, T. H., J. S. Collins, and D. L. Perlman. 1981. A wavelength discrimination function for the hummingbird *Archilochus alexandri*. *Journal of Comparative Physiology* 143:103–110.

Goldsmith, T. H., and K. M. Goldsmith. 1979. Discrimination of colors by the Black-chinned Hummingbird, *Archilochus alexandri*. *Journal of Comparative Physiology* 130:209–220.

Goller, B., B. F. Blackwell, T. L. DeVault, P. E. Baumhardt, and E. Fernández-Juricic. 2018. Assessing bird avoidance of high-contrast lights using a choice test approach: Implications for reducing human-induced avian mortality. *PeerJ* 6: Article e5404.

Gómez-Robledo, L., E. M. Valero, R. Huertas, M. A. Martínez-Domingo, and J. Hernández-Andrés. 2018. Do EnChroma glasses improve color vision for colorblind subjects? *Optics Express* 26: Article 28693.

Gruson, H., C. Andraud, W. D. D. Marcillac, S. Berthier, M. Elias, and D. Gomez. 2019. Quantitative characterization of iridescent colours in biological studies: A novel method using optical theory. *Journal of the Royal Society Interface Focus* 9: Article 20180049.

Gruson, H., M. Elias, C. Andraud, C. Djediat, S. Berthier, C. Doutrelant, and D. Gomez. 2019. Hummingbird iridescence: An unsuspected structural diversity influences colouration at multiple scales. *bioRxiv*: Article 699744.

Haines, C. D., E. M. Rose, K. J. Odom, and K. E. Omland. 2020. The role of diversity in science: A case study of women advancing female birdsong research. *Animal Behaviour* 168: 19–24.

Hamchand, R., D. Hanley, R. O. Prum, and C. Brückner. 2020. Expanding the eggshell colour gamut: Uroerythrin and bilirubin from tinamou (Tinamidae) eggshells. *Scientific Reports* 10: Article 11264.

Hanley, D., and S. M. Doucet. 2012. Does environmental contamination influence egg coloration? A long-term study in herring gulls. *Journal of Applied Ecology* 49:1055–1063.

Hanley, D., T. Grim, P. Cassey, and M. E. Hauber. 2015. Not so colourful after all: Eggshell pigments constrain avian eggshell colour space. *Biology Letters* 11: Article 20150087.

Hart, N. S., and D. M. Hunt. 2007. Avian visual pigments: Characteristics, spectral tuning, and evolution. *The American Naturalist* 169:S7–S26.

Håstad, O., and A. Ödeen. 2014. A vision physiological estimation of ultraviolet window marking visibility to birds. *PeerJ* 2: Article e621.

He, Y., Z. K. Varley, L. O. Nouri, C. J. A. Moody, M. D. Jardine, S. Maddock, G. H. Thomas, and C. R. Cooney. 2022. Deep learning image segmentation reveals patterns of UV reflectance evolution in passerine birds. *Nature Communications* 13: Article 5068.

Hidalgo, M., C. Curantz, N. Quenech'Du, J. Neguer, S. Beck, A. Mohammad, and M. Manceau. 2022. A conserved molecular template underlies color pattern diversity in estrildid finches. *Science Advances* 8: Article eabm5800.

Hill, G. E. 2018. Color and ornamentation. Pages 309–332 in M. L. Morrison, A. D. Rodewald, G. Voelker, M. R. Colón, and J. F. Prather, eds. *Ornithology: Foundation, Analysis, and Application*. Johns Hopkins University Press, Baltimore, MD.

G. E. Hill and K. J. McGraw, eds. 2006a. *Bird Coloration: Volume 1. Mechanisms and Measurements*. Harvard University Press, Cambridge, MA.

G. E. Hill and K. J. McGraw, eds. 2006b. *Bird Coloration: Volume 2. Function and Evolution*. Harvard University Press, Cambridge, MA

Hogan, B. G., and M. C. Stoddard. 2018. Synchronization of speed, sound and iridescent color in a hummingbird aerial courtship dive. *Nature Communications* 9:1–8.

Hogan, B. G., and M. C. Stoddard. 2024. Hyperspectral imaging in animal coloration research: A user-friendly pipeline for image generation, analysis, and integration with 3D modeling. *PLoS Biology* 22(12): Article e3002867.

Hu, D., J. A. Clarke, C. M. Eliason, R. Qiu, Q. Li, M. D. Shawkey, C. Zhao, L. D'Alba, J. Jiang, and X. Xu. 2018. A bony-crested Jurassic dinosaur with evidence of iridescent plumage highlights complexity in early paravian evolution. *Nature Communications* 9: Article 217.

Hubbard, J. K., J. A. C. Uy, M. E. Hauber, H. E. Hoekstra, and R. J. Safran. 2010. Vertebrate pigmentation: From underlying genes to adaptive function. *Trends in Genetics* 26:231–239.

Hutton, P., B. M. Seymoure, K. J. McGraw, R. A. Ligon, and R. K. Simpson. 2015. Dynamic color communication. *Current Opinion in Behavioral Sciences* 6:41–49.

Inaba, M., and C.-M. Chuong. 2020. Avian pigment pattern formation: Developmental control of macro- (across the body) and micro- (within a feather) level of pigment patterns. *Frontiers in Cell and Developmental Biology* 8: Article 620.

Iverson, E. N. K., and J. Karubian. 2017. The role of bare parts in avian signaling. *The Auk* 134:587–611.

Jacobs, G. H. 2018. Photopigments and the dimensionality of animal color vision. *Neuroscience and Biobehavioral Reviews* 86:108–130.

Jacquin, L., P. Lenouvel, C. Haussy, S. Ducatez, and J. Gasparini. 2011. Melanin-based coloration is related to parasite intensity and cellular immune response in an urban free living bird: the feral pigeon *Columba livia*. *Journal of Avian Biology* 42:11–15.

Jagannath, A., R. F. Shore, L. A. Walker, P. N. Ferns, and A. G. Gosler. 2007. Eggshell pigmentation indicates pesticide contamination. *Journal of Applied Ecology* 45:133–140.

Jetz, W., G. H. Thomas, J. B. Joy, K. Hartmann, and A. O. Mooers. 2012. The global diversity of birds in space and time. *Nature* 491:444–448.

Jones, T. M., A. D. Rodewald, and D. P. Shustack. 2010. Variation in plumage coloration of Northern Cardinals in urbanizing landscapes. *Wilson Journal of Ornithology* 122:326–333.

Justyn, N. M., and R. J. Weaver. 2022. Painting the bunting: Carotenoids and structural elements combine to produce the feather coloration of the male Painted Bunting. *Ornithology* 140: Article ukac052.

Karell, P., K. Ahola, T. Karstinen, J. Valkama, and J. E. Brommer. 2011. Climate change drives microevolution in a wild bird. *Nature Communications* 2: Article 208.

Kelber, A. 2016. Colour in the eye of the beholder: Receptor sensitivities and neural circuits underlying colour opponency and colour perception. *Current Opinion in Neurobiology* 41:106–112.

Kelber, A. 2019. Bird colour vision—from cones to perception. *Current Opinion in Behavioral Sciences* 30:34–40.

Kelber, A., M. Vorobyev, and D. Osorio. 2003. Animal colour vision—behavioural tests and physiological concepts. *Biological Reviews of the Cambridge Philosophical Society* 78:81–118.

Khalil, S., J. F. Welklin, K. J. McGraw, J. Boersma, H. Schwabl, M. S. Webster, and J. Karubian. 2020. Testosterone regulates CYP2J19-linked carotenoid signal expression in male red-backed fairywrens (*Malurus melanocephalus*). *Proceedings of the Royal Society B: Biological Sciences* 287: Article 20201687.

Khudiyev, T., T. Dogan, and M. Bayindir. 2014. Biomimicry of multifunctional nanostructures in the neck feathers of mallard (*Anas platyrhynchos* L.) drakes. *Scientific Reports* 4: Article 4718.

Kilner, R. M. 2006. The evolution of egg colour and patterning in birds. *Biological Reviews* 81:383–406.

Kim, M. H., T. A. Harvey, D. S. Kittle, and H. Rushmeier. 2012. 3D imaging spectroscopy for measuring hyperspectral patterns on solid objects. *ACM Transactions on Graphics* 31: Article 38.

Kjernsmo, K., H. M. Whitney, N. E. Scott-Samuel, J. R. Hall, H. Knowles, L. Talas, and I. C. Cuthill. 2020. Iridescence as camouflage. *Current Biology* 30:551–555.

Klem, D., and P. G. Saenger. 2013. Evaluating the effectiveness of select visual signals to prevent bird-window collisions. *Wilson Journal of Ornithology* 125:406–411.

Kolmann, M. A., M. Kalacska, O. Lucanus, L. Sousa, D. Wainwright, J. P. Arroyo-Mora, and M. C. Andrade. 2021. Hyperspectral data as a biodiversity screening tool can differentiate among diverse Neotropical fishes. *Scientific Reports* 11: Article 16157.

Koskenpato, K., A. Lehikoinen, C. Lindstedt, and P. Karell. 2020. Gray plumage color is more cryptic than brown in snowy landscapes in a resident color polymorphic bird. *Ecology and Evolution* 10:1751–1761.

Krinsky, N. I., and E. J. Johnson. 2005. Carotenoid actions and their relation to health and disease. *Molecular Aspects of Medicine* 26:459–516.

LaFountain, A. M., R. O. Prum, and H. A. Frank. 2015. Diversity, physiology, and evolution of avian plumage carotenoids and the role of carotenoid–protein interactions in plumage color appearance. *Archives of Biochemistry and Biophysics* 572:201–212.

Langi, P., S. Kiokias, T. Varzakas, and C. Proestos. 2018. Microbial carotenoids: Methods and protocols. *Methods in Molecular Biology* 1852:57–71.

Leveau, L. 2021. United colours of the city: A review about urbanisation impact on animal colours. *Austral Ecology* 46:670–679.

Li, Q., K.-Q. Gao, Q. Meng, J. A. Clarke, M. D. Shawkey, L. D'Alba, R. Pei, M. Ellison, M. A. Norell, and J. Vinther. 2012. Reconstruction of microraptor and the evolution of iridescent plumage. *Science* 335:1215–1219.

Lišková, S., and D. Frynta. 2013. What determines bird beauty in human eyes? *Anthrozoos* 26:27–41.

Lišková, S., E. Landová, and D. Frynta. 2015. Human preferences for colorful birds: Vivid colors or pattern? *Evolutionary Psychology* 13:339–359.

Liu, H., J. Hu, Z. Guo, W. Fan, Y. Xu, S. Liang, D. Liu, Y. Zhang, M. Xie, J. Tang, W. Huang, Q. Zhang, Y. Xi, Y. Li, L. Wang, S. Ma, Y. Jiang, Y. Feng, Y. Wu, J. Cao, Z. Zhou, and S. Hou. 2020. A single nucleotide polymorphism variant located in the cis-regulatory region of the ABCG2 gene is associated with mallard egg color. *Molecular Ecology* 30:1477–1491.

Londoño, G. A., D. A. García, and M. A. S. Martínez. 2015. Morphological and behavioral evidence of Batesian mimicry in nestlings of a lowland Amazonian bird. *The American Naturalist* 185:135–141.

Lopes, R. J., J. D. Johnson, M. B. Toomey, M. S. Ferreira, P. M. Araujo, J. Melo-Ferreira, L. Andersson, G. E. Hill, J. C. Corbo, and M. Carneiro. 2016. Genetic basis for red coloration in birds. *Current Biology* 26:1427–1434.

Loss, S. R., T. Will, and P. P. Marra. 2014. Direct mortality of birds from anthropogenic causes. *Annual Review of Ecology, Evolution, and Systematics* 46:1–22.

Maclary, E. T., B. Phillips, R. Wauer, E. F. Boer, R. Bruders, T. Gilvarry, C. Holt, M. Yandell, and M. D. Shapiro. 2021. Two genomic loci control three eye colors in the domestic pigeon (*Columba livia*). *Molecular Biology and Evolution* 38:5376–5390.

Maia, R., D. R. Rubenstein, and M. D. Shawkey. 2013. Key ornamental innovations facilitate diversification in an avian radiation. *Proceedings of the National Academy of Sciences of the USA* 110:10687–10692.

Marcondes, R. S., and R. T. Brumfield. 2019. Fifty shades of brown: Macroevolution of plumage brightness in the Furnariida, a large clade of drab Neotropical passerines. *Evolution* 73: 704–719.

Mason, N. A., and R. C. K. Bowie. 2020. Plumage patterns: Ecological functions, evolutionary origins, and advances in quantification. *The Auk* 137: Article ukaa060.

Mayani-Parás, F., R. M. Kilner, M. C. Stoddard, C. Rodríguez, and H. Drummond. 2015. Behaviorally induced camouflage: A new mechanism of avian egg protection. *The American Naturalist* 186:E91–E97.

McCoy, D. E., T. Feo, T. A. Harvey, and R. O. Prum. 2018. Structural absorption by barbule microstructures of super black bird of paradise feathers. *Nature Communications* 9: Article 1.

McCoy, D. E., and R. O. Prum. 2019. Convergent evolution of super black plumage near bright color in 15 bird families. *Journal of Experimental Biology* 222: Article jeb208140.

McCoy, D. E., A. J. Shultz, C. Vidoudez, E. V. D. Heide, J. E. Dall, S. A. Trauger, and D. Haig. 2021. Microstructures amplify carotenoid plumage signals in tanagers. *Scientific Reports* 11: Article 8582.

McDougal, A., B. Miller, M. Singh, and M. Kolle. 2019. Biological growth and synthetic fabrication of structurally colored materials. *Journal of Optics* 21: Article 073001.

McGraw, K. J. 2006. Mechanics of uncommon colors: Pterins, porphyrins, and psittacofulvins. Pages 354–398 in G. E. Hill and K. J. McGraw, eds. *Bird Coloration: Volume 1. Mechanisms and Measurements*. Harvard University Press, Cambridge, MA.

McNamara, M. E., V. Rossi, T. S. Slater, C. S. Rogers, A. L. Ducrest, S. Dubey, and A. Roulin. 2021. Decoding the evolution of melanin in vertebrates. *Trends in Ecology & Evolution* 36:430–443.

Meadows, M. G., N. I. Morehouse, R. L. Rutowski, J. M. Douglas, and K. J. McGraw. 2011. Quantifying iridescent coloration in animals: A method for improving repeatability. *Behavioral Ecology and Sociobiology* 65:1317–1327.

Medina, I., E. Newton, M. R. Kearney, R. A. Mulder, W. P. Porter, and D. Stuart-Fox. 2018. Reflection of near-infrared light confers thermal protection in birds. *Nature Communications* **9**: Article 3610.

Medina, J. J., J. M. Maley, S. Sannapareddy, N. N. Medina, C. M. Gilman, and J. E. McCormack. 2020. A rapid and cost-effective pipeline for digitization of museum specimens with 3D photogrammetry. *PLoS One* 15: Article e0236417.

Mendes-Pinto, M. M., A. M. LaFountain, M. C. Stoddard, R. O. Prum, H. A. Frank, and B. Robert. 2012. Variation in carotenoid–protein interaction in bird feathers produces novel plumage coloration. *Journal of the Royal Society: Interface* 9:3338–3350.

Miller, A. E., B. G. Hogan, and M. C. Stoddard. 2022. Color in motion: Generating 3-dimensional multispectral models to study dynamic visual signals in animals. *Frontiers in Ecology and Evolution* 10: Article 983369.

Miller, E. T., G. M. Leighton, B. G. Freeman, A. C. Lees, and R. A. Ligon. 2019. Ecological and geographical overlap drive plumage evolution and mimicry in woodpeckers. *Nature Communications* 10: Article 1602.

Mills, L. S., E. V. Bragina, A. V. Kumar, M. Zimova, D. J. R. Lafferty, J. Feltner, B. M. Davis, K. Hackländer, P. C. Alves, J. M. Good, J. Melo-Ferreira, A. Dietz, A. V. Abramov, N. Lopatina, and K. Fay. 2018. Winter color polymorphisms identify global hot spots for evolutionary rescue from climate change. *Science* 359:1033–1036.

Morrison, E. S., and A. V. Badyaev. 2018. Structure versus time in the evolutionary diversification of avian carotenoid metabolic networks. *Journal of Evolutionary Biology* 31:764–772.

Mundy, N. I. 2005. A window on the genetics of evolution: MC1R and plumage colouration in birds. *Proceedings of the Royal Society B: Biological Sciences* 272:1633–1640.

Mundy, N. I. 2018. Colouration genetics: Pretty polymorphic parrots. *Current Biology* 28: R113–R114.

Mundy, N. I., J. Stapley, C. Bennison, R. Tucker, H. Twyman, K.-W. Kim, T. Burke, Tim R. Birkhead, S. Andersson, and J. Slate. 2016. Red carotenoid coloration in the zebra finch is controlled by a cytochrome P450 gene cluster. *Current Biology* 26:1435–1440.

Ng, C. S., and W.-H. Li. 2018. Genetic and molecular basis of feather diversity in birds. *Genome Biology and Evolution* 10:2572–2586.

Nicolaï, M. P. J., M. D. Shawkey, S. Porchetta, R. Claus, and L. D'Alba. 2020. Exposure to UV radiance predicts repeated evolution of concealed black skin in birds. *Nature Communications* 11: Article 2414.

Nordén, K. K., C. M. Eliason, and M. C. Stoddard. 2021. Evolution of brilliant iridescent feather nanostructures. *eLife* 10: Article e71179.

Nordén, K. K., J. W. Faber, F. Babarović, T. L. Stubbs, T. Selly, J. D. Schiffbauer, P. P. Štefanić, G. Mayr, F. M. Smithwick, and J. Vinther. 2018. Melanosome diversity and convergence in the evolution of iridescent avian feathers: Implications for paleocolor reconstruction. Evolution 73: 15–27.

Nowik, K., E. Langwińska-Wośko, P. Skopiński, K. E. Nowik, and J. P. Szaflik. 2020. Bionic eye review: An update. *Journal of Clinical Neuroscience* 78:8–19.

Obukhova, N. Y. 2011. Dynamics of balanced polymorphism morphs in blue rock pigeon *Columbia livia*. *Russian Journal of Genetics* 47:83–89.

Oliphant, L. W., J. Hudon, and J. T. Bagnara. 1992. Pigment cell refugia in homeotherms: The unique evolutionary position of the iris. *Pigment Cell Research* 5:367–371.

Orteu, A., and C. D. Jiggins. 2020. The genomics of coloration provides insights into adaptive evolution. *Nature Reviews Genetics* 21:461–475.

Piselli, D. 2020. How to prevent millions of unnecessary bird deaths from collisions with windows. *Proceedings of the Institution of Civil Engineers – Civil Engineering* 173:53–53.

Price, T. D., M. C. Stoddard, S. K. Shevell, and N. I. Bloch. 2019. Understanding how neural responses contribute to the diversity of avian colour vision. *Animal Behaviour* 155:297–305.

Price-Waldman, R. M., J. R. Ali, A. J. Shultz, B. G. Hogan, and M. C. Stoddard. 2025. Hidden white and black feather layers enhance plumage coloration in tanagers and other songbirds. *Science Advances* 11: Article eadw5857.

Price-Waldman, R., and M. C. Stoddard. 2021. Avian coloration genetics: Recent advances and emerging questions. *Journal of Heredity* 112:395–416.

Price-Waldman, R. M., A. J. Shultz, and K. J. Burns. 2020. Speciation rates are correlated with changes in plumage color complexity in the largest family of songbirds. *Evolution* 74:1155–1169.

Prum, R. 2006. Anatomy, physics, and evolution of structural colors. Pages 295–353 in G. E. Hill and K. J. McGraw, eds. *Bird Coloration: Volume 1. Mechanisms and Measurements.* Harvard University Press, Cambridge, MA.

Prum, R. O., J. S. Berv, A. Dornburg, D. J. Field, J. P. Townsend, E. M. Lemmon, and A. R. Lemmon. 2015. A comprehensive phylogeny of birds (Aves) using targeted next-generation DNA sequencing. *Nature* 526:569–573.

Prum, R. O., E. R. Dufresne, T. Quinn, and K. Waters. 2009. Development of colour-producing β-keratin nanostructures in avian feather barbs. *Journal of the Royal Society: Interface* 6:S253–S265.

Prum, R. O., and R. Torres. 2003. Structural colouration of avian skin: Convergent evolution of coherently scattering dermal collagen arrays. *Journal of Experimental Biology* 206:2409–2429.

Renoult, J. P., A. Kelber, and H. M. Schaefer. 2017. Colour spaces in ecology and evolutionary biology. *Biological Reviews* 92:292–315.

Rocha, F. A. F., C. A. Saito, L. C. L. Silveira, J. M. D. Souza, and D. F. Ventura. 2008. Twelve chromatically opponent ganglion cell types in turtle retina. *Visual Neuroscience* 25:307–315.

Rogalla, S., A. Patil, A. Dhinojwala, M. D. Shawkey, and L. D'Alba. 2021. Enhanced photothermal absorption in iridescent feathers. *Journal of the Royal Society: Interface* 18: Article 20210252.

Rogalla, S., M. D. Shawkey, and L. D'Alba. 2022. Thermal effects of plumage coloration. *Ibis* 164:933–948.

Romano, A., R. Séchaud, A. H. Hirzel, and A. Roulin. 2019. Climate-driven convergent evolution of plumage colour in a cosmopolitan bird. *Global Ecology and Biogeography* 28:496–507.

Rosenthal, G. G. 2007. Spatiotemporal dimensions of visual signals in animal communication. *Annual Review of Ecology, Evolution, and Systematics* 38:155–178.

Roulin, A., and A.-L. Ducrest. 2013. Genetics of colouration in birds. *Seminars in Cell & Developmental Biology* 24:594–608.

Roy, A., M. Pittman, E. T. Saitta, T. G. Kaye, and X. Xu. 2020. Recent advances in amniote palaeocolour reconstruction and a framework for future research. *Biological Reviews of the Cambridge Philosophical Society* 95:22–50.

Rubenstein, D. R., A. Corvelo, M. D. MacManes, R. Maia, G. Narzisi, A. Rousaki, P. Vandenabeele, M. D. Shawkey, and J. Solomon. 2021. Feather gene expression elucidates the developmental basis of plumage iridescence in African starlings. *Journal of Heredity* 112:417–429.

Russell, B. J., and H. M. Dierssen. 2015. Use of hyperspectral imagery to assess cryptic color matching in *Sargassum* associated crabs. *PLoS One* 10: Article e0136260.

Saranathan, V., and C. Finet. 2021. Cellular and developmental basis of avian structural coloration. *Current Opinion in Genetics & Development* 69:56–64.

Saranathan, V., S. Narayanan, A. Sandy, E. R. Dufresne, and R. O. Prum. 2021. Evolution of single gyroid photonic crystals in bird feathers. *Proceedings of the National Academy of Sciences of the USA* 118: Article e2101357118.

Shawkey, M. D., and L. D'Alba. 2017. Interactions between colour-producing mechanisms and their effects on the integumentary colour palette. *Philosophical Transactions of the Royal Society B: Biological Sciences* 372: Article 20160536.

Shawkey, M. D., and L. D'Alba. 2019. Egg pigmentation probably has an early Archosaurian origin. *Nature* 570:E43–E45.

Shawkey, M. D., and G. E. Hill. 2006. Significance of a basal melanin layer to production of non-iridescent structural plumage color: Evidence from an amelanotic Steller's jay (*Cyanocitta stelleri*). *Journal of Experimental Biology* 209:1245–1250.

Shawkey, M. D., N. I. Morehouse, and P. Vukusic. 2009. A protean palette: Colour materials and mixing in birds and butterflies. *Journal of the Royal Society: Interface* -6:S221–S231.

Sheppard, C. D. 2019. Evaluating the relative effectiveness of patterns on glass as deterrents of bird collisions with glass. *Global Ecology and Conservation* 20: Article e00795.

Si, S., X. Xu, Y. Zhuang, X. Gao, H. Zhang, Z. Zou, and S.-J. Luo. 2021. The genetics and evolution of eye color in domestic pigeons (*Columba livia*). *PLoS Genetics* 17: Article e1009770.

Sicher, A., R. Ganz, A. Menzel, D. Messmer, G. Panzarasa, M. Feofilova, R. O. Prum, R. W. Style, V. Saranathan, R. M. Rossi, and E. R. Dufresne. 2021. Structural color from solid-state polymerization-induced phase separation. *Soft Matter* 17:5772–5779.

Simpson, R. K., and K. J. McGraw. 2018a. It's not just what you have, but how you use it: Solar-positional and behavioural effects on hummingbird colour appearance during courtship. *Ecology Letters* 21:1413–1422.

Simpson, R. K., and K. J. McGraw. 2018b. Two ways to display: Male hummingbirds show different color-display tactics based on sun orientation. *Behavioral Ecology* 29:637–648.

Simpson, R. K., and K. J. McGraw. 2019. Interspecific covariation in courtship displays, iridescent plumage, solar orientation, and their interactions in hummingbirds. *The American Naturalist* 194:441–454.

Sin, S. Y. W., L. Lu, and S. V. Edwards. 2020. De novo assembly of the Northern Cardinal (*Cardinalis cardinalis*) genome reveals candidate regulatory regions for sexually dichromatic red plumage coloration. *G3* 10:3541–3548.

Solonen, T. 2021a. Does plumage colour signal fitness in the tawny owl *Strix aluco*? *Journal of Avian Biology* 52: Article 02470.

Solonen, T. 2021b. Significance of plumage colour for winter survival in the Tawny Owl (*Strix aluco*): Revisiting the camouflage hypothesis. *Ibis* 163:1437–1442.

Sparks, N. H. C. 2011. Eggshell pigments from formation to deposition. *Avian Biology Research* 4:162–167.

Stavenga, D. G., H. L. Leertouwer, N. J. Marshall, and D. Osorio. 2011. Dramatic colour changes in a bird of paradise caused by uniquely structured breast feather barbules. *Proceedings of the Royal Society B: Biological Sciences* 278:2098–2104.

Stevens, M., C. A. Párraga, I. C. Cuthill, J. C. Partridge, and T. S. Troscianko. 2007. Using digital photography to study animal coloration. *Biological Journal of the Linnean Society* 90: 211–237.

Stoddard, M. C., H. N. Eyster, B. G. Hogan, D. H. Morris, E. R. Soucy, and D. W. Inouye. 2020. Wild hummingbirds discriminate nonspectral colors. *Proceedings of the National Academy of Sciences of the USA* 117:15112–15122.

Stoddard, M. C., A. E. Miller, H. N. Eyster, and D. Akkaynak. 2019. I see your false colours: How artificial stimuli appear to different animal viewers. *Interface Focus* 9: Article 20180053.

Stoddard, M. C., and D. Osorio. 2019. Animal coloration patterns: Linking spatial vision to quantitative analysis. *The American Naturalist* 193:164–186.

Stoddard, M. C., and R. O. Prum. 2011. How colorful are birds? Evolution of the avian plumage color gamut. *Behavioral Ecology* 22:1042–1052.

Stowers, J. R., M. Hofbauer, R. Bastien, J. Griessner, P. Higgins, S. Farooqui, R. M. Fischer, K. Nowikovsky, W. Haubensak, I. D. Couzin, K. Tessmar-Raible, and A. D. Straw. 2017. Virtual reality for freely moving animals. *Nature Methods* 14:995–1002.

Stuart-Fox, D., L. Ospina-Rozo, L. Ng, and A. M. Franklin. 2021. The paradox of iridescent signals. *Trends in Ecology & Evolution* 36:187–195.

Sumasgutner, P., M. Adrion, and A. Gamauf. 2018. Carotenoid coloration and health status of urban Eurasian kestrels (*Falco tinnunculus*). *PLoS One* 13: Article e0191956.

Swaddle, J. P., L. C. Emerson, R. G. Thady, and T. J. Boycott. 2020. Ultraviolet-reflective film applied to windows reduces the likelihood of collisions for two species of songbird. *PeerJ* 8: Article e9926.

Tavares, D. C., J. F. Moura, and A. Merico. 2019. Anthropogenic debris accumulated in nests of seabirds in an uninhabited island in West Africa. *Biological Conservation* 236:586–592.

Tay, E. J., J. E. Barnsley, D. B. Thomas, and K. C. Gordon. 2021. Elucidating the resonance Raman spectra of psittacofulvins. *Spectrochimica Acta Part A: Molecular and Biomolecular Spectroscopy* 262: Article 120146.

Terrill, R. S., and A. J. Schultz. 2023. Feather function and the evolution of birds. *Biological Reviews* 98:540–566.

Thomas, D. B., M. E. Hauber, and D. Hanley. 2015. Analysing avian eggshell pigments with Raman spectroscopy. *Journal of Experimental Biology* 218:2670–2674.

Thomas, D. B., C. M. McGoverin, K. J. McGraw, H. F. James, and O. Madden. 2013. Vibrational spectroscopic analyses of unique yellow feather pigments (spheniscins) in penguins. *Journal of the Royal Society: Interface* 10: Article 20121065.

Thomas, D. B., K. J. McGraw, M. W. Butler, M. T. Carrano, O. Madden, and H. F. James. 2014. Ancient origins and multiple appearances of carotenoid-pigmented feathers in birds. *Proceedings of the Royal Society B: Biological Sciences* 281: Article 20140806.

Thomas, D. B., K. J. McGraw, H. F. James, and O. Madden. 2013. Non-destructive descriptions of carotenoids in feathers using Raman spectroscopy. *Analytical Methods* 6:1301–1308.

Tian, L., and M. J. Benton. 2020a. Predicting biotic responses to future climate warming with classic ecogeographic rules. *Current Biology* 30:R744–R749.

Tian, L., and M. J. Benton. 2020b. Response to Delhey et al. *Current Biology* 30:R1408.

Tidemann, S. C., and A. Gosler, eds. 2010. *Ethno-ornithology*. Earthscan, New York.

Toews, D. P. L., N. R. Hofmeister, and S. A. Taylor. 2017. The evolution and genetics of carotenoid processing in animals. *Trends in Genetics* 33:171–182.

Toews, D. P. L., S. A. Taylor, R. Vallender, A. Brelsford, B. G. Butcher, P. W. Messer, and I. J. Lovette. 2016. Plumage genes and little else distinguish the genomes of hybridizing warblers. *Current Biology* 26:2313–2318.

Toomey, M. B., A. M. Collins, R. Frederiksen, M. C. Cornwall, J. A. Timlin, and J. C. Corbo. 2015. A complex carotenoid palette tunes avian colour vision. *Journal of the Royal Society: Interface* 12: Article 20150563.

Toomey, M. B., and J. C. Corbo. 2017. Evolution, development and function of vertebrate cone oil droplets. *Frontiers in Neural Circuits* 11:793–717.

Toomey, M. B., O. Lind, R. Frederiksen, R. W. Curley, K. M. Riedl, D. Wilby, S. J. Schwartz, C. C. Witt, E. H. Harrison, N. W. Roberts, M. Vorobyev, K. J. McGraw, M. C. Cornwall, A. Kelber, and J. C. Corbo. 2016. Complementary shifts in photoreceptor spectral tuning unlock the full adaptive potential of ultraviolet vision in birds. *eLife* 5: Article e15675.

Toomey, M. B., C. I. Marques, P. M. Araújo, D. Huang, S. Zhong, Y. Liu, G. D. Schreiner, C. A. Myers, P. Pereira, S. Afonso, P. Andrade, M. A. Gazda, R. J. Lopes, I. Viegas, R. E. Koch, M. E. Haynes, D. J. Smith, Y. Ogawa, D. Murphy, R. E. Kopec, D. M. Parichy, M. Carneiro, and J. C. Corbo. 2022. A mechanism for red coloration in vertebrates. *Current Biology* 32:4201–4214.

Toomey, M. B., and K. L. Ronald. 2021. Avian color expression and perception: Is there a carotenoid link? *Journal of Experimental Biology* 224: Article jeb203844.

Troscianko, J., and M. Stevens. 2015. Image calibration and analysis toolbox: A free software suite for objectively measuring reflectance, colour and pattern. *Methods in Ecology and Evolution* 6:1320–1331.

Twyman, H., S. Andersson, and N. I. Mundy. 2018. Evolution of CYP2J19, a gene involved in colour vision and red coloration in birds: Positive selection in the face of conservation and pleiotropy. *BMC Evolutionary Biology* 18: Article 22.

Twyman, H., M. Prager, N. I. Mundy, and S. Andersson. 2018. Expression of a carotenoid-modifying gene and evolution of red coloration in weaverbirds (Ploceidae). *Molecular Ecology* 27:449–458.

Twyman, H., N. Valenzuela, R. Literman, S. Andersson, and N. I. Mundy. 2016. Seeing red to being red: Conserved genetic mechanism for red cone oil droplets and co-option for red coloration in birds and turtles. *Proceedings of the Royal Society B: Biological Sciences* 283: Article 20161208.

Velando, A., R. Beamonte-Barrientos, and R. Torres. 2006. Pigment-based skin colour in the blue-footed booby: An honest signal of current condition used by females to adjust reproductive investment. *Oecologia* 149:535–542.

Venable, G. X., K. Gahm, and R. O. Prum. 2022. Hummingbird plumage color diversity exceeds the known gamut of all other birds. *Communications Biology* 5: Article 576.

Vinther, J. 2020. Reconstructing vertebrate paleocolor. *Annual Review of Earth and Planetary Sciences* 48:345–375.

Wang, S., S. Rohwer, D. R. Zwaan, D. P. L. Toews, I. J. Lovette, J. Mackenzie, and D. Irwin. 2020. Selection on a small genomic region underpins differentiation in multiple color traits between two warbler species. *Evolution Letters* 4:502–515.

Wang, Z., L. Qu, J. Yao, X. Yang, G. Li, Y. Zhang, J. Li, X. Wang, J. Bai, G. Xu, X. Deng, N. Yang, and C. Wu. 2013. An EAV-HP insertion in 5′ flanking region of SLCO1B3 causes blue eggshell in the chicken. *PLoS Genetics* 9: Article e1003183.

Wiemann, J., T.-R. Yang, and M. A. Norell. 2018. Dinosaur egg colour had a single evolutionary origin. *Nature* 563:555–558.

Wiemann, J., T.-R. Yang, and M. A. Norell. 2019. Reply to: Egg pigmentation probably has an Archosaurian origin. *Nature* 570:E46–E50.

Wiemann, J., T. R. Yang, P. N. Sander, M. Schneider, M. Engeser, S. Kath-Schorr, C. E. Muller, and P. M. Sander. 2017. Dinosaur origin of egg color: Oviraptors laid blue-green eggs. *PeerJ* 5: Article e3706.

Wisocki, P. A., P. Kennelly, I. R. Rivera, P. Cassey, M. L. Burkey, and D. Hanley. 2019. The global distribution of avian eggshell colours suggest a thermoregulatory benefit of darker pigmentation. *Nature Ecology & Evolution* 4:148–155.

Wragg, D., J. M. Mwacharo, J. A. Alcalde, C. Wang, J.-L. Han, J. Gongora, D. Gourichon, M. Tixier-Boichard, and O. Hanotte. 2013. Endogenous retrovirus EAV-HP linked to blue egg phenotype in Mapuche fowl. *PLoS One* 8: Article e71393.

Xiao, M., Y. Li, M. C. Allen, D. D. Deheyn, X. Yue, J. Zhao, N. C. Gianneschi, M. D. Shawkey, and A. Dhinojwala. 2015. Bio-inspired structural colors produced via self-assembly of synthetic melanin nanoparticles. *ACS Nano* 9:5454–5460.

Xiao, M., M. D. Shawkey, and A. Dhinojwala. 2020. Bioinspired melanin-based optically active materials. *Advanced Optical Materials* 8: Article 2000932.

Zeuss, D., R. Brandl, M. Brändle, C. Rahbek, and S. Brunzel. 2014. Global warming favours light-coloured insects in Europe. *Nature Communications* 5: Article 3874.

Zimova, M., K. Hackländer, J. M. Good, J. Melo-Ferreira, P. C. Alves, and L. S. Mills. 2018. Function and underlying mechanisms of seasonal colour moulting in mammals and birds: What keeps them changing in a warming world? *Biological Reviews* 93:1478–1498.

Zipple, M. N., E. M. Caves, P. A. Green, S. Peters, S. Johnsen, and S. Nowicki. 2019. Categorical colour perception occurs in both signalling and non-signalling colour ranges in a songbird. *Proceedings of the Royal Society B: Biological Sciences* 286: Article 20190524.

16
Bird Migration in the Age of Rapid Technological Advance and Global Change

Kristen C. Ruegg, Dmitry Kishkinev, and Barbara Helm

Questions about where birds migrate and how they find their way have fascinated scientists and natural history enthusiasts for centuries. We have come a long way in our understanding of where and how birds migrate since early natural historians such as John James Audubon reportedly tied "silver threads" to the legs of Eastern Phoebes in efforts to document their potential for natal philopatry (Audubon 1834; but see Montgomerie 2018). In the past several decades, multiple books have been written on the topic of bird migration, covering everything from methods for studying migration to how birds orient and navigate and theories on why birds migrate (Gauthreaux 1980, Alerstam 1993, Berthold 1993, Newton 2010, Hansson and Åkesson 2014, McKinnon and Love 2018). However, the field of migratory science is changing fast, and rather than summarize previous reviews here, we focus on highlighting novel insights gained in recent decades. Specifically, we are currently living in the midst of the fourth industrial revolution, with rapid advances in the fields of artificial intelligence, robotics, genetic engineering, and quantum computing, and all of these advances are having an impact on our ability to answer fundamental questions in the field of migration science (La Sorte et al. 2018, Shamoun-Baranes et al. 2019, Nathan et al. 2022). At the same time, the natural world is changing rapidly, with habitat alteration, climate change, and increasingly frequent extreme climate events all having an impact on migratory birds. In this chapter, we address how new technology is facilitating a more synthetic view of migratory bird behavior and how these behaviors are being influenced by a rapidly changing world.

Migratory Patterns

The migratory patterns of birds are as diverse as the patterns of abundance in the resources upon which they depend for their survival and reproduction. For simplicity, previous authors have often categorized migratory birds into groups based on the distance of the migratory journey and the regularity of those journeys (Newton 2010). These groups include traditional obligate migrants that undertake predictable annual migrations between breeding and nonbreeding grounds each year, facultative migrants that vary year-to-year in the extent of migration, partial migrants where only part of the population migrates, and nomadic migrants that move less

Kristen C. Ruegg, Dmitry Kishkinev, and Barbara Helm, *Bird Migration in the Age of Rapid Technological Advance and Global Change*. In: *New Perspectives in Ornithology*. Edited by: Scott V. Edwards and J. Michael Reed,
Oxford University Press. © Oxford University Press (2025). DOI: 10.1093/oso/9780197787670.003.0016

predictably from one breeding ground to the next depending on resource availability. While recent technological advances in high-performance computing, data science, genomics, and device miniaturization have confirmed the existence of species within each of these categories, they have also helped broaden our understanding of the diversity and complexity of migratory movements, as well as deepen our understanding of the connection between these movements and the environmental conditions that help shape them (Figure 16.1).

One major way that the field of information technology has propelled the field of avian migration science is by facilitating the development of community science-based methods for documenting migratory movements such as eBird, eurobirdportal, biolovision, or FeederWatch (Louv and Fitzpatrick 2012, La Sorte et al. 2018). In particular, as one of the world's largest community science projects, eBird helps connect more than 100 million bird sightings a year with information on timing and habitat in order to document how birds move across landscapes. For example, the application of community science data into macroecological studies has led to important insights into the vulnerability of certain groups of birds to future global change. For example, La Sorte and Graham (2021) used eBird data in conjunction with remotely sensed landscape information to show that species across guilds synchronize seasonal migration with vegetative greenness. These results highlight the potential for phenological mismatches as vegetation phenology responds to climate change. Similarly, in a separate study, Youngflesh et al. (2021) used eBird data to demonstrate that earlier arriving, shorter distance migrants are more easily able to adjust their migratory timing to sync with climate change–induced phenological shifts.

Analysis of eBird data sets has also improved our knowledge of migratory pathways across taxa (Laughlin et al. 2013, Klipp et al. 2018), including revealing new insights into the frequency and evolutionary origins of migration patterns that are difficult to assess with other tracking tools, such as irruptive and altitudinal movements (Strong et al. 2015, Tsai et al. 2021). For example, integration of eBird data with ringing/banding, tracking, and isotope data has led to the discovery of important stopover areas in previously understudied species (Heim et al. 2020) or new migratory routes that may otherwise go undetected (Dufour et al. 2021). In one study, Heim et al. (2020) used eBird occurrence data in combination with geolocator and ring recovery data to identify new population-specific migratory corridors used by Barn Swallows (*Hirundo rustica*) and Siberian Rubythroats (*Calliope calliope*) over the eastern mainland of China and along a chain of islands in the Pacific Ocean. Similarly, tracking data were combined with citizen science bird data and climatic niche modeling to document an extraordinary new longitudinal migration route in Siberian-breeding Richard's Pipit (*Anthus richardi*; Dufour et al. 2021). As citizen science-based methods for recording individual observations continue to grow in popularity, so too will our ability to document and understand the movement patterns of birds over time and space.

Another area in which high-performance computing and data science are rapidly advancing migration research is in facilitating the development of increasingly sophisticated external tracking devices (Nathan et al. 2022). Specifically, the miniaturization of electronic components has facilitated the development of light-level, radio, and satellite-based devices that can be used to track highly detailed individual movements of birds at ever finer timescales (from seconds to years). Such advances are

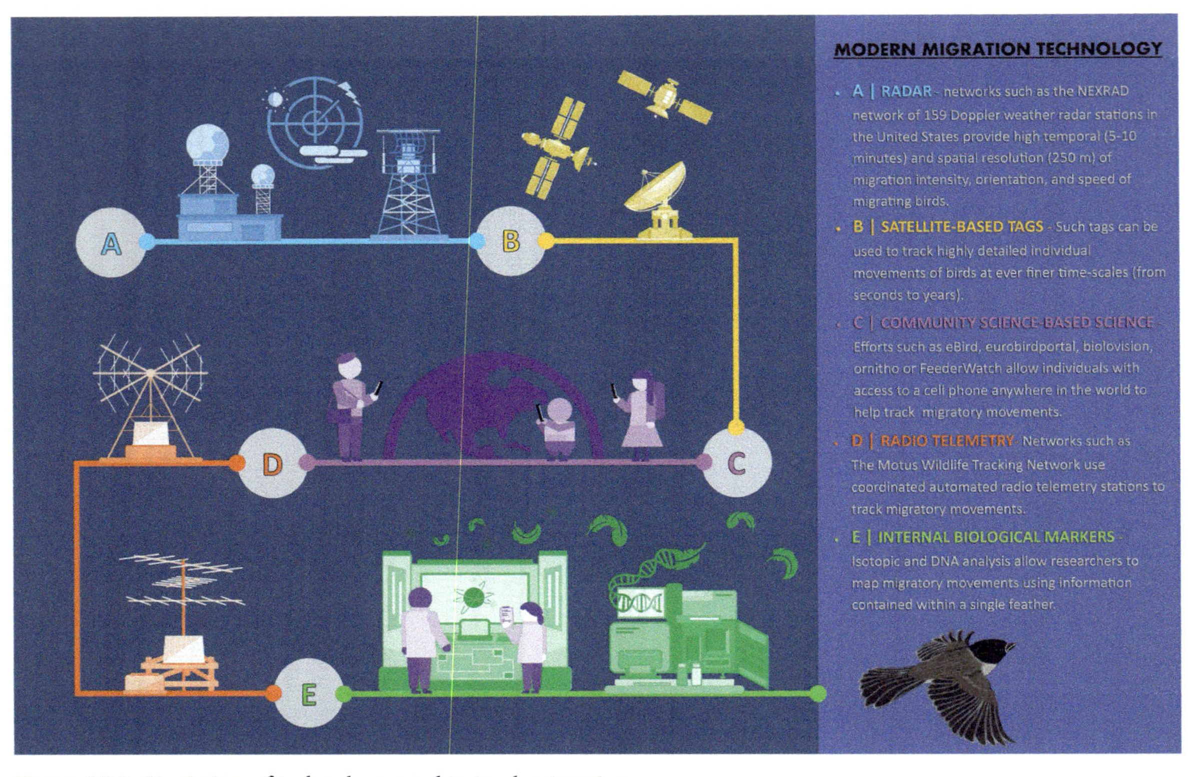

Figure 16.1 Depiction of technology used to track migration.
Illustration by Eden Vo, Colorado State University.

expanding our knowledge not only of specific migratory pathways that birds take but also of the altitudes they reach, the stopovers they select, and the timing of their journeys. One of the most spectacular insights gained from satellite tracking devices in recent years is the discovery of the Bar-Tailed Godwits' more than 10,000-km, 9-day, nonstop flight over the Pacific Ocean (Gill et al. 2009). The duration and distance of the Bar-Tailed Godwits' nonstop flight are nearly twice those of its nearest known competitor, the Eastern Curlew (*Numenius madagascariensis*), leading researchers to reconsider existing paradigms around the physiological capabilities of vertebrates and the long-held view that oceans represent substantial barriers to migration. Separate tracking studies using multisensor data loggers have further demonstrated that when traditionally nocturnally migrating birds are forced to migrate during the day in order to cross major barriers, they increase their flight altitude significantly above previously known extremes (Lindström et al. 2021, Sjöberg et al. 2021). Such findings are leading researchers to hypothesize that nocturnal migration may have evolved as a strategy for avoiding increased temperatures associated with solar radiation. Last, perhaps one of the most spectacular revelations from tracking technology is confirmation of the hitherto suspected, almost completely aerial life of some species. Specifically, it is now known that the aerial feeding Common Swift (*Apus apus*) remains airborne year-round with the exception of a 2-month breeding season (Hedenström et al. 2016). Based on accelerometer data, over their lifetime, the birds may cover the distance to the moon several times, leaving humans dazzled by their ability to nonetheless obtain sufficient sleep.

A technological advance that neatly combines community-based science with advances in tracking is the Motus Wildlife Tracking Network, an international collaborative research network that uses coordinated automated radiotelemetry to track migratory movements (Taylor et al. 2017). Some of the most exciting research made possible by Motus radiotelemetry arrays is helping to deepen our understanding of the role of winter habitat quality and pesticide exposure on migratory movements, with important implications for migrant conservation. For example, tracking data from radiotelemetry arrays across Columbia and the northeastern United States were used in combination with stable isotope analysis to demonstrate that Swainson's Thrushes (*Catharus ustulatus*) wintering in high-quality native forest depart later than migrants wintering in shade coffee plantations (González et al. 2020). Such results run counter to existing paradigms around earlier departure from higher quality habitat and suggest that delaying departure to take advantage of high-quality habitat may benefit migrants. Separate studies using the Motus network are helping advance our understanding of the effect of pesticides on migratory birds. In particular, by combining controlled dosing of wild-caught White-Crowned Sparrows (*Zonotrichia leucophrys*) with Motus tracking, researchers recently demonstrated that birds exposed to pesticides exhibited an approximately 3.5-day delay in departure from northern Ontario stopover sites while they were recovering from intoxication (Eng et al. 2019). Given that migration delays can negatively impact migrant survival and fitness, this work highlights the critical role of high-quality stopover habitats and the need to consider all stages of the annual cycle in migrant conservation efforts.

Although radar has been used to study migration for decades, the ability to utilize large-scale weather radar networks to quantify and forecast broadscale migratory movements has been facilitated recently via advances in high-performance computing (Van Doren et al. 2017, 2021; Horton et al. 2020). Specifically, the NEXRAD network of 159 Doppler weather radar stations in the United States provides high temporal (5–10 minutes) and spatial resolution (250 m) of migration intensity, orientation, and speed of migrating birds. These developments have allowed biologists to pursue a remarkable number of scientific endeavors and are well-suited to identify "hot spots" of migratory activity. For instance, the utilization of such data has emerged as a powerful approach for highlighting the significance of tropical wintering regions to the survival of long-distance Neotropical migrants (Dokter et al. 2018) while also capturing the dependence of migrants on weather conditions that may undergo shifts due to climate change (Horton et al. 2020). Furthermore, radar studies have also greatly contributed to revealing the sheer magnitude of harm to birds from light pollution and associated collisions with buildings (Van Doren et al. 2021). It is estimated that nearly 1 billion birds in the United States alone die in this way, of which most are migrants. As a result of these studies, urgent action to mitigate such effects is currently underway. This includes the implementation of mitigation measures that use radar-based forecasts of numbers of birds migrating in a given area on a given day to encourage lights to be reduced in those areas when heavy migration is expected. Overall, the combined use of real-time forecasting with lights-out programs provides an inspiring application of a technology-driven solution to conservation challenges facing migratory birds.

In addition to improvements in tracking technology, we have also seen advances in the use of internal biological markers to track migratory movements, including in isotopic and DNA analysis. Samples obtained through collaborations with bird banding stations or by accessing specimens in museum collections offer an excellent source of isotopic and DNA material (Smith et al. 2003). Specifically, stable isotope analysis first relies on creating an isoscape, or map of how isotopic elements vary across geographic space, and then uses animal tissues such as feathers collected outside of the breeding range to trace the molting origin for birds (Hobson et al. 2010). Although the use of isotopes has propelled avian science on numerous fronts, including furthering our understanding of migrant speciation (Bearhop et al. 2005; Turbek et al. 2018, 2022), migratory connectivity (Rundel et al. 2013, González-Prieto et al. 2017, Ruegg et al. 2017, Jiguet et al. 2019), and the transmission pathways of avian parasites (Pulgarín-R et al. 2019), some of the most innovative isotope studies in recent decades have been done in combination with other tracking technologies to reveal entirely new patterns of migration. For example, isotopes and an analysis of feather growth patterns were combined to demonstrate that increasingly warmer winters brought about by climate change have resulted in a rapid reduction in migration distance in Barn Swallows (Møller et al. 2021). In a separate study on Western Hemisphere Barn Swallows, a combination of isotopes and light-level geolocators was used to describe what is among the most dramatic reversals in migratory behavior ever documented (Areta et al. 2021). Specifically, researchers were able to show that a small population of barn swallows that formerly bred in North America and migrated to Argentina for the winter underwent a complete hemispheric switch in their behavior so that the new population wintered in Argentina and bred no further north than northern South America (Garcia-Perez et al. 2013, Winkler et al. 2017). Overall, isotopic analysis has proven an especially powerful tool for tracking migratory movements in cases in which it may be cost prohibitive or logistically too complex to attach external tracking devices to many hundreds of individuals.

Similar to isotope approaches, DNA analysis relies on first assessing how genetic variation is distributed across the breeding ground of a species and then using DNA from a bird captured anywhere outside of its breeding range to trace its breeding population of origin (Rundel et al. 2013, Ruegg et al. 2017). Although the ability to analyze DNA sequence data in non-model organisms has been around since the late 20th century, the advent of next-generation sequencing has led to a rapid increase in the amount of sequence data that can be generated and to a corresponding decrease in the per base pair cost. This explosion has resulted in an increase in the precision of resulting assignments of birds to breeding regions. Similar to isoscapes, the map of how population genetic variation is distributed across geographic space is called a genoscape. The overall advantage of isoscape- and genoscape-based approaches is that thousands of individuals can be screened using biological information found in feather, toepad, muscle, blood, or skin samples. Furthermore, the information from both methods is often complementary. When birds breed in the same place that they molt their feathers, the two sources of information can be combined to yield higher quality information then either method used in isolation (Clegg et al. 2003, Kelly et al. 2005, Garcia-Perez et al. 2013, Ruegg et al. 2017). Alternatively, when birds molt outside of the breeding area, isotopes can be used to identify migratory stopover or wintering areas, and genetics can be used to identify breeding sites. A powerful recent demonstration of such combined approaches is an investigation of the evolution of migration around a major mountain barrier, the Karakoram range (Turbek et al. 2022). In this study on Barn Swallows, stable isotope approaches and geolocators revealed a continental migration divide of Chinese breeding birds, whereby one population overwinters in East Africa and the other in south India. DNA sequencing indicated strong divergence between these populations and suggested behavioral mechanisms that drive it.

One of the advantages of genoscape methods in comparison to other techniques is the ability to first identify locally adapted populations across the breeding range and then understand how these populations are connected across the annual cycle (Bay et al. 2018, 2021; Ruegg et al. 2018, 2021). When genome-wide genomic data are combined with environmental data, it is possible to assess how signatures of selection and niche use compare on the breeding and wintering grounds. This, in turn, opens up the possibility of identifying whether populations are exposed to parallel or contrasting selective pressures on breeding and wintering areas. Initial work by Ruegg et al. (2021) using Willow Flycatchers (*Empidonax traillii*) as a model system found that southwestern U.S. populations occupied similar climatic niches on their tropical wintering grounds but that genetically distinct populations breeding in the east switch climatic niche between breeding and wintering areas. Interestingly, when the results were compared to patterns of population decline based on Breeding Bird Survey data, the researchers found that the degree of niche tracking was correlated with population vulnerability. Thus, highly specialized niche trackers are either already endangered or show steep declines, and more general niche switching populations have increased or remained stable. Further work by Bay et al. (2021) on a separate species, the Yellow Warbler (*Setophaga petechia*), built on previous population-level analysis to demonstrate how genetic markers can be used to assign wintering individuals to specific breeding regions. When combined with climatic data from the site of capture on the wintering grounds and the predicted origin on the breeding grounds, the research determined that Yellow Warblers seem to track similar precipitation regimes across the annual cycle. Both of these studies lay the groundwork for future investigations of the process of natural selection across the annual cycle.

Recent technological innovations in the use of external tracking devices and community science have also allowed researchers to revisit classic work on the genetic basis of migratory timing with a new, more integrative lens. This area of research started with centuries-old observations that even if caged, migrants show migration-related changes in behavior and physiology at the right times of year (e.g., zugunruhe, directed movement, and super-athletes; reviewed in Berthold 2001). In particular, birds have been shown to possess genetic migration programs that set the timing and direction of their movement in interaction with environmental information (Åkesson and Helm 2020). These programs help birds track and anticipate environmental conditions in often faraway locations and find goal areas at the right time. Timing is mechanistically based on dial (circadian), annual (circannual), and possibly lunar clocks (Norevik et al. 2019), which are all highly sensitive to ambient light. The use of long-term community-based science demonstrated that these clocks seem to be at least to some extent malleable to evolution in our rapidly changing world. For example, in Pied Flycatchers (*Ficedula hypoleuca*), a replicated captivity experiment combined with analysis of long-term field-based ringing data showed that the advancement of spring activities recorded in the field was likely based on a change in the timing program (Helm et al. 2019). This work suggests that in populations with enough standing genetic variation, the timing of long-distance migration in birds may have a greater potential to respond to changing climate conditions than previously thought. Alternatively, a recent study using geolocators documented a hemispheric switch in American hirudines (Winkler et al. 2017). In this example, the birds switched their migration programs to breed in the Southern Hemisphere and winter to the north, and then proceeded to behave in ways predicted from a reversed program (Helm and Muheim 2021). Overall, the combination of citizen science and tracking has provided valuable insights into migratory timing, revealing that certain aspects of timing can be flexible, whereas others are influenced by intrinsic components of the migratory program.

Technological advances in genomic sequencing are also contributing to a more holistic understanding of the genetic basis of migration timing by synthesizing information from the molecular structure of genes to the gene networks and pathways involved (Liedvogel et al. 2011). For example, many studies focused on the identification of highly significant outliers between closely related populations of birds with divergent migratory phenotypes have succeeded in identifying a variety of genes and genomic regions putatively important to migration timing (Lundberg et al. 2013, 2017; Franchini et al. 2017; Gu et al. 2021). Although much of this work in birds has fallen short in demonstrating functional links with actual clock pathways, recent work suggests that this may, in part, be due to the focus on core clock genes rather than genes that entrain and modulate the clock pathway and/or to limitations in detection methods that focus only on highly significant outlier loci, thus ignoring genes of small effect (Bossu et al. 2022). Specifically, work by Bossu et al. (2022) using relaxed detection thresholds supports the idea that intraspecific variation in migratory timing in American Kestrels (*Falco sparverius*) is in part due to genetic variation in a suite of metabolic and light input pathway genes that help modulate biological clocks. Future work focused on re-examining migration genetic data sets in other taxa with relaxed thresholds may also help identify whether or not there is a consistent pattern regarding the importance of clock-linked genes in regulating migration timing.

Research on the systems that birds use to orient and navigate during migration has also become increasingly technology-driven. In recent decades, such work often involves sophisticated instrumentation and collaboration with engineers and biophysicists to manipulate and measure natural cues and quantify animal movement in great detail, which benefits from rapidly developing wildlife tracking technologies (Taylor et al. 2017, Nathan et al. 2022). Although it is well established that stars, sun, and Earth's magnetic field underlie the compass sense, the hierarchical relationship between compasses remains poorly understood (Chernetsov 2017). The importance of different compass systems varies from study to study, with some studies suggesting simple dominance of one compass (Chernetsov et al. 2011, Schmaljohann et al. 2012) and other studies suggesting the magnetic compass is calibrated by the sunset cues (Cochran et al. 2004, Muheim et al. 2006) or vice versa (Wiltschko et al. 2001). Further investigations are clearly needed to disentangle the ecological or taxonomical factors explaining how and why different species use different combinations of compass systems.

The role of the magnetic compass remains perhaps the most enigmatic of all the senses thought to be involved in navigation. Its existence in birds is beyond reasonable doubt due to multiple studies demonstrating behavioral responses (usually in round arenas). Yet, the primary magnetosensory cells along with their exact locus in the body, ultracellular structure, perceptive principle, and functional characteristics remain elusive, making it "sense without a sensor" (Nordmann et al. 2017). Currently, at least three hypotheses attempt to indirectly explain how magnetic sensors may promote avian migration. The radical pair hypothesis (Ritz et al. 2000, Hore and Mouritsen 2016) suggests that magnetic field may be perceived visually by specialized photoreceptors of the retina of the eye, and this perception is mediated by magnetically sensitive chemical reactions involving light-sensitive proteins known of as cryptochromes. Experiments on European robins with a lesioned Cluster N (a forebrain region processing visual information from the eye's retina) found impaired use of the magnetic field for choosing migratory direction, but this did not affect their star and sun compasses and overall vision, suggesting the magnetic compass is associated with vision (Zapka et al. 2009). Alternatively, the trigeminal nerve hypothesis assumes the existence of magnetoreceptor cells associated with the endings of the trigeminal nerve (V1 nerve). Studies suggested that the nerve transmits magnetic information to the brain in some migratory birds (Heyers et al. 2010, Elbers et al. 2017), but its function is not fully understood. Integrity of the V1 nerve is crucial for navigational responses in Common Reed Warblers (Kishkinev et al. 2013, Pakhomov et al. 2018), but it appears to be unnecessary for navigation in a long-distance migrating gull species (Wikelski et al. 2015). Last, the electromagnetic induction hypothesis (known since the mid-19th century) recently received experimental support in nonmigratory birds (pigeons). The hypothesis proposes the electrosensitive hair cells of the semicircular canals of the inner ear as magnetoreceptors (Nimpf et al. 2019), but it has not yet been tested in any migratory species. All the hypotheses mentioned coexist, and it is commonly assumed that a single species may use more than one mechanism, although the functional implications of that remain unknown.

A growing number of recent studies combine various sensory manipulations impairing olfactory or magnetic senses with long-distance physical translocations and wildlife tracking (free-ranging animals) or orientation tests (captive birds). So far, this approach has obtained mixed results, with some studies supporting the role of magnetic sense associated with the trigeminal nerve for navigation (Kishkinev et al. 2013), whereas others lend support to the sense of smell (Gagliardo et al. 2013, Pollonara et al. 2015, Wikelski et al. 2015) or even show no effect of these treatments (Kishkinev et al. 2016). Satellite tracking and aerial observations have clearly demonstrated the role of olfactory cues for finding foraging patches across short distance (tens of miles)—for example, in white storks feeding on patches of freshly cut grass or in albatrosses finding sea surface spots with high abundance of food (Nevitt et al. 2008, Wikelski et al. 2021). In summary, studies suggest that the navigation system of birds is based on multiple cues, and their role depends on scale, environment, and taxon. Future studies will be technology-driven and should help unravel the complex relationships between navigational systems of different taxa and their evolutionary history, neurophysiological adaptations, and ecological factors.

Conclusion

It is an exciting time to be a migration scientist, with technological advances facilitating opportunities for the synthesis of information across fields in ways that were unimaginable only a couple of decades ago. Although migratory feats have always held a place of wonder for scientists and natural history enthusiasts, it is clear that recent technological advances have allowed us to appreciate that birds are migrating farther, higher, and faster than we previously thought possible. It is also becoming increasingly apparent that the systems and sensors that birds rely on to undertake such migratory journeys are more complex and varied than previously realized. As we continue to rapidly gather information about where birds migrate and how they find their way, we are also able to more easily document the ways in which anthropogenic change is influencing their future survival. The challenge for the next generation will be to continue to push the boundaries of our knowledge regarding avian migration while simultaneously harnessing those same innovations to catalyze avian conservation efforts at an equal or greater pace.

References

Åkesson, S., and B. Helm. 2020. Endogenous programs and flexibility in bird migration. *Frontiers in Ecology and Evolution* 8: Article 78.

Alerstam, T. 1993. *Bird Migration*. Cambridge University Press, Cambridge, UK.

Areta, J. I., S. A. Salvador, F. A. Gandoy, E. S. Bridge, F. C. Gorleri, T. M. Pegan, E. R. Gulson-Castillo, K. A. Hobson, and D. W. Winkler. 2021. Rapid adjustments of migration and life history in hemisphere-switching cliff swallows. *Current Biology* 31:2914–2919.

Audubon, J. J. 1834. *Ornithological Biography: Or An Account of the Habits of the Birds of the United States of America; Accompanied by Descriptions of the Objects Represented in the Work Entitled The Birds of America, and Interspersed With Delineations of American Scenery and Manners.* 5 Vol. Il. Q. Vol. IA: Black, London; Vols. 2–5, A. & C. Black, London.

Bay, R. A., R. J. Harrigan, V. L. Underwood, H. L. Gibbs, T. B. Smith, and K. Ruegg. 2018. Genomic signals of selection predict climate-driven population declines in a migratory bird. *Science* 359:83–86.

Bay, R. A., D. S. Karp, J. F. Saracco, W. R. L. Anderegg, L. O. Frishkoff, D. Wiedenfeld, T. B. Smith, and K. Ruegg. 2021. Genetic variation reveals individual-level climate tracking across the annual cycle of a migratory bird. *Ecology Letters* 24:819–828.

Bearhop, S., W. Fiedler, R. W. Furness, S. C. Votier, S. Waldron, J. Newton, G. J. Bowen, P. Berthold, and K. Farnsworth. 2005. Assortative mating as a mechanism for rapid evolution of a migratory divide. *Science* 310:502–504.

Berthold, P. 1993. *Bird Migration: A General Survey.* Oxford University Press, New York.

Berthold, P. 2001. *Bird Migration: A General Survey.* 2nd ed. Oxford University Press, New York.

Bossu, C. M., J. A. Heath, G. S. Kaltenecker, B. Helm, and K. C. Ruegg. 2022. Clock-linked genes underlie seasonal migratory timing in a diurnal raptor. *Proceedings of the Royal Society B: Biological Sciences* 289: Article 20212507.

Chernetsov, N. 2017. Compass systems. *Journal of Comparative Physiology A* 203:447–453.

Chernetsov, N., D. Kishkinev, V. Kosarev, and C. V. Bolshakov. 2011. Not all songbirds calibrate their magnetic compass from twilight cues: A telemetry study. *Journal of Experimental Biology* 214:2540–2543.

Clegg, S. M., J. F. Kelly, M. Kimura, and T. B. Smith. 2003. Combining genetic markers and stable isotopes to reveal population connectivity and migration patterns in a Neotropical migrant, Wilson's warbler (*Wilsonia pusilla*). *Molecular Ecology* 12:819–830.

Cochran, W. W., H. Mouritsen, and M. Wikelski. 2004. Migrating songbirds recalibrate their magnetic compass daily from twilight cues. *Science* 304:405–408.

Dokter, A. M., A. Farnsworth, D. Fink, V. Ruiz-Gutierrez, W. M. Hochachka, F. A. La Sorte, O. J. Robinson, K. V. Rosenberg, and S. Kelling. 2018. Seasonal abundance and survival of North America's migratory avifauna determined by weather radar. *Nature Ecology & Evolution* 2:1603–1609.

Dufour, P., C. de Franceschi, P. Doniol-Valcroze, F. Jiguet, M. Guéguen, J. Renaud, S. Lavergne, and P.-A. Crochet. 2021. A new westward migration route in an Asian passerine bird. *Current Biology* 31:5590–5596.

Elbers, D., M. Bulte, F. Bairlein, H. Mouritsen, and D. Heyers. 2017. Magnetic activation in the brain of the migratory northern wheatear (*Oenanthe oenanthe*). *Journal of Comparative Physiology A* 203:591–600.

Eng, M. L., B. J. M. Stutchbury, and C. A. Morrissey. 2019. A neonicotinoid insecticide reduces fueling and delays migration in songbirds. *Science* 365:1177–1180.

Franchini, P., I. Irisarri, A. Fudickar, A. Schmidt, A. Meyer, M. Wikelski, and J. Partecke. 2017. Animal tracking meets migration genomics: Transcriptomic analysis of a partially migratory bird species. *Molecular Ecology* 26:3204–3216.

Gagliardo, A., J. Bried, P. Lambardi, P. Luschi, M. Wikelski, and F. Bonadonna. 2013. Oceanic navigation in Cory's shearwaters: Evidence for a crucial role of olfactory cues for homing after displacement. *Journal of Experimental Biology* 216:2798–2805.

Garcia-Perez, B., K. A. Hobson, R. L. Powell, C. J. Still, and G. H. Huber. 2013. Switching hemispheres: A new migration strategy for the disjunct Argentinean breeding population of Barn Swallow (*Hirundo rustica*). *PLoS One* 8: Article e55654.

Gauthreaux, S. A. 1980. Preface. Pages xi–xii in S. A. Gauthreaux, ed. *Animal Migration, Orientation and Navigation.* Academic Press, New York.

Gill, R. E., T. L. Tibbitts, D. C. Douglas, C. M. Handel, D. M. Mulcahy, J. C. Gottschalck, N. Warnock, B. J. McCaffery, P. F. Battley, and T. Piersma. 2009. Extreme endurance flights by landbirds crossing the Pacific Ocean: Ecological corridor rather than barrier? *Proceedings of the Royal Society B: Biological Sciences* 276:447–457.

González, A. M., N. J. Bayly, and K. A. Hobson. 2020. Earlier and slower or later and faster: Spring migration pace linked to departure time in a Neotropical migrant songbird. *Journal of Animal Ecology* 89:2840–2851.

González-Prieto, A. M., N. J. Bayly, G. J. Colorado, and K. A. Hobson. 2017. Topography of the Andes Mountains shapes the wintering distribution of a migratory bird. *Diversity and Distributions* 23:118–129.

Gu, Z., S. Pan, Z. Lin, L. Hu, X. Dai, J. Chang, Y. Xue, H. Su, J. Long, M. Sun, S. Ganusevich, V. Sokolov, A. Sokolov, I. Pokrovsky, F. Ji, M. W. Bruford, A. Dixon, and X. Zhan. 2021. Climate-driven flyway changes and memory-based long-distance migration. *Nature* 591:259–264.

Hansson, L.-A., and S. Åkesson. 2014. *Animal Movement Across Scales.* Oxford University Press, New York.

Hedenström, A., G. Norevik, K. Warfvinge, A. Andersson, J. Bäckman, and S. Åkesson. 2016. Annual 10-month aerial life phase in the common swift *Apus apus. Current Biology* 26:3066–3070.

Heim, W., R. J. Heim, I. Beermann, O. A. Burkovskiy, Y. Gerasimov, P. Ktitorov, K. Ozaki, I. Panov, M. M. Sander, S. Sjöberg, S. M. Smirenski, A. Thomas, A. P. Tøttrup, I. M. Tiunov, M. Willemoes, N. Hölzel, K. Thorup, and J. Kamp. 2020. Using geolocator tracking data and ringing archives to validate citizen-science based seasonal predictions of bird distribution in a data-poor region. *Global Ecology and Conservation* 24: Article e01215.

Helm, B., and R. Muheim. 2021. Bird migration: Clock and compass facilitate hemisphere switching. *Current Biology* 31:R1058–R1061.

Helm, B., B. M. Van Doren, D. Hoffmann, and U. Hoffmann. 2019. Evolutionary response to climate change in migratory pied flycatchers. *Current Biology* 29:3714–3719.

Heyers, D., M. Zapka, M. Hoffmeister, J. M. Wild, and H. Mouritsen. 2010. Magnetic field changes activate the trigeminal brainstem complex in a migratory bird. *Proceedings of the National Academy of Sciences of the USA* 107:9394–9399.

Hobson, K. A., R. Barnett-Johnson, and T. Cerling. 2010. Using Isoscapes to Track Animal Migration. Pages 273–298 in J. West, G. Bowen, T. Dawson, and K. Tu, eds. *Isoscapes.* Springer, Dordrecht, the Netherlands.

Hore, P. J., and H. Mouritsen. 2016. The radical-pair mechanism of magnetoreception. *Annual Review of Biophysics* 45:299–344.

Horton, K. G., F. A. La Sorte, D. Sheldon, T.-Y. Lin, K. Winner, G. Bernstein, S. Maji, W. M. Hochachka, and A. Farnsworth. 2020. Phenology of nocturnal avian migration has shifted at the continental scale. *Nature Climate Change* 10:63–68.

Jiguet, F., A. Robert, R. Lorrillière, K. A. Hobson, K. J. Kardynal, R. Arlettaz, F. Bairlein, V. Belik, P. Bernardy, J. L. Copete, M. A. Czajkowski, S. Dale, V. Dombrovski, D. Ducros, R. Efrat, J. Elts, Y. Ferrand, R. Marja, S. Minkevicius, P. Olsson, M. Pérez, M. Piha, M. Raković, H. Schmaljohann, T. Seimola, G. Selstam, J.-P. Siblet, M. Skierczyǹski, A. Sokolov, J. Sondell, and C. Moussy. 2019. Unravelling migration connectivity reveals unsustainable hunting of the declining ortolan bunting. *Science Advances* 5: Article eaau2642.

Kelly, J. F., K. C. Ruegg, and T. B. Smith. 2005. Combining isotopic and genetic markers to identify breeding origins of migrant birds. *Ecological Applications* 15:1487–1494.

Kishkinev, D., N. Chernetsov, D. Heyers, and H. Mouritsen. 2013. Migratory Reed Warblers need intact trigeminal nerves to correct for a 1,000 km eastward displacement. *PLoS One* 8: Article e65847.

Kishkinev, D., D. Heyers, B. K. Woodworth, G. W. Mitchell, K. A. Hobson, and D. R. Norris. 2016. Experienced migratory songbirds do not display goal-ward orientation after release following a cross-continental displacement: An automated telemetry study. *Scientific Reports* 6: Article 37326.

Klipp, J. C., M. P. Cruz, M. E. Iezzi, D. Varela, and U. Balza. 2018. Determining the wintering range of Broad-Winged Hawk (*Buteo platypterus*) in South America using citizen-science database. *Ornitología Neotropica* 29:337–342.

La Sorte, F. A., and C. H. Graham. 2021. Phenological synchronization of seasonal bird migration with vegetation greenness across dietary guilds. *Journal of Animal Ecology* 90:343–355.

La Sorte, F. A., C. A. Lepczyk, J. L. Burnett, A. H. Hurlbert, M. W. Tingley, and B. Zuckerberg. 2018. Opportunities and challenges for big data ornithology. *The Condor* 120:414–426.

Laughlin, A. J., C. M. Taylor, D. W. Bradley, D. LeClair, R. G. Clark, R. D. Dawson, P. O. Dunn, A. Horn, M. Leonard, D. R. Sheldon, D. Shutler, L. A. Whittingham, D. W. Winkler, and D. R. Norris. 2013. Integrating information from geolocators, weather radar, and citizen science to uncover a key stopover area of an aerial insectivore. *The Auk* 130:230–239.

Liedvogel, M., S. Åkesson, and S. Bensch. 2011. The genetics of migration on the move. *Trends in Ecology & Evolution* 26:561–569.

Lindström, Å., T. Alerstam, A. Andersson, J. Bäckman, P. Bahlenberg, R. Bom, R. Ekblom, R. H. G. Klaassen, M. Korniluk, S. Sjöberg, and J. K. M. Weber. 2021. Extreme altitude changes between night and day during marathon flights of great snipes. *Current Biology* 31:3433–3439.

Louv, R., and J. W. Fitzpatrick. 2012. *Citizen Science*. Cornell University Press, Ithaca, NY.

Lundberg, M., J. Boss, B. Canbäck, M. Liedvogel, K. W. Larson, M. Grahn, S. Åkesson, S. Bensch, and A. Wright. 2013. Characterisation of a transcriptome to find sequence differences between two differentially migrating subspecies of the willow warbler *Phylloscopus trochilus*. *BMC Genomics* 14: Article 330.

Lundberg, M., M. Liedvogel, K. Larson, H. Sigeman, M. Grahn, A. Wright, S. Akesson, and S. Bensch. 2017. Genetic differences between willow warbler migratory phenotypes are few and cluster in large haplotype blocks. *Evolution Letters* 1:155–168.

McKinnon, E. A., and O. P. Love. 2018. Ten years tracking the migrations of small landbirds: Lessons learned in the golden age of bio-logging. *The Auk* 135:834–856.

Møller, A. P., T. van Nus, and K. A. Hobson. 2021. Rapid reduction in migration distance in relation to climate in a long-distance migratory bird. *Current Zoology* 68:233–235.

Montgomerie, B. 2018. Audubon's legendary experiments. American Ornithological Society. https://americanornithology.org/audubons-legendary-experiments.

Muheim, R., J. B. Phillips, and S. Åkesson. 2006. Polarized light cues underlie compass calibration in migratory songbirds. *Science* 313:837–839.

Nathan, R., C. T. Monk, R. Arlinghaus, T. Adam, J. Alós, M. Assaf, H. Baktoft, C. E. Beardsworth, M. G. Bertram, A. I. Bijleveld, T. Brodin, J. L. Brooks, A. Campos-Candela, S. J. Cooke, K. Ø. Gjelland, P. R. Gupte, R. Harel, G. Hellström, F. Jeltsch, S. S. Killen, T. Klefoth, R. Langrock, R. J. Lennox, E. Lourie, J. R. Madden, Y. Orchan, I. S. Pauwels, M. Říha, M. Roeleke, U. E. Schlägel, D. Shohami, J. Signer, S. Toledo, O. Vilk, S. Westrelin, M. A. Whiteside, and I. Jarić. 2022. Big-data approaches lead to an increased understanding of the ecology of animal movement. *Science* 375: Article eabg1780.

Nevitt, G. A., M. Losekoot, and H. Weimerskirch. 2008. Evidence for olfactory search in wandering albatross, *Diomedea exulans*. *Proceedings of the National Academy of Sciences of the USA* 105:4576–4581.

Newton, I. 2010. *The Migration Ecology of Birds*. Elsevier, New York.

Nimpf, S., G. C. Nordmann, D. Kagerbauer, E. P. Malkemper, L. Landler, A. Papadaki-Anastasopoulou, L. Ushakova, A. Wenninger-Weinzierl, M. Novatchkova, P. Vincent, T. Lendl, M. Colombini, M. J. Mason, and D. A. Keays. 2019. A putative mechanism for magnetoreception by electromagnetic induction in the pigeon inner ear. *Current Biology* 29:4052–4059.

Nordmann, G. C., T. Hochstoeger, and D. A. Keays. 2017. Magnetoreception: A sense without a receptor. *PLoS Biology* 15: Article e2003234.

Norevik, G., S. Åkesson, A. Andersson, J. Bäckman, and A. Hedenström. 2019. The lunar cycle drives migration of a nocturnal bird. *PLoS Biology* 17: Article e3000456.

Pakhomov, A., A. Anashina, D. Heyers, D. Kobylkov, H. Mouritsen, and N. Chernetsov. 2018. Magnetic map navigation in a migratory songbird requires trigeminal input. *Scientific Reports* 8: Article 11975.

Pollonara, E., P. Luschi, T. Guilford, M. Wikelski, F. Bonadonna, and A. Gagliardo. 2015. Olfaction and topography, but not magnetic cues, control navigation in a pelagic seabird: Displacements with shearwaters in the Mediterranean Sea. *Scientific Reports* 5: Article 16486.

Pulgarín-R, P. C., C. Gómez, N. J. Bayly, S. Bensch, A. M. FitzGerald, N. Starkloff, J. J. Kirchman, A. M. González-Prieto, K. A. Hobson, J. Ungvari-Martin, H. Skeen, M. I. Castaño, and C. D. Cadena. 2019. Migratory birds as vehicles for parasite dispersal? Infection by avian haemosporidians over the year and throughout the range of a long-distance migrant. *Journal of Biogeography* 46:83–96.

Ritz, T., S. Adem, and K. Schulten. 2000. A model for photoreceptor-based magnetoreception in birds. *Biophysical Journal* 78:707–718.

Ruegg, K., E. C. Anderson, M. Somveille, R. A. Bay, M. Whitfield, E. H. Paxton, and T. B. Smith. 2021. Linking climate niches across seasons to assess population vulnerability in a migratory bird. *Global Change Biology* 27:3519–3531.

Ruegg, K., R. A. Bay, E. C. Anderson, J. F. Saracco, R. J. Harrigan, M. Whitfield, E. H. Paxton, and T. B. Smith. 2018. Ecological genomics predicts climate vulnerability in an endangered southwestern songbird. *Ecology Letters* 21:1085–1096.

Ruegg, K. C., E. C. Anderson, R. J. Harrigan, K. L. Paxton, J. F. Kelly, F. Moore, and T. B. Smith. 2017. Genetic assignment with isotopes and habitat suitability (GAIAH), a migratory bird case study. *Methods in Ecology and Evolution* 8:1241–1252.

Rundel, C. W., M. B. Wunder, A. H. Alvarado, K. C. Ruegg, R. Harrigan, A. Schuh, J. F. Kelly, R. B. Siegel, D. F. DeSante, T. B. Smith, and J. Novembre. 2013. Novel statistical methods for integrating genetic and stable isotope data to infer individual-level migratory connectivity. *Molecular Ecology* 22:4163–4176.

Schmaljohann, H., T. Rautenberg, R. Muheim, B. Naef-Daenzer, and F. Bairlein. 2012. Response of a free-flying songbird to an experimental shift of the light polarization pattern around sunset. *Journal of Experimental Biology* 216:1381–1387.

Shamoun-Baranes, J., C. Nilsson, S. Bauer, and J. Chapman. 2019. Taking radar aeroecology into the 21st century. *Ecography* 42:847–851.

Sjöberg, S., G. Malmiga, A. Nord, A. Andersson, J. Bäckman, M. Tarka, M. Willemoes, K. Thorup, B. Hansson, T. Alerstam, and D. Hasselquist. 2021. Extreme altitudes during diurnal flights in a nocturnal songbird migrant. *Science* 372:646–648.

Smith, T. B., P. P. Marra, M. S. Webster, I. Lovette, H. L. Gibbs, R. T. Holmes, K. A. Hobson, and S. Rohwer. 2003. A call for feather sampling. *The Auk* 120:218–221.

Strong, C., B. Zuckerberg, J. L. Betancourt, and W. D. Koenig. 2015. Climatic dipoles drive two principal modes of North American boreal bird irruption. *Proceedings of the National Academy of Sciences of the USA* 112:E2795–E2802.

Taylor, P. D., T. L. Crewe, S. A. Mackenzie, D. Lepage, Y. Aubry, Z. Crysler, G. Finney, C. M. Francis, C. G. Guglielmo, D. J. Hamilton, R. L. Holberton, P. H. Loring, G. W. Mitchell, D. R. Norris, J. Paquet, R. A. Ronconi, J. R. Smetzer, P. A. Smith, L. J. Welch, and B. K. Woodworth. 2017. The Motus Wildlife Tracking System: A collaborative research network to enhance the understanding of wildlife movement. *Avian Conservation and Ecology* 12: Article 8.

Tsai, P. Y., C. J. Ko, S. Y. Chia, Y. J. Lu, and M. N. Tuanmu. 2021. New insights into the patterns and drivers of avian altitudinal migration from a growing crowdsourcing data source. *Ecography* 44:75–86.

Turbek, S. P., D. R. Schield, E. S. C. Scordato, A. Contina, X.-W. Da, Y. Liu, Y. Liu, E. Pagani-Núñez, Q.-M. Ren, C. C. R. Smith, C. A. Stricker, M. Wunder, D. M. Zonana, and R. J. Safran. 2022. A migratory divide spanning two continents is associated with genomic and ecological divergence. *Evolution* 76:722–736.

Turbek, S. P., E. S. C. Scordato, and R. J. Safran. 2018. The role of seasonal migration in population divergence and reproductive isolation. *Trends in Ecology & Evolution* 33:164–175.

Van Doren, B. M., K. G. Horton, A. M. Dokter, H. Klinck, S. B. Elbin, and A. Farnsworth. 2017. High-intensity urban light installation dramatically alters nocturnal bird migration. *Proceedings of the National Academy of Sciences of the USA* 114:11175–11180.

Van Doren, B. M., D. E. Willard, M. Hennen, K. G. Horton, E. F. Stuber, D. Sheldon, A. H. Sivakumar, J. Wang, A. Farnsworth, and B. M. Winger. 2021. Drivers of fatal bird collisions in an urban center. *Proceedings of the National Academy of Sciences of the USA* 118: Article e2101666118.

Wikelski, M., E. Arriero, A. Gagliardo, R. A. Holland, M. J. Huttunen, R. Juvaste, I. Mueller, G. Tertitski, K. Thorup, M. Wild, M. Alanko, F. Bairlein, A. Cherenkov, A. Cameron, R. Flatz, J. Hannila, O. Huppop, M. Kangasniemi, B. Kranstauber, M. L. Penttinen, K. Safi, V. Semashko, H. Schmid, and R. Wistbacka. 2015. True navigation in migrating gulls requires intact olfactory nerves. *Scientific Reports* 5: Article 17061.

Wikelski, M., M. Quetting, Y. Cheng, W. Fiedler, A. Flack, A. Gagliardo, R. Salas, N. Zannoni, and J. Williams. 2021. Smell of green leaf volatiles attracts white storks to freshly cut meadows. *Scientific Reports* 11: Article 12912.

Wiltschko, R., U. Munro, H. Ford, and W. Wiltschko. 2001. Orientation in migratory birds: Time-associated relearning of celestial cues. *Animal Behaviour* 62:245–250.

Winkler, D. W., F. A. Gandoy, J. I. Areta, M. J. Iliff, E. Rakhimberdiev, K. J. Kardynal, and K. A. Hobson. 2017. Long-distance range expansion and rapid adjustment of migration in a newly established population of barn swallows breeding in Argentina. *Current Biology* 27:1080–1084.

Youngflesh, C., J. Socolar, B. R. Amaral, A. Arab, R. P. Guralnick, A. H. Hurlbert, R. LaFrance, S. J. Mayor, D. A. W. Miller, and M. W. Tingley. 2021. Migratory strategy drives species-level variation in bird sensitivity to vegetation green-up. *Nature Ecology & Evolution* 5:987–994.

Zapka, M., D. Heyers, C. M. Hein, S. Engels, N.-L. Schneider, J. Hans, S. Weiler, D. Dreyer, D. Kishkinev, J. M. Wild, and H. Mouritsen. 2009. Visual but not trigeminal mediation of magnetic compass information in a migratory bird. *Nature* 461:1274–1277.

SECTION IV

DATABASES AND CITIZEN SCIENCE

17

Raising Cassandra's Voice

A New Dawn for Social Justice and Reliable Knowledge in Social–Ornithological Systems

David C. Pavlacky, Jr.

From oral–poetic traditions of Greek and Roman mythology (Schmitz 1867), Cassandra was a daughter of Trojan royalty Priam and Hecabe. As a child, Cassandra was left at the temple of Apollo, where attendant serpents whispered into her ears, giving her understanding of divine sounds of nature and voices of birds, and thereby the ability to predict the future. Later in life, she defied Apollo, and the god cursed her so that none would believe her true prophecies. In a tragic twist of fate, Cassandra warned the Trojan people that Achaean warriors were hiding in the Wooden Horse, but disbelieving her, the Trojans demeaned her and silenced her voice. The Achaean army triumphed, and the great citadel of Troy fell.

The ancient tale of Cassandra's Curse provides several points of access to the following chapters in this section on citizen science and avian databases: Chapter 18, "Recent Advances in Avian Occurrence and Movement Databases," by Frank A. La Sorte and Adriaan M. Dokter; Chapter 19, "Participatory Science in the Field of Ornithology," by Bradley Allf, Deja Perkins, Jin Bai, and Caren Cooper; and Chapter 20, "Expanding the Scope of Ornithology by Engaging Diverse Communities," by Sebastian Moreno, and Paige S. Warren. Considering the injustice faced by Cassandra when her voice was silenced within the deceptive plan of the Trojan Horse, "expanding the scope of ornithology" may well consider extending recommendations to include social objectives aimed at reversing historical injustice from systemic oppression of marginalized identities (Foggin et al. 2021). Interestingly, the silencing of Cassandra's voice and downfall of Troy were inextricably linked within a complicit plan of conquest, which parallels a need for "participatory science" to redefine the discipline as a social–ecological system in which both biodiversity declines and social injustices are often created by the same policies and societal norms of dominant social groups (Schell et al. 2020). The serpent attendants of Apollo gifted Cassandra with augury, or the ability to predict the future by observing bird behavior, but the god then cursed her so that none would believe her prophecies. Despite purported predictive performance of modeling efforts in Chapters 18 and 19, Cassandra's Curse is invoked when sampling and observation biases and reliance on untested model assumptions elevate the meaning and impact of model output beyond levels of belief supported by underlying data (Diffendorfer and Doherty 2004).

An increasing recognition of tight linkages between social and natural systems has led to the development of transdisciplinary programs to integrate social

David C. Pavlacky, Jr. *Raising Cassandra's Voice*. In: *New Perspectives in Ornithology*. Edited by: Scott V. Edwards and J. Michael Reed, Oxford University Press. © Oxford University Press (2025). DOI: 10.1093/oso/9780197787670.003.0017 and ecological sciences (Kareiva and Marvier 2012). Applied social science for environmental and conservation education often seeks to better understand target audiences and evaluate the effectiveness of education and outreach (Bennett et al. 2017). However, processes for imagining desirable futures and identifying actions to improve social outcomes are poorly developed in the social sciences (Bennett et al. 2017). Often, engagement with society is conducted as an end in itself rather than a means to achieving substantive goals (Haarstad et al. 2018). Standard definitions of public engagement as "the process of encouraging people to be interested in the work of an organization" often fail to acknowledge plurality in public sectors representing multiple voices and perspectives, and they often do not ensure objectives of affected parties are heard, understood, and respected (McShane et al. 2011, p. 967). Chapter 20 discussed several objectives for broadening engagement that can be classified as (1) improving diversity and inclusion in science, (2) enhancing participant science and nature learning goals, (3) advancing science and research methods, and (4) furthering bird conservation. Although engagement to build relationships, develop trust, and understand diverse perspectives is unquestionably vital for ornithology (Saunders et al. 2021), clearly articulating objectives to address social vulnerability and injustice provides the means to affect tangible and meaningful social change (Crausbay et al. 2022). Engagement strategies to make progress toward accountability and sustainability ultimately require developing measurable social objectives that reflect desirable futures for well-being, values, agency, and equality (Hicks et al. 2016). Another way to pursue the intersection between ornithology and social justice is to cultivate partnerships with organizations led by underrepresented groups, such as Birdability, Feminist Bird Club, and Latino Outdoors, and founders of Black Birders Week, Black AF in STEM. Our professional societies are increasingly becoming a forum for embracing social change, including Out to Innovate's LGBTQ+ partnership with the Rainbow Lorikeets of the American Ornithological Society (AOS) and Native American Fish and Wildlife Society's partnership with the Native Peoples' Wildlife Management Working Group of the Wildlife Society. Because social injustice and biodiversity declines are often caused by the same underlying social norms and policies (Schell et al. 2020), advancing social objectives for well-being, agency, and equality of marginalized people will likely also benefit biodiversity.

When considering the growth and development of ornithological citizen science, it is important to recognize the extent to which colonialism (Trisos et al. 2021), inequality (Crosman et al. 2022), and systemic racism (Schell et al. 2020) have shaped and continue to shape the discipline. The disheartening magnitude of social injustice occurring at the dawn of citizen science and first wave of environmentalism (late 1800s and early 1900s) can be readily experienced by anyone attempting an internet search for historical contributions of Indigenous people, people of color, LGBTQ+, or women in the field of ornithology. By looking closely, we can see glimpses of a forgotten history of ornithology, including contributions from Black Walden (1780–1820); York from the Corps of Discovery Expedition (1804–1806); and Harriet Tubman (1822–1913), the American abolitionist, social activist, and naturalist (Chavis et al. 2021). Graceanna Lewis (1821–1912) was a pioneer female ornithologist and activist in the anti-slavery, temperance, and women's suffrage movements. Charles Henry Turner (1867–1923) was one of the first African Americans to earn a PhD from the

University of Chicago and the first black scientist with a focus on animal behavior to publish avian research in the journal *Science* (Turner 1892). Robert Alexander Gilbert (1870–1942) was an ornithologist and employee of the Museum of Comparative Zoology, Harvard University (Mitchell 2006).

The millennia-long intellectual traditions of traditional ecological knowledge (TEK) rivals the annals of Western science, yet has been historically marginalized by the scientific community (Kimmerer 2002). Interestingly, the act of looking backward in history for significant contributions to ornithology represents a departure in the way Western scientists and native peoples contemplate their place in history. Western European immigrants to North America tend to look backward and forward in time for their sense of place in history, whereas native peoples of North America look to connections around them for their sense of place in history (Pierotti and Wildcat 2000). Nevertheless, it is important to recognize that traditional ornithological knowledge was sampled and extracted by Western European colonists over the early centuries, informing the development of what we now recognize as modern ornithology (Kimmerer 2002). The recognition that TEK is the foundation of modern ornithology should encourage scientists to build authentic relationships with Indigenous people, take action to support communities, and acknowledge how displacement from ancestral land affects Indigenous cultural identities.

Rapid growth in citizen science and the second wave of environmentalism (1960s) corresponded to the emergence of social movements for civil rights, coinciding with adoption of the International Covenant on Civil and Political Rights. It was not until then that society throughout the world began to acknowledge contributions of Indigenous people, people of color, women, and other underrepresented identities in the field of ornithology. And yet, dark shadows of social imbalance operating at the dawn of environmentalism and citizen science persist and are painfully evident in the current composition of citizen scientists and professional societies in the United States. For example, Chapter 19 highlights a National Science Foundation–funded study in 1974 that found half of ornithological society members were amateurs and called for societies to increase public participation in ornithology. However, in 1971, the Sierra Club surveyed members for interest in addressing issues of social justice and found 41% strongly disagreed and only 15% strongly agreed the club should be actively involved in conservation problems involving urban poor and ethnic identities (Gottlieb 2005). Chapter 19 primarily focuses on science objectives for contributory citizen science and suggests that grassroots community science has primarily involved nest monitoring and provisioning of nesting sites. However, the collaboration and co-production aspects of community science hold promise for addressing a wide variety of ecological questions as well as affecting social empowerment, agency, and collective action (Charles et al. 2020). Community science involves collaboration among stakeholders and scientists, who mutually define the problem, research questions, methods, and outputs, as well as make scientific inferences and decide on appropriate use of science. The recognition that participatory citizen science has failed to serve all segments of society equitably has led to calls for increased funding for community science, the subset of citizen science capable of addressing the interests, concerns, and needs of historically and currently underserved members of society (Cooper et al. 2021). Much of the previous discussion involves critical social science, recognizing subjects of justice, harms that constitute injustice, mechanisms that produce injustices, and responses to alleviate these (Martin et al. 2016) in a way that allows ecologists to properly locate themselves as a part of the social–ecological system they study (Trisos et al. 2021).

The rapid emergence of big data from crowdsourcing and technology has driven the development of a new paradigm to organize large quantities of data, employ computer algorithms to discover complex patterns, and then develop hypotheses about processes generating the patterns (Kelling et al. 2009). However, the new paradigm depends on the retrospective analysis of observational data known to make weaker inferences than prospective analysis of observational data guided by a priori hypotheses (Nichols et al. 2012). For example, spatial patterns are often used to make inferences about dynamic processes of movement and habitat quality that may have produced them, but the reliability of such inferences is questionable because a multitude of post hoc explanations can be invoked to describe most ecological patterns (MacKenzie et al. 2018). Considering the rapid proliferation of modeling methodology, reliable knowledge from ecological research depends on understanding how sampling and data analysis affect inference (Williams and Brown 2019). Overall, lack of attention to fundamentals of statistical inference has contributed to imprecise concepts of sampling and observation bias in citizen science. Reliable knowledge from large-scale biodiversity monitoring requires addressing two important sources of variation in raw data: geographic sampling and incomplete detection (Nichols et al. 2012).

Geographic sampling describes the process of making inference to unsurveyed locations, and it involves aggregating information across locations and estimating spatial variation of ecological processes (Royle and Dorazio 2008). It is important to recognize that sampling bias affects both the accuracy and precision of ecological estimates. Design-based inference to unsampled locations proceeds through probabilistic sampling designs, whereas model-based inference to unsampled locations proceeds via covariate predictions and distributional assumptions (Williams and Brown 2019). Model-based approaches have the advantage of making inference from nonrandom sampling, but reliable inference depends on the validity of several assumptions, including that the sample adequately represents the population from which it was obtained (Williams and Brown 2019). Although sampling bias in the citizen science literature is often discussed in terms of high variability of data, spatiotemporal clustering in tandem with very large sample sizes are likely to underestimate variance and bias estimates, resulting in very precise, wrong answers. Detection refers to an aspect of most ecological data in which not all individuals of a species present at a location are detected (Nichols et al. 2012). The observation process involves a probabilistic model of detections that approximates the process producing the observed data (Royle and Dorazio 2008), and reliable inference often involves allowing the observation process to vary over space and time. However, the observation process is often misrepresented in the citizen science literature by low precision concepts related to the effort expended by observers at locations. The effort expended at a location can be used as a covariate for estimating incomplete availability or detection (Nichols et al. 2009), but survey effort on its own is not sufficient to approximate the observation process. For example, time to event methods require observation data on time to detect each species or individual rather than overall survey duration.

Data standardization, filtering, and correction procedures are routinely recommended to reduce sources of bias and error in citizen science data, but simulations show sampling and observer bias persist, indicating simple filtering methods are inadequate for improving inference (Isaac et al. 2014). At best, encounter histories generated by citizen science programs represent a confounding between observation and ecological processes that cannot be resolved without data collection protocols to estimate the observation process (Nichols et al. 2012). The classification of citizen science into unstructured, semistructured, and structured designs represents the very lowest end of the inferential gradient from retrospective observational studies to true experiments (Nichols et al. 2012). This is an important distinction because monitoring programs with defensible sampling designs, such as Integrated Monitoring in Bird Conservation Regions (IMBCR; Pavlacky et al. 2017) and several programs in the Pan-European Common Bird Monitoring Scheme (PECBMS; Brlik et al. 2021), provide much stronger inferences than so-called structured citizen science programs. Even the most structured citizen science programs, such as the North American Breeding Bird Survey, employ convenience sampling on roadways and lack data collection protocols for estimating the observation process. In contrast, IMBCR and many PECBMS programs feature probabilistic sampling and data collection protocols for estimating incomplete detection, leading to reliable knowledge of geographic variation and population parameters. It is important to recognize that remedial measures involving filtering, standardization, sophisticated modeling, and untested assumptions do not necessarily elevate strength of inference beyond levels of belief supported by underlying data (Blanco et al. 2012).

Chapter 18 suggests that advances in innovation and rigor are being driven by data integration, yet most of the cited examples do not take advantage of the best available science for model-based integration to accommodate the strengths and potential biases of each data source (Isaac et al. 2020). Approaches involving simple data pooling or merging fail to address biases and share components of variation among data sources (Fletcher et al. 2019), and using these approaches to integrate unstructured citizen science and movement databases will likely propagate sampling and observation bias discussed above. Although data integration at macro-ecological scales faces significant challenges (Zipkin et al. 2021), the ability to make reliable inference from data contaminated by sampling and observation biases can be improved by integrating programs with defensible sampling designs (Fletcher et al. 2019, Zhao et al. 2024). The IMBCR and PECBMS databases mentioned above are accessible through data sharing agreements with parent organizations (Bird Conservancy of the Rockies, PECBMS), and many other monitoring programs in North America can be accessed through the Avian Knowledge Network (Iliff et al. 2009). The emergence of the Motus Wildlife Tracking System (Birds Canada 2019) is expected to play an important role in integrating individual-level movement and population-level parameters (McClintock et al. 2022). Finally, Chapter 18 refers to eBird as community science, but this misappropriates a term that has long referred to distinct, grassroots science with far greater capacities for collaboration and co-production among those underserved by science, and thus is not synonymous with citizen science (Cooper et al. 2021). Interestingly, shifting from academic-focused to community-focused science may provide greater ability to develop systems of reliable knowledge than programs discussed in Chapter 19 because initial investment in a defensible sampling design will yield stronger inferences with much lower end-user modeling expertise than is required for large unstructured data sets (Pavlacky et al. 2017).

Biodiversity conservation in increasingly human-modified ecosystems must eventually contend with the realization that natural systems can no longer be considered separate from human systems (Kareiva and Marvier 2012). The New Conservation Debate centers around positions on a continuum from anthropocentric to biocentric values (Hunter et al. 2014). The anthropocentric conservationists suggest preserving biodiversity and ecosystem processes must be reframed as societal problems that are created by, cause harm to, and can only be solved by humans (Hackmann et al. 2014). Because biological conservation is not yet fully integrated into economic and social planning, the ultimate drivers of biodiversity decline often continue unchecked while conservationists address proximate drivers with limited success (Johnson et al. 2017). The biocentric conservationists suggest that adopting an anthropogenic perspective and managing nature for human benefit will promote irreparable attrition and accelerate the current trajectory toward a biologically impoverished future (Doak et al. 2014). Moreover, the ongoing biodiversity crisis cannot be solved using the same ideology that created the problem in the first place; conservation strategies for saving species from extinction and maintaining ecosystem processes have become paramount for a human-modified planet (Miller et al. 2014). Going beyond the dichotomy of the New Conservation Debate, there is an increased recognition of a need to reconcile these schools of thought within cooperative systems that embrace social justice and biodiversity conservation (Shoreman-Ouimet and Kopnina 2015). Transformative change to counter biodiversity loss, climate change, and social injustice (Fougères et al. 2022) must resist the siren song of win–win solutions and instead embrace difficult choices among complex trade-offs that exist between human well-being and biodiversity conservation goals (McShane et al. 2011). Deeper integration across biological and social sciences is imperative for developing applied solutions that promote environmental justice, equity, and sustainability (Schell et al. 2020).

Advances in avian databases and participatory science to monitor avian populations over decadal and continental scales have greatly increased the ability of ornithologists to address research questions for range dynamics, climate change, and migration ecology across the full annual cycle. The development of a global observational network for migratory birds and early warning system of population declines may require extending research questions for ecological transformation to investigate amplifying and dampening mechanisms, rates of change, and the effectiveness of resilience strategies on bird populations and movement (Crausbay et al. 2022). Employing citizen science and avian databases to develop dynamic predictions of species' daily distributions during migration will likely play an important role in understanding how press–pulse processes influence bird population trajectories for climate change–driven transformations (Crausbay et al. 2022). However, although machine learning models trained on data with sampling and observation biases often demonstrate high predictive performance (Lozier et al. 2009), and applying analytical accounting techniques to low-quality data sets is often justified by crisis mentality to conserve rapidly declining biodiversity, relying on untested model assumptions often elevates the meaning and impact of model output beyond levels of belief supported by data (Diffendorfer and Doherty 2004, Blanco et al. 2012). Research applications of citizen science data can improve rigor by implementing the best available science, adhering to the scientific method to evaluate competing hypotheses, comparing predictions to observed data, identifying model assumptions, and explaining relevant uncertainties (Tear et al. 2005). Demonstrating scientific rigor is particularly important when setting conservation objectives and evaluating the success of conservation actions, especially when consequences of making wrong decisions are considered a high risk to decision-makers (Tear et al. 2005).

The noteworthy conclusions of Chapters 18, 19, and 20 represent outstanding social and scientific advancements and are to be commended and applauded. Increasing diversity and inclusion in science, involvement of lay bird enthusiasts in citizen science, providing learning opportunities for participants, and employing crowdsourcing and technology to address broad-scale research questions represent considerable progress toward meeting social, ecological, and conservation goals within the discipline of ornithology. However, by pursuing ornithological research within the silos of our subject areas, we run the risk of working on symptoms or downstream proximate factors of social and biological problems while the ultimate factors underlying injustice and biodiversity loss continue unchecked (Hackmann et al. 2014, Johnson et al. 2017). Sustaining diverse and just futures for life on Earth will require collective action within a transdisciplinary research agenda to understand and realize desired futures for social–ecological systems (Wyborn et al. 2021).

Transformative approaches to nature conservation provide systems to integrate biodiversity conservation, ecosystem services for human well-being, as well as social equity and justice (Foggin et al. 2021). The wider lens of thinking like a mountain (Leopold 1949) provides a template for reimagining the discipline of ornithology within a framework to promote and protect all human rights, including a greater acceptance of Indigenous peoples and traditional ways (Foggin et al. 2021). Participatory citizen science can look to transformative approaches as a model for collaboration and co-production that go beyond rebranding in name toward the realization of community science (Cooper et al. 2021). Decision science may ultimately provide the nexus for ornithology and transformation as the workhorse for balancing complex trade-offs between ecological and social objectives (Robinson et al. 2019). The future of transformative decision science is a promising direction for extending participatory science to empower collaboration and co-production in community science (Cooper et al. 2021) and for expanding the scope of ornithology through engaging diverse communities to develop measurable objectives for social well-being, values, agency, and equality (Hicks et al. 2016). Together with social objectives for well-being and justice, increasing the rigor of avian databases by integrating data with defensible sampling designs (Fletcher et al. 2019) will be important for developing ecological scenarios to imagine desired futures for social–ecological transformation (Crausbay et al. 2022).

Birds speak to us in the language of diversity: In subtropical rainforest of eastern Australia, early morning mimicry of the Albert's Lyrebird (*Menura alberti*) celebrates diversity of the local bird community; in the high plains of North America, dawn flight calls of the Mountain Plover (*Charadrius montanus*) proclaim life-giving heterogeneity of the prairie dog colony; and in lowland rainforest of the Amazon Basin, descending whistles of the Black-Spotted Bare-Eye (*Phlegopsis nigromaculata*) pronounce the ant-following principle of mutual dependence. The following is an important question for ornithologists and bird enthusiasts to ask: How satisfied would you be if your efforts contributed to solving the avian biodiversity crisis, but systematic oppression of underrepresented groups continued to be prevalent in society? Considering the historical subjugation of marginalized identities, and current composition of Western White, educated, industrialized, rich, and democratic (WEIRD) participants of science (Crosman et al. 2022), ethical use of citizen science data may involve asking who has been harmed and who is benefitting from the system of data collection (Trisos et al. 2021). This is an important edition. Lilla Watson does not want to be recognized as the sole source of the quote. She would prefer the quote referenced by 'Aboriginal activists group, Queensland, 1970s'. The Production contact and landed on the sensu reference, but the note about the aboriginal activist's group attribution was lost in the shuffle. Originating from an aboriginal activist's group in Queensland during the 1970s, Lilla Watson, an Indigenous Australian visual artist, activist, and academic is known to say (sensu Watson 2004, para. 7), "If you have come here to help me, you are wasting your time. But if you have come because your liberation is bound up with mine, then let us work together." The recognition that scientific endeavors in ornithology are inextricably linked to social systems provides an opportunity for scientists and lay bird enthusiasts to properly locate themselves within the context of larger societal and ecological problems, which is fundamental to working together as part of the solution.

In closing, I share a personal reflection of an encounter with a young Alaska Native in Katmai National Park following the Anchorage AOS meeting in 2019. Standing in an ancestral house depression, Sam related a story told by his grandfather of "How Raven Stole the Sun." The story touched me emotionally and moved me to tears. There was something intimate about a grandchild relating a grandparent's story at an ancestral home site; something sorrowful about a history of oppression and displacement; and something beautiful about inseparability of the people, land, wildlife, and spirituality. Here, the story is recounted by Charlotte Alstrom and the Marshall Cultural Atlas (Alstrom 1995):

> There once lived a very powerful and rich chief who had a beautiful young daughter. Somehow, the chief got the sun and the moon and he hung them up in his house. Because he had the sun and the moon, it became dark everywhere. Because of the darkness, the people could not hunt or fish. When they went out to find wood to burn in their fires, they had to crawl around in the forest feeling with their hands until they found something which might be wood. Then they would bite it to make certain that it was indeed firewood. Raven learned that the great chief had taken the sun and moon, so he went to his house to take it back. He asked the chief if he would return the sun and moon, but he would not. So the smart black bird devised a plan. He saw how the chief's daughter went to a small stream to get water every morning, so he hid near there and waited for her to return. When he saw her coming down the

> trail, he turned himself into a fingerling, a tiny fish, and jumped into the water. After the girl arrived, she filled a bucket with water. Then she dipped her drinking cup into the stream and Raven, disguised as a fingerling, quickly swam into it. She did not see Raven and drank the water. Inside her body, Raven turned into a baby and so the girl became pregnant. After a short time the daughter gave birth to a baby boy which was really Raven. The baby grew fast and was soon a young boy. The grandfather was very fond of his grandson and would do anything for him. One day the boy began crying for something. The chief asked him, "What do you want, grandson?" The boy pointed to the sun and moon hanging from the ceiling. The chief decided to let him play with them if it would make him stop crying. So the boy took them outside and played with them for a while, but then he threw them high into the air. When the old chief ran out to see what had happened, Raven became himself again and flew away. Since that time there has been light.

At the risk of overinterpreting, with respect for traditional cultural messages, Raven gives us a template for a smart and just plan of transformational action: Working from the inside out to free oppressed people and biodiversity from the selfishness of dominant social groups, with a view that both social and ecological well-being are profoundly intertwined. Integrating social objectives and best available ornithological science can give light to the truth of this precious world—the wonderful diversity of people and birds.

References

Alstrom, C., trans. 1995. Raven steals the light. In *Marshall Cultural Atlas*. Alaska Native Knowledge Network, University of Alaska, Fairbanks, AK. Accessed January 22, 2023, from http://www.ankn.uaf.edu/NPE/CulturalAtlases/Yupiaq/Marshall/raven/RavenStealsTheLight.html.

Bennett, N. J., R. Roth, S. C. Klain, K. Chan, P. Christie, D. A. Clark, G. Cullman, D. Curran, T. J. Durbin, G. Epstein, A. Greenberg, M. P. Nelson, J. Sandlos, R. Stedman, T. L. Teel, R. Thomas, D. Verissimo, and C. Wyborn. 2017. Conservation social science: Understanding and integrating human dimensions to improve conservation. *Biological Conservation* 205:93–108.

Birds Canada. 2019. Motus Wildlife Tracking System. Birds Canada, Port Rowan, Ontario, Canaday. Accessed December 24, 2022, from http://www.motus.org.

Blanco, G., F. Sergio, J. A. Sanchez-Zapata, J. M. Perez-Garcia, F. Botella, F. Martinez, I. Zuberogoitia, O. Frias, F. Roviralta, J. E. Martinez, and F. Hiraldo. 2012. Safety in numbers? Supplanting data quality with fanciful models in wildlife monitoring and conservation. *Biodiversity and Conservation* 21:3269–3276.

Brlik, V., E. Šilarová, Š. J., H. Alonso, M. Anton, A. Aunins, Z. Benkö, G. Biver, M. Busch, T. Chodkiewicz, P. Chylarecki, D. Coombes, E. de Carli, J. C. Del Moral, A. Derouaux, V. Escandell, D. P. Eskildsen, B. Fontaine, R. P. B. Foppen, A. Gamero, R. D. Gregory, S. Harris, S. Herrando, I. Hristov, M. Husby, C. Ieronymidou, F. Jiquet, J. A. Kålås, J. Kamp, P. Kmecl, K. P., A. Lehikoinen, L. Lewis, Å. Lindström, A. Manolopoulos, D. Marti, D. Massimino, C. Moshøj, R. Nellis, D. Noble, A. Paquet, J. Y. Paquet, D. Portolou, I. Ramírez, C. Redel, J. Reif, J. Ridzoň, H. Schmid, B. Seaman, L. Silva, L. Soldaat, S. Spasov, A. Staneva, T. Szep, G. T. Florenzano, N. Teufelbauer, S. Trautmann, T. van der Meij, A. van Strien, C. van Turnhout, G. Vermeersch, Z. Vermouzek, T. Vikstrøm, V. P., A. Weiserbs, and A. Klvaňová. 2021. Long-term and large-scale multispecies dataset tracking population changes of common European breeding birds. *Scientific Data* 8: Article 21.

Charles, A., L. Loucks, F. Berkes, and D. Armitage. 2020. Community science: A typology and its implications for governance of social-ecological systems. *Environmental Science & Policy* 106:77–86.

Chavis, C., T. Gallo, L. Brannan, and A. Rountree. 2021. The enslaved naturalist. George Mason University, John Mitchell, Jr. Program for History, Justice and Race, Digital Museum, Arlington, VA. Accessed December 16, 2022, from https://jmjp.gmu.edu/exhibits/enslaved-naturalist-landing.

Cooper, C. B., C. L. Hawn, L. R. Larson, J. K. Parrish, G. Bowser, D. Cavalier, R. R. Dunn, M. Haklay, K. K. Gupta, N. O. Jelks, V. A. Johnson, M. Katti, Z. Leggett, O. R. Wilson, and S. Wilson. 2021. Inclusion in citizen science: The conundrum of rebranding. *Science* 372:1386–1388.

Crausbay, S. D., H. R. Sofaer, A. E. Cravens, B. C. Chaffin, K. R. Clifford, J. E. Gross, C. N. Knapp, D. J. Lawrence, D. R. Magness, A. J. Miller-Rushing, G. W. Schuurman, and C. S. Stevens-Rumann. 2022. A science agenda to inform natural resource management decisions in an era of ecological transformation. *Bioscience* 72:71–90.

Crosman, K. M., I. Jurcevic, C. Van Holmes, C. C. Hall, and E. H. Allison. 2022. An equity lens on behavioral science for conservation. *Conservation Letters* 15: Article e12885.

Diffendorfer, J., and P. Doherty. 2004. Lifting Cassandra's curse. *Conservation Biology* 18:600.

Doak, D. F., V. J. Bakker, B. E. Goldstein, and B. Hale. 2014. What is the future of conservation? *Trends in Ecology & Evolution* 29:77–81.

Fletcher, R. J., T. J. Hefley, E. P. Robertson, B. Zuckerberg, R. A. McCleery, and R. M. Dorazio. 2019. A practical guide for combining data to model species distributions. Ecology 100: Article e02710.

Foggin, J. M., D. Brombal, and A. Razmkhah. 2021. Thinking like a mountain: Exploring the potential of relational approaches for transformative nature conservation. Sustainability 13: Article 12884.

Fougères, D., M. Jones, P. D. McElwee, A. Andrade, and S. R. Edwards. 2022. Transformative conservation of ecosystems. *Global Sustainability* 5: Article e5.

Gottlieb, R. 2005. *Forcing the Spring: The Transformation of the American Environmental Movement.* Island Press, Washington, DC.

Haarstad, H., S. Sareen, T. I. Wanvik, J. Grandin, K. Kjaeras, S. E. Oseland, H. Kvamsas, K. Lillevold, and M. Wathne. 2018. Transformative social science? Modes of engagement in climate and energy solutions. *Energy Research & Social Science* 42:193–197.

Hackmann, H., S. C. Moser, and A. L. St. Clair. 2014. The social heart of global environmental change. *Nature Climate Change* 4:653–655.

Hicks, C. C., A. Levine, A. Agrawal, X. Basurto, S. J. Breslow, C. Carothers, S. Charnley, S. Coulthard, N. Dolsak, J. Donatuto, C. Garcia-Quijano, M. B. Mascia, K. Norman, M. R. Poe, T. Satterfield, K. S. Martin, and P. S. Levin. 2016. Engage key social concepts for sustainability. *Science* 352:38–40.

Hunter, M. L., K. H. Redford, and D. B. Lindenmayer. 2014. The complementary niches of anthropocentric and biocentric conservationists. *Conservation Biology* 28:641–645.

Iliff, M., L. Salas, E. R. Inzunza, G. Ballard, D. Lepage, and S. Kelling. 2009. The Avian Knowledge Network: A partnership to organize, analyze, and visualize bird observation data for education, conservation, research, and land management. Pages 365–373 in T. D. Rich, C. Arizmendi, D. Demarest, and C. Thompson, eds. *Tundra to tropics: Connecting birds, habitats and people. Proceedings of the 4th International Partners in Flight Conference, 13–16 February 2008.* Partners in Flight, McAllen, TX. Accessed December 24, 2022, from https://www.fs.usda.gov/psw/publications/4251/psw_2009_salas(iliff)001.pdf.

Isaac, N. J. B., M. A. Jarzyna, P. Keil, L. I. Dambly, P. H. Boersch-Supan, E. Browning, S. N. Freeman, N. Golding, G. Guillera-Arroita, P. A. Henrys, S. Jarvis, J. Lahoz-Monfort, J. Pagel, O. L. Pescott, R. Schmucki, E. G. Simmonds, and R. B. O'Hara. 2020. Data integration for large-scale models of species distributions. *Trends in Ecology & Evolution* 35:56–67.

Isaac, N. J. B., A. J. van Strien, T. A. August, M. P. de Zeeuw, and D. B. Roy. 2014. Statistics for citizen science: Extracting signals of change from noisy ecological data. *Methods in Ecology and Evolution* 5:1052–1060.

Johnson, C. N., A. Balmford, B. W. Brook, J. C. Buettel, M. Galetti, G. C. Lei, and J. M. Wilmshurst. 2017. Biodiversity losses and conservation responses in the Anthropocene. *Science* 356:270–274.

Kareiva, P., and M. Marvier. 2012. What is conservation science? *Bioscience* 62:962–969.

Kelling, S., W. M. Hochachka, D. Fink, M. Riedewald, R. Caruana, G. Ballard, and G. Hooker. 2009. Data-intensive science: A new paradigm for biodiversity studies. *Bioscience* 59:613–620.

Kimmerer, R. W. 2002. Weaving traditional ecological knowledge into biological education: A call to action. *Bioscience* 52:432–438.

Leopold, A. 1949. *A Sand County Almanac: And Sketches Here and There.* Oxford University Press, New York.

Lozier, J. D., P. Aniello, and M. J. Hickerson. 2009. Predicting the distribution of Sasquatch in western North America: Anything goes with ecological niche modelling. *Journal of Biogeography* 36:1623–1627.

MacKenzie, D. I., J. D. Nichols, J. A. Royle, K. H. Pollock, L. L. Bailey, and J. E. Hines. 2018. Introduction. Pages 3–26 in *Occupancy Estimation and Modeling: Inferring Patterns and Dynamics of Species Occurrence.* Academic Press, London.

Martin, A., B. Coolsaet, E. Corbera, N. M. Dawson, J. A. Fraser, I. Lehmann, and I. Rodriguez. 2016. Justice and conservation: The need to incorporate recognition. *Biological Conservation* 197:254–261.

McClintock, B. T., B. Abrahms, R. B. Chandler, P. B. Conn, S. J. Converse, R. L. Emmet, B. Gardner, N. J. Hostetter, and D. S. Johnson. 2022. An integrated path for spatial capture–recapture and animal movement modeling. *Ecology* 103: Article e3473.

McShane, T. O., P. D. Hirsch, T. C. Trung, A. N. Songorwa, A. Kinzig, B. Monteferri, D. Mutekanga, H. V. Thang, J. L. Dammert, M. Pulgar-Vidal, M. Welch-Devine, J. P. Brosius, P. Coppolillo, and S. O'Connor. 2011. Hard choices: Making trade-offs between biodiversity conservation and human well-being. *Biological Conservation* 144:966–972.

Miller, B., M. E. Soule, and J. Terborgh. 2014. "New conservation" or surrender to development? *Animal Conservation* 17:509–515.

Mitchell, J. H. 2006. *Looking for Mr. Gilbert: The Reimagined life of an African American.* Shoemaker & Hoard, Washington, DC.

Nichols, J. D., E. G. Cooch, J. M. Nichols, and J. R. Sauer. 2012. Studying biodiversity: Is a new paradigm really needed? *Bioscience* 62:497–502.

Nichols, J. D., L. Thomas, and P. B. Conn. 2009. Inferences about landbird abundance from count data: Recent advances and future directions. Pages 201–235 in D. L. Thomson, E. G. Cooch, and M. J. Conroy, eds. *Modeling Demographic Processes in Marked Populations.* Springer, New York.

Pavlacky, D. C., Jr., P. M. Lukacs, J. A. Blakesley, R. C. Skorkowsky, D. S. Klute, B. A. Hahn, V. J. Dreitz, T. L. George, and D. J. Hanni. 2017. A statistically rigorous sampling design to integrate avian monitoring and management within Bird Conservation Regions. *PLos One* 12: Article e0185924.

Pierotti, R., and D. Wildcat. 2000. Traditional ecological knowledge: The third alternative [Commentary]. *Ecological Applications* 10:1333–1340.

Robinson, K. F., A. K. Fuller, R. C. Stedman, W. F. Siemer, and D. J. Decker. 2019. Integration of social and ecological sciences for natural resource decision making: Challenges and opportunities. *Environmental Management* 63:565–573.

Royle, J. A., and R. M. Dorazio. 2008. Conceptual and philosophical considerations in ecology and statistics. Pages 1–26 in *Hierarchical Modeling and Inference in Ecology: The Analysis of Data From Populations, Metapopulations and Communities.* Academic Press, San Diego.

Saunders, S. P., J. X. Wu, E. A. Gow, E. Adams, B. L. Bateman, T. Bayard, S. Beilke, A. A. Dayer, A. M. V. Fournier, K. Fox, P. Heglund, S. B. Lerman, N. L. Michel, E. H. Paxton, Ç. H. Şekercioğlu, M. A. Smith, W. Thogmartin, M. S. Woodrey, and C. van Riper III. 2021. Bridging the research–implementation gap in avian conservation with translational ecology. *Ornithological Applications* 123:1–13.

Schell, C. J., K. Dyson, T. L. Fuentes, S. Des Roches, N. C. Harris, D. S. Miller, C. A. Woelfle-Erskine, and M. R. Lambert. 2020. The ecological and evolutionary consequences of systemic racism in urban environments. *Science* 369: Article eaay4497.

Schmitz, L. 1867. Cassandra. Page 621 in W. Smith, ed. *A Dictionary of Greek and Roman Biography and Mythology.* Little, Brown, Boston. Accessed January 2, 2023, from https://quod.lib.umich.edu/m/moa/acl3129.0001.001/636?rgn=full±text;view=image;q1=Cassandra.

Shoreman-Ouimet, E., and H. Kopnina. 2015. Reconciling ecological and social justice to promote biodiversity conservation. *Biological Conservation* 184:320–326.

Tear, T. H., P. Kareiva, P. L. Angermeier, P. Comer, B. Czech, R. Kautz, L. Landon, D. Mehlman, K. Murphy, M. Ruckelshaus, J. M. Scott, and G. Wilhere. 2005. How much is enough? The recurrent problem of setting measurable objectives in conservation. *Bioscience* 55:835–849.

Trisos, C. H., J. Auerbach, and M. Katti. 2021. Decoloniality and anti-oppressive practices for a more ethical ecology. *Nature Ecology & Evolution* 5:1205–1212.

Turner, C. H. 1892. A few characteristics of the avian brain. *Science* 19:6–17.

Watson, L. 2004. Recognition of indigenous terms of reference. Keynote address. A Contribution to Change: Cooperation Out of Conflict Conference: Celebrating Difference, Embracing Equality, Center for Action and Contemplation, 21–24 September 2004, Hobart, Tasmania, AUS. Retrieved June 4, 2025, from https://uniting.church/lilla-watson-let-us-work-together.

Williams, B. K., and E. D. Brown. 2019. Sampling and analysis frameworks for inference in ecology. *Methods in Ecology and Evolution* 10:1832–1842.

Wyborn, C., J. Montana, N. Kalas, S. Clement, F. Davila, N. Knowles, E. Louder, M. Balan, J. Chambers, L. Christel, T. Forsyth, G. Henderson, S. I. Tort, M. Lim, M. J. Martinez-Harms, J. Mercon, E. Nuesiri, L. Pereira, V. Pilbeam, E. Turnhout, S. Wood, and M. Ryan. 2021. An agenda for research and action toward diverse and just futures for life on Earth. *Conservation Biology* 35:1086–1097.

Zhao, Q., Q. S. Latif, B. L. Nuse, D. C. Pavlacky Jr., C. L. Kilner, T. B. Ryder, and C. E. Latimer. 2024. Integrating counts from rigorous surveys and participatory science to better understand spatiotemporal variation in population processes. *Methods in Ecology and Evolution* 15:1380–1393.

Zipkin, E. F., E. R. Zylstra, A. D. Wright, S. P. Saunders, A. O. Finley, M. C. Dietze, M. S. Itter, and M. W. Tingley. 2021. Addressing data integration challenges to link ecological processes across scales. *Frontiers in Ecology and the Environment* 19:30–37.

18

Recent Advances in Avian Occurrence and Movement Databases

Frank A. La Sorte and Adriaan M. Dokter

Birds are a highly successful group of animals whose species occur across all the world's continents and oceans (Newton 2003). The roughly 11,000 bird species that have been described are characterized by a remarkable diversity of form and function (Jetz et al. 2012, 2014), and the potential remains for more species to be discovered in the future (Barrowclough et al. 2016). The ease of observing individual birds through sight and sound has created unique opportunities for data acquisition and scientific study. Avian occurrence and movement databases are currently experiencing a renaissance driven by technological advancements and changes in how occurrence and movement data are acquired, stored, and retrieved. The rapid emergence of these new data sources has motivated the development of specialized statistical and computational methods for their analyses. In some cases, the scientific application of these data has required the development and application of new conceptual perspectives that account for the unique characteristics of the data. As these methods and perspectives are tested and refined, researchers can better understand what natural phenomena are being measured by the data and what questions or hypotheses can be most readily addressed. These developments are occurring during a period of global environmental change, creating novel opportunities for scientists to address the ecological implications for bird populations.

History of Avian Occurrence and Movement Data

Museum Specimens

Many of the first avian occurrence and movement databases started through the collection of specimens. Here, individual birds are collected in the field and brought back to the lab, where their skins are prepared and preserved. These efforts were often initiated by local naturalists or artists who were interested in the color and arrangement of feathers. These efforts expanded during the age of European discovery in which expeditions were tasked with collecting bird specimens from increasingly remote regions of the world. This includes the 1831–1836 expedition of the *H. M. S. Beagle* to survey the southern coast of South America, where Charles Darwin collected 468 bird skins (Steinheimer 2004). Many of these early collections now reside in natural

Frank A. La Sorte and Adriaan M. Dokter, *Recent Advances in Avian Occurrence and Movement Databases*. In: *New Perspectives in Ornithology*. Edited by: Scott V. Edwards and J. Michael Reed, Oxford University Press.
 DOI: 10.1093/oso/9780197787670.003.0018

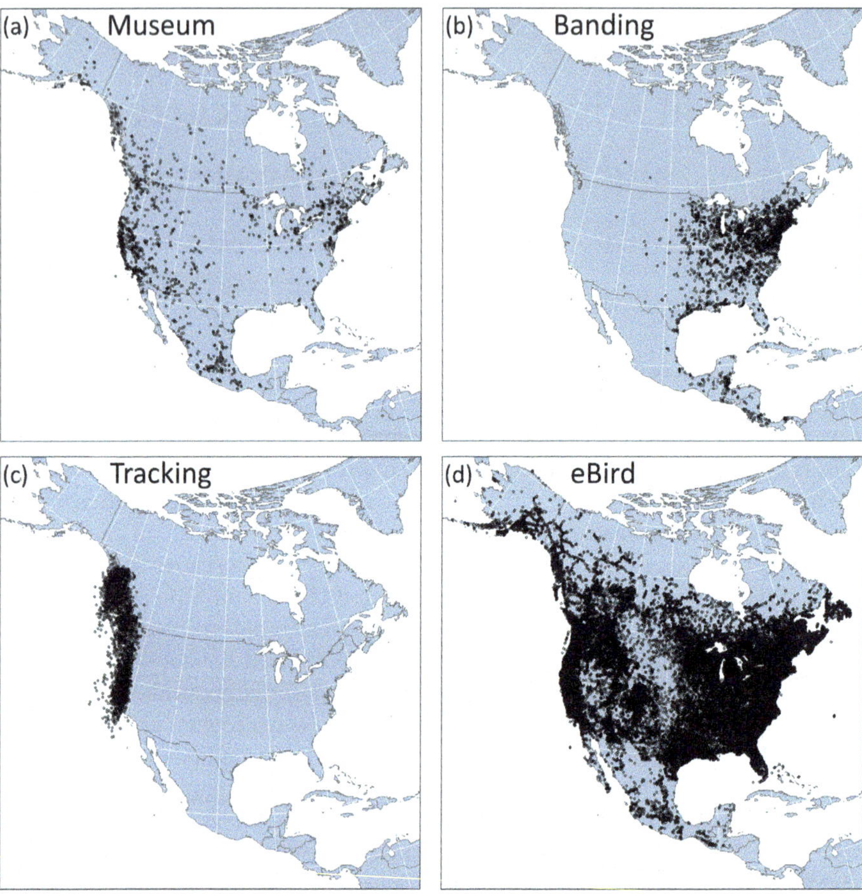

Figure 18.1 The geographic distributions of unique locations for the Hermit Thrush (*Catharus guttatus*) in North America based on four data sources: (a) museum specimen locations compiled from 1818 to 2020, (b) banding locations with related encounter locations (0.63%) compiled from 1960 to 2021, (c) tracking locations compiled from 2013 to 2014 for 16 individuals (Nelson et al. 2016), and (d) eBird locations compiled from 2002 to 2021.

history museums, where they continue to be augmented by the addition of modern specimens and continue to provide a rich source of information for research and education (Figure 18.1a). The relatively deep historical record contained in many of these collections comprises unique scientific value, especially as an ecological baseline when assessing the developing implications of global change (Winker 2005). When using specimens in scientific research, potential sources of bias in the collection process (N. Cooper, Bond, et al. 2019, Meineke and Daru 2021) and the accuracy of the spatial and temporal information accompanying the specimens must be carefully evaluated (Boakes et al. 2010). Under the current extinction crisis, the practice of removing individuals from populations has come under increasing scrutiny, especially for rare or endangered species (Winker et al. 2010, Rocha et al. 2014). These issues add complexities to the acquisition and use of specimen data. However, the physical time series of the specimens contains a unique and rich source of biological information whose scientific value is only likely to increase as global change progresses.

Banding

Avian databases that do not require removing individuals from the populations started to increase in popularity during the beginning of the 20th century. One of the first consisted of banding birds with unique identifying tags that allowed individual identity to be chronicled over time (Figure 18.1b). The practice of bird banding dates back several centuries and was initially used to denote ownership (Harold 1945). Over time, it became clear that the information from the recovery or resighting of banded birds could provide unique insights into bird behavior and movement. In Europe, systematic banding became popular after 1899 through the efforts of Christian Mortensen, who wanted to determine where the birds he observed in Viborg, Denmark, went during the winter (Harold 1945). These efforts started with the European Starling (*Sturnus vulgaris*), followed by storks, ducks, and raptors, eventually culminating in the formation of the first banding stations in Europe. Systematic banding in North America started with the banding of Black-Crowned Night Herons (*Nycticorax nycticorax*) by the Smithsonian Institution's Paul Bartsch in 1902. These efforts were followed by the founding of the American Bird Banding Association and the establishment of the federal Bird Banding Laboratory in 1920 to support the collection, archiving, management, and dissemination of information from banded birds (Davis et al. 2009). A key weakness of banding data is that the banded individual must be recaptured, collected, or resighted at some point in the future. In most cases, banded birds are never observed again after they are released. This outcome is especially common with fledglings whose chance of survival during the first year is very low. Capturing and handling birds during the banding process can also be stressful for birds, raising ethical considerations (Huber et al. 2021). However, band recoveries can provide insights into avian behavior and ecology that cannot be acquired from specimens, and the banding process allows for the collection of ancillary data (e.g., body mass and blood) that cannot be easily acquired from preserved specimens. The genetic information acquired from these efforts can be used to address questions in migratory connectivity, allowing researchers to determine the population and geographical origins of individual birds (Webster et al. 2002). Band recoveries can also be used to address demographic questions. For example, the Monitoring Avian Productivity and Survivorship (MAPS) program coordinated by the Institute for Bird Populations (IBP) collects breeding season data on vital rates and body condition from more than 1,200 MAPS stations located throughout North America. These data are then used to estimate productivity, recruitment, and survival of individual bird species (DeSante et al. 2015). Another IBP initiative that collects nonbreeding season data on vital rates and body condition is the MoSI program (Monitoreo de Sobrevivencia Invernal = Monitoring Overwintering Survival). MoSI is an international network of bird monitoring stations located throughout the northern Neotropics that compiles data on resident birds with a special focus on Neotropical migrant landbirds that breed in the United States and Canada. In combination, data from MAPS and MoSI have allowed researchers to estimate adult apparent survival rates and to explore linkages between breeding and wintering populations (Saracco et al. 2008).

Tracking Technology

Another avian database that is related to banding is the use of tracking technology. Here, after a bird is captured and banded, an electronic tracking device is attached to the bird, creating opportunities for researchers to acquire data from birds in the wild (Figure 18.1c). Tracking technology was first used as a scientific tool in the 1960s to monitor the respiration of the Mallard Duck (*Anas platyrhynchos*) during flight (Lord et al. 1962). The technology of tracking devices has changed dramatically over time through reductions in mass, smaller dimensions, greater capability, and increased affordability (Bridge et al. 2011, Guilford et al. 2011). The first tracking devices emitted a radio signal that could be detected by a receiver, providing information on direction or location through triangulation. These initial devices were very effective when studying short-distance movements but became increasingly impractical as distances increased beyond a few kilometers. The development of solar geolocation, the Global Positioning System, and cellular networks allowed tracking devices to provide location information over much longer distances (Bridge et al. 2011, Guilford et al. 2011). Specialized technologies were developed in unison with tracking technology that allow ancillary data to be recorded on avian physiology, motion, and characteristics of the environment (Bridge et al. 2011, Guilford et al. 2011). A key limitation of tracking devices is that the mass, attachment type, and position can adversely affect behavior, which can interfere with survival, reproduction, or parental care (Bodey et al. 2018, Geen et al. 2019). A key advantage is that measurements and samples can be collected from individuals when they are first captured, and tracking devices can generate data remotely over an extended period of time without further interference. Tracking devices therefore provided some of the first full annual cycle research perspectives in ornithology, generating unique insights into the patterns and drivers of bird movements within and between seasons, including those associated with migratory connectivity (Webster et al. 2002).

Bird-Watching

A third avian database that does not require capturing and handling birds is based on the practice of bird watching. Here, human observers identify bird species in the field using distinguishing morphological characteristics, behaviors, songs, or calls. This approach represents a transition from taking or capturing birds to one based on nonintrusive observations by a skilled observer. The term "bird watching" first emerged early in the 20th century as bird-watching increased in popularity among hobbyist and scientists (Selous 1901). The potential of bird-watching to acquire large amounts of high-quality observations on bird occurrence, abundance, and behavior was quickly realized by the scientific community. An early example is the North

American Christmas Bird Count, which was started in 1900 by Frank M. Chapman as an alternative to the common practice of Christmas "side hunts." The first Christmas Bird Count took place in 1900 and consisted of 27 volunteers distributed across 25 locations in North America. Today, the Christmas Bird Count consists of tens of thousands of volunteers compiling bird observations from approximately 2,100 sites throughout the Americas. The data from the Christmas Bird Count provide an extended time series that can be used to address basic and applied questions on the ecology and behavior of winter bird communities (Bock and Root 1981, Butcher et al. 1990, Soykan et al. 2016). An early example in which bird-watching was used to compile data outside of North America is Max Nicholson's 1928 heronry census, which utilized the labor of hundreds of bird-watchers to map heron breeding colonies across Great Britain (Guida 2019). These early bird-watching initiatives led to the formation of breeding bird surveys in North America (1966), Great Britain (1994), and France (2001). The data generated by these breeding bird surveys have seen broad applications in avian research, conservation, and management.

Radar Technology

A final avian database that does not require capturing or handling birds or the presence of skilled bird watchers is based on the application of radar (radio detection and ranging) technology. Radar was first developed as a tool to detect and track enemy aircraft during World War II. Immediately after the war, the application of radar as a meteorological tool for storm detection and tracking was explored and developed (Maynard 1945). It also became clear during this time that radar could be used to detect and track flying organisms (Brooks 1945, Lack and Varley 1945). These insights led to the use of radar to document flight behavior and estimate the biomass of organisms in the atmosphere (Bonham and Blake 1956), leading to the use of radar to study bird migration (Eastwood 1967; Bruderer 1997a, 1997b). The advent of large networks of weather surveillance radar stations in the contiguous United States (Figure 18.2) and Europe (Figure 18.3) has allowed scientists to document migration patterns and associations within the atmosphere at increasingly larger spatial extents (Gauthreaux and Belser 1998). A key limitation of radar is that species composition usually cannot be ascertained, and migrants cannot be tracked continuously along their entire migration route. Nevertheless, radar is the best tool for measuring aerial biomass flows, and radar networks are providing the basis for the rapidly emerging field of aeroecology (Kunz et al. 2008, Shamoun-Baranes et al. 2019). The number of applications of radar data in avian ecology (Bauer et al. 2019, Becciu et al. 2019) and conservation (Hüppop et al. 2019) continues to grow as scientists explore the wealth of information contained in the data.

Sampling Perspectives

A key characteristic that can be used to classify the occurrence and movement databases summarized above is their ability to retain individual-level information and

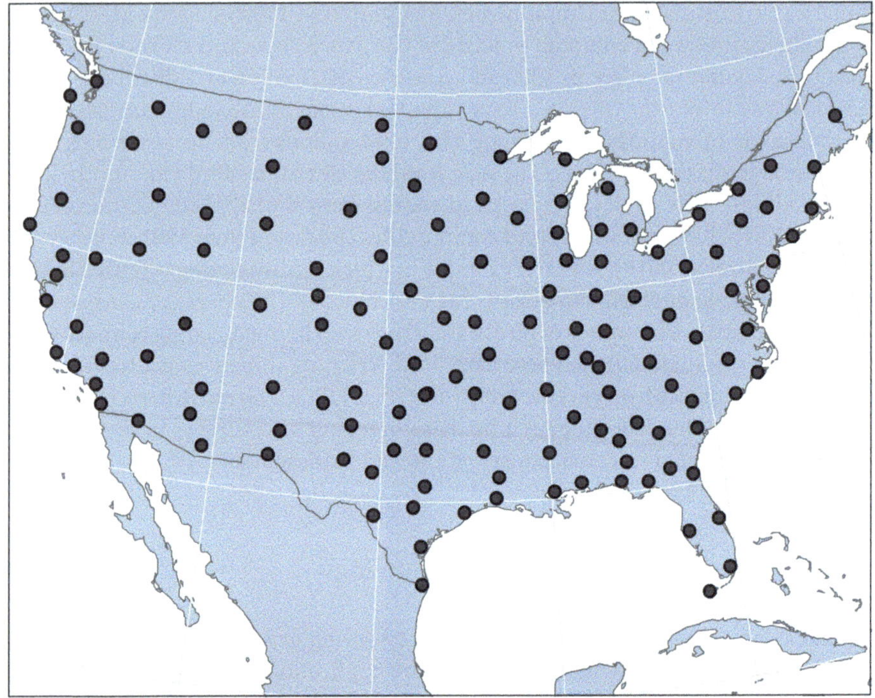

Figure 18.2 The location of the National Oceanic and Atmospheric Administration's NEXRAD network of 144 S-band Doppler weather surveillance radars stations (WSR-88D) in the contiguous United States (Crum and Alberty 1993).

species-level information across space and time. Three general perspectives can be defined. The first is the individual-level perspective, which is the traditional perspective in ornithology. Here, observations are made on the same individual bird across time and space. The primary example is banding and tracking data, where birds are captured and tagged and then monitored with varying levels of spatial and temporal precision (Figures 18.1b and 18.1c). The second is the population-level perspective. Here, species-level information is recorded by an observer with varying levels of spatial, temporal, and taxonomic precision, but individual-level information cannot be retained. The primary example is citizen-science initiatives such as eBird (Sullivan et al. 2014), in which individual birds are identified by human observers at a given location and time and then not tracked over time (Figure 18.1d). The third is the aggregate-level perspective. Here, a cumulative measure of bird density or biomass is acquired, where individual identity and species identity are integrated into a general measure of birds aloft. The primary example is radar data (Figures 18.2 and 18.3).

When using these avian databases to study bird occurrence or movements, a researcher must carefully consider which perspective has the greatest capacity to address the question or hypothesis of interest (Supp et al. 2021). If one perspective is not sufficient, there is also the option of combining perspectives across data sources. For example, population-level data (eBird) can be combined with individual-level

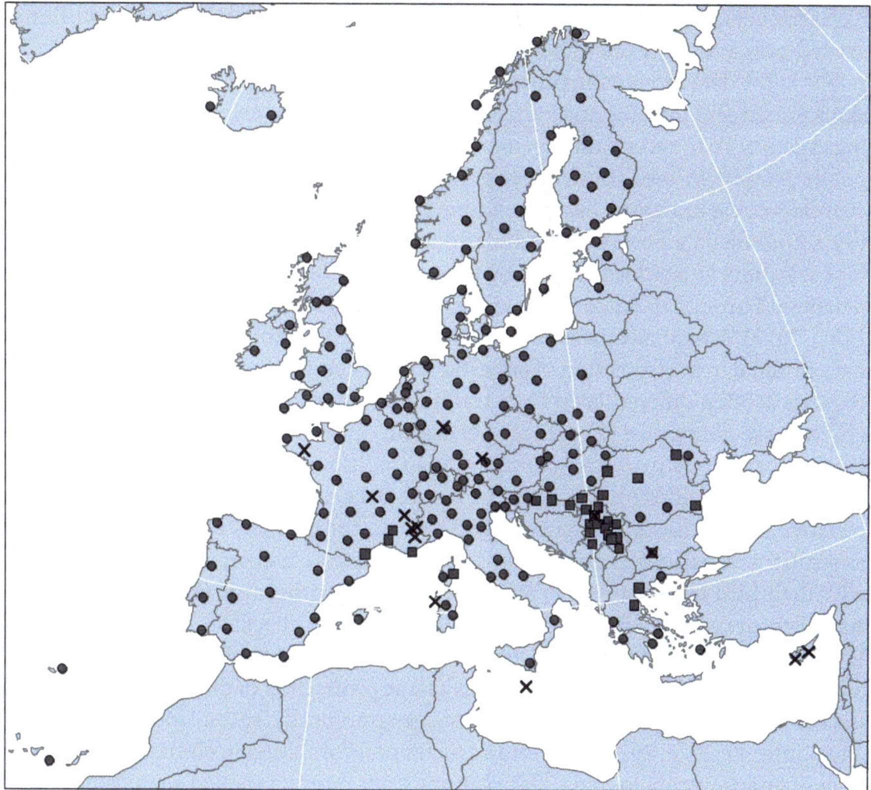

Figure 18.3 The location of 230 weather surveillance radar (WSR) stations in the Network of European Meteorological Services' (EUMETNET) Operational Program for Exchange of Weather Radar Information (OPERA) (Köck et al. 2000). The WSR stations include 184 C-band (circles), 31 S-band (squares), and 14 X-band (×).

data (tracking) to enhance estimates of species' distributions and habitat associations (N. Cooper, Ewert, et al. 2019). Population-level data (eBird) can also be combined with aggregate-level data (radar) to estimate the occurrence of individual migratory bird species (Laughlin et al. 2013, 2016) or estimate characteristics of the migratory bird assemblages aloft (La Sorte, Hochachka, et al. 2015; La Sorte, Hochachka, Farnsworth, et al. 2015; Horton et al. 2018). As these examples suggest, combining data and perspectives can generate more detailed and comprehensive information for scientific study, resulting in more refined inferences.

eBird Citizen Science Program

Avian databases compiled through bird-watching activities represent the beginning of crowdsourcing in ornithology. Here, the collection of bird observations in the field is not tasked to a few trained ornithologists but, rather, to many volunteer bird-watchers. In fact, the growth of bird-watching in North America and Europe started primarily as a scientific enterprise in which the objective was to engage the public in the data acquisition process and tap into their natural history knowledge and skills to increase the efficiency and scope of the scientific enterprise while minimizing expense. With recent advancements in information technology and the increased public interest in bird-watching (e.g., Ma et al. 2013), the potential of these early crowdsourcing initiatives has grown tremendously. With the development of sophisticated online data portals and cellular devices and networks, observations made by bird-watchers at any location at any time can now be readily entered into a central database. Through these developments, ornithology has entered the age of "big data" (Tien 2013), where advancements in information technology and data acquisition have resulted in the formation of large digital data sets (La Sorte, Lepczyk, et al. 2018). Big data is often differentiated from traditional data sources based on three features: (1) Data are numerous; (2) data are generated, captured, and processed rapidly; and (3) data cannot be readily organized into a traditional relational database (Hashem et al. 2015). The eBird program (Sullivan et al. 2014) is a primary example of a big data resource in ornithology (Figure 18.4). The data compiled by eBird can be classified as semistructured, where participants have the option to enter ancillary information with their bird observations on sampling effort and procedure (Sullivan et al. 2014). As of February 2022, eBird had compiled 85.7 million checklists containing 1.15 billion bird observations acquired by 756,300 bird-watchers. These observations are acquired across the year and encompass a large proportion of the terrestrial surface of the Earth (see Figure 18.4), including from geographically isolated locations (La Sorte and Somveille 2020, 2021). The bird occurrence information in eBird can also be used to guide participants to enter observations from poorly sampled regions or regions containing enhanced research and conservation needs (Freeman and Peterson 2019, Callaghan et al. 2021).

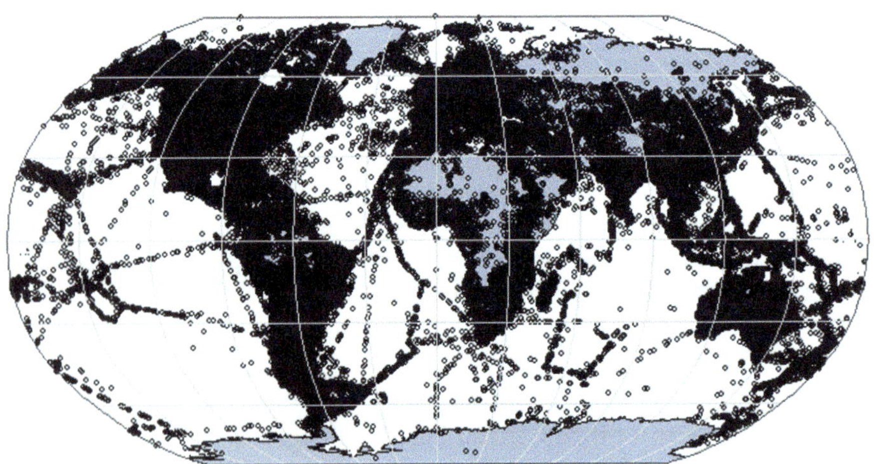

Figure 18.4 The geographic distribution of 11,942,843 unique locations where eBird checklists were submitted from 2002 to 2021.

The application of data from citizen science initiatives requires the development of statistical and analytical methods that account for the unique features of the data. For example, eBird participants have the option of selecting a general sampling protocol that best describes their bird-watching activity and entering related information on sampling effort. The lack of a highly structured sampling protocol in eBird can introduce biases into the data that must be carefully addressed in their scientific application (Johnston, Moran, et al. 2020; Tang et al. 2021). The sampling protocol and effort information available in eBird can be used to filter the data to minimize biases, and it can be included as covariates in statistical models to control for variation in sampling effort (Johnston et al. 2021). The large quantity of observations with variable sampling structures requires the applications of statistical methods that can generate rigorous results consistently across space and time (Fink et al. 2010, 2020). These methods continue to be refined through statistical innovations, advancements in computational resources, the increased availability of relevant covariates, and advancements in artificial intelligence technologies (McClure et al. 2020). Another avenue for enhancing the rigor of these methods is to combine eBird data with other avian data resources (Fletcher et al. 2019). For example, model quality has been improved by integrating eBird data with expert survey data (Robinson et al. 2020), planned-survey data (Zulian et al. 2021), distribution-wide standardized aerial survey data (Tanner et al. 2020), and standardized but sparse aerial survey data (Adde et al. 2021). Another approach is to use banding and tracking data to validate eBird-based estimates in regions that are poorly sampled by eBird (Heim et al. 2020). In total, these efforts are advancing the application of eBird data, transforming how we study and understand the patterns and drivers of bird occurrence across the annual cycle.

Seasonal Migration Patterns and Drivers

A major asset of eBird data is that they provide occurrence information across the full annual cycle. Documenting patterns and associations across the full annual cycle is critical when working with migratory bird species that have seasonally dynamic geographical distributions (Marra et al. 2015). This is also relevant when assessing the implications of global change for species that have geographically dynamic annual life cycles. Another strength of eBird is that it can provide population-level information for species that have traditionally been very difficult to study at the individual level. Examples include small-bodied species such as hummingbirds (Supp et al. 2015); species that are elusive and difficult to observe, such as pelagic seabirds (Martín et al. 2020); and species that conduct irruptive migration where the timing and extent of seasonal movements are irregular, such as the Red-Breasted Nuthatch (*Sitta canadensis*; Freeman et al. 2019). Another asset of eBird is that migratory behaviors that have been challenging to study at the individual level, such as altitudinal (Tsai et al. 2021), intratropical (Dias et al. 2021), or hemispheric-level movements (La Sorte, Fink, et al. 2016), can now be examined in detail at the population level. eBird data can also be used to develop and test theories of bird migration related to the factors that drive seasonal movements and geographic redistributions (Somveille et al. 2021). eBird has also created opportunities to document habitat associations of migratory birds during poorly studied phases of their annual life cycles. Examples include habitat associations during migration for forest passerines in eastern North America (Zuckerberg et al. 2016) and during the nonbreeding season for the Sprague's Pipits (*Anthus spragueii*) in southern North America (Muller et al. 2018).

Migration Phenology

eBird data have been used to assess the geographic and environmental factors that affect migration phenology (Smith et al. 2020, Powers et al. 2021). This includes seasonal patterns of vegetation green-up in the spring and senescence in the autumn (La Sorte and Graham 2021) and the role of migration strategy in determining how migration phenology is affected by changes in vegetation green-up in the spring (Youngflesh et al. 2021). For species that roost in large numbers during migration, such as the Tree Swallow (*Tachycineta bicolor*), eBird data have been combined with radar data and tracking data to estimate seasonal occupancy dynamics (Laughlin et al. 2013, 2016).

Introductions, Range Expansion, and Invasions

Birds have provided many illustrations of range expansion through human introductions or through other non-anthropogenic mechanisms. The population-level perspective provided by eBird has played a pivotal role in promoting the scientific study of these phenomena. For example, eBird data have been used to study the northward range expansion of the Anna's Hummingbird (*Calypte anna*) in western North America (Battey 2019), the southern range expansion of the Eastern Fox Sparrow (*Passerella iliaca iliaca*) in eastern North America (Lloyd 2018), and the northward range expansion of Cory's Shearwater (*Calonectris diomedea*) into the North Atlantic (Gjerdrum et al. 2018). eBird has also been used to assess the patterns and correlates of vagrancy in migratory birds. Examples include the winter vagrancy of Purple Gallinules (*Porphyrio martinicus*) in the North Atlantic (Farnsworth et al. 2015) and the breeding season vagrancy of Ash-Throated Flycatchers (*Myiarchus cinerascens*) along the east coast of North America (Farnsworth et al. 2015).

In addition to documenting range expansions, eBird data have been used to assess invasion risk of introduced bird species. Examples include estimates of invasion risk based on the global trade network of wild-caught birds (Reino et al. 2017). Parrots (Psittaciformes) are often classified as one of the most widely traded and invasive bird taxa in the world, and eBird has played an important role in studying parrot introductions and invasions. For example, eBird data have been used to study parrot introductions and the formation of self-sustaining breeding populations of parrots in Puerto Rico (Falcón and Tremblay 2018), the United States (Uehling et al. 2019), and globally (Calzada Preston and Pruett-Jones 2021). eBird has also been used to assess how regional bans on the wild-bird trade have affected invasion risk of parrots worldwide (Cardador et al. 2017).

Urban Ecology

For bird-watchers who live in cities, the nearest bird-watching opportunities are often located in urban parks or green spaces. As a result, eBird provides a rich source of information on the occurrence of birds in urban environments throughout the world. This has led to the increased application of eBird as a scientific resource in urban ecology. For example, eBird data have been used to assess how humans interact with wild bird populations in Chicago (Lopez et al. 2020). eBird data have also been used to identify avian traits that are associated with urban tolerance in Los Angeles (D. Cooper et al. 2020); to study the suitability and persistence of the California Quail (*Callipepla californica*) in urban parks (Iknayan et al. 2022); and to explore how the size, shape, and location of urban green spaces in New York City affect the seasonal occurrence of birds (La Sorte et al. 2020). In general, as the popularity of eBird increases, its potential as a global resource for studying urban bird assemblages across the annual cycle is likely to grow.

Protected Areas

Another popular destination for bird-watchers is protected areas, where the abundance and diversity of birds tend to be highest. Consequently, eBird provides unique opportunities for studying the bird species that occur in the world's protected areas. For example, eBird data have been used to assess the effectiveness of protected areas in conserving tropical forest birds across three continents (Cazalis et al. 2020) and the effectiveness of protected areas within the Western Hemisphere in conserving sensitive bird species during the breeding season (Cazalis et al. 2021). eBird models have been used to document how stewardship responsibilities on federal protected lands within the contiguous United States are defined across the annual cycle for migratory birds (La Sorte, Fink, et al. 2015). eBird data have also been used to evaluate the efficacy of protected areas for vulnerable waterbird species across the Gulf of Mexico and Atlantic coasts of the United States (Michel et al. 2021). Last, eBird data have been used to identify priority stopover sites in North America during migration and the extent to which these sites are protected now and into the future (H. Lin et al. 2020).

Conservation Planning, Policy, and Regulation

eBird data and models have played a critical role in conservation planning, policy, and regulation, especially regarding migratory birds in North America. For example, eBird data have been used to assess the co-occurrence of mining concessions with migratory bird populations during the nonbreeding season (Rodewald et al. 2019). eBird data have also been used to inform policy decisions for bird species of conservation concern (Long et al. 2019), to support bird conservation within an environmental regulatory context (Young et al. 2019), and to explore how conservation efforts applied to migratory birds in the Neotropics can benefit threatened resident vertebrate species (A. Wilson et al. 2021). In addition, eBird data have been used to contrast the use of range maps with modeled estimates of abundance in conservation planning for migratory birds (Johnston, Auer, et al. 2020). eBird-based estimates of abundance have also been combined with population trend data and land-use change projections to develop conservation plans for migratory bird species (S. Wilson et al. 2019).

Global Climate Change

Due in part to data quantity, quality, and availability, birds have been a popular taxon for studying or modeling the current or future implications of climate change for natural systems (McCarty 2001). These opportunities have continued to expand with the growth of eBird. For example, eBird data have been used to assess how the migration phenology of wading birds within the Afro-Palearctic and the Nearctic migration flyways has changed under climate change (Mondain-Monval et al. 2021). eBird data have also been used to model how bird distributions are projected to change under global warming (Freeman et al. 2019) and to examine how migratory bird populations are projected to be associated with novel climates across their annual life cycles (La Sorte, Fink, et al. 2018, 2019). In addition to global warming, another important consequence of climate change is the increasing frequency, duration, and intensity of climate extremes. eBird data have been used to document how the timing of spring migration is affected by spring heat wave in North America (La Sorte, Hochachka, et al. 2016). eBird data have also been used to explore how extreme heat, drought, and rainfall events have affected bird occurrence across temporal scales (J. Cohen et al. 2020) and how extreme heat and cold events have affected bird occurrence and abundance during the nonbreeding season in North America (J. Cohen et al. 2021). Another aspect of global change that is occurring in unison with climate change is land-use change. eBird-based estimates of occurrence and abundance have been used to assess how projected changes in climate and land-use in combination could affect migratory bird populations that winter in Central America (La Sorte, Fink, Blancher, et al. 2017).

Environmental Pollution

Birds have long been used as sentinels of environmental pollution (Morrison 1986, Hollamby et al. 2006). Three sources of environmental pollution that are especially detrimental to birds are light pollution, air pollution, and noise pollution. It has long been known that light pollution disorients nocturnally migrating birds (Gauthreaux and Belser 2006). eBird data have been used to show that the abundance of nocturnally migrating birds are skewed toward urban sources of light pollution during migration (La Sorte, Fink, Buler, et al. 2017). eBird data have also been used to show that the species richness of nocturnally migrating birds is negatively correlated with urban sources of light pollution during the winter and summer and positively correlated with urban sources of light pollution during spring and autumn migration (La Sorte and Horton 2021).

A second important source of environmental pollution for birds is air pollution. The unique structural and functional refinements of the avian respiratory system (Maina 2000) enhance the toxicity of air pollution, especially when individuals are undertaking rigorous exercise that requires increased breathing rates in the open air, such as migratory flight (Brown et al. 1997). eBird data have been used to show that improvements in air quality in the United States have benefited the region's bird populations (Liang et al. 2020).

Another important source of environmental pollution for birds is noise pollution. Noise pollution is becoming more pervasive throughout the world through urbanization, the growth of transportation networks, industrial agriculture, and resource extraction. Birds have played a leading role in noise pollution research (Francis and Barber 2013, Shannon et al. 2016). Vocal communication serves a variety of important functions for birds associated with mating activities, territorial defense, and predator avoidance, making birds especially sensitive to the presence of noise pollution (Francis et al. 2011, Francis 2015). eBird-based estimates of occurrence and abundance have been used to assess the association between birds and noise pollution during the breeding season in North America (Klingbeil et al. 2020).

Future Directions

As the breadth and depth of the eBird database expand and as statistical methods and computational resources improve, many novel and innovative applications of eBird data are emerging. These advances are being driven in part through data integration. Combining eBird data with environmental data or different sources of avian data is enhancing inferential quality and breadth. Exciting examples include the use of eBird data to detect biological invasives (Larson et al. 2020), estimate rates of hybridization (Justyn et al. 2020, Ottenburghs and Slager 2020), estimate the strength of migratory connectivity (Fournier et al. 2017, Reese et al. 2019), estimate predator–prey dynamics during migration using eDNA (Bourbour et al. 2024), model and predict the transmission of avian diseases (Kain and Bolker 2019), model the threat of bird strikes at commercial airports (Nilsson et al. 2021), quantify the cultural niche of birds (Schuetz and Johnston 2019), measure recolonization after extreme events (Lee et al. 2023), explore how abundance is structured within geographic ranges (Dallas et al. 2017), and estimate population trends (Fink et al. 2020). These efforts portend an exciting period of scientific innovation and discovery as the scope of the eBird database expands and our ability to work effectively with the data improves.

Radar Technology

Twice a year, seasonal bird migration leads to one of the largest rearrangements of animal biomass throughout the world. More than 80% of temperate breeding bird species are partially or fully migratory (Sheard et al. 2020), and seasonal migration involves billions of individuals moving each spring and autumn between their breeding and wintering areas. Despite the global pervasiveness of bird migration, studying these mass avifaunal movements has remained elusive. Most species migrate under the cover of darkness, and their movements are only given away by faint call notes that can be detected in the night sky by trained observers (Farnsworth 2005). It is the enormous magnitude of bird migration and the vast spatial scales involved that have made it difficult to obtain a bird's-eye view of this phenomena and to document these mass movements in their entirety.

Since its discovery as a tool for observing birds in flight after World War II, radar has become an invaluable tool to observe many fundamental characteristics of bird migration (Bauer et al. 2019). It remains the only tool by which migrating birds can be observed in active migration within the atmosphere over large distances in a fully noninvasive fashion. Because species identity cannot typically be determined from radar, the emerging field of radar ornithology has focused on questions that are general to the migration process. Most of the early studies were performed using portable military radars, leading to a wealth of natural history information on the altitude, speed, and intensity of bird migration. The detail and spatial scope of these early assessments expanded dramatically as the number, strength, and quality of radar systems grew to meet the needs of air traffic control, air defense, and weather forecasting (Gurbuz et al. 2015).

Since the turn of the 21st century, a renewed interest in radar technology as an empirical tool in ornithology has emerged (Shamoun-Baranes et al. 2019). This has led to a renaissance in radar ornithology, which can be explained as a top-down coalescence of new technologies in combination with the emergence of multiple research needs driven in part by the need to assess the implications of global change for migratory birds. Recent advancements in information technology and the resulting big data revolution (Hampton et al. 2013) have driven many innovations in radar ornithology. For decades, storing, accessing, and analyzing the enormous volumes of data produced by radar systems substantially hindered scientific progress. These constraints changed dramatically as data storage capacity increased and as storage and computational costs decreased. Simultaneously, new methodologies were developed that could analyze the vast quantities of radar data automatically in a rigorous and efficient fashion without the need for manual data screening (Dokter et al. 2019, T. Lin et al. 2019), which was a common practice in the early years of radar ornithology (Eastwood 1967). In addition, small-scale radars with enhanced high-frequency sampling and machine classification on wingbeat signals have increased the taxonomic resolution of radars, and the development of affordable portable radars has met the need for bird monitoring near airports and wind farms, broadening the community of researchers with access to radar systems (Nilsson, Dokter, et al. 2018).

In the early 1970s, researchers in North America pioneered the use of weather surveillance radar networks as a data source for monitoring bird migration (Gauthreaux 1970). The unlocking of meteorological radar archives, combined with the development of automated methods that could extract bird migration information (and remove weather) from these images, transformed the field. Currently, the weather surveillance radar archives that were established in the mid-1990s in the

United States (see Figure 18.2) and Europe (see Figure 18.3) represent the largest and most detailed databases on avian movement in the world, unique in their range of spatial (kilometers to continents) and temporal (minutes to decades) scales.

Simultaneously, new questions and applications have emerged to meet the increased availability of radar data and methods. This includes questions on avian behavior and migration strategies within the atmosphere, the choice of stopover sites during migration, the current and future implications of environmental pollution and climate change, and the need to better understand the timing and location of large-scale bird migration events to mitigate wildlife conflicts. The increased need to mitigate aerial conflicts has also led to the appreciation that the atmosphere is a habitat in its own right containing distinct ecological features with unique conservation needs. Through the influence of macroecology, interest has also grown in using radar to understand continental-level patterns of avian movement at a system level, focused on estimating biomass flow and macro-demographic parameters (Dokter et al. 2018).

Avian Behavior and Migration Ecology

Radar data are increasingly being used to address questions related to avian behavior and migration ecology. This is occurring at the individual level for specific species that display unique behaviors which can be easily detected by radar and at an aggregate level for all species combined, which in most cases can be readily detected by radar. For species that display strong communal roosting behaviors, the connection between the occurrence of a species on the ground and their detection in the atmosphere by radar can be quite strong. For example, radar data can be used to document the morning and evening flights (Russell and Gauthreaux 1998) and the spatial and temporal dynamics and habitat associations of Purple Martin (*Progne subis*) roosts (Russell et al. 1998, Russell and Gauthreaux 1999, Bridge et al. 2016). Radar data can also be used to document the twilight ascents of Common Swift (*Apus apus*) roosts (Dokter, Åkesson, et al. 2013), to estimate their flight speeds during nocturnal migration and during nocturnal roosting flights (Henningsson et al. 2009), and to study how wind speed and direction affect their orientation and flight behavior during nocturnal roosting flights (Bäckman and Alerstam 2001, 2002) and nocturnal migration (Karlsson et al. 2010). In addition, radar data can be used to explore the behavioral drivers of Tree Swallow (*Tachycineta bicolor*) roosts (Laughlin et al. 2014) and to estimate the population size of American Robin (*Turdus migratorius*) winter roosts (Van den Broeke 2019).

In addition to detecting activities associated with the large communal roosts of specific species, radar data can be used to study at an aggregate level the behavior and ecology of all nocturnally migrating birds aloft. Examples include the use of radar data to document seasonal differences in ground speed and air speed of migrating birds (Karlsson et al. 2012, Horton et al. 2016a). Radar data have been used to locate and characterize stopover sites of nocturnally migrating birds (Buler and Dawson 2014, Cabrera-Cruz et al. 2020, E. Cohen et al. 2021) and to explore how nocturnally migrating birds navigate around or across ecological barrier to migration, such as mountain ranges (Aschwanden et al. 2020) and large water bodies (Clipp et al. 2021). Radar data have also been used to study where and when nocturnally migrating birds compensate for wind drift (McLaren et al. 2012, Horton et al. 2016c, Newcombe et al. 2019), the correspondence between flight altitude and wind assistance (Dokter, Shamoun-Baranes, et al. 2013; Kemp et al. 2013; Horton et al. 2016b), the causes of reverse migration (Nilsson and Sjöberg 2016), the effects of fog on diurnal soaring bird migration (Panuccio et al. 2019), and the effects of a solar eclipse on flying birds (Nilsson, Horton, et al. 2018).

Typically, species identity cannot be acquired from the aggregate measures of bird migration acquired from radar systems. However, species composition in some cases can be inferred at a coarse level using diurnal bird observations compiled within the vicinity of the radar stations. For example, when combined with radar data, the bird observations provided by eBird can be used to explore the atmospheric, environmental, morphological, and behavioral correlates of migration timing (La Sorte, Hochachka, et al. 2015). Radar and eBird data in combination can also be used to document how variation in migration altitudes is related to seasonal variation in atmospheric conditions and species composition (La Sorte, Hochachka, Farnsworth, et al. 2015) and where and when nocturnally migrating birds compensate for wind drift based on body mass (Horton et al. 2018),

Avian Conservation

A key application of radar data is to address issues associated with bird conservation, especially those arising for nocturnally migrating birds. Migrating birds face an array of unique risks and challenges during migration, many of which are exacerbated by human activities. Some of the factors can be studied very effectively using radar data, including the risk of colliding with military or domestic flights (van Gasteren et al. 2019, Nilsson et al. 2021), wind turbines (Liechti et al. 2013, Cabrera-Cruz and Villegas-Patraca 2016, Welcker et al. 2017), or windows (Elmore et al. 2021). A key factor that enhances the risks and challenges of nocturnal migration is the disorienting effects of artificial light at night on migratory behavior (Gauthreaux and Belser 2006). Light pollution can increase time and energy expenditures as migratory birds are drawn away from their intended migration routes (Larkin and Frase 1988, Van Doren et al. 2017). Light pollution can also enhance the risk of colliding with illuminated structures such as buildings or communications towers (Lao et al. 2020, Van Doren et al. 2021). Radar data can be used to study how nocturnally migrating birds are attracted to artificial light at night (Van Doren et al. 2017) and the implications of light pollution on the selection of stopover habitat (McLaren et al. 2018). In addition to light pollution, another key challenge human activities have created for migrating birds is global climate change (Zurell et al. 2018). Radar data have been used to assess the current effects of climate change on broad-scale migratory behavior, such as its effects on migration phenology (Horton et al. 2019, 2020), and to estimate the implications of projected changes in wind speed and direction under climate change for nocturnally migrating bird populations (La Sorte, Horton, et al. 2019).

Future Directions

Radar data contain many exciting opportunities to advance avian research and inform conservation efforts. Recent examples include using radar data to model and predict day-to-day variation in the occurrence of migratory birds in the atmosphere (Van Doren and Horton 2018, Van Doren et al. 2021). Other important innovations include the use of electromagnetic modeling techniques (Mirkovic et al. 2016) and fluid dynamics models (Nussbaumer et al. 2021) to estimate the density and taxonomy of migrating birds aloft. As these methods mature, we expect the integration of radar data with other avian databases to become increasingly important as a source of innovation. As demonstrated by several recent examples (La Sorte, Hochachka, et al. 2015; La Sorte, Hochachka, Farnsworth, et al. 2015; Horton et al. 2018), radar and eBird data are highly complementary, and the two data sources in combination contain significant potential to drive scientific innovation. eBird provides detailed information on species occurrence across broad geographic regions during the day after migration. Radar, in contrast, provides an aggregate measure at a high temporal resolution of birds aloft during nocturnal migration. These two data sources are naturally complementary, and their integrated analyses may lead to dynamic predictions of species' daily distributions during migration. This integration could be enhanced through the addition of acoustic data on nocturnal flight calls, which will connect species-level information to the aggregate measurements of bird migration (Farnsworth et al. 2004).

There are many national radar systems, networks, and archives throughout the world that remain largely untapped by ornithologists, including in Canada, Central America, China, India, and Russia (Shamoun-Baranes et al. 2019). It is conceivable that these resources could support a global bird migration monitoring system in which the seasonal concentrations of migrating birds in the atmosphere can be tracked and monitored from one year to the next. As global change progresses, this information could allow researchers to assess the current and future implications of global change for migratory bird populations and the best strategies to mitigate population declines. Many important hurdles, however, need to be overcome, including developing data infrastructure, harmonization of data collection, addressing variation in data quality, and the promotion of open-source policies for scientific research. Similarly, as the spatial breadth and depth of eBird data expand worldwide (La Sorte and Somveille 2020), additional opportunities for data integration will likely emerge. This will lead to an increasingly global observational network for the study and monitoring of migratory birds, providing opportunities to develop and address novel research questions, including the creation of an early warning system of population declines.

References

Adde, A., C. Casabona I Amat, M. J. Mazerolle, M. Darveau, S. G. Cumming, and R. B. O'Hara. 2021. Integrated modeling of waterfowl distribution in western Canada using aerial survey and citizen science (eBird) data. *Ecosphere* 12: Article e03790.

Aschwanden, J., M. Schmidt, G. Wichmann, H. Stark, D. Peter, T. Steuri, and F. Liechti. 2020. Barrier effects of mountain ranges for broad-front bird migration. *Journal of Ornithology* 161:59–71.

Bäckman, J., and T. Alerstam. 2001. Confronting the winds: Orientation and flight behaviour of roosting swifts, *Apus apus*. *Proceedings of the Royal Society B: Biological Sciences* 268:1081–1087.

Bäckman, J., and T. Alerstam. 2002. Harmonic oscillatory orientation relative to the wind in nocturnal roosting flights of the swift *Apus apus*. *Journal of Experimental Biology* 205:905–910.

Barrowclough, G. F., J. Cracraft, J. Klicka, and R. M. Zink. 2016. How many kinds of birds are there and why does it matter? *PLoS One* 11: Article e0166307.

Battey, C. J. 2019. Ecological release of the Anna's Hummingbird during a northern range expansion. *The American Naturalist* 194:306–315.

Bauer, S., J. Shamoun-Baranes, C. Nilsson, A. Farnsworth, J. F. Kelly, D. R. Reynolds, A. M. Dokter, J. F. Krauel, L. B. Petterson, K. G. Horton, and J. W. Chapman. 2019. The grand challenges of migration ecology that radar aeroecology can help answer. *Ecography* 42:861–875.

Becciu, P., M. H. M. Menz, A. Aurbach, S. A. Cabrera-Cruz, C. E. Wainwright, M. M. Scacco, M. Ciach, L. B. Pettersson, I. Maggini, G. M. Arroyo, J. J. Buler, D. R. Reynolds, and N. Sapir. 2019. Environmental effects on flying migrants revealed by radar. *Ecography* 42:942–955.

Boakes, E. H., P. J. K. McGowan, R. A. Fuller, D. Chang-qing, N. E. Clark, K. O'Connor, and G. M. Mace. 2010. Distorted views of biodiversity: Spatial and temporal bias in species occurrence data. *PLoS Biology* 8: Article e1000385.

Bock, C. E., and T. L. Root. 1981. The Christmas Bird Count and avian ecology. *Studies in Avian Biology* 6:17–23.

Bodey, T. W., I. R. Cleasby, F. Bell, N. Parr, A. Schultz, S. C. Votier, and S. Bearhop. 2018. A phylogenetically controlled meta-analysis of biologging device effects on birds: Deleterious effects and a call for more standardized reporting of study data. *Methods in Ecology and Evolution* 9:946–955.

Bonham, L. L., and L. V. Blake. 1956. Radar echoes from birds and insects. *Scientific Monthly* 82:204–209.

Bourbour, R. P., C. M. Aylward, T. D. Meehan, B. L. Martinico, M. E. Badger, A. M. Goodbla, A. M. Fish, T. E. Ely, C. W. Briggs, and E. M. Hull. 2024. Feeding *en route*: Prey availability and traits influence prey selection by an avian predator on migration. *Journal of Animal Ecology* 93:1176–1191.

Bridge, E. S., S. M. Pletschet, T. Fagin, P. B. Chilson, K. G. Horton, K. R. Broadfoot, and J. F. Kelly. 2016. Persistence and habitat associations of Purple Martin roosts quantified via weather surveillance radar. *Landscape Ecology* 31:43–53.

Bridge, E. S., K. Thorup, M. S. Bowlin, P. B. Chilson, R. H. Diehl, R. W. Fléron, P. Hartl, R. Kays, J. F. Kelly, W. D. Robinson, and M. Wikelski. 2011. Technology on the move: Recent and forthcoming innovations for tracking migratory birds. *BioScience* 61:689–698.

Brooks, M. 1945. Electronics as a possible aid in the study of bird flight and migration. *Science* 101:329–329.

Brown, R. E., J. D. Brain, and N. Wang. 1997. The avian respiratory system: A unique model for studies of respiratory toxicosis and for monitoring air quality. *Environmental Health Perspectives* 105:188–200.

Bruderer, B. 1997a. The study of bird migration by radar Part 1: The technical basis. *Naturwissenschaften* 84:1–8.

Bruderer, B. 1997b. The study of bird migration by radar Part 2: Major achievements. *Naturwissenschaften* 84:45–54.

Buler, J. J., and D. K. Dawson. 2014. Radar analysis of fall bird migration stopover sites in the northeastern U.S. *The Condor* 116:357–370.

Butcher, G. S., M. R. Fuller, L. S. McAllister, and P. H. Geissler. 1990. An evaluation of the Christmas Bird Count for monitoring population trends of selected species. *Wildlife Society Bulletin* 18:129–134.

Cabrera-Cruz, S. A., E. B. Cohen, J. A. Smolinsky, and J. J. Buler. 2020. Artificial light at night is related to broad-scale stopover distributions of nocturnally migrating landbirds along the Yucatan Peninsula, Mexico. *Remote Sensing* 12: Article 395.

Cabrera-Cruz, S. A., and R. Villegas-Patraca. 2016. Response of migrating raptors to an increasing number of wind farms. *Journal of Applied Ecology* 53:1667–1675.

Callaghan, C. T., J. E. M. Watson, M. B. Lyons, W. K. Cornwell, and R. A. Fuller. 2021. Conservation birding: A quantitative conceptual framework for prioritizing citizen science observations. *Biological Conservation* 253: Article 108912.

Calzada Preston, C. E., and S. Pruett-Jones. 2021. The number and distribution of introduced and naturalized parrots. *Diversity* 13: Article 412.

Cardador, L., M. Lattuada, D. Strubbe, J. L. Tella, L. Reino, R. Figueira, and M. Carrete. 2017. Regional bans on wild-bird trade modify invasion risks at a global scale. *Conservation Letters* 10: 717–725.

Cazalis, V., M. D. Barnes, A. Johnston, J. E. M. Watson, C. H. Sekercioglu, and A. S. L. Rodrigues. 2021. Mismatch between bird species sensitivity and the protection of intact habitats across the Americas. *Ecology Letters* 24:2394–2405.

Cazalis, V., K. Princé, J.-B. Mihoub, J. Kelly, S. H. M. Butchart, and A. S. L. Rodrigues. 2020. Effectiveness of protected areas in conserving tropical forest birds. *Nature Communications* 11: Article 4461.

Clipp, H. L., J. J. Buler, J. A. Smolinsky, K. G. Horton, A. Farnsworth, and E. B. Cohen. 2021. Winds aloft over three water bodies influence spring stopover distributions of migrating birds along the Gulf of Mexico coast. *Ornithology* 138: Article ukab051.

Cohen, E. B., K. G. Horton, P. P. Marra, H. L. Clipp, A. Farnsworth, J. A. Smolinsky, D. Sheldon, and J. J. Buler. 2021. A place to land: Spatiotemporal drivers of stopover habitat use by migrating birds. *Ecology Letters* 24:38–49.

Cohen, J. M., D. Fink, and B. Zuckerberg. 2020. Avian responses to extreme weather across functional traits and temporal scales. *Global Change Biology* 26:4240–4250.

Cohen, J. M., D. Fink, and B. Zuckerberg. 2021. Extreme winter weather disrupts bird occurrence and abundance patterns at geographic scales. *Ecography* 44:1143–1155.

Cooper, D. S., A. J. Shultz, and D. T. Blumstein. 2020. Temporally separated data sets reveal similar traits of birds persisting in a United States megacity. *Frontiers in Ecology and Evolution* 8: Article 251.

Cooper, N., A. L. Bond, J. L. Davis, R. Portela Miguez, L. Tomsett, and K. M. Helgen. 2019. Sex biases in bird and mammal natural history collections. *Proceedings of the Royal Society B: Biological Sciences* 286: Article 20192025.

Cooper, N. W., D. N. Ewert, J. M. Wunderle, Jr., E. H. Helmer, and P. P. Marra. 2019. Revising the wintering distribution and habitat use of the Kirtland's warbler using playback surveys, citizen scientists, and geolocators. *Endangered Species Research* 38:79–89.

Crum, T. D., and R. L. Alberty. 1993. The WSR-88D and the WSR-88D operational support facility. *Bulleting of the American Meteorological Society* 74:1669–1687.

Dallas, T., R. R. Decker, and A. Hastings. 2017. Species are not most abundant in the centre of their geographic range or climatic niche. *Ecology Letters* 20:1526–1533.

Davis, W. E., J. A. Jackson, and J. Tautin. 2009. *Bird Banding in North America: The First Hundred Years.* Nuttall Ornithological Club, Cambridge, MA.

DeSante, D. F., D. R. Kaschube, and J. F. Saracco. 2015. Vital rates of North American landbirds. The Institute for Bird Populations. https://www.VitalRatesOfNorthAmericanLandbirds.org.

Dias, R. I., V. R. Vianna, and R. Araujo. 2021. Annual movements of four South American tyrant flycatchers: Uncovering intratropical migration patterns. *Bird Study* 68:76–89.

Dokter, A. M., S. Åkesson, H. Beekhuis, W. Bouten, L. Buurma, H. van Gasteren, and I. Holleman. 2013. Twilight ascents by common swifts, *Apus apus*, at dawn and dusk: Acquisition of orientation cues? *Animal Behaviour* 85:545–552.

Dokter, A. M., P. Desmet, J. H. Spaaks, S. van Hoey, L. Veen, L. Verlinden, C. Nilsson, G. Haase, H. Leijnse, A. Farnsworth, W. Bouten, and J. Shamoun-Baranes. 2019. bioRad: biological analysis and visualization of weather radar data. *Ecography* 42:852–860.

Dokter, A. M., A. Farnsworth, D. Fink, V. Ruiz-Gutierrez, W. M. Hochachka, F. A. La Sorte, O. J. Robinson, K. V. Rosenberg, and S. Kelling. 2018. Seasonal abundance and survival of North America's migratory avifauna determined by weather radar. *Nature Ecology & Evolution* 2:1603–1609.

Dokter, A. M., J. Shamoun-Baranes, M. U. Kemp, S. Tijm, and I. Holleman. 2013. High altitude bird migration at temperate latitudes: A synoptic perspective on wind assistance. *PLoS One* 8: Article e52300.

Eastwood, E. 1967. *Radar Ornithology*. Methuen, London.

Elmore, J. A., C. S. Riding, K. G. Horton, T. J. O'Connell, A. Farnsworth, and S. R. Loss. 2021. Predicting bird-window collisions with weather radar. *Journal of Applied Ecology* 58:1593–1601.

Falcón, W., and R. L. Tremblay. 2018. From the cage to the wild: Introductions of Psittaciformes to Puerto Rico. *PeerJ* 6: Article e5669.

Farnsworth, A. 2005. Flight calls and their value for future ornithological studies and conservation research. *The Auk* 122:733–746.

Farnsworth, A., S. A. Gauthreaux, and D. Blaricom. 2004. A comparison of nocturnal call counts of migrating birds and reflectivity measurements on Doppler radar. *Journal of Avian Biology* 35:365–369.

Farnsworth, A., F. A. La Sorte, and M. J. Iliff. 2015. Warmer summers and drier winters correlate with more winter vagrant Purple Gallinules (*Porphyrio martinicus*) in the North Atlantic region. *Wilson Journal of Ornithology* 127:582–592.

Fink, D., T. Auer, A. Johnston, V. Ruiz-Gutierrez, W. M. Hochachka, and S. Kelling. 2020. Modeling avian full annual cycle distribution and population trends with citizen science data. *Ecological Applications* 30: Article e02056.

Fink, D., W. M. Hochachka, B. Zuckerberg, D. W. Winkler, B. Shaby, M. A. Munson, G. Hooker, M. Riedewald, D. Sheldon, and S. Kelling. 2010. Spatiotemporal exploratory models for broad-scale survey data. *Ecological Applications* 20:2131–2147.

Fletcher, R. J., T. J. Hefley, E. P. Robertson, B. Zuckerberg, R. A. McCleery, and R. M. Dorazio. 2019. A practical guide for combining data to model species distributions. *Ecology* 100: Article e02710.

Fournier, A. M. V., A. R. Sullivan, J. K. Bump, M. Perkins, M. C. Shieldcastle, and S. L. King. 2017. Combining citizen science species distribution models and stable isotopes reveals migratory connectivity in the secretive Virginia rail. *Journal of Applied Ecology* 54:618–627.

Francis, C. D. 2015. Vocal traits and diet explain avian sensitivities to anthropogenic noise. *Global Change Biology* 21:1809–1820.

Francis, C. D., and J. R. Barber. 2013. A framework for understanding noise impacts on wildlife: An urgent conservation priority. *Frontiers in Ecology and the Environment* 11:305–313.

Francis, C. D., C. P. Ortega, and A. Cruz. 2011. Noise pollution filters bird communities based on vocal frequency. *PLoS One* 6: Article e27052.

Freeman, B., and A. T. Peterson. 2019. Completeness of digital accessible knowledge of the birds of western Africa: Priorities for survey. *The Condor* 121: Article duz035.

Freeman, B., J. Sunnarborg, and A. T. Peterson. 2019. Effects of climate change on the distributional potential of three range-restricted West African bird species. *The Condor* 121: Article duz012.

Gauthreaux, S. A., Jr. 1970. Weather radar quantification of bird migration. *BioScience* 20:17–20.

Gauthreaux, S. A., Jr., and C. G. Belser. 1998. Displays of bird movements on the WSR-88D: Patterns and quantification. *Weather Forecast* 13:453–464.

Gauthreaux, S. A., Jr., and C. G. Belser. 2006. Effects of artificial night lighting on migrating birds. Pages 67–93 in C. Rich and T. Longcore, eds. *Ecological Consequences of Artificial Night Lighting*. Island Press, Washington, DC.

Geen, G. R., R. A. Robinson, and S. R. Baillie. 2019. Effects of tracking devices on individual birds: A review of the evidence. *Journal of Avian Biology* 50.

Gjerdrum, C., J. Loch, and D. A. Fifield. 2018. The recent invasion of Cory's Shearwaters into Atlantic Canada. *Northeastern Naturalist* 25:532–544.

Guida, M. 2019. 1928. Popular bird-watching becomes scientific: The first national bird census in Britain. *Public Understanding of Science* 28:622–627.

Guilford, T., S. Åkesson, A. Gagliardo, R. A. Holland, H. Mouritsen, R. Muheim, R. Wiltschko, W. Wiltschko, and V. P. Bingman. 2011. Migratory navigation in birds: New opportunities in an era of fast-developing tracking technology. *Journal of Experimental Biology* 214:3705–3712.

Gurbuz, S. Z., D. R. Reynolds, J. Koistinen, F. Liechti, H. Leijnse, J. Shamoun-Baranes, A. M. Dokter, J. Kelly, and J. W. Chapman. 2015. Exploring the skies: Technological challenges in radar aeroecology. Pages 817–822 in *2015 IEEE Radar Conference (RadarCon 2015)*. Institute of Electrical and Electronics Engineers.

Hampton, S. E., C. A. Strasser, J. J. Tewksbury, W. K. Gram, A. E. Budden, A. L. Batcheller, C. S. Duke, and J. H. Porter. 2013. Big data and the future of ecology. *Frontiers in Ecology and the Environment* 11:156–162.

Harold, B. W. 1945. The history of bird banding. *The Auk* 62:256–265.

Hashem, I. A. T., I. Yaqoob, N. B. Anuar, S. Mokhtar, A. Gani, and S. Ullah Khan. 2015. The rise of "big data" on cloud computing: Review and open research issues. *Information Systems* 47:98–115.

Heim, W., R. J. Heim, I. Beermann, O. A. Burkovskiy, Y. Gerasimov, P. Ktitorov, K. Ozaki, I. Panov, M. M. Sander, S. Sjöberg, S. M. Smirenski, A. Thomas, A. P. Tøttrup, I. M. Tiunov, M. Willemoes, N. Hölzel, K. Thorup, and J. Kamp. 2020. Using geolocator tracking data and ringing archives to validate citizen-science based seasonal predictions of bird distribution in a data-poor region. *Global Ecology and Conservation* 24: Article e01215.

Henningsson, P., H. Karlsson, J. Bäckman, T. Alerstam, and A. Hedenström. 2009. Flight speeds of swifts (*Apus apus*): Seasonal differences smaller than expected. *Proceedings of the Royal Society B: Biological Sciences* 276:2395–2401.

Hollamby, S., J. Afema-Azikuru, S. Waigo, K. Cameron, A. Rae Gandolf, A. Norris, and J. G. Sikarskie. 2006. Suggested guidelines for use of avian species as biomonitors. *Environmental Monitoring and Assessment* 118:13–20.

Horton, K. G., F. A. La Sorte, D. Sheldon, T.-Y. Lin, K. Winner, G. Bernstein, S. Maji, W. M. Hochachka, and A. Farnsworth. 2020. Phenology of nocturnal avian migration has shifted at the continental scale. *Nature Climate Change* 10:63–68.

Horton, K. G., B. M. Van Doren, F. A. La Sorte, E. B. Cohen, H. L. Clipp, J. J. Buler, D. Fink, J. F. Kelly, and A. Farnsworth. 2019. Holding steady: Little change in intensity or timing of bird migration over the Gulf of Mexico. *Global Change Biology* 25:1106–1118.

Horton, K. G., B. M. Van Doren, F. A. La Sorte, D. Fink, D. Sheldon, A. Farnsworth, and J. F. Kelly. 2018. Navigating north: How body mass and winds shape avian flight behaviours across a North American migratory flyway. *Ecology Letters* 21:1055–1064.

Horton, K. G., B. M. Van Doren, P. M. Stepanian, A. Farnsworth, and J. F. Kelly. 2016a. Seasonal differences in landbird migration strategies. *The Auk* 133:761–769.

Horton, K. G., B. M. Van Doren, P. M. Stepanian, A. Farnsworth, and J. F. Kelly. 2016b. Where in the air? Aerial habitat use of nocturnally migrating birds. *Biology Letters* 12.

Horton, K. G., B. M. Van Doren, P. M. Stepanian, W. M. Hochachka, A. Farnsworth, and J. F. Kelly. 2016c. Nocturnally migrating songbirds drift when they can and compensate when they must. *Scientific Reports* 6: Article 21249.

Huber, N., K. Mahr, Z. Tóth, E. Z. Szarka, Y. U. Çınar, P. Salmón, and Á. Z. Lendvai. 2021. The stressed bird in the hand: Influence of sampling design on the physiological stress response in a free-living songbird. *Physiology & Behavior* 238: Article 113488.

Hüppop, O., M. Ciach, R. Diehl, D. R. Reynolds, P. M. Stepanian, and M. H. M. Menz. 2019. Perspectives and challenges for the use of radar in biological conservation. *Ecography* 42:912–930.

Iknayan, K. J., M. M. Wheeler, S. M. Safran, J. S. Young, and E. N. Spotswood. 2022. What makes urban parks good for California quail? Evaluating park suitability, species persistence, and the potential for reintroduction into a large urban national park. *Journal of Applied Ecology* 59:199–209.

Jetz, W., G. H. Thomas, J. B. Joy, K. Hartmann, and A. O. Mooers. 2012. The global diversity of birds in space and time. *Nature* 491:444–448.

Jetz, W., G. H. Thomas, J. B. Joy, D. W. Redding, K. Hartmann, and A. O. Mooers. 2014. Global distribution and conservation of evolutionary distinctness in birds. *Current Biology* 24:919–930.

Johnston, A., T. Auer, D. Fink, M. Strimas-Mackey, M. Iliff, K. V. Rosenberg, S. Brown, R. Lanctot, A. D. Rodewald, and S. Kelling. 2020. Comparing abundance distributions and range maps in spatial conservation planning for migratory species. *Ecological Applications* 30: Article e02058.

Johnston, A., W. M. Hochachka, M. E. Strimas-Mackey, V. R. Gutierrez, O. J. Robinson, E. T. Miller, T. Auer, S. T. Kelling, and D. Fink. 2021. Analytical guidelines to increase the value of community science data: An example using eBird data to estimate species distributions. *Diversity and Distributions* 27:1265–1277.

Johnston, A., N. Moran, A. Musgrove, D. Fink, and S. R. Baillie. 2020. Estimating species distributions from spatially biased citizen science data. *Ecological Modelling* 422: Article 108927.

Justyn, N. M., C. T. Callaghan, and G. E. Hill. 2020. Birds rarely hybridize: A citizen science approach to estimating rates of hybridization in the wild. *Evolution* 74:1216–1223.

Kain, M. P., and B. M. Bolker. 2019. Predicting West Nile virus transmission in North American bird communities using phylogenetic mixed effects models and eBird citizen science data. *Parasites & Vectors* 12: Article 395.

Karlsson, H., P. Henningsson, J. Bäckman, A. Hedenström, and T. Alerstam. 2010. Compensation for wind drift by migrating swifts. *Animal Behaviour* 80:399–404.

Karlsson, H., C. Nilsson, J. Bäckman, and T. Alerstam. 2012. Nocturnal passerine migrants fly faster in spring than in autumn: A test of the time minimization hypothesis. *Animal Behaviour* 83:87–93.

Kemp, M. U., J. Shamoun-Baranes, A. M. Dokter, E. van Loon, and W. Bouten. 2013. The influence of weather on the flight altitude of nocturnal migrants in mid-latitudes. *Ibis* 155:734–749.

Klingbeil, B. T., F. A. La Sorte, C. A. Lepczyk, D. Fink, and C. H. Flather. 2020. Geographical associations with anthropogenic noise pollution for North American breeding birds. *Global Ecology and Biogeography* 29:148–158.

Köck, K., T. Leltne, W. L. Randeu, M. Divjak, and K. J. Schrelber. 2000. OPERA: Operational Programme for the Exchange of Weather Radar Information. First results and outlook for the future. *Physics and Chemistry of the Earth, Part B: Hydrology, Oceans and Atmosphere* 25:1147–1151.

Kunz, T. H., S. A. Gauthreaux, Jr., N. I. Hristov, J. W. Horn, G. Jones, E. K. V. Kalko, R. P. Larkin, G. F. McCracken, S. M. Swartz, R. B. Srygley, R. Dudley, J. K. Westbrook, and M. Wikelski. 2008. Aeroecology: Probing and modeling the aerosphere. *Integrative and Comparative Biology* 48:1–11.

La Sorte, F. A., M. F. J. Aronson, C. A. Lepczyk, and K. G. Horton. 2020. Area is the primary correlate of annual and seasonal patterns of avian species richness in urban green spaces. *Landscape and Urban Planning* 203: Article 103892.

La Sorte, F. A., D. Fink, P. J. Blancher, A. D. Rodewald, V. Ruiz-Gutierrez, K. V. Rosenberg, W. M. Hochachka, P. H. Verburg, and S. Kelling. 2017. Global change and the distributional dynamics of migratory bird populations wintering in Central America. *Global Change Biology* 23: 5284–5296.

La Sorte, F. A., D. Fink, J. J. Buler, A. Farnsworth, and S. A. Cabrera-Cruz. 2017. Seasonal associations with urban light pollution for nocturnally migrating bird populations. *Global Change Biology* 23:4609–4619.

La Sorte, F. A., D. Fink, W. M. Hochachka, J. L. Aycrigg, K. V. Rosenberg, A. D. Rodewald, N. E. Bruns, A. Farnsworth, B. L. Sullivan, C. Wood, and S. Kelling. 2015. Documenting stewardship responsibilities across the annual cycle for birds on U.S. public lands. *Ecological Applications* 25:39–51.

La Sorte, F. A., D. Fink, W. M. Hochachka, and S. Kelling. 2016. Convergence of broad-scale migration strategies in terrestrial birds. *Proceedings of the Royal Society B: Biological Sciences* 283: Article 20152588.

La Sorte, F. A., D. Fink, and A. Johnston. 2018. Seasonal associations with novel climates for North American migratory bird populations. *Ecology Letters* 21:845–856.

La Sorte, F. A., D. Fink, and A. Johnston. 2019. Time of emergence of novel climates for North American migratory bird populations. *Ecography* 42:1079–1091.

La Sorte, F. A., and C. H. Graham. 2021. Phenological synchronization of seasonal bird migration with vegetation greenness across dietary guilds. *Journal of Animal Ecology* 90:343–355.

La Sorte, F. A., W. M. Hochachka, A. Farnsworth, A. A. Dhondt, and D. Sheldon. 2016. The implications of mid-latitude climate extremes for North American migratory bird populations. *Ecosphere* 7: Article e01261.

La Sorte, F. A., W. M. Hochachka, A. Farnsworth, D. Sheldon, D. Fink, J. Geevarghese, K. Winner, B. M. Van Doren, and S. Kelling. 2015. Migration timing and its determinants for nocturnal migratory birds during autumn migration. *Journal of Animal Ecology* 84:1202–1212.

La Sorte, F. A., W. M. Hochachka, A. Farnsworth, D. Sheldon, B. M. Van Doren, D. Fink, and S. Kelling. 2015. Seasonal changes in the altitudinal distribution of nocturnally migrating birds during autumn migration. *Royal Society Open Science* 2: Article 150347.

La Sorte, F. A., and K. G. Horton. 2021. Seasonal variation in the effects of artificial light at night on the occurrence of nocturnally migrating birds in urban areas. *Environmental Pollution* 270: Article 116085.

La Sorte, F. A., K. G. Horton, C. Nilsson, and A. M. Dokter. 2019. Projected changes in wind assistance under climate change for nocturnally migrating bird populations. *Global Change Biology* 25:589–601.

La Sorte, F. A., C. A. Lepczyk, J. L. Burnett, A. H. Hurlbert, M. W. Tingley, and B. Zuckerberg. 2018. Opportunities and challenges for big data ornithology. *The Condor* 120:414–426.

La Sorte, F. A., and M. Somveille. 2020. Survey completeness of a global citizen-science database of bird occurrence. *Ecography* 43:34–43.

La Sorte, F. A., and M. Somveille. 2021. The island biogeography of the eBird citizen-science programme. *Journal of Biogeography* 48:628–638.

Lack, D., and G. C. Varley. 1945. Detection of birds by radar. *Nature* 156:446.

Lao, S., B. A. Robertson, A. W. Anderson, R. B. Blair, J. W. Eckles, R. J. Turner, and S. R. Loss. 2020. The influence of artificial light at night and polarized light on bird-building collisions. *Biological Conservation* 241: Article 108358.

Larkin, R. P., and B. A. Frase. 1988. Circular paths of birds flying near a broadcasting tower in cloud. *Journal of Comparative Psychology* 102:90–93.

Larson, E. R., B. M. Graham, R. Achury, J. J. Coon, M. K. Daniels, D. K. Gambrell, K. L. Jonasen, G. D. King, N. LaRacuente, T. I. N. Perrin-Stowe, E. M. Reed, C. J. Rice, S. A. Ruzi, M. W. Thairu, J. C. Wilson, and A. V. Suarez. 2020. From eDNA to citizen science: Emerging tools for the early detection of invasive species. *Frontiers in Ecology and the Environment* 18:194–202.

Laughlin, A. J., D. R. Sheldon, D. W. Winkler, and C. M. Taylor. 2014. Behavioral drivers of communal roosting in a songbird: A combined theoretical and empirical approach. *Behavioral Ecology* 25:734–743.

Laughlin, A. J., D. R. Sheldon, D. W. Winkler, and C. M. Taylor. 2016. Quantifying non-breeding season occupancy patterns and the timing and drivers of autumn migration for a migratory songbird using Doppler radar. *Ecography* 39:1017–1024.

Laughlin, A. J., C. M. Taylor, D. W. Bradley, D. LeClair, R. G. Clark, R. D. Dawson, P. O. Dunn, A. Horn, M. Leonard, D. R. Sheldon, D. Shutler, L. A. Whittingham, D. W. Winkler, and D. R. Norris. 2013. Integrating information from geolocators, weather radar, and citizen science to uncover a key stopover area of an aerial insectivore. *The Auk* 130:230–239.

Lee, J. S., C. T. Callaghan, and W. K. Cornwell. 2023. Using citizen science to measure recolonisation of birds after the Australian 2019–2020 mega-fires. *Austral Ecology* 48:31–40.

Liang, Y., I. Rudik, E. Y. Zou, A. Johnston, A. D. Rodewald, and C. L. Kling. 2020. Conservation cobenefits from air pollution regulation: Evidence from birds. *Proceedings of the National Academy of Sciences of the USA* 117: Article 30900.

Liechti, F., J. Guélat, and S. Komenda-Zehnder. 2013. Modelling the spatial concentrations of bird migration to assess conflicts with wind turbines. *Biological Conservation* 162:24–32.

Lin, H.-Y., R. Schuster, S. Wilson, S. J. Cooke, A. D. Rodewald, and J. R. Bennett. 2020. Integrating season-specific needs of migratory and resident birds in conservation planning. *Biological Conservation* 252: Article 108826.

Lin, T.-Y., K. Winner, G. Bernstein, A. Mittal, A. M. Dokter, K. G. Horton, C. Nilsson, B. M. Van Doren, A. Farnsworth, F. A. La Sorte, S. Maji, and D. Sheldon. 2019. MistNet: Measuring historical bird migration in the US using archived weather radar data and convolutional neural networks. *Methods in Ecology and Evolution* 10:1908–1922.

Lloyd, J. D. 2018. The recent expansion of Fox Sparrow (*Passerella iliaca iliaca*) breeding range into the northeastern United States. *PeerJ* 6: Article e6087.

Long, A. M., B. L. Pierce, A. D. Anderson, K. L. Skow, A. Smith, and R. R. Lopez. 2019. Integrating citizen science and remotely sensed data to help inform time-sensitive policy decisions for species of conservation concern. *Biological Conservation* 237:463–469.

Lopez, B., E. Minor, and A. Crooks. 2020. Insights into human–wildlife interactions in cities from bird sightings recorded online. *Landscape and Urban Planning* 196: Article 103742.
Lord, R. D., F. C. Bellrose, and W. W. Cochran. 1962. Radiotelemetry of the respiration of a flying duck. *Science* 137:39–40.
Ma, Z., Y. Cheng, J. Wang, and X. Fu. 2013. The rapid development of birdwatching in mainland China: A new force for bird study and conservation. *Bird Conservation International* 23:259–269.
Maina, J. N. 2000. What it takes to fly: The structural and functional respiratory refinements in birds and bats. *Journal of Experimental Biology* 203:3045–3064.
Marra, P. P., E. B. Cohen, S. R. Loss, J. E. Rutter, and C. M. Tonra. 2015. A call for full annual cycle research in animal ecology. *Biology Letters* 11: Article 20150552.
Martín, B., A. Onrubia, J. González-Arias, and J. A. Vicente-Vírseda. 2020. Citizen science for predicting spatio-temporal patterns in seabird abundance during migration. *PLoS One* 15: Article e0236631.
Maynard, R. H. 1945. Radar and weather. *Journal of Meteorology* 2:214–226.
McCarty, J. P. 2001. Ecological consequences of recent climate change. *Conservation Biology* 15:320–331.
McClure, E. C., M. Sievers, C. J. Brown, C. A. Buelow, E. M. Ditria, M. A. Hayes, R. M. Pearson, V. J. D. Tulloch, R. K. F. Unsworth, and R. M. Connolly. 2020. Artificial intelligence meets citizen science to supercharge ecological monitoring. *Patterns* 1:100109.
McLaren, J. D., J. J. Buler, T. Schreckengost, J. A. Smolinsky, M. Boone, E. Emiel van Loon, D. K. Dawson, and E. L. Walters. 2018. Artificial light at night confounds broad-scale habitat use by migrating birds. *Ecology Letters* 21:356–364.
McLaren, J. D., J. Shamoun-Baranes, and W. Bouten. 2012. Wind selectivity and partial compensation for wind drift among nocturnally migrating passerines. *Behavioral Ecology* 23:1089–1101.
Meineke, E. K., and B. H. Daru. 2021. Bias assessments to expand research harnessing biological collections. *Trends in Ecology & Evolution* 36:1071–1082.
Michel, N. L., S. P. Saunders, T. D. Meehan, and C. B. Wilsey. 2021. Effects of stewardship on protected area effectiveness for coastal birds. *Conservation Biology* 35:1484–1495.
Mirkovic, D., P. M. Stepanian, J. F. Kelly, and P. B. Chilson. 2016. Electromagnetic model reliably predicts radar scattering characteristics of airborne organisms. *Scientific Reports* 6: Article 35637.
Mondain-Monval, T. O., M. Amos, J. L. Chapman, A. MacColl, and S. P. Sharp. 2021. Flyway-scale analysis reveals that the timing of migration in wading birds is becoming later. *Ecology and Evolution* 11:14135–14145.
Morrison, M. L. 1986. Bird populations as indicators of environmental change. Pages 429–451 in R. F. Johnston, ed. *Current Ornithology*. Vol. 3. Springer, Boston.
Muller, J. A., J. A. Veech, and R. M. Kostecke. 2018. Landscape-scale habitat associations of Sprague's Pipits wintering in the southern United States. *Journal of Field Ornithology* 89:326–336.
Nelson, A. R., R. L. Cormier, D. L. Humple, J. C. Scullen, R. Sehgal, and N. E. Seavy. 2016. Migration patterns of San Francisco Bay Area Hermit Thrushes differ across a fine spatial scale. *Animal Migration* 3:1–13.
Newcombe, P. B., C. Nilsson, T.-Y. Lin, K. Winner, G. Bernstein, S. Maji, D. Sheldon, A. Farnsworth, and K. G. Horton. 2019. Migratory flight on the Pacific Flyway: Strategies and tendencies of wind drift compensation. *Biology Letters* 15: Article 20190383.
Newton, I. 2003. *The Speciation and Biogeography of Birds*. Academic Press, London.
Nilsson, C., A. M. Dokter, B. Schmid, M. Scacco, L. Verlinden, J. Bäckman, G. Haase, G. Dell'Omo, J. W. Chapman, H. Leijnse, and F. Liechti. 2018. Field validation of radar systems for monitoring bird migration. *Journal of Applied Ecology* 55:2552–2564.
Nilsson, C., K. G. Horton, A. M. Dokter, B. M. V. Doren, and A. Farnsworth. 2018. Aeroecology of a solar eclipse. *Biology Letters* 14: Article 20180485.
Nilsson, C., F. A. La Sorte, A. Dokter, K. Horton, B. M. Van Doren, J. J. Kolodzinski, J. Shamoun-Baranes, and A. Farnsworth. 2021. Bird strikes at commercial airports explained by citizen science and weather radar data. *Journal of Applied Ecology* 58:2029–2039.
Nilsson, C., and S. Sjöberg. 2016. Causes and characteristics of reverse bird migration: An analysis based on radar, radio tracking and ringing at Falsterbo, Sweden. *Journal of Avian Biology* 47:354–362.

Nussbaumer, R., S. Bauer, L. Benoit, G. Mariethoz, F. Liechti, and B. Schmid. 2021. Quantifying year-round nocturnal bird migration with a fluid dynamics model. *Journal of the Royal Society: Interface* 18: Article 20210194.

Ottenburghs, J., and D. L. Slager. 2020. How common is avian hybridization on an individual level? *Evolution* 74:1228–1229.

Panuccio, M., G. Dell'Omo, G. Bogliani, C. Catoni, and N. Sapir. 2019. Migrating birds avoid flying through fog and low clouds. *International Journal of Biometeorology* 63:231–239.

Powers, B. F., J. M. Winiarski, J. M. Requena-Mullor, and J. A. Heath. 2021. Intra-specific variation in migration phenology of American Kestrels (*Falco sparverius*) in response to spring temperatures. *Ibis* 163:1448–1456.

Reese, J., C. Viverette, C. M. Tonra, N. J. Bayly, T. J. Boves, E. Johnson, M. Johnson, P. Marra, E. M. Ames, A. Caguazango, M. DeSaix, A. Matthews, A. Molina, K. Percy, M. C. Slevin, and L. Bulluck. 2019. Using stable isotopes to estimate migratory connectivity for a patchily distributed, wetland-associated Neotropical migrant. *The Condor* 121:1–15.

Reino, L., R. Figueira, P. Beja, M. B. Araújo, C. Capinha, and D. Strubbe. 2017. Networks of global bird invasion altered by regional trade ban. *Science Advances* 3: Article e1700783.

Robinson, O. J., V. Ruiz-Gutierrez, M. D. Reynolds, G. H. Golet, M. Strimas-Mackey, and D. Fink. 2020. Integrating citizen science data with expert surveys increases accuracy and spatial extent of species distribution models. *Diversity and Distributions* 26:976–986.

Rocha, L. A., A. Aleixo, G. Allen, F. Almeda, C. C. Baldwin, M. V. L. Barclay, J. M. Bates, A. M. Bauer, F. Benzoni, C. M. Berns, M. L. Berumen, D. C. Blackburn, S. Blum, F. Bolaños, R. C. K. Bowie, R. Britz, R. M. Brown, C. D. Cadena, K. Carpenter, L. M. Ceríaco, P. Chakrabarty, G. Chaves, J. H. Choat, K. D. Clements, B. B. Collette, A. Collins, J. Coyne, J. Cracraft, T. Daniel, M. R. de Carvalho, K. de Queiroz, F. Di Dario, R. Drewes, J. P. Dumbacher, A. Engilis, M. V. Erdmann, W. Eschmeyer, C. R. Feldman, B. L. Fisher, J. Fjeldså, P. W. Fritsch, J. Fuchs, A. Getahun, A. Gill, M. Gomon, T. Gosliner, G. R. Graves, C. E. Griswold, R. Guralnick, K. Hartel, K. M. Helgen, H. Ho, D. T. Iskandar, T. Iwamoto, Z. Jaafar, H. F. James, D. Johnson, D. Kavanaugh, N. Knowlton, E. Lacey, H. K. Larson, P. Last, J. M. Leis, H. Lessios, J. Liebherr, M. Lowman, D. L. Mahler, V. Mamonekene, K. Matsuura, G. C. Mayer, H. Mays, J. McCosker, R. W. McDiarmid, J. McGuire, M. J. Miller, R. Mooi, R. D. Mooi, C. Moritz, P. Myers, M. W. Nachman, R. A. Nussbaum, D. Ó. Foighil, L. R. Parenti, J. F. Parham, E. Paul, G. Paulay, J. Pérez-Emán, A. Pérez-Matus, S. Poe, J. Pogonoski, D. L. Rabosky, J. E. Randall, J. D. Reimer, D. R. Robertson, M. O. Rödel, M. T. Rodrigues, P. Roopnarine, L. Rüber, M. J. Ryan, F. Sheldon, G. Shinohara, A. Short, W. B. Simison, W. F. Smith-Vaniz, V. G. Springer, M. Stiassny, J. G. Tello, C. W. Thompson, T. Trnski, P. Tucker, T. Valqui, M. Vecchione, E. Verheyen, P. C. Wainwright, T. A. Wheeler, W. T. White, K. Will, J. T. Williams, G. Williams, E. O. Wilson, K. Winker, R. Winterbottom, and C. C. Witt. 2014. Specimen collection: An essential tool. *Science* 344:814–815.

Rodewald, A. D., M. Strimas-Mackey, R. Schuster, and P. Arcese. 2019. Beyond canaries in coal mines: Co-occurrence of Andean mining concessions and migratory birds. *Perspectives in Ecology and Conservation* 17:151–156.

Russell, K. R., and S. A. Gauthreaux. 1998. Use of weather radar to characterize movements of roosting purple martins. *Wildlife Society Bulletin* 26:5–16.

Russell, K. R., and S. A. Gauthreaux. 1999. Spatial and temporal dynamics of a Purple Martin premigratory roost. *Wilson Bulletin* 111:354–362.

Russell, K. R., D. S. Mizrahi, and S. A. Gauthreaux. 1998. Large-scale mapping of Purple Martin premigratory roosts using WSR-88D weather surveillance radar. *Journal of Field Ornithology* 69:316–325.

Saracco, J., D. Desante, M. Nott, and D. Kaschube. 2008. Using the MAPS and MoSI programs to monitor landbirds and inform conservation. Pages 651–658 in T. D. Rich, C. D. Thompson, D. Demarest, and C. Arizmendi, eds. *Proceedings of the Fourth International Partners in Flight Conference: Tundra to Tropics*. University of Texas-Pan American Press, Edinburg, TX.

Schuetz, J. G., and A. Johnston. 2019. Characterizing the cultural niches of North American birds. *Proceedings of the National Academy of Sciences of the USA* 116:10868–10873.

Selous, E. 1901. *Bird Watching*. Dent, London.

Shamoun-Baranes, J., C. Nilsson, S. Bauer, and J. Chapman. 2019. Taking radar aeroecology into the 21st century. *Ecography* 42:847–851.

Shannon, G., M. F. McKenna, L. M. Angeloni, K. R. Crooks, K. M. Fristrup, E. Brown, K. A. Warner, M. D. Nelson, C. White, J. Briggs, S. McFarland, and G. Wittemyer. 2016. A synthesis of two decades of research documenting the effects of noise on wildlife. *Biological Reviews* 91: 982–1005.

Sheard, C., M. H. Neate-Clegg, N. Alioravainen, S. E. Jones, C. Vincent, H. E. MacGregor, T. P. Bregman, S. Claramunt, and J. A. Tobias. 2020. Ecological drivers of global gradients in avian dispersal inferred from wing morphology. *Nature Communications* 11: Article 2463.

Smith, J. A. M., K. Regan, N. W. Cooper, L. Johnson, E. Olson, A. Green, J. Tash, D. C. Evers, and P. P. Marra. 2020. A green wave of saltmarsh productivity predicts the timing of the annual cycle in a long-distance migratory shorebird. *Scientific Reports* 10: Article 20658.

Somveille, M., R. A. Bay, T. B. Smith, P. P. Marra, and K. C. Ruegg. 2021. A general theory of avian migratory connectivity. *Ecology Letters* 24:1848–1858.

Soykan, C. U., J. Sauer, J. G. Schuetz, G. S. LeBaron, K. Dale, and G. M. Langham. 2016. Population trends for North American winter birds based on hierarchical models. *Ecosphere* 7: Article e01351.

Steinheimer, F. D. 2004. Charles Darwin's bird collection and ornithological knowledge during the voyage of H.M.S. "Beagle," 1831–1836. *Journal of Ornithology* 145:300–320.

Sullivan, B. L., J. L. Aycrigg, J. H. Barry, R. E. Bonney, N. Bruns, C. B. Cooper, T. Damoulas, A. A. Dhondt, T. Dietterich, A. Farnsworth, D. Fink, J. W. Fitzpatrick, T. Fredericks, J. Gerbracht, C. Gomes, W. M. Hochachka, M. J. Iliff, C. Lagoze, F. A. La Sorte, M. Merrifield, W. Morris, T. B. Phillips, M. Reynolds, A. D. Rodewald, K. V. Rosenberg, N. M. Trautmann, A. Wiggins, D. W. Winkler, W.-K. Wong, C. L. Wood, J. Yu, and S. Kelling. 2014. The eBird enterprise: An integrated approach to development and application of citizen science. *Biological Conservation* 169:31–40.

Supp, S. R., G. Bohrer, J. Fieberg, and F. A. La Sorte. 2021. Estimating the movements of terrestrial animal populations Using broad-scale occurrence data. *Movement Ecology* 9: Article 60.

Supp, S. R., F. A. La Sorte, T. A. Cormier, M. C. W. Lim, D. R. Powers, S. M. Wethington, S. Goetz, and C. H. Graham. 2015. Citizen-science data provides new insight into annual and seasonal variation in migration patterns. *Ecosphere* 6: Article 5.

Tang, B., J. S. Clark, and A. E. Gelfand. 2021. Modeling spatially biased citizen science effort through the eBird database. *Environmental and Ecological Statistics* 28:609–630.

Tanner, A. M., E. P. Tanner, M. Papeş, S. D. Fuhlendorf, R. D. Elmore, and C. A. Davis. 2020. Using aerial surveys and citizen science to create species distribution models for an imperiled grouse. *Biodiversity and Conservation* 29:967–986.

Tien, J. M. 2013. Big data: Unleashing information. *Journal of Systems Science and Systems Engineering* 22:127–151.

Tsai, P. Y., C. J. Ko, S. Y. Chia, Y. J. Lu, and M. N. Tuanmu. 2021. New insights into the patterns and drivers of avian altitudinal migration from a growing crowdsourcing data source. *Ecography* 44:75–86.

Uehling, J. J., J. Tallant, and S. Pruett-Jones. 2019. Status of naturalized parrots in the United States. *Journal of Ornithology* 160:907–921.

Van den Broeke, M. S. 2019. Radar quantification, temporal analysis and influence of atmospheric conditions on a roost of American Robins (*Turdus migratorius*) in Oklahoma. *Remote Sensing in Ecology and Conservation* 5:193–204.

Van Doren, B. M., and K. G. Horton. 2018. A continental system for forecasting bird migration. *Science* 361:1115–1118.

Van Doren, B. M., K. G. Horton, A. M. Dokter, H. Klinck, S. B. Elbin, and A. Farnsworth. 2017. High-intensity urban light installation dramatically alters nocturnal bird migration. *Proceedings of the National Academy of Sciences of the USA* 114:11175–11180.

Van Doren, B. M., D. E. Willard, M. Hennen, K. G. Horton, E. F. Stuber, D. Sheldon, A. H. Sivakumar, J. Wang, A. Farnsworth, and B. M. Winger. 2021. Drivers of fatal bird collisions in an urban center. *Proceedings of the National Academy of Sciences of the USA* 118: Article e2101666118.

van Gasteren, H., K. L. Krijgsveld, N. Klauke, Y. Leshem, I. C. Metz, M. Skakuj, S. Sorbi, I. Schekler, and J. Shamoun-Baranes. 2019. Aeroecology meets aviation safety: Early warning systems in Europe and the Middle East prevent collisions between birds and aircraft. *Ecography* 42:899–911.

Webster, M. S., P. P. Marra, S. M. Haig, S. Bensch, and R. T. Holmes. 2002. Links between worlds: Unraveling migratory connectivity. *Trends in Ecology & Evolution* 17:76–83.

Welcker, J., M. Liesenjohann, J. Blew, G. Nehls, and T. Grünkorn. 2017. Nocturnal migrants do not incur higher collision risk at wind turbines than diurnally active species. *Ibis* 159:366–373.

Wilson, A. A., M. A. Ditmer, J. R. Barber, N. H. Carter, E. T. Miller, L. P. Tyrrell, and C. D. Francis. 2021. Artificial night light and anthropogenic noise interact to influence bird abundance over a continental scale. *Global Change Biology* 27:3987–4004.

Wilson, S., R. Schuster, A. D. Rodewald, J. R. Bennett, A. C. Smith, F. A. La Sorte, P. H. Verburg, and P. Arcese. 2019. Prioritize diversity or declining species? Trade-offs and synergies in spatial planning for the conservation of migratory birds in the face of land cover change. *Biological Conservation* 239: Article 108285.

Winker, K. 2005. Bird collections: Development and use of a scientific resource. *The Auk* 122:966–971.

Winker, K., J. M. Reed, P. Escalante, R. A. Askins, C. Cicero, G. E. Hough, and J. Bates. 2010. The importance, effects, and ethics of bird collecting. *The Auk* 127:690–695.

Young, B. E., N. Dodge, P. D. Hunt, M. Ormes, M. D. Schlesinger, and H. Y. Shaw. 2019. Using citizen science data to support conservation in environmental regulatory contexts. *Biological Conservation* 237:57–62.

Youngflesh, C., J. Socolar, B. R. Amaral, A. Arab, R. P. Guralnick, A. H. Hurlbert, R. LaFrance, S. J. Mayor, D. A. W. Miller, and M. W. Tingley. 2021. Migratory strategy drives species-level variation in bird sensitivity to vegetation green-up. *Nature Ecology & Evolution* 5:987–994.

Zuckerberg, B., D. Fink, F. A. La Sorte, W. M. Hochachka, and S. Kelling. 2016. Novel seasonal land cover associations for eastern North American forest birds identified through dynamic species distribution modelling. *Diversity and Distributions* 22:717–730.

Zulian, V., D. A. W. Miller, and G. Ferraz. 2021. Integrating citizen-science and planned-survey data improves species distribution estimates. *Diversity and Distributions* 27:2498–2509.

Zurell, D., C. H. Graham, L. Gallien, W. Thuiller, and N. E. Zimmermann. 2018. Long-distance migratory birds threatened by multiple independent risks from global change. *Nature Climate Change* 8:992–996.

19
Participatory Science in the Field of Ornithology

Bradley Allf, Deja Perkins, Jin Bai, and Caren Cooper

Use of the term *citizen science* to describe the centralization of observations by the lay public for scientific purposes was coined in the United States in 1995 at the Cornell Lab of Ornithology. By 2014, the Citizen Science Association formed as the first professional association of practitioners who make citizen science happen by designing, facilitating, evaluating, and examining citizen science and advancing the theory and practice across the discipline. Despite the relatively recent formalization of citizen science as a distinct field of practice (Jordan et al. 2015), and evolving terminology (Table 19.1), the participation of bird-watchers in science activities began long before.

The first wave of environmentalism in the United States was prompted in part by declines in birds, particularly due to overhunting for the millinery (hat-making) industry (Taylor 2016). In the United States during the late 1800s and early 1900s, socially advantaged people with interests in birds formed societies that specifically focused on avian conservation, such as the Audubon Society, the Nuttall Ornithological Society, the American Ornithological Union (AOU), the Wilson Ornithological Society, and the Cooper Ornithological Society, mirroring societies that already existed in Europe such as the British Ornithologists' Union. The power of these societies is evident in the legislation that was passed around the turn of the century at their urging: the Lacey Act and the Migratory Bird Treaty Act, which heavily restricted the killing of birds. This legislation not only restricted bird hunting for hats and food but also put an end to the killing and collection of birds, as well as their nests, feathers, and eggs, by amateur ornithologists—the primary means of ornithological study up until this period. At the same time, binoculars became a more affordable purchase, which allowed for non-extractive bird study. By the late 1800s, it was relatively common for individuals, particularly those who were wealthy, educated, and with leisurely interests in birds, to maintain state- and county-level lists of bird species they had seen (Barrow 2000, Shane 2012).

Over time, these ornithological societies not only advocated for the conservation of birds but also were organizing to study them. For instance, the AOU recruited lighthouse keepers to record the birds they found dead or that struck the lighthouse. The effort failed in part because of a lack of consensus on bird names, with keepers using local colloquial names for birds and AOU scientists unable to figure out what they meant (Merriam 1885, Droege 2007). Around this time, a schoolteacher named W. W. Cooke organized tens of thousands of volunteers to track the timing of bird migrations on standardized note cards. This effort continued into the mid-1900s, and these cards

Bradley Allf et al., *Participatory Science in the Field of Ornithology*. In: *New Perspectives in Ornithology*.
Edited by: Scott V. Edwards and J. Michael Reed, Oxford University Press. © Oxford University Press (2025).
DOI: 10.1093/oso/9780197787670.003.0019

Table 19.1 Definition of Terms[a]

Term	Definition
Participatory science(s)	A descriptive phrase referring to a range of approaches to scientific activities that involve those without scientific credentials. In this chapter, we refer to *participatory science* as a term that includes both citizen science and community science projects.
Citizen science	Projects in which participants share observations or specimens with a central entity for a shared purpose. These projects are most commonly initiated and administered by large institutions such as the Cornell Lab of Ornithology and the National Audubon Society. This chapter primarily focuses on citizen science. In some contexts, such as in U.S. federal law (i.e., the Crowdsourcing and Citizen Science Act of 2016, H.R. 6414), the term *citizen science* has a broader definition, referring to a range of approaches to scientific activities that involve those without scientific credentials.
Community science	Grassroots, community-driven participatory sciences. In this chapter, we use the term *community science* to refer to projects initiated, administered, and/or coordinated by grassroots efforts, such as the North American Bluebird Society and the Purple Martin Conservation Association. These examples also involve sharing data to a central entity, but participants are often involved in more self-directed activities for the study and conservation of birds.

[a] In recognition of evolving terminology, we define terms in relation to their recent historical use.

were digitized for decades through a project called the North American Bird Phenology Program, organized by the U.S. Geological Survey (Greenwood 2007, Mayer 2010). The Christmas Bird Count began in 1900 as a nationwide survey of birds to replace the earlier tradition of a Christmas bird hunt and was later administered by the Audubon Society. Bird-watchers collated observations among the thousands of participating members and circulated lists in society periodicals (Stewart 1954). The Christmas Bird Count was the first successful and longest running example of citizen science for avian research.

Before the emergence of ornithology as a profession, studies were conducted by people who today we might refer to as amateurs because their primary employment did not involve avian research (Miller-Rushing et al. 2012). By the late 1800s, a distinction began to emerge between paid avian researchers ("professionals") and unpaid bird hobbyists ("amateurs"; Figure 19.1A). As this distinction grew, the research interests of these two groups began to split, with amateurs focused more on local natural history studies, borne of informal education and an interest in local ecology (Vetter 2011).

Amateurs led high-quality studies and made important contributions to ornithology. An amateur birder named J. H. Gurney conducted the first worldwide census of

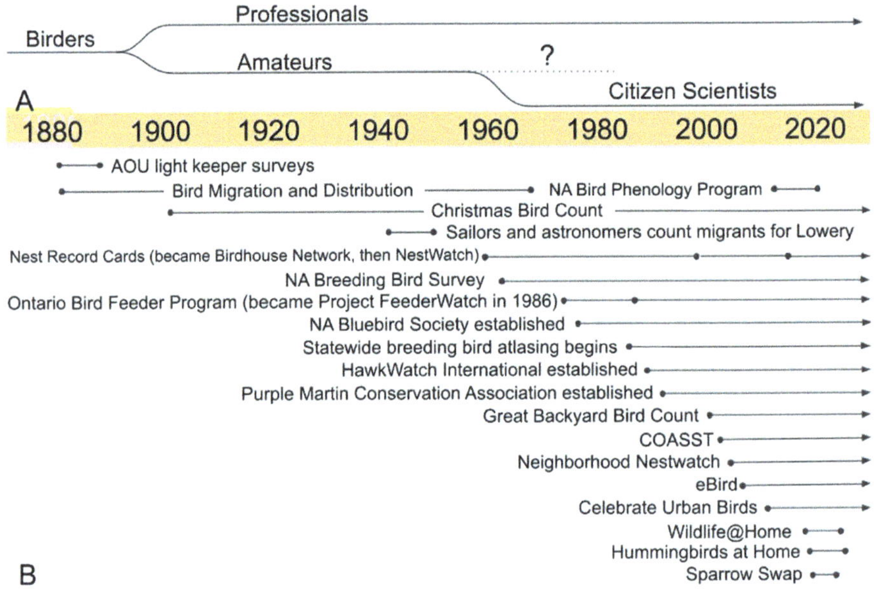

Figure 19.1 Timeline of trends in ornithology and related citizen science activity focused on birds within North America. (A) Bird enthusiasts began separating into professionals and amateurs around 1900. By the 1960s, an offshoot of amateurs became citizen science data collectors, while amateurs leading scientific studies became much less common. (B) A timeline of the emergence of popular avian participatory science activities and organizations in the United States since 1880. Lines ending in a circle indicate that a project ended; lines ending in an arrow indicate a project is still active. Note that the Bird Migration and Distribution program started in 1880 by Wells Cooke stopped in 1970 but then restarted in the 2000s as the North American Bird Phenology Program, whose goal was to digitize the earlier migration records. "Lowery" refers to George Lowery, ornithologist at Louisiana State University. AOU, American Ornithological Union; COASST, Coastal Observation and Seabird Survey Team; NA, North American.

the Atlantic Gannett in 1913, for instance, and a businessman named Arthur C. Bent published 22 volumes of his *Life Histories of North American Birds* series starting in 1919, despite never receiving a salary for his research and lacking formal training in ornithology (Barrow 2000, Greenwood 2007). Other pioneering amateurs of the early 20th century include the Michigan dentist Lawrence Walkinshaw, who wrote an authoritative account of field sparrows based on observations made on an abandoned farm near his home (Mayfield 1991), and Harold Mayfield, a businessman who pioneered new techniques for measuring nest success and was later elected president of the American Ornithologists' Union (Tramer 2007). Indeed, between the 1950s and 1970s, a quarter of all researchers who received the AOU's Brewster medal were amateurs (King and Bock 1978).

Rachel Carson's *Silent Spring* (1962) played a significant role in the emergence of the second wave of environmentalism in the United States, and once again, bird conservation was at the fore of this movement. Carson's book brought to light the impact of pesticides on ecosystems, and her central premise—that untargeted pesticides might lead to springtime soundscapes without bird or insect songs—had a direct influence on public interest in bird conservation and the subsequent development of the U.S. Breeding Bird Survey (standardized avian surveys along roadsides) and the North American Nest Record Card Program (standardized cards for monitoring nests), which were both launched in the 1960s (Pettingill 1966, Sauer et al. 2017). A decade before Carson's book, however, a retired banker and amateur birder in Florida named Charles Broley, who banded thousands of bald eagles, may have been the first person to suggest a link between the pesticide DDT and egg mortality in eagles and other raptors (Barrow 2000).

In 1974, the National Science Foundation funded a study on the state of ornithology that culminated with a white paper (King and Bock 1978). In the white paper, King and Bock found that half of the members of ornithological societies were amateurs, and the authors encouraged ornithological societies to be prime movers of increasing public participation in ornithology. The societies clearly recognized amateurs as a valuable resource that should be supported, yet the continued professionalization of ornithology coincided with an apparent steady decrease of amateurs in societies. Instead of professional associations taking on the role of promoting and sustaining amateur engagement, the role of nonprofessionals in ornithology was transformed from amateurs producing natural history papers to the engagement of a broader array of bird enthusiasts in data collection through citizen science projects organized by institutions such as federal agencies, the National Audubon Society, and the Cornell Lab of Ornithology.

Web 2.0 technologies emerged in the early 2000s and allowed users of online spaces to add content, making it easier for scientists to create citizen science projects that can accept observation via the internet. At the same time, the rise of social media fostered a culture of online sharing. In this atmosphere, the number of citizen science projects continued to grow, including fully online projects in which volunteers assisted in transcribing text and annotating photographs and videos (Figure 19.1B).

The history of participatory science creates a legacy of data in ornithology. For example, in large part because of well-kept records of bird clubs, nest monitoring projects, and banders, half of what is understood about migratory birds and climate change is based on studies that leveraged these types of crowdsourced data (Cooper 2014). These studies involved participants contributing bird checklists, banding/resighting birds, and monitoring nesting birds. Although these three activities form the basis of most avian citizen science, Web 2.0 technologies have enabled other ways for bird enthusiasts to contribute to ornithological research. With the rise of the internet, the smartphone, and big data visualization, an even more expansive vision for ornithological citizen science has emerged.

Successes of Ornithological Citizen Science

The greatest value of avian citizen science is in the broadscale data sets it generates, which would be otherwise logistically or financially unfeasible for scientists to generate on their own. Whereas some citizen science projects are structured to test specific hypotheses, many collect baseline information on many aspects of avian species, such as population size, range, migration phenology, reproductive success, and mortality rates and causes.

Volunteer-generated data have high variability (e.g., measurement error) and bias (e.g., spatiotemporal clustering). Projects recognize and address data quality issues by, for example, providing volunteer training, data standardization, validation and filtering procedures, and other actions to reduce sources of error and bias before, during, and after data are collected (Wiggins et al. 2011, Bonter and Cooper 2012). For instance, the Cornell Lab of Ornithology's Bird Academy provides trainings in citizen science practices and bird identification that can improve citizen scientists' data quality, and the Lab also has automated processes that use algorithms to flag potentially inaccurate data records submitted to their projects (Sullivan et al. 2014). All data have both strengths and limitations that are important to recognize (including data collected by scientists), and the types of limitations associated with citizen science data can be specific to the volunteer activity and type of data collected. We review some areas of success associated with different participant activities.

Specimen Collection

During the European Renaissance, it was fashionable for wealthy European men to collect birds, as well as other animals, and classify, catalog, and preserve them in "cabinets of curiosities." Some of these efforts were participatory. For instance, in the 1500s, Pierre Belon encouraged people to send him bird specimens from throughout the world, from which he produced a book of woodcut bird illustrations (Chansigaud 2010).

The colonial roots of ornithology are apparent as the practice of collecting specimens grew and evolved during the 17th and 18th centuries, with governments in Western Europe sponsoring voyages around the world to lay claim to new territories and extract resources. Museum and university collections grew with thousands of specimens that explorers collected (Chansigaud 2010). For example, in the early 1800s, on the Lewis and Clark expedition, explorers made significant efforts to describe the birds they saw and collect bird specimens (Davis and Stevenson 1934). In the 1850s, Spencer Fullerton Baird, as the first curator of the Smithsonian Institution, recruited doctors in the U.S. Army to collect birds on their journeys (Shane 2012).

Collections have been used for significant findings, such as the role of DDT in eggshell thinning (Hickey and Anderson 1968). Today, specimen collection is a fairly uncommon citizen science activity, partly restricted by the Migratory Bird Treaty Act and other laws protecting wildlife. Collections of skins, nests, and shells provide useful biological vouchers for DNA, contaminant data, and stable isotope measurements. For example, Full Cycle Phenology is a project that relies on the volunteer collection of feathers to better understand kestrel migration (Joray 2020). Sparrow Swap was a project in which volunteers mailed the eggs of House Sparrows (a non-native species not protected by the law in the United States) to the North Carolina Museum of Natural Sciences for analysis of heavy metals and egg color patterns (Hartley 2019). Feather

Map of Australia involves volunteer collection of feathers and lab analyses of stable isotopes (Brandis et al. 2021).

Bird-Watching Checklists

One of the most prominent and straightforward types of volunteer data contribution is the checklist of wild living birds encountered. The most popular checklist project is eBird, by the Cornell Lab of Ornithology. Volunteers contribute to eBird using a mobile app to record the number of birds of different species they see in a given area and for a given amount of time. As of 2022, more than 800,000 citizen scientists had contributed more than 1 billion bird observations and 70 million checklists across more than 200 countries. Few, if any, citizen science projects, from any discipline, approach the magnitude of data generated by eBird, and the scope of this project is a noteworthy development in the history of science. In 2021, eBird volunteers submitted more than 1 million records of Red-Tailed Hawks alone—a species whose total global population is estimated to be approximately 3 million (Partners in Flight 2020). Acknowledging the fact that some of these records are of the same individual bird, it is still remarkable that data are being collected from a substantial fraction of all members of an entire species.

Data from eBird have been used in more than 100 peer-reviewed publications and many management plans. For instance, La Sorte et al. (2017) used eBird data to examine large-scale bird migration patterns, specifically whether nocturnal migratory birds are attracted to artificial lights at night in urban areas. In addition to eBird, other projects use various forms of checklists, such as the Cornell Lab's Project FeederWatch, which tasks volunteers with counting birds that visit feeders; Audubon's Climate Watch, in which volunteers count 12 species at a specific location twice per year; and the U.S. Geological Survey's North American Breeding Bird Survey, in which volunteers count birds along randomly established roadside routes. An offshoot of Project FeederWatch, the House Finch Disease Survey, tasked volunteers with maintaining checklists of House Finches with visible conjunctivitis to track the spread of avian disease (Hochachka and Dhondt 2000). Checklist data are most often used to better understand avian population dynamics, the distribution of invasive species, endangered species management, and how birds respond to climate change and urbanization.

Checklists provide three main types of data about living birds (Bird et al. 2014): presence-only observations (Is a species present in an area?), presence *and* absence observations (Is a species present in an area and not in another?), and counts (the number of individuals of each species observed). In addition, checklist data can be structured, unstructured, or semistructured—a distinction that applies to any citizen science data. Unstructured data are opportunistic and can be collected at any time and place, such as bird observations submitted to the project iNaturalist (Callaghan et al. 2019). Unstructured projects lower barriers to participation because they require little or no training and can thereby involve more volunteers, resulting in a higher likelihood of observing rare events or species. However, the nature of unstructured data results in spatial, taxonomic, and other biases, such as concentrations of observations in urban areas, overrepresentation of large or interesting species, and underrepresentation of common species (Kamp et al. 2016, Callaghan et al. 2019). Structured projects, on the other hand, such as the North American Breeding Bird Survey and Breeding Bird Atlas programs, rely on formal protocols for data collection, which can result in a more robust understanding of distributions but at the cost of engaging fewer volunteers due to the higher degree of volunteer training required (Bowler et al. 2022). Semistructured projects, such as eBird, utilize flexible protocols, such as allowing volunteers to choose whether they will submit a robust checklist of every individual seen or a more casual checklist of the individuals or species of interest to the contributor (Kelling et al. 2019). Measuring survey efforts, standardizing data collection protocols, and accounting for observers' biases can improve data quality and contribute to more robust estimates of avian diversity and abundance from checklists. In addition, because volunteers often participate in multiple projects, citizen science projects can be scaffolded such that volunteers first gain foundational skills in simpler unstructured projects before moving on to contributing high-quality data to more complicated structured projects (Allf et al. 2022).

Nest Monitoring

Whereas checklisting is heavily derived from British colonial culture, interest in managing and monitoring nesting has broader cultural roots. For instance, the Chickasaw and Choctaw, and other Indigenous groups in the eastern areas of what is now the United States, influenced the colonial breeding habits of Purple Martins—specifically their breeding in clusters of hollow gourds provided by people—long before the first European settlers (Doughty and Fergus 2002).

Although bird enthusiasts can potentially find and share data about any nesting species, the most commonly monitored are those using nesting sites provided by people. Most instances of grassroots community science focused on birds have been (and continue to be) related to the provisioning of nesting sites. Grassroots community science has supported bird conservation, particularly of target species commensal with humans. There are three species that now mostly depend exclusively on human structures for breeding in the United States—the Purple Martin, Chimney Swift, and Barn Swallow—and community science has played a significant role in the management of at least two of them (Chimney Swifts and Purple Martins). For example, in 1987, the Purple Martin Conservation Association (PMCA) was formed in order to better organize conservation efforts related to martins. The PMCA later created Project Martin Watch, the Purple Martin Survey, and the Scout-Arrival Study, each of which engages volunteers (particularly landowners with artificial Purple Martin roosts) with collecting data about martin nesting success, migration phenology, and other information. The PMCA also sells roosts and shares educational information about Purple Martins with volunteers.

Another group of species that relies heavily, but not exclusively, on human structures is bluebirds. There has been significant community science and citizen science organized around each of the three species of bluebirds in the United States. Volunteer-generated bluebird data support studies of population trends, population dynamics, phenology, seasonal and geographic patterns, habitat comparisons, and more. For instance, data from bluebird monitors in The Birdhouse Network (predecessor to the Cornell Lab's NestWatch project) were used to understand differences in clutch size in the Eastern Bluebird across latitudes (Cooper et al. 2005). Nest monitoring data have also been influential in understanding phenological responses to climate change (Crick and Sparks 1999). In fact, the work of Crick and Sparks with the British Trust for Ornithology's Nest Record Scheme was cited to urge policymakers to ratify the Kyoto Protocol, an international treaty that extended the United Nations Framework Convention on Climate Change. NestWatch observers have documented patterns of egg incubation (Nord and Cooper 2020), twinning in wild bird eggs (Bailey and Clark 2014), and the influence of sensory pollutants (noise and light) on breeding phenology and reproductive fitness across the United States (Senzaki et al. 2020). Cameras mounted on nests have expanded public engagement in nest monitoring—2 million people every month watch the Cornell Lab's nest cameras alone (Saulnier 2021)—and scientific insights, such as the development of new proximate mechanisms for determining clutch size (Cooper et al. 2009). CitSciGrid was a platform that hosted several citizen science projects, including Wildlife@Home, which engaged participants in the processing of images and videos of nesting birds (Desell et al. 2015). Like other forms of data collection, nest monitoring data contain bias, such as increased frequency of observation on weekends, which can be resolved with analytical techniques such as backdating (Cooper 2014).

Banding

Bird banding in the United States dates back to at least the early 1800s, with John James Audubon's early attempts to band Eastern Phoebes with silver wire (Jackson et al. 2008). Currently, the U.S. Fish and Wildlife Service administers a program of bird banding permits, with a system for sub-permittees. In some citizen science projects, participants can work as sub-permittees, such as through the Monitoring Avian Productivity and Survivorship (MAPS) project (DeSante 1992). MAPS, an initiative of the Institute for Bird Populations, started in 1989 and produces avian productivity indices based on ratios of young and adult birds captured in mist nets, and it estimates adult apparent survival and population growth rates based on band-recapture statistical methods (DeSante et al. 1995). In other projects, participants serve as location hosts for bird banding on their property, such as with the Smithsonian Institute's Neighborhood NestWatch (Evans et al. 2005). In addition, well-established banding stations, such as Braddock Bay Bird Observatory in New York, Point Reyes Bird Observatory in California, and Powdermill Nature Preserve in Pennsylvania, have programs for volunteers. Banding data support studies of migration rates, migration phenology, adult survival rates, and related population studies.

Dead Bird Monitoring

Citizen science efforts play a critical role in monitoring avian mortality. Perhaps the best known of these projects is the Coastal Observation and Seabird Survey Team (COASST), which relies on volunteers to monitor the occurrence of beached birds—that is, birds found dead on beaches. COASST was started in 1999 to document baseline levels of beached birds and subsequent departures from baseline, as well as to investigate the causes of mortality events and to create predictive models of mortality events. COASST participants follow standardized protocols that volunteers helped develop (Parrish et al. 2017). COASST engages approximately 800 monthly collectors across 450 sites between northern California and Utqiagvik, Alaska, and west through the Aleutians (Parrish et al. 2022). COASST baseline data have served as a sort of scientific anchor, allowing the research community to understand the severity and geographic and taxonomic breadth of multiple marine bird mass mortality events collectively resulting in the death of several million birds. COASST makes contributions to ornithology; for example, it has submitted 60 significant bird records to the Washington Bird Records Committee of the Washington Ornithological Society, including four species with no previous record of occurrence in Washington (Least Auklet, Wedge-Tailed Shearwater, Red-Footed Booby, and Purple Gallinule), and it has reclassified the Horned Puffin as a regularly occurring species rather than a rare species (Julia Parrish, University of Washington, personal communication). COASST contributes to management, as evident from the hundreds of data requests it receives from federal agencies, academic groups, nongovernmental organizations, and state wildlife agencies. For instance, the National Oceanic and Atmospheric Association uses COASST baseline data in its annual ecosystem assessments in California and Alaska (Julia Parrish, personal communication).

Another type of dead bird monitoring focuses on collecting and analyzing observations of bird collisions with built structures. The earliest documented record of a bird–window collision in North America was by Nuttall (1832), who observed a Sharp-Shinned Hawk (*Accipiter striatus*) collide with windows while pursuing prey. Today, a variety of citizen science projects collect observations of bird–window collisions, such as Minnesota Project BirdSafe and the Fatal Light Awareness Program (Cusa et al. 2015, Nichols et al. 2018). Science museums also play an important role in keeping bird–window collision records because many bird carcasses collected by surveyors are brought to museums to be added to ornithological collections. As more people have become aware of the collision issue, the number of collision monitoring projects and the number of published articles on bird–window collisions have increased, and these projects have been at the forefront of Lights Out campaigns to reduce light pollution that causes nocturnal migrants to collide with buildings, as well as campaigns to retrofit high-collision-risk windows with films and stickers that make building windows more visible to birds (Basilio et al. 2020).

Bird–window collision projects can be structured, using standardized surveys of targeted buildings over a period of time, or unstructured, making use of opportunistic observations, such as the Bird–Window Collisions project on iNaturalist (Winton et al. 2018). One benefit of bird–window collision projects is that they are often easy to participate in; collecting these observations does not require the bird identification skills needed to participate in eBird, for instance. Whereas most of these projects collect contemporary data, some studies have used memory-based surveys that ask volunteers if they have seen birds strike windows in the past (Kummer et al. 2016).

Experiments

Most ornithological citizen science projects focus on observational data such as bird counts, but there are some projects that engage citizen scientists in conducting ecological experiments. One study recruited volunteers who were enrolled in Project FeederWatch to participate in a hawk playback experiment in people's yards (Zuckerberg et al. 2022). A randomized control-treatment design was used to ask participants to broadcast either hawk or goldfinch calls to backyard birds. Participants recorded birds' anti-predator behavior before, during, and after playback, and researchers used these data to explore how urbanization influences birds' behavior. In a different study, in 1993 the Cornell Lab recruited volunteers to put white millet, red milo, or black-oil sunflower seeds on separate pieces of cardboard and then test which bird species fed on which seeds (Trumbull et al. 2000). In general, experiment-based citizen science projects are uncommon due to the complexity of the tasks that volunteers are required to carry out.

Impacts for Participant Learning

Avian citizen science has benefits beyond its ecological value. An emerging research practice focuses on evaluating the "learning outcomes" of participation in these projects—that is, how experiences in avian citizen science lead to an increased understanding of science, ecology, and/or conservation. For instance, one study found that participants in The Birdhouse Network increased their knowledge of bird biology, according to comparisons of surveys taken before and after volunteers participated in the project (Brossard et al. 2005). In addition, some participants in this project set up their own experiments related to The Birdhouse Network project or engaged in more "pro-bird" behaviors such as erecting nest boxes, indicating the potential for avian citizen science to inspire volunteers to more deeply pursue scientific inquiry and to change their conservation behaviors, respectively (Brossard et al. 2005, Bonney et al. 2009). Other studies have found that volunteers with the Smithsonian Institute's Neighborhood Nestwatch program increased their ecological literacy and their sense of the value of backyard habitats for birds (Evans et al. 2005), that kindergarten through grade 12 participants in the curriculum-based Cornell Lab project BirdSleuth increased their knowledge of bird biology and identification (Thompson 2007), and that nest box monitors are more likely to manage for invasive bird species than people who do not do nest monitoring (Phillips et al. 2021). In general, citizen science is often shown to increase volunteers' scientific or ecological knowledge; changes in higher level cognitive processes such as pro-environmental attitudes, pro-environmental behaviors, and conceptualizations of science and scientific epistemologies are rarer, due in part to the fact that many citizen science volunteers enter into projects already highly interested in, and engaged with, science and nature (Brossard et al. 2005, Pandya and Dibner 2018). Designing project protocols to achieve learning goals requires intentionality; projects designed solely to accomplish scientific goals without considering volunteers' needs are generally poorly equipped to also have significant value for learning. In addition, documenting changes in volunteers' knowledge or beliefs inherently requires engaging volunteers who enter into projects with "room to grow" in terms of their baseline knowledge or beliefs about birds and conservation.

Room for Improvement

Neither professional ornithology nor citizen science has produced research representative of the distribution of birds. For example, the ornithological literature is dominated by studies from the Global North (Europe, the United States, and Canada) despite the fact that more than one-third of all bird species live in the Neotropics (E. Blake 1977, Hedblom and Murgui 2017). Similarly, although approximately 84% of the Earth's terrestrial surface contains bird occurrence information, these data are concentrated in North America, Europe, India, Australia, and New Zealand, and more than half of eBird checklists come from the United States alone (La Sorte and Somveille 2020). Gaps in data on bird occurrences are most prevalent in central South America, northern and central Africa, and northern Asia (La Sorte and Somveille 2020). These trends reflect the birding legacy of British colonization. In addition, ornithology is part of the hegemonic structure of science, with dominant narratives in the English language and research priorities dictated by those in the Global North. Citizen science counters the hierarchical systems of Eurocentric science by valuing the observations of those without scientific credentials. Nevertheless, citizen science reinforces colonial Western reductionist approaches to science and mimics the extractive practices of imperial scientists taking specimens to European museums. In addition, citizen science typically does not situate other forms of knowledge about birds from different cultures and perspectives into ornithology.

Although approximately one-third of adults in the United States go bird-watching, at least on occasion (Cordell and Herbert 2002), participant demographics are not representative of the general population. In the United States, this demographic skew is particularly prominent in terms of race and ethnicity. For example, in a survey with more than 3,000 respondents, 96% of Christmas Bird Count participants identified as White and non-Hispanic (Allf et al. 2022). Citizen science across disciplines shows similar trends (C. Blake et al. 2020, Pateman et al. 2021, Mahmoudi et al. 2022). Citizen science initiatives such as the Cornell Lab's Celebrate Urban Birds project are attempting to foster greater diversity in avian citizen science by collaborating with existing local organizations to engage diverse communities in avian study (Purcell et al. 2012). Celebrate Urban Birds makes program materials available in both English and Spanish and offers mini-grants to enable partner organizations to use project resources to create their own bird-related events. Although not citizen science per se,

Table 19.2 Types of Bird-Related Recreation Activities[a]

Bird-Related Activity	% Male (*n*)	% Female (*n*)
Supportive (least competitive)	46 (508,113)	54 (587,233)
Participatory (moderately competitive)	55 (45,312)	45 (37,800)
Competitive (highly competitive)	97 (6,815)	2 (118)
Authoritative (most competitive)	90 (230)	10 (26)

[a] Women were much less likely to participate in the more competitive activity types.
Source: Extracted from Cooper and Smith (2010).

Black Birders Week, a week-long event series meant to increase the visibility of Black birders in response to the harassment of a Black birder in New York City, is another initiative meant to create space for more diverse bird-watching communities.

Gender patterns are another important consideration in the design of inclusive ornithological citizen science. For instance, competitiveness in birding culture affects gender disparities in engagement and the design of citizen science projects (Table 19.2) (Cooper and Smith 2010). Specifically, women are less likely to be competitive birders (e.g., participate in the World Series of Birding). More generally, women are also less likely to be in positions of authority in avian citizen science (e.g., eBird reviewers; Cooper and Smith 2010).

Physical accessibility barriers also exist that can limit participation in ornithological citizen science. Indeed, 35% of wildlife viewers report that they experience accessibility-related challenges, including those related to mobility, vision, hearing, mental or chronic illness, intellectual or developmental disabilities, or neurodiversity—which means a sizable portion of citizen science participants likely experience these challenges as well (Sinkular et al. 2022). Disabilities also increase with age, and many avian citizen science volunteers are middle-aged or older (Allf et al. 2022). Birdability, a nonprofit with the mission of sharing the joys of birding with people who have disabilities, publishes an international map of birding locations that are accessible to people needing special accommodations, such as places to sit, paved terrain, tactile or Braille signs, and even the dimensions of bathroom sinks at birding locations. As was made clear during the COVID-19 pandemic, group size and whether or not an activity is indoors are important considerations for people with compromised immune systems and are important considerations in inclusive design. This is not to say that all birding activities must take place outdoors in small groups; rather, projects should be clear about the accommodations that are afforded to participants and what volunteers with different needs can expect from an activity. Making ornithological citizen science more accessible can also involve providing special equipment such as wheelchair-mounted scopes, trail chairs, table "digiscoping" adaptors for scopes, binocular mounts, winged eyecups for binoculars, outdoor microphones, hearing-assistance devices, car window mounts, and more.

Financial barriers to participating in avian monitoring also exist. Some ornithology projects require significant costs (e.g., Earthwatch expeditions) or low costs (e.g., Project FeederWatch), and any costs can create a barrier to entry and participation.

Wealth-based barriers can occur without direct fees, such as the cost of purchasing binoculars or spotting scopes; costs involved with setting up feeders, nest boxes, or providing supplemental food; and even costs associated with time commitments or required flexibility at particular times of day. Each of these factors can skew participation. Even among people who can afford the financial commitment of a citizen science project, those with lower incomes may simply weigh the cost more heavily when considering to what degree to pursue a project than people with higher incomes.

For two main reasons, avian citizen science would be improved by broadening the diversity of volunteers who participate in projects. First, participating in citizen science confers benefits on participants, such as gaining new skills, new social connections (e.g., with other volunteers or with scientists), or a stronger connection to nature in their community. If most people who do citizen science are White, educated, able-bodied, and/or wealthy, then the practice is concentrating these benefits among people who are already well-resourced in our society. Broadening the diversity of participants in citizen science can also simply result in higher quality data (C. Blake et al. 2020, Mahmoudi et al. 2022). For instance, projects that oversample wealthy neighborhoods compared to poorer minority communities are generating biased data. Minority communities are often disproportionately burdened with air and water pollution and lower tree canopy cover that may affect local bird populations. Thus, an urban ornithological project might give an unrealistically optimistic portrayal of a city's bird populations if it does not collect data from regions where bird populations are suffering due to environmental contamination and habitat degradation. Broadening volunteer diversity starts with reaching out to and connecting with new communities, but it cannot end there; these relationships must be nurtured through sustained engagement with the goals and interests of volunteers and the public. This might involve gamification of project protocols, giving volunteers more autonomy to pursue their own research questions, and/or providing more opportunities for social interactions among volunteers.

Conclusion

Throughout the history of ornithology, amateurs have made important contributions to avian science. Although the professionalization of ornithology and the rise of citizen science have coincided with a reduction in amateurs leading full-blown research studies, amateur participation in the study of birds remains high, and the amount of data generated by these volunteers is unrivaled by the efforts of amateurs working in any other scientific discipline. Indeed, more than half of all records of any species in the Global Biodiversity Information Facility are of birds, despite the minuscule fraction of global species that are birds. These data have resulted in hundreds of publications and new understandings of avian migration, feeding, breeding, and mortality. Avian citizen science has also enabled millions of people to connect with the natural world and contribute to science.

Citizen science data, like all data, have inherent biases, and avian citizen science can be highly spatially biased, in particular. To remedy this patchiness, two approaches have been suggested. In one approach, incentive systems, including gamified approaches, encourage current participants to make the most meaningful contributions by sampling in rural and remote undersampled areas (Xue et al. 2016, Alexandrino et al. 2019). In another approach, participatory science projects are encouraged to pursue inclusive and equitable methods for engaging *new* participants living in, near, or with attachment to undersampled areas (Cooper et al. 2021). Apps such as Merlin, which uses artificial intelligence to assist novices with bird identification, have the potential to improve the skill level of volunteers who enter into projects with less birding experience. The great successes experienced by ornithological citizen science thus far could be magnified further if projects broaden participation.

References

Alexandrino, E. R., A. B. Navarro, V. F. Paulete, M. Camolesi, V. G. R. Lima, A. Green, T. D. Conto, M. P. M. D. B. Ferraz, Ç. H. Şekercioğlu, and H. T. Z. Couto. 2019. Challenges in engaging birdwatchers in bird monitoring in a forest patch: Lessons for future citizen science projects in agricultural landscapes. *Citizen Science: Theory and Practice* 4: Article 4.

Allf, B. C., C. B. Cooper, L. R. Larson, R. R. Dunn, S. E. Futch, M. Sharova, and D. Cavalier. 2022. Citizen science as an ecosystem of engagement: Implications for learning and broadening participation. *BioScience* 72:651–663.

Bailey, R. L., and G. E. Clark. 2014. Occurrence of twin embryos in the eastern bluebird. *PeerJ* 2: Article e273.

Barrow, M. 2000. *A Passion for Birds: American Ornithology After Audubon*. Princeton University Press, Princeton, NJ.

Basilio, L. G., D. J. Moreno, and A. J. Piratelli. 2020. Main causes of bird-window collisions: A review. *Anais da Academia Brasileira de Ciências* 92: Article e20180745.

Bird, T. J., A. E. Bates, J. S. Lefcheck, N. A. Hill, R. J. Thomson, G. J. Edgar, R. D. Stuart-Smith, S. Wotherspoon, M. Krkosek, J. F. Stuart-Smith, G. T. Pecl, N. Barrett, and S. Frusher. 2014. Statistical solutions for error and bias in global citizen science datasets. *Biological Conservation* 173:144–154.

Blake, C., A. Rhanor, and C. Pajic. 2020. The demographics of citizen science participation and its implications for data quality and environmental justice. *Citizen Science: Theory and Practice* 5: Article 21.

Blake, E. R. 1977. *Manual of Neotropical Birds*. University of Chicago Press, Chicago.

Bonney, R., H. Ballard, R. Jordan, E. McCallie, T. Phillips, J. Shirk, and C. C. Wilderman. 2009. Public participation in scientific research: Defining the field and assessing its potential for informal science education. A CAISE inquiry group report. CAISE, Washington, DC.

Bonter, D. N., and C. B. Cooper. 2012. Data validation in citizen science: A case study from Project FeederWatch. *Frontiers in Ecology and the Environment* 10:305–307.

Bowler, D. E., N. Bhandari, L. Repke, C. Beuthner, C. T. Callaghan, D. Eichenberg, K. Henle, R. Klenke, A. Richter, F. Jansen, H. Bruelheide, and A. Bonn. 2022. Decision-making of citizen scientists when recording species observations. *Scientific Reports* 12: Article 11069.

Brandis, K. J., D. Mazumder, P. Gadd, B. Ji, R. T. Kingsford, and D. Ramp. 2021. Using feathers to map continental-scale movements of waterbirds and wetland importance. *Conservation Letters* 14: Article e12798.

Brossard, D., B. Lewenstein, and R. Bonney. 2005. Scientific knowledge and attitude change: The impact of a citizen science project. *International Journal of Science Education* 27:1099–1121.

Callaghan, C. T., J. J. L. Rowley, W. K. Cornwell, A. G. B. Poore, and R. E. Major. 2019. Improving big citizen science data: Moving beyond haphazard sampling. *PLoS Biology* 17: Article e3000357.

Carson, R. 1962. *Silent Spring*. Houghton Mifflin, Boston, MA.

Chansigaud, V. 2010. *All About Birds: A Short Illustrated History of Ornithology*. Princeton University Press, Princeton, NJ.

Cooper, C. B. 2014. Is there a weekend bias in clutch-initiation dates from citizen science? Implications for studies of avian breeding phenology. *International Journal of Biometeorology* 58:1415–1419.

Cooper, C. B., C. L. Hawn, L. R. Larson, J. K. Parrish, G. Bowser, D. Cavalier, R. R. Dunn, M. Haklay, K. K. Gupta, N. O. Jelks, V. A. Johnson, M. Katti, Z. Leggett, O. R. Wilson, and S. Wilson. 2021. Inclusion in citizen science: The conundrum of rebranding. *Science* 372:1386–1388.

Cooper, C. B., W. M. Hochachka, G. Butcher, and A. A. Dhondt. 2005. Seasonal and latitudinal trends in clutch size: Thermal constraints during laying and incubation. *Ecology* 86:2018–2031.

Cooper, C. B., and J. A. Smith. 2010. Gender patterns in bird-related recreation in the USA and UK. *Ecology and Society* 15: Article 4.

Cooper, C. B., M. A. Voss, and B. Zivkovic. 2009. Extended laying interval of ultimate eggs of the Eastern Bluebird. *The Condor* 111:752–755.

Cordell, H. K., and N. G. Herbert. 2002. The popularity of birding is still growing. *Birding* 34:54–61.

Crick, H. Q. P., and T. H. Sparks. 1999. Climate change related to egg-laying trends. *Nature* 399:423–423.

Cusa, M., D. A. Jackson, and M. Mesure. 2015. Window collisions by migratory bird species: Urban geographical patterns and habitat associations. *Urban Ecosystems* 18:1427–1446.

Davis, W. B., and J. Stevenson. 1934. The type localities of three birds collected by Lewis and Clark in 1806. *The Condor* 36:161–163.

DeSante, D. F. 1992. Monitoring Avian Productivity and Survivorship (MAPS): A sharp, rather than blunt, tool for monitoring and assessing landbird populations. Pages 511–521 in D. R. McCullough and R. H. Barrett, eds. *Wildlife 2001: Populations*. Springer, Dordrecht, the Netherlands.

DeSante, D. F., K. M. Burton, J. F. Saracco, and B. L. Walker. 1995. Productivity indices and survival rate estimates from MAPS, a continent-wide programme of constant-effort mist-netting in North America. *Journal of Applied Statistics* 22:935–948.

Desell, T., K. Goehner, A. Andes, R. Eckroad, and S. Ellis-Felege. 2015. On the effectiveness of crowd sourcing avian nesting video analysis at Wildlife@Home. *Procedia Computer Science* 51:384–393.

Doughty, R. W., and R. Fergus. 2002. *The Purple Martin*. University of Texas Press, Austin, TX.

Droege, S. 2007. Just because you paid them doesn't mean their data are better. Pages 13–26 in C. MeEver, R. Bonney, J. Dickinson, S. Kelling, K. V. Rosenberg, and J. Shirk, eds. *Citizen Science Toolkit Conference*. Cornell Lab of Ornithology, Ithaca, NY.

Evans, C., E. Abrams, R. Reitsma, K. Roux, L. Salmonsen, and P. P. Marra. 2005. The neighborhood Nestwatch program: Participant outcomes of a citizen-science ecological research project. *Conservation Biology* 19:589–594.

Greenwood, J. J. D. 2007. Citizens, science and bird conservation. *Journal of Ornithology* 148:77–124.

Hartley, S. M. 2019. Sparrow swap: Testing management strategies for house sparrows and exploring the use of their eggshells for monitoring heavy metal pollution [Master's thesis]. North Carolina State University, Raleigh, NC.

Hedblom, M., and E. Murgui. 2017. Urban bird research in a global perspective. Pages 3–10 in E. Murgui and M. Hedblom, eds. *Ecology and Conservation of Birds in Urban Environments*. Springer, Cham, Switzerland.

Hickey, J. J., and D. W. Anderson. 1968. Chlorinated hydrocarbons and eggshell changes in raptorial and fish-eating birds. *Science* 162:271–273.

Hochachka, W. M., and A. A. Dhondt. 2000. Density-dependent decline of host abundance resulting from a new infectious disease. *Proceedings of the National Academy of Sciences of the USA* 97:5303–5306.

Jackson, J. A., W. E. Davis, and J. Tautin. 2008. *Bird Banding in North America: The First 100 Years*. Nuttall Ornithological Club, Cambridge, MA.

Joray, T. P. 2020. Determination of the summer origins of American Kestrels (*Falco sparverius*) wintering in northern and central Illinois [Master's thesis]. Illinois State University, Normal, IL.

Jordan, R., A. Crall, S. Gray, T. Phillips, and D. Mellor. 2015. Citizen science as a distinct field of inquiry. *BioScience* 65:208–211.

Kamp, J., S. Oppel, H. Heldbjerg, T. Nyegaard, and P. F. Donald. 2016. Unstructured citizen science data fail to detect long-term population declines of common birds in Denmark. *Diversity and Distributions* 22:1024–1035.

Kelling, S., A. Johnston, A. Bonn, D. Fink, V. Ruiz-Gutierrez, R. Bonney, M. Fernandez, W. M. Hochachka, R. Julliard, R. Kraemer, and R. Guralnick. 2019. Using semistructured surveys to improve citizen science data for monitoring biodiversity. *BioScience* 69:170–179.

King, J. R., and W. J. Bock. 1978. Workshop on a national plan for ornithology, final report. National Science Foundation and the Council of the American Ornithologists' Union, Washington, DC.

Kummer, J. A., E. M. Bayne, and C. S. Machtans. 2016. Use of citizen science to identify factors affecting bird–window collision risk at houses. *The Condor* 118:624–639.

La Sorte, F. A., D. Fink, J. J. Buler, A. Farnsworth, and S. A. Cabrera-Cruz. 2017. Seasonal associations with urban light pollution for nocturnally migrating bird populations. *Global Change Biology* 23:4609–4619.

La Sorte, F. A., and M. Somveille. 2020. Survey completeness of a global citizen-science database of bird occurrence. *Ecography* 43:34–43.

Mahmoudi, D., C. L. Hawn, E. H. Henry, D. J. Perkins, C. B. Cooper, and S. M. Wilson. 2022. Mapping for whom? Communities of color and the citizen science gap. ACME 21:372–388.

Mayer, A. 2010. Phenology and citizen science: Volunteers have documented seasonal events for more than a century, and scientific studies are benefiting from the data. *BioScience* 60: 172–175.

Mayfield, H. 1991. The amateur: Finding a niche in ornithology. *Nebraska Bird Review* 59:39–42.

Merriam, C. H. 1885. Preliminary report of the Committee on Bird Migration. *The Auk* 2:53–65.

Miller-Rushing, A., R. Primack, and R. Bonney. 2012. The history of public participation in ecological research. *Frontiers in Ecology and the Environment* 10:285–290.

Nichols, K. S., T. Homayoun, J. Eckles, and R. B. Blair. 2018. Bird-building collision risk: An assessment of the collision risk of birds with buildings by phylogeny and behavior using two citizen-science datasets. *PLoS One* 13: Article e0201558.

Nord, A., and C. B. Cooper. 2020. Night conditions affect morning incubation behaviour differently across a latitudinal gradient. *Ibis* 162:827–835.

Nuttall, T. 1832. *A Manual of the Ornithology of the United States and of Canada*. Hilliard & Brown, Cambridge, MA.

Pandya, R. E., and K. A. Dibner. 2018. *Learning Through Citizen Science: Enhancing Opportunities by Design*. National Academies Press, Washington, DC.

Parrish, J. K., H. Burgess, J. Lindsey, L. Divine, R. Kaler, S. Pearson, and J. Dolliver. 2022. Partnering with the public: The Coastal Observation and Seabird Survey Team. Pages 87–108 in G. Auad and F. K. Wiese, eds. *Partnerships in Marine Research*. Elsevier, New York.

Parrish, J. K., K. Litle, J. Dolliver, T. Hass, H. K. Burgess, E. Frost, C. W. Wright, and T. Jones. 2017. *The Coastal Observation and Seabird Survey Team (COASST): Citizen Science for Coastal and Marine Conservation*. Routledge, Abingdon, UK.

Partners in Flight. 2020. Population Estimates Database, Version 3.1. https://pif.birdconservancy.org/population-estimates-database.

Pateman, R., A. Dyke, and S. West. 2021. The diversity of participants in environmental citizen science. *Citizen Science: Theory and Practice* 6: Article 9.

Pettingill, O. S., Jr. 1966. The North American Nest-Record Card Program for 1966. *Wilson Bulletin* 78:136.

Phillips, T. B., R. L. Bailey, V. Martin, H. Faulkner-Grant, and D. N. Bonter. 2021. The role of citizen science in management of invasive avian species: What people think, know, and do. *Journal of Environmental Management* 280: Article 111709.

Purcell, K., C. Garibay, and J. L. Dickinson. 2012. 13. A gateway to science for all: Celebrate urban birds. Pages 191–200 in J. L. Dickinson and R. Bonney, eds. Citizen Science: Public Participation in Environmental Research. Cornell University Press, Ithaca, NY.

Sauer, J. R., K. L. Pardieck, D. J. Ziolkowski, Jr., A. C. Smith, M.-A. R. Hudson, V. Rodriguez, H. Berlanga, D. K. Niven, and W. A. Link. 2017. The first 50 years of the North American Breeding Bird Survey. *The Condor* 119:576–593.

Saulnier, B. 2021. Birdcams offer up-close views of avian life. *Cornellians.* https://alumni.cornell.edu/cornellians/birdcams.

Senzaki, M., J. R. Barber, J. N. Phillips, N. H. Carter, C. B. Cooper, M. A. Ditmer, K. M. Fristrup, C. J. W. McClure, D. J. Mennitt, L. P. Tyrrell, J. Vukomanovic, A. A. Wilson, and C. D. Francis. 2020. Sensory pollutants alter bird phenology and fitness across a continent. *Nature* 587:605–609.

Shane, T. 2012. *A Two-Hundred Year History of Ornithology, Avian Biology, Bird Watching, and Birding in Kansas (1810–2010).* Zea E-Books, Lincoln, NE.

Sinkular, E. N., P. C. Pototsky, and A. A. Dayer. 2022. New Mexico results of the Wildlife Viewer Survey: Enhancing relevancy and engaging support from a broader constituency [Report]. Virginia Tech, Blacksburg, VA.

Stewart, P. A. 1954. The value of the Christmas Bird Counts. *Wilson Bulletin* 66:184–195.

Sullivan, B. L., J. L. Aycrigg, J. H. Barry, R. E. Bonney, N. Bruns, C. B. Cooper, T. Damoulas, A. A. Dhondt, T. Dietterich, A. Farnsworth, D. Fink, J. W. Fitzpatrick, T. Fredericks, J. Gerbracht, C. Gomes, W. M. Hochachka, M. J. Iliff, C. Lagoze, F. A. La Sorte, M. Merrifield, W. Morris, T. B. Phillips, M. Reynolds, A. D. Rodewald, K. V. Rosenberg, N. M. Trautmann, A. Wiggins, D. W. Winkler, W.-K. Wong, C. L. Wood, J. Yu, and S. Kelling. 2014. The eBird enterprise: An integrated approach to development and application of citizen science. *Biological Conservation* 169:31–40.

Taylor, D. E. 2016. *The Rise of the American Conservation Movement: Power, Privilege, and Environmental Protection.* Duke University Press, Durham, NC.

Thompson, S. 2007. BirdSleuth: Final evaluation report. Seavoss Associates, Ithaca, NY.

Tramer, E. J. 2007. In memoriam: Harold F. Mayfield, 1911–2007. *The Auk* 124:1453–1455.

Trumbull, D. J., R. Bonney, D. Bascom, and A. Cabral. 2000. Thinking scientifically during participation in a citizen-science project. *Science Education* 84:265–275.

Vetter, J. 2011. Introduction: Lay participation in the history of scientific observation. *Science in Context* 24:127–141.

Wiggins, A., G. Newman, R. D. Stevenson, and K. Crowston. 2011. Mechanisms for data quality and validation in citizen science. Pages 14–19 in *2011 IEEE Seventh International Conference on e-Science Workshops.* IEEE, Piscataway, NJ.

Winton, R. S., N. Ocampo-Peñuela, and N. Cagle. 2018. Geo-referencing bird–window collisions for targeted mitigation. *PeerJ* 6: Article e4215.

Xue, Y., I. Davies, D. Fink, C. Wood, and C. P. Gomes. 2016. Avicaching: A two stage game for bias reduction in citizen science. Pages 776–785 in *Proceedings of the 15th International Conference on Autonomous Agents and Multiagent Systems (AAMAS 2016).* International Foundation for Autonomous Agents and Multiagent Systems, Taiwan.

Zuckerberg, B., J. D. McCabe, and N. A. Gilbert. 2022. Antipredator behaviors in urban settings: Ecological experimentation powered by citizen science. *Ecology and Evolution* 12: Article e9269.

20
Expanding the Scope of Ornithology by Engaging Diverse Communities

Sebastian Moreno and Paige S. Warren

A question scientists, managers, and birders often ask is how they, their team, or their project can engage with diverse communities in ornithological research, management, or outreach events. This is a critical question to grapple with because the conservation of birds is intricately linked with experiences with nature (Gaston and Soga 2020). Our fascination with birds spans throughout human history. From cave paintings to religious symbols, their prominence in mythology, and modern song and poetry, birds have been a mainstay—whether it is Prince singing about doves crying (Prince 1984), Bob Dylan pining about robins as the harbingers of spring in the song "If Not for You" (Dylan 1970), or Emily Dickinson conjuring a sense of optimism in one of her most famous poems, "Hope Is the Thing With Feathers" (Dickinson 1891). No doubt, a bird's ability to fly, both short and epic distances, can conjure feelings of freedom. Furthermore, they sing intricate songs; exhibit dazzling colors and outrageous feather patterns; are mostly diurnal; and are found in our bustling city centers, neighborhood parks, backyards, and wild places. Birds surround us and can strengthen our sense of place (Luck et al. 2009), and their familiarity can increase the accessibility of the science we share and inspire conservation action (Prévot et al. 2018). Thus, birds represent an ideal platform for engaging with the public. To effectively answer the question posed above requires additional information about who we mean to engage and why it matters, the motivation and objectives for engagement, and how to expand the scope of ornithology—essentially the who, what, why, and how. In this chapter, we provide an entry point for exploring these issues, and we also provide some examples of action that can be taken, drawn from the literature and our professional experiences.

What Are "Diverse Communities"?

According to the Cambridge Dictionary, *diversity* is "the condition or fact of being different or varied; variety." From a biological perspective, (bio)diversity refers to the variety of all life forms, ranging from within species (i.e., genetic) to between species, and between ecosystems (United Nations 1992). Biodiversity plays a critical role in supporting a functional ecosystem and the associated services provided to people (Cardinale et al. 2011, 2012). Specifically, birds disperse seeds, pollinate plants,

Sebastian Moreno and Paige S. Warren, *Expanding the Scope of Ornithology by Engaging Diverse Communities*.
In: *New Perspectives in Ornithology*. Edited by: Scott V. Edwards and J. Michael Reed, Oxford University Press.
 DOI: 10.1093/oso/9780197787670.003.0020

scavenge carrion, consume pests, and cycle nutrients (Şekercioğlu et al. 2004, Luck et al. 2009, Cardinale et al. 2012). These ecosystem functions have tremendous implications for humanity by increasing economic yields in agriculture, promoting forest growth and plant dispersal, providing pest control of insect outbreaks and rodents, as well as providing cultural services that enhance aesthetic experiences or sense of place (Luck et al. 2009). In a similar way, human diversity is essential for the "functioning" of the field of ornithology.

When describing diverse communities from a people perspective, diversity often implies the ways we identify, including gender, race, ethnicity, class, sexuality, religion, political beliefs, and other social identities. Several identities can intersect and have served as advantages for some and barriers for others for participating in ornithology (Crenshaw 1989, Cronin et al. 2021). For this chapter, we use *communities* to encompass the members of the public involved in outreach and engagement activities and the communities of ornithologists engaged in scientific practice. Many groups marginalized in the broader society (e.g., Black people, Indigenous people, and people of color) have also historically been excluded from fully participating in practicing ornithology. Likewise, public engagement programming also has well-known historic biases with respect to factors such as race, ethnicity, income, and disability status (Cooper et al. 2021). Thus, increasing participation of diverse communities means creating more inclusive spaces, identifying opportunities to enhance engagement events, and revising research programs so that more voices can participate in ornithology.

Why Diversity Matters in Ornithology

Ornithologists spend careers researching, teaching about, and protecting bird diversity. However, scientific knowledge might be limited because of a lack of diversity among ornithologists. Like a functioning ecosystem, advancing science is intricately related to the scientists doing the work. Although we aim for objectivity, the very nature of the (Western) scientific method commences with a subjective action: We make an observation. And biases, both conscious and unconscious, inherently affect the observer and, consequently, the data (Sauer et al. 1994). For example, ornithologists had generally characterized birdsong as a male trait, with only males having complex vocalizations. More recent research has documented the breadth and ancestral nature of female birdsong (Odom et al. 2014). Haines and colleagues investigated the shift in understanding and concluded that ornithologists who identify as women were more often leading research on female birdsong (Haines et al. 2020). With more women in the science, technology, engineering, and mathematics (STEM) fields in general and ornithology in particular, these new perspectives drove research programs and discovery.

Many ornithologists get exposed to a career in ornithology while conducting fieldwork. Indeed, these experiences can clarify biological concepts while strengthening our connection with our study species or place. However, the simple act of "going to the field" can be dangerous, unwelcoming, and/or logistically challenging in ways that are connected to identity (e.g., minority identities related to race, ethnicity, gender identity, religion, disability, and sexual orientation; Demery and Pipkin 2021) as well as for parents or guardians with young children or other dependent individuals (Lerman, Pejchar, et al. 2021). Removing some of the barriers faced by these and other groups becomes imperative to create an inclusive ornithological community. For example, when supervisors, department heads, and other leaders within an organization take the initiative and develop safety guidelines and opportunities that specifically address identity-based concerns, it reduces the burden for those with marginalized identities and allows them to focus on the fieldwork (Demery and Pipkin 2021). Hiring extra field technicians can increase safety while also providing some flexibility for parents at various stages who are unable to conduct or limited in conducting a full day of fieldwork (e.g., pregnant, lactating, or constrained by school schedules; Lerman, Pejchar, et al. 2021). These and other efforts can advance a more inclusive ornithological community.

Why, How, and Where We Engage With the Public

Returning to the Cambridge Dictionary, *engagement* is "the process of encouraging people to be interested in the work of an organization." Engagement provides an opportunity for ornithologists to co-create scientific knowledge and convey the importance of science for advancing understanding and the relevance to society (Pham 2016, Struminger et al. 2018). We engage to increase scientific literacy, regain public trust in science, and provide an opportunity for scientists to identify research needs and new lines of inquiry (Pham 2016, Charles et al. 2020). Engagement, like other aspects of our research and management, requires thoughtful planning, an understanding of the audience, and a level of professionalism (Pham 2016). Often, poorly delivered engagement activities or products (e.g., media) can backfire and decrease support for and understanding of scientific research (Bucchi and Neresini 2002), undermining the goals of the engagement activity. Engaging with the public is an effective conservation tool because it allows for increasing conservation partners and sharing the love of birds, with the hope that these engagement activities will develop a deep appreciation for birds and ultimately inspire actions that benefit birds (Soga and Gaston 2016). Engagement activities can also provide a spark for the next generation of ornithologists, particularly when personal narratives are shared regarding career paths and when engagement provides mentoring opportunities. The most effective engagement activities focus on optimism and hope (McAfee et al. 2019). Although we are in a biodiversity crisis, with the loss of billions of birds (Rosenberg et al. 2019), perpetuating a doom-and-gloom message when engaging with the public can lead to alienation, a sense of hopelessness, and disengagement in conservation (McAfee et al. 2019).

Engagement with the public occurs through specific programs or organized activities often geared toward targeted audiences (Struminger et al. 2018). The activities can provide place-based and hands-on experiences. Engaging with the public provides a vehicle for informal environmental education and can address the broader impacts scientific research has beyond the scientific community. Indeed, the National

Science Foundation requires grantees to articulate how their research will serve society beyond the intellectual merit (National Science Board 2011). Understanding the needs and cares of the engaged community and its members' learning goals and interests can increase the audience experience in engagement activities (Struminger et al. 2018). When engagement occurs in familiar places and includes interactive activities, it increases knowledge about birds (Evans et al. 2005, Struminger et al. 2018).

Engagement often takes place in traditional settings, such as field stations and nature centers, and these events invite the public to scientists' spaces (i.e., they bring the people to the birds). The mission of many field stations includes providing educational experiences to the public (Billick et al. 2013). These experiences can range from delivering workshops to kindergarten through grade 12 teachers to providing tours of the research station to give a "behind-the-scenes" look and scheduling programs that highlight the research activities (Billick et al. 2013). Bird outings sponsored by nature centers offer interactions between local birding experts and the public (all ages). These events provide knowledge exchange and can include how to use binoculars, what features to focus on when the bird and binoculars match up, and other tips on how to observe and enjoy birds. Nature centers also sponsor public talks, with topics ranging from trip reports to local research projects. These events can occur in person and online, increasing opportunities to engage with diverse audiences.

An underutilized yet effective engagement strategy is to actively partner with local schools and design environmental education activities within the schoolyard or local green spaces. By serving as an "expert guest," classroom visits can reduce some of the barriers to students for traveling to wild places, particularly for students in urban areas (Awasthy et al. 2012). In a study from Lower Hutt City, New Zealand, the authors found that students who participated in a radiotelemetry activity with Kereru (New Zealand pigeon; *Hemiphaga novaeseelandiae*: Columbidae) in a local green space had a greater level of awareness of wildlife compared with students who did not participate in the hands-on experience (Awasthy et al. 2012). This study demonstrated the importance of designing environmental education programs and engaging students in ornithological research (e.g., radiotelemetry) within their local surroundings for providing a critical link between awareness and pro-environmental behavior. The research further demonstrates the importance of urban green spaces, particularly to children during formative ages (e.g., ages 10–13 years; Kellert 1985), for increasing cognitive and factual understanding of wildlife, which can ultimately help build values that center wildlife conservation (Awasthy et al. 2012). In addition to increasing pro-environmental behaviors and knowledge about birds, when ornithologists engage with students, they can also serve as role models (particularly when they have shared identities with the students) and demonstrate to students the possibility of pursuing a career in ornithology (Dasgupta and Asgari 2004).

Broadening Engagement

An important goal of engagement is making ornithology broadly relevant and accessible. But the activities described in the previous section have frequently embedded biases and messaging that can unconsciously limit their audience. For example, ornithologists and experienced birders often spend time studying or chasing rare or elusive species. The thrill of seeing a new species can inspire many predawn starts, and catching a fleeting glimpse of a warbler as it briefly flits about in the upper canopy can be a satisfying experience. However, this might not resonate with novice bird-watchers or the general public. When ornithologists celebrate and appreciate some of our most common species—American Robins (*Turdus migratorius*), Rock Pigeons (*Columba livia*). and House Sparrows (*Passer domesticus*)—it demonstrates to the public that interactions with these species have value. Indeed, experiences with common species might be the only opportunity many people have with birds, given that the majority of people now live in urban and urbanizing environments (Dunn et al. 2006).

Understanding which species people value and what it is about these species they value can help advance conservation efforts because these and other species can provide an entry point into the world of ornithology. For example, Schuetz and Johnston (2019) ranked 621 North American species based on public interest scores by integrating Google searches with eBird data. By establishing "cultural niches," the authors identified additional ways to evaluate birds, by centering the observer rather than the conservation status or biological trait (Schuetz and Johnston 2019). Further research in Phoenix, Arizona, has delved even deeper by assessing perceptions of cultural (e.g., aesthetics and song quality) and ecological traits (e.g., diet and nesting strata) of birds that people experience close to home. Distinctive features, such as colorfulness, and desert species were associated with positive attitudes, whereas some of the most iconic urban species, such as Rock Pigeons, were associated with negative attitudes (Andrade et al. 2022). Furthermore, Rock Pigeons, European Starlings, and House Sparrows increased their abundance in predominantly Latine[1] neighborhoods (Lerman and Warren 2011, Andrade et al. 2022), likely due to different landscaping features and development patterns. These patterns and explanations highlight the context on the complexity of human–bird interactions. Nonetheless, these and other studies demonstrate how and where we interact with birds and nature and can shape programs that celebrate all birds.

Expanding locations beyond nature centers and field stations can provide additional engagement opportunities. Delivering programs in neighborhood parks and other public places set the stage for ornithologists to bring science to the people in familiar places and spaces. Understanding the complex relationships people have with nearby nature can help inform engagement activities that are relevant to the intended audience. Providing binoculars and access to field guides and a scope can diminish some barriers for participating. Although early morning starts correlate with the most bird activity, scheduling walks later in the morning could expand participation by catering to different ages, personal schedules, and physical abilities.

[1] Here, we use *Latine* (pronounced la · ti·· ne) as a gender-neutral form of the word *Latino*, created by LGBTQIA+, gender nonbinary, and feminist communities in Spanish-speaking countries. The objective of the term Latine is to remove gender from the Spanish word Latino by replacing it with the gender-neutral Spanish letter "e." This language shift is native to the Spanish language and can be seen in many gender-neutral words such as "estudiante" (student).

Expanding the Scope of Ornithology

Just as important as who conducts the ornithological research, expanding where ornithological research and birding take place can further welcome new voices to the field. Understanding the natural history of many North American birds provided the cornerstone of American ornithology, and in the early years of the American Ornithologists' Union, many publications in *The Auk* reported on field notes and observations in wild places, away from people and developed areas, both in the United States and abroad (e.g., Brewster 1884, Bendire 1889, Lawrence 1889). These expeditions set the standard for the discipline, and the following seven decades followed suit. The expansion of what consists of bird habitat, of what is worthy of investigation, built momentum in the early 1970s with two key publications: "Invite Wildlife to Your Backyard" (Thomas et al. 1973) and research on the Tucson, Arizona bird community (Emlen 1974). These and other publications not only inspired a new focus on urban areas in ecological research but also increased the accessibility of nature and ornithology. No longer was it required to travel to far-flung places to make scientific discoveries, nor were there limitations in creating and protecting bird habitat close to home. Although birds and their habitat had existed in urban and suburban spaces long before these publications, they represented a shift in how scholars defined what constitutes ornithology. By diversifying where ornithology happens, opportunities arise for conducting science and engagement activities in spaces that resonate with most of the human population (Soga and Gaston 2016, Lerman et al. 2023).

Celebrating and researching urban birds and conserving their habitat have the potential to reach broader audiences—to expand the focus to match the human demographics. Programs such as the U.S. Fish and Wildlife Service's Urban Bird Treaty Program (https://www.fws.gov/program/urban-bird-treaty) provide grants to support urban communities in creating bird-friendly cities. A key goal of the program is providing historically excluded groups increased access to nature through bird-focused activities. A National Wildlife Federation program certifies yards as wildlife habitat. Inspired by the publication by Thomas et al. (1973), the program guides householders on how to create wildlife habitat. These and other programs expand our notion of habitat and also promote more opportunities for people to interact with nearby birds. Indeed, research shows that the efficacy of certified yards in supporting birds is strong (Lerman, Narango, et al. 2021). Furthermore, yards also support more popular species as per Schuetz and Johnston (2019), lending additional support for celebrating common birds.

How We Do Ornithology

In addition to who and where is how. Most ornithological training occurs within the constraints of Western science: guided by the scientific method and assessed by peers through a peer-review process (Kimmerer 2002). Students receive rigorous training in statistics and other quantitative tools, scientific writing, reductionism, objectivity, and designing experimental and/or observational approaches that follow strict protocols

(Berkes 1993). However, science does not happen in a vacuum, and more recently, there has been an increased interest in complementing Western science with traditional ecological knowledge (TEK; Kimmerer 2002), a way of knowing that has been practiced for millennia. Within TEK, the accumulation of knowledge and beliefs is shared through generations and is about the interconnected relationships of all living beings, including humans and their environment (Berkes 1993). One of the primary ways TEK differs from Western science is that TEK is integrated within social and spiritual contexts of the local culture (Kimmerer 2002). TEK is often value-based and takes on a holistic understanding of the natural world. Indeed, the two types of science complement each other, with TEK providing in-depth understanding of a place and the cultural framework the knowledge sits within (Kimmerer 2002), which is vital for solving some of the most pressing problems within conservation.

Glimpses of a TEK approach to the early days of modern-day ornithology (ca. 1880s—1940s) are evident. When perusing historical ornithological publications, the lyrical writing and emotions of observing birds in the field portray the subjective nature of the author. For example, it is easy to picture Joseph Grinnell hunkering down in the California foothills, watching California Thrashers (*Toxostoma redivivum*) rustling around in the chaparral (Grinnell 1917). The astute description of the habitat relationships and the natural history of the species is not embedded in sophisticated statistical models, rendering his science and that of his contemporaries more accessible to the nonscientist. This style of ornithology more closely aligns with TEK in that the intimacy Grinnell conveys is reminiscent of a spiritual awakening in addition to scientific discovery, although it falls short in fully articulating how humans are part of the system (Kimmerer 2002). Much of the current ornithological literature is anchored by statistical models, written in the third person, objective, and the reported research is often conducted by outsiders (Berkes 1993). Although publications might inspire action (e.g., Rosenberg et al. [2019], leading to the Bring Back Birds campaign: https://www.3billionbirds.org), the writing lacks the lyricism and poetic musings once evident in the ornithological literature from the late 19th and early 20th centuries. Embracing both Western science and TEK, acknowledging the mutual benefits each approach provides, can lead to more robust understanding of the ornithological world (Rattling Leaf Sr. 2022).

In addition to acknowledging how human values and the individual scientist inherently shape science, recognizing the damage "parachute science" has caused to local communities can further inform how to improve the practice of ornithology. Parachute science, also called "colonial science," happens when affluent Western scientists "drop down" into another country—often a low-income nation—gather data, and return to their home countries. The Western scientist often does not acknowledge local knowledge, engage with local communities, or invest in local infrastructure, rendering the local community dependent on the Western scientist for expertise. This ultimately has negative implications for conservation efforts (de Vos 2020). One only needs to explore the millions of museum skins in North America and Europe to fully appreciate the extent of extraction that has taken place (Trisos et al. 2021). The collection label tells an important story, including where the specimen was collected and by whom. However, the tides appear to be changing. With a renewed focus on how systemic racism and coloniality have guided Western science, ornithologists are taking small steps to reverse the extractive nature of the past.

In the early 20th century, prominent ornithologist Frank Chapman led a research team sponsored by the American Museum of Natural History (AMNH) to Colombia to catalog the avifauna. Over the course of 5 years, he and five other Americans collected close to 16,000 specimens, which he brought back to the United States and cataloged at the AMNH. The extraction and exportation of scientific discovery and the exploitive nature of the exhibition alienated locals from involvement in research, and the research team failed to share the exciting findings with stewards of the land (Taylor 2022). Although Chapman acknowledged the infrastructural support he received by the Colombian government and expressed his appreciation for housing and mules provided by locals, the expedition exemplified parachute science as described above (Chapman 1917). However, the story does not end with 100+-year-old specimens sitting in museum collections in New York City. In 2019, Colombian scientists launched the Colombia Resurvey Project (Gomez et al. 2022) to document changes in the distribution and diversity of birds as documented by Chapman (1917). Unlike the previous exhibitions, this one is directed, led, and researched by Colombians. The data will remain in Colombia and be archived to share with local communities (e.g., digitizing collections for online sharing), and the project has involved local communities from the start. Initial interactions included informing locals about the nature of the research, sharing information about the need to collect some individuals, and training locals on how to share their observations with the project (Gomez et al. 2022). The project exemplifies how to decolonize science and serves as a model for future ornithological research.

The Power of Citizen Science for Engaging With the Public

Ornithological-based citizen science programs create opportunities to invite the public to participate in science. Citizen science can also further increase the relevancy of ornithology to the public by exposing the public to the various steps within the scientific process (Cooper et al. 2007). Citizen science has a lengthy history with ornithology. One of the oldest continuous bird-centered citizen science projects, the National Audubon's Christmas Bird Count (CBC; established in 1901), began as a move to redirect away from the traditional Christmas hunts with the end goal of killing as many birds encountered and toward (nonlethally) counting the total number of birds seen. Data gathered over more than 120 years by the CBC have informed roughly 350 scientific papers and advanced our understanding of population trends for North American birds (Butcher and Niven 2007). Furthermore, incorporating citizen science methodology provides a powerful tool for improving equitable access to science and can deliver to the participant the associated benefits resulting from the research (Schuttler et al. 2018). However, the level of benefit to the participant depends on the type of citizen science program and where the data are collected.

Citizen science programs are classified into three categories based on the level of public engagement: contributory, collaborative, and co-created (community science;

Charles et al. 2020). Contributory science programs include eBird (https://ebird.org), a global repository for bird sightings (Sullivan et al. 2014), and, more recently, iNaturalist (https://www.inaturalist.org), a similar citizen science program whose scope extends beyond birds. Birders across large geographic ranges (i.e., global) contribute their individual bird sightings, and collectively these checklists have contributed to scientific knowledge (Sullivan et al. 2014). eBird, iNaturalist, and other large-scale projects have a top-down approach in which scientists design the protocol and develop research questions based on the contributed data. In addition to collating an extensive database, contributory citizen science makes science relevant to birders, and participating also creates a sense of community, although this is not unique to contributory projects (Price and Lee 2013).

Contributory citizen science projects have provided valuable information on bird conservation (La Sorte et al. 2017, Cazalis et al. 2021) and have advanced understanding of habitat relationships, particularly in urban settings (Callaghan et al. 2019, Iknayan et al. 2022). However, these data, although spatially extensive, might not fully reflect bird distribution patterns and habitat relationships due to spatial gaps in data collection. In a recent study, Perkins (2020) demonstrated a distribution bias in eBird checklists. By comparing eBird checklists from Tucson, Arizona, Fresno, California, and Raleigh–Durham, North Carolina, with systematic point count surveys in these same cities, Perkins found an overrepresentation of more middle to upper income neighborhoods and underrepresentation in predominantly lower income neighborhoods. The systematic point count survey data more evenly sampled these cities, particularly including lower income neighborhoods. These results have implications for better assessing whether and how social inequalities drive urban bird distribution (Perkins 2020, Schell et al. 2020). Similar eBird sampling biases have been observed in three other cities (Buffalo, New York; Boston; and Phoenix, Arizona; Walker et al. 2017, Grade et al. 2022), and in Chicago, bird observations were more commonly posted to eBird, iNaturalist, and Flickr from open spaces than residential areas (Lopez et al. 2020). Modeling demonstrates that sampling biases can lead to drawing incorrect inferences from these data, particularly when the bias is associated with the variable of interest in the analysis, such as race or income (Grade et al. 2022). Engaging broader participation in citizen science can improve scientific understanding, potentially reducing these sampling biases.

When citizen scientists collaborate (collect, analyze, and disseminate data) and co-create (participant-based research questions) with researchers beyond the contribution of data collection, the dynamics can shift from academic-focused to community-focused science (Charles et al. 2020). This bottom-up approach leads to the leveraging of resources among the researchers and organization to better encourage a successful project. Because the citizen scientist drives the research question, these types of citizen science projects are typically smaller in scale (Charles et al. 2020, Cooper et al. 2021). In addition, the primary motivation for designing the research might lack grounding in ornithological theory yet have important consequences for bird conservation. For example, a proposed finfish aquaculture operation in Nova Scotia, Canada, threatened "safe havens" for lobster traps, potentially destabilizing the local fishing community and economy. The fishing community organized meetings and began documenting how the ecosystem at large and lobsters might react to the fish farm. Scientists attended these meetings and asked how the system functioned, acknowledging the deep amount of local knowledge held by the fishing community. The fisherfolk and scientists co-developed several collaborative studies, integrating their respective scientific and natural history strengths, to test whether waste from the fish farm might contaminate the broader ecosystem. Studies documented elevated concentrations of copper (Loucks et al. 2012), which likely contributed to the loss of other species, including Piping Plovers (*Charadrius melodus*), because they nested within the vicinity of the fish farm (Loucks et al. 2014). Ultimately, the collective action halted the approval of new aquaculture operations in the bay (Charles et al. 2020), likely improving conditions for Piping Plovers as well. The process demonstrates how a systems approach, particularly when the affected community drives the research, can have far-reaching benefits. Collaborative and co-created citizen science empowers citizen scientists by providing a voice in the scientific process. Here, the citizen scientist can guide research in a direction that addresses environmental and social justice questions that address concerns of the community.

Whether the goal is acquisition of new ecological knowledge, strengthening a sense of place, and/or increasing scientific literacy, citizen science projects can set up a feedback loop of co-production of science, benefiting the scientist, the citizen scientist, and the community at large (Dickinson et al. 2012). Although the scientist teaches the participant how to be a scientist (e.g., sharing information on the scientific method and protocols), the citizen scientist not only contributes data but also shares their experiences and knowledge with the study organism or system, their understanding of the project goals, and their additional observations with the scientist, which can help refine or create new research questions. As with the example of women ornithologists studying female birdsong, diverse voices from beyond the academy can thus reshape and refresh ornithology. Given the benefits of citizen science programs to science discovery, conservation, and the participant (Dickinson et al. 2012, Schuttler et al. 2018), ensuring equitable access to participating in programs is an important consideration, particularly within the lens of diversifying science.

Cultivating and Expanding Partnerships

More broadly, cultivating authentic partnerships can expand who has access to ornithology and whether this happens during engagement events, citizen science programs, or building a pathway toward a career in ornithology. When partnerships extend beyond the usual suspects (e.g., nature centers and bird clubs), more people have opportunities to participate in ornithology. And creatively linking ornithological activities with other aspects of our lives can foster a more inclusive atmosphere. With the increased focus on building a more welcoming scientific community (Cronin et al. 2021), several groups and events have emerged to bridge social justice and ornithology.

Highlighting and celebrating specific identities, affinity groups and events can demonstrate the importance of representation and the diverse voices within the ornithological community. In response to recent hate crimes experienced by Black people in outdoor spaces (e.g., Ahmaud Arbery and Christian Cooper), Black Birders Week (https://www.audubon.org/black-birders-week) was created to highlight Black nature enthusiasts and increase the visibility of Black birders who face unique challenges and dangers when engaging in outdoor activities. Organizations such as Latino Outdoors create partnerships with existing groups embedded in communities to engage Latine people in enjoying nature through sharing and celebrating stories, knowledge, and culture. Similarly, Birdability (https://www.birdability.org) and the Feminist Bird Club (https://www.feministbirdclub.org) are nonprofits that ensure the birding community and outdoors are welcoming, inclusive, safe, and accessible for everyone. Birdability focuses on people with mobility challenges, blindness or low vision, chronic illness, intellectual or developmental disabilities, or mental illness and those who are neurodivergent, deaf or hard of hearing, or who have other health concerns. The Feminist Bird Club aims to make birding and outdoors inclusive and affirming to people who may not have safe access to it (Black, Indigenous, and people of color; LGTBQIA+; and women). These organizations advocate for social justice and lasting social change for those who may lack safe access to the outdoors. Addressing underlying safety issues and accessibility can help reduce some barriers for participating in ornithology.

Developing community partnerships and programs that address local interests can provide further opportunities for authentic collaborations with larger, ornithological organizations. For example, Mapping Migraciones (https://www.audubon.org/mapping-migraciones) is a recent (2021) community project between Audubon California, the National Audubon Society, and Latino Outdoors. The interactive map integrates bird species, their natural history, migration routes, and personal narratives from the Latine community. In addition, the project hosts monthly webinars and events that spotlight sense of place, inclusivity, political borders, and birding. The project provides a fuller picture of how birds and people are connected through geography and culture.

The Community Congress on English Bird Names, sponsored by the Diversity and Inclusion Committee of the American Ornithological Society, provides another example of how forging partnerships can build a more inclusive ornithological community (Liu et al. 2024). The event focused on eponymous bird names (i.e., species named for people) and the challenges and opportunities for changing these names. The crux of the issue revolved around the harm eponymous names may cause. For example, McCown's Longspur (*Rhyncophanes mccownii*) was originally named for amateur ornithologist John McCown, who was a Confederate general, led ranks during the Civil War to preserve slavery, and played a role in persecuting Indigenous tribes (Driver and Bond 2021). A bird name represents one of our first access points for engaging with an individual bird and provides a gateway to more information about its life history, ecology, and conservation status. The Congress sparked conversations on who decides common bird names (historically, the North American Classification Committee) and how to include a broader group of bird name users in these decisions. These and other conversations are identifying concrete actions to create a more welcoming ornithological and birding community.

Conclusion

Several recent high-impact publications have reignited conversations about how systemic racism has created barriers for some marginalized groups to participate in science and how ornithology, science, and society can be more inclusive (e.g., Schell et al. 2020, Cronin et al. 2021, Trisos et al. 2021). Many of these publications reference how to change practices within academia and research groups (e.g., Ten Simple Rules; Chaudhary and Berhe 2020, Golden et al. 2021). In this chapter, we focused on engagement, although within a more holistic framework, ornithologists can work to create a more inclusive scientific society. We highlighted the importance of expanding beyond the question of who to engage with to include where to engage and how to engage more people in ornithology. Whether it is through citizen science programs or forging new partnerships, pursuing these interactions with care and authenticity is key. Making space for diverse voices and promoting new leadership in ornithology can lead to new insights, innovative approaches to doing ornithology (Gibbs 2014), and, ultimately, stronger relationships between birds and people. This last point is essential for the conservation of birds and can advance human well-being.

References

Andrade, R., K. L. Larson, J. Franklin, S. B. Lerman, H. L. Bateman, and P. S. Warren. 2022. Species traits explain public perceptions of human–bird interactions. *Ecological Applications* 32: Article e2676.

Awasthy, M., A. Z. Popovic, and W. L. Linklater. 2012. Experience in local urban wildlife research enhances a conservation education programme with school children. *Pacific Conservation Biology* 18:41–46.

Bendire, C. E. 1889. Notes on the habits, nests, and eggs of *Dendragapus obscurus fuliginosus*, the Sooty Grouse. *The Auk* 6:32–39.

Berkes, F. 1993. Traditional ecological knowledge in perspective. Pages 1–9 in J. T. Inglis, eds. *Traditional Ecological Knowledge: Concepts and Cases.* Canadian Museum of Nature, Ottawa, Ontario, Canada.

Billick, I., I. Babb, B. Kloeppel, J. Leong, J. Hodder, J. Sanders, and H. Swain. 2013. Field stations and marine laboratories of the future: A strategic vision. National Association of Marine Laboratories and Organization of Biological Field Stations. https://obfs.org/wp-content/uploads/2024/10/fsml_final_report.pdf

Brewster, W. 1884. Notes on the summer birds of Berkshire County, Massachusetts. *The Auk* 1:5–16.

Bucchi, M., and F. Neresini. 2002. Biotech remains unloved by the more informed. *Nature* 416:261.

Butcher, G. S., and D. K. Niven. 2007. Combining data from the Christmas Bird Count and the Breeding Bird Survey to determine the continental status and trends of North America birds. National Audubon Society.

Callaghan, C. T., G. Bino, R. E. Major, J. M. Martin, M. B. Lyons, and R. T. Kingsford. 2019. Heterogeneous urban green areas are bird diversity hotspots: Insights using continental-scale citizen science data. *Landscape Ecology* 34:1231–1246.

Cardinale, B. J., J. E. Duffy, A. Gonzalez, D. U. Hooper, C. Perrings, P. Venail, A. Narwani, G. M. Mace, D. Tilman, and D. A. Wardle. 2012. Biodiversity loss and its impact on humanity. *Nature* 486:59–67.

Cardinale, B. J., K. L. Matulich, D. U. Hooper, J. E. Byrnes, E. Duffy, L. Gamfeldt, P. Balvanera, M. I. O'Connor, and A. Gonzalez. 2011. The functional role of producer diversity in ecosystems. *American Journal of Botany* 98:572–592.

Cazalis, V., M. D. Barnes, A. Johnston, J. E. M. Watson, C. H. Sekercioglu, and A. S. L. Rodrigues. 2021. Mismatch between bird species sensitivity and the protection of intact habitats across the Americas. *Ecology Letters* 24:2394–2405.

Chapman, F. M. 1917. The distribution of bird-life in Colombia. *Bulletin of the American Museum of Natural History* 36.

Charles, A., L. Loucks, F. Berkes, and D. Armitage. 2020. Community science: A typology and its implications for governance of social-ecological systems. *Environmental Science & Policy* 106:77–86.

Chaudhary, V. B., and A. A. Berhe. 2020. Ten simple rules for building an antiracist lab. *PLoS Computational Biology* 16: Article e1008210.

Cooper, C. B., J. Dickinson, T. Phillips, and R. Bonney. 2007. Citizen science as a tool for conservation in residential ecosystems. *Ecology and Society* 12: Article 11.

Cooper, C. B., C. L. Hawn, L. R. Larson, J. K. Parrish, G. Bowser, D. Cavalier, R. R. Dunn, M. Haklay, K. K. Gupta, N. O. Jelks, V. A. Johnson, M. Katti, Z. Leggett, O. R. Wilson, and S. Wilson. 2021. Inclusion in citizen science: The conundrum of rebranding. *Science* 372:1386–1388.

Crenshaw, K. 1989. Demarginalizing the intersection of race and sex: A Black feminist critique of antidiscrimination doctrine, feminist theory and antiracist politics. *University of Chicago Legal Forum* 1989:139–167.

Cronin, M. R., S. H. Alonzo, S. K. Adamczak, D. N. Baker, R. S. Beltran, A. L. Borker, A. B. Favilla, R. Gatins, L. C. Goetz, N. Hack, J. G. Harenčár, E. A. Howard, M. C. Kustra, R. Maguiña, L. Martinez-Estevez, R. S. Mehta, I. M. Parker, K. Reid, M. B. Roberts, S. B. Shirazi, T.-A. M. Tatom-Naecker, K. M. Voss, E. Willis-Norton, B. Vadakan, A. M. Valenzuela-Toro, and E. S. Zavaleta. 2021. Anti-racist interventions to transform ecology, evolution and conservation biology departments. *Nature Ecology & Evolution* 5:1213–1223.

Dasgupta, N., and S. Asgari. 2004. Seeing is believing: Exposure to counterstereotypic women leaders and its effect on the malleability of automatic gender stereotyping. *Journal of Experimental Social Psychology* 40:642–658.

de Vos, A. 2020. The problem of "colonial science." *Scientific American.*

Demery, A.-J. C., and M. A. Pipkin. 2021. Safe fieldwork strategies for at-risk individuals, their supervisors and institutions. *Nature Ecology & Evolution* 5:5–9.

Dickinson, E. 1891. *Poems by Emily Dickinson.* Roberts Brothers, Boston, MA.

Dickinson, J. L., J. Shirk, D. Bonter, R. Bonney, R. L. Crain, J. Martin, T. Phillips, and K. Purcell. 2012. The current state of citizen science as a tool for ecological research and public engagement. *Frontiers in Ecology and the Environment* 10:291–297.

Driver, R. J., and A. L. Bond. 2021. Towards redressing inaccurate, offensive and inappropriate common bird names. *Ibis* 163:1492–1499.

Dunn, R. R., M. C. Gavin, M. C. Sanchez, and J. N. Solomon. 2006. The pigeon paradox: Dependence of global conservation on urban nature. *Conservation Biology* 20:1814–1816.

Dylan, B. 1970. If not for you. *New Morning.* Columbia Records.

Emlen, J. T. 1974. An urban bird community in Tucson, Arizona: Derivation, structure, regulation. *The Condor* 76:184–197.

Evans, C., E. Abrams, R. Reitsma, K. Roux, L. Salmonsen, and P. P. Marra. 2005. The Neighborhood Nestwatch program: Participant outcomes of a citizen-science ecological research project. *Conservation Biology* 19:589–594.

Gaston, K. J., and M. Soga. 2020. Extinction of experience: The need to be more specific. *People and Nature* 2:575–581.

Gibbs, K., Jr. 2014. Diversity in STEM: What it is and why it matters. *Scientific American.*

Golden, N., K. Devarajan, C. Balantic, J. Drake, M. T. Hallworth, and T. L. Morelli. 2021. Ten simple rules for productive lab meetings. *PLoS Computational Biology* 17: Article e1008953.

Gomez, C., C. D. Cadena, A. M. Cuervo, J. Díaz-Cárdenas, F. García-Cardona, N. Niño-Rodríguez, N. Ocampo-Peñuela, D. Ocampo, G. Seeholzer, A. Sierra-Ricaurte, and J. Soto-Patiño. 2022. Reexpedición Colombia: Entender el pasado para empoderar acciones que fortalezcan el conocimiento y conservación de las aves. *Biota Colombiana* 23: Article e984. https://doi.org/10.21068/2539200x.984.

Grade, A. M., N. W. Chan, P. Gajbhiye, D. J. Perkins, and P. S. Warren. 2022. Evaluating the use of semi-structured crowdsourced data to quantify inequitable access to urban biodiversity: A case study with eBird. *PLoS One* 17: Article e0277223.

Grinnell, J. 1917. The niche-relationships of the California Thrasher. *The Auk* 34:427–433.

Haines, C. D., E. M. Rose, K. J. Odom, and K. E. Omland. 2020. The role of diversity in science: A case study of women advancing female birdsong research. *Animal Behaviour* 168:19–24.

Iknayan, K. J., M. M. Wheeler, S. M. Safran, J. S. Young, and E. N. Spotswood. 2022. What makes urban parks good for California quail? Evaluating park suitability, species persistence, and the potential for reintroduction into a large urban national park. *Journal of Applied Ecology* 59:199–209.

Kellert, S. R. 1985. Birdwatching in American society. *Leisure Sciences* 7:343–360.

Kimmerer, R. W. 2002. Weaving traditional ecological knowledge into biological education: A call to action. *BioScience* 52:432–438.

La Sorte, F. A., D. Fink, P. J. Blancher, A. D. Rodewald, V. Ruiz-Gutierrez, K. V. Rosenberg, W. M. Hochachka, P. H. Verburg, and S. Kelling. 2017. Global change and the distributional dynamics of migratory bird populations wintering in Central America. *Global Change Biology* 23:5284–5296.

Lawrence, N. G. 1889. An account of the breeding habits of *Puffinus auduboni* in the island of Grenada, West Indies, with a note on *Zenaida rubripes*. *The Auk* 6:19–21.

Lerman, S. B., K. L. Larson, D. L. Narango, M. A. Goddard, and P. P. Marra. 2023. Humanity for habitat: Residential yards as an opportunity for biodiversity conservation. *BioScience* 73:671–689.

Lerman, S. B., D. L. Narango, M. L. Avolio, A. R. Bratt, J. M. Engebretson, P. M. Groffman, S. J. Hall, J. B. Heffernan, S. E. Hobbie, K. L. Larson, D. H. Locke, C. Neill, K. C. Nelson, J. Padullés Cubino, and T. L. E. Trammell. 2021. Residential yard management and landscape cover affect urban bird community diversity across the continental USA. *Ecological Applications* 31: Article e02455.

Lerman, S. B., L. Pejchar, L. Benedict, K. M. Covino, J. L. Dickinson, J. E. Fantle-Lepczyk, A. D. Rodewald, and C. Vleck. 2021. Juggling parenthood and ornithology: A full lifecycle approach to supporting mothers through the American Ornithological Society. *Ornithological Applications* 123:1–9.

Lerman, S. B., and P. S. Warren. 2011. The conservation value of residential yards: Linking birds and people. *Ecological Applications* 21:1327–1339.

Liu, I. A., E. R. Gulson-Castillo, J. X. Wu, A.-J. C. Demery, N. Cortes-Rodriguez, K. M. Covino, S. B. Lerman, S. A. Gill, and V. Ruiz Gutierrez. 2024. Building bridges in the conversation on eponymous common names of North American birds. *Ibis* 166:1092–1102.

Lopez, B., E. Minor, and A. Crooks. 2020. Insights into human–wildlife interactions in cities from bird sightings recorded online. *Landscape and Urban Planning* 196: Article 103742.

Loucks, R. H., R. E. Smith, C. V. Fisher, and E. Brian Fisher. 2012. Copper in the sediment and sea surface microlayer near a fallowed, open-net fish farm. *Marine Pollution Bulletin* 64:1970–1973.

Loucks, R. H., R. E. Smith, and E. B. Fisher. 2014. Interactions between finfish aquaculture and lobster catches in a sheltered bay. *Marine Pollution Bulletin* 88:255–259.

Luck, G. W., R. Harrington, P. A. Harrison, C. Kremen, P. M. Berry, R. Bugter, T. P. Dawson, F. de Bello, S. Díaz, C. K. Feld, J. R. Haslett, D. Hering, A. Kontogianni, S. Lavorel, M. Rounsevell, M. J. Samways, L. Sandin, J. Settele, M. T. Sykes, S. van den Hove, M. Vandewalle, and M. Zobel. 2009. Quantifying the contribution of organisms to the provision of ecosystem services. *BioScience* 59:223–235.

McAfee, D., Z. A. Doubleday, N. Geiger, and S. D. Connell. 2019. Everyone loves a success story: Optimism inspires conservation engagement. *BioScience* 69:274–281.

National Science Board. 2011. National Science Foundation's merit review criteria: Review and revisions. NSB/MR-11-22.

Odom, K. J., M. L. Hall, K. Riebel, K. E. Omland, and N. E. Langmore. 2014. Female song is widespread and ancestral in songbirds. *Nature Communications* 5: Article 3379.

Perkins, D. J. 2020. Blind spots in citizen science data: Implications of volunteer biases in eBird data [Master's thesis]. North Carolina State University, Raleigh, NC.

Pham, D. 2016. Public engagement is key for the future of science research. *npj Science of Learning* 1: Article 16010.

Prévot, A.-C., H. Cheval, R. Raymond, and A. Cosquer. 2018. Routine experiences of nature in cities can increase personal commitment toward biodiversity conservation. *Biological Conservation* 226:1–8.

Price, C. A., and H.-S. Lee. 2013. Changes in participants' scientific attitudes and epistemological beliefs during an astronomical citizen science project. *Journal of Research in Science Teaching* 50:773–801.

Prince. 1984. When doves cry. *Purple Rain*. Warner Brothers Records.

Rattling Leaf, J., Sr. 2022. What is traditional ecological knowledge and why does it matter? *Frontiers in Ecology and the Environment* 20:3.

Rosenberg, K. V., A. M. Dokter, P. J. Blancher, J. R. Sauer, A. C. Smith, P. A. Smith, J. C. Stanton, A. Panjabi, L. Helft, M. Parr, and P. P. Marra. 2019. Decline of the North American avifauna. *Science* 366:120–124.

Sauer, J. R., B. G. Peterjohn, and W. A. Link. 1994. Observer differences in the North American Breeding Bird Survey. *The Auk* 111:50–62.

Schell, C. J., K. Dyson, T. L. Fuentes, S. Des Roches, N. C. Harris, D. S. Miller, C. A. Woelfle-Erskine, and M. R. Lambert. 2020. The ecological and evolutionary consequences of systemic racism in urban environments. *Science* 369: Article eaay4497.

Schuetz, J. G., and A. Johnston. 2019. Characterizing the cultural niches of North American birds. *Proceedings of the National Academy of Sciences of the USA* 116:10868–10873.

Schuttler, S. G., A. E. Sorensen, R. C. Jordan, C. Cooper, and A. Shwartz. 2018. Bridging the nature gap: Can citizen science reverse the extinction of experience? *Frontiers in Ecology and the Environment* 16:405–411.

Şekercioğlu, Ç. H., G. C. Daily, and P. R. Ehrlich. 2004. Ecosystem consequences of bird declines. *Proceedings of the National Academy of Sciences of the USA* 101:18042–18047.

Soga, M., and K. J. Gaston. 2016. Extinction of experience: The loss of human–nature interactions. *Frontiers in Ecology and the Environment* 14:94–101.

Struminger, R., J. Zarestky, R. A. Short, and A. M. Lawing. 2018. A framework for informal STEM education outreach at field stations. *BioScience* 68:969–978.

Sullivan, B. L., J. L. Aycrigg, J. H. Barry, R. E. Bonney, N. Bruns, C. B. Cooper, T. Damoulas, A. A. Dhondt, T. Dietterich, A. Farnsworth, D. Fink, J. W. Fitzpatrick, T. Fredericks, J. Gerbracht, C. Gomes, W. M. Hochachka, M. J. Iliff, C. Lagoze, F. A. La Sorte, M. Merrifield, W. Morris, T. B. Phillips, M. Reynolds, A. D. Rodewald, K. V. Rosenberg, N. M. Trautmann, A. Wiggins, D. W. Winkler, W.-K. Wong, C. L. Wood, J. Yu, and S. Kelling. 2014. The eBird enterprise: An integrated approach to development and application of citizen science. *Biological Conservation* 169:31–40.

Taylor, L. 2022. Landmark Colombian bird study repeated to right colonial-era wrongs. *Nature* 601:178–179.

Thomas, J. W., R. O. Brush, and R. M. DeGraaf. 1973. Invite wildlife to your backyard. *National Wildlife* 11:5–16.

Trisos, C. H., J. Auerbach, and M. Katti. 2021. Decoloniality and anti-oppressive practices for a more ethical ecology. *Nature Ecology & Evolution* 5:1205–1212.

United Nations. 1992. United Nations Convention on Biological Diversity, Rio de Janeiro, Brazil. United Nation, New York.

Walker, C. M., K. Colton Flynn, G. A. Ovando-Montejo, E. A. Ellis, and A. E. Frazier. 2017. Does demolition improve biodiversity? Linking urban green space and socioeconomic characteristics to avian richness in a shrinking city. *Urban Ecosystems* 20:1191–1202.

SECTION V
CONSERVATION AND MANAGEMENT

21
Commentary on Conservation and Management

Alberto Yanosky

I am honored to introduce the section on conservation and management for this book. Most conservation biologists involved in the professional activity of ornithology are faced with conservation challenges that in many cases require active management. This is the case for many grassland birds (Douglas et al. 2023), where induced and natural disturbances force us to find solutions to enable the species to survive (Vickery et al. 2000). In addition, ornithology has also developed management techniques regardless of conservation matters, such as the case of parasitism (Fiorini et al. 2019). All that we know and the decisions we make to conserve and manage birds both in the wild and in captivity rely on diverse sources of data, as well as scientific and traditional knowledge, and the syntheses we produce from these sources to help advance the science of ornithology.

This section presents the following topics with regard to avian conservation:

- A useful overview of the geography of bird conservation
- Climate change and its consequences for bird conservation, a topic of global concern
- A focus on the Neotropics, with significant knowledge gaps and the challenges of diverse anthropogenic threats
- A more management-oriented perspective from sub-Saharan Africa, on the value of chicken breeds adapted to the tropics
- What is happening to African bird conservation initiatives
- Addressing the ways to do conservation within the climate change scenario

The following chapters in this section together make for a timely review of the current state of knowledge of avian conservation, as well as specific gaps that should be prioritized in the years to come to address conservation and management problems in ornithology.

Birds are amazingly diverse, and their diversity varies over space and time; this geography is clearly affected by human societies. The geographic aspects of conservation help focus actions implemented to mitigate threats and promote the desired sustainability of society with nature. Conservation geography relies on data to understand the status of bird species globally, and this information is influenced by the specific methods employed to detect and count birds in their habitats and provide estimates of their abundance to understand trends in bird populations. This

Alberto Yanosky, *Commentary on Conservation and Management*. In: *New Perspectives in Ornithology*.
Edited by: Scott V. Edwards and J. Michael Reed, Oxford University Press. © Oxford University Press (2025).
DOI: 10.1093/oso/9780197787670.003.0021

information is used and periodically revised by organizations such as the International Union for the Conservation of Nature to assess extinction risks. Extinction risks are closely associated with different aspects of avian ecology, especially for migratory species or species that congregate in specific areas or ecological gradients, such as elevational range or behavior. The combination of species that may be threatened and the added complication of some ranges overlapping imposes challenges, and the difficulty of addressing these factors is also related to intraspecific or genetic diversity. The practice of red-listing bird species is extensively analyzed, and it is noted that the robustness of information generated at global and subglobal levels differs. Bird status throughout the world varies, and this is related not only to human influences but also to ornithological evolutionary ecology and biogeography. Citizen science and local participation is of crucial importance to understand where the species are (Yanosky et al. 2023). The collection and participation of local people including natural history aspects are needed especially where limited knowledge about bird species exist (Marcot et al. 2022). Human influences have been analyzed for agricultural and aquacultural systems, including biological resource usage; invasive species and diseases; infrastructure development; natural events (e.g., those related to fire and water); climate change and variability, including severe weather conditions; pollution; energy and mining; transportation; human disturbances (tourism and incidentals); and geological events. Cardador and Blackburn (2020) have shown that the expansion of human niches involved colonization of more disturbed environments by species not responding positively to human influence in their native ranges.

Chapter 22, on conservation geography, clearly demonstrates that although the threats are geographically known, the development of actions for those geographies as a predictive science is still underway. Habitat protection and management and species management, in terms of their geographies, are related to education and awareness building, law and policy, as well as incentives for conservation. In Chapter 22, Thomas M. Brooks notes that although much information is available to address geographical aspects of bird conservation, more data are needed to reinforce conservation actions, especially outside the protected areas scheme, such as local and Indigenous knowledge, avitourism, the implementation of policy mechanisms, ecosystem restoration, and financial safeguards. This chapter also highlights the importance of societal participation and citizen science.

The geography of conservation varies depending on whether it is analyzed globally or on a subglobal level; however, the planet faces events related to human-induced climate change that are global, with impacts at the local level. Once one understands the geography of conservation, one may consider one of the largest threats of which humans are aware: climate change. Climate change is being addressed on both mitigation and adaptation fronts, to reduce greenhouse gas emissions and find ways to reduce the vulnerability and increase the resilience of humans and ecosystems to change. Bird conservation is one of the areas in which climate change needs to be better understood. Ramírez et al. (2018) present examples from waterbirds as a "flagship community of species" from which positive and negative impacts can be better understood. Chapter 22 analyses how this threat affects birds and discusses appropriate responses if global trends of climate conditions continue to worsen. Also, climate affects not only birds but also other species, affecting relationships and the whole ecosystem, in some cases being beneficial and in others detrimental to bird species.

A related aspect is linked to societal solutions for the climate crisis, which may have implications for our work in bird conservation. Chapter 22 notes that birds are exposed to different threats, and climate change, although general, may alter the frequency and impact of other current threats; this forces us to think about these threats, which at geographical scales may differ and have implications for coordinating and planning bird conservation. The interaction of climate change with other, better understood threats is still poorly addressed. Conservation requires securing suitable habitats, and the larger and more representative they are, the higher the chance that suitable climatic conditions may be found in areas that birds can move into. This could mean that places that are secure habitats today may not be so in the future. This view, and related considerations for protected areas, may affect our discourse and actions because sites secured for conservation may not be secure for species or a set of species. Diseases and climate change also interact, which can be detrimental to birds. Ecological networks may be altered by climate change, not only positively or negatively affecting different species but also affecting the distribution of species. If our knowledge of the network of dependencies among species and their habitats is fragmentary, the predictions about the impacts of climate change on these networks may be less precise. At a broader scale, ecosystem-wide changes could be affected by other, more chemical- and physical-oriented elements in the environment, such as the carbon, mineral, and nutrient cycles.

Chapter 22 reviews the importance of society's role in fighting climate change. It has the potential to greatly affect human societies, and the attention given to reducing the vulnerability of people and their livelihoods to climate change may also benefit birds. The climate change effects of moving agricultural patterns and the agroecological production zones will also have implications for birds and biodiversity. Tirivangasi et al. (2022) provide a local analysis in South Africa to better understand the effects of changing weather patterns on human livelihood, demonstrating a decline in a subsistence economy, a lack of water, and exposure to dust and wind.

Chapter 22 details the existing knowledge and all the areas that still must be better understood in terms of climate change effects and biodiversity and how scientists and managers must deal with these problems. These challenges will not be easy because climate change is happening at a faster pace than our ability to acquire knowledge and effectively address them. This chapter also notes the importance of flexibility, not only in terms of making decisions but also in terms of acting to change circumstances; the effects of changes in climate are upon us, urging us to act even if knowledge is insufficient to advance with science-based solutions. Planning takes a predominant role in these situations, and conservation plans, well designed in advance, will make our lives easier when we need to act because the effects of climate change are in front of us.

The geography of bird conservation and the effects of climate change on birds are a global focus, including one of the richest biodiversity regions of the planet, the Neotropics, to understand how knowledge of birds and their ecological functions in the tropics have helped advance conservation efforts. These conservation efforts have been demonstrated to be better represented by site protection and their networks, the protected area systems. Protected areas in the Neotropics, not only those officially declared but also those of voluntary creation, such as conservation areas, cover approximately one-fourth of the land area and represent a slightly higher percentage of the total species distribution. Chapter 23, by David A. Prieto-Torres et al., notes that for many species, little of their distributional ranges is protected, and with a coverage increase of 30%, we could increase approximately 36% of all species' ranges and more than 42% of threatened species' ranges. This analysis is very opportune because the post-2020 biodiversity targets aim toward a nature-positive situation by 2030 (Yanosky 2024). Especially considering the consensus that current protected areas are insufficient by far, it is generally agreed that there is a need to ensure the long-term persistence of bird populations and ecosystem functionalities given the knowledge on land-use change, habitat fragmentation and degradation, climate change, and other threats.

Decision-makers have the huge challenge to make decisions based on the information available, desired outcomes, and the need to act in the landscape and at site levels. Given all these threats and their trends, Chapter 23 focuses on the identification of safe places where species richness and wildernesses coincide, or where human interventions are not expected to occur, and the need to go beyond commonly known and used indicators such as species richness: Other aspects, such as invasive species, soil management, and ecological interactions, are also part of our information basis for effective conservation actions. There is a need for much stronger and better communication among all parties, including the public and private sectors, civil society, and academia. These interactions, and a collective understanding, are required to effectively create bridges that will allow us to identify and promote sustainable-use strategies. The combination of bird conservation priorities for the protection of species and their networks and the presence of developing countries—in which financial contributions may not be optimal—poses another threat, and the need for more protection at the site level and the landscape level and effective monitoring of actions that must go beyond jurisdictions and political boundaries exacerbates this situation.

Chapter 24, by Abdulmojeed Yakubu et al., focuses on Indigenous chicken production in Africa. This addresses an important aspect of husbandry and management, for which ample knowledge of the species of interest, and the adaptive management of this bird based on the domestication process, is required. For Indigenous chickens, not only the ecological and biological management of the species but also the economic and cultural aspects of bird species management must be addressed; this strategy involves innovation, technology, and approaches to increase smallholder poultry production to grow rural economies and increase diversified and equitable livelihoods. Artificial, as opposed to natural, selection has occurred in tropically adapted birds, and different breeds have been produced—this is described for different African countries in terms of egg and meat production attributes. Morphobiometric traits and systems of production are at the center of the analysis, including a variety of systems ranging from scavenging to intensive conditions. Nutrition plays a major role in the management of genetically improved chickens. In general, addressing poultry production from the perspective of the chicken value chain requires also address marketing strategies, credits and incentives, extension services, community-based management, and infrastructure development to facilitate the success of the poultry production initiatives, biosecurity measures, breed registration, and farmers associations for enhancing information and sharing experience.

Chapter 24 highlights the importance of the Convention on Biodiversity's Nagoya Protocol, in which the use of genetic and biological resources as bioresources for innovation and the generation and sharing of benefits in a fair and equitable way are addressed.

This section on conservation and management not only discusses the current knowledge in terms of bird species and their conservation status but also discusses the local, regional, and global efforts to understand the relevant conservation geography, the effects of climate change and other threats on species, conservation areas in the Neotropics, and Indigenous chicken productivity and its implications for the conservation of genetic resources. The general aspects discussed, the state of the art of management and conservation, and the dynamics of knowledge from global to local and vice versa with the interaction of current and more studied threats to bird conservation raise a myriad of complex issues that are applicable to different conditions regardless of where birds exist—and hardly any place on Earth is birdless. The metric related to bird ecology and conservation in population and community ecology levels needs to involve more social and humanistic approaches with society participation and citizen science, with legal implications and policymaking.

In Chapter 25, Hazell Shokellu Thompson and Samuel Ivande provide a detailed history and milestones for African ornithology, exploring the multifaceted challenges and opportunities surrounding bird conservation, which apply globally but particularly in Africa. This chapter highlights the importance of building local ornithological expertise and engaging citizen scientists in conservation efforts, and it is focused on two pillars—capacity development and citizen science. There is a detailed development of actions toward bird conservation in Africa during the past 30 years; nevertheless, African birds have sharply declined in this period. Knowledge on bird ecology and conservation has greatly increased, including threats to bird survival, although actions to address these threats and lower key barriers are beginning to take effect. South Africa has greatly contributed with practitioners and with scientific knowledge, demonstrating a continental disbalance in terms of capacity. The Pan-African Ornithological Congress has also played a key role.

Chapter 25 emphasizes the need for collaborative approaches to address threats to bird populations and promote sustainable practices for their long-term survival, and it also describes the "Indigenous conservation leadership vacuum" and how it is changing with a combination of continental flavor and international support. Publications and resources for publishing are addressed as critical barriers, as is the African linguistic diversity, which necessitates scientific materials being translated into local languages to facilitate bird conservation.

The importance of information on where the birds are for their conservation and protection is discussed, as is the role of developing atlases and encouraging citizen science initiatives with crucial platforms to capture and increase local awareness and local capacity for a meaningful engagement.

Chapter 26, by Chris S. Elphick, highlights that climate change differs from other threats in its rate and global scale. This chapter explores four ideas and what they mean to ornithologists in their bird conservation present and future work. The chapter analyzes the unique challenges that climate change poses to bird conservation, and it discusses climate change's rapid pace and global scale, which are causing near-simultaneous, widespread disruptions across species' ranges, impacting physiological processes and potentially creating irreversible changes. Elphick indicates that this necessitates a shift from reversing conditions to mitigation strategies.

A highlighted aspect is the synergistic effect of climate change with other threats, as it intensifies land-cover change; reduces habitat suitability; and increases disease risk, such as the case related to avian malaria. Some species might benefit from range expansion, and this depends on both habitat and species mobility, often hindered by slow vegetation changes, whereas land protection strategies must adapt, potentially involving changing the legal protection status of less valuable protected areas.

Climate change appears to disrupt ecological networks through species losses and colonizations, affecting food webs, creating phenological mismatches, and exacerbating insect declines, thus affecting prey availability and with cascading effects on bird populations. This happens in an uncertain scenario in which our understanding of species interactions and trophic dynamics is still poor.

Societal responses to climate change present further challenges, especially because interventions require careful planning and mitigation, such as those in renewable energy development, agricultural changes, and geoengineering, which can negatively impact bird species and their populations and interactions. Elphick advocates for a holistic approach in which benefits and costs are weighed in actions based on population consequences and not just using mortality rates, which are the most commonly available data.

Chapter 26 also notes that climate change, compared to other current threats that affect more bird species, requires significant, proactive, and globally coordinated attention due to its unique and far-reaching consequences. Climate change threatens approximately 1 in 10 bird species, and it combines impacts with other existing threats, which will eventually impact nearly all living organisms and thus trigger societal responses affecting bird conservation. This threat is happening fast and is of global scale, requiring faster, larger scale conservation actions in combination with proactive planning and improved coordination among different sectors—the public, academia, and organizations.

I have no doubt about the importance of this section and the wisely selected and invited chapters, in addition to the relevance of this section for a broad audience from those in science and research—including an update of the scientific information and areas on which to focus future research—but also to the practitioners and those who are taking decisions for bird conservation, and the students—future researchers and practitioners who will have the enormous challenge of conserving birds and their populations in an apparently changing planet, not only environmentally but also economically and socially.

References

Cardador, L., and T. M. Blackburn. 2020. A global assessment of human influence on niche shifts and risk predictions of bird invasions. *Global Ecology and Biogeography* 29:1956–1966. https://doi.org/10.1111/geb.13166.

Douglas, D. J., J. Waldinger, Z. Buckmire, K. Gibb, J. P. Medina, L. Sutcliffe, C. Beckmann, N. J. Collar, R. Jansen, J. Kamp, I. Little, R. Sheldon, A. Yanosky, and N. Koper. 2023. A global review identifies agriculture as the main threat to declining grassland birds. *Ibis* 165:1107–1128.

Fiorini, V. D., M. C. De Mársico, C. A. Ursino, and J. C. Reboreda. 2019. Obligate brood parasitism on Neotropical birds. Pages 103–131 in J. C. Reboreda, V. D. Fiorini, and D. Tuero, eds. Behavioral *Ecology* of Neotropical *Birds*. Springer, Cham, Switzerland.

Marcot, B. G., D. E. Gawlik, A. Yanosky, J. Anderson, A. Gupta, and G. Sundar. 2022. The value and necessity of natural history studies of waterbirds. *Waterbirds* 45:iv–ix. https://bioone.org/journals/Waterbirds.

Ramírez, F., C. Rodríguez, J. Seoane, J. Figuerola, and J. Bustamante. 2018. How will climate change affect endangered Mediterranean waterbirds? *PLoS One* 13: Article 0192702. https://doi.org/10.1371/journal.pone.0192702.

Tirivangasi, H. M., S. A. Rankoana, and S. S. Mugambiwa. 2022. Community perceptions on the effects of climate change on socio-economic and health conditions of Dikgale community, Limpopo province South Africa. *African Journal of Development Studies* 12:183–200. https://doi.org/10.31920/2634-3649/2022/v12n4a9.

Vickery, P. D., J. R. Herkert, F. L. Knopf, J. Ruth, and C. E. Keller. 2000. Grassland birds: An overview of threats and recommended management strategies. Pages 74–77 in R. Bonney, D. N. Pashley, R. Cooper, and L. Niles, eds. *Strategies for Bird Conservation: The Partners in Flight Planning Process.* U.S. Forest Service, Rocky Mountain Research Station, Ogden, UT.

Yanosky, A. 2024. Will the new global biodiversity agreement be another failed effort? Pages 57–60 in M. Mastrangelo, A. Attanasio, and I. Torres, eds. *Biodiversity in Latin America and the Caribbean.* IUCN, Quito, Ecuador.

Yanosky, A., R. Irala, and T. Galluppi. 2023. Citizen science updates on the occurrence and conservation of herons in Paraguay. *Journal of Heron Biology and Conservation* 8:1–8.

22
Bird Conservation Geography

Thomas M. Brooks

It is now well-documented and widely accepted that human activities are driving a crisis of life on Earth (Intergovernmental Science-Policy Platform on Biodiversity and Ecosystem Services 2019), with species extinction rates hundreds of times higher than those typical of the planet's history (Pimm et al. 2014). Given that the approximately 11,000 species of the Class Aves comprise the most easily observable and thus best-known component of biodiversity (Bibby 1999), with estimates of total numbers of individuals ranging from 50 billion (Callaghan et al. 2021) to 200–400 billion (Gaston and Blackburn 1997), much of the evidence documenting this crisis comes from birds. However, if the first "rule" of biogeography is that biodiversity varies enormously over space and time, the first "rule" of conservation science is that human impacts on living nature are equally variable (Brooks 2010). In this chapter, I review the multiple dimensions of this variation. My focus is on the global scale.

My approach is that of "conservation geography," defined by Di Minin et al. (2022) as studying "where, when, and what conservation actions should be implemented in order to mitigate threats and promote sustainable people–nature interactions" (p. 42). Specifically, I draw from the classic state–pressure–response framework to structure the chapter. In turn, I therefore consider the variation over space and time of the state of the world's birds, pressures on the world's birds, and responses to conserve the world's birds; a similar approach was taken in the article by Lees et al. (2022), published while the current chapter was under peer review. I restrict consideration to wild birds, and so exclude the world's approximately 20 billion chickens and other domestics (Robinson et al. 2014).

Conservation Geography of the State of the World's Birds

Three primary mechanisms for data generation and synthesis underlie our understanding of the conservation geography of the state of the world's birds. First and most obviously are counts of the number of individuals of specific species in particular places. The tendency of some bird species to congregate—for example, wetland birds spending the nonbreeding station on lakes and marshlands (Boere et al. 2006), shorebirds and raptors concentrating at bottleneck sites on migration (Bildstein 2006), and seabird nesting colonies (Croxall et al. 1984)—means that for these we have for many decades been able to track the number of individuals for some hundreds of bird species with considerable precision. The advent of satellite telemetry now means

Thomas M. Brooks, *Bird Conservation Geography*. In: *New Perspectives in Ornithology*.
Edited by: Scott V. Edwards and J. Michael Reed, Oxford University Press.
DOI: 10.1093/oso/9780197787670.003.0022

that we can even track the position of individuals within such species all the time (López-López 2016). The limitation of this approach is one of sampling: For species distributed across the landscape overall, and species which are difficult to detect, counting all individuals is typically impractical (Bibby et al. 2000).

Second, and at a higher level of abstraction, are standardized survey programs designed to overcome sampling limitations through measurement of relative abundance rather than absolute counts. Examples include bird atlases, dedicating a set level of effort in survey across a set (typically equal area) spatial unit (Donald and Fuller 1998), and common bird surveys such as the Pan European Common Bird Monitoring Scheme (Brlík et al. 2021) and the North American Breeding Bird Survey (Sauer et al. 2013). The primary shortcoming of such standardized surveys is that they are labor-intensive and so have not been implemented at scale in regions where volunteer capacity is scarce (e.g., in much of the megadiverse tropics) and where human population density is low (e.g., in the boreal regions) (Gregory and van Strien 2010). Semistructured citizen science platforms such as eBird (Neate-Clegg et al. 2020) are helping mobilize data from remote regions but nevertheless face similar challenges in uneven data coverage (see the section titled "Education and Awareness").

Third, and heavily informed by data emerging from counts and atlases where available, are protocols for the assessment of species extinction risk through Red Listing. Such approaches date back to the 1960s (Hilton-Taylor 2014), but they were revolutionized in the 1990s with the development of quantitative categories and criteria as the basis for the International Union for Conservation of Nature (IUCN) Red List of Threatened Species (Rodrigues et al. 2006, IUCN 2012). These encompass three threatened categories (Critically Endangered [CR], Endangered [EN], and Vulnerable [VU]), two extinct categories (Extinct [EX] and Extinct in the Wild [EW]), in addition to Near Threatened (NT), Least Concern (LC), Data Deficient (DD), and Not Evaluated (NE) (Mace et al. 2008). BirdLife International hosts the Red List Authority for bird species, with 11,158 species now assessed, including 1,481 (13%) threatened and a further 1,001 (9%) NT, in addition to 164 (1.5%) EX or EW (IUCN 2021b). Although there is some spatial bias in the assessment of species as DD (i.e., where there is inadequate information to assess the species' risk of extinction), with, for example, the largest numbers in Indonesia and Papua New Guinea (Butchart and Bird 2010), only 52 bird species (<0.5%) are now assessed as DD.

Crucially, the IUCN Red List of Threatened Species incorporates, for all species, not only assessment of extinction risk but also required documentation on, for example, the species' habitats, threats, and distribution (IUCN 2013). Documentation on distribution can incorporate a range of data types (IUCN 2021c); for birds, a range map is provided for all species (BirdLife International 2019). This maps the current limits of distribution of a species, accounting for all known, inferred or projected sites of occurrence, and can be refined in turn using data on habitat and elevation to generate Area of Habitat maps (Brooks et al. 2019), which can be validated through harnessing occurrence data, for example, from citizen science (Huang et al. 2021).

Spatial Patterns in State of the World's Birds

Given the limitations of sampling that face counts and atlas approaches, assessments of birds from the IUCN Red List of Threatened Species are the only available source for documentation of the state of birds at the global level. However, different insights can be obtained by presenting different dimensions of these data, as well as from using different mapping techniques. Overlay of range maps across grid cells for all threatened bird species reveals dramatic variation in extinction risk throughout the world (Figure 22.1A). The ranges of one or two threatened species overlap most areas of the land's surface. However, this overlap rises to peaks of several dozen co-occurring threatened species in the world's tropical forest hot spots (Myers et al. 2000, Mittermeier et al. 2004), especially in the Caribbean, Tropical Andes, Cerrado, Atlantic Forest, East Africa, Madagascar, the Himalayas, IndoBurma, Sundaland, and the Philippines.

Species richness maps such as Figure 22.1A suffer from the serious shortcoming that they are heavily driven by the distributions of species with large ranges, which are typically those of least relevance to conservation (Lennon et al. 2004, Orme et al. 2005). This is problematic for birds, for which 25% of all species have global distributions of less than 50,000 km^2 (Stattersfield et al. 1998). The problem can be overcome by weighting in inverse proportion to range size, which when combined with weighting by extinction risk (based on the IUCN Red List categories) yields a Species Threat Abatement and Restoration (STAR) metric of the potential contribution of any given area toward global reduction of extinction risk to background levels (Mair et al. 2021). STAR is an additive metric, and so can be mapped across any spatial units, including grid cells (Mair et al. 2021) but also, importantly, political units, to shed light on national responsibilities toward bird conservation. Figure 22.1B therefore maps STAR threat-abatement values for the world's bird species across countries. This again shows the highest values for countries such as Colombia, Brazil, and Indonesia, which encompass multiple tropical forest hot spots. It also reveals intermediate values in large, higher latitude countries such as Australia, China, Russia, and the United States, which encompass large proportions of the distributions of broader-ranged species, and lowest values in small, high-latitude nations, especially in Europe.

The distribution of recently extinct bird species is rather different, with the great majority (89%) of known extinctions since 1500 having occurred on islands (Figure 22.1C), despite the fact that most bird species overall (>80%) live on continents (Butchart et al. 2006). Particularly severe concentrations of recent bird extinctions have occurred in Hawai'i (27), Mauritius (18), New Zealand (14), Réunion (11), and St. Helena (9).

This wide variation in extinction risk is also revealed in studies considering other spatial and biological dimensions of bird conservation geography, such as elevational range or migratory behavior. For example, Manne et al. (1999) showed that for the passerine species of the Americas, lowland continental species face the highest prevalence of extinction risk when controlling for range size. More generally, White and Bennett (2015) found a strong negative relationship between elevational range, midpoint elevation, and maximum elevation and avian extinction risk.

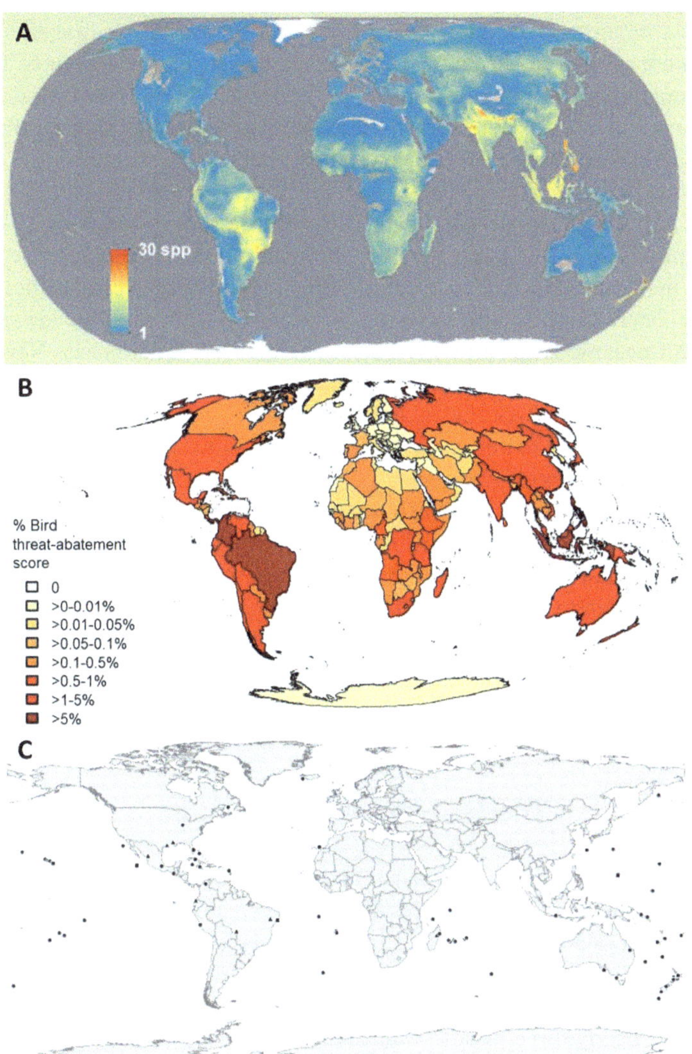

Figure 22.1 (A) Overlay of the range maps across grid cells of globally threatened terrestrial bird species assessed by BirdLife International for the IUCN Red List of Threatened Species, based on 2014 data for a total of 1,308 species. (B) STAR scores for threat abatement for terrestrial birds, by country, as the percentage of the total global STAR threat-abatement score. (C) Last localities for species assessed as Extinct (squares; 131 species), Extinct in the Wild (circles; 4 species), and Critically Endangered (possibly extinct) (triangles; 15 species).

(A) Reprinted with permission from Pimm et al. (2014). (B) Reprinted with permission from Mair et al. (2021). (C) Reprinted with permission from Butchart et al. (2006).

Somveille et al. (2013) restricted consideration to migratory bird species, finding that threatened migrant species have breeding concentrations in Central Asia and non-breeding concentrations in South and Southeast Asia. This is explained by the fact that although there are similar numbers of threatened migrant bird species between the Western and Eastern Hemispheres, those in Asia have larger ranges and so overlap across larger areas. Reed (1999) reviewed other behavioral characteristics associated with elevated extinction risk in birds.

At the intraspecific level, more severely threatened bird species have lower genetic (Evans and Sheldon 2008) and genomic (Brüniche-Olsen et al. 2021) diversity. Considering the extinction risk to avian evolutionary history—through, for example, mapping genera in which all species are threatened—again reveals wide spatial variation, with oceanic island hot spots such as Polynesia–Micronesia, New Zealand, Madagascar and the Indian Ocean Islands, Wallacea, and the Caribbean holding the greatest concentrations (Brooks et al. 2005). Mapping the evolutionary distinctiveness rarity (a measure of how isolated a species is within a phylogenetic tree divided by its range size) yields similar results (Jetz et al. 2014). Although extinction risk facing bird functional diversity has not yet been mapped at a fine geographic resolution, ecologically rare species are disproportionately threatened (Loiseau et al. 2020), and there is again substantial variation across biogeographic realms, with impacts on functional diversity projected to be greatest in the Indo-Malay realm (Toussaint et al. 2021). The loss of birds on islands has had disproportionate impacts on both phylogenetic diversity (Baiser et al. 2018) and functional diversity (Sayol et al. 2021), uncompensated for by subsequently introduced species (Sobral et al. 2016).

Trends in State of the World's Birds

The same sampling limitations that restrict global documentation of bird conservation geography over space also restrict documentation over time (Fraixedas et al. 2020). As such, repeated IUCN Red List assessments provide the basis for the only robust global indicators of the state of the world's birds. The concept of deriving an indicator of aggregate change in extinction risk from repeat Red List assessments was derived nearly three decades ago (Smith et al. 1993), but it was plagued from the outset by the fact that most Red List category changes in between assessments result from changing knowledge rather than actual changing extinction risk (Cuarón 1993). This challenge was finally overcome by Butchart et al. (2004, 2007) in establishing a structure to document reasons for change between assessments, and thus factor out non-genuine change to yield an authoritative Red List Index. In the Red List Index, a value of 0 indicates that all species are EX, a value of 1 indicates that all species are LC, and thus slope indicates the aggregate deterioration in extinction risk over the time periods covered. It is widely applied in policy, for example, to track progress toward Sustainable Development Goal target 15.5 and targets under multilateral environmental agreements such as the Conservation on Biological Diversity (Brooks et al. 2015), and likely to also be used under the emerging Post-2020 Global Biodiversity Framework (B. Williams et al. 2020). The global Red List Index for the world's birds, based on eight full global assessments of all the world's bird species since the first global assessment by Collar and Andrew (1988), shows a persistent deterioration in extinction risk during the past three decades (Birdlife International 2018b; Figure 22.2A).

The major limitation of the Red List Index is its low temporal resolution (Brooks and Kennedy 2004). The IUCN Red List categories comprise broad bands of

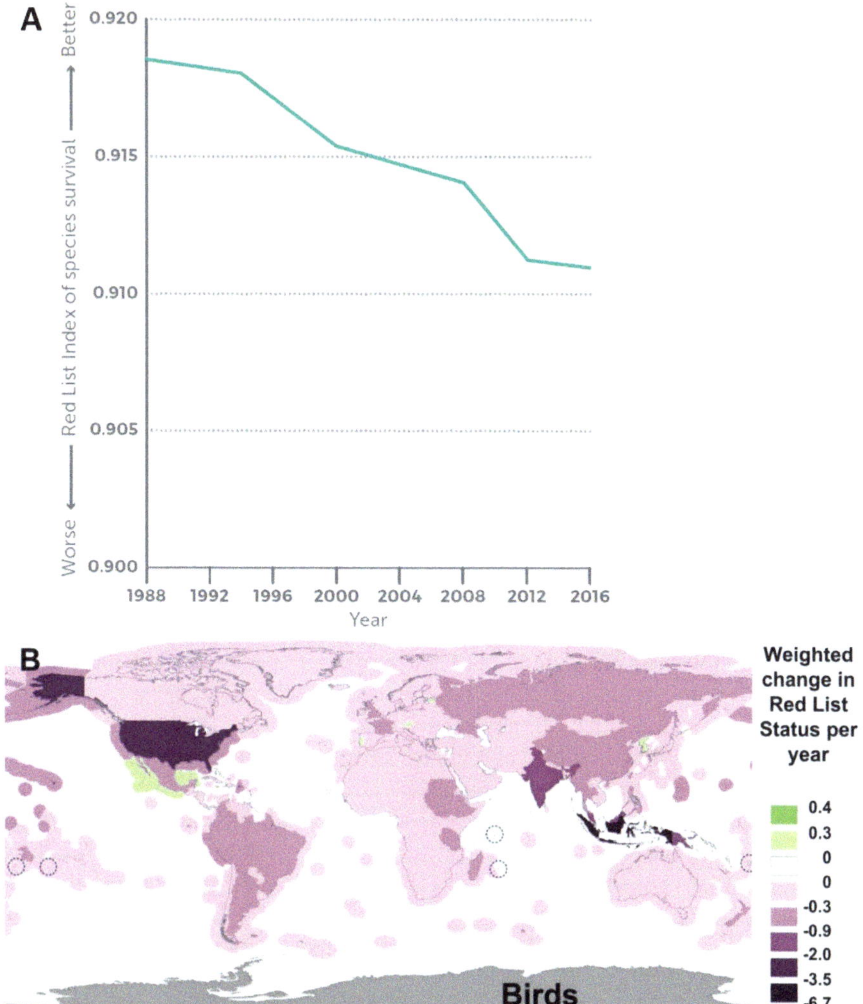

Figure 22.2 (A) The Red List Index for the world's birds, for 10,903 non–Data Deficient, extant species. (B) National disaggregation of the global Red List Index for the world's birds, mapped as the variation across countries in the weighted change in Red List status per year. The Cook Islands, Fiji, Mauritius, the Seychelles, and Tonga are indicated by dashed circles.

(A) Reprinted with permission from BirdLife International (2018a). (B) Reprinted with permission from Rodrigues et al. (2014).

extinction risk, and thus genuine transitions between them are profound events, unlikely to be reflected over short time periods in depauperate biotas. As such, disaggregation of the global Red List Index for birds down to finer spatial resolution (e.g., to countries) yields flat lines for many time periods. Nevertheless, national disaggregations of the Red List Index for birds (Rodrigues et al. 2014; Figure 22.2B) reveal patterns such as rapid deteriorations in Indonesia (largely due to rapid conversion of Sundaic lowland forests) and the United States (largely due to the impacts of invasive species and diseases in Hawai'i) but also improvements in countries such as the Cook Islands, Fiji, Mauritius, Seychelles, and Tonga (resulting from successful conservation action; see the section titled "Conservation Geography of Pressure on the World's Birds").

Do population-level indicators allow insight into the changing state of the world's birds? Gaston et al. (2003) derived a coarse estimate that the total number of individual birds in the world may have been reduced by 20–25% over human history. More recently, the Living Planet Index has been developed, seeking to aggregate unstructured vertebrate population monitoring time series into a global index (Loh et al. 2005). Within this, approximately one-third of both population time series (6,666 of 20,811) and species (1,586 of 4,392) are for birds (World Wildlife Fund 2020), although the index is not taxonomically disaggregated. The great majority of these time series are from Europe and North America, and so the index calculation is weighted based on the total numbers of species within biogeographic realms (McRae et al. 2017). Even with these corrections, however, the Living Planet Index is very sensitive to outliers, with the removal of just 3% of rapidly declining populations apparently switching the overall index from a global decline to an increase (Leung et al. 2020, 2022).

At subglobal levels, however, population indicators of bird status are increasingly robust. The best developed of these are Wild Bird Indices (Gregory et al. 2003). For Europe, these reveal general stability across all species during the past five decades, but with dramatic declines for farmland species (Gregory and van Strien 2010; see the section titled "Biological Resource Use") and an overall decline of 17–19% in overall breeding bird abundance since 1980 (Burns et al. 2021). By contrast, K. Rosenberg et al. (2019) found a 29% overall decline in the North America avifauna since 1970, based largely on Breeding Bird Survey and migration monitoring data. The BioTIME database (https://biotime.st-andrews.ac.uk) includes 41 population time series for birds (as of October 2022), but nearly all of these are from temperate latitudes. No continent-wide bird population indicators are available for the tropics, but systematic local studies provide evidence of some declines (e.g., Şekercioğlu et al. 2019).

Conservation Geography of Pressure on the World's Birds

As revealed in the previous section, the state of birds varies widely throughout the world and over recent decades. This variation is the result of interplay of underlying evolutionary ecology and biogeography (e.g., in determining species' body size,

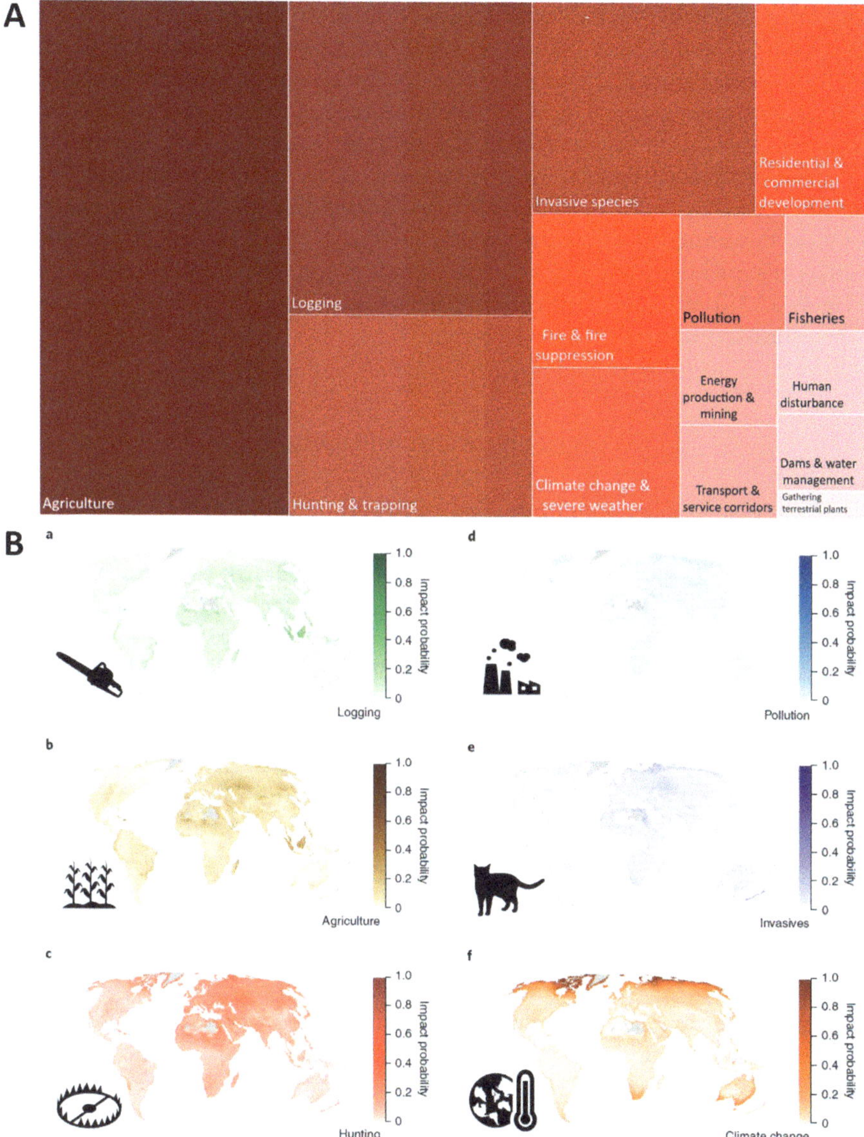

Figure 22.3 (A) Visualization of the relative importance of threats documented to incur medium and high impacts on globally threatened bird species based on the number of species affected; many species are impacted by more than one threat. (B) Probability that a randomly selected bird in a given 50 km^2 cell is impacted by (a) logging, (b) agriculture, (c), hunting, (d) pollution, (e) invasive species, and (f) climate change.

(A) Reprinted with permission from BirdLife International (2021a). (B) Reprinted with permission from Harfoot et al. (2021).

fecundity, and other factors associated with increased vulnerability; Bennett and Owens 1997) and human influences. Given that overall extinction risk of birds is increasing (see Figure 22.2A) and that bird populations are decreasing in many places, it is apparent that many of these human influences represent negative pressures. To better understand these, Salafsky et al. (2008) developed a standard, hierarchical typology of threats to species, which is required documentation in all IUCN Red List assessments (IUCN 2013). BirdLife International (2021a) visualized the prevalence of these threats across all threatened bird species (Figure 22.3A), and Harfoot et al. (2021) mapped the likelihood of impact of six of the most severe threats on birds (Figure 22.3B). What does the conservation geography of the world's birds look like when considering these threats one by one?

Agriculture and Aquaculture

The relationship of agriculture to bird conservation geography is complex, given that, on the one hand, agriculture is the single most prevalent threat to threatened bird species (1,143 species; all numbers from IUCN 2021b), whereas on the other hand, agricultural systems provide important habitat for many "farmland" species (474 species). The major driver of threat from agriculture to birds is the conversion of natural habitats, with regions holding multiple small-ranged bird species disproportionately converted for agriculture in the present, past, and future (Scharlemann et al. 2004). Forestry and agroforestry plantations also typically have negative impacts (Bohada-Murillo et al. 2019). Impacts are particularly severe in Southeast Asia (Figure 22.3B.b), with the conversion of lowland forests to oil palm plantations, for example, threatening numerous species, such as Gurney's Pitta *Pitta gurneyi* CR (Donald et al. 2009). Extinction risk due to agriculture is also severe in the Caribbean, the Andes, the Brazilian Atlantic Forest, eastern Madagascar, and parts of the Sahel and of Central and East Asia (see Figure 22.3B.b). Much of this extinction risk is embodied in global trade (typically from the Global South to the Global North): For example, Irwin et al. (2022) found that for birds, approximately 45% of the STAR scores embodied in trade flows are due to the food and beverage and agricultural sectors alone. Perhaps surprisingly, agriculture does not appear to reduce avian functional diversity (Matuoka et al. 2020), although evidence from Costa Rica reveals reduction in phylogenetic diversity with increasing agricultural intensity (Frishkoff et al. 2014).

Even in regions such as Europe with disproportionately many farmland species, agricultural intensification is driving severe population declines (Donald et al. 2001), at much faster rates than those for forest species in the same region (Gregory et al. 2019). This may suggest that "wildlife-friendly farming" approaches could be more effective in temperate latitudes, and "sustainable intensification" approaches that increase agricultural yields per unit area and thus leave more natural habitat unconverted could be more effective in the tropics (i.e., "land sharing" compared to "land sparing"; Phalan et al. 2011). However, evidence not only from the tropics (Phalan et al. 2011) but also from the United Kingdom (Finch et al. 2020) indicates that land sparing may result in lower impacts on birds for a given level of agricultural production across the board.

Biological Resource Use

Biological resource use encompasses both logging and hunting, as well as fishing and gathering plants (Salafsky et al. 2008), with the first two of these particularly prevalent threats to birds (see Figure 22.3A), impacting 773 and 600 species, respectively (IUCN 2021b), especially in Southeast Asia for logging (Figure 22.3B.a) and in Eurasia and North Africa for hunting (Figure 22.3B.c). In a meta-analysis of impacts of anthropogenic tropical forest disturbances, Gibson et al. (2011) found birds to be the most severely impacted taxon from among all species groups considered. Among ecological groups, forest fragmentation disproportionately impacts insectivores and large frugivores (Bregman et al. 2014). More generally, logging appears to reduce functional diversity of birds (Matuoka et al. 2020).

Hunting encompasses wild bird harvest for pets, food, and a range of other uses, and although extinction risk has been increasing during the past three decades at approximately the same rate for both utilized and non-utilized species, the aggregate extinction risk of the latter is more severe (Butchart 2008), perhaps suggesting that sustainable harvesting can reduce extinction risk (see the section titled "Species Management"). However, much harvest is clearly unsustainable, with Benítez-López et al. (2017), for example, showing through meta-analysis that hunting drives declines of hunted tropical forest birds by 58% in hunted compared to nonhunted areas. Perhaps the most serious impacts of unsustainable harvest, however, are those on individual species perceived as being particularly valuable or prestigious. Examples include the pet trade for Southeast Asian songbirds (Eaton et al. 2015); illegal killing of migratory species in the Mediterranean (Brochet et al. 2016); and, most dramatically, 85–95% declines from 1980 to 2013 in the Yellow-Breasted Bunting *Emberiza aureola* CR due to trapping in China for food (Kamp et al. 2015).

Invasive and Other Problematic Species, Genes, and Diseases

Invasive alien species are another highly prevalent threat to birds (see Figure 22.3A), impacting 584 threatened species (IUCN 2021b). This is especially the case for species with small ranges and species living in island systems (Clavero et al. 2009), such as New Zealand (Figure 22.3B.e). In such systems, invasive species, especially mammalian predators, have also constituted a major driver of bird extinctions (Blackburn et al. 2004). The impacts of cats (Medina et al. 2011) and rodents (Jones et al. 2008) have been particularly severe, exemplified by the intense pressure on the Tristan Albatross *Diomedea dabbenena* CR from introduced House Mice *Mus musculus* on Gough Island (Cuthbert and Hilton 2004). On some islands, invasive diseases are also causing catastrophic impacts on threatened bird species—for example, in the case of avian malaria on Hawai'i (Van Riper III et al. 1986, LaPointe et al. 2012).

Butchart (2008) disaggregated the Red List Index to consider changes in aggregate extinction risk to birds driven by changes in the threat from invasive alien species—either from new invasions and novel impacts or from efforts to control these. He showed that although invasives are indeed a driver of ongoing increases in extinction risk, these deteriorations are less rapid than those driven by other causes, because efforts to control invasive species have themselves had substantial positive impacts (see the section titled "Land and Water Management"). It is also worth noting that a few alien bird species are threatened in their native ranges, and thus their introduced populations provide an important buffer against extinction; Yellow-Crested Cockatoo *Cacatua sulphurea* CR, severely threatened in its native range in Indonesia but increasing as an introduced species in Hong Kong, is an example (Gibson and Yong 2017).

Residential and Commercial Development

Urbanization is a threat of moderate prevalence to bird species (see Figure 22.3A), impacting 405 threatened species worldwide, although only 52 threatened species inhabit urban areas (IUCN 2021b). Across 54 cities, urban areas are documented to hold only 8% of the density (species per square kilometer) of that estimated in equivalent non-urban areas (Aronson et al. 2014). Although urbanization appears to have little effect on many macroecological relationships for birds, cities remove the positive relationships between productivity and species richness and evenness found in natural environments (Pautasso et al. 2011). Urbanization also has negative impacts on bird functional and phylogenetic diversity (Morelli et al. 2016, La Sorte et al. 2018, Matuoka et al. 2020, Sol et al. 2020), but apparently not on genetic diversity (Schmidt et al. 2020). The primary mechanism for these impacts is the loss of habitats for specialist bird species through urbanization (Chace and Walsh 2006), exacerbated by the effects of noise and light pollution (Marín-Gómez and MacGregor-Fors 2021). The changing nature of urbanization also impacts some species, such as the Chimney Swift *Chaetura pelagica* VU, for which declines of 67% from 1970 to 2014 have been at least partially attributed to loss of nesting sites as old buildings and chimneys have been demolished (BirdLife International 2018a).

Natural System Modifications

Fire and fire suppression, dams and water management, and other natural system modifications are another threat of moderate prevalence (see Figure 22.3A), impacting 419 threatened bird species globally (IUCN 2021b). Fire regimes are changing dramatically worldwide under both direct anthropogenic modification and climate change (Krawchuk et al. 2009), with complex and idiosyncratic effects across species, but generally increasing impacts with increasing fire severity (Smucker et al. 2005). Similarly, dam construction and other water management mean that most of the world's rivers are now heavily modified (Nilsson et al. 2005). This imposes direct impacts onto a fairly small number of waterbird species, such as the Indian Skimmer *Rynchops albicollis* EN (Debata et al. 2019), as well as severe impacts on freshwater biodiversity in general (Harrison et al. 2018). It also drives local impacts on terrestrial species through flooding and fragmentation (Nasruddin-Roshidi et al. 2021), which can drive global extinction risk where the species in question have small range sizes in susceptible habitats, such as the Marsh Tapaculo *Scytalopus iraiensis* EN (Vasconcelos et al. 2008).

Climate Change and Severe Weather

In total, 512 threatened bird species are known to be impacted by climate change (IUCN 2021b), rendering this as another threat of intermediate prevalence (see Figure 22.3A), at least over the time frames of 10 years or three generations (whichever is longer) encompassed by the Red List (Trull et al. 2018). As would be anticipated, there is strong geographic patterning in this prevalence, with great concentration at high latitudes (Figure 22.3B.f), likely driving, for example, declines and redistributions in the Steller's Eider *Polysticta stelleri* VU (Aarvak et al. 2013). Climate change is driving a poleward shift of the ranges of north-temperate breeding species overall, consistently across both North America and Europe (Stephens et al. 2016). Similar trends are apparent for wintering assemblages (Princé and Zuckerberg 2015). Models also suggest upslope range shifts (Şekercioğlu et al. 2008), with evidence of such an "escalator to extinction" from the Peruvian Andes (Freeman et al. 2018). Although Important Bird Areas overall (see the section titled "Conservation Geography of Responses to Conserve the World's Birds") may retain their importance under climate change (e.g., see Hole et al. [2009] for Africa continent-wide), they may suffer a net loss of suitability at higher latitudes (e.g., see Coetzee et al. [2009] for South Africa specifically). Climate change also drives phenological changes, with, for example, Pied Flycatchers *Ficedula hypoleuca* LC advancing laying date under warmer temperatures but not able to advance arrival dates on their breeding grounds (Both and Visser 2001). It also decreases migration distance, especially for species breeding in open dry areas (Visser et al. 2009). In the Northern Hemisphere overall, more southerly and resident or short-distance migrant species tend to increase, whereas more northerly and longer distance migrant species decrease (Pearce-Higgins et al. 2015), with, for example, 90% of North American shorebirds predicted to suffer increasing vulnerability under climate change.

Pollution

Despite locally severe impacts, overall pollution is a low-prevalence threat to threatened bird species (see Figure 22.3A), impacting 240 species (IUCN 2021b), with little geographic pattern (see Figure 22.3B.d). The specific drivers are diverse across pollution types, species, and regions. At sea, oil (Wilhelm et al. 2009) and plastics (Wilcox et al. 2015) are perhaps the most prevalent pollution threats impacting birds. On land, pollution includes mercury (Seewagen 2020), pesticides (Li et al. 2020), and light and noise (Senzaki et al. 2020). Incorporation of anthropogenic debris into bird nests is pervasive and increases with human footprint (Jagiello et al. 2019). Perhaps most dramatic, the impact of the anti-inflammatory veterinary drug diclofenac in causing renal failure has driven declines of greater than 95% in three CR South Asian vulture species (Indian *Gyps indicus*, Slender-Billed *G. tenuirostris*, and White-Rumped *G. bengalensis*) since the 1990s (Oaks et al. 2004).

Energy Production and Mining

Energy production and extractive industries are again a rather low-prevalence threat (see Figure 22.3A), impacting 263 threatened bird species overall (IUCN 2021b). Clearly, extractive industries have intense local impacts, which can drive global extinction risk where these coincide with the distributions of small-ranged tropical birds. An example is the Sierra Leone Prinia *Schistolais leontica* EN, for which the two most important sites (Mont Nimba and Pic de Fon, in the Upper Guinea forests of West Africa) are severely threatened by iron ore mining (BirdLife International 2017). There is also growing concern regarding the threat posed by renewable energy development, especially wind farms, given the risk of collision imposed on migrating birds (Marques et al. 2014).

Transportation and Service Corridors

Linear infrastructure, again despite local impacts, has a generally overall low prevalence as a threat to threatened birds (see Figure 22.3A), impacting 223 species globally. In some cases, these can be severe, with the construction of power line corridors, for instance, an important factor threatening the Great Indian Bustard *Ardeotis nigriceps* CR (Collar et al. 2015). At the local level, however, roads certainly drive substantial changes to bird communities: Cooke et al. (2020), for example, found significant impacts up to 700 m alongside roads (an area covering 70% of the country) on populations of 58 of 75 common species in the United Kingdom, with approximately half of these being positive and the other half negative. These impacts are driven by a range of factors, including direct habitat loss; vehicle-caused mortality; chemical, noise, and light pollution; movement barriers; and edge effects (Kociolek et al. 2011).

Human Intrusions and Disturbance

Tourism, warfare, and other incidental human disturbance are a low-prevalence threat to threatened birds (see Figure 22.3A), impacting 190 species globally (IUCN 2021b). Tourism drives threat to 99 of these, with waterbirds and seabirds disproportionately likely to be threatened (Steven and Castley 2013), but also severe impacts on landbirds such as the Grenada Dove *Leptotila wellsi* CR, for which multinational hotel development is a major threat (J. Rosenberg 2018). Warfare, by contrast, is documented as threatening just 15 bird species, perhaps because despite disproportionate likelihood of conflicts occurring within the ranges of threatened birds (Mendiratta et al. 2021), most conflicts during the past 30 years have occurred outside of the specific sites important for birds and biodiversity (IUCN 2021a).

Geological Events

The final threat class documented by Salafsky et al. (2008), comprising natural extreme events including volcanoes, earthquakes, tsunamis, avalanches, and landslides, threats just 30 threatened bird species globally (IUCN 2021b) and so only merits mention to underscore that the vast majority of threats to birds are anthropogenic. A rare example in which geological events are an important contributing factor to bird extinction risk is that of the Montserrat Oriole *Icterus oberi* VU, declining at least partly due to the impact of volcanic eruptions on its tiny range (Dalsgaard et al. 2007).

Conservation Geography of Responses to Conserve the World's Birds

The conservation geography of bird conservation actions as a predictive science is much less advanced than is the conservation geography of their threats (Luther et al. 2016), for example, in considering impacts as binary outcomes rather than examining and seeking to explain change over time and patterns over space. Nevertheless, there is growing recognition of the importance of documenting evidence of the outcomes of conservation interventions (D. Williams et al. 2013) and of assessing impacts relative to counterfactual scenarios (Rodrigues et al. 2006). Thus, for example, Hoffmann et al. (2010) compared the Red List Index 1988–2008 for birds with that anticipated in the absence of conservation interventions, finding that actions slowed the deterioration of survival probability by approximately one-fifth. Disaggregating the Red List Index to national levels reveals a number of small island nations where the impacts of habitat protection and invasive species management have been sufficient to "bend the curve" on extinction risk (Mace et al. 2018), such that they now have positive Red List Indices (Rodrigues et al. 2014; see Figure 22.2B). Bolam et al. (2020) extended the approach, finding that 21–32 bird extinctions have been averted from 1993 to 2020 due to conservation action; in comparison to the 10 extinctions that did occur during this period, this represents a two-thirds to three-fourths reduction in extinction.

However, although conservation has had substantial impact in preventing movement from the Critically Endangered category to the Extinct category, it may have been less effective in slowing deteriorations across other Red List categories (Brooke et al. 2008), and in Europe at least, Wild Bird Indices reveal that the rarest species tend to be increasing while many numerically common species are decreasing (Inger et al. 2015). Akçakaya et al. (2018) developed a framework for systematic assessment of progress toward species recovery, which was applied to 35 bird species by

Grace et al. (2021). Applying such Green Status assessments more broadly will help disentangle the geography of bird conservation impact, as well as support the delivery of a species goal under the Post-2020 Global Biodiversity Framework (B. Williams et al. 2020).

Land and Water Protection

Land and water protection is the most frequently documented conservation action needed for threatened birds, applying to 981 species (66%; IUCN 2021b). As such, it is unsurprising that more attention has been focused on the conservation geography of protected areas for birds than on any other action. For this, the comprehensive identification of more than 13,000 Important Bird and Biodiversity Areas (IBAs), against standard criteria (Donald et al. 2019) and within a comprehensive monitoring framework (Mwangi et al. 2010), has been fundamental both in guiding action and in assessing its effectiveness. IBAs have been validated against formal calculations of irreplaceability (Di Marco et al. 2016), and thence form the subset of Key Biodiversity Areas (KBAs) contributing significantly to the global persistence of bird biodiversity (IUCN 2016, BirdLife International 2021b). Trends over time in the coverage of KBAs by protected areas are used to generate the official indicators for the Sustainable Development Goal targets 14.5, 15.1, and 15.4 (United Nations [UN] 2021). The bird subset of this indicator shows a steady increase in protected area coverage since 1980, but even now only one-fifth of IBAs are fully covered, and mean coverage is less than half (Butchart et al. 2015; Figure 22.4A). For migrants, IBAs are fully protected across all sites required seasonally for only 3% of species (Runge et al. 2015).

Analysis of the IBA data set, especially relative to protected areas data, allows numerous other insights regarding the geography of bird conservation action. Butchart et al. (2012) showed that deteriorations in Red List Indices 1988–2008 were twice as severe for bird species, with more than 50% of their IBAs covered by protected areas, compared to those with 50% or less of their IBAs covered by protected areas (Figure 22.4B). Ricketts et al. (2005) showed that 217 CR and EN species have their global populations restricted to effectively one single site ("Alliance for Zero Extinction" species); by 2018, successful conservation had allowed the removal of 21 species from such assessment (Luther et al. 2021). McCarthy et al. (2012) modeled costs of managing IBAs to estimate that $7.1 billion is invested annually in current protection and that an additional $57.8 billion would be needed to safeguard the entire network.

Evidence for protected area impacts on bird conservation is mixed when considering data sets beyond IBAs. Barnes et al. (2016) used the Living Planet Index data set to examine correlates of population change for 1,068 bird populations within protected areas, finding that protected areas maintained populations overall, with more positive trends for larger-bodied species and for protected areas in countries with higher development scores. At a finer scale, Jellesmark et al. (2021) compared population trends from 1994 to 2018 for 5 species of birds of conservation concern across 47 lowland wet grassland Royal Society for the Protection of Birds reserves in the United Kingdom against counterfactual trends from Breeding Bird Survey observations in equivalent unprotected sites, revealing positive impacts for 4 species.

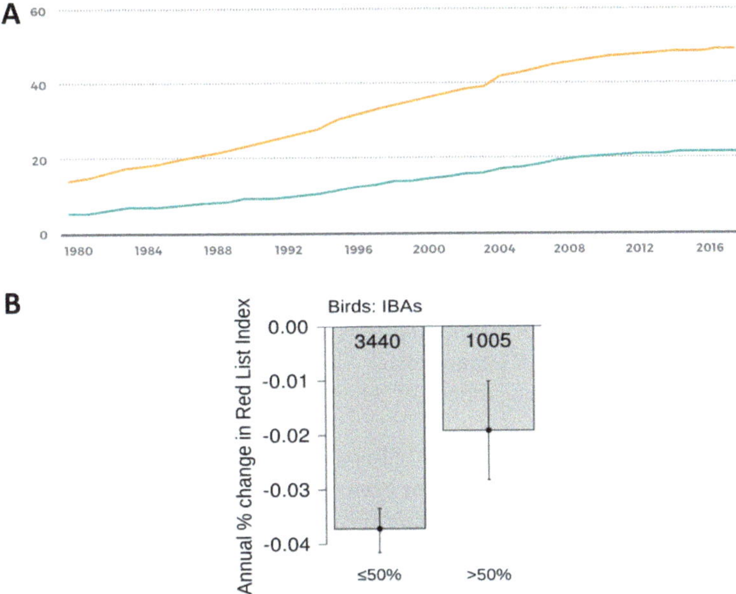

Figure 22.4 (A) Coverage of Important Bird and Biodiversity Areas (IBAs) by protected areas, showing mean percentage of area of IBAs covered by protected areas in orange and percentage of IBAs completely covered in blue. (B) Annual percentage decline in Red List Index 1988–2008 for sets of bird species with ≤50% or >50% of IBAs completely protected. Numbers within each bar refer to the number of species. Error bars show 95% confidence intervals based on uncertainty around the estimated value that is introduced by Data Deficient species.

(A) Reprinted with permission from BirdLife International (2018). (B) Reprinted with permission from Butchart et al. (2012).

There is also evidence that protected areas can be moderately effective in buffering bird communities from climate change, from both France (Gaüzère et al. 2016) and Finland (Santangeli et al. 2017). By contrast, Rayner et al. (2013) found that population trends from 2000 to 2010 for 60 species across 58 woodland reserves in the Australian Capital Territory were not more positive than those in equivalent unprotected woodland. Moreover, comparing mist-netting surveys of the 96-acre Hutcheson Memorial Forest in New Jersey, from 1960 to 1967 with those from 2009 to 2015 revealed that one-third of species had been lost from the protected area during this period (Brown et al. 2019).

Land and Water Management

Land and water management encompasses actions at broad scales as well as within sites, including invasive species management and habitat restoration, and is documented as needed for 875 threatened bird species (59%; IUCN 2021b). Overall, 19% of threatened birds require broadscale management actions, either because they are area-demanding themselves or because they are dependent on broadscale ecological processes (Boyd et al. 2008). Lens et al. (2002) showed that patch occupancy of Kenyan forest fragments was jointly predicted by species mobility and tolerance to habitat deterioration, highlighting the importance of conservation across landscapes as well as within sites.

Regarding invasive species management, Jones et al. (2016) assessed the impact of 251 eradications of invasive mammals from 181 islands, finding that 162 bird species benefitted, whereas only 7 species were negatively impacted, and predicting that at least 67 threatened bird species across 161 populations have likely benefitted. For three of these, impacts have been so positive as to trigger reduction in extinction risk category: Seychelles Magpie-Robin *Copsychus sechellarum* (CR to EN), Cook's Petrel *Pterodroma cookii* (EN to VU), and Black-Vented Shearwater *Puffinus opisthomelas* (VU to NT). Equivalent assessment of other invasive species management impacts has yet to be undertaken.

Given the profile of the UN Decade of Restoration (Mrema et al. 2020), it is perhaps surprising that more attention has not been given to the conservation geography of restoration actions for bird conservation specifically. This could be because bird data have tended to be merged with data for other tetrapod species in relevant analyses (e.g., Luther et al. 2020, Strassburg et al. 2020). Specific consideration of the impacts and implications of restoration actions for bird conservation is a research priority.

Species Management

Single-species management actions are documented as needed for 401 threatened bird species (27%) and are the only conservation actions currently possible for 5 EW bird species (IUCN 2021b). Overall, 257 threatened, EW, and NT species are documented as needing ex situ conservation, with no populations yet in captivity for 8–21% of these (Collar and Butchart 2014). Perhaps the most iconic example is California Condor *Gymnogyps californianus* CR, which was saved from extinction by bringing all individuals into captivity in 1987, and for which reintroduction has now produced a wild population of more than 200 individuals (Walters et al. 2010).

The other major species management subclass concerns in situ management, especially concerning harvest management for sustainable use. Although the literature regarding unsustainable harvest is extensive (see the section titled "Biological Resource Use"), analysis of sustainable use as a bird conservation action is sparse beyond the case of waterfowl hunting in North America (Bolen 2000). Recently, Marsh et al. (2022) found that 77% of bird species documented as being used are assessed as LC on the IUCN Red List, and 43% of used species are LC with stable or increasing population trends, suggesting that sustainable use is indeed quite extensive.

Education and Awareness

Education and awareness are documented as needed for 530 threatened bird species (36%) (IUCN 2021b). A number of studies have assessed the geography of bird conservation awareness from different perspectives. Most comprehensive are a number of analyses of the tools, extent, and utility of birding citizen science (Şekercioğlu 2003). The predominant example is eBird (Sullivan et al. 2009), which is specifically targeted to engage birders in data generation (Wood et al. 2011) and now encompasses more than 60 million checklists from more than 720,000 birders worldwide (https://ebird.org), of which greater than 80% of both checklists and observers are from North America. However, national bird citizen science platforms remain the dominant approach in Europe (https://www.eurobirdportal.org); whether and how integration emerges between these and eBird will be a key question in the global value of bird citizen science over coming years.

Another important dimension of connecting people to birds comes through garden bird feeding, which reaches prevalence of greater than 50% of households in Europe, North America, and Australia, with massive but as yet little explored benefits for awareness, as well as impacts on bird communities (Reynolds et al. 2017). Similarly, visitations to bird reserves and other important sites represent excellent opportunities for assessment of bird conservation education and awareness. Thus, for example, Steven et al. (2013) analyzed the contributions of tourism revenue toward the conservation of CR and EN species, finding that they are highest in Latin America and Africa. Social media represents a powerful emerging tool for exploring visitations, with Hausmann et al. (2019), for example, finding more than 65 million posts from more than 10 million twitter users during 2016 and 2017 from within IBAs. Finally, improved documentation and application of Indigenous and local knowledge of birds (Diamond 1966) serves an important role in strengthening the contributions of traditional land custodians toward conservation overall (Garnett et al. 2018).

Law and Policy

IUCN (2021b) documents species-specific law and policy actions as needed for 399 threatened bird species (27%; nonspecific law and policy interventions are clearly needed for all threatened species), but the impacts of such broadscale actions are notoriously difficult to measure. A rare example of robust evaluation of bird conservation policy impacts comes from the European Union's Birds Directive, for which Donald et al. (2007) showed that species listed on Annex I have more positive trends than do species not on the Annex, since introduction of the policy, and compared to adjacent countries outside of the European Union. IBAs serve an important role in guiding a number of other policy commitments, for example, under the Ramsar Convention on Wetlands and under the Convention on Migratory Species (Waliczky et al. 2019). At a broader scale still, Amano et al. (2018) modeled population time series for 461 waterbird species across 25,769 wetland sites globally, finding national governance effectiveness to be the strongest predictor of positive impacts of conservation efforts. Most comprehensive of all, BirdLife International (2020) used bird conservation data to evaluate progress toward the 20 Aichi Targets of the Strategic Plan for Biodiversity 2011–2020, agreed under the Convention on Biological Diversity, finding that none of its targets had been met in full.

Livelihood, Economic, and Other Incentives

The final category of conservation actions concerns economic incentives and is documented by IUCN (2021b) as needed for 157 threatened birds (11%). Much potential lies in the concept of "other effective area-based conservation measures"—that is, sites with primary objectives other than nature but that nevertheless contribute effectively to the maintenance of the biodiversity for which they are important (Alves-Pinto et al. 2021). Donald et al. (2019) assessed 740 unprotected KBAs across 10 countries, finding that three-fourths of these are likely being managed in ways consistent with this definition. Another promising line of research lies in economic evaluation of the ecosystem services provided by birds (Wenny et al. 2011). Among these, regulating and supporting services include predation, pollination, scavenging, seed dispersal and predation, and ecosystem engineering (Whelan et al. 2008), whereas provisioning services mainly encompass wild harvest (see the sections titled "Biological Resource Use" and "Species Management") and cultural services are manifest as public engagement (see the section titled "Education and Awareness"). Perhaps most profound is the potential to influence flows of capital away from investment in activities and places that would compromise bird conservation; Waliczky et al. (2019) explain how the International Finance Corporation's Performance Standard 6 (http://www.ifc.org/ps6) has advanced this through the inclusion of IBA data (and KBA data overall) in the Integrated Biodiversity Assessment Tool (https://www.ibat-alliance.org).

Conclusion

What general insights emerge from this broad overview of the state of the science of bird conservation geography? First, despite the fact that data for birds are more extensive than for any other taxonomic group, it is clear that severe gaps and constraints remain on the mobilization of underpinning data. There are no robust population bird monitoring programs beyond North America, Europe, Australasia, and South Africa. Citizen science has the potential to fill these gaps, but coverage remains low in the megadiverse tropics, and programs are fragmented between North America and Europe. Although data synthesis through application of standards for assessment of extinction risk and identification of important sites are essential in allowing for global coverage, work to maintain the IUCN Red List and the assessment of KBAs is severely under-resourced, with current investment only approximately half that necessary to maintain these fundamental data sets for bird conservation geography (Juffe-Bignoli et al. 2016).

Thematically, whereas the conservation geography of the state of the world's birds is now quite well-known, despite the large gaps in population time-series data for tropical countries, threats have been less comprehensively analyzed, and the predictive science of conservation actions remains in its infancy, especially beyond protected areas. Application of the Green Status of Species more broadly across birds will help advance this. Particular themes for which knowledge gaps are apparent concern the distribution and impact on birds of ecosystem restoration, sustainable harvest,

Indigenous and local knowledge, garden bird feeding, avitourism, policy instruments, other effective area-based conservation measures, ecosystem service measurement, and financial safeguards.

Bird conservation geography also serves a fundamental role in advancing the frontiers of biodiversity conservation science across the board (Brooks et al. 2008); methods and workflows developed for bird conservation are often later extended to other species groups, when data for these become available. Citizen science is one example, in which advances for birds have triggered the development of platforms for documentation of unstructured observations of species from other taxonomic groups through platforms such as iNaturalist (Unger et al. 2020; http://www.inaturalist.org). Similarly, comprehensive reassessment of extinction risk of all the world's birds has been followed by equivalent reassessment of other taxa, including mammals, amphibians, corals, and cycads (Hoffmann et al. 2010), and identification of IBAs broadened with the development of the global KBA standard to encompass other species groups as well as biodiversity at ecosystem and genetic levels (IUCN 2016).

Finally, there is a crucial urgency to the narrative of hope provided by bird conservation geography, given the daunting cost of failure. That 10 bird species have likely been driven extinct by people over recent decades (Bolam et al. 2020), including one Alagoas Foliage-Gleaner *Philydor novaesi* EN within the past 10 years (Lees et al. 2014), threatens to make the challenge appear overwhelming. Better understanding not just the threats facing the world's birds but also how conservation actions can and do make a difference to abate and reverse these is a fundamental contribution not just to the science but also to the psychology of the conservation endeavor overall.

Acknowledgments

Many thanks to Scott Edwards and Michael Reed for the invitation to contribute this chapter and to an anonymous reviewer for their helpful comments on the manuscript. The views expressed here do not necessarily reflect those of IUCN, and the presentation of geographical entities does not imply any opinion of IUCN concerning the legal status or boundaries of any country or territory.

References

Aarvak, T., I. Jostein Øien, Y. V. Krasnov, M. V. Gavrilo, and A. A. Shavykin. 2013. The European wintering population of Steller's Eider *Polysticta stelleri* reassessed. *Bird Conservation International* 23:337–343.

Akçakaya, H. R., E. L. Bennett, T. M. Brooks, M. K. Grace, A. Heath, S. Hedges, C. Hilton-Taylor, M. Hoffmann, D. A. Keith, B. Long, D. P. Mallon, E. Meijaard, E. J. Milner-Gulland, A. S. L. Rodrigues, J. P. Rodriguez, P. J. Stephenson, S. N. Stuart, and R. P. Young. 2018. Quantifying species recovery and conservation success to develop an IUCN Green List of Species. *Conservation Biology* 32:1128–1138.

Alves-Pinto, H., J. Geldmann, H. Jonas, V. Maioli, A. Balmford, A. Ewa Latawiec, R. Crouzeilles, and B. Strassburg. 2021. Opportunities and challenges of other effective area-based conservation measures (OECMs) for biodiversity conservation. *Perspectives in Ecology and Conservation* 19:115–120.

Amano, T., T. Székely, B. Sandel, S. Nagy, T. Mundkur, T. Langendoen, D. Blanco, C. U. Soykan, and W. J. Sutherland. 2018. Successful conservation of global waterbird populations depends on effective governance. *Nature* 553:199–202.

Aronson, M. F. J., F. A. La Sorte, C. H. Nilon, M. Katti, M. A. Goddard, C. A. Lepczyk, P. S. Warren, N. S. G. Williams, S. Cilliers, B. Clarkson, C. Dobbs, R. Dolan, M. Hedblom, S. Klotz, J. L. Kooijmans, I. Kühn, I. MacGregor-Fors, M. McDonnell, U. Mörtberg, P. Pyšek, S. Siebert, J. Sushinsky, P. Werner, and M. Winter. 2014. A global analysis of the impacts of urbanization on bird and plant diversity reveals key anthropogenic drivers. *Proceedings of the Royal Society B: Biological Sciences* 281: Article 20133330.

Baiser, B., D. Valle, Z. Zelazny, and J. G. Burleigh. 2018. Non-random patterns of invasion and extinction reduce phylogenetic diversity in island bird assemblages. *Ecography* 41:361–374.

Barnes, M. D., I. D. Craigie, L. B. Harrison, J. Geldmann, B. Collen, S. Whitmee, A. Balmford, N. D. Burgess, T. Brooks, M. Hockings, and S. Woodley. 2016. Wildlife population trends in protected areas predicted by national socio-economic metrics and body size. *Nature Communications* 7: Article 12747.

Benítez-López, A., R. Alkemade, A. M. Schipper, D. J. Ingram, P. A. Verweij, J. A. J. Eikelboom, and M. A. J. Huijbregts. 2017. The impact of hunting on tropical mammal and bird populations. *Science* 356:180–183.

Bennett, P. M., and I. P. F. Owens. 1997. Variation in extinction risk among birds: Chance or evolutionary predisposition? *Proceedings of the Royal Society B: Biological Sciences* 264:401–408.

Bibby, C. J. 1999. Making the most of birds as environmental indicators. *Ostrich* 70:81–88.

Bibby, C. J., N. D. Burgess, D. M. Hillis, D. A. Hill, and S. Mustoe. 2000. *Bird Census Techniques.* Academic Press, London.

Bildstein, K. L. 2006. *Migrating Raptors of the World: Their Ecology & Conservation.* Cornell University Press, Ithaca, NY.

BirdLife International. 2017. *Schistolais leontica*: IUCN Red List of Threatened Species. International Union for Conservation of Nature.

BirdLife International. 2018a. *Chaetura pelagica*: IUCN Red List of Threatened Species. International Union for Conservation of Nature.

Birdlife International. 2018b. *State of the World's Birds: Taking the Pulse of the Planet.* BirdLife International, Cambridge, UK.

BirdLife International. 2019. *Bird Species Distribution Maps of the World: Version 2019.1.* BirdLife International, Cambridge, UK.

BirdLife International. 2020. *Birds and Biodiversity Targets.* BirdLife International, Cambridge, UK.

BirdLife International. 2021a. *State of the World's Birds: 2021 Annual Update.* BirdLife International, Cambridge, UK.

BirdLife International. 2021b. The World Database of Key Biodiversity Areas. Developed by the KBA Partnership: BirdLife International, International Union for the Conservation of Nature, Amphibian Survival Alliance, Conservation International, Critical Ecosystem Partnership Fund, Global Environment Facility, Global Wildlife Conservation, NatureServe, Rainforest Trust, Royal Society for the Protection of Birds, Wildlife Conservation Society and World Wildlife Fund.

Blackburn, T. M., P. Cassey, R. P. Duncan, K. L. Evans, and K. J. Gaston. 2004. Avian extinction and mammalian introductions on oceanic islands. *Science* 305:1955–1958.

Boere, G. C., C. A. Galbraith, and D. A. Stroud. 2006. *Waterbirds Around the World.* The Stationery Office, London.

Bohada-Murillo, M., G. J. Castaño-Villa, and F. E. Fontúrbel. 2019. The effects of forestry and agroforestry plantations on bird diversity: A global synthesis. *Land Degradation & Development* 31:646–654.

Bolam, F. C., L. Mair, M. Angelico, T. M. Brooks, M. Burgman, C. Hermes, M. Hoffmann, R. W. Martin, P. J. K. McGowan, A. S. L. Rodrigues, C. Rondinini, J. R. S. Westrip, H. Wheatley, Y. Bedolla-Guzmán, J. Calzada, M. F. Child, P. A. Cranswick, C. R. Dickman, B. Fessl, D. O. Fisher,

S. T. Garnett, J. J. Groombridge, C. N. Johnson, R. J. Kennerley, S. R. B. King, J. F. Lamoreux, A. C. Lees, L. Lens, S. P. Mahood, D. P. Mallon, E. Meijaard, F. Méndez-Sánchez, A. R. Percequillo, T. J. Regan, L. M. Renjifo, M. C. Rivers, N. S. Roach, L. Roxburgh, R. J. Safford, P. Salaman, T. Squires, E. Vázquez-Domínguez, P. Visconti, J. C. Z. Woinarski, R. P. Young, and S. H. M. Butchart. 2020. How many bird and mammal extinctions has recent conservation action prevented? *Conservation Letters* 14: Article e12762.

Bolen, E. G. 2000. Waterfowl management: Yesterday and tomorrow. *Journal of Wildlife Management* 64:323–335.

Both, C., and M. E. Visser. 2001. Adjustment to climate change is constrained by arrival date in a long-distance migrant bird. *Nature* 411:296–298.

Boyd, C., T. M. Brooks, S. H. M. Butchart, G. J. Edgar, G. A. B. Da Fonseca, F. Hawkins, M. Hoffmann, W. Sechrest, S. N. Stuart, and P. P. Van Dijk. 2008. Spatial scale and the conservation of threatened species. *Conservation Letters* 1:37–43.

Bregman, T. P., C. H. Sekercioglu, and J. A. Tobias. 2014. Global patterns and predictors of bird species responses to forest fragmentation: Implications for ecosystem function and conservation. *Biological Conservation* 169:372–383.

Brlík, V., E. Šilarová, J. Škorpilová, H. Alonso, M. Anton, A. Aunins, Z. Benkö, G. Biver, M. Busch, T. Chodkiewicz, P. Chylarecki, D. Coombes, E. de Carli, J. C. Del Moral, A. Derouaux, V. Escandell, D. P. Eskildsen, B. Fontaine, R. P. B. Foppen, A. Gamero, R. D. Gregory, S. Harris, S. Herrando, I. Hristov, M. Husby, C. Ieronymidou, F. Jiquet, J. A. Kålås, J. Kamp, P. Kmecl, P. Kurlavičius, A. Lehikoinen, L. Lewis, Å. Lindström, A. Manolopoulos, D. Martí, D. Massimino, C. Moshøj, R. Nellis, D. Noble, A. Paquet, J.-Y. Paquet, D. Portolou, I. Ramírez, C. Redel, J. Reif, J. Ridzoň, H. Schmid, B. Seaman, L. Silva, L. Soldaat, S. Spasov, A. Staneva, T. Szép, G. T. Florenzano, N. Teufelbauer, S. Trautmann, T. van der Meij, A. van Strien, C. van Turnhout, G. Vermeersch, Z. Vermouzek, T. Vikstrøm, P. Voříšek, A. Weiserbs, and A. Klvaňová. 2021. Long-term and large-scale multi-species dataset tracking population changes of common European breeding birds. *Scientific Data* 8: Article 21.

Brochet, A.-L., W. Van Den Bossche, S. Jbour, P. K. Ndang'Ang'A, V. R. Jones, W. A. L. I. Abdou, A. R. Al- Hmoud, N. G. Asswad, J. C. Atienza, I. Atrash, N. Barbara, K. Bensusan, T. Bino, C. Celada, S. I. Cherkaoui, J. Costa, B. Deceuninck, K. S. Etayeb, C. Feltrup-Azafzaf, J. Figelj, M. Gustin, P. Kmecl, V. Kocevski, M. Korbeti, D. Kotrošan, J. Mula Laguna, M. Lattuada, D. Leitão, P. Lopes, N. López-jiménez, V. Lucić, T. Micol, A. Moali, Y. Perlman, N. Piludu, D. Portolou, K. Putilin, G. Quaintenne, G. Ramadan-Jaradi, M. Ružić, A. Sandor, N. Sarajli, D. Saveljić, R. D. Sheldon, T. Shialis, N. Tsiopelas, F. Vargas, C. Thompson, A. Brunner, R. Grimmett, and S. H. M. Butchart. 2016. Preliminary assessment of the scope and scale of illegal killing and taking of birds in the Mediterranean. *Bird Conservation International* 26:1–28.

Brooke, M. d. L., S. H. M. Butchart, S. T. Garnett, G. M. Crowley, N. B. Mantilla-Beniers, and A. J. Stattersfield. 2008. Rates of movement of threatened bird species between IUCN Red List categories and toward extinction. *Conservation Biology* 22:417–427.

Brooks, T. 2010. Conservation planning and priorities. Pages 199–219 in N. S. Sodhi and P. R. Ehrlich eds. *Conservation Biology for All.* Oxford University Press, New York.

Brooks, T., and E. Kennedy. 2004. Biodiversity barometers. *Nature* 431:1046–1047.

Brooks, T. M., S. H. M. Butchart, N. A. Cox, M. Heath, C. Hilton-Taylor, M. Hoffmann, N. Kingston, J. P. Rodríguez, S. N. Stuart, and J. Smart. 2015. Harnessing biodiversity and conservation knowledge products to track the Aichi Targets and Sustainable Development Goals. *Biodiversity* 16: 157–174.

Brooks, T. M., N. J. Collar, R. E. Green, S. J. Marsden, and D. J. Pain. 2008. The science of bird conservation. *Bird Conservation International* 18:S2–S12.

Brooks, T. M., J. D. Pilgrim, A. S. L. Rodrigues, and G. A. B. Fonseca. 2005. Conservation status and geographic distribution of avian evolutionary history. Pages 267–294 in A. Purvis, J. L. Gittleman, and T. M. Brooks, eds. *Phylogeny and Conservation.* Cambridge University Press, Cambridge, UK.

Brooks, T. M., S. L. Pimm, H. R. Akçakaya, G. M. Buchanan, S. H. M. Butchart, W. Foden, C. Hilton-Taylor, M. Hoffmann, C. N. Jenkins, L. Joppa, B. V. Li, V. Menon, N. Ocampo-Peñuela, and C. Rondinini. 2019. Measuring terrestrial area of habitat (AOH) and its utility for the IUCN Red List. *Trends in Ecology & Evolution* 34:977–986.

Brown, J. A., J. L. Lockwood, J. D. Avery, J. Curtis Burkhalter, K. Aagaard, and K. H. Fenn. 2019. Evaluating the long-term effectiveness of terrestrial protected areas: A 40-year look at forest bird diversity. *Biodiversity and Conservation* 28:811–826.

Brüniche-Olsen, A., K. F. Kellner, J. L. Belant, and J. A. DeWoody. 2021. Life-history traits and habitat availability shape genomic diversity in birds: Implications for conservation. *Proceedings of the Royal Society B: Biological Sciences* 288: Article 20211441.

Burns, F., M. A. Eaton, I. J. Burfield, A. Klvaňová, E. Šilarová, A. Staneva, and R. Gregory. 2021. Abundance decline in the avifauna of the European Union conceals complex patterns of biodiversity change. Authorea.

Butchart, S. H. M. 2008. Red List Indices to measure the sustainability of species use and impacts of invasive alien species. *Bird Conservation International* 18:S245–S262.

Butchart, S. H. M., and J. P. Bird. 2010. Data Deficient birds on the IUCN Red List: What don't we know and why does it matter? *Biological Conservation* 143:239–247.

Butchart, S. H. M., M. Clarke, R. J. Smith, R. E. Sykes, J. P. W. Scharlemann, M. Harfoot, G. M. Buchanan, A. Angulo, A. Balmford, B. Bertzky, T. M. Brooks, K. E. Carpenter, M. T. Comeros-Raynal, J. Cornell, G. F. Ficetola, L. D. C. Fishpool, R. A. Fuller, J. Geldmann, H. Harwell, C. Hilton-Taylor, M. Hoffmann, A. Joolia, L. Joppa, N. Kingston, I. May, A. Milam, B. Polidoro, G. Ralph, N. Richman, C. Rondinini, D. B. Segan, B. Skolnik, M. D. Spalding, S. N. Stuart, A. Symes, J. Taylor, P. Visconti, J. E. M. Watson, L. Wood, and N. D. Burgess. 2015. Shortfalls and solutions for meeting national and global conservation area targets. *Conservation Letters* 8:329–337.

Butchart, S. H. M., H. Resit Akçakaya, J. Chanson, J. E. M. Baillie, B. Collen, S. Quader, W. R. Turner, R. Amin, S. N. Stuart, and C. Hilton-Taylor. 2007. Improvements to the Red List Index. *PLoS One* 2: Article e140.

Butchart, S. H. M., J. P. W. Scharlemann, M. I. Evans, S. Quader, S. Aricò, J. Arinaitwe, M. Balman, L. A. Bennun, B. Bertzky, C. Besançon, T. M. Boucher, T. M. Brooks, I. J. Burfield, N. D. Burgess, S. Chan, R. P. Clay, M. J. Crosby, N. C. Davidson, N. De Silva, C. Devenish, G. C. L. Dutson, D. F. D. Z. Fernández, L. D. C. Fishpool, C. Fitzgerald, M. Foster, M. F. Heath, M. Hockings, M. Hoffmann, D. Knox, F. W. Larsen, J. F. Lamoreux, C. Loucks, I. May, J. Millett, D. Molloy, P. Morling, M. Parr, T. H. Ricketts, N. Seddon, B. Skolnik, S. N. Stuart, A. Upgren, and S. Woodley. 2012. Protecting important sites for biodiversity contributes to meeting global conservation targets. *PLoS One* 7: Article e32529.

Butchart, S. H. M., A. J. Stattersfield, L. A. Bennun, S. M. Shutes, H. R. Akçakaya, J. E. M. Baillie, S. N. Stuart, C. Hilton-Taylor, and G. M. Mace. 2004. Measuring global trends in the status of biodiversity: Red List Indices for birds. *PLoS Biology* 2: Article e383.

Butchart, S. H. M., A. J. Stattersfield, and T. M. Brooks. 2006. Going or gone: Defining "Possibly Extinct" species to give a truer picture of recent extinctions. *Bulletin of the British Ornithologists' Club* 126A:7–24.

Callaghan, C. T., S. Nakagawa, and W. K. Cornwell. 2021. Global abundance estimates for 9,700 bird species. *Proceedings of the National Academy of Sciences of the USA* 118: Article e2023170118.

Chace, J. F., and J. J. Walsh. 2006. Urban effects on native avifauna: A review. *Landscape and Urban Planning* 74:46–69.

Clavero, M., L. Brotons, P. Pons, and D. Sol. 2009. Prominent role of invasive species in avian biodiversity loss. *Biological Conservation* 142:2043–2049.

Coetzee, B. W. T., M. P. Robertson, B. F. N. Erasmus, B. J. van Rensburg, and W. Thuiller. 2009. Ensemble models predict Important Bird Areas in southern Africa will become less effective for conserving endemic birds under climate change. *Global Ecology and Biogeography* 18:701–710.

Collar, N., and S. Butchart. 2014. Conservation breeding and avian diversity: Chances and challenges. *International Zoo Yearbook* 48:7–28.

Collar, N., P. Patil, and G. Bhardwaj. 2015. What can save the Great Indian Bustard? *BirdingAsia* 23:15–24.

Collar, N. J., and P. Andrew. 1988. *Birds to Watch.* International Council for Bird Preservation, Cambridge, UK.

Cooke, S. C., A. Balmford, P. F. Donald, S. E. Newson, and A. Johnston. 2020. Roads as a contributor to landscape-scale variation in bird communities. *Nature Communications* 11: Article 3125.

Croxall, J. P., P. G. H. Evans, and E. A. Schreiber. 1984. *Status and Conservation of the World's Seabirds*. Paston Press, UK.

Cuarón, A. D. 1993. Extinction rate estimates. *Nature* 366:118.

Cuthbert, R., and G. Hilton. 2004. Introduced house mice *Mus musculus*: A significant predator of threatened and endemic birds on Gough Island, South Atlantic Ocean? *Biological Conservation* 117:483–489.

Dalsgaard, B., G. M. Hilton, G. A. L. Gray, L. Aymer, J. Boatswain, J. Daley, C. Fenton, J. Martin, L. Martin, P. Murrain, W. J. Arendt, D. W. Gibbons, and J. M. Olesen. 2007. Impacts of a volcanic eruption on the forest bird community of Montserrat, Lesser Antilles . *Ibis* 149:298–312.

Debata, S., T. Kar, H. S. Palei, and K. K. Swain. 2019. Breeding ecology and causes of nest failure in the Indian Skimmer *Rynchops albicollis*. *Bird Study* 66:243–250.

Di Marco, M., T. Brooks, A. Cuttelod, L. D. C. Fishpool, C. Rondinini, R. J. Smith, L. Bennun, S. H. M. Butchart, S. Ferrier, R. P. B. Foppen, L. Joppa, D. Juffe-Bignoli, A. T. Knight, J. F. Lamoreux, P. F. Langhammer, I. May, H. P. Possingham, P. Visconti, J. E. M. Watson, and S. Woodley. 2016. Quantifying the relative irreplaceability of important bird and biodiversity areas. *Conservation Biology* 30:392–402.

Di Minin, E., R. A. Correia, and T. Toivonen. 2022. Quantitative conservation geography. *Trends in Ecology & Evolution* 37:42–52.

Diamond, J. M. 1966. Zoological classification system of a primitive people. *Science* 151:1102–1104.

Donald, P. F., S. Aratrakorn, T. Win Htun, J. C. Eames, H. Hla, S. Thunhikorn, K. Sribua-Rod, P. Tinun, S. M. Aung, S. M. Zaw, and G. M. Buchanan. 2009. Population, distribution, habitat use and breeding of Gurney's Pitta *Pitta gurneyi* in Myanmar and Thailand. *Bird Conservation International* 19:353–366.

Donald, P. F., L. D. C. Fishpool, A. Ajagbe, L. A. Bennun, G. Bunting, I. J. Burfield, S. H. M. Butchart, S. Capellan, M. J. Crosby, M. P. Dias, D. Diaz, M. I. Evans, R. Grimmett, M. Heath, V. R. Jones, B. G. Lascelles, J. C. Merriman, M. O'Brien, I. RamÍRez, Z. Waliczky, and D. C. Wege. 2019. Important Bird and Biodiversity Areas (IBAs): The development and characteristics of a global inventory of key sites for biodiversity. *Bird Conservation International* 29:177–198.

Donald, P. F., and R. J. Fuller. 1998. Ornithological atlas data: A review of uses and limitations. *Bird Study* 45:129–145.

Donald, P. F., R. Green, and M. Heath. 2001. Agricultural intensification and the collapse of Europe's farmland bird populations. *Proceedings of the Royal Society B: Biological Sciences* 268:25–29.

Donald, P. F., F. J. Sanderson, I. J. Burfield, S. M. Bierman, R. D. Gregory, and Z. Waliczky. 2007. International conservation policy delivers benefits for birds in Europe. *Science* 317:810–813.

Eaton, J., C. Shepherd, F. Rheindt, J. Harris, S. Van Balen, D. Wilcove, and N. Collar. 2015. Trade-driven extinctions and near-extinctions of avian taxa in Sundaic Indonesia. *Forktail* 31:1–12.

Evans, S. R., and B. C. Sheldon. 2008. Interspecific patterns of genetic diversity in birds: Correlations with extinction risk. *Conservation Biology* 22:1016–1025.

Finch, T., B. H. Day, D. Massimino, J. W. Redhead, R. H. Field, A. Balmford, R. E. Green, and W. J. Peach. 2020. Evaluating spatially explicit sharing-sparing scenarios for multiple environmental outcomes. *Journal of Applied Ecology* 58:655–666.

Fraixedas, S., A. Lindén, M. Piha, M. Cabeza, R. Gregory, and A. Lehikoinen. 2020. A state-of-the-art review on birds as indicators of biodiversity: Advances, challenges, and future directions. *Ecological Indicators* 118: Article 106728.

Freeman, B. G., M. N. Scholer, V. Ruiz-Gutierrez, and J. W. Fitzpatrick. 2018. Climate change causes upslope shifts and mountaintop extirpations in a tropical bird community. *Proceedings of the National Academy of Sciences of the USA* 115:11982–11987.

Frishkoff, L. O., D. S. Karp, L. K. M'Gonigle, C. D. Mendenhall, J. Zook, C. Kremen, E. A. Hadly, and G. C. Daily. 2014. Loss of avian phylogenetic diversity in neotropical agricultural systems. *Science* 345:1343–1346.

Garnett, S. T., N. D. Burgess, J. E. Fa, Á. Fernández-Llamazares, Z. Molnár, C. J. Robinson, J. E. M. Watson, K. K. Zander, B. Austin, E. S. Brondizio, N. F. Collier, T. Duncan, E. Ellis, H. Geyle, M. V. Jackson, H. Jonas, P. Malmer, B. McGowan, A. Sivongxay, and I. Leiper. 2018. A spatial overview of the global importance of Indigenous lands for conservation. *Nature Sustainability* 1:369–374.

Gaston, K. J., and T. M. Blackburn. 1997. How many birds are there? *Biodiversity and Conservation* 6:615–625.
Gaston, K. J., T. M. Blackburn, and K. Klein Goldewijk. 2003. Habitat conversion and global avian biodiversity loss. *Proceedings of the Royal Society B: Biological Sciences* 270:1293–1300.
Gaüzère, P., F. Jiguet, and V. Devictor. 2016. Can protected areas mitigate the impacts of climate change on bird's species and communities? *Diversity and Distributions* 22:625–637.
Gibson, L., T. M. Lee, L. P. Koh, B. W. Brook, T. A. Gardner, J. Barlow, C. A. Peres, C. J. A. Bradshaw, W. F. Laurance, T. E. Lovejoy, and N. S. Sodhi. 2011. Primary forests are irreplaceable for sustaining tropical biodiversity. *Nature* 478:378–381.
Gibson, L., and D. L. Yong. 2017. Saving two birds with one stone: Solving the quandary of introduced, threatened species. *Frontiers in Ecology and the Environment* 15:35–41.
Grace, M. K., H. R. Akçakaya, E. L. Bennett, T. M. Brooks, A. Heath, S. Hedges, C. Hilton-Taylor, M. Hoffmann, A. Hochkirch, R. Jenkins, D. A. Keith, B. Long, D. P. Mallon, E. Meijaard, E. J. Milner-Gulland, J. P. Rodriguez, P. J. Stephenson, S. N. Stuart, R. P. Young, P. Acebes, J. Alfaro-Shigueto, S. Alvarez-Clare, R. R. Andriantsimanarilafy, M. Arbetman, C. Azat, G. Bacchetta, R. Badola, L. M. D. Barcelos, J. P. Barreiros, S. Basak, D. J. Berger, S. Bhattacharyya, G. Bino, P. A. V. Borges, R. K. Boughton, H. J. Brockmann, H. L. Buckley, I. J. Burfield, J. Burton, T. Camacho-Badani, L. S. Cano-Alonso, R. H. Carmichael, C. Carrero, J. P. Carroll, G. Catsadorakis, D. G. Chapple, G. Chapron, G. W. Chowdhury, L. Claassens, D. Cogoni, R. Constantine, C. A. Craig, A. A. Cunningham, N. Dahal, J. C. Daltry, G. C. Das, N. Dasgupta, A. Davey, K. Davies, P. Develey, V. Elangovan, D. Fairclough, M. D. Febbraro, G. Fenu, F. M. Fernandes, E. P. Fernandez, B. Finucci, R. Földesi, C. M. Foley, M. Ford, M. R. J. Forstner, N. García, R. Garcia-Sandoval, P. C. Gardner, R. Garibay-Orijel, M. Gatan-Balbas, I. Gauto, M. G. U. Ghazi, S. S. Godfrey, M. Gollock, B. A. González, T. D. Grant, T. Gray, A. J. Gregory, R. H. A. van Grunsven, M. Gryzenhout, N. C. Guernsey, G. Gupta, C. Hagen, C. A. Hagen, M. B. Hall, E. Hallerman, K. Hare, T. Hart, R. Hartdegen, Y. Harvey-Brown, R. Hatfield, T. Hawke, C. Hermes, R. Hitchmough, P. M. Hoffmann, C. Howarth, M. A. Hudson, S. A. Hussain, C. Huveneers, H. Jacques, D. Jorgensen, S. Katdare, L. K. D. Katsis, R. Kaul, B. Kaunda-Arara, L. Keith-Diagne, D. T. Kraus, T. M. de Lima, K. Lindeman, J. Linsky, E. Louis, A. Loy, E. N. Lughadha, J. C. Mangel, P. E. Marinari, G. M. Martin, G. Martinelli, P. J. K. McGowan, A. McInnes, E. Teles Barbosa Mendes, M. J. Millard, C. Mirande, D. Money, J. M. Monks, C. L. Morales, N. N. Mumu, R. Negrao, A. H. Nguyen, M. N. H. Niloy, G. L. Norbury, C. Nordmeyer, D. Norris, M. O'Brien, G. A. Oda, S. Orsenigo, M. E. Outerbridge, S. Pasachnik, J. C. Pérez-Jiménez, C. Pike, F. Pilkington, G. Plumb, R. d. C. Q. Portela, A. Prohaska, M. G. Quintana, E. F. Rakotondrasoa, D. H. Ranglack, H. Rankou, A. P. Rawat, J. T. Reardon, M. L. Rheingantz, S. C. Richter, M. C. Rivers, L. R. Rogers, P. da Rosa, P. Rose, E. Royer, C. Ryan, Y. J. S. de Mitcheson, L. Salmon, C. H. Salvador, M. J. Samways, T. Sanjuan, A. Souza Dos Santos, H. Sasaki, E. Schutz, H. A. Scott, R. M. Scott, F. Serena, S. P. Sharma, J. A. Shuey, C. J. P. Silva, J. P. Simaika, D. R. Smith, J. L. Y. Spaet, S. Sultana, B. K. Talukdar, V. Tatayah, P. Thomas, A. Tringali, H. Trinh-Dinh, C. Tuboi, A. A. Usmani, A. M. Vasco-Palacios, J. C. Vié, E. Virens, A. Walker, B. Wallace, L. J. Waller, H. Wang, O. R. Wearn, M. van Weerd, S. Weigmann, D. Willcox, J. Woinarski, J. W. H. Yong, and S. Young. 2021. Testing a global standard for quantifying species recovery and assessing conservation impact. *Conservation Biology* 35:1833–1849.
Gregory, R. D., D. Noble, R. Field, J. Marchant, M. Raven, and D. Gibbons. 2003. Using birds as indicators of biodiversity. *Ornis Hungarica* 12:11–24.
Gregory, R. D., J. Skorpilova, P. Vorisek, and S. Butler. 2019. An analysis of trends, uncertainty and species selection shows contrasting trends of widespread forest and farmland birds in Europe. *Ecological Indicators* 103:676–687.
Gregory, R. D., and A. van Strien. 2010. Wild bird indicators: Using composite population trends of birds as measures of environmental health. *Ornithological Science* 9:3–22.
Harfoot, M. B. J., A. Johnston, A. Balmford, N. D. Burgess, S. H. M. Butchart, M. P. Dias, C. Hazin, C. Hilton-Taylor, M. Hoffmann, N. J. B. Isaac, L. L. Iversen, C. L. Outhwaite, P. Visconti, and J. Geldmann. 2021. Using the IUCN Red List to map threats to terrestrial vertebrates at global scale. *Nature Ecology & Evolution* 5:1510–1519.
Harrison, I., R. Abell, W. Darwall, M. L. Thieme, D. Tickner, and I. Timboe. 2018. The freshwater biodiversity crisis. *Science* 362: Article 1369.

Hausmann, A., T. Toivonen, C. Fink, V. Heikinheimo, H. Tenkanen, S. H. M. Butchart, T. M. Brooks, and E. Di Minin. 2019. Assessing global popularity and threats to Important Bird and Biodiversity Areas using social media data. *Science of the Total Environment* 683:617–623.
Hilton-Taylor, C. 2014. A history of the IUCN Red List. Pages 9–27 in J. Smart, C. Hilton-Taylor, and R. A. Mittermeier, eds. *The IUCN Red List: 50 Years of Conservation.* Earth in Focus, Washington, DC.
Hoffmann, M., C. Hilton-Taylor, A. Angulo, M. Böhm, T. M. Brooks, S. H. M. Butchart, K. E. Carpenter, J. Chanson, B. Collen, N. A. Cox, W. R. T. Darwall, N. K. Dulvy, L. R. Harrison, V. Katariya, C. M. Pollock, S. Quader, N. I. Richman, A. S. L. Rodrigues, M. F. Tognelli, J.-C. Vié, J. M. Aguiar, D. J. Allen, G. R. Allen, G. Amori, N. B. Ananjeva, F. Andreone, P. Andrew, A. L. A. Ortiz, J. E. M. Baillie, R. Baldi, B. D. Bell, S. D. Biju, J. P. Bird, P. Black-Decima, J. J. Blanc, F. Bolaños, W. Bolivar-G, I. J. Burfield, J. A. Burton, D. R. Capper, F. Castro, G. Catullo, R. D. Cavanagh, A. Channing, N. L. Chao, A. M. Chenery, F. Chiozza, V. Clausnitzer, N. J. Collar, L. C. Collett, B. B. Collette, C. F. C. Fernandez, M. T. Craig, M. J. Crosby, N. Cumberlidge, A. Cuttelod, A. E. Derocher, A. C. Diesmos, J. S. Donaldson, J. W. Duckworth, G. Dutson, S. K. Dutta, R. H. Emslie, A. Farjon, S. Fowler, J. Freyhof, D. L. Garshelis, J. Gerlach, D. J. Gower, T. D. Grant, G. A. Hammerson, R. B. Harris, L. R. Heaney, S. B. Hedges, J.-M. Hero, B. Hughes, S. A. Hussain, J. Icochea M, R. F. Inger, N. Ishii, D. T. Iskandar, R. K. B. Jenkins, Y. Kaneko, M. Kottelat, K. M. Kovacs, S. L. Kuzmin, E. La Marca, J. F. Lamoreux, M. W. N. Lau, E. O. Lavilla, K. Leus, R. L. Lewison, G. Lichtenstein, S. R. Livingstone, V. Lukoschek, D. P. Mallon, P. J. K. McGowan, A. McIvor, P. D. Moehlman, S. Molur, A. M. Alonso, J. A. Musick, K. Nowell, R. A. Nussbaum, W. Olech, N. L. Orlov, T. J. Papenfuss, G. Parra-Olea, W. F. Perrin, B. A. Polidoro, M. Pourkazemi, P. A. Racey, J. S. Ragle, M. Ram, G. Rathbun, R. P. Reynolds, A. G. J. Rhodin, S. J. Richards, L. O. Rodríguez, S. R. Ron, C. Rondinini, A. B. Rylands, Y. Sadovy de Mitcheson, J. C. Sanciangco, K. L. Sanders, G. Santos-Barrera, J. Schipper, C. Self-Sullivan, Y. Shi, A. Shoemaker, F. T. Short, C. Sillero-Zubiri, D. L. Silvano, K. G. Smith, A. T. Smith, J. Snoeks, A. J. Stattersfield, A. J. Symes, A. B. Taber, B. K. Talukdar, H. J. Temple, R. Timmins, J. A. Tobias, K. Tsytsulina, D. Tweddle, C. Ubeda, S. V. Valenti, P. Paul van Dijk, L. M. Veiga, A. Veloso, D. C. Wege, M. Wilkinson, E. A. Williamson, F. Xie, B. E. Young, H. R. Akçakaya, L. Bennun, T. M. Blackburn, L. Boitani, H. T. Dublin, G. A. B. da Fonseca, C. Gascon, T. E. Lacher, G. M. Mace, S. A. Mainka, J. A. McNeely, R. A. Mittermeier, G. M. Reid, J. P. Rodriguez, A. A. Rosenberg, M. J. Samways, J. Smart, B. A. Stein, and S. N. Stuart. 2010. The impact of conservation on the status of the world's vertebrates. *Science* 330:1503–1509.
Hole, D. G., S. G. Willis, D. J. Pain, L. D. Fishpool, S. H. M. Butchart, Y. C. Collingham, C. Rahbek, and B. Huntley. 2009. Projected impacts of climate change on a continent-wide protected area network. *Ecology Letters* 12:420–431.
Huang, R. M., W. Medina, T. M. Brooks, S. H. M. Butchart, J. W. Fitzpatrick, C. Hermes, C. N. Jenkins, A. Johnston, D. J. Lebbin, B. V. Li, N. Ocampo-Peñuela, M. Parr, H. Wheatley, D. A. Wiedenfeld, C. Wood, and S. L. Pimm. 2021. Batch-produced, GIS-informed range maps for birds based on provenanced, crowd-sourced data inform conservation assessments. *PLoS One* 16: Article e0259299.
Inger, R., R. Gregory, J. P. Duffy, I. Stott, P. Voříšek, and K. J. Gaston. 2015. Common European birds are declining rapidly while less abundant species' numbers are rising. *Ecology Letters* 18:28–36.
Intergovernmental Science-Policy Platform on Biodiversity and Ecosystem Services. 2019. *The IPBES Global Assessment Report on Biodiversity and Ecosystem Services.* Secretariat of the Intergovernmental Science-Policy Platform on Biodiversity and Ecosystem Services, Bonn, Germany.
International Union for Conservation of Nature. 2012. IUCN Red List categories and criteria.
International Union for Conservation of Nature. 2013. Documentation standards and consistency checks for IUCN Red List assessments and species accounts: Version 2.
International Union for Conservation of Nature. 2016. A global standard for the identification of key biodiversity areas: Version 1.0.
International Union for Conservation of Nature. 2021a. Conflict and conservation: Nature in a globalised world. Report No. 1.
International Union for Conservation of Nature. 2021b. The IUCN Red List of Threatened Species.
International Union for Conservation of Nature. 2021c. Mapping standards and data quality for the IUCN Red List categories and criteria: Version 1.19.

Irwin, A., A. Geschke, T. M. Brooks, J. Siikamaki, L. Mair, and B. B. N. Strassburg. 2022. Quantifying and categorising national extinction-risk footprints. *Scientific Reports* 12: Article 5861.

Jagiello, Z., Ł. Dylewski, M. Tobolka, and J. I. Aguirre. 2019. Life in a polluted world: A global review of anthropogenic materials in bird nests. *Environmental Pollution* 251:717–722.

Jellesmark, S., M. Ausden, T. M. Blackburn, R. D. Gregory, M. Hoffmann, D. Massimino, L. McRae, and P. Visconti. 2021. A counterfactual approach to measure the impact of wet grassland conservation on U.K. breeding bird populations. *Conservation Biology* 35:1575–1585.

Jetz, W., G. H. Thomas, J. B. Joy, D. W. Redding, K. Hartmann, and A. O. Mooers. 2014. Global distribution and conservation of evolutionary distinctness in birds. *Current Biology* 24: 919–930.

Jones, H. P., N. D. Holmes, S. H. M. Butchart, B. R. Tershy, P. J. Kappes, I. Corkery, A. Aguirre-Muñoz, D. P. Armstrong, E. Bonnaud, A. A. Burbidge, K. Campbell, F. Courchamp, P. E. Cowan, R. J. Cuthbert, S. Ebbert, P. Genovesi, G. R. Howald, B. S. Keitt, S. W. Kress, C. M. Miskelly, S. Oppel, S. Poncet, M. J. Rauzon, G. Rocamora, J. C. Russell, A. Samaniego-Herrera, P. J. Seddon, D. R. Spatz, D. R. Towns, and D. A. Croll. 2016. Invasive mammal eradication on islands results in substantial conservation gains. *Proceedings of the National Academy of Sciences of the USA* 113:4033–4038.

Jones, H. P., B. R. Tershy, E. S. Zavaleta, D. A. Croll, B. S. Keitt, M. E. Finkelstein, and G. R. Howald. 2008. Severity of the effects of invasive rats on seabirds: A global review. *Conservation Biology* 22:16–26.

Juffe-Bignoli, D., T. M. Brooks, S. H. M. Butchart, R. B. Jenkins, K. Boe, M. Hoffmann, A. Angulo, S. Bachman, M. Böhm, N. Brummitt, K. E. Carpenter, P. J. Comer, N. Cox, A. Cuttelod, W. R. T. Darwall, M. Di Marco, L. D. C. Fishpool, B. Goettsch, M. Heath, C. Hilton-Taylor, J. Hutton, T. Johnson, A. Joolia, D. A. Keith, P. F. Langhammer, J. Luedtke, E. Nic Lughadha, M. Lutz, I. May, R. M. Miller, M. A. Oliveira-Miranda, M. Parr, C. M. Pollock, G. Ralph, J. P. Rodríguez, C. Rondinini, J. Smart, S. Stuart, A. Symes, A. W. Tordoff, S. Woodley, B. Young, and N. Kingston. 2016. Assessing the cost of global biodiversity and conservation knowledge. *PLoS One* 11: Article e0160640.

Kamp, J., S. Oppel, A. A. Ananin, Y. A. Durnev, S. N. Gashev, N. Hölzel, A. L. Mishchenko, J. Pessa, S. M. Smirenski, E. G. Strelnikov, S. Timonen, K. Wolanska, and S. Chan. 2015. Global population collapse in a superabundant migratory bird and illegal trapping in China. *Conservation Biology* 29:1684–1694.

Kociolek, A. V., A. P. Clevenger, C. C. St. Clair, and D. S. Proppe. 2011. Effects of road networks on bird populations. *Conservation Biology* 25:241–249.

Krawchuk, M. A., M. A. Moritz, M.-A. Parisien, J. Van Dorn, and K. Hayhoe. 2009. Global pyrogeography: The current and future distribution of wildfire. *PLoS One* 4: Article e5102.

La Sorte, F. A., C. A. Lepczyk, M. F. J. Aronson, M. A. Goddard, M. Hedblom, M. Katti, I. MacGregor-Fors, U. Mörtberg, C. H. Nilon, P. S. Warren, N. S. G. Williams, and J. Yang. 2018. The phylogenetic and functional diversity of regional breeding bird assemblages is reduced and constricted through urbanization. *Diversity and Distributions* 24:928–938.

LaPointe, D. A., C. T. Atkinson, and M. D. Samuel. 2012. Ecology and conservation biology of avian malaria. *Annals of the New York Academy of Sciences* 1249:211–226.

Lees, A. C., C. Albano, G. M. Kirwan, J. F. Pacheco, and A. Whittaker. 2014. The end of hope for Alagoas Foliage-Gleaner *Philydor novaesi*? *Neotropical Birding* 14:20–28.

Lees, A. C., L. Haskell, T. Allinson, S. B. Bezeng, I. J. Burfield, L. M. Renjifo, K. V. Rosenberg, A. Viswanathan, and S. H. M. Butchart. 2022. State of the world's birds. *Annual Review of Environment and Resources* 47:231–260.

Lennon, J. J., P. Koleff, J. J. D. Greenwood, and K. J. Gaston. 2004. Contribution of rarity and commonness to patterns of species richness. *Ecology Letters* 7:81–87.

Lens, L., S. Van Dongen, K. Norris, M. Githiru, and E. Matthysen. 2002. Avian persistence in fragmented rainforest. *Science* 298:1236–1238.

Leung, B., A. L. Hargreaves, D. A. Greenberg, B. McGill, M. Dornelas, and R. Freeman. 2020. Clustered versus catastrophic global vertebrate declines. *Nature* 588:267–271.

Leung, B., A. L. Hargreaves, D. A. Greenberg, B. McGill, M. Dornelas, and R. Freeman. 2022. Reply to: Emphasizing declining populations in the Living Planet Report. *Nature* 601:E25–E26.

Li, Y., R. Miao, and M. Khanna. 2020. Neonicotinoids and decline in bird biodiversity in the United States. *Nature Sustainability* 3:1027–1035.

Loh, J., R. E. Green, T. Ricketts, J. Lamoreux, M. Jenkins, V. Kapos, and J. Randers. 2005. The Living Planet Index: Using species population time series to track trends in biodiversity. *Philosophical Transactions of the Royal Society B: Biological Sciences* 360:289–295.

Loiseau, N., N. Mouquet, N. Casajus, M. Grenié, M. Guéguen, B. Maitner, D. Mouillot, A. Ostling, J. Renaud, C. Tucker, L. Velez, W. Thuiller, and C. Violle. 2020. Global distribution and conservation status of ecologically rare mammal and bird species. *Nature Communications* 11: Article 5071.

López-López, P. 2016. Individual-based tracking systems in ornithology: Welcome to the era of big data. *Ardeola* 63:103–136.

Luther, D., C. R. Beatty, J. Cooper, N. Cox, S. Farinelli, M. Foster, J. Lamoreux, P. J. Stephenson, and T. M. Brooks. 2020. Global assessment of critical forest and landscape restoration needs for threatened terrestrial vertebrate species. *Global Ecology and Conservation* 24: Article e01359.

Luther, D., W. J. Cooper, J. Wong, M. Walker, S. Farinelli, I. Visseren-Hamakers, I. J. Burfield, A. Simkins, G. Bunting, T. M. Brooks, K. Dicks, J. Scott, J. R. S. Westrip, J. Lamoreux, M. Parr, N. de Silva, M. Foster, A. Upgren, and S. H. M. Butchart. 2021. Conservation actions benefit the most threatened species: A 13-year assessment of Alliance for Zero Extinction species. *Conservation Science and Practice* 3: Article e510.

Luther, D. A., T. M. Brooks, S. H. M. Butchart, M. W. Hayward, M. E. Kester, J. Lamoreux, and A. Upgren. 2016. Determinants of bird conservation-action implementation and associated population trends of threatened species. *Conservation Biology* 30:1338–1346.

Mace, G. M., M. Barrett, N. D. Burgess, S. E. Cornell, R. Freeman, M. Grooten, and A. Purvis. 2018. Aiming higher to bend the curve of biodiversity loss. *Nature Sustainability* 1:448–451.

Mace, G. M., N. J. Collar, K. J. Gaston, C. Hilton-Taylor, H. R. AkÇAkaya, N. Leader-Williams, E. J. Milner-Gulland, and S. N. Stuart. 2008. Quantification of extinction risk: IUCN's system for classifying threatened species. *Conservation Biology* 22:1424–1442.

Mair, L., L. A. Bennun, T. M. Brooks, S. H. M. Butchart, F. C. Bolam, N. D. Burgess, J. M. M. Ekstrom, E. J. Milner-Gulland, M. Hoffmann, K. Ma, N. B. W. Macfarlane, D. C. Raimondo, A. S. L. Rodrigues, X. Shen, B. B. N. Strassburg, C. R. Beatty, C. Gómez-Creutzberg, A. Iribarrem, M. Irmadhiany, E. Lacerda, B. C. Mattos, K. Parakkasi, M. F. Tognelli, E. L. Bennett, C. Bryan, G. Carbone, A. Chaudhary, M. Eiselin, da Fonseca, G.A.B., R. Galt, A. Geschke, L. Glew, R. Goedicke, J. M. H. Green, R. D. Gregory, S. L. L. Hill, D. G. Hole, J. Hughes, J. Hutton, M. P. W. Keijzer, L. M. Navarro, E. Nic Lughadha, A. J. Plumptre, P. Puydarrieux, H. P. Possingham, A. Rankovic, E. C. Regan, C. Rondinini, J. D. Schneck, J. Siikamäki, C. Sendashonga, G. Seutin, S. Sinclair, A. L. Skowno, C. A. Soto-Navarro, S. N. Stuart, H. J. Temple, A. Vallier, F. Verones, V, L. R. Viana, J. Watson, S. Bezeng, M. Böhm, I. J. Burfield, V. Clausnitzer, C. Clubbe, N. A. Cox, J. Freyhof, L. R. Gerber, C. Hilton-Taylor, R. Jenkins, A. Joolia, L. N. Joppa, L. P. Koh, T. E. Lacher Jr., P. F. Langhammer, B. Long, D. Mallon, M. Pacifici, B. A. Polidoro, C. M. Pollock, M. C. Rivers, N. S. Roach, J. P. Rodríguez, J. Smart, B. E. Young, F. Hawkins, and P. J. K. McGowan. 2021. Measuring spatially-explicit contributions to science-based species targets. *Nature Ecology & Evolution* 5:836–844.

Manne, L. L., T. M. Brooks, and S. L. Pimm. 1999. Relative risk of extinction of passerine birds on continents and islands. *Nature* 399:258–261.

Marín-Gómez, O. H., and I. MacGregor-Fors. 2021. A global synthesis of the impacts of urbanization on bird dawn choruses. *Ibis* 163:1133–1154.

Marques, A. T., H. Batalha, S. Rodrigues, H. Costa, M. J. R. Pereira, C. Fonseca, M. Mascarenhas, and J. Bernardino. 2014. Understanding bird collisions at wind farms: An updated review on the causes and possible mitigation strategies. *Biological Conservation* 179:40–52.

Marsh, S. M., M. Hoffmann, N. D. Burgess, T. M. Brooks, D. W. Challender, P. J. Cremona, C. Hilton-Taylor, F. L. de Micheaux, G. Lichtenstein, and D. Roe. 2022. Prevalence of sustainable and unsustainable use of wild species inferred from the IUCN Red List of Threatened Species. *Conservation Biology* 36: Article e13844.

Matuoka, M. A., M. Benchimol, J. M. D. Almeida-Rocha, and J. C. Morante-Filho. 2020. Effects of anthropogenic disturbances on bird functional diversity: A global meta-analysis. *Ecological Indicators* 116: Article 106471.

McCarthy, D. P., P. F. Donald, J. P. W. Scharlemann, G. M. Buchanan, A. Balmford, J. M. H. Green, L. A. Bennun, N. D. Burgess, L. D. C. Fishpool, S. T. Garnett, D. L. Leonard, R. F. Maloney, P.

Morling, H. M. Schaefer, A. Symes, D. A. Wiedenfeld, and S. H. M. Butchart. 2012. Financial costs of meeting global biodiversity conservation targets: Current spending and unmet needs. *Science* 338:946–949.

McRae, L., S. Deinet, and R. Freeman. 2017. The diversity-weighted Living Planet Index: Controlling for taxonomic bias in a global biodiversity indicator. *PLoS One* 12: Article e0169156.

Medina, F. M., E. Bonnaud, E. Vidal, B. R. Tershy, E. S. Zavaleta, C. Josh Donlan, B. S. Keitt, M. Corre, S. V. Horwath, and M. Nogales. 2011. A global review of the impacts of invasive cats on island endangered vertebrates. *Global Change Biology* 17:3503–3510.

Mendiratta, U., A. M. Osuri, S. J. Shetty, and A. Harihar. 2021. Mammal and bird species ranges overlap with armed conflicts and associated conservation threats. *Conservation Letters* 14: Article e12815.

Mittermeier, R. A., G. P. Robles, M. Hoffmann, J. Pilgrim, T. Brooks, C. G. Mittermeier, J. Lamoreux, and G. A. B. Fonseca. 2004. *Hotspots Revisited: Earth's Biologically Richest and Most Endangered Ecoregions*. CEMEX, Mexico City, Mexico.

Morelli, F., Y. Benedetti, J. D. Ibáñez-Álamo, J. Jokimäki, R. Mänd, P. Tryjanowski, and A. P. Møller. 2016. Evidence of evolutionary homogenization of bird communities in urban environments across Europe. *Global Ecology and Biogeography* 25:1284–1293.

Mrema, E. M., T. W. Crowther, W. Mathai, T. Brooks, S. Archibald, P. Brancalion, E. B. Barbier, D. Obura, V. Reyes-García, W. J. Bond, L. P. Koh, T. Christophersen, and E. S. Brondizio. 2020. Ten years to restore a planet. *One Earth* 3:647–652.

Mwangi, M. A. K., S. H. M. Butchart, F. B. Munyekenye, L. A. Bennun, M. I. Evans, L. D. C. Fishpool, E. Kanyanya, I. Madindou, J. Machekele, P. Matiku, R. Mulwa, A. Ngari, J. Siele, and A. J. Stattersfield. 2010. Tracking trends in key sites for biodiversity: A case study using Important Bird Areas in Kenya. *Bird Conservation International* 20:215–230.

Myers, N., R. A. Mittermeier, C. G. Mittermeier, G. A. B. da Fonseca, and J. Kent. 2000. Biodiversity hotspots for conservation priorities. *Nature* 403:853–858.

Nasruddin-Roshidi, A., M. S. Mansor, N. A. Ismail, E. Ngadi, M. Izzat-Husna, S. M. Husin, F. S. Mohd-Taib, R. Illias, and S. M. Nor. 2021. Recovery of bird communities following the construction of a large-scale hydroelectric dam. *Ecological Processes* 10: Article 30.

Neate-Clegg, M. H. C., J. J. Horns, F. R. Adler, M. Ç. Kemahlı Aytekin, and Ç. H. Şekercioğlu. 2020. Monitoring the world's bird populations with community science data. *Biological Conservation* 248: Article 108653.

Nilsson, C., C. A. Reidy, M. Dynesius, and C. Revenga. 2005. Fragmentation and flow regulation of the world's large river systems. *Science* 308:405–408.

Oaks, J. L., M. Gilbert, M. Z. Virani, R. T. Watson, C. U. Meteyer, B. A. Rideout, H. L. Shivaprasad, S. Ahmed, M. J. Iqbal Chaudhry, M. Arshad, S. Mahmood, A. Ali, and A. Ahmed Khan. 2004. Diclofenac residues as the cause of vulture population decline in Pakistan. *Nature* 427:630–633.

Orme, C. D. L., R. G. Davies, M. Burgess, F. Eigenbrod, N. Pickup, V. A. Olson, A. J. Webster, T.-S. Ding, P. C. Rasmussen, R. S. Ridgely, A. J. Stattersfield, P. M. Bennett, T. M. Blackburn, K. J. Gaston, and I. P. F. Owens. 2005. Global hotspots of species richness are not congruent with endemism or threat. *Nature* 436:1016–1019.

Pautasso, M., K. Böhning-Gaese, P. Clergeau, V. R. Cueto, M. Dinetti, E. Fernández-Juricic, M.-L. Kaisanlahti-Jokimäki, J. Jokimäki, M. L. McKinney, N. S. Sodhi, D. Storch, L. Tomialojc, P. J. Weisberg, J. Woinarski, R. A. Fuller, and E. Cantarello. 2011. Global macroecology of bird assemblages in urbanized and semi-natural ecosystems. *Global Ecology and Biogeography* 20:426–436.

Pearce-Higgins, J. W., S. M. Eglington, B. Martay, and D. E. Chamberlain. 2015. Drivers of climate change impacts on bird communities. *Journal of Animal Ecology* 84:943–954.

Phalan, B., M. Onial, A. Balmford, and R. E. Green. 2011. Reconciling food production and biodiversity conservation: Land sharing and land sparing compared. *Science* 333:1289–1291.

Pimm, S. L., C. N. Jenkins, R. Abell, T. M. Brooks, J. L. Gittleman, L. N. Joppa, P. H. Raven, C. M. Roberts, and J. O. Sexton. 2014. The biodiversity of species and their rates of extinction, distribution, and protection. *Science* 344: Article 1246752.

Princé, K., and B. Zuckerberg. 2015. Climate change in our backyards: The reshuffling of North America's winter bird communities. *Global Change Biology* 21:572–585.

Rayner, L., D. B. Lindenmayer, J. T. Wood, P. Gibbons, and A. D. Manning. 2013. Are protected areas maintaining bird diversity? *Ecography* 37:43–53.

Reed, J. M. 1999. The role of behavior in recent avian extinctions and endangerments. *Conservation Biology* 13:232–241.

Reynolds, S. J., J. A. Galbraith, J. A. Smith, and D. N. Jones. 2017. Garden bird feeding: Insights and prospects from a north–south comparison of this global urban phenomenon. *Frontiers in Ecology and Evolution* 5: Article 24.

Ricketts, T. H., E. Dinerstein, T. Boucher, T. M. Brooks, S. H. M. Butchart, M. Hoffmann, J. F. Lamoreux, J. Morrison, M. Parr, J. D. Pilgrim, A. S. L. Rodrigues, W. Sechrest, G. E. Wallace, K. Berlin, J. Bielby, N. D. Burgess, D. R. Church, N. Cox, D. Knox, C. Loucks, G. W. Luck, L. L. Master, R. Moore, R. Naidoo, R. Ridgely, G. E. Schatz, G. Shire, H. Strand, W. Wettengel, and E. Wikramanayake. 2005. Pinpointing and preventing imminent extinctions. *Proceedings of the National Academy of Sciences of the USA* 102:18497–18501.

Robinson, T. P., G. R. W. Wint, G. Conchedda, T. P. Van Boeckel, V. Ercoli, E. Palamara, G. Cinardi, L. D'Aietti, S. I. Hay, and M. Gilbert. 2014. Mapping the global distribution of livestock. *PLoS One* 9: Article e96084.

Rodrigues, A., J. Pilgrim, J. Lamoreux, M. Hoffmann, and T. Brooks. 2006. The value of the IUCN Red List for conservation. *Trends in Ecology & Evolution* 21:71–76.

Rodrigues, A. S. L., T. M. Brooks, S. H. M. Butchart, J. Chanson, N. Cox, M. Hoffmann, and S. N. Stuart. 2014. Spatially explicit trends in the global conservation status of vertebrates. *PLoS One* 9: Article e113934.

Rosenberg, J. 2018. Transnational advocacy and the politics of sustainable development in a small island developing state: An uncertain future for the Grenada Dove. *Journal of Environment & Development* 27:236–261.

Rosenberg, K. V., A. M. Dokter, P. J. Blancher, J. R. Sauer, A. C. Smith, P. A. Smith, J. C. Stanton, A. Panjabi, L. Helft, M. Parr, and P. P. Marra. 2019. Decline of the North American avifauna. *Science* 366:120–124.

Runge, C. A., J. E. M. Watson, S. H. M. Butchart, J. O. Hanson, H. P. Possingham, and R. A. Fuller. 2015. Protected areas and global conservation of migratory birds. *Science* 350:1255–1258.

Salafsky, N., D. Salzer, A. J. Stattersfield, C. Hilton-Taylor, R. Neugarten, S. H. M. Butchart, B. E. N. Collen, N. Cox, L. L. Master, S. O'Connor, and D. Wilkie. 2008. A standard lexicon for biodiversity conservation: Unified classifications of threats and actions. *Conservation Biology* 22:897–911.

Santangeli, A., A. Rajasärkkä, and A. Lehikoinen. 2017. Effects of high latitude protected areas on bird communities under rapid climate change. *Global Change Biology* 23:2241–2249.

Sauer, J. R., W. A. Link, J. E. Fallon, K. L. Pardieck, and D. J. Ziolkowski. 2013. The North American Breeding Bird Survey 1966–2011: Summary analysis and species accounts. *North American Fauna* 79:1–32.

Sayol, F., R. S. C. Cooke, A. L. Pigot, T. M. Blackburn, J. A. Tobias, M. J. Steinbauer, A. Antonelli, and S. Faurby. 2021. Loss of functional diversity through anthropogenic extinctions of island birds is not offset by biotic invasions. *Science Advances* 7: Article eabj5790.

Scharlemann, J. P. W., R. E. Green, and A. Balmford. 2004. Land-use trends in Endemic Bird Areas: Global expansion of agriculture in areas of high conservation value. *Global Change Biology* 10:2046–2051.

Schmidt, C., M. Domaratzki, R. P. Kinnunen, J. Bowman, and C. J. Garroway. 2020. Continent-wide effects of urbanization on bird and mammal genetic diversity. *Proceedings of the Royal Society B: Biological Sciences* 287: Article 20192497.

Seewagen, C. L. 2020. The threat of global mercury pollution to bird migration: Potential mechanisms and current evidence. *Ecotoxicology* 29:1254–1267.

Şekercioğlu, Ç. H. 2003. Birding economics: Conservation through commodification. *Birding* 35:394–402.

Şekercioğlu, Ç. H., C. D. Mendenhall, F. Oviedo-Brenes, J. J. Horns, P. R. Ehrlich, and G. C. Daily. 2019. Long-term declines in bird populations in tropical agricultural countryside. *Proceedings of the National Academy of Sciences of the USA* 116:9903–9912.

Şekercioğlu, Ç. H., S. H. Schneider, J. P. Fay, and S. R. Loarie. 2008. Climate change, elevational range shifts, and bird extinctions. *Conservation Biology* 22:140–150.

Senzaki, M., J. R. Barber, J. N. Phillips, N. H. Carter, C. B. Cooper, M. A. Ditmer, K. M. Fristrup, C. J. W. McClure, D. J. Mennitt, L. P. Tyrrell, J. Vukomanovic, A. A. Wilson, and C. D. Francis. 2020. Sensory pollutants alter bird phenology and fitness across a continent. *Nature* 587:605–609.

Smith, F. D. M., R. M. May, R. Pellew, T. H. Johnson, and K. S. Walter. 1993. Estimating extinction rates. *Nature* 364:494–496.

Smucker, K. M., R. L. Hutto, and B. M. Steele. 2005. Changes in bird abundance after wildfire: Importance of fire severity and time since fire. *Ecological Applications* 15:1535–1549.

Sobral, F. L., A. C. Lees, and M. V. Cianciaruso. 2016. Introductions do not compensate for functional and phylogenetic losses following extinctions in insular bird assemblages. *Ecology Letters* 19:1091–1100.

Sol, D., C. Trisos, C. Múrria, A. Jeliazkov, C. González-Lagos, A. L. Pigot, C. Ricotta, C. M. Swan, J. A. Tobias, and S. Pavoine. 2020. The worldwide impact of urbanisation on avian functional diversity. *Ecology Letters* 23:962–972.

Somveille, M., A. Manica, S. H. M. Butchart, and A. S. L. Rodrigues. 2013. Mapping global diversity patterns for migratory birds. *PLoS One* 8: Article e70907.

Stattersfield, A., M. Crosby, A. Long, and D. C. Wege. 1998. Endemic bird areas of the world: Priorities for biodiversity conservation. *The Auk* 115:1089–1091.

Stephens, P. A., L. R. Mason, R. E. Green, R. D. Gregory, J. R. Sauer, J. Alison, A. Aunins, L. Brotons, S. H. M. Butchart, T. Campedelli, T. Chodkiewicz, P. Chylarecki, O. Crowe, J. Elts, V. Escandell, R. P. B. Foppen, H. Heldbjerg, S. Herrando, M. Husby, F. Jiguet, A. Lehikoinen, A. Lindstrom, D. G. Noble, J. Y. Paquet, J. Reif, T. Sattler, T. Szep, N. Teufelbauer, S. Trautmann, A. J. van Strien, C. A. M. van Turnhout, P. Vorisek, and S. G. Willis. 2016. Consistent response of bird populations to climate change on two continents. *Science* 352:84–87.

Steven, R., and J. G. Castley. 2013. Tourism as a threat to critically endangered and endangered birds: Global patterns and trends in conservation hotspots. *Biodiversity and Conservation* 22:1063–1082.

Steven, R., J. G. Castley, and R. Buckley. 2013. Tourism revenue as a conservation tool for threatened birds in protected areas. *PLoS One* 8: Article e62598.

Strassburg, B. B., A. Iribarrem, H. L. Beyer, C. L. Cordeiro, R. Crouzeilles, C. C. Jakovac, A. Braga Junqueira, E. Lacerda, A. E. Latawiec, and A. Balmford. 2020. Global priority areas for ecosystem restoration. *Nature* 586:724–729.

Sullivan, B. L., C. L. Wood, M. J. Iliff, R. E. Bonney, D. Fink, and S. Kelling. 2009. eBird: A citizen-based bird observation network in the biological sciences. *Biological Conservation* 142:2282–2292.

Toussaint, A., S. Brosse, C. G. Bueno, M. Pärtel, R. Tamme, and C. P. Carmona. 2021. Extinction of threatened vertebrates will lead to idiosyncratic changes in functional diversity across the world. *Nature Communications* 12: Article 5162.

Trull, N., M. Böhm, and J. Carr. 2018. Patterns and biases of climate change threats in the IUCN Red List. *Conservation Biology* 32:135–147.

Unger, S., M. Rollins, A. Tietz, and H. Dumais. 2020. iNaturalist as an engaging tool for identifying organisms in outdoor activities. *Journal of Biological Education* 55:537–547.

United Nations. 2021. The sustainable development goals report 2021.

Van Riper, C., III, S. G. van Riper, M. L. Goff, and M. Laird. 1986. The epizootiology and ecological significance of malaria in Hawaiian land birds. *Ecological Monographs* 56:327–344.

Vasconcelos, M., G. N. Maurício, G. M. Kirwan, and L. F. Silveira. 2008. Range extension for Marsh Tapaculo *Scytalopus iraiensis* to the highlands of Minas Gerais, Brazil, with an overview of the species' distribution. *Bulletin of the British Ornithologists' Club* 128:101–106.

Visser, M. E., A. C. Perdeck, J. H. van Balen, and C. Both. 2009. Climate change leads to decreasing bird migration distances. *Global Change Biology* 15:1859–1865.

Waliczky, Z., L. D. C. Fishpool, S. H. M. Butchart, D. Thomas, M. F. Heath, C. Hazin, P. F. Donald, A. Kowalska, M. P. Dias, and T. S. M. Allinson. 2019. Important Bird and Biodiversity Areas (IBAs): Their impact on conservation policy, advocacy and action. *Bird Conservation International* 29:199–215.

Walters, J. R., S. R. Derrickson, D. Michael Fry, S. M. Haig, J. M. Marzluff, and J. M. Wunderle. 2010. Status of the California Condor (*Gymnogyps californianus*) and efforts to achieve its recovery. *The Auk* 127:969–1001.

Wenny, D. G., T. L. DeVault, M. D. Johnson, D. Kelly, C. H. Sekercioglu, D. F. Tomback, and C. J. Whelan. 2011. The need to quantify ecosystem services provided by birds. *The Auk* 128:1–14.

Whelan, C. J., D. G. Wenny, and R. J. Marquis. 2008. Ecosystem services provided by birds. *Annals of the New York Academy of Sciences* 1134:25–60.

White, R. L., and P. M. Bennett. 2015. Elevational distribution and extinction risk in birds. *PLoS One* 10: Article e0121849.

Wilcox, C., E. Van Sebille, and B. D. Hardesty. 2015. Threat of plastic pollution to seabirds is global, pervasive, and increasing. *Proceedings of the National Academy of Sciences of the USA* 112:11899–11904.

Wilhelm, S. I., G. J. Robertson, P. C. Ryan, S. F. Tobin, and R. D. Elliot. 2009. Re-evaluating the use of beached bird oiling rates to assess long-term trends in chronic oil pollution. *Marine Pollution Bulletin* 58:249–255.

Williams, B. A., J. E. M. Watson, S. H. M. Butchart, M. Ward, T. M. Brooks, N. Butt, F. C. Bolam, S. N. Stuart, L. Mair, P. J. K. McGowan, R. Gregory, C. Hilton-Taylor, D. Mallon, I. Harrison, and J. S. Simmonds. 2020. A robust goal is needed for species in the Post-2020 Global Biodiversity Framework. *Conservation Letters* 14: Article e12778.

Williams, D. R., R. G. Pople, D. A. Showler, L. V. Dicks, M. F. Child, E. K. Zu Ermgassen, and W. J. Sutherland. 2013. *Bird Conservation: Global Evidence for the Effects of Interventions.* Pelagic, Exeter, UK.

Wood, C., B. Sullivan, M. Iliff, D. Fink, and S. Kelling. 2011. eBird: Engaging birders in science and conservation. *PLoS Biology* 9: Article e1001220.

World Wildlife Fund. 2020. Living planet report 2020: A deep dive into the Living Planet Index.

23

Bird Conservation in the Neotropics

Gaps and Challenges in the Face of Anthropocene Threats

David A. Prieto-Torres, Luis A. Sánchez-González, Octavio R. Rojas-Soto, María del Coro Arizmendi, and Adolfo G. Navarro-Sigüenza

During the past 50 years, numerous studies have drawn attention to the magnitude and consequences of human impacts on global biodiversity. Although multiple pressures act in combination, deforestation and anthropogenic climate change are currently the most important drivers of the biodiversity crisis (Intergovernmental Science-Policy Platform on Biodiversity and Ecosystem Services [IPBES] 2019, Lovejoy and Hannah 2019). These drivers have a profound and strong impact not only on biodiversity, from genetic to ecosystem (functional traits) levels (Etard et al. 2022), but also on human life quality (Díaz et al. 2019). Despite our growing understanding of these problems, decision-making criteria about resource management have failed to keep up with these global drivers (Xu et al. 2021). Species extinction rates due to these anthropogenic disturbances are today 1,000 times higher than rates due to natural causes (Ceballos and Ehrlich 2018), and in this century, predictions are not optimistic due to population declines and important habitat reductions as a consequence of human impacts (De Vos et al. 2015).

Protected areas (PAs) are a fundamental tool for preservation of biodiversity (Rodrigues et al. 2004, Leroux et al. 2010). These areas are defined as "a geographical space clearly delimited and especially dedicated to the protection and maintenance of biological diversity, and of natural and associated cultural resources, and managed through legal or other effective means" (Dudley 2008, p. 10). As a cornerstone of biodiversity conservation, PAs are essential for maintaining wildlife populations at sustainable levels and ensuring the long-term maintenance of human well-being services (Gray et al. 2016, Díaz et al. 2019). However, it is well known that historically the allocation of PAs has been mainly opportunistic and highly influenced by economic activities (logging and agriculture, mining, tourism, etc.). Therefore, the global PA network is far from complete (Rodrigues et al. 2004, Cantú-Salazar and Gaston 2010), showing important gaps in megadiverse countries such as Brazil, Costa Rica, Peru, and Mexico (Teixeira et al. 2018, Vieira et al. 2019, Godínez-Gómez et al. 2021). Furthermore, the lack of funds for maintenance, infrastructure, surveillance, and implementation of both management and monitoring plans is an important limitation in PAs' functioning (Zafra-Calvo and Geldmann 2020, Powlen et al. 2021). Debate is ongoing on their overall effectiveness to the long-term conservation of biodiversity under global change scenarios (Asamoah et al. 2021),

David A. Prieto-Torres et al., *Bird Conservation in the Neotropics*. In: *New Perspectives in Ornithology*.
Edited by: Scott V. Edwards and J. Michael Reed, Oxford University Press. © Oxford University Press (2025).
DOI: 10.1093/oso/9780197787670.003.0023

especially considering the high species richness and endemism of the Neotropics (Pearson et al. 2019, Antonelli 2022).

Neotropical birds are not the exception in these critical scenarios. Today, more than 75% of all threatened birds are affected by anthropogenic impacts, including several species on the brink of extinction that have lost more than 99% of their original ranges (Buckton 2001, BirdLife International 2018, Bird et al. 2020). Although birds are a charismatic and well-known group of vertebrates, awareness of spatiotemporal biodiversity patterns (in both alpha and beta levels) is still incomplete and patchily distributed across the region (e.g., Etard et al. 2022). Significant gaps in vulnerability and extinction risk for these taxa (and their ecological interactions) in the forthcoming decades have been reported (Şekercioğlu et al. 2012, Heinen et al. 2020). One important conclusion of these studies is that in most cases, the current PA network may not be conservation-effective for distributions of individual taxa and bird communities (e.g., Navarro-Sigüenza et al. 2011, Prieto-Torres et al. 2018, Ramírez-Albores et al. 2021). Consequently, the Neotropics are losing bird species at unprecedented rates, often without properly identifying and describing their biodiversity (Ceballos and Ehrlich 2018, Lees et al. 2020). Therefore, expansion of the Neotropical PA network is urgent. Still, further research is needed, particularly for determining which species and regions are more susceptible, as well as trying to identify priorities for conservation (Pearson et al. 2019, Jennings et al. 2020).

In this chapter, we review the already established conservation areas for Neotropical birds to detect coverage in species representativeness and identify which high-species-richness areas require immediate conservation action. Herein, we adopt a Neotropics' definition that includes a vast range of biomes and habitat types extending from the lowlands of northern Mexico to Argentina and Brazil, including the Caribbean (Morrone 2013, Dinerstein et al. 2017; see Figure 23.1). Specifically, we present a general characterization of Neotropical avian conservation efforts by addressing the following questions: (1) What are the current representativeness values of the existing PAs for species distribution? (2) Which bird species and groups currently represent important conservation gaps within PAs? and (3) Where must future conservation areas (complementary to the current PA network) be established to maximize species representativeness to reach the Post-2020 Biodiversity Framework targets efficiently? Finally, we discuss current and future challenges and opportunities for research. All these results represent a critical step in informing policymakers and conservation planners where to focus local and large-scale efforts to protect these vulnerable and important taxa in the context of anthropic pressures.

Birds as Surrogates of Biodiversity

The Neotropics, with an unparalleled diversity of avian species (*N* > 3,800; Stotz et al. 1996, BirdLife International 2018), is a bird-watcher's paradise that includes seven of the most bird-rich countries (Colombia, Peru, Brazil, Ecuador, Venezuela, Bolivia, and Mexico) and nine with the highest number of restricted-range species globally. Birds are important indicators of landscape conditions due to their strong vulnerability to environmental changes (e.g., Browder et al. 2002, Sherry 2021), and they play an important role in tropical ecosystem functioning (e.g., seed dispersion, pollination, and plant reproduction; Sekercioglu 2006). For example, hummingbirds are specialized nectarivores that play an important role by pollinating nearly 15% of the plant species in North and South America (Able 2000, Buzato et al. 2000). The loss of these key ecological interactions can have detrimental impacts on ecosystems and even cause their collapse, mainly because many species cannot complete their life cycles without their interaction partners (Jordano 2016). Moreover, we have more extensive knowledge (including taxonomy, distribution, and ecology) of birds than of many other groups (Peterson and Navarro-Sigüenza 2017). For all these reasons, they frequently serve as biodiversity indicators and surrogates. Birds have become a crucial element in highlighting and promoting conservation strategies across the region (e.g., the Important Bird Area program; see https://www.birdlife.org). Therefore, delineating areas that are important for bird conservation (based on vulnerability and irreplaceability criteria) provides benefits to other poorly known taxa and less charismatic taxa (the "umbrella" effect; Caro 2010).

Examining Conservation Gaps and Representativeness in Protected Areas

Providing accurate information to decision-makers about geographical areas to prioritize the design of conservation areas is urgent. During the past two decades, different conservation planning schemes have been developed to prioritize the most important sites for conservation and that are also compatible with sustainable human use (e.g., Ciarleglio et al. 2009, Sarkar and Illoldi-Rangel 2010, Navarro-Sigüenza et al. 2011, Moilanen et al. 2014, Brum et al. 2017). These approaches are based on the distribution of key biodiversity features and anthropic variables, promoting a representative and connected network of PAs that contributes to the viability of biodiversity and ecosystem functioning.

We selected 3,671 Neotropical species—belonging to 34 orders, 122 families, and 996 genera—and defined their potential ranges using digital range maps (extent of occurrence maps with a resolution of 0.08333°) available at the International Union for Conservation of Nature (IUCN) database (https://www.iucnredlist.org; accessed May 7, 2022). We summarized the diversity patterns (Figure 23.1) using a per-site species richness (S) approach and a Sørensen-based multiple-site index (i.e., species turnover [β]) to determine the variation in species composition for the entire assembly across the Neotropics (Baselga and Orme 2012). Then, using Zonation (Moilanen et al. 2014), and based on two different removal rules (core-area zonation [CAZ], which prioritizes areas containing rare and/or highly weighted species, and additive benefit function [ABF], which prioritizes high species richness), we examined what would be the resulting priority areas if we (1) assumed that existing PAs could be redrawn (i.e., starting from scratch) and (2) expanded existing PAs (to reach Post-2020 Biodiversity Framework targets; Woodley et al. 2019). To do the latter, we included all IUCN categories for PA (IUCN and United Nations Environment Programme World Conservation Monitoring Centre [UNEP-WCMC] 2022): (I) strict protection (i.e., strict nature reserves and wilderness areas), (II) ecosystem

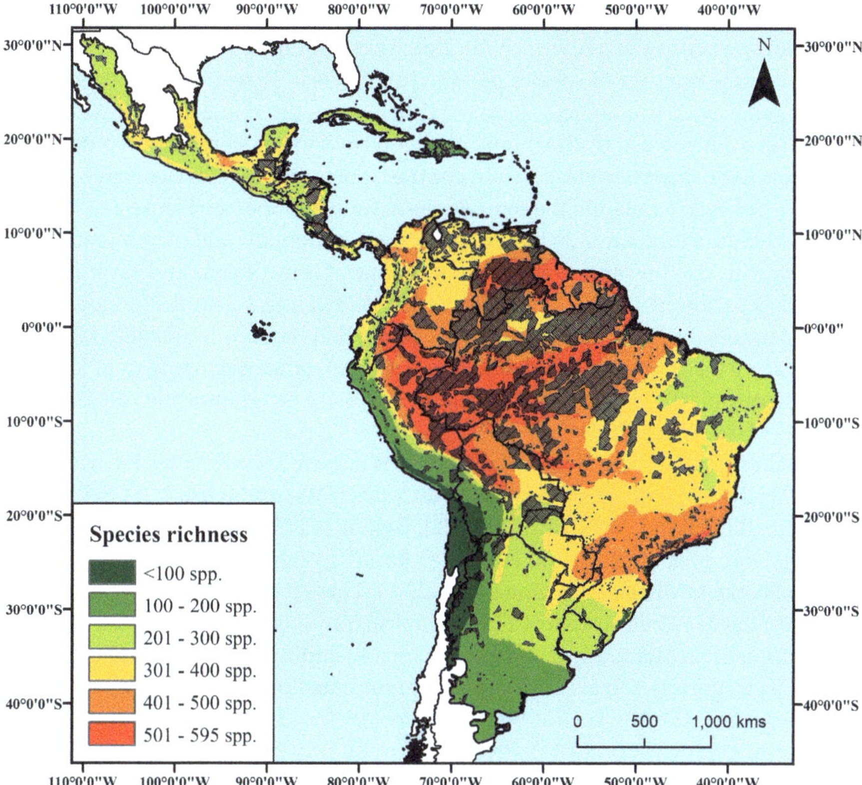

Figure 23.1 Spatial patterns of species richness for 3,671 bird species throughout the Neotropics. Hatched black-and-gray areas correspond to the current protected areas (both officially and voluntarily designated terrestrial conservation areas) according to the World Database of Protected Areas (IUCN and UNEP-WCMC 2022).

Source: Map created using ArcMap 10.8.2 (ID subscription: 5783807033).

conservation and protection (i.e., national parks), (III) conservation of natural features—natural monument, (IV) conservation through active management (i.e., habitat/species management areas), (V) landscape conservation and recreation (i.e., protected landscapes), and (VI) sustainable use of natural resources (i.e., managed resource protected areas).

Given that most bird species cannot adequately be protected within highly modified areas (Pimm et al. 2014), the analyses were performed explicitly considering anthropic pressures (sites with >50% cover loss [European Space Agency Climate Change Initiative 2017] and high human influence [see Global Terrestrial Human Footprint map; Venter et al. 2016, 2018]). This last step prevented Zonation from assigning high conservation values to such areas. In addition, to define those with highest priority (highly weighted) species, we used their current conservation status (Butchart et al. 2004): Least Concern [LC] = 1, Near Threatened [NT] = 2, Vulnerable [VU] and Data Deficient [DD] = 3, Endangered [EN] = 4, and Critically Endangered

[CR] = 5. Thus, species that are highly threatened and geographically restricted (i.e., endemic) were considered a priority in Zonation solutions.

The average number of species per site (one pixel = 10 km^2) was 350 ± 127 spp. Overall, sites with the highest species richness were distributed in the Amazonian region (including western Brazil and the eastern Andean slope from southwestern Colombia to northwestern Bolivia). In contrast, the lowest species richness was found in the Caribbean islands and the southwestern Andean slope (see Figure 23.1). Species richness tends to be greatest in humid areas, such as montane and tropical rainforests, and lowest in dry lowland forests (e.g., Caatinga, Gran Chaco, and savanna). Bird assemblages showed a mean β diversity of 0.70, indicating a clear distinction in composition across the Neotropical region. We found that only approximately 4.2% of the species are widely distributed (defined by a species' range occurring in at least 50% of the Neotropics), whereas approximately 75.6% are geographically restricted (i.e., occur in <10% of this region).

A total of 24.0% of the Neotropical region is currently covered by PAs (including both officially and voluntarily designated terrestrial PAs; see Figure 23.1). Seven countries (Brazil [55.03%], Venezuela [11.08%], Bolivia [5.80%], Peru [5.63%], Argentina [5.11%], Colombia [4.13%], and Mexico [3.08%]) account for approximately 90% of current PA extent within the region (Table 23.1). These PAs contain on average 27.20 ± 19.14% of the distribution area of all Neotropical bird species and 27.8 ± 24.65% of the distribution of threatened species (n = 357 spp.; including VU, EN, and CR according to IUCN categories). On average, the best-represented orders within PAs were those including some seabirds, such as Phaethontiformes (n = 2 spp.; 61.44 ± 13.44% of their distributional range within PAs), Procellariiformes (n = 23 spp.; 44.64 ± 24.96%), and Sphenisciformes (n = 5 spp.; 41.38 ± 22.33%). Taxa with lowest representativeness values were Cariamiformes (n = 2 spp.; 9.52 ± 1.26%), Anseriformes (n = 62 spp.; 15.05 ± 12.45%), and Podicipediformes (n = 9 spp.; 16.42 ± 13.17%). A high number of species (n = 558; 15.2%) had less than 10% of their range protected, whereas most of the species (35.74%) had 20–40% of their ranges within PAs. Only 365 spp. (9.94%) possessed at least 50% of their distribution within the PAs' boundaries (Figure 23.2).

Identifying Priority Conservation Areas: Are Protected Areas Located Optimally?

According to our prioritization analyses, protecting an additional 6% of the total area in the Neotropics (Figures 23.2 and 23.3) would substantially increase the species representativeness in the PA network, covering between 35.96 ± 21.91% (CAZ rule) and 51.55 ± 28.61% (ABF rule) when all species are considered. Protecting the identified priority areas, regardless of the removal rule, would increase the proportion of species' ranges covered by PAs: Species for which less than 10% of their distribution range is currently protected would drop to 4.79% (see Figure 23.2). In comparison, 23.40% (CAZ) to 42.55% (ABF) of species could have at least 50% of their distribution under protection. Moreover, these potential conservation areas would include the

Table 23.1 Current Extension by Countries for the Neotropics' Surface Under Protection and the Proportions of the Additional Complementary Conservation Areas Estimated Herein for Each Country to Increase Coverage to Match the Post-2020 Biodiversity Framework (30%)[a]

Country	Current PA Surface (km^2)	% of Total PA	Priority Conservation Areas Analyses					
			CAZ		ABF		Consensus Areas	
			km^2	%	km^2	%	km^2	%
Peru	304,400	5.63	256,400	19.00	253,600	18.79	76,200	28.74
Mexico	166,700	3.08	337,700	25.02	55,000	4.07	51,400	19.39
Colombia	223,000	4.13	238,500	17.67	73,200	5.42	36,400	13.73
Ecuador	68,600	1.27	139,600	10.34	46,700	3.46	25,800	9.73
Brazil	2,975,000	55.03	101,200	7.50	405,400	30.03	24,000	9.05
Bolivia	313,100	5.79	56,100	4.16	66,400	4.92	16,000	6.04
Guyana	20,600	0.38	10,100	0.75	78,800	5.84	9,000	3.39
Panama	26,800	0.50	24,600	1.82	6,900	0.51	6,500	2.45
Chile	18,500	0.34	10,300	0.76	20,200	1.50	3,900	1.47
Venezuela	598,700	11.07	14,500	1.07	66,100	4.90	2,800	1.06
Others	690,500	12.77	160,800	11.91	277,500	20.56	13,100	4.94
Total	5,405,600	100	1,349,800	100	1,349,800	100	265,100	100

[a] Results were estimated based on the 10 most important countries found.
ABF, additive benefit function; CAZ, core-area zonation; PA, protected area.

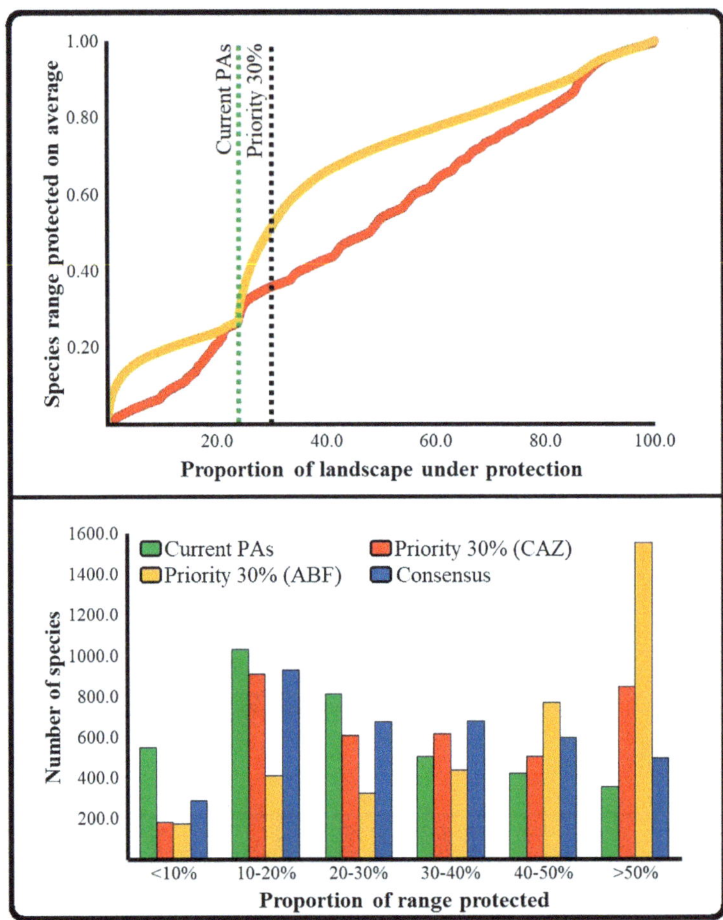

Figure 23.2 Protection for the Neotropical avifauna considering the current protected areas (PAs) and the 30% of high-priority conservation sites. (Top) Proportion of available grid cells that are protected and their corresponding average species range protection. (Bottom) Histogram showing the average percentage of geographic distribution and the number of bird species found within the current PAs and the identified priority areas. ABF, additive benefit function; CAZ, core-area zonation.

distributional range of 42.45 ± 28.44% (CAZ) to 66.79 ± 33.04% (ABF) of threatened species. Both prioritization analyses showed a spatial consensus pattern of 17.61% for the priority areas selected (see Figure 23.3). When these consensus priority conservation areas are combined with current PAs, an average of 33.52 ± 33.04% of the ranges for all the bird species and an average of 40.33 ± 27.79% for threatened species are covered.

We determined that the consensuses of priority conservation areas covered broad areas adjacent to currently designated PAs, mostly across Peru (with 28.74% of the identified areas), Mexico (19.39%), Colombia (13.73%), Ecuador (9.73%), and Brazil (9.05%) (see Table 23.1). We highlight four key areas that substantially increase the

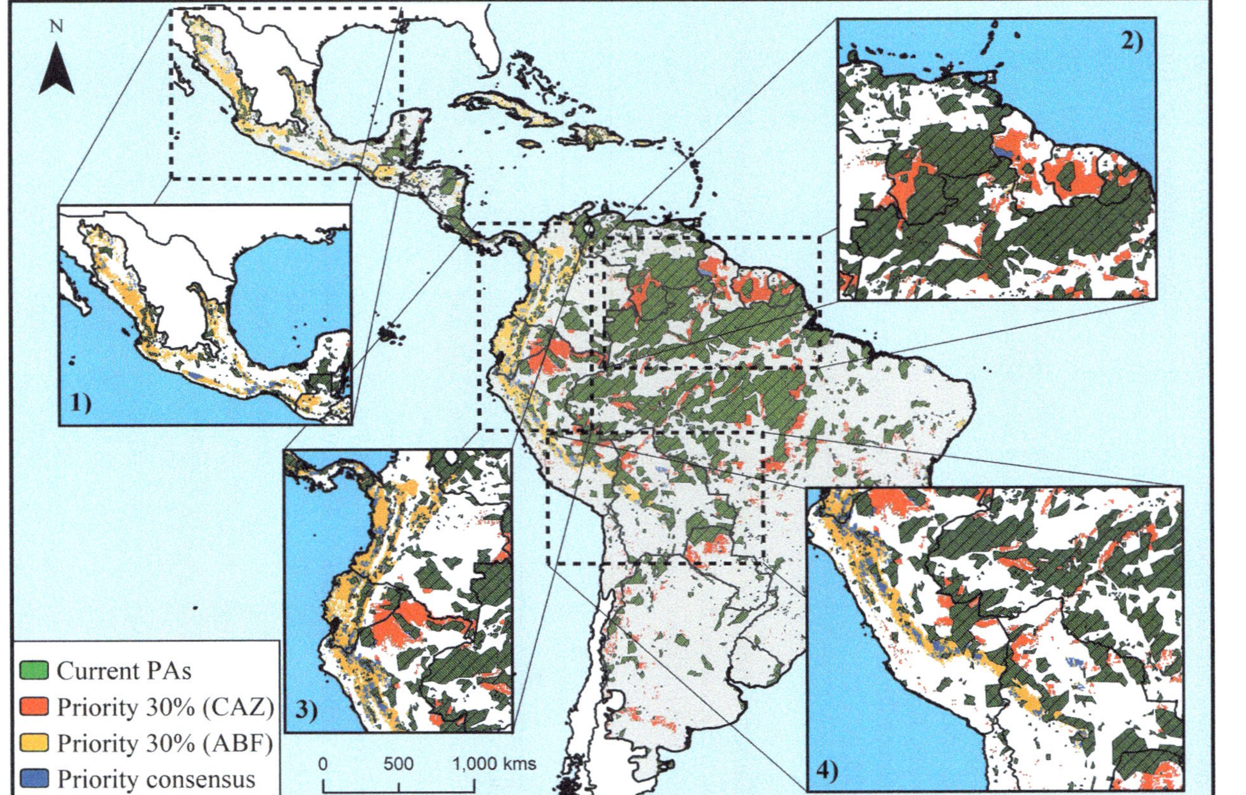

Figure 23.3 Map showing existing protected areas (PAs) of the region (green). Insets show the potential expansion of areas identified in our spatial prioritization analysis for each analysis (additive benefit function [ABF] and core-area zonation [CAZ] removal rules) and the areas where they overlap (consensus).

Source: Map created using ArcMap 10.8.2 (ID subscription: 5783807033).

connectivity among high-species-richness areas in the Neotropics (see Figure 23.3): (1) the dry lowland and pine-oak forests of the western Pacific slope from northwestern Mexico to northwestern Central America (Guatemala, Honduras, and El Salvador); (2) the Amazon region and the Guiana Shield; (3) the northern Andean slope, from northwestern Colombia–Venezuela to western Peru (covered mainly with montane, dry, and moist forests); and (4) the central-southern Andean slope, from northern Peru to northwestern Bolivia and western Argentina (covered with montane, wet, and Yunga forests). The combined current PAs and the identified priority consensus areas showed significantly higher species richness per site (408 ± 113 spp.) and outside them (331 ± 126 spp.). At the same time, however, bird assemblages within PAs showed a distinctive composition, with a β diversity of 0.65. It is important to note that these consensus conservation areas matched 64.28% with those defined by Nori et al. (2020) as priority areas for the conservation and research of terrestrial vertebrates. Consensus areas identified herein were distributed mainly in 15 ecoregions (sensu Dinerstein et al. 2017), covering more than 60% of the estimated area (Table 23.2).

Table 23.2 Current Extension by Terrestrial Ecosystems Throughout Neotropics for the Current Protected Areas and the Additional-Complementary Conservation Areas Estimated Herein to Increase Coverage to Match the Post-2020 Biodiversity Framework (30%)[a]

Ecosystem	Current PAs		Consensus Area	
	km	%	km	%
Peruvian Yungas	35,800	0.66	30,500	11.51
Eastern Cordillera real montane forests	26,700	0.49	23,800	8.98
Choco-Darien moist forests	13,300	0.25	14,500	5.47
Northwestern Andean montane forests	15,400	0.28	13,700	5.17
Sierra Madre Occidental pine-oak forests	29,900	0.55	11,100	4.19
Sierra Madre del Sur pine-oak forests	2,100	0.04	9,400	3.55
Ucayali moist forests	29,900	0.55	9,200	3.47
Central Andean wet puna	22,500	0.42	7,100	2.68
Guianan Highlands moist forests	130,600	2.42	7,000	2.64
Beni savanna	23,400	0.43	6,800	2.57
Bahia coastal forests	10,600	0.20	6,600	2.49
Southwest Amazon moist forests	335,400	6.20	5,600	2.11
Bolivian montane dry forests	13,700	0.25	5,200	1.96
Cordillera Central *páramo*	2,300	0.04	4,400	1.66
Southern Pacific dry forests	1,900	0.04	4,400	1.66
Others	4,712,400	87.17	105,800	39.91
Total	5,405,900	100	265,100	100

[a] Results were estimated based on the 15 most important ecosystems found.
PAs, protected areas.

Finally, when starting from scratch, priority areas identified by Zonation showed medium overlap values with the current PA network (34.80% [ABF] and 46.85% [CAZ]). These overlapping areas occur in a relatively continuous stretch throughout the Amazonian region and involve mainly (40.81% [CAZ] and 46.94% [ABF]) the PAs categorized as type V and VI by IUCN. In addition, these areas match 19.92% (CAZ) to 23.91% (ABF) with Neotropical Important Bird Areas (BirdLife International 2021).

Future Challenges and Opportunities for Conservation Planning

Historically, detecting important conservation areas for the avifauna and proposing measures for their protection have been common topics in the literature at local (Levey et al. 2021), national (Arizmendi and Márquez-Valdelamar 2000, Íñigo-Elías and Enkerlin-Hoeflich 2003, Prieto-Torres et al. 2021c), regional (Sánchez-Ramos et al. 2018, Ramírez-Albores et al. 2021), and continental scales (Donald et al. 2018, Prieto-Torres et al. 2018). However, and despite an increase in the extent of conservation areas during the past decades, the existence of critical sites outside PAs or alternative conservation programs across the Neotropics is very striking. In fact, the current level of protection of bird faunas is still woefully inadequate, and several globally threatened, ecologically restricted, and endemic species, as well as species assemblages, remain unprotected (e.g., Arizmendi et al. 2016; Prieto-Torres, Nuñez Rosas, et al. 2021; Prieto-Torres, Sánchez-González et al. 2021; Ramírez-Albores et al. 2021). Thus, current PAs are insufficient to ensure the long-term persistence of populations and functional ecosystems considering past and current trends in land use change, habitat fragmentation and degradation, and also future climate change (Hannah et al. 2007, IPBES 2019, Lovejoy and Hannah 2019, Asamoah et al. 2021). Most important, many of the priority areas predicted to be highly resilient to current and future scenarios of global changes have been frequently reported outside of current PAs (e.g., Rojas-Soto et al. 2012; Prieto-Torres, Nori, et al. 2021; Prieto-Torres, Sánchez-González et al. 2021; Prieto-Torres et al. 2022; Ramírez-Albores et al. 2021). In this context, it is important that decision-makers define desired outcomes clearly and spatially target interventions accordingly.

Previous analyses have highlighted the importance of different regions throughout the Neotropics as important bird areas (e.g., Arizmendi and Márquez-Valdelamar 2000, Devenish et al. 2009), highly endemic areas (Stattersfield et al. 1998), or areas that contain attribute of evolutionary importance (e.g., Álvarez-Mondragón and Morrone 2004, García-Trejo and Navarro 2004). In fact, many of the areas identified as a priority for conservation have been recognized as biodiversity hot spots and centers of species richness and endemic vertebrates (Myers et al. 2000). Failure to protect this megadiverse avifauna may result in major losses of a huge and unique evolutionary history globally. However, as the influence of the Anthropocene progresses, the ratio of species' ranges will likely decrease substantially even in those contained within PAs. Therefore, future conservation policies must be planned differently if efforts are to succeed under different anthropogenic practices (Baldi 2020).

The identification of "safe places," where high species richness coincides with areas in which human-induced changes are not expected in the near future, is of particular interest in this pursuit (Hannah et al. 2007; Pearson et al. 2019; Prieto-Torres, Nori, et al. 2021). This latter is not a minor detail; in fact, they would probably be the better cost-efficient areas to implement conservation and restoration actions of endangered Neotropical ecosystems and, therefore, undoubtedly must be urgently protected. However, it is not enough to protect a large area of land; policies and strategies must be crafted that address the myriad of threats to biota. For example, in western Mexico's lowlands and the Andean slope's highlands, several sites have been identified as diversification hot spots with historical climatic stability for biotas (e.g., Hazzi et al. 2018, Castillo-Chora et al. 2021). However, despite having PAs, these same areas are subject to high levels of habitat transformation and destruction by agricultural activities and climate change due to anthropogenic activities (e.g., Prieto-Torres et al. 2016; Prieto-Torres, Nori, et al. 2021; Rodríguez-Echeverry and Leiton 2021). Therefore, actions (aimed at increasing the development of sustainable use of the territories) should be taken now, especially in places such as Peru, Mexico, and Colombia, where our analyses indicated a conservation debt regarding species and area protected. Because these countries hold a remarkable amount of natural resources and biodiversity, they are responsible for biodiversity conservation (Barr et al. 2011, Baldi 2020).

To truly maximize the performance of the PA network and biodiversity conservation, we need to account for criteria other than the traditional species richness, especially considering that factors such as biotic interactions, eco-friendly soil management, and invasive species management in PAs are usually ignored. This is critical because most of the publications on Neotropical birds address a single dimension of information (taxonomic) at local scales, whereas fewer studies consider ecological interactions with conservation purposes (Heinen et al. 2020). There are still large literature gaps in this area (Pearson et al. 2019), but predictions of the impact of global changes on bird–plant networks are not optimistic (e.g., Remolina-Figueroa et al. 2022, Sonne et al. 2022, Nuñez Landa et al. 2023). In addition, recent evidence indicates that invasive species are present within PAs (Liu et al. 2020, Rico-Sánchez et al. 2020), which may threaten the biota and ecosystem integrity directly and/or indirectly. All this represents knowledge gaps that restrict our understanding of the future consequences and threats to avifaunal and biodiversity (Haines-Young 2009, Zabel et al. 2019). Therefore, more studies are needed to address this issue. Monitoring and evaluation can increase the ability of PA management to adapt to specific threats by providing sufficient information for decision-making, thereby increasing the likelihood of achieving specific goals (Powlen et al. 2021).

We encourage future researchers to include detailed assessments on how anthropization may promote changes in both alpha and beta birds' diversities, as well as ecosystem functioning within PAs (Salgado-Negret and Paz 2015). The upsurge of interest in developing long-term biomonitoring programs offers relevant baseline information for both in-depth ecological studies on ecosystem dynamics and conservation decision-making for long-term protection (Navarro et al. 2017). Unfortunately, evidence suggests that these biomonitoring programs have been gradually reduced or even cut in recent decades due to financial considerations (Ferreira and Klütsch

2021). For most Neotropical countries, financial resources for conservation are limited and overshadowed by competing economic interests (e.g., Lessmann et al. 2016, Pringle 2017). Thus, PAs will benefit only by strengthening the capacity of local governments to integrate voluntary conservation and development projects (Chazdon et al. 2009), as well as by strengthening environmental legislation and increasing surveillance activities (Rico-Sánchez et al. 2020).

Despite that, our results are valuable to guide future conservation strategies. We argue that it is time for an integrative viewpoint which includes measurements and careful assessments of empirically quantified trends. National assessments are the next natural step in connecting science and decision-makers to make better decisions on biodiversity management and long-term conservation planning (Pearson et al. 2019). For example, in 2005, a major effort of this type was launched in Mexico: El Capital Natural de México (Sarukhán 2008). This was a country-level assessment on the state of knowledge, the status of the components, and the function and use of biodiversity, but also to develop approaches to its conservation and management. Lessons learned from this assessment can be applied to many other megadiverse developing countries. Different studies suggest that it is possible to reduce conflicts with other land uses by integrating socioeconomic, legal, and political actions with strategies to conserve biodiversity and sustain rural livelihoods in agricultural landscapes (Harvey et al. 2008). To succeed, it is important to promote and financially support their long-term maintenance as well as restoration (Naime et al. 2020, Sánchez-Romero et al. 2021).

Forest restoration is now a global priority in recovering ecosystem health and to achieve Sustainable Development Goals (Minnermayer et al. 2011, Crouzeilles et al. 2016). This is particularly important considering that degradation, fragmentation, and loss of tropical forests have exponentially increased in the past few decades, leading to unprecedented rates of species extinction and loss of ecosystem functions and services (Ceballos and Ehrlich 2018). Several studies have shown significant reductions in bird species' abundance resulting from forest patch edge effects (even in unfragmented-forest interior birds) due to area reduction, isolation of patches, and possibly climate change. All these also involve a decrease in quality foraging opportunities (Stouffer 2020, Sherry 2021). Therefore, identifying priority areas and strategies for forest restoration for multiple objectives (i.e., promote ecosystem services provision, increase landscape connectivity, and minimize impacts on local livelihoods) is key for successful conservation outcomes (e.g., Melo et al. 2020, Develey 2021). Building strong partnerships to implement on-the-ground strategies that balance biodiversity conservation and human development is a priority to produce tangible outcomes that both benefit humans and protect biodiversity (e.g., Fitzpatrick and Rodeald 2016, Morán-Ordóñez et al. 2022). Thus, future conservation, restoration, and management actions could be possible if there is also political will.

In this regard, integrative conservation actions across political boundaries are crucial for proper landscape-scale planning and to achieve a connected PA network (Tobón et al. 2017, Saura et al. 2018, Godínez-Gómez et al. 2021). Thus, initiatives aimed at establishing biological corridors would benefit from our results. This information will facilitate the definition of management strategies to stop and reverse the negative effects of human influence on natural landscapes and their bird diversity.

To achieve this, additional efforts involving interdisciplinary and complementary programs such as vegetation restoration, especially to avoid the loss of ecosystem services, are crucial. There is an urgent need to assess the long-term implementation of restoration programs, including habitat traits such as natural vegetation coverage, the complexity of the vegetation, shrub and tree stratum density, resource availability, and habitat connectivity, that favor native bird diversity (Ortega-Alvarez and Lindig-Cisneros 2012, Ramírez-Bastida et al. 2018, Zúñiga-Vega et al. 2019, Sherry 2021).

All of these tasks are not easy because they require the integration of academia, nongovernmental organizations, governments, and decision-makers and local communities, which often have dissimilar interests. Therefore, we need to build bridges among stakeholders to identify and promote sustainable-use strategies. This is an important point to address because most priority areas and biomes/ecosystems are in developing countries. It will also be important to promote financial strategies (e.g., payment for ecosystem services, tax relief, ecotourism, etc.) at local, national, and regional levels (Banks-Leite et al. 2014). Fortunately, positive and effective experiences have been reported for the region (Naime et al. 2020, Sánchez-Romero et al. 2021), especially across priority areas categorized by the IUCN as types V and VI.

We recognize that conservation challenges that we face today are more intractable than they were only a few decades ago. We have acquired a new vision: "Conservation biodiversity requires taking bold steps beyond the protection of areas minimally impacted by past or present human activities" (Chazdon et al. 2009, p. 148). Although the current perception is that much remains to be done, our analysis aimed to serve as a baseline toward future actions to establish conservation programs, taking into account what the birds, their diversity, and the environments demand of us. A new conservation paradigm must incorporate human-modified landscapes in the assessment of biodiversity and ecosystem services, planning of corridors and buffer zones, and restoration of degraded lands (Karp et al. 2013). Management effectiveness is expected to lead to better conservation outcomes as more strategic planning, better monitoring capacity, improved accountability and transparency, and sufficient human and financial resources should theoretically result in more effective enforcement and governance (Powlen et al. 2021). This is perhaps the most important legacy that this conservation work can leave today.

Conclusion

In this chapter, we explored bird distribution and diversity as a proxy to fill conservation gaps, as well as our current viewpoint on the need for the implementation of ornithological knowledge into conservation practice. Several taxa were considered as gap species, but we also identified many important areas and ecoregions poorly represented in PAs. This marks the need for Neotropical countries to increase their PAs, implement long-term effective biomonitoring and conservation programs, as well as promote transboundary conservation agreements. Our findings represent a valuable step to guide the future establishment of new and efficient conservation areas across the region, conveying important recommendations about which areas require attention. These results could be used to convince relevant governments to establish new efforts for efficient conservation planning. This is a good moment to ensure that protected and conservation areas are increasingly strategic and in compliance with the Post-2020 Global Biodiversity Framework. Decision-makers should also translate broad public support for bird conservation to identify and promote sustainable-use strategies across political boundaries, including preserving biodiversity and ecosystem services. This is imperative to move toward sustainability in the use of natural ecosystems in a continent privileged by its diversity of natural resources inhabited by a huge and growing human population.

Acknowledgments

We thank the editors of the book for inviting us to write this chapter. We also thank the Dirección General de Asuntos del Personal Académico from Universidad Nacional Autónoma de México (DGAPA-UNAM; PAPIIT projects IN215818, IN221920, IN214621, and IA202822), the Programa de Investigación en Cambio Climático (PINCC-UNAM [grant to DAP-T and MCA]), and the Rufford Foundation (grants 16017-1, 20284-2, and 28502B to DAP-T) for the financial support that allowed us to perform the research whose results are included in this chapter.

References

Able, K. P. 2000. *Handbook of the Birds of the World, Volume 5, Barn-Owls to Hummingbirds. The Auk* 117:532–534.

Álvarez-Mondragón, E., and J. J. Morrone. 2004. Propuesta de áreas para la conservación de aves de México, empleando herramientas panbiogeográficas e índices de complementariedad. *Interciencia* 29:112–120.

Antonelli, A. 2022. The rise and fall of Neotropical biodiversity. *Botanical Journal of the Linnean Society* 199:8–24.

Arizmendi, M. C., H. Berlanga, C. Rodríguez-Flores, V. Vargas-Canales, L. Montes-Leyva, and R. Lira. 2016. Hummingbird conservation in Mexico: The natural protected areas system. *Natural Areas Journal* 36:366–376.

Arizmendi, M. C., and L. Márquez-Valdelamar. 2000. *Áreas de Importancia Para la Conservación de las Aves en México*. CIPAMEX-CONABIO-CCA, México City, México.

Asamoah, E. F., L. J. Beaumont, and J. M. Maina. 2021. Climate and land-use changes reduce the benefits of terrestrial protected areas. *Nature Climate Change* 11:1105–1110.

Baldi, G. 2020. Nature protection across countries: Do size and power matter? *Journal for Nature Conservation* 56: Article 125860.

Banks-Leite, C., R. Pardini, L. R. Tambosi, W. D. Pearse, A. A. Bueno, R. T. Bruscagin, T. H. Condez, M. Dixo, A. T. Igari, A. C. Martensen, and J. P. Metzger. 2014. Using ecological thresholds to evaluate the costs and benefits of set-asides in a biodiversity hotspot. *Science* 345:1041–1045.

Barr, L. M., R. L. Pressey, R. A. Fuller, D. B. Segan, E. McDonald-Madden, and H. P. Possingham. 2011. A new way to measure the world's protected area coverage. *PLoS One* 6: Article e24707.

Baselga, A., and C. D. L. Orme. 2012. betapart: An R package for the study of beta diversity. *Methods in Ecology and Evolution* 3:808–812.

Bird, J. P., R. Martin, H. R. Akçakaya, J. Gilroy, I. J. Burfield, S. T. Garnett, A. Symes, J. Taylor, Ç. H. Şekercioğlu, and S. H. M. Butchart. 2020. Generation lengths of the world's birds and their implications for extinction risk. *Conservation Biology* 34:1252–1261.

BirdLife International. 2018. *IUCN Red List for Birds: July 2022 Version*. BirdLife International, Cambridge, UK.

BirdLife International. 2021. *Important Bird and Biodiversity Area (IBA) Digital Boundaries: September 2021 Version*. BirdLife International, Cambridge, UK.

Browder, S. F., D. H. Johnson, and I. J. Ball. 2002. Assemblages of breeding birds as indicators of grassland condition. *Ecological Indicators* 2:257–270.

Brum, F. T., C. H. Graham, G. C. Costa, S. B. Hedges, C. Penone, V. C. Radeloff, C. Rondinini, R. Loyola, and A. D. Davidson. 2017. Global priorities for conservation across multiple dimensions of mammalian diversity. *Proceedings of the National Academy of Sciences of the USA* ***114***: 7641–7646.

Buckton, S. E. B. 2001. Threatened Birds of the World. Birdlife International (2000). Barcelona and Cambridge, UK: Lynx Edicions and BirdLife International. 852 pages, £70. *Bird Conservation International* 11:71–75.

Butchart, S. H. M., A. J. Stattersfield, L. A. Bennun, S. M. Shutes, H. R. Akçakaya, J. E. M. Baillie, S. N. Stuart, C. Hilton-Taylor, and G. M. Mace. 2004. Measuring global trends in the status of biodiversity: Red List indices for birds. *PLoS Biology* 2: Article e383.

Buzato, S., M. Sazima, and I. Sazima. 2000. Hummingbird-pollinated floras at three Atlantic forest sites. *Biotropica* 32:824–841.

Cantú-Salazar, L., and K. J. Gaston. 2010. Very large protected areas and their contribution to terrestrial biological conservation. *BioScience* 60:808–818.

Caro, T. 2010. *Conservation by Proxy: Indicator, Umbrella, Keystone, Flagship, and Other Surrogate Species*. Island Press, Washington, DC.

Castillo-Chora, V. D. J., L. A. Sánchez-González, A. Mastretta-Yanes, D. A. Prieto-Torres, and A. G. Navarro-Sigüenza. 2021. Insights into the importance of areas of climatic stability in the evolution and maintenance of avian diversity in the Mesoamerican dry forests. *Biological Journal of the Linnean Society* 132:741–758.

Ceballos, G., and P. R. Ehrlich. 2018. The misunderstood sixth mass extinction. *Science* 360:1080–1081.

Chazdon, R. L., C. A. Harvey, O. Komar, D. M. Griffith, B. G. Ferguson, M. Martínez-Ramos, H. Morales, R. Nigh, L. Soto-Pinto, M. van Breugel, and S. M. Philpott. 2009. Beyond reserves: A research agenda for conserving biodiversity in human-modified tropical landscapes. *Biotropica* 41:142–153.

Ciarleglio, M., J. Wesley Barnes, and S. Sarkar. 2009. ConsNet: New software for the selection of conservation area networks with spatial and multi-criteria analyses. *Ecography* 32: 205–209.

Crouzeilles, R., M. Curran, M. S. Ferreira, D. B. Lindenmayer, C. E. V. Grelle, and J. M. Rey Benayas. 2016. A global meta-analysis on the ecological drivers of forest restoration success. *Nature Communications* 7: Article 11666.

De Vos, J. M., L. N. Joppa, J. L. Gittleman, P. R. Stephens, and S. L. Pimm. 2015. Estimating the normal background rate of species extinction. *Conservation Biology* 29:452–462.

Develey, P. F. 2021. Bird conservation in Brazil: Challenges and practical solutions for a key megadiverse country. *Perspectives in Ecology and Conservation* 19:171–178.

Devenish, C., D. F. Díaz-Fernández, R. P. Clay, I. Davidson, and Í. Yépez-Zabala, eds. 2009. *Important Bird Areas Americas: Priority Sites for Biodiversity Conservation*. BirdLife Conservation Series No. 16. BirdLife International, Quito, Ecuador.

Díaz, S., J. Settele, E. S. Brondízio, H. T. Ngo, J. Agard, A. Arneth, P. Balvanera, K. A. Brauman, S. H. Butchart, and K. M. Chan. 2019. Pervasive human-driven decline of life on Earth points to the need for transformative change. *Science* 366: Article eaax3100.

Dinerstein, E., D. Olson, A. Joshi, C. Vynne, N. D. Burgess, E. Wikramanayake, N. Hahn, S. Palminteri, P. Hedao, R. Noss, M. Hansen, H. Locke, E. C. Ellis, B. Jones, C. V. Barber, R. Hayes, C. Kormos, V. Martin, E. Crist, W. Sechrest, L. Price, J. E. M. Baillie, D. Weeden, K. Suckling, C. Davis, N. Sizer, R. Moore, D. Thau, T. Birch, P. Potapov, S. Turubanova, A. Tyukavina, N. de

Souza, L. Pintea, J. C. Brito, O. A. Llewellyn, A. G. Miller, A. Patzelt, S. A. Ghazanfar, J. Timberlake, H. Klöser, Y. Shennan-Farpón, R. Kindt, J.-P. B. Lillesø, P. van Breugel, L. Graudal, M. Voge, K. F. Al-Shammari, and M. Saleem. 2017. An ecoregion-based approach to protecting half the terrestrial realm. *BioScience* 67:534–545.

Donald, P. F., L. D. C. Fishpool, A. Ajagbe, L. A. Bennun, G. Bunting, I. J. Burfield, S. H. M. Butchart, S. Capellan, M. J. Crosby, M. P. Dias, D. Diaz, M. I. Evans, R. Grimmett, M. Heath, V. R. Jones, B. G. Lascelles, J. C. Merriman, M. O'Brien, I. RamÍRez, Z. Waliczky, and D. C. Wege. 2018. Important Bird and Biodiversity Areas (IBAs): The development and characteristics of a global inventory of key sites for biodiversity. *Bird Conservation International* 29:177–198.

Dudley, N. 2008. *Directrices Para la Aplicación de las Categorías de Gestión de Áreas Protegidas.* Unión Internacional para la Conservación de la Naturaleza, St. Gallen, Switzerland.

Etard, A., A. L. Pigot, and T. Newbold. 2022. Intensive human land uses negatively affect vertebrate functional diversity. *Ecology Letters* 25:330–343.

European Space Agency Climate Change Initiative. 2017. 300m annual global land cover time series from 1992 to 2015. Retrieved May 19, 2022, from https://www.esa-landcover-cci.org.

Ferreira, C. C., and C. Klütsch, eds. 2021. *Closing the Knowledge-Implementation Gap in Conservation Science.* Springer, Cham, Switzerland.

Fitzpatrick, J. W., and A. D. Rodeald. 2016. *Handbook of Bird Biology.* Cornell Lab of Ornithology, Ithaca, NY.

García-Trejo, E. A., and A. G. Navarro. 2004. Patrones biogeográficos de la riqueza de especies y el endemismo de la avifauna en el oeste de México. *Acta Zoológica Mexicana* 20:167–185.

Godínez-Gómez, O., T. Urquiza-Haas, P. Koleff, C. A. Correa, and L. S. Castillo. 2021. ¿Qué tan conectados están los sistemas nacionales de áreas protegidas terrestres en latinoamérica y el Caribe? Pages 5–28 in M. Álvarez Malvido, C. Lázaro, X. De Lamo, D. Juffe-Bignoli, R. Cao, P. Bueno, C. Sofrony, C. Maretti, and F. Guerra, eds. *Informe Planeta Protegido 2020: Latinoamérica y el Caribe.* International Union for Conservation of Nature and Natural Resources, Gland, Switzerland.

Gray, C. L., S. L. L. Hill, T. Newbold, L. N. Hudson, L. Börger, S. Contu, A. J. Hoskins, S. Ferrier, A. Purvis, and J. P. W. Scharlemann. 2016. Local biodiversity is higher inside than outside terrestrial protected areas worldwide. *Nature Communications* 7: Article 12306.

Haines-Young, R. 2009. Land use and biodiversity relationships. *Land Use Policy* 26:S178–S186.

Hannah, L., G. Midgley, S. Andelman, M. Araújo, G. Hughes, E. Martinez-Meyer, R. Pearson, and P. Williams. 2007. Protected area needs in a changing climate. *Frontiers in Ecology and the Environment* 5:131–138.

Harvey, C. A., O. Komar, R. Chazdon, B. G. Ferguson, B. Finegan, D. M. Griffith, M. MartÍNez-Ramos, H. Morales, R. Nigh, L. Soto-Pinto, M. Van Breugel, and M. Wishnie. 2008. Integrating agricultural landscapes with biodiversity conservation in the Mesoamerican hotspot. *Conservation Biology* 22:8–15.

Hazzi, N. A., J. S. Moreno, C. Ortiz-Movliav, and R. D. Palacio. 2018. Biogeographic regions and events of isolation and diversification of the endemic biota of the tropical Andes. *Proceedings of the National Academy of Sciences of the USA* 115:7985–7990.

Heinen, J. H., C. Rahbek, and M. K. Borregaard. 2020. Conservation of species interactions to achieve self-sustaining ecosystems. *Ecography* 43:1603–1611.

Íñigo-Elías, E. E., and E. C. Enkerlin-Hoeflich. 2003. Amenazas, estrategias e instrumentos para la conservación de las aves. Pages 86–132 in H. Gómez de Silva and A. Oliveras de Ita, eds. *Conservación de Aves, Experiencias en México.* CIPAMEX, Morelia, México.

Intergovernmental Science-Policy Platform on Biodiversity and Ecosystem Services. 2019. *The IPBES Global Assessment on Biodiversity and Ecosystem Services.* Secretariat of the Intergovernmental Science-Policy Platform on Biodiversity and Ecosystem Services, Bonn, Germany.

International Union for Conservation of Nature and United Nations Environment Programme World Conservation Monitoring Centre. 2022. Protected Planet: The latest initiative harnessing the world database on protected areas. World Database on Protected Areas (WDPA).

Jennings, M., E. Haeuser, D. Foote, R. Lewison, and E. Conlisk. 2020. Planning for dynamic connectivity: Operationalizing robust decision-making and prioritization across landscapes experiencing climate and land-use change. *Land* 9: Article 341.

Jordano, P. 2016. Chasing ecological interactions. *PLoS Biology* 14: Article e1002559.
Karp, D. S., C. D. Mendenhall, R. F. Sandí, N. Chaumont, P. R. Ehrlich, E. A. Hadly, and G. C. Daily. 2013. Forest bolsters bird abundance, pest control and coffee yield. *Ecology Letters* 16:1339–1347.
Lees, A. C., S. Attwood, J. Barlow, and B. Phalan. 2020. Biodiversity scientists must fight the creeping rise of extinction denial. *Nature Ecology & Evolution* 4:1440–1443.
Leroux, S. J., M. A. Krawchuk, F. Schmiegelow, S. G. Cumming, K. Lisgo, L. G. Anderson, and M. Petkova. 2010. Global protected areas and IUCN designations: Do the categories match the conditions? *Biological Conservation* 143:609–616.
Lessmann, J., J. Fajardo, J. Muñoz, and E. Bonaccorso. 2016. Large expansion of oil industry in the Ecuadorian Amazon: Biodiversity vulnerability and conservation alternatives. *Ecology and Evolution* 6:4997–5012.
Levey, D. R., A. Estrada, P. L. Enríquez, and A. G. Navarro-Sigüenza. 2021. The importance of forest–nonforest transition zones for avian conservation in a vegetation disturbance gradient in the northern Neotropics. *Tropical Conservation Science* 14: Article 194008292110080.
Liu, X., T. M. Blackburn, T. Song, X. Wang, C. Huang, and Y. Li. 2020. Animal invaders threaten protected areas worldwide. *Nature Communications* 11: Article 2892.
Lovejoy, T. E., and L. J. Hannah. 2019. *Biodiversity and climate change: Transforming the biosphere.* Yale University Press, New Haven, CT.
Melo, M. A., M. A. G. D. A. Silva, and A. J. Piratelli. 2020. Improvement of vegetation structure enhances bird functional traits and habitat resilience in an area of ongoing restoration in the Atlantic Forest. *Anais da Academia Brasileira de Ciências* 92: Article e20191241.
Minnermayer, S., L. Laestadius, and N. Sizer. 2011. *A World of Opportunity.* World Resources Institute, Washington, DC.
Moilanen, A., A. Arponen, J. Leppänen, L. Meller, and H. Kujala. 2014. *Spatial Conservation Planning Methods and Software Zonation Version 3.1 User Manual.* Biodiversity Conservation Informatics Group, Department of Biosciences, University of Helsinki, Helsinki, Finland.
Morán-Ordóñez, A., V. Hermoso, and A. Martínez-Salinas. 2022. Multi-objective forest restoration planning in Costa Rica: Balancing landscape connectivity and ecosystem service provisioning with sustainable development. *Journal of Environmental Management* 310: Article 114717.
Morrone, J. J. 2013. Cladistic biogeography of the Neotropical region: Identifying the main events in the diversification of the terrestrial biota. *Cladistics* 30:202–214.
Myers, N., R. A. Mittermeier, C. G. Mittermeier, G. A. B. da Fonseca, and J. Kent. 2000. Biodiversity hotspots for conservation priorities. *Nature* 403:853–858.
Naime, J., F. Mora, M. Sánchez-Martínez, F. Arreola, and P. Balvanera. 2020. Economic valuation of ecosystem services from secondary tropical forests: Trade-offs and implications for policy making. *Forest Ecology and Management* 473: Article 118294.
Navarro, L. M., N. Fernández, C. Guerra, R. Guralnick, W. D. Kissling, M. C. Londoño, F. Muller-Karger, E. Turak, P. Balvanera, M. J. Costello, A. Delavaud, G. Y. El Serafy, S. Ferrier, I. Geijzendorffer, G. N. Geller, W. Jetz, E.-S. Kim, H. Kim, C. S. Martin, M. A. McGeoch, T. H. Mwampamba, J. L. Nel, E. Nicholson, N. Pettorelli, M. E. Schaepman, A. Skidmore, I. Sousa Pinto, S. Vergara, P. Vihervaara, H. Xu, T. Yahara, M. Gill, and H. M. Pereira. 2017. Monitoring biodiversity change through effective global coordination. *Current Opinion in Environmental Sustainability* 29:158–169.
Navarro-Sigüenza, A. G., A. Lira-Noriega, M. D. C. Arizmendi, P. Koleff, J. García-Moreno, A. T. Peterson, P. Koleff, and T. Urquiza-Haas. 2011. Áreas de conservación para las aves: hacia la integración de criterios de priorización. Pages 109–129 in P. Koleff and T. Urquiza-Haas, eds. *Planeación Para la Conservación de la Biodiversidad Terrestre en México: Retos en un País Megadiverso.* Comisión Nacional Para el Conocimiento y Uso de la Biodiversidad and Comisión Nacional de Áreas Naturales Protegidas, México City, México.
Nori, J., R. Loyola, and F. Villalobos. 2020. Priority areas for conservation of and research focused on terrestrial vertebrates. *Conservation Biology* 34:1281–1291.
Nuñez Landa, M. d. L., J. C. Montero Castro, T. C. Monterrubio-Rico, S. I. Lara-Cabrera, and D. A. Prieto-Torres. 2023. Predicting co-distribution patterns of parrots and woody plants under global changes: The case of the Lilac-Crowned Amazon and Neotropical dry forests. *Journal for Nature Conservation* 71: Article 126323.

Ortega-Alvarez, R., and R. Lindig-Cisneros. 2012. Feathering the scene: The effects of ecological restoration on birds and the role birds play in evaluating restoration outcomes. *Ecological Restoration* 30:116–127.

Pearson, R. G., E. Martínez-Meyer, M. A. Velázquez, M. Caron, R. O. Corona-Núñez, K. Davis, A. P. Durán, R. García-Morales, T. D. Hackett, D. J. Ingram, R. L. Díaz, J. Lescano, A. Lira-Noriega, Y. López-Maldonado, D. Manuschevich, A. Mendoza, B. Milligan, S. C. Mills, D. Moreira-Arce, L. F. Nava, V. Oostra, N. Owen, D. Prieto-Torres, C. R. Soto, T. Smith, A. J. Suggitt, C. T. Haristoy, J. Velásquez-Tibatá, S. Díaz, and P. A. Marquet. 2019. Research priorities for maintaining biodiversity's contributions to people in Latin America. *UCL Open Environment* 1: Article e002.

Peterson, A. T., and A. G. Navarro-Sigüenza. 2017. What bird specimens can reveal about species-level distributional ecology. Pages 111–126 in M. S. Webster, ed. *The Extended Specimen: Emerging Frontiers in Collections-Based Ornithological Research*. CRC Press, Boca Raton, FL.

Pimm, S. L., C. N. Jenkins, R. Abell, T. M. Brooks, J. L. Gittleman, L. N. Joppa, P. H. Raven, C. M. Roberts, and J. O. Sexton. 2014. The biodiversity of species and their rates of extinction, distribution, and protection. *Science* 344: Article 1246752.

Powlen, K. A., M. C. Gavin, and K. W. Jones. 2021. Management effectiveness positively influences forest conservation outcomes in protected areas. *Biological Conservation* 260: Article 109192.

Prieto-Torres, D. A., S. Díaz, J. M. Cordier, R. Torres, M. Caron, and J. Nori. 2022. Analyzing individual drivers of global changes promotes inaccurate long-term policies in deforestation hotspots: The case of Gran Chaco. *Biological Conservation* 269: Article 109536.

Prieto-Torres, D. A., A. G. Navarro-Sigüenza, D. Santiago-Alarcon, and O. R. Rojas-Soto. 2016. Response of the endangered tropical dry forests to climate change and the role of Mexican Protected Areas for their conservation. *Global Change Biology* 22:364–379.

Prieto-Torres, D. A., J. Nori, and O. R. Rojas-Soto. 2018. Identifying priority conservation areas for birds associated to endangered Neotropical dry forests. *Biological Conservation* 228:205–214.

Prieto-Torres, D. A., J. Nori, O. R. Rojas-Soto, and A. G. Navarro-Sigüenza. 2021. Challenges and opportunities in planning for the conservation of Neotropical seasonally dry forests into the future. *Biological Conservation* 257: Article 109083.

Prieto-Torres, D. A., L. E. Nuñez Rosas, D. Remolina Figueroa, and M. D. C. Arizmendi. 2021. Most Mexican hummingbirds lose under climate and land-use change: Long-term conservation implications. *Perspectives in Ecology and Conservation* 19:487–499.

Prieto-Torres, D. A., L. A. Sánchez-González, M. F. Ortiz-Ramírez, J. E. Ramírez-Albores, E. A. García-Trejo, and A. G. Navarro-Sigüenza. 2021. Climate warming affects spatio-temporal biodiversity patterns of a highly vulnerable Neotropical avifauna. *Climatic Change* 165: Article 57.

Pringle, R. M. 2017. Upgrading protected areas to conserve wild biodiversity. *Nature* 546:91–99.

Ramírez-Albores, J. E., D. A. Prieto-Torres, A. Gordillo-Martínez, L. E. Sánchez-Ramos, and A. G. Navarro-Sigüenza. 2021. Insights for protection of high species richness areas for the conservation of Mesoamerican endemic birds. *Diversity and Distributions* 27:18–33.

Ramírez-Bastida, P., A. Meléndez-Herrada, E. G. López-Saut, S. Saldaña-Martínez, A. Ruiz-Rodríguez, M. Vargas-Gómez, A. R. Cruz-Nava, M. I. Dávalos-Fong, U. D. García, and C. Sánchez-Sánchez. 2018. Importancia de los ambientes acuáticos urbanos para las aves nativas: el caso de la zona metropolitana de la Ciudad de México. Pages 5–28 in A. Ramírez-Bautista and R. Pineda-López, eds. *Ecología y Conservación de Fauna en Ambientes Antropizados*. REFAMA-CONACyT-UAQ, Querétaro, México.

Remolina-Figueroa, D., D. A. Prieto-Torres, W. Dáttilo, E. Salgado Díaz, L. E. Nuñez Rosas, C. Rodríguez-Flores, A. G. Navarro-Sigüenza, and M. D. C. Arizmendi. 2022. Together forever? Hummingbird–plant relationships in the face of climate warming. *Climatic Change* 175: Article 2.

Rico-Sánchez, A. E., A. Sundermann, E. López-López, M. J. Torres-Olvera, S. A. Mueller, and P. J. Haubrock. 2020. Biological diversity in protected areas: Not yet known but already threatened. *Global Ecology and Conservation* 22: Article e01006.

Rodrigues, A. S. L., S. J. Andelman, M. I. Bakarr, L. Boitani, T. M. Brooks, R. M. Cowling, L. D. C. Fishpool, G. A. B. da Fonseca, K. J. Gaston, M. Hoffmann, J. S. Long, P. A. Marquet, J. D. Pilgrim,

R. L. Pressey, J. Schipper, W. Sechrest, S. N. Stuart, L. G. Underhill, R. W. Waller, M. E. J. Watts, and X. Yan. 2004. Effectiveness of the global protected area network in representing species diversity. *Nature* 428:640–643.

Rodríguez-Echeverry, J., and M. Leiton. 2021. State of the landscape and dynamics of loss and fragmentation of forest critically endangered in the tropical Andes hotspot: Implications for conservation planning. *Journal of Landscape Ecology* 14:73–91.

Rojas-Soto, O. R., V. Sosa, and J. F. Ornelas. 2012. Forecasting cloud forest in eastern and southern Mexico: Conservation insights under future climate change scenarios. *Biodiversity and Conservation* 21:2671–2690.

Salgado-Negret, B., and H. Paz. 2015. Escalando de los rasgos funcionales a procesos poblacionales, comunitarios y ecosistémicos. Pages 12–35 in B. Salgado-Negret, ed. *La Ecología Funcional Como Aproximación al Estudio, Manejo y Conservación de la Biodiversidad: Protocolos y Aplicaciones.* Instituto de Investigación de Recursos Biológicos Alexander von Humboldt, Bogota, Columbia.

Sánchez-Ramos, L. E., A. Gordillo-Martínez, C. R. Gutiérrez-Arellano, T. Kobelkowsky-Vidrio, C. A. Ríos-Muñoz, and A. G. Navarro-Sigüenza. 2018. Bird diversity patterns in the nuclear Central American highlands: A conservation priority in the northern Neotropics. *Tropical Conservation Science* 11: Article 194008291881907.

Sánchez-Romero, R., P. Balvanera, A. Castillo, F. Mora, L. E. García-Barrios, and C. E. González-Esquivel. 2021. Management strategies, silvopastoral practices and socioecological drivers in traditional livestock systems in tropical dry forests: An integrated analysis. *Forest Ecology and Management* 479: Article 118506.

Sarkar, S., and P. Illoldi-Rangel. 2010. Systematic conservation planning: An updated protocol. *Natureza & Conservação* 8:19–26.

Sarukhán, J. 2008. *Capital Natural y Bienestar Social.* Comisión Nacional Para el Conocimiento y Uso de la Biodiversidad, México City, México.

Saura, S., B. Bertzky, L. Bastin, L. Battistella, A. Mandrici, and G. Dubois. 2018. Protected area connectivity: Shortfalls in global targets and country-level priorities. *Biological Conservation* 219:53–67.

Sekercioglu, C. 2006. Increasing awareness of avian ecological function. *Trends in Ecology & Evolution* 21:464–471.

Şekercioğlu, Ç. H., R. B. Primack, and J. Wormworth. 2012. The effects of climate change on tropical birds. *Biological Conservation* 148:1–18.

Sherry, T. W. 2021. Sensitivity of tropical insectivorous birds to the Anthropocene: A review of multiple mechanisms and conservation implications. *Frontiers in Ecology and Evolution* 9: Article 662873.

Sonne, J., B. Dalsgaard, M. K. Borregaard, J. Kennedy, J. Fjeldså, and C. Rahbek. 2022. Biodiversity cradles and museums segregating within hotspots of endemism. *Proceedings of the Royal Society B: Biological Sciences* 289: Article 20221102.

Stattersfield, A., M. Crosby, A. Long, and D. Wege. 1998. Endemic bird areas of the world: Priorities for biodiversity conservation. *The Auk* 115:1089–1091.

Stotz, D. F., J. W. Fitzpatrick, T. A. I. Parker, and D. K. Moskovits. 1996. *Neotropical Birds: Ecology and Conservation.* University of Chicago Press, Chicago.

Stouffer, P. C. 2020. Birds in fragmented Amazonian rainforest: Lessons from 40 years at the Biological Dynamics of Forest Fragments Project. *The Condor* 122: Article duaa005.

Teixeira, E. C., J. S. Santos, M. R. G. Silva, A. C. M. Malhado, R. J. Ladle, and V. S. Batista. 2018. Scientific research in Neotropical protected areas: Themes and gaps. *PeerJ* Preprints 6: Article e27086v1.

Tobón, W., T. Urquiza-Haas, P. Koleff, M. Schröter, R. Ortega-Álvarez, J. Campo, R. Lindig-Cisneros, J. Sarukhán, and A. Bonn. 2017. Restoration planning to guide Aichi targets in a megadiverse country. *Conservation Biology* 31:1086–1097.

Venter, O., E. W. Sanderson, A. Magrach, J. R. Allan, J. Beher, K. R. Jones, H. P. Possingham, W. F. Laurance, P. Wood, and B. M. Fekete. 2018. Last of the Wild Project, Version 3 (LWP-3): 2009 human footprint, 2018 release. NASA Socioeconomic Data and Applications Center (SEDAC), Palisades, NY.

Venter, O., E. W. Sanderson, A. Magrach, J. R. Allan, J. Beher, K. R. Jones, H. P. Possingham, W. F. Laurance, P. Wood, B. M. Fekete, M. A. Levy, and J. E. M. Watson. 2016. Sixteen years of change in the global terrestrial human footprint and implications for biodiversity conservation. *Nature Communications* 7: Article 12558.

Vieira, R. R. S., R. L. Pressey, and R. Loyola. 2019. The residual nature of protected areas in Brazil. *Biological Conservation* 233:152–161.

Woodley, S., H. Locke, D. Laffoley, K. MacKinnon, T. Sandwith, and J. Smart. 2019. A review of evidence for area-based conservation targets for the Post-2020 Global Biodiversity Framework. *Parks* 25:31–46.

Xu, H., Y. Cao, D. Yu, M. Cao, Y. He, M. Gill, and H. M. Pereira. 2021. Ensuring effective implementation of the post-2020 global biodiversity targets. *Nature Ecology & Evolution* 5:411–418.

Zabel, F., R. Delzeit, J. M. Schneider, R. Seppelt, W. Mauser, and T. Václavík. 2019. Global impacts of future cropland expansion and intensification on agricultural markets and biodiversity. *Nature Communications* 10: Article 2844.

Zafra-Calvo, N., and J. Geldmann. 2020. Protected areas to deliver biodiversity need management effectiveness and equity. *Global Ecology and Conservation* 22: Article e01026.

Zúñiga-Vega, J. J., I. Solano-Zavaleta, M. F. Sáenz-Escobar, and G. A. Ramírez-Cruz. 2019. Habitat traits that increase the probability of occupancy of migratory birds in an urban ecological reserve. *Acta Oecologica* 101: Article 103480.

24
The Introduction, Management, and Utilization of Tropically Adapted Chickens in Sub-Saharan Africa

Abdulmojeed Yakubu, Oladeji Bamidele, and Oludayo Michael Akinsola

Chickens, the world's most numerous and widely used farm animals, have a significant genetic diversity (Tiemann et al. 2020, Hata et al. 2021, Lawal and Hanotte 2021). Indigenous chickens (ICs) of small-scale holdings constitute 90% of total poultry production in some low- and middle-income countries (LMICs) and contribute substantially to rural households through enhanced diet, nutritional diversity, improved livelihoods, and food security (Hoffmann 2005, Wong et al. 2017, Mathiu et al. 2021, Mujyambere et al. 2021, Rajkumar et al. 2021). Also, IC has contributed significantly to fulfilling social and cultural obligations (Manyelo et al. 2020, Banda and Tanganyika 2021). The current state of knowledge on IC genetic resources of Africa with regard to domestication, distribution, importance, and performance has been reviewed (Desta 2021a, Pius et al. 2021).

Despite the multifunctional roles of the indigenous birds in Africa, they have low levels of production and productivity (Yakubu 2010, Desta 2021b, Ogbu 2021). This is worrisome because approximately 1.3 billion people in the world depend on livestock for their livelihoods, and the demand for animal-source foods (meat, eggs, and milk) is expected to triple in Africa by 2050, while its population is projected to almost double, reaching approximately 2.5 billion people in 2050 (Latino et al. 2020). The 2019 EAT-*Lancet* planetary health report suggested that "a diet rich in plant-based foods and with fewer animal source foods confers both improved health and environmental benefits" (EAT 2019). However, many populations, especially in LMICs, continue to face significant burdens of undernutrition, and obtaining adequate quantities of micronutrients from plant source foods alone can be difficult. Therefore, there is a need for new innovations, technologies, and approaches to increase smallholder poultry production to complement diets from plant-based sources. It is in line with one of the strategies of the United Nations' Sustainable Development Goals (SDGs) (Food and Agriculture Organization [FAO] 2019) to increase the potential of food systems to grow rural economies and generate diversified and equitable livelihoods (Independent Group of Scientists 2019, International Fund for Agricultural Development 2021). In this direction, smallholder farmers in Africa require hybrid, improved, dual-purpose chickens of high productivity, disease resistance, and climate resilience that can produce high-quality meat and eggs at earlier ages with better

Abdulmojeed Yakubu et al., *The Introduction, Management, and Utilization of Tropically Adapted Chickens in Sub-Saharan Africa*. In: *New Perspectives in Ornithology*. Edited by: Scott V. Edwards and J. Michael Reed, Oxford University Press.
 DOI: 10.1093/oso/9780197787670.003.0024

economic returns than their indigenous counterparts. Such superior, low-input–high-output, improved tropically adapted breeds (iTABs) of chickens have recently been tested on-farm and on-station in some countries in sub-Saharan Africa. These include ShikaBrown, FUNAAB Alpha, Noiler, Potchefstroom Koekoek, Sasso, and Kuroiler breeds (Yakubu and Ari 2018, Abegaz et al. 2019, Bamidele et al. 2020, Yakubu et al. 2020, Akinsola et al. 2021). This chapter addresses the distribution, production potential, adaptive capability, sustainable utilization, trait preferences, genetic improvement, and strategies for continuous availability of the six introduced iTABs in Africa.

Origin and Distribution of Improved Tropically Adapted Chickens in Sub-Saharan Africa

Origin

The improved tropically adapted birds have resulted from artificial selection (Bosse 2019). The Potchefstroom Koekoek is a dual-purpose breed that was developed in the 1950s at Potchefstroom Agricultural College in South Africa (Dessie and Getachew 2016, Hlokoe et al. 2022). Sasso dual-purpose birds came into being in 1978 through Hendrix Genetics, a private company based in France (Hendrix Genetics 2022). Kuroiler hybrid dual-purpose chicken was created by Keggs Farms in India in the early 1990s (Dessie and Getachew 2016). ShikaBrown was developed by the National Animal Production Research Institute (NAPRI), Shika, Zaria, and registered as the first Nigerian improved layer breed for commercialization in 1998 (NAPRI 1998). Noiler, a dual-purpose breed that was introduced into the Nigerian market in 2014, is a product of Amo Farm Sieberer Hatchery, a pure line breeding company in Awe, Oyo State, Nigeria. FUNAAB Alpha originated from the Federal University of Agriculture, Abeokuta, Nigeria, and was registered as the second improved chicken breed (dual-purpose) in Nigeria on July 26, 2018, by the National Centre for Genetic Resources and Biotechnology, an agency under the Ministry of Science and Technology of the Federal Republic of Nigeria.

Distribution

Although there is no official census indicating the populations of the six iTABs, a report (Oduntan 2020) shows that 29 million Noiler day-old chicks were supplied to smallholder farmers in West Africa from 2014 to 2020. From this figure, 3,648 mother units and 778,492 smallholder farmers benefitted. Sasso is widely distributed in more than 50 countries across the continents of Africa, North America, South America, Asia, and Europe (Hendrix Genetics 2022). Kuroiler is popular among farmers in Uganda, where it was first introduced to Africa and neighboring countries such as Kenya, Malawi, and Rwanda (Vernooij et al. 2018, Department for International Development 2019). Kuroiler is also being extended to Ghana (Dessie 2021).

On-farm and on-station performance testing of the six improved tropically adapted chicken breeds was recently performed in Ethiopia, Nigeria, and Tanzania under the African Chicken Genetic Gains project (2015–2019) along with six indigenous strains (Abegaz et al. 2019). The on-farm studies were carried out in five zones of each country in 6,348 households under a smallholder semi-scavenging production system.

Utilization of Improved Tropically Adapted Chickens

The tropically adapted breeds of chickens have played a significant role in increasing the meat yield and egg number in a sustainable manner under smallholder production systems (Bamidele et al. 2020). More income is also derived from the sale of the eggs, live and dressed birds, and their byproducts for human consumption; in some markets, IC meat attracts a premium price for its flavor. They are also important in upgrading smallholder poultry to small-scale profit-oriented systems (Alemu et al. 2021), which is central to the empowerment of women and youth (although men tend to dominate commercialized value chains; Chawala et al. 2022) who are directly involved in poultry production. This has been reported to impact positively on food security; improve nutrition in resource-poor settings; and have positive effects on vulnerable groups, health, poverty alleviation, and livelihoods of the farming populace within the context of rural and peri-urban development (O. O. Alabi et al. 2020, Passarelli et al. 2020, Geremew and Dessie 2021). Associated benefits include increased activities with respect to input supply, processing, packaging, and distribution, thereby expanding the business opportunities in the chicken value chain (Birhanu, Esatu, et al. 2021; Birhanu, Geremew, et al. 2021). Generally, the benefits derived from small-scale poultry-keeping have direct relevance to the first three United Nations' SDGs: no poverty (SDG1), zero hunger (SDG2), and good health and well-being (SDG3) (FAO 2018, 2019, 2021).

Morphobiometric Traits

Phenotypic characterization is a significant tool in breeding programs that allows the preservation of animal biodiversity and supports consumer demands (Brito et al. 2021). In poultry, genetic differences between populations can be determined by body size and skeletal dimensions (Mushi et al. 2020, Shoyombo et al. 2021). Therefore, standardization and classification of iTAB populations based on qualitative (morphological) and quantitative (morphometric) traits are required for future genetic improvement strategies.

Kuroiler was larger in 6-week body weight (450.86 vs. 416.82 vs. 228.66 g), neck circumference (6.11 vs. 5.82 vs. 4.03 cm), thigh circumference (6.92 vs. 6.15 vs. 4.55 cm), and shank circumference (3.87 vs. 3.42 vs. 2.98 cm) and longer in back length (19.25 vs. 17.68 vs. 13.28 cm), wing length (13.02 vs. 12.48 vs. 10.80 cm), thigh length (7.76 vs. 7.52 vs. 6.34 cm), and shank length (4.93 vs. 4.66 vs. 3.46 cm) than its Sasso and indigenous Fulani counterparts in Nigeria, respectively (Table 24.1). However,

Table 24.1 Strain Effect on Body Weight and Linear Body Measurements of 42-Day-Old Chickens of Equal Sexes of Different Strains

		Strain		
		Sasso	**Kuroiler**	**Fulani**
Trait	*n*	Mean ± SE	Mean ± SE	Mean ± SE
Body weight (g)	150	416.82 ± 6.49	450.86 ± 6.49	228.66 ± 6.49
Breast girth (cm)	150	17.72 ± 0.15	18.116 ± 0.15	12.66 ± 0.15
Neck circumference (cm)	150	5.82 ± 0.07	6.11 ± 0.07	4.03 ± 0.07
Back length (cm)	150	17.68 ± 0.20	19.25 ± 0.20	13.28 ± 0.20
Wing length (cm)	150	12.48 ± 0.11	13.02 ± 0.11	10.80 ± 0.11
Thigh length (cm)	150	7.52 ± 0.07	7.76 ± 0.07	6.34 ± 0.07
Thigh circumference (cm)	150	6.15 ± 0.10	6.92 ± 0.10	4.55 ± 0.10
Shank length (cm)	150	4.66 ±0.07	4.93±0.07	3.46 ±0.07
Shank circumference (cm)	150	3.42 ±0.04	3.87 ±0.04	2.98 ±0.04

Source: Yakubu and Ari (2018).

Kuroiler and Sasso were similar but had larger chest circumferences than the Fulani birds (18.12 vs. 17.72 vs. 21.66 cm, respectively) (Yakubu and Ari 2018).

In a 20-week on-station trial in Nigeria, Kuroiler and Sasso had higher body height (31.5 vs. 30.3 vs. 28.7 vs. 28.6 vs. 28.1 cm) and shank length (13.2 vs. 12.5 vs. 10.0 vs. 9.4 vs. 10.2 cm) than Noiler, FUNAAB Alpha, and ShikaBrown, respectively, but they had similar body length as Noiler (45.8 vs. 45.1 vs. 44.5 cm, respectively). Lowest values were recorded for the indigenous Fulani in most of the traits (Oyewale et al. 2021). In an on-farm study in Ethiopia, there was agroecology × genotype interaction in mature body weight and some linear body measurements, although Sasso appeared to be superior to Koekoek, which in turn was larger in size than the local birds (Assefa et al. 2018). The following ranges were obtained in the lowland and midland zones, respectively, for the three genotypes: body weight (2.05–3.38; 1.98–3.19 kg), body length (34.2–42.3; 33.9–40.0 cm), wing span (30.4–38.2; 33.4–36.4 cm), chest width (25.4–34.1; 26.5–33.3 cm), shank length (6.71–7.69; 7.29–7.71 cm), and shank circumference (4.38–5.77; 4.46–5.92 cm). For all measures and in all genotypes, male birds were larger. O. J. Alabi et al. (2012) also found sex differences in all traits and reported the following 22-week body parameters for Koekoek cocks and hens, respectively, in South Africa: body weight (2.51 vs. 1.70 kg), body length (43.33 vs. 39.80 cm), chest circumference (46.67 vs. 36.00 cm), wing length (19.67 vs. 18.90 cm), shank thickness (5.50 vs. 4.20 cm), and shank length (29.33 vs. 25.60 cm). The larger size of the iTABs could be attributed to their higher genetic potential compared to their IC counterparts. Sexual differences in body traits could be as a result of between-sex differential hormonal action (Baéza et al. 2001).

There is virtually no comprehensive work in the literature regarding within- and between-breed comparison of the qualitative traits of iTABs such as color patterns, comb types, comb size, head shape, body shape, feather morphology, feather growth, and feather distribution. This calls for future investigation because information on qualitative traits will aid the identification and classification of iTABs with regard to their adaptability to varying environmental conditions.

Production Systems

Poultry production systems are generally classified into three sectors based on several characteristics, among which biosecurity, housing, product marketing, and breed type are primary (FAO 2004, 2007c). Based on these characteristics, poultry production systems are categorized as industrial/integrated, commercial, and village or backyard. In practice, the rearing of iTABs is predominantly within the village or backyard sector. This sector is characterized by family poultry (i.e., keeping of chickens by households based on purpose and use of family labor) and can be further subdivided into four categories based on the level, scale, and intensity of management (Table 24.2).

Studies across East Africa and West Africa have shown that most smallholder farmers raise iTABs under the semi-scavenging production system of the village or backyard poultry (Moges et al. 2010, Guteta and Ameha 2020, Kassa et al. 2021). The adoption of the semi-scavenging production system for rearing of improved chickens hinges on its suitability for providing high output with minimal inputs. In a recent survey (N = 350) conducted in Nigeria, a larger percentage of the farmers (49.2%) practiced semi-scavenging compared to semi-intensive (37.1%), scavenging (13.7%), and intensive (0%) production systems (Bamidele, Amole, et al. 2022).

Nutrition

The nutritional requirements of improved chickens are based on the premise that improved genetics require improved feeds and feeding for improved productivity and performance. The nutritional profiles, as evidenced by the feed conversion ratio (FCR) and feed/dozen eggs, of the iTABs under on-station, intensive production system demonstrated the full genetic potentials of the improved tropically adapted chickens for dual-purpose (meat and egg) functions under the smallholder poultry production systems. Table 24.3 shows that the FCR (0–20 weeks for both sexes) and feed/dozen eggs (kg) of the iTABs raised in Nigeria ranged from 5.4 (Noiler) to 8.9 (FUNAAB Alpha) and 3.0 kg (ShikaBrown) to 8.5 kg (Sasso), respectively. In Ethiopia, FCR for the ITABS ranged from 5.3 (Kuroiler) to 7.4 (Horro), whereas feed/dozen eggs ranged from 1.9 kg (Sasso-RIR) to 3.2 kg (Horro). The FCR (6–20 weeks) and feed/dozen eggs of Sasso and Kuroiler in Tanzania were 4.8 and 3.8 and 5.8 and 3.9 kg, respectively. Overall, the FCR and feed/dozen eggs of the improved breeds were superior compared to those previously reported for local chickens raised under similar management (intensive) conditions across Africa (Nigeria, Ghana, Ethiopia, Kenya, and Tanzania) (Tadelle et al. 2000, Ajayi 2010, Osei-Amponsah et al. 2015, Chimenem-Amadi et al. 2021, Mujyambere et al. 2021).

Under on-farm conditions, the nutritional challenge is with the availability of scavenging feed resources as well as the scavenging ability of the improved chickens. The scavenging feed resource base available for the production of improved chickens

Table 24.2 Characteristics of the Village or Backyard Poultry Sector

Characteristics	Village or Backyard Sector			
	Scavenging	Semi-Scavenging	Semi-Intensive	Intensive
Environment	Rural	Rural	Rural, urban, and peri-urban	Urban, peri-urban
Movement	Unrestricted (free-range)	Unrestricted (free-range)	Confined to an enclosed area (runs)	Restricted (battery cages, pens)
Shelter/housing	None	Nighttime	Nighttime	All day
Feeds and feeding	No supplementary feeding and water	Supplementary feeding (household waste, farm residues) and water	Supplementary feeding (household waste, farm residues, proprietary feed) and water	Recommended daily feeding (proprietary/compounded feed) and water
Breed type	Local	Local, and improved	Local, improved, hybrid	Improved, hybrid, and exotic
Breed function	Dual-purpose	Dual-purpose	Dual-purpose	Dual-purpose, single function
Source of chicks/restocking	Natural incubation by broody hens	Natural incubation by broody hens or purchase of day-old chicks/brooded chicks	Natural incubation by broody hens or purchase of day-old chicks/brooded chicks	Purchase of day-old chicks or brooded chicks
Purpose	Subsistence	Subsistence and profit-oriented (minimal)	Profit-oriented (moderate)	Profit-oriented (high)
Markets	Rarely	Hawking at roadsides/highways, village markets	Hawking at roadsides/highways, village/city markets, access to offtakers and aggregators	City markets, access to offtakers and aggregators
Vaccination	None	Prevaccinated chicks of the improved breeds against Newcastle, Mareks, and Gumboro disease	Prevaccinated chicks of the improved breeds and intermittent vaccination for other age groups and bird type	Prevaccinated chicks and routine vaccination schedule for all age groups and bird type
Input–output	Low input–low output	Minimal input–high output	Moderate input–high output	High input–high output

Adapted from FAO (2004, 2007c, 2014).

Table 24.3 Feed Conversion Ratio and Feed/Dozen Eggs of the Improved, Tropically Adapted Breeds Under On-Station Conditions

Country	iTABs	FCR (Mean ± SD)	Feed/Dozen Eggs (kg; Mean ± SD)
Ethiopia	Horro	7.4 ± 3.6	3.2 ± 4.2
	Koekoek	5.7 ± 1.4	2.4 ± 1.6
	Kuroiler	5.3 ± 1.1	2.8 ± 0.8
	Sasso-RIR	6.3 ± 2.5	1.9 ± 1.0
Nigeria	FUNAAB Alpha	8.9 ± 1.9	3.1 ± 1.0
	Noiler	5.4 ± 1.3	3.6 ± 1.6
	ShikaBrown	8.7 ± 2.1	3.0 ± 0.7
	Kuroiler	6.9 ± 3.0	4.6 ± 1.9
	Sasso	6.4 ± 2.8	8.5 ± 0.4
Tanzania	Sasso	4.8 ± 0.1	3.8 ± 1.3
	Kuroiler	5.8 ± 0.1	3.9 ± 0.8

FCR, feed conversion ratio; iTABs, improved, tropically adapted breeds.

in rural and peri-urban communities is dependent on several agroecological factors (season and biodiversity) and anthropogenic activities (food waste, garbage disposal, and gardening) as well as the stocking density around the households. The provision of supplementary feeding and water by farmers has been adopted as an important strategy to combat the nutritional challenge facing the production of iTABs under smallholder poultry production systems. In addition to the scavenging feed resource base, supplementary feeding is an important characteristic of the semi-scavenging system of production practiced by most smallholder farmers, and it has contributed to the superior growth and laying performance of the improved breeds introduced into different agroecologies in sub-Saharan Africa.

Production Attributes

Meat Production

The production performance of improved chickens, relative to local chickens, has been tested across multiple agroecologies (within country and between countries) under various conditions (on-station and on-farm) and production systems. The growth performance is defined by the liveweights of the male birds at 18–20 weeks (on-station and on-farm). These liveweights form the basis of the dressing percentages (DPs) for the chickens. Figure 24.1 shows the male liveweights of some iTABs under on-station conditions in selected African countries. The increase in body weight over the local chickens in Ethiopia (698 g), Nigeria (975 g), and Tanzania (1,609 g) was 231–280%, 136–304%, and 144–168%, respectively. The DPs observed for these breeds in Nigeria and Tanzania ranged from 60.9% (Fulani) to 69.3%

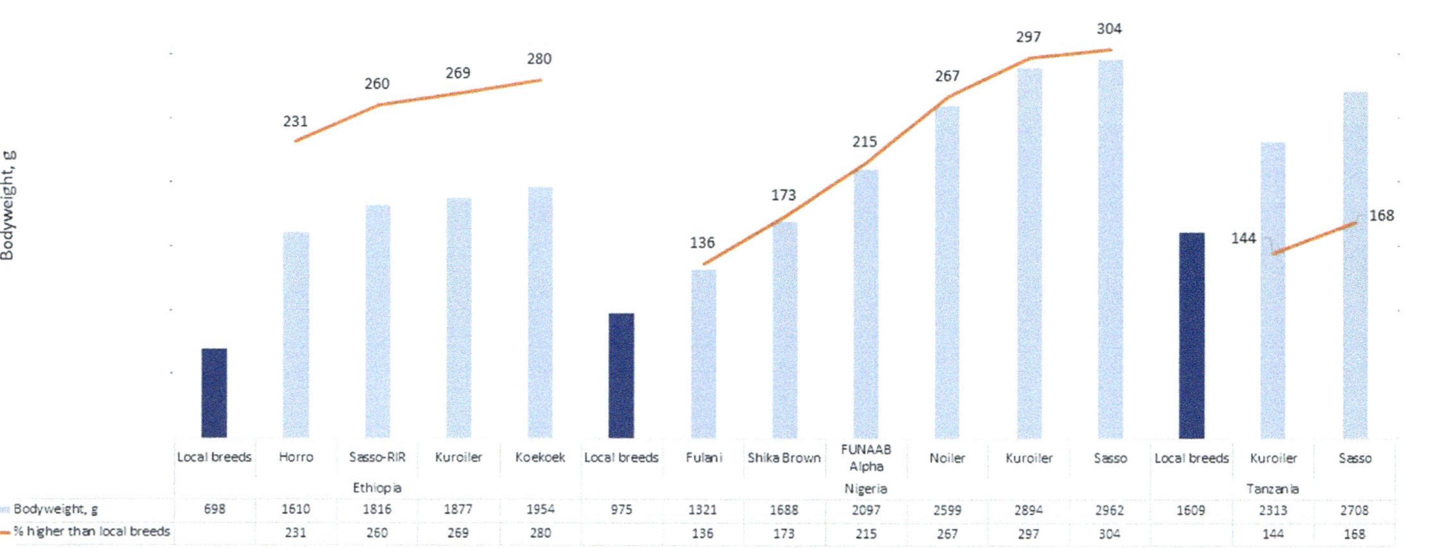

Figure 24.1 Live weights of male birds (20 weeks) of the iTABs in Ethiopia, Nigeria, and Tanzania under on-station conditions.

Sources: Bamidele et al. (2019); and Guni Mbaga, Katule, and Goromela (2021).

(Noiler) and from 69.5% (Kuroiler) to 72.1% (Sasso), respectively (Guni, Mbaga, Katule, and Goromela 2021; Oyewale et al. 2021). The DPs of the improved chickens in Tanzania were higher than that previously reported for the local chickens in Tanzania (66.2%) (Sanka and Mbaga 2014), and the DPs of the improved chickens in Nigeria were lower than the percentages (70.6–77.2%) reported for exotic chickens reared under similar conditions in Nigeria (Fadare et al. 2020). The growth performances of the improved breeds under on-farm conditions in the selected countries were 128–155% (Ethiopia), 120–215% (Nigeria), and 149–171% (Tanzania) higher than those of the local breeds in the respective countries: These values were higher than those of the local chickens raised under similar village or backyard production systems in the respective countries (Figure 24.2). Higher liveweights of iTABs are suggestive of higher dressing percentages and meat production compared to the local, unimproved chickens. Figure 24.3 shows the differences in the on-station and on-farm liveweights of both the iTABs and local breeds across the three countries.

Egg Production

The age at first egg, total number of eggs/hen/year, and the average egg weight are parameters used in defining the laying performances of the iTABs. Table 24.4 shows the average laying performances for the improved chicken breeds under on-station and on-farm conditions. Overall, the improved chicken breeds had superior performance compared to the local chickens at age at first egg, average number of eggs/hen/year, and egg weight. The average egg number/hen/year for the improved chickens was 152% (on-station) and 142% (on-farm) higher than that of the local chickens. Compared with local chickens, the average egg weight of the improved chickens was higher by 125% (on-station) and 143% (on-farm).

The high growth and laying performance of the improved chickens, relative to the local chickens, supports the use of the improved tropically adapted chickens as low-input–high-output, dual-purpose breeds (meat and eggs) for the development of smallholder poultry production and improvement in the livelihoods and food security of farmers.

Reproduction

A major characteristic of improved tropically adapted chickens is the lack of broodiness, which has resulted in increased egg production for household consumption and sales. The implication of this is that at the household level, bird multiplication is achieved primarily through the purchase of day-old-chicks (DOCs) and brooded chicks from hatcheries and mother units, respectively. This ensures that the DOCs and the brooded chicks are prevaccinated against Newcastle disease, Mareks disease, and infectious bursal disease at day-old (hatcheries) and against fowl pox during brooding (mother units). This model is being introduced into sub-Saharan African countries, having been successfully demonstrated in other LMICs (Bangladesh and India). However, a major challenge to this model is that of supply and logistics,

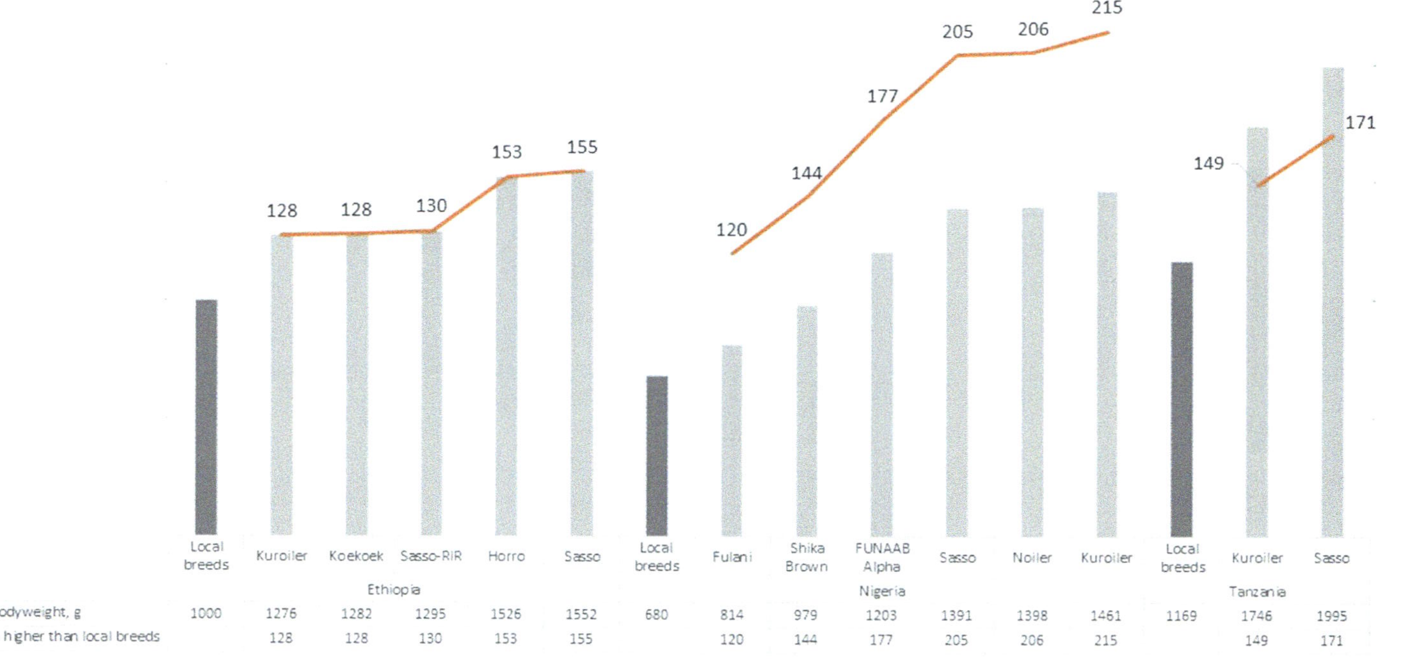

Figure 24.2 Live weights of male birds (20 weeks) of the iTABs in Ethiopia, Nigeria, and Tanzania under on-farm conditions.

Sources: Ajayi et al. (2020); Abegaz et al. (2019); Tadese (2017); Guni, Mbaga, Katule, and Goromela (2021); Kassa et al. (2021); and Fekede et al. (2021).

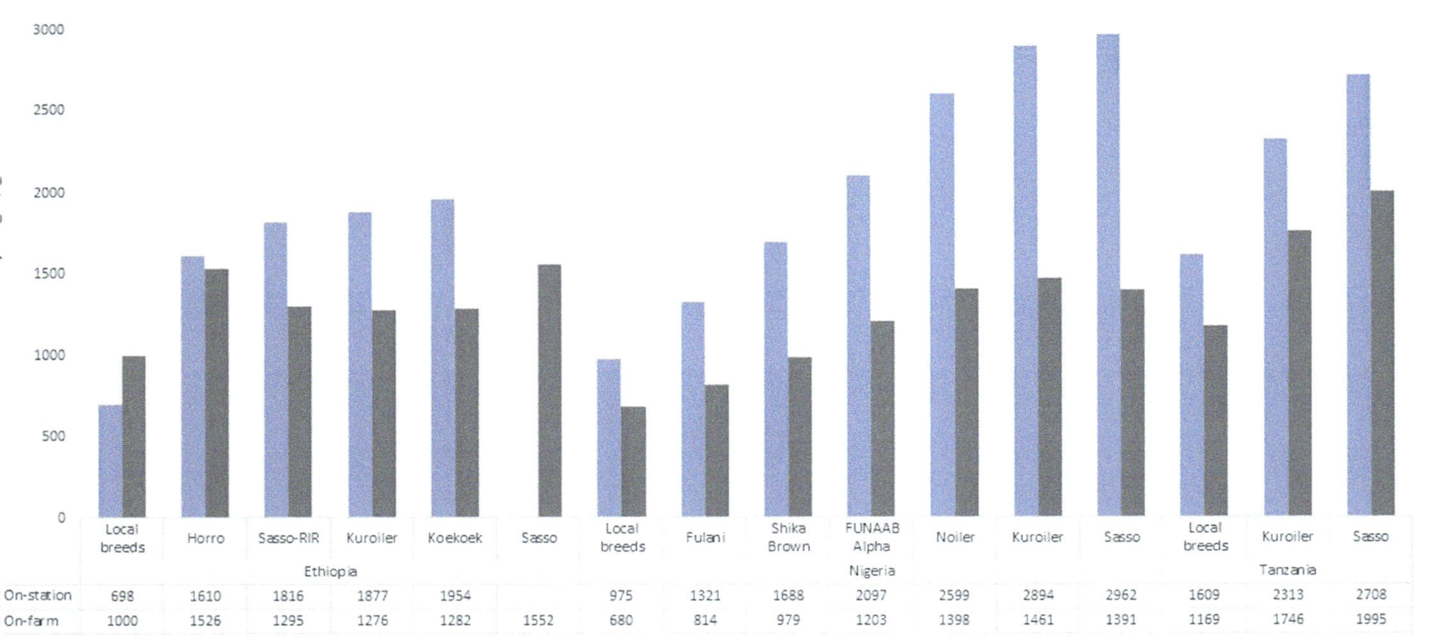

Figure 24.3 On-station and on-farm live weights of male birds (20 weeks) of the iTABs in Ethiopia, Nigeria, and Tanzania. Note that for Ethiopia, no data were available for Sasso under on-station condition.

Sources: Ajayi et al. (2020); Abegaz et al. (2019); Tadese (2017); Guni, Mbaga, Katule, and Goromela (2021); Kassa et al. (2021); and Fekede et al. (2021).

Table 24.4 Laying Performances of the Improved Tropically Adapted Breeds in Ethiopia, Nigeria, and Tanzania

Characteristics	Ethiopia[a]		Nigeria[b]		Tanzania[c]		Overall	
	Improved (Mean ± SD)	Local (Mean ± SD)	Improved (Mean ± SD)	Local (Mean ± SD)	Improved (Mean ± SD)	Local (Mean ± SD)	Improved (Mean ± SD)	Local (Mean ± SD)
Management Condition—On-Station								
Age at first egg (days)	130.5 ± 9.8	171.5 ± 37.5	124.7 ± 5.9	160.7 ± 4.0	153.4 ± 1.7	170.6 ± 3.7	136.2 ± 15.2	167.6 ± 6.0
Average no. of eggs/hen/year	173.8 ± 20.2	82.0 ± 15.8	140.8 ± 51.7	128.0 ± 12.3	108.3 ± 4.3	68.0 ± 5.7	140.9 ± 328	92.7 ± 31.4
Egg weight (g)	53.0 ± 3.7	46.5 ± 1.0	51.6 ± 6.4	37.5 ± 2.5	56.0 ± 2.3	43.7 ± 1.8	53.5 ± 2.3	42.6 ± 4.6
Management Condition—On-Farm								
Age at first egg (days)	182.1 ± 3.6	215.8 ± 16.7	154 ± 2.1	192.5 ± 15.9	179.1 ± 1.2	224.8 ± 17.8	171.7 ± 15.4	211.0 ± 16.7
Average no. of eggs/hen/year	103.0 ± 9.1	63.9 ± 13.0	44.5 ± 3.6	30.0 ± 10.0	50.5 ± 1.2	45.2 ± 5.3	66.0 ± 32.2	46.4 ± 16.9
Egg weight (g)	52.3 ± 6.6	39.4 ± 3.5	51.9 ± 4.7	30.0 ± 5.0	57.5 ± 3.5	43.9 ± 1.1	53.9 ± 3.1	37.8 ± 7.1

[a] Data from Markos et al. (2015), Abegaz et al. (2019), Bamidele et al. (2019), Fekede et al. (2021), and Kassa et al. (2021).
[b] Data from Adedokun and Sonaiya (2001), Odubote (2015), Bamidele et al. (2019), and Ajayi et al. (2020).
[c] Data from Lwelamira et al. (2008); Guni et al. (2013); Magonka et al. (2016); Bamidele et al. (2019); and Guni, Mbaga, and Katule (2021).

Table 24.5 Hatchability Parameters of Improved, Tropically Adapted Breeds in Three Sub-Saharan African Countries

Country	Improved Breed	Hatchability % of Eggs Set	Hatchability % of Fertile Eggs
Ethiopia	Horro	34.9	51.8
	Koekoek	52.1	50.0
	Kuroiler	52.0	64.3
	Sasso-RIR	64.7	78.9
	Overall	55.4	63.8
Nigeria	Fulani	54.8	72.2
	FUNAAB Alpha	55.3	81.5
	Kuroiler	67.0	79.8
	Noiler	83.5	89.0
	Sasso	70.8	84.0
	ShikaBrown	77.1	85.2
	Overall	69.7	83.3
Tanzania	Kuroiler	60.0	71.0
	Sasso	92.0	96.0
	Overall	83.7	90.1

Source: Bamidele et al. (2019).

particularly to farmers living in rural areas. This challenge was evident during the lockdowns occasioned by the COVID-19 pandemic, and it negatively impacted farmers' ability to restock their flock (Bamidele and Amole 2021). During that period, farmers resorted to the use of local hens to brood and hatch new chicks. Integration of a community-based breeding program has been proposed as a viable improvement to the current model (Kanyama et al. 2020, Bamidele and Amole 2021).

Table 24.5 summarizes the hatchability parameters for the improved chickens in three selected countries. The hatchability percentages were within the range (52–92%) previously reported for local, cross-bred and dual-purpose breeds (Ministry of Agriculture, Livestock Development and Marketing 1993, Marwa et al. 2018, Bamidele et al. 2020, Wolde et al. 2021). The wide range in the hatchability parameters may have been influenced by breed differences and environmental factors (storage and handling, hatchery operations, and incubator conditions) (Bamidele et al. 2020).

Whole Genome Sequencing and Genome-Wide Association Studies

Most of the genetic and genomics studies on improved chickens have focused on genetic admixture and diversity, through variations in microsatellite markers

(Bakare et al. 2021) and nuclear and mitochondria DNA (Malomane et al. 2019, Okani-Onyejiaka et al. 2022). Also, several studies have applied the candidate gene approach in investigating genetic variations (single nucleotide polymorphisms [SNPs]—an SNP is a DNA sequence variation that occurs when a single nucleotide [adenine, thymine, cytosine, or guanine] in the genome sequence is altered and the particular alteration is present in at least 1% of the population—deletions, and insertions) in some quantitative traits (growth, carcass, and morphometrics) (Wheto et al. 2017, Okafor et al. 2019). Some of the genes that have been investigated include the ghrelin gene (Sanda et al. 2021), chicken growth hormone (Okafor et al. 2019), and insulin-like growth factor (Wheto et al. 2016, 2020). There is a dearth of information on whole genome sequencing and the application of genome-wide association studies on economically important traits such as disease resistance, heat stress, body size, and egg production of the improved chickens. This may be due to the recent introduction of the iTABs within poultry research in general, as well as the emerging awareness and gradual adoption of the breeds by researchers. Therefore, this gap presents a major opportunity for the application of genomics in the continuous improvement and development of the improved breeds.

Genetic Improvements

Trait Preferences and Choice of Genotypes

Stated preference-based valuation is widely used in identifying preferred traits of livestock and economic valuation of animal genetic resources (Terfa et al. 2019) for the development of appropriate breeding programs, especially in smallholder flocks (Tilahun et al. 2022). Farmers' specific preferences for traits and iTABs of chickens differ in environmental gradients within and between countries in East Africa and West Africa. In a post–on-farm study to determine chicken genotypes of choice and traits preference in chicken by smallholder farmers in Nigeria, data were obtained from a total of 2,063 farmers using structured questionnaires in five agroecological zones (Kwara, Rivers, Imo, Nasarawa, and Kebbi) of the country. In the preceding on-farm study (using a randomized complete block design of 420 farmers per agroecology), the six iTABs were randomly allocated to the farmers, and each farmer received one breed of 30 birds aged 6 weeks (Yakubu et al. 2020). The preference for a chicken genotype by farmers was influenced by agroecological zone, with the exception of Shika Brown (the percentage of farmers who liked this genotype ranged from 52.6% to 72.3%) (Yakubu et al. 2020). FUNAAB Alpha was well-liked in Rivers (90.9%), Nasarawa (89.6%), and Kebbi (87.5%), respectively. Kuroiler was most like in Imo (88.1%), Rivers (83.1%), and Kwara (81.0%), whereas Sasso's likeness was 91.7% (Imo), 88.0% (Rivers), 79.8% (Nasarawa), and 73.8% (Kwara). Similarly, 88.1% (Nasarawa), 86.9% (Imo), and 79.8% (both Kwara and Kebbi) of farmers showed a remarkable likeness for Noiler birds. However, the indigenous Fulani birds were least liked by farmers across zones (5.6–48.5%). Overall, FUNAAB Alpha, Sasso, and Noiler were jointly rated highest, followed by Kuroiler, ShikaBrown, and Fulani. Within each zone, traits of preference for selection of breeding stock tended toward body size, egg number, egg size, and meat taste. Preferences for FUNAAB Alpha, Sasso, and Noiler birds could have been influenced mainly by their body size and egg number (Yakubu et al. 2020).

The Sasso chicken breed was rated higher than the Ethiopian IC breeds in terms of egg production, body size, feed efficiency, and good physical appearance. However, the indigenous birds were noted for higher adaptability (Getiso et al. 2017). Farmers in Uganda preferred Kuroiler chickens rather than their indigenous counterparts in terms of egg size (92.0 vs. 3.0%), meat taste (78.9 vs. 3.9%), and meat texture (75.0 vs. 5.3%) (Sharma et al. 2015). However, Kyarisiima et al. (2011), using a systematic sampling survey in Uganda, reported that approximately 80% of chicken consumers preferred IC meat compared to exotic chicken strains. Using latent segmentation in Ghana, Addo (2020) found that consumers had a high preference for domestic chicken and were willing to pay a much higher premium for IC. Similarly, consumers of chickens in Kenya exhibited high preferences for IC and willingness to pay a premium price (Bett et al. 2011). A study that employed a contingent valuation experiment involving farmers in Kenya estimated that consumers are willing to pay 23.26% per kilogram more for IC meat and 41.53% more for eggs compared to exotic chicken (Bett et al. 2013). Using a multistage sampling procedure, it was found that smallholder farmers of Makueni and Kakamega counties in Kenya preferred improved chickens compared to ICs for survivability (69.86 vs. 67.55%), size of eggs (2.97 vs. 2.80 cm), and age at first egg lay (21.27 vs. 21.99 weeks) (Christopher et al. 2018). This is buttressed by the ability of ICs to survive under harsh weather conditions (hardy) and withstand feed fluctuation.

At the traits level, smallholder farmers in Tanzania preferred Kuroiler and Sasso due to their fast growth rate, large body size, increased egg production, ability to adapt, bigger egg size, and easy sales at good prices, although these varied between the two agroecological zones of Mwanza and Mbeya. This is associated with the decision of most of the farmers (72.7%) to adopt iTABs compared to the ICs (27.3%) in Tanzania (Wolfgang 2020).

Crossbreeding

Crossbreeding using improved chickens is of practical relevance in genetic improvement of ICs to increase their production and productivity in a community-based poultry production system. The ICs in Africa are known for their ability to survive in harsh environmental conditions (Bettridge et al. 2018, El-Tahawy 2020, Soliman et al. 2020). Therefore, there is a need for an easy to adopt and sustainable breeding program to explore the matrix effect of introgressing (i.e., the introduction of new genetic variation into a population, which can cause traits to be shared among the introgressing breeds) genes of iTABs into the gene pool of their IC counterparts under a low-input production system. Such future breeding program should also take into consideration the interests of marketers and consumers because of the probable rejection of chicken/chicken products that do not include the traits of preference of some critical stakeholders in the chicken value chain (Reithmayer and Mußhoff 2019, Terfa et al. 2019). The expected outcome by farmers/poultry breeders of crossing iTABs with IC is the production of more resilient dual-purpose birds with higher meat yield and egg production, including adaptation to different environmental hot spots and better acceptability.

Crossbreeding work in Africa using iTABs and ICs is still in its infancy. However, an on-station study on Normal feathered local (LL), Sasso-RIR (SRSR), and their F1-cross (LSR) chickens has been performed in Ethiopia (Wolde et al. 2021). It was found that next to SRSR chickens, LSR performed higher than LL chickens in terms of egg number per hen (96.8 vs. 75.6), egg weight (46.9 vs. 44.4 g/egg), body weight (1,882 vs. 1,449 g/hen), albumen height (5.97 vs. 5.67 mm), albumen weight (27.2 vs. 25.3 g/egg), and albumen weight ratio (56.8 vs. 55.6).

Strategies for Sustainable Production and Utilization of iTABs

Considering the important roles of iTABs, there is a need to devise means to scale up the production of these birds on a sustainable basis to empower and improve the livelihoods of the rural poor households. This will stimulate population growth as entrenched in the action plan of the Africa Sustainable Livestock 2050 (FAO 2017, Latino et al. 2020). Approaches for sustainable production and utilization of iTABs may depend on locality (Bettridge et al. 2018). Some of the strategies that can be adopted are highlighted here.

Transformation of the Chicken Value Chain

Poultry value chains link the actors and activities involved in the delivery of poultry and poultry products to the final consumers, with increasing product value at every stage (FAO 2022). Globally, the chains are increasingly dualistic, with integrated market leaders coexisting with comparatively disadvantaged (in terms of economies of scale) small producers, who are less competitive within the value chain (Ngenoh et al. 2019, Garcia-Dorado et al. 2021). The constraints and challenges faced by smallholder poultry farmers within the poultry value chain limit their opportunities for growth. Improving access to growth opportunities for smallholder farmers requires a re-mapping of the entire poultry value chain to allow for a smallholder poultry-specific value chain, where activities involved in the production and marketing of high-value products of smallholder poultry production systems are conducted without competitive interference from large commercial farm enterprises (Bamidele and Amole 2021). In view of this, transformation of the poultry value chain for specificity to smallholder poultry enterprises and its use of iTABs can be achieved through the means discussed next.

Provision of Prevaccinated Day-Old Chicks by Breeding Companies to Smallholder Farmers Through Mother Units

One of the problems confronting smallholder farmers is difficulty in getting access to improved genetics. It has been established that access to improved low-input and dual-purpose chicken to supplement the local indigenous breeds could transform the rural and peri-urban poultry enterprise (Chaiban et al. 2020, Baldinger and Bussemas 2021). Therefore, establishing a link between certified breeding companies and the smallholder farmers through the mother units will facilitate access to high-quality day-old chicks. The mother unit model has been trialed in some African countries, including Burkina Faso, Ethiopia, Nigeria, Tanzania, and Uganda, where investments were made to support the African Poultry Multiplication Initiative or other comparable structures (Duijvesteijn and Perrault 2019). Generally, the model operates through capable local private breeding companies to establish a parent stock and hatchery operation for the production of day-old chicks of improved dual-purpose chicken breeds for smallholder farmers.

The day-old chicks from the parent stock farm are transported to the so-called mother units (equivalent to "brooder units" or "ambassadors"), which specialize in the brooding, feeding, and proper vaccination process for the first 30–40 days of the chicks' lives (Duijvesteijn and Perrault 2019). Such mother units are typically located in areas close to the smallholder farmers for easy patronage. This model is also able to reduce inbreeding among the iTABs. This is because farmers, especially those under the extensive and semi-intensive management systems, may be discouraged to propagate the chicks produced from the uncontrolled matings of these birds due to the fact that they can easily purchase directly from the mother units or mobile vendors (Wong et al. 2017). Because birds from mother units receive adequate nutrition, vaccinations, and medications before the farmers obtain the chicks, there is the tendency that they will perform well under the smallholder production systems (Yitayih et al. 2019).

Improved Production Practices

The housing system for ICs is mainly free-range, and feed sources include scavenging and kitchen byproducts (Aslam et al. 2020). However, improved housing in the form of semi-scavenging systems has been advocated for iTABs (Bettridge et al. 2018, Guteta and Ameha 2020, Kassa et al. 2021). Despite the fact that iTABs can scavenge, their diets should be supplemented with high-quality formulated feed to realize their genetic potential. In this sense, farmers, through guidance from extension workers, can formulate their own feeds using locally available materials (Oyewale et al. 2020) or buy feeds in small package bags from commercial feed producers or retailers (Garcia-Dorado et al. 2021). For poultry-keeping households not producing crops, establishing a link with crop farmers is imperative for continued access to cereal and leguminous crops. According to A. Wilson et al. (2021), chicken feed is an important entry point for integrating the crop–chicken value chains because maize and soybean meal are key sources of energy and protein for chickens. The use of local animal health service providers (including community vaccinators) along with ethnoveterinary medicine will be important to facilitate access to drugs and effective vaccines. Enahoro, Mensah, et al. (2021) were emphatic regarding the need to improve the supply of good quality drugs and vaccines in rural areas for effective poultry health care delivery. The adoption by farmers of the enterprise development model is suggested. In this model, an input supplier organizes the backyard farmers into small groups and supports them to set up and develop an enterprise that covers production and marketing (Beesabathuni et al. 2018). Support by the input supplier to farmers includes the provision of the input package with credit, training, and access to the markets for their poultry products (Beesabathuni et al. 2018).

Improved Marketing Strategies

Establishment of links with aggregators (market aggregation mechanisms) that focus especially on live sales can help support small and independent producers, especially women and youth (Dumas et al. 2016; Ramalingia et al. 2017; FAO 2018; Enahoro, Mensah, et al. 2021). Participation in well-organized collective marketing will lead to superior bargaining powers of smallholder poultry farmers in the value chain and reduce transaction costs through cost-sharing (Kiprop et al. 2020). There should be institutional architecture on the part of governments to reduce transportation costs of birds through reduction in taxes paid by transporters to local park authorities (Pica-Ciamarra 2005) and to ensure improved security for humans and animals. This will lead to more profits from the sale of chickens and chicken products, thereby stimulating interest in the production of iTABs. Governments should also control marketing places and informal traders and set standards. It is also imperative to establish an organized forum where producers, traders, and consumers can interact to collectively make the poultry enterprise a viable venture (Birhanu, Esatu, et al. 2021; Birhanu, Geremew, et al. 2021).

Provision of Credit Facilities to Smallholder Farmers

Credit facilities are essential for increasing the capital available to finance and invest in chicken meat and egg production (Isack 2019, Gill et al. 2021). There should be a policy thrust of governments in terms of formal credit, specifically targeting smallholder farmers. This could be in the form of low interest rates for the purchase of inputs or no-interest loans (Pica-Ciamarra 2005). Such interventions should take into account the economic circumstances (economic performance of the farms) and sociocultural (traditional beliefs and practices) context in which the farmers live (R. Wilson 2021).

Improved Extension Services

There should be a free flow of innovations and technologies from research institutions to farmers through government-approved extension agents and accredited nongovernmental organizations. There is also a need for feedback mechanisms from the farmers to the research institutions to solve problems using a bottom-top approach (Sankhyan et al. 2013). In this approach, farmers identify their own research problems/needs and pass such information to research institutions, which then attempt to find solutions. Extension could also be an avenue to equip farmers with the necessary technical and business skills through regular training (Dumas et al. 2016; U.S. Agency for International Development 2021a, 2021b). This will provide them the opportunity to improve the production and productivity of their birds, including marketing of the final products. It is in line with the recommendation of Kamau et al. (2018) that policies should strengthen farmers' access to adequate extension services for optimal production. Other ways of supporting access of the farmers to information include the use of phones and the internet.

Infrastructure Development

Quality infrastructure, such as good road networks and electricity and telecommunication services, is critical to the success of poultry enterprises (Liverpool-Tasie et al. 2016) because it leads to reduction in the cost of production (Bah and Gajigo 2019). Policies should therefore promote investment in rural infrastructure so that smallholder farmers can have reliable access to inputs (DOCs, feeds, and vaccines) and deliver their products easily to the markets for increased sales (Vos and Cattaneo 2020; Enahoro, Galiè, et al. 2021).

Ensuring Farm Biosecurity

The FAO defines *biosecurity* as the implementation of measures to reduce the risk of the introduction and spread of disease agents (FAO 2007a). Although there is no standardized way of classifying biosecurity measures, they can be grouped into bioexclusion (preventing diseases from entering the farm) and biocontainment (preventing diseases from leaving the farm) (Petrovan et al. 2021). The prevailing tropical climate in Africa offers a conducive environment for many pests, diseases, and vectors; hence, outbreaks of livestock diseases have led to high levels of flock mortality and morbidity (African Union–Interafrican Bureau for Animal Resources 2019, Isack 2019, Attree et al. 2021, Otiang et al. 2021). This may have significant impacts on geographically concentrated breed populations (World Intellectual Property Organization [WIPO] 2014). The smallholder poultry production systems have also been identified as a hot spot for the development of antimicrobial resistance (Bamidele, Joseph, et al. 2022), with potential health implications that could reduce chicken populations. Therefore, smallholder farmers producing iTABs should factor biosecurity into their decisions for infectious disease prevention and control (Conan et al. 2012, FAO 2018, Tasie et al. 2020).

Where risks of disease spread impact on the wider community and generate significant externalities, backyard poultry-keepers must be supported by wider societal actions (Otte et al. 2021). In light of this, industry stakeholders and policymakers should adapt their current disease programs and contingency plans to the reality of the health and biosecurity status of backyard poultry (Correia-Gomes and Sparks 2020). Other options available to minimize risks of disease transfer from backyard poultry flocks include the initiation of outreach programs on disease prevention and biosecurity practices, along with prophylactic campaigns, by governments at all levels (Fagrach et al. 2023). With the introduction of iTABs into the backyard poultry system, such engagements with smallholder farmers will help in disseminating reliable biosecurity and public health information.

Breed Registration

There is a need for concerted efforts at registering the iTABs in individual countries as unique breeds, with relevant information on them deposited in established and accessible data banks. It will make them more valuable to the farmers, consumers, government and nongovernment organizations, and local and foreign investors. This is in compliance with the Convention on Biological Diversity (CBD; WIPO 2014) to protect animal genetic diversity and exploit the sustainable utilization of animals through enhanced national, regional, and world livestock databases (FAO 2007d, 2018; Bolatito and Aladele 2019).

Formation of Breeders' and Farmers' Associations

There is a need for the formation of breeders' and farmers' associations. A breeders' association is a group of individuals that focuses on activities centered on a single, specific breed of a particular species of domesticated animal, whereas a poultry farmers' association is a group that focuses on the poultry farming business only (buying and selling of poultry products and services). The formation of breeders' associations or farmers' groups will enhance information sharing on iTABs (e.g., input supply, credit facilities, research and extension, and marketing and government policies), thereby aiding the production of these unique genetics (FAO 2013, Kamau et al. 2018). Group formation may also assist in improving the financial capacity of the farmers through cooperative activities. Candemir et al. (2021) reported that cooperatives play an important role in farm economic sustainability.

Compliance With Nagoya Protocol

African countries should follow the principles of the Nagoya Protocol on Access to Genetic Resources and the Fair and Equitable Sharing of Benefits Arising From Their Utilization (CBD 2011). It was introduced in 2010 as a supplementary agreement to the 1992 CBD. All member states should introduce practical mechanisms for its application in their own legal and regulatory systems (WIPO 2014). The intent was to use livestock genetic/biological resources for innovation, which is expected to lead to the establishment of intellectual property rights with benefits to be shared through the National Competent Authority or for research purposes (Martyniuk and Haska 2021). The significance of multi-access intellectual property protocols under smallholder production systems was highlighted by Sartas et al. (2021). This will also assist in the sustainable production and utilization of iTABs.

Monitoring of the Genetic Populations of iTABs

The Global Plan of Action for Animal Genetic Resources (FAO 2007b) identifies the need for country-based strategies to ensure that inventory and monitoring activities can be linked to and coordinated with action plans such as agricultural censuses or livestock population surveys (FAO 2011). Monitoring requires the regular checking of population status and the evaluation of trends in the size and structure of breeds/populations, their geographical distribution, production systems, and genetic diversity. Such monitoring is expected to contribute to the planning of national development policies and other local and international interventions (Pym 2010), which may be useful to iTABs.

Development of a Community-Based Breeding Program

The availability and sustained distribution of the iTABs in rural and remote communities remain a challenge. Accessibility of these breeds was made particularly difficult with the outbreak of COVID-19 pandemic, partly due to logistical challenges (Bamidele and Amole 2021). For instance, the average monthly income of the respondents before and during the COVID-19 pandemic was $62.70 and $38.10, respectively (Bamidele and Amole 2021). In the post-COVID-19 era, this gap in the supply of improved chickens requires appropriate interventions in order to sustain the gains (production and impact) made prior to the pandemic: In addition to the positive impact on household nutrition and income, there was an increased capacity of farmers and increased empowerment of women toward initiating several business ventures along the smallholder poultry value chain (O. O. Alabi et al. 2020). The development and implementation of a community-based breeding strategy is a viable option for ensuring continuous and sustainable delivery of the iTABs (Bamidele and Amole 2021). Similar approaches have been demonstrated for ICs in Bangladesh, Egypt, and Malawi (Jensen 2000, Kolstand and Abdou 2000, FAO 2003). However, implementing the community-based breeding strategy for iTABs requires considerable adjustments to the overarching breeding options available in poultry breeding programs. Such adjustments may be necessary to protect the interests of the breeding companies while improving the access and affordability of the improved breeds to poor farmers living in the difficult terrains of sub-Saharan Africa.

Conclusion

This chapter highlighted the potential of the six newly introduced iTABs (ShikaBrown, FUNAAB Alpha, Noiler, Potchefstroom Koekoek, Sasso, and Kuroiler) in complementing ICs within small-scale poultry production systems in Africa. The greater productivity of iTABs has potential to enhance benefits accruing from ICs, such as women and youth empowerment, increased food security, improvement in nutrition and health, poverty alleviation, and improved livelihoods of the rural and peri-urban dwellers. The birds are more suitable for the semi-scavenging system and superior to their IC counterparts in biometrics, productive and reproductive capacities, as well as rankings. Some of the strategic interventions to produce and utilize iTABs on a sustainable basis require the involvement of governments and the private sector in the areas of chicken value chain transformation, rural and peri-urban development, conducive rearing environment, breed registration and formation of farmers' groups, compliance with institutional regulations and policies on animal biodiversity, and development of a community-based breeding program.

Acknowledgment

This work was partly funded by the CGIAR Research Program for Agriculture for Nutrition and Health (A4NH) as part of the CGIAR COVID-19 Hub project in Nigeria.

References

Abegaz, S., W. Esatu, G. Assefa, E. H. Goromela, E. B. Sonaiya, S. H. Mbaga, O. Adebambo, O. Bamidele, A. Teressa, J. Bruno, J. Poole, F. Getachew, H. Kasaye, and T. Dessie. 2019. On-farm performance testing of tropically adaptable chicken strains under small holder management in three countries of sub-Saharan Africa. Paper presented at the Seventh All Africa Conference on Animal Agriculture, Accra, Ghana, July 29–August 2.

Addo, A. C. 2020. Analysis of consumer attitudes, preferences, and demand for poultry. Doctoral dissertation, Georg-August-University, Goettingen, Germany.

Adedokun, S. A., and Sonaiya, E. B. 2001. Comparison of the performance of Nigerian indigenous chickens from three agro-ecological zones. *Livestock Research for Rural Development* 13:15.

African Union–Interafrican Bureau for Animal Resources. 2019. *The State of Farm Animal Genetic Resources in Africa.*

Ajayi, F. O. 2010. Nigerian indigenous chicken: A valuable genetic resource for meat and egg production. *Asian Journal of Poultry Science* 4:164–172.

Ajayi, F. O., O. Bamidele, W. A. Hassan, U. Ogundu, A. Yakubu, O. O. Alabi, O. M. Akinsola, E. B. Sonaiya, and O. A. Adebambo. 2020. Production performance and survivability of six dual-purpose breeds of chicken under smallholder farmers' management practices in Nigeria. *Archives Animal Breeding* 63:387–408.

Akinsola, O. M., E. B. Sonaiya, O. Bamidele, W. A. Hassan, A. Yakubu, F. O. Ajayi, U. Ogundu, O. O. Alabi, and O. A. Adebambo. 2021. Comparison of five mathematical models that describe growth in tropically adapted dual-purpose breeds of chicken. *Journal of Applied Animal Research* 49:158–166.

Alabi, O. J., J. Ngambi, D. Norris, and S. Egena. 2012. Comparative study of three indigenous chicken breeds of South Africa: Body weight and linear body measurements. *Agricultural Journal* 7:220–225.

Alabi, O. O., F. O. Ajayi, O. Bamidele, A. Yakubu, E. U. Ogundu, E. B. Sonaiya, M. A. Ojo, W. A. Hassan, and O. A. Adebambo. 2020. Impact assessment of improved chicken genetics on livelihoods and food security of smallholder poultry farmers in Nigeria. *Livestock Research for Rural Development* 32.

Alemu, S. W., O. Hanotte, F. G. Kebede, W. Esatu, S. Abegaz, J. E. Bruno, B. Abrar, T. Alemayehu, R. Mrode, and T. Dessie. 2021. Evaluation of live-body weight and the number of eggs produced for introduced and local chickens in Ethiopia. *Acta Agriculturae Scandinavica Section A: Animal Science* 70:71–77.

Aslam, H. B., P. Alarcon, T. Yaqub, M. Iqbal, and B. Häsler. 2020. A value chain approach to characterize the chicken sub-sector in Pakistan. *Frontiers in Veterinary Science* 7: Article 361.

Assefa, S., A. Melesse, and S. Banerjee. 2018. Egg production and linear body measurement traits of local and three exotic chicken genotypes reared under two agroecological zones. *International Journal of Ecology and Ecosolution* 5:18–23.

Attree, E., G. Sanchez-Arsuaga, M. Jones, D. Xia, V. Marugan-Hernandez, D. Blake, and F. Tomley. 2021. Controlling the causative agents of coccidiosis in domestic chickens: An eye on the past and considerations for the future. *CABI Agriculture and Bioscience* 2: Article 37.

Baéza, E., N. Williams, D. Guemene, and M. J. Duclos. 2001. Sexual dimorphism for growth in Muscovy duck and changes in insulin-like growth factor I (IGF-I), growth hormone (GH) and triiodothyronine (T_3) plasma levels. *Reproduction, Nutrition, Development* 41:173–179.

Bah, E.-h., and O. Gajigo. 2019. *Improving the Poultry Value Chain in Mozambique.* African Development Bank, Abidjan, Ivory Coast.

Bakare, I. O., B. M. Ilori, M. Wheto, L. T. Egbeyale, A. J. Sanda, and O. Olowofeso. 2021. Genetic diversity and gene flow among three chicken populations in Nigeria using microsatellite markers. *Agriculturae Conspectus Scientificus* 86:173–181.

Baldinger, L., and R. Bussemas. 2021. Dual-purpose production of eggs and meat—Part 2: Hens of crosses between layer and meat breeds show moderate laying performance but choose feed with less protein than a layer hybrid, indicating the potential to reduce protein in diets. *Organic Agriculture* 11:73–87.

Bamidele, O., and T. A. Amole. 2021. Impact of COVID-19 on smallholder poultry farmers in Nigeria. *Sustainability* 13: Article 11475.

Bamidele, O., T. A. Amole, O. A. Oyewale, O. O. Bamidele, A. Yakubu, U. E. Ogundu, F. O. Ajayi, and W. A. Hassan. 2022. Antimicrobial usage in smallholder poultry production in Nigeria. Veterinary Medicine International 2022: Article 7746144.

Bamidele, O., E. B. Joseph, and A. Yakubu. 2022. Characterising antibiotic resistance of bacterial isolates from smallholder poultry droppings in selected locations in Nasarawa State, Nigeria. Pages 360–364 in Y. P. Mancha, D. J. U. Kalla, T. T. Akpensuen, T. T. Igila, J. S. Luka, and U. Okpanachi, eds. *Proceedings of the Nigerian Society for Animal Production (NSAP) 47th Annual Conference.* Nigerian Society for Animal Production.

Bamidele, O., E. B. Sonaiya, O. Adebambo, G. Assefa, S. Abegaz, W. Esatu, E. H. Goromela, S. H. Mbaga, and T. Dessie. 2019. On-station performance evaluation of improved tropically adapted chicken strains for smallholder poultry production systems in sub-Saharan Africa. Paper presented at the Seventh All Africa Conference on Animal Agriculture.

Bamidele, O., E. B. Sonaiya, O. A. Adebambo, and T. Dessie. 2020. On-station performance evaluation of improved tropically adapted chicken breeds for smallholder poultry production systems in Nigeria. *Tropical Animal Health and Production* 52:1541–1548.

Banda, L. J., and J. Tanganyika. 2021. Livestock provide more than food in smallholder production systems of developing countries. *Animal Frontiers* 11: Article 2.

Beesabathuni, K., S. Lingala, and K. Kraemer. 2018. Increasing egg availability through smallholder business models in East Africa and India. *Maternal & Child Nutrition* 14: Article e12667.

Bett, H. K., K. J. Peters, and W. Bokelmann. 2011. Hedonic price analysis to guide in breeding and production of indigenous chicken in Kenya. *Livestock Research for Rural Development* 23: Article 142.

Bett, H. K., K. J. Peters, U. M. Nwankwo, and W. Bokelmann. 2013. Estimating consumer preferences and willingness to pay for the underutilised indigenous chicken products. *Food Policy* 41: 218–225.

Bettridge, J. M., A. Psifidi, Z. G. Terfa, T. T. Desta, M. Lozano-Jaramillo, T. Dessie, P. Kaiser, P. Wigley, O. Hanotte, and R. M. Christley. 2018. The role of local adaptation in sustainable production of village chickens. *Nature Sustainability* 1:574–582.

Birhanu, M. Y., W. Esatu, K. Geremew, T. Yemane, O. A. Oyewale, A. Adesina, and T. Dessie. 2021a. *Economic and Marketing Performance of Chicken Value Chain Actors in Nigeria: Challenges and Business Opportunities for Sustainable Livelihood.* International Livestock Research Institute, Nairobi, Kenya.

Birhanu, M. Y., K. Geremew, W. Esatu, T. Girma, F. Getachew, S. Worku, and T. Dessie. 2021b. *Economic and Marketing Performance of Chicken Value Chain Actors in Ethiopia: Challenges and Business Opportunities for Sustainable Livelihoods.* International Livestock Research Institute, Nairobi, Kenya.

Bolatito, O., and S. Aladele. 2019. Preliminary information on the status of sustainable use, characterization and inventory of animal genetic resources in Nigeria. *Nigerian Journal of Animal Science* 21:9–17.

Bosse, M. 2019. No "doom" in chicken domestication? *PLoS Genetics* 15: Article 1008089.

Brito, N. V., J. C. Lopes, V. Ribeiro, R. Dantas, and J. V. Leite. 2021. Biometric characterization of the Portuguese autochthonous hens breeds. *Animals* 11: Article 498.

Candemir, A., S. Duvaleix, and L. Latruffe. 2021. Agricultural cooperatives and farm sustainability: A literature review. *Journal of Economic Surveys* 35:1118–1144.

Chaiban, C., T. P. Robinson, E. M. Fèvre, J. Ogola, J. Akoko, M. Gilbert, and S. Vanwambeke. 2020. Early intensification of backyard poultry systems in the tropics: A case study. *Animal* 14:2387–2396.

Chawala, M., B. Mwiya, J. Tembo, and G. Kabwe. 2022. Examining gender differences in indigenous chicken commercialisation intent: Evidence from north-western Zambia. *Cogent Business & Management* 9: Article 2007745.

Chimenem-Amadi, S. N., V. U. Oleforuh-Okoleh, B. O. Agaviezor, and H. H. Gunn. 2021. Comparative study of body weight and some performance traits of improved Nigerian indigenous chickens raised in the south–south region of Nigeria. *Nigerian Journal of Animal Production* 48:14–22.

Christopher, N. K., W. Lucy, E. K. Kabuage, and H. Bett. 2018. Impact of improved indigenous chicken breeds on productivity. The case of smallholder farmers in Makueni and Kakamega counties, Kenya. *Cogent Food & Agriculture* 4: Article 1477232.

Conan, A., F. L. Goutard, S. Sorn, and S. Vong. 2012. Biosecurity measures for backyard poultry in developing countries: A systematic review. *BMC Veterinary Research* 8: Article 240.

Convention on Biological Diversity. 2011. Nagoya Protocol on Access to Genetic Resources and the Fair and Equitable Sharing of Benefits Arising From Their Utilization to the Convention on Biological Diversity. Secretariat of the Convention on Biological Diversity, Montreal, Canada.

Correia-Gomes, C., and N. Sparks. 2020. Exploring the attitudes of backyard poultry keepers to health and biosecurity. *Preventive Veterinary Medicine* 174: Article 104812.

Dessie, T. 2021. Tropical Poultry Genetic Solutions (TPGS): Delivering farmer preferred, productive and ecologically adapted poultry to smallholders. Paper presented at the CTLGH Virtual Development Meeting. International Livestock Research Institute.

Dessie, T., and F. Getachew. 2016. African chicken genetic gains: Factsheet 3. The Potchefstroom Koekoek Breed. International Livestock Research Institute, Nairobi, Kenya.

Desta, T. T. 2021a. Indigenous village chicken production: A tool for poverty alleviation, the empowerment of women, and rural development. *Tropical Animal Health and Production* 53: Article 1.

Desta, T. T. 2021b. Sustainable intensification of indigenous village chicken production system: Matching the genotype with the environment. *Tropical Animal Health and Production* 53: Article 337.

Department for International Development. 2019. Poultry Sector Study.

Duijvesteijn, N., and L. Perrault. 2019. How dual-purpose chickens can help African smallholder farmers. *Journal of Animal Science* 97:182.

Dumas, S. E., L. Lungu, N. Mulambya, W. Daka, E. McDonald, E. Steubing, T. Lewis, K. Backel, J. Jange, B. Lucio-Martinez, D. Lewis, and A. J. Travis. 2016. Sustainable smallholder poultry interventions to promote food security and social, agricultural, and ecological resilience in the Luangwa Valley, Zambia. *Food Security* 8:507–520.

EAT. 2019. Summary report of the EAT–*Lancet* Commission. An adapted summary of the Commission Food in The Anthropocene: The EAT–*Lancet* Commission on Healthy Diets From Sustainable Food Systems.

El-Tahawy, W. 2020. Analysis of heterotic components in a cross bred between two Egyptian local chicken strains. *Egyptian Poultry Science Journal* 40:525–535.

Enahoro, D., A. Galiè, Y. Abukari, G. H. Chiwanga, T. R. Kelly, J. Kahamba, F. A. Massawe, F. Mapunda, H. Jumba, and C. Weber. 2021. Strategies to upgrade animal health delivery in village poultry systems: Perspectives of stakeholders from northern Ghana and central zones in Tanzania. *Frontiers in Veterinary Science* 8: Article 611357.

Enahoro, D., C. Mensah, G. Cooper, and K. M. Richa. 2021. Upgrading village chicken value chains in Ghana: An application of spatial group model building. Paper presented at Tropentag 2021, hybrid conference (Conference on International Research on Food Security, Natural Resource Management and Rural Development) organized by the University of Hohenheim.

Fadare, A., T. Dawodu, and J. Ilufoye. 2020. Variations in the carcass traits of three strains of broiler chickens. *Nigerian Journal of Animal Science* 22:7–12.

Fagrach, A., S. Fellahi, M. K. Challioui, O. Arbani, I. El Zirani, F. Kichou, and M. Bouslikhane. 2023. Backyard poultry flocks in Morocco: Demographic characteristics. husbandry practices, and disease and biosecurity management. *Animals* 13: Article 202.

Fekede, G., Y. Tadesse, W. Esatu, and T. Dessie. 2021. On-farm comparative production and reproduction performance evaluation of Sasso, Sasso-RIR, Koekoek and improved local chicken breeds in Bako Tibe and Dano Districts of Western Oromia, Ethiopia. *Livestock Research for Rural Development* 33: Article 15.
Food and Agriculture Organization. 2003. Community-based management of animal genetic resources.
Food and Agriculture Organization. 2004. Small scale poultry production.
Food and Agriculture Organization. 2007a. *FAO Biosecurity Toolkit.* Food and Agriculture Organization, Rome.
Food and Agriculture Organization. 2007b. *Global Plan of Action for Animal Genetic Resources and the Interlaken Declaration.* Food and Agriculture Organization, Rome.
Food and Agriculture Organization. 2007c. *Scale and Structures of the Poultry Sector and Factors Inducing Change: Intercountry Differences and Expected Trends.* Food and Agriculture Organization, Rome.
Food and Agriculture Organization. 2007d. *The State of the World's Animal Genetic Resources for Food and Agriculture.* Food and Agriculture Organization, Rome.
Food and Agriculture Organization. 2011. *Surveying and Monitoring of Animal Genetic Resources: FAO Animal Production and Health Guidelines.* Food and Agriculture Organization, Rome
Food and Agriculture Organization. 2013. *In Vivo Conservation of Animal Genetic Resources: FAO Animal Production and Health Guidelines.* Food and Agriculture Organization, Rome.
Food and Agriculture Organization. 2014. *Decision Tools for Family Poultry Development: FAO Animal Production and Health Guidelines.* Food and Agriculture Organization, Rome.
Food and Agriculture Organization. 2017. *Thinking Beyond LIVESTOCK Today for the People of TOMORROW: Africa Sustainable Livestock 2050 (ASL2050).* Food and Agriculture Organization, Rome.
Food and Agriculture Organization. 2018. *World Livestock: Transforming the Livestock Sector Through the Sustainable Development Goals.* Food and Agriculture Organization, Rome.
Food and Agriculture Organization. 2019. *Business Development Portfolio: Opportunities to Invest in Sustainable Development (2019–2020 Cycle).* Food and Agriculture Organization, Rome.
Food and Agriculture Organization. 2021. *Backyard Farming and Slaughtering: Keeping Tradition Safe: Food Safety Technical Toolkit for Asia and the Pacific.* Food and Agriculture Organization, Rome.
Food and Agriculture Organization. 2022. The poultry chain. In *Gateway to Poultry Production and Products.* Food and Agriculture Organization, Rome.
Garcia-Dorado, S. C., K. Queenan, B. Shankar, B. Häsler, T. Mabhaudhi, G. Cooper, and R. Slotow. 2021. Using qualitative system dynamics analysis to promote inclusive livestock value chains: A case study of the South African broiler value chain. *Frontiers in Sustainable Food Systems* 5: Article 670756.
Geremew, K., and T. Dessie. 2021. Human nutrition trials around chicken meat and egg consumed in Ethiopia and Tanzania: A technical report. International Livestock Research Institute, Nairobi, Kenya.
Getiso, A., A. Jimma, M. Asrat, H. Giorgis, B. Zeleke, and T. Birhanu. 2017. Management practices and productive performances of Sasso chickens breed under village production system in SNNPR, Ethiopia. *Journal of Biology, Agriculture and Healthcare* 7:120–135.
Gill, T., R. Nisengwe, H. Goertz, D. Ader, K. McGehee, R. Nshuti, A. Gumisiriza, M. Smith, and E. Urban. 2021. Strengthening smallholder engagement and integration in the Rwandan commercial broiler value chain. *World's Poultry Science Journal* 77:1059–1078.
Guni, F. S., A. M. Katule, and P. A. A. Mwakilembe. 2013. Characterization of local chickens in selected districts of the Southern Highlands of Tanzania: II. Production and morphometric traits. *Livestock Research for Rural Development* 25: Article 190.
Guni, F. S., S. Mbaga, A. Katule, and E. Goromela. 2021. Evaluation of carcass characteristics and egg quality traits of Kuroiler and Sasso chickens reared under on-station and on-farm management conditions in Tanzania. Preprint.

Guni, F. S., S. H. Mbaga, and A. M. Katule. 2021. Performance evaluation of Kuroiler and Sasso chicken breeds reared under on-farm and on-station management conditions in Tanzania. *European Journal of Agriculture and Food Sciences* 3:53–59.

Guteta, A., and N. Ameha. 2020. Characterization of scavenging and intensive chicken production system in Lume District, East Showa Zone, Oromia Regional State, Ethiopia. *International Journal of Livestock Production* 11:8–20.

Hata, A., M. Nunome, T. Suwanasopee, P. Duengkae, S. Chaiwatana, W. Chamchumroon, T. Suzuki, S. Koonawootrittriron, Y. Matsuda, and K. Srikulnath. 2021. Origin and evolutionary history of domestic chickens inferred from a large population study of Thai red junglefowl and indigenous chickens. *Scientific Reports* 11: Article 2035.

Hendrix Genetics. 2022. The story of Sasso.

Hlokoe, V., T. Tyasi, and B. Gunya. 2022. Chicken ovarian follicles morphology and *growth differentiation factor 9* gene expression in chicken ovarian follicles. *Heliyon* 8: Article e08742.

Hoffmann, I. 2005. Research and investment in poultry genetic resources: Challenges and options for sustainable use. *World's Poultry Science Journal* 61:57–70.

Independent Group of Scientists appointed by the Secretary-General. 2019. *Global Sustainable Development Report 2019: The Future Is Now: Science for Achieving Sustainable Development.* United Nations, New York.

International Fund for Agricultural Development. 2021. Transforming food systems for rural prosperity. International Fund for Agricultural Development, Rome, Italy.

Isack, N. 2019. Value chain analysis of the tropically adapted improved chicken in Lindi rural and Masasi districts. Doctoral dissertation, University of Agriculture, Morogoro, Tanzania.

Jensen, H. A. 2000. Structures for improving smallholder chickens in Bangladesh: Breeding strategy. Paper presented at the Workshop on Developing Breeding Strategies for Lower Input Production Environment.

Kamau, C. N. K., L. W, and E. K. Bett. 2018. Impact of improved indigenous chicken breeds on productivity. The case of smallholder farmers in Makueni and Kakamega Counties, Kenya. *Cogent Food & Agriculture* 4: Article 1.

Kanyama, C. M., A. F. Moss, and T. M. Crowley. 2020. Smallholder farmer community-based breeding program of indigenous chickens key to in-situ genetic resource conservation and enhancing rural livelihoods in Zambia. Pages 23–24 in *2020 PSA Annual Meeting Abstract Book.* Poultry Science Association. https://www.poultryscience.org/viewdocument/2020-psa-annual-meeting-abstract-bo.

Kassa, B., Y. Tadesse, W. Esatu, T. Dessie, M. Lakew, and A. Tesfa. 2021. On-farm performance evaluation of tropically adapted chicken strains under semi-scavenging production system in western Amhara region, Ethiopia. *Ethiopian Journal of Applied Science and Technology* 12:32–42.

Kiprop, E., C. Okinda, S. Wamuyu, and X. Geng. 2020. Factors influencing smallholder farmers participation in collective marketing and the extent of participation in improved indigenous chicken markets in Baringo, Kenya. *Asian Journal of Agricultural Extension, Economics & Sociology* 37:1–12.

Kolstand, N., and F. A. Abdou. 2000. NORFA: The Norwegian–Egyptian Project for Improving Local Breeds of Laying Hens in Egypt. Paper presented at the Workshop on Developing Breeding Strategies for Lower Input Production Environment.

Kyarisiima, C. C., F. A. Nagujja, H. Magala, H. Kwizera, D. R. Kugonza, and J. Bonabana- Wabbi. 2011. Perceived tastes and preferences of chicken meat in Uganda. Livestock Research for Rural Development 23.

Latino, L., U. Pica-Ciamarra, and D. Wisser. 2020. Africa: The livestock revolution urbanizes. *Global Food Security* 26: Article 100399.

Lawal, R. A., and O. Hanotte. 2021. Domestic chicken diversity: Origin, distribution, and adaptation. *Animal Genetics* 52:385–394.

Liverpool-Tasie, S., B. Omonona, A. Sanou, W. Ogunleye, S. Padilla, and T. Reardon. 2016. Growth and transformation of chicken & eggs value chains in Nigeria. Feed the Future Innovation Lab for Food Security Policy Research Paper 22.

Lwelamira, J., G. Kifaro, and P. Gwakisa. 2008. On station and on-farm evaluation of two Tanzania chicken ecotypes for body weights at different ages and for egg production. *African Journal of Agricultural Research* 3:843–851.

Magonka, J. M., D. S. Sendalo, E. H. Goromela, P. B. Malingila, and E. Daniel. 2016. Production performance of indigenous #chicken under semi-intensive management conditions in central Tanzania. *Huria Journal of the Open University of Tanzania* 22.

Malomane, D. K., H. Simianer, A. Weigend, C. Reimer, A. O. Schmitt, and S. Weigend. 2019. The SYNBREED chicken diversity panel: A global resource to assess chicken diversity at high genomic resolution. *BMC Genomics* 20:1–15.

Manyelo, T. G., L. Selaledi, Z. M. Hassan, and M. Mabelebele. 2020. Local chicken breeds of Africa: Their description, uses and conservation methods. *Animals* 10: Article 2257.

Markos, S., B. Belay, and T. Dessie. 2015. On farm performance evaluation of three local chicken ecotypes in western zone of Tigray, northern Ethiopia. *Journal of Biology, Agriculture and Healthcare* 5:158–169.

Martyniuk, E., and A. Haska. 2021. Implementation of the Nagoya Protocol in livestock sector: What have we learnt so far? *Animals* 11: Article 2354.

Marwa, L. J., S. H. Mbaga, S. K. Mutayoba, and B. Lukuyu. 2018. The productivity and management systems of free range local chickens in rural areas of Babati District, Tanzania. *Livestock Research for Rural Development* 30: Article 134.

Mathiu, E. M., S. N. Ndirangu, and S. C. Mwangi. 2021. Production of indigenous poultry among smallholder farmers in Tigania West Meru County, Kenya. *African Journal of Agricultural Research* 17:705–713.

Ministry of Agriculture, Livestock Development and Marketing. 1993. Annual report.

Moges, F., A. Tegegne, and T. Dessie. 2010. Indigenous chicken production and marketing systems in Ethiopia: Characteristics and opportunities for market-oriented development. IPMS (Improving Productivity and Market Success) of Ethiopian Farmers Project Working Paper 24. International Livestock Research Institute, Nairobi, Kenya.

Mujyambere, V., K. Adomako, S. O. Olympio, M. Ntawubizi, L. Nyinawamwiza, J. Mahoro, and A. Conroy. 2021. Local chickens in East African region: Their production and potential. *Poultry Science* 101: Article 101547.

Mushi, J. R., G. H. Chiwanga, E. N. Amuzu-Aweh, M. Walugembe, R. A. Max, S. J. Lamont, T. R. Kelly, E. L. Mollel, P. L. Msoffe, and J. Dekkers. 2020. Phenotypic variability and population structure analysis of Tanzanian free-range local chickens. *BMC Veterinary Research* 16: Article 360.

National Animal Production Research Institute. 1998. *A Handbook for ShikaBrown parents and Commercial Layer.* National Institute for Animal Production Research Institute (NAPRI), Ahmadu Bello University, Zaria, Nigeria.

Ngenoh, E., B. K. Kurgat, H. K. Bett, S. W. Kebede, and W. Bokelmann. 2019. Determinants of the competitiveness of smallholder African indigenous vegetable farmers in high-value agro-food chains in Kenya: A multivariate probit regression analysis. *Agricultural and Food Economics* 7:1–17.

Odubote, I. K. 2015. The local chickens of Nigeria: A review. Department of Animal Science, Obafemi Awolowo University, Ile-Ife, Osun state, Nigeria.

Oduntan, A. 2020. Noiler: Research and innovation improve nutrition, profit and social impact in West African backyards. Keynote paper presented at the 2019–2020 Carter Virtual Conference hosted by the Center for African Studies and Feed the Future Innovation Lab for Livestock Systems.

Ogbu, C. C. 2021. Utilization and conservation of landrace chickens of Nigeria: Physical and performance characteristics, issues and concerns. In A. Elkelish, ed. *Landraces: Traditional Variety and Natural Breed.* IntechOpen, London.

Okafor, O., V. Okoro, C. Mbajiorgu, I. Okoli, I. Ogbuewu, and U. Ogundu. 2019. Influence of chicken growth hormone (cGH) SNP genotypes on morphometric and growth traits of three chicken breeds in Nigeria. *Indian Journal of Animal Research* 53:1559–1565.

Okani-Onyejiaka, M. C., U. E. Ogundu, S. F. Boudali, O. Bamidele, I. P. Ogbuewu, and N. O. Aladi. 2022. Genetic diversity of mitochondrial DNA (mtDNA) D-loop sequences in six improved tropically adapted chicken breeds (iTABs) in Imo State, Nigeria. *Genetic and Biodiversity Journal* 6:199–219.

Osei-Amponsah, R., B. Kayang, A. Naazie, M. Tiexier-Boichard, and X. Rognon. 2015. Phenotypic characterization of local Ghanaian chickens: Egg-laying performance under improved management conditions. *Animal Genetic Resources* 56:29–35.

Otiang, E., T. S. M, Z. A. Campbell, L. W. Njagi, P. N. Nyaga, and G. H. Palmer. 2021. Impact of routine Newcastle disease vaccination on chicken flock size in smallholder farms in western Kenya. *PLoS One* 16: Article 0248596.

Otte, J., J. Rushton, E. Rukambile, and R. G. Alders. 2021. Biosecurity in village and other free-range poultry: Trying to square the circle? *Frontiers in Veterinary Science* 8: Article 678419.

Oyewale, O., O. Bamidele, J. Oyedele, and E. Sonaiya. 2021. Morphometrics and carcass characteristics of males of six dual purpose chicken breeds under intensive management. *Nigerian Journal of Animal Production* 48:1–7.

Oyewale, O. A., O. O. Ojebiyi, F. A. Adedeji, O. Bamidele, and E. B. Sonaiya. 2020. A smallholder poultry feed app: Development and field test. *International Journal of Poultry Science* 19:176–185.

Passarelli, S., R. Ambikapathi, N. S. Gunaratna, I. Madzorera, C. R. Canavan, A. R. Noor, A. Worku, Y. Berhane, S. Abdelmenan, S. Sibanda, B. Munthali, T. Madzivhandila, L. M. Sibanda, K. Geremew, T. Dessie, S. Abegaz, G. Assefa, C. Sudfeld, M. McConnell, K. Davison, and W. Fawzi. 2020. A chicken production intervention and additional nutrition behavior change component increased child growth in Ethiopia: A cluster-randomized trial. *Journal of Nutrition* 150:2806–2817.

Petrovan, S. O., D. C. Aldridge, H. Bartlett, A. J. Bladon, H. Booth, S. Broad, D. M. Broom, N. D. Burgess, S. Cleaveland, A. A. Cunningham, M. Ferri, A. Hinsley, F. Hua, A. C. Hughes, K. Jones, M. Kelly, G. Mayes, M. Radakovic, C. A. Ugwu, N. Uddin, D. Veríssimo, C. Walzer, T. B. White, J. L. Wood, and W. J. Sutherland. 2021. Post COVID-19: A solution scan of options for preventing future zoonotic epidemics. *Biological Reviews* 96:2694–2715.

Pica-Ciamarra, U. 2005. Livestock policies for poverty alleviation: Theory and practical evidence from Africa, Asia and Latin America. Pro-Poor Livestock Policy Initiative Working Paper No.61.

Pius, L. O., P. Strausz, and S. Kusza. 2021. Overview of poultry management as a key factor for solving food and nutritional security with a special focus on chicken breeding in East African countries. *Biology* 10: Article 810.

Pym, R. 2010. Poultry genetics and breeding in developing countries: Genetic diversity and conservation of genetic resources. *Food and Agriculture Organization of the United Nations Poultry Development Review* 1–3.

Rajkumar, U., S. V. Rama Rao, M. V. L. N. Raju, and J. N. Chatterjee. 2021. Backyard poultry farming for sustained production and enhanced nutritional and livelihood security with special reference to India: A review. *Tropical Animal Health and Production* 53: Article 176.

Ramalingia, L., S. Kamardi, J. Vankayala, H. Narasimhamurthy, H. Upendra, and Shilpa Shree, J. 2017. Marketing pattern of coloured broiler birds in Karnataka. *International Journal of Livestock Research* 7:220–227.

Reithmayer, C., and O. Mußhoff. 2019. Consumer preferences for alternatives to chick culling in Germany. *Poultry Science* 98:4539–4548.

Sanda, A., M. Bemji, M. Wheto, A. Oso, M. Sanda, and O. Olowofeso. 2021. Ghrelin (GHRL) gene polymorphism and its association with growth and body size parameters in three Nigerian chicken breeds. *Bulletin of the University of Agricultural Sciences & Veterinary Medicine Cluj-Napoca: Animal Science & Biotechnologies* 78.

Sanka, Y., and S. Mbaga. 2014. Evaluation of Tanzanian local chicken reared under intensive and semi-intensive systems: I. Growth performance and carcass characteristics. *Livestock Research for Rural Development* 26:1–7.

Sankhyan, V., S. Katoch, Y. P. Thakur, K. Dinesh, S. Patial, and N. Bhardwaj. 2013. Analysis of characteristics and improvement strategies of rural poultry farming in north western Himalayan state of Himachal Pradesh, India. *Livestock Research for Rural Development* 25.

Sartas, M., E. Kang'ethe, and I. Dror. 2021. *Complete Scaling Readiness Study of Tropical Poultry Genetic Solutions Strategy in Ethiopia, Tanzania and Nigeria*. International Livestock Research Institute, Nairobi, Kenya.

Sharma, J., J. Xie, M. Boggess, E. Galukande, D. Semambo, and S. Sharma. 2015. Higher weight gain by Kuroiler chickens than indigenous chickens raised under scavenging conditions by rural households in Uganda. *Livestock Research for Rural Development* 27: Article 178.

Shoyombo, A., A. Yakubu, A. Adebambo, M. Popoola, O. Olafadehan, M. Wheto, O. Alabi, H. Osaiyuwu, C. Ukim, and A. Olayanju. 2021. Characterisation of indigenous helmeted guinea fowls in Nigeria for meat and egg production. *World's Poultry Science Journal* 77:1037–1058.

Soliman, M. A., M. H. Khalil, K. El-Sabrout, and M. K. Shebl. 2020. Crossing effect for improving egg production traits in chickens involving local and commercial strains. *Veterinary World* 13:407–412.

Tadelle, D., Y. Alemu, and K. J. Peters. 2000. Indigenous chicken in Ethiopia: Genetic potential and attempts at improvement. *World's Poultry Science Journal* 56:45–54.

Tadese, B. 2017. The production performance of the Ethiopian indigenous chickens. GRIN Verlag, Munich, Germany. https://www.grin.com/document/384386.

Tasie, C., G. Wilcox, and A. Kalio. 2020. Adoption of biosecurity for disease prevention and control by poultry farmers in Imo State, Nigeria. *Journal of Agriculture and Food Sciences* 18:85–97.

Terfa, Z. G., S. Garikipati, G. T. Kassie, T. Dessie, and R. M. Christley. 2019. Understanding farmers' preference for traits of chickens in rural Ethiopia. *Agricultural Economics* 50:451–463.

Tiemann, I., S. Hillemacher, and M. Wittmann. 2020. Are dual-purpose chickens twice as good? Measuring performance and animal welfare throughout the fattening period. *Animals* 10: Article 1980.

Tilahun, M., M. Mitiku, and W. Ayalew. 2022. Agroecology is affecting village chicken producers' breeding objective in Ethiopia. *Scientifica* 2022: Article 9492912.

U.S. Agency for International Development. 2021a. Feed the Future Rwanda Orora Wihaze Activity Fact Sheet.

U.S. Agency for International Development. 2021b. Agribusiness development activity annual report.

Vernooij, A., M. N. Masaki, and D. Meijer-Willems. 2018. Regionalisation in poultry development in Eastern Africa. Wageningen Livestock Research Report No. 1121.

Vos, R., and A. Cattaneo. 2020. Smallholders and rural people: Making food system value chains inclusive. Pages. 14–27 in *2020 Global Food Policy Report*. International Food Policy Research Institute, Washington, DC.

Wheto, M., M. Adeleke, A. Ogundero, N. Tor, S. Amusan, O. Ogunpaimo, H. Ojoawo, A. Adebambo, and O. Adebambo. 2020. Polymorphism and diversity studies of insulin-like growth factor I gene among indigenous and FUNAAB Alpha chicken breeds in Nigeria. *Nigerian Journal of Animal Production* 47:33–40.

Wheto, M., A. Adenaike, A. Sanda, B. Ilori, K. Akano, T. Sanni, O. Olowofeso, C. Ikeobi, and O. Adebambo. 2017. Association between insulin like growth factor-1 (IGF-1) gene polymorphism and carcass traits in improved Nigerian indigenous chickens. *Nigerian Journal of Biotechnology* 33:125–130.

Wheto, M., B. M. Ilori, K. Akano, Adebambo, O. A. Ojoawo, H. O. Durosaro, A. J. Sanda, and A. O. Adebambo. 2016. Effect of IGF1 gene polymorphism on carcass traits in improved Nigerian indigenous chicken. Page 6 in *41st Annual Conference of Nigerian Society for Animal Production*. Federal University of Agriculture, Abeokuta, Nigeria.

Wilson, A. A., M. A. Ditmer, J. R. Barber, N. H. Carter, E. T. Miller, L. P. Tyrrell, and C. D. Francis. 2021. Artificial night light and anthropogenic noise interact to influence bird abundance over a continental scale. *Global Change Biology* 27:3987–4004.

Wilson, R. T. 2021. An overview of traditional small-scale poultry production in low-income, food-deficit countries. *Annals of Agricultural and Crop Sciences* 6: Article 1077.

Wolde, S., T. Mirkena, A. Melesse, T. Dessie, and S. Abegaz. 2021. Hatchability and growth performances of normal feathered local, Sasso-RIR and their F1-cross chickens managed under on-station condition in southern Ethiopia. *Tropical Animal Health and Production* 53:1–8.

Wolfgang, G. 2020. Farmers' preferences for tropically adapted improved chicken breeds in selected agro-ecological zones in Tanzania. Doctoral Dissertation Sokoine University of Agriculture, Morogoro, Tanzania.

Wong, J. T., J. Bruyn, B. Bagnol, H. Grieve, M. Li, R. Pym, and R. G. Alders. 2017. Small-scale poultry and food security in resource-poor settings: A review. *Global Food Security* 15:43–52.

World Intellectual Property Organization. 2014. Patent landscape report on animal genetic resources. World Intellectual Property Organization in cooperation with the Food and Agriculture Organization of the United Nations.

Yakubu, A. 2010. Indigenous chicken flocks of Nasarawa State, north central Nigeria: Their characteristics, husbandry and productivity. *Tropical and Subtropical Agroecosystems* 12:69–76.

Yakubu, A., and M. M. Ari. 2018. Principal component and discriminant analyses of body weight and conformation traits of Sasso, Kuroiler and indigenous Fulani chickens in Nigeria. *Journal of Animal and Plant Sciences* 28:46–55.

Yakubu, A., O. Bamidele, W. Hassan, F. Ajayi, U. Ogundu, O. Alabi, E. Sonaiya, and O. Adebambo. 2020. Farmers' choice of genotypes and trait preferences in tropically adapted chickens in five agro-ecological zones in Nigeria. *Tropical Animal Health and Production* 52:95–107.

Yitayih, M., W. Esatu, K. Geremew, T. Yemane, and T. Dessie. 2019. Economic contribution of tropically adapted and more productive breeds based village chicken production in sub-Saharan Africa: ACGG model. Paper presented at the Seventh All Africa Conference on Animal Agriculture, Accra, Ghana, July 29–August 2, 2019.

25

Trends in Bird Conservation in Africa

Hazell Shokellu Thompson and Samuel Tertese Ivande

In recent years, ornithology in Africa has faced a quandary. On one hand, it is viewed as an elite pursuit, with highly educated ornithologists advancing knowledge of African birds and ecosystems while benefiting financially from their work. On the other hand, African ornithologists report marginalization in global contexts, particularly in terms of funding, data access, publication, and recognition (Mwampamba et al. 2022, Tuyisenge et al. 2023). There are also views that ornithology in Africa is influenced by non-African cultures, rooted in colonial history, lacks Indigenous capacity (Brito and Oprea 2009, Cresswell 2018), and is sometimes complicit in "parachute science" (Miller et al. 2023, Mwampamba et al. 2022). Moreover, some consider African ornithology irrelevant to pressing regional issues such as poverty and health (T. Brooks and Thompson 2001, Kilpatrick et al. 2017), and it has sometimes struggled to engage with disadvantaged communities.

This chapter argues that African ornithology and bird conservation are undergoing a renaissance that is addressing some of these long-standing challenges. We focus on two key areas—capacity development and citizen science—noting the global relevance of these issues, given the interconnected nature of ornithology (e.g. bird migration). Finally, we explore future challenges and opportunities for African ornithology.

Africa's Avian Biodiversity

Africa is home to approximately 2500 bird species, representing 22% of the global total, with nearly 60% being endemic (BirdLife International 2018, 2024b). One in nine African bird species face the threat of global extinction (BirdLife International 2024b). Iconic and formerly widespread species, such as Vultures and Parrots, are now in decline (Annorbah et al. 2016; Ogada and Buij 2011; Thiollay 2007, 2017; Williams et al. 2021). In addition, Palearctic migrants to Africa are declining faster than other bird groups (Vickery et al. 2023). However, it is not all doom and gloom. Although Africa holds 30 critically endangered species, such as the Liben Lark (*Heteromirafra archeri*) and White-Winged Flufftail (*Sarothrura ayresi*), there have not been any confirmed bird extinctions in sub-Saharan Africa in the past 300 years. Moreover, some species, such as the Seychelles Magpie-Robin (*Copsychus sechellarum*) and the Mauritius Kestrel (*Falco punctatus*), have been rescued from the brink of extinction in the past three decades (Burt et al. 2016, Tatayah 2022).

Hazell Shokellu Thompson and Samuel Tertese Ivande, *Trends in Bird Conservation in Africa*. In: *New Perspectives in Ornithology*. Edited by: Scott V. Edwards and J. Michael Reed, Oxford University Press. © Oxford University Press (2025). DOI: 10.1093/oso/9780197787670.003.0025

Despite some positive developments, the overall conservation status of African birds has sharply declined during the past 30 years (BirdLife International 2023, Shaw et al. 2024). The main drivers of this decline include habitat loss and degradation (due to poorly planned agriculture, urbanization, and logging), climate change (which affects migration, breeding, food availability, and habitat suitability), hunting and poaching (including the wild bird trade), and invasive species (Beresford et al. 2019; BirdLife International 2018, 2024b; De Vos et al. 2024; Shaw et al. 2024; Thiollay 2006; Zwarts et al. 2018). Urgent conservation efforts and ornithological research linked to advances in ornithological capacity are needed to address these threats. Encouragingly, as we show in this chapter, efforts to lower some key barriers to sustainable bird conservation and research in Africa are starting to take effect. The following sections explore how capacity development and citizen science initiatives are contributing to progress in African ornithology and bird conservation.

Capacity Development

According to the United Nations Development Programme (UNDP 2017, p6) Capacity Development is about transformations that empower individuals, leaders, organizations and societies. If something does not lead to change that is generated, guided and sustained by those whom it is meant to benefit, then it cannot be said to have enhanced capacity, even if it has served a valid development purpose.

This perspective stresses empowerment and transformation, highlighting that capacity development is an ongoing process, which considers local contexts and promotes resilience and adaptation to changes (Global Partnership for Effective Development Co-operation 2016, UNDP 2017).

For ornithology and bird conservation to progress in Africa, strong ornithological capacity is obviously needed. As Beale (2018, p. 100) states, "If ornithology is to thrive in Africa, practitioners must come from a broad diversity of backgrounds, be widely established in Indigenous institutions and publish science of the highest calibre." Also, according to several authors, developing Indigenous ornithological capacity at the systemic, institutional, and individual level is probably the only way to ensure tangible, long-term, and sustainable bird research and conservation impact in Africa (A. P. Leventis Ornithological Research Institute [APLORI] 2024, Awoyemi and Bown 2019, T. Brooks and Thompson 2001, Cresswell 2018, Gichuki et al. 1999, Tende et al. 2024). These approaches influence the way in which capacity development efforts have been described below.

Ornithological Capacity in Africa: Numbers and Distribution

Until recently, there have been relatively few practicing ornithologists in Africa outside South Africa, and capacity for ornithology in Africa has been viewed as limited (Brito and Oprea 2009, Cresswell 2018, O'Connell et al. 2019).

There are some indications that this situation is changing. Ornithological capacity in Africa, measured by peer-reviewed conservation research output, has increased significantly during the past 30 years (Pototsky and Cresswell 2021). In 1987, only three ornithological articles were published by national authors working in institutions within sub-Saharan African countries—these were from Zimbabwe, South Africa, and Ethiopia. By 2017, 12,701 papers had been published, with 38% of 41 sub-Saharan countries contributing (Pototsky and Cresswell 2021). Similarly, Tuyisenge et al. (2023) reported increased African authorship in conservation literature originating from Africa during the period 2015–2021. More broadly, Africa's overall research output has steadily risen, with its share of global scientific publications doubling from 1.5% in 2005 to 3.2% in 2016. In this context, ornithology (irrespective of the country of origin of authors) ranked sixth in Africa among 273 Web of Science subject categories in its contribution to global scientific output (Beaudry et al. 2018).

However, South Africa has dominated the ornithological publishing landscape in Africa: 11 of the top 15 most productive sub-Saharan institutions in terms of published ornithological papers in 2017 were South African, and the remaining institutions produced less than 1% of papers with national primary authors (Pototsky and Cresswell 2021). Therefore, although ornithological capacity has been increasing, this seems, until recently, mainly only in a few countries. This situation is also slowly changing, and some of the changes are outlined below.

The Pan-African Ornithological Congress

A second key sign of progress in African ornithology is the growing participation in the Pan-African Ornithological Congress (PAOC). The PAOC (which takes place every four years) has existed since 1957 and has been a major platform for discussion, research, and presentation of work on the birds of Africa (Beale 2018, Louette and Urban 2007, Thompson 2001). Composition of attendance according to participants' country of origin has evolved considerably since its inception, with the most noteworthy trend being a steady and significant increase in attendance of ornithologists from areas of Africa outside Southern Africa (where the congress originated). Notably, and regrettably, the inaugural congress had no Indigenous African attendees (Louette and Urban 2007), likely due to the political climate at the time. However, Indigenous African participation has steadily increased, reaching more than 60% by 2004 (Louette and Urban 2007) and now typically surpasses international attendance. The most recent congress, PAOC 15 in 2022, drew up to 300 delegates from 60 nations, including 33 African countries (Slater and Mundava 2024). These trends underscore the PAOC's growing role in fostering inclusive, geographically diverse scientific dialogue on African ornithology, and they reflect the rapidly strengthening nature of the African ornithological community.

Ornithological Institutions in Sub-Saharan Africa

Until 2002, the only dedicated ornithological institution in sub-Saharan Africa was the Fitzpatrick Institute of African Ornithology based at the University of Cape Town in South Africa. Since its establishment in 1960, to date, "Fitztitute" researchers have published nearly 8,000 peer-reviewed scientific and public intellectual papers and books. More than 80% of its nearly 500 MSc, PhD, and postdoctoral students have found relevant employment at, and have become leaders within, scores of academic/public/private conservation and research organizations. In terms of demographics, 21% of graduates have been "black" and 52% female. In all, they hail from 46 countries, of which 23 are in Africa. In recognition of its achievements and future potential, in 2004 the Fitztitute was selected as one of the first six South African Centres of Excellence by the South African Department of Science and Technology and National Research Foundation (Crowe and Ryan 2024).

Despite the good work of the Fitzpatrick Institute, a single ornithological research institution serving the whole of the African continent was clearly not an ideal situation. That situation changed in 2002 when the APLORI was set up in Nigeria to address the dearth of dedicated ornithological capacity development opportunities in West Africa (Vickery and Jones 2002).

Capacity development for ornithology at APLORI is achieved primarily by running a master's degree in conservation biology granted through the University of Jos, Nigeria. Teaching is by invited international academics and staff of the University of Jos, with exams moderated by staff of the University of St. Andrews and external examiners from international universities throughout Africa and the world, such as Edinburgh, St. Andrews, Ghent, Groningen, and Oxford.

Since 2004, when the first set of eight students completed their program, APLORI has trained over 146 young West Africans at the master's level in ornithology and conservation biology. More than 25% of these graduates have gone on to complete PhDs, and another approximately 24% are currently studying for PhDs. Approximately 90% of APLORI graduates remain active contributors to biodiversity conservation efforts as researchers, lecturers, teachers, government officials, and conservation nongovernmental organization (NGO) founders and workers (APLORI 2024).

Although APLORI's capacity-building work started in Nigeria, this has extended across the West African subregion. Since 2012, when the first international student from Liberia was admitted to the program, other young West Africans have participated in the APLORI MSc conservation biology program. Through regional partnerships with the University of Cape Coast in Ghana, Fauna and Flora—Liberia, and universities in Sierra Leone, 10 Ghanaian, 4 Liberian, and 3 Sierra Leonean students have participated in the APLORI MSc program (Tende et al. 2024). In 2023, the first student outside West Africa was admitted to the APLORI MSc program from Zimbabwe, and with the first student from Kenya joining in 2025, the program now includes participants from West, Southern, and Eastern Africa. This regional expansion is expected to continue and increase in the coming years.

After 23 years of quality capacity-building work, some APLORI graduates are now in leadership positions in Africa and globally, where they are driving efforts for the conservation of biodiversity at frontline biodiversity conservation organizations. These leaders are working at organizations such as BirdLife International, Fauna and Flora, the Wildlife Conservation Society, the Nigerian Conservation Foundation, the Leventis Foundation, The Nature Conservancy (TNC), and the Indianapolis Zoo Global Center for Species Survival. Several APLORI graduates, especially those who have gained advanced degrees, have also taken up teaching and research positions across different universities and academic institutions, bringing hitherto limited capacity to help introduce ornithology and conservation training across these institutions and to further strengthen ornithological capacity at more foundational (undergraduate) levels. Some other APLORI graduates have founded and are leading biodiversity conservation efforts at NGOs, both locally and internationally—for example, the EDEN Creation Care Initiative and the Better Earth Foundation, both in Nigeria (APLORI 2024). This aspect of APLORI's work has had a significant positive impact on the issue of "leadership models" for ornithology in Africa (see the section titled "African Leadership and Mentorship in Conservation").

In addition to capacity-building work, APLORI also conducts and facilitates research in tropical ornithology and ecology. Through APLORI's work, the capacity of researchers in West Africa to study the environment and investigate ecological processes, especially in tropical habitats in Africa, has increased. From an average of three published articles from APLORI per year during the first five years of establishment, researchers from APLORI alone are now contributing an annual average of 15 peer-reviewed scientific articles to improve understanding of ecological processes in tropical environments in West Africa. If this rate of publication is maintained or increased in the coming years, a minimum of 450 papers will be produced from APLORI alone during the next 30 years. Compare this to the fact that overall, only 129 papers with West African authors were produced from 1987 to 2016 (Cresswell 2018). Because the number of papers produced by local authors in West Africa approximately doubled each decade in the 30 years prior to 2016, it is not unrealistic to predict that the number of papers produced by 2046 would be close to or exceed 1,000.

Note that both dedicated ornithological research institutions in Africa are based in English-speaking countries—South Africa and Nigeria—and the intake has come mainly from English-speaking countries. There are ongoing efforts to establish a third ornithological research station in a French-speaking country—Senegal. This should further boost capacity development for ornithology (but see the section titled "Do Barriers to Ornithological Capacity Development Still Exist in Africa?").

African Leadership and Mentorship in Conservation

A side effect of the domination of ornithological research output in Africa by expatriate and South African institutions during the past 30–40 years (Beale 2018, Brito and Oprea 2009, Cresswell 2018, O'Connell et al. 2019, Pototsky and Cresswell 2021, Tuyisenge et al. 2023) has been a paucity of African conservation leaders promoting ornithology and conservation on the African continent outside Southern Africa and acting as role models/mentors for young African ornithologists. The limited number of African scientists publishing their work has also reduced the availability of potential role models and mentors (Maas et al. 2021), and this in turn may have slowed the growth of ornithological capacity in Africa (Li et al. 2019, Tuyisenge et al. 2023).

However, this picture of "an Indigenous conservation leadership vacuum" is changing. During the past 30 years, a dynamic network of national bird and nature conservation organizations has developed across Africa. The strongest manifestation of this has been the emergence and growth of the BirdLife Africa Partnership. This is a growing network of 27 national nature conservation organizations in 27 African countries with a combined total of more than 500 staff and 87,000 national members (BirdLife International 2024a).

The BirdLife Africa Partnership currently comprises the following organizations: BirdLife Botswana, NATURAMA (Burkina Faso), Association Burundaise pour la Protection de la Nature (Burundi), Biosfera (Cabo Verde), SOS-Forêts (Côte d'Ivoire), Nature Conservation Egypt, Ethiopian Wildlife and Natural History Society, Ghana Wildlife Society, Guinée-Ecologie (Guinea), Nature Kenya—The East Africa Natural History Society, The Society for the Conservation of Nature of Liberia, Asity Madagascar, Wildlife and Environmental Society of Malawi, Nature Mauritanie (Mauritania), Mauritian Wildlife Foundation, GREPOM/BirdLife Maroc (Morocco), Nigerian Conservation Foundation, Nature Rwanda, Nature-Communautés-Développement (Senegal), Nature Seychelles, The Conservation Society of Sierra Leone (CSSL), BirdLife South Africa, Nature Tanzania, Association "Les Amis des Oiseaux" (Tunisia), NatureUganda, BirdWatch Zambia, and BirdLife Zimbabwe (BirdLife International 2024a).

These organizations provide important conduits for the capacity development of young African ornithologists that did not exist 30–40 years ago. In effect, they provide a system in which conservation leadership and mentorship are being provided.

In addition to leading national bird conservation organizations, in the past 30 years, Africans have emerged on the global conservation stage as notable global or regional leaders. In the past decade, these have included leadership positions at global nature conservation organizations such as BirdLife International (interim chief executive officer, global directors, and regional directors), TNC (Africa Regional Director), and the Indianapolis Zoo Global Center for Species Survival (Conservation Coordinator (Birds)—IUCN SSC Species Specialist Groups). Consequently, role models for Africans in global ornithology are now more visible, which is an encouraging development that was not present before the turn of the 21st century.

Do Barriers to Ornithological Capacity Development Still Exist in Africa?

Given the progress and development noted so far, do barriers to ornithological capacity development still exist in Africa? The answer is undoubtedly yes. A significant issue in this regard may be broadly categorized as "barriers to publication."

Several authors have highlighted significant barriers to research and publication for African ornithologists and Global South scientists, including acceptance bias, funding limitations, restricted data access, and language challenges (Maas et al. 2021, Negret et al. 2022, Tuyisenge et al. 2023). Each of these issues is briefly examined below.

We use the term *acceptance bias* to refer to perceived inequalities in the way scientific manuscripts are accepted by and shared with the wider scientific community. For example, publications originating from sub-Saharan Africa have been found to face lower acceptance rates and tend to be frequently published in lower impact journals (Mammides et al. 2016, Tuyisenge et al. 2023). Several factors could be responsible for this, including editorial ignorance of the African context and inadequate mentoring and training support (Beaudry et al. 2018). We argue that it is incumbent on publishers, editors, and senior scientists to review their publication records and seek ways and means to redress any perceived bias (see also Li et al. 2019). Some specific policies that journal editors and senior scientists can implement to prevent or redress acceptance bias include having regular and authentic collaborations with young ornithologists, installing regional journal editors for global publications, providing editing assistance, waiving publication fees, and seeking locally focused studies from which globally relevant strategies and lessons can be drawn (Li et al. 2019, Mammides et al. 2016).

Unsurprisingly, funding is a significant barrier to publishing for African ornithologists. In addition to the well-documented lack of funding for higher education, research, and conservation in Africa (Beaudry et al. 2018, Mammides et al. 2016), individual researchers also face personal financial constraints. Limited resources and limited time for article writing and mentoring are common in African academic environments (Beaudry et al. 2018). Many researchers take on multiple jobs to supplement their income, reducing time for publication efforts, and some either leave conservation science or publish lower quality articles to disseminate their work quickly (Tuyisenge et al. 2023).

The issue of funding is closely linked to one of the oldest and most significant barriers to publication in Africa (and globally): the subscription or pay-per-view paywall that restricts access to much of the ornithological and scientific literature. This paywall limits the availability of published information for scientists, including African ornithologists (Bolick et al. 2017, Fuller et al. 2014, Harnad et al. 2008, Taylor et al. 2008). In Africa, the problem is worsened by the fact that the paywall is often not offset by institutional access (Sunderland et al. 2009) and also by the relatively high publication costs compared to personal incomes. In addition, comparatively low internet penetration further limits access to scientific data and the ability to publish.

However, significant progress is being made in improving data and publication access. For example, Research4Life provides free or discounted access to scientific content from major journal publishers to institutions in developing countries (Research4Life 2024). In addition, the number of journals offering open access (OA) publishing has increased rapidly, particularly through the hybrid model, in which authors can choose to publish their papers as OA by paying a fee or keep them behind a subscription firewall. This growth is driven by the increasing demand from the scientific community for accessible scientific information and requirements from European and U.S. funding agencies for research to be published in OA sources (Buchanan and Pruett-Jones 2023).

Many established conservation scientists now publish in OA journals, ensuring that the license permits reuse with proper attribution, or they opt for per-article OA payments (Fuller et al. 2014). Others are self-archiving their articles on preprint servers or institutional repositories, making them freely available online. Moreover, more international journals are allowing author self-archiving (Harnad et al. 2008). These trends should be encouraged, and senior scientists should mentor early career ornithologists in identifying and utilizing OA options for accessing and publishing research.

To conclude this section on barriers to publication in Africa, it is important to examine the role of language. Ornithological research and teaching are typically conducted in colonial or dominant languages such as English, French, or Portuguese, and most ornithological literature is available primarily in these languages. This creates barriers for local researchers, students, and community members who speak Indigenous languages (Negret et al. 2022). The language divide restricts the dissemination of ornithological findings and hampers collaboration between local communities, African ornithologists unfamiliar with local languages, and international researchers. In addition, Western terminology and concepts (e.g., "spring migration") can alienate students and communities. Multilingual communication strategies are vital to bridging these gaps and ensuring diverse perspectives are included in ornithological research.

In multilingual Africa (as elsewhere), Indigenous languages, such as Kiswahili in East Africa, are associated with many bird species. These languages could be key to conservation, research, policies, and practices. However, scientific information that is available only in non-English languages is usually omitted from the scientific literature. The omission of such non-English-language information can bias inferences and cause gaps in the spatial and taxonomic coverage of scientific evidence that is useful for ornithological research and conservation. It also limits the accessibility of information on birds for local communities and governments, many of which may work primarily in Indigenous languages (Negret et al. 2022). Translation and localization of ornithological reports are important for ensuring that research influences conservation policies effectively.

Africa's linguistic diversity means that ornithological research often involves collaboration across languages and cultures. International research teams, local communities, and governmental bodies must communicate across language barriers. Translating scientific materials into local languages and making technical terms accessible are crucial to ensuring that all stakeholders understand the research goals and findings.

African ornithologists are actively addressing language and cultural biases in the field by promoting the use of Indigenous languages in ornithological materials. For instance, *The Common Birds of Sierra Leone: An Indigenous Language Guide* (Thompson et al. 2024) is a handbook that describes common bird species in Sierra Leone, listing bird names in three of the country's main Indigenous languages—Mende, Temne, and Krio. It also highlights cultural connections where they exist, aiming to foster greater interest in birds and bridge gaps between ornithologists and local communities.

Similarly, in South Africa, the South African Names for South African Birds Working Group is developing a list of bird names in traditional South African languages (BirdLife International 2024c). This initiative is intended to make bird conservation more accessible and inclusive, helping a wider range of South Africans engage with and protect their avian heritage.

Addressing the underrepresentation of non-English knowledge and research in the ornithological and scientific literature as well as in policy implementation will require multiple strategies, including increased use of technological advancements (e.g., artificial intelligence–driven tools such as Google Translate), institutional support (governments may fund localization efforts to promote access in local languages), and a shift in academic culture (e.g., routine inclusion of translation and localization costs as part of original grant proposals for research projects).

Conclusions and Recommendations – Ornithological Capacity

Ornithological capacity is increasing across Africa with some very positive developments, the most notable of which include the following:

- The establishment of the APLORI in Nigeria in 2002 to address the lack of ornithological capacity in West Africa. Since its inception, APLORI has trained ornithologists from across the region and beyond, complementing the long-standing contributions of the FitzPatrick Institute of African Ornithology in South Africa and further boosting research and publication rates throughout the continent.
- The emergence of a growing network of national bird and nature conservation organizations throughout Africa, the strongest manifestation of which is the BirdLife Africa Partnership. These organizations provide important conduits for the development of young African ornithologists that did not exist 30 years ago.
- The appearance and development of African leaders in ornithology and conservation at national, regional, and global levels, serving as role models for aspiring young African ornithologists. By mentoring young scientists and advocating for more inclusive, regionally focused research, these leaders are helping cultivate the next generation of African ornithologists and conservationists, reinforcing the growing role of Africa in global ornithological research.
- The evolution of the PAOC into an inclusive, geographically diverse, thriving platform for scientific dialogue on African ornithology, reflecting the rapidly strengthening nature of the African ornithological community.

To further improve capacity development in ornithology in Africa, we recommend the following:

- Development of continent-wide "mentoring" schemes for young African ornithologists through the International Ornithological Union (IOU) or the PAOC. Recent discussions on advancing African ornithology at PAOC 15 (S. Banda, personal communication, September 24, 2022) and the recent approval by the IOU Council to form a Working Group for African Ornithology as a platform for networking and collaboration for African ornithologists are positive steps in this direction (E. Yohannes, personal communication, April 26, 2023).
- Changes in institutional policies at universities, government agencies, NGOs, and other agencies, with increased allocation of resources for article writing and publication and also for mentorship and institutional capacity development.

- Encouragement of greater use of OA publications by individuals and institutions, and further development of funding mechanisms to support these publications.
- Support for ongoing efforts to establish a francophone ornithological research institute.

Citizen Science

A very urgent data requirement for African ornithology and bird conservation is distributional information: "We cannot study or protect the continent's birds if we do not know where they are found" (T. Brooks and Thompson 2001). Until approximately 40 years ago, most countries in Africa lacked information on the distribution and abundance of the birds within their national borders, and large areas of the continent remained almost unexplored by ornithologists. Outside South Africa, new data on bird distribution and abundance were typically being provided by expatriate expeditions from outside Africa and/or cross-border initiatives (e.g., Dowsett and Dowsett-Lemaire 1989, Ryan et al. 1999). Although quite valuable, many of these ornithological initiatives were project-based, time-bound, and expensive, and some of them had little input from local ornithologists. A recent alternative but complementary approach on the African continent has been the use of the bird atlas concept and citizen science to obtain bird distributional information in a cost-effective manner (Tende et al. 2024).

The atlas concept was first developed in the late 1950s to map the distribution of the British flora (Perring and Walters 1962). It was adopted to map the distribution of British birds in the late 1960s (Lord and Munns 1970, Sharrock 1976). Since the 1960s, the concept has spread rapidly throughout the world. As of 2021, there were more than 600 bird atlas projects following standardized protocols in 93 countries, involving more than 380,000 participants worldwide (Shan 2024). A crucial element of the concept is that it is underpinned by citizen science—the engagement and contribution of data by members of the public through field surveys (Kobori et al. 2016, Kullenberg and Kasperowski 2016, Tende et al. 2024).

The first bird atlas project in Africa, based on a standardized protocol (BirdMap), started in 1986 (Harrison et al. 2008, Lee et al. 2022). This project (SABAP1) covered six countries—Botswana, Lesotho, Namibia, South Africa, Swaziland, and Zimbabwe—and resulted in the publication of the atlas of southern African birds (Harrison et al. 1997). It showcased the significance of citizen science in providing knowledge useful for promoting the conservation of African avifauna. A second bird atlas project (SABAP2) was started in 2007 to build on the successes recorded previously and was expanded to cover Kenya through the Kenyan Bird Map (KBM) in 2013 and Nigeria through the Nigerian Bird Atlas Project (NiBAP) in late 2015 (Lee et al. 2022, Tende et al. 2024). The initiative has morphed into the Africa Bird Atlas Project (ABAP), which is probably the single largest citizen science project in Africa. It holds more than 18 million bird records from 47 countries in Africa, with more than 1.9 million records submitted every year. The following 15 countries have formal projects with registered project participants: South Africa, eSwatini, Lesotho, Namibia, Zimbabwe, Zambia, Mozambique, Botswana, Malawi, Nigeria, Kenya, Sierra Leone, Liberia, Ghana, and Uganda (M. Brooks et al. 2022).

Bird atlas data are typically submitted to the ABAP central database and associated BirdMap projects across Africa by means of a mobile BirdLasser app. The app and its integrated gamification network have experienced a high adoption rate by atlassers, bird-watchers, and conservationists, contributing to causes, creating life lists, and taking part in events. The app has been associated with the recruitment of new participants and has been credited with increased submission of data (Lee and Nel 2020). Recent collaborative work between ABAP and eBird, a globally popular platform, will allow approximately 3 million bird records previously collected in Africa to become available in the ABAP database. Importantly, the eBird app can now be used to collect data for the ABAP project, considerably widening the scope of atlassing in Africa (U. Ottosson, personal communication, October 31, 2024).

All BirdMap data can be accessed through the ABAP website (http://www.birdmap.africa). In addition to South Africa, where the project started, Nigeria and Kenya have dedicated websites for data curation. The KBM project, which is collaboratively managed by the Ornithology Section of the National Museums of Kenya, A Rocha Kenya, the Tropical Biology Association, and the Kenya Nature Bird Committee, focuses on collecting data across Kenya (http://kenya.birdmap.africa), and NiBAP (http://nigeria.birdmap.africa), in conjunction with APLORI, manages the project in Nigeria. In addition, data collected by BirdMap protocols are stored and made publicly available for almost all sub-Saharan countries in Africa and its neighboring islands (M. Brooks et al. 2022).

As a case study from outside South Africa, the development of NiBAP clearly shows the potential of citizen science for bird data collection and the promotion of biodiversity conservation at a relatively low cost in Africa. The project undertook a foundational process of identifying, mapping, and convening pre-existing local capacity for ornithology and bird-watching in Nigeria. It then recruited and supported the capacity development and growth of citizen scientists in Nigeria through bird clubs and, finally, assessed the impacts of this growing bird-watching and citizen science community on data generation and survey coverage for the NiBAP (Tende et al. 2024). The results have been striking. From just two expatriate-dominated bird clubs in Nigeria, the NiBAP has founded and established 20 bird clubs with more than 1,000 mostly Indigenous members. From three active atlassers in 2015, there are now more than 827 citizen scientists contributing to the Nigerian bird atlas as a live and active initiative. These citizen scientists have helped survey 8,709 (79%) of the 11,080 pentads needed for a complete survey coverage of Nigeria (Tende et al. 2024). At the inception of the initiative in December 2015, only seven pentads had been covered. The pentad is the spatial unit of the grid system used by the ABAP to collect data on bird distribution. It is based on the latitude/longitude coordinates system and measures approximately 5 × 5 geographical coordinate minutes; i.e., approximately 9 × 9 km in Africa.

Ornithological surveys have been accompanied by a vigorous education and awareness campaign. The NiBAP Facebook group has approximately 5,000 people participating in the general discourse about birds and biodiversity conservation in Nigeria (APLORI 2024).

The initiative has also trained potential local experts, who are urgently needed in Africa (Awoyemi and Ibáñez-Álamo 2023). Some members have advanced in their scientific endeavors, analyzing and publishing several bird-related papers. The approach has shown the potential for converting enthusiasts into citizen scientists through knowledge transfer from ornithologists (Tende et al. 2024).

At the other end of the spectrum, the progress of a similar initiative in Sierra Leone shows that there are still considerable challenges in rolling out citizen science and bird atlassing across the African continent. In Sierra Leone, five bird clubs were established between the inception of a national bird atlas initiative in 2018 and 2024 (at the time of writing). There are now 100 bird club members throughout the country. Atlassing coverage increased from just 1 atlassed pentad in 2018 to 50 atlassed pentads in 2024, and bird record submission increased from 185 to 3,036 records in BirdLasser (ABAP 2024). Current pentad coverage represents 6% of the coverage needed to cover the country. The hosting organization, Conservation Society of Sierra Leone (CSSL), has more than 500 members, and its education and awareness campaigns reach approximately 2,000 people on the CSSL Facebook page alone. These figures represent progress (albeit quite slow) and show the value of inter-African collaboration—participants in the Sierra Leone initiative were trained by NiBAP members at an inception workshop in 2018. However, they also illustrate the scale of the capacity development efforts required in citizen science initiatives. In Sierra Leone, ownership of the smartphones required for use of the BirdLasser app has been an issue, and some lay experts have struggled with the technology.

Conclusions and Recommendations – Citizen Science

Clearly, the continuing development of citizen science in Africa is turning out to be a relatively low-cost and more sustainable approach to generating much needed information about the distribution and natural history of birds across the continent. Citizen science is also providing a crucial platform to foster and increase local awareness and Indigenous capacity for meaningful engagement and participation in bird and biodiversity monitoring and conservation in Africa.

As citizen science continues to grow and local capacity to analyze and generate relevant output from collected data increases, we can expect its immense potential to influence and promote effective conservation decision-making across the region to begin to be unlocked.

Despite the remarkable development and growth of citizen science across the continent evidenced by the establishment of national bird atlas projects and submitted records, this development also reflects the current variation in ornithological capacity throughout Africa, further underscoring the need for continued and more widespread capacity development across the continent.

To fast-track and consolidate the effective use of citizen science throughout Africa, we recommend the following:

- Creation of complementary platforms (or updating current platforms with pathways) that allow the incorporation of Indigenous knowledge and processes to make local participation and engagement more meaningful, genuine, and relevant.
- Adaptation and replication of successful citizen science models such as in South Africa and Nigeria across the continent, emphasizing the creation of bird clubs that encourage skills and knowledge transfer between trained ornithologists and nature enthusiasts through social learning and other less formal capacity-building processes.
- Investigation and encouragement of sustainable ways to incentivize and increase participation of young people and Indigenous communities in citizen science and other similar community conservation programs at scale across the continent.
- More regional, national, and local government support, including through the allocation of resources, to support the growth and expansion of citizen science programs across the continent and the adoption and integration of results and outcomes in biodiversity management and decision-making processes. South Africa's citizen science initiatives in biodiversity monitoring, particularly through government-supported programs such as the South African National Biodiversity Institute, could provide a possible model here (SANBI 2024).

Future Challenges and Opportunities for African Ornithology

The future of bird conservation and ornithology in Africa will face a combination of challenges and opportunities, many of which will be shaped by emerging socioeconomic and environmental factors on a macro scale (see Thompson 2001).

A major challenge will continue to be human population growth across the continent. Although rates of increase have slowed, sub-Saharan Africa has the highest rate of population growth in the world (United Nations 2024a). The growing population will intensify pressures on birds, habitats, and biodiversity through activities such as urban expansion, deforestation, and agricultural intensification. The situation will be exacerbated by climate change.

However, as the 21st century advances, improvements in living standards and education in Africa may help mitigate the challenges posed by population growth. Poverty and lack of education are generally acknowledged as contributors to population increase, which drives environmental degradation (United Nations Department of Economic and Social Affairs 2020). Because economic development and education are top priorities for African governments and international investment, a key challenge—and opportunity—for ornithologists and conservationists will be to more effectively integrate bird and biodiversity conservation into economic development and population education strategies. Development and conservation have often been viewed as mutually exclusive, but as living standards and education improve, population growth rates are likely to decline, benefiting birds and biodiversity. Economic development may also result in increases in the funding resources available for higher education, research, and conservation in Africa, which constitute one of the key existing barriers to ornithological capacity development in the region.

Although population growth may lead to habitat loss for birds and wildlife, Africa's expanding population also presents opportunities. With 70% of sub-Saharan Africa's population younger than age 30 years, the region has the youngest demographic globally (United Nations 2024b). This youth surge offers potential for both economic growth and bird conservation, provided these younger generation members are empowered to reach their full potential. If strategically engaged, Africa's youth could be a key resource for conservation. With the right education, outreach, and funding, this large, youthful population can be mobilized as citizen scientists, advocates, and future conservation professionals. Increased support for ornithological training, coupled with local and international funding and mentoring, could help unlock this potential, equipping the next generation to address biodiversity challenges. In addition, we anticipate further expansion of the civil society conservation organization sector in Africa (e.g., the BirdLife Africa Partnership). With its emphasis on transformative capacity development, especially leadership and mentorship, Africa's youth should benefit.

Potential roadmaps for accelerating future conservation capacity development in Africa could include (1) scaling up the Africa component of the Conservation Leadership Programme, which has already directed project funding and training to more than 3,100 early career leaders from developing countries who are tackling priority conservation challenges (Conservation Leadership Programme 2024) and/or (2) using a similar framework as The Tony Elumelu Foundation Entrepreneurship Programme. Since the launch of the program in 2015, the foundation has trained more than 1.5 million young Africans on its digital hub, TEFConnect, and disbursed nearly $100 million in direct funding to more than 19,000 African women and men, who have collectively created more than 400,000 direct and indirect jobs. Funding has come from African philanthropists, entrepreneurs, and innovative partnerships with the private sector, governments, and development agencies throughout the world (The Tony Elumelu Foundation 2024).

Another advantage of Africa's young and tech-savvy population is the speed with which it is adapting to the emerging technologies that are revolutionizing ornithological research and conservation, such as mobile data collection platforms, audio monitors, drones and Global Positioning System tags for wildlife monitoring, and e-DNA and artificial intelligence for species identification. With appropriate training and support, young conservationists can conduct surveys of large and inaccessible areas, monitor bird populations, track migratory patterns, and assess habitat changes with unprecedented accuracy. This could also help reduce the costs and time needed for traditional fieldwork and make conservation efforts more efficient.

With increasing access to smartphones and internet connectivity, global mobile data collection apps such as eBird and iNaturalist (iNaturalist 2024, Sullivan et al. 2009) are empowering young African citizen scientists, researchers, and conservationists to record bird sightings and biodiversity data in real time, often from remote and previously understudied areas. Several apps are also currently being developed on the African continent (e.g., BirdLasser and BirdPlus; Kumdet et al. 2023, Lee and Nel 2020) that reflect relevant local needs and incorporate Indigenous knowledge systems to further accelerate adoption and engagement across the continent. This democratization of data collection will continue and should be encouraged, expanding the scope of ornithological research, filling in knowledge gaps, and allowing for more localized and timely conservation responses as local capacity for the analyses of these data also continues to grow.

However, a youthful population also brings challenges. Many young people in Africa face underemployment and limited opportunities—exacerbated by lack of internet access, which restricts access to information about available prospects. If unaddressed, these limitations can lead to "brain drain," in which Africa loses its most talented ornithologists to better opportunities abroad, leaving gaps in local capacity and leadership. However, as socioeconomic and professional opportunities improve, this trend may reverse, leading to a "brain gain." Returning ornithologists bring valuable skills and knowledge, and with the rise of capable conservation professionals within Africa, the continent's role in global conservation will continue to strengthen.

As African ornithologists gain leadership experience, they can play a more central role in shaping regional and global conservation strategies, including advancement of the Global Biodiversity Framework targets (Abu-Bakarr et al. 2022, Convention on Biological Diversity 2020). African countries, supported by this local expertise, may have better prospects for meeting these biodiversity goals while also influencing international policy through more equitable partnerships with the Global North (Rodríguez et al. 2022). Opportunities to further improve African ornithologists' skills may emerge through national and regional capacity-building initiatives, enhanced access to international networks, and collaborative research platforms that promote equity in knowledge sharing and publication. In this regard, barriers to capacity development for ornithology in Africa that have significant contributions from the Global North will probably be further lowered with OA facilitating publication and information flow and with parachute science becoming obsolete.

Thompson (2001) predicted that "the cadre of African ornithologists will continue to increase and become technically stronger." Twenty plus years later, this chapter has shown this prediction to be true, and we can confidently predict that this improvement in numbers and technical knowledge will continue into the foreseeable future. The PAOC and existing ornithological research institutions—the Fitzpatrick Institute of African Ornithology in South Africa and APLORI—will continue to play pivotal roles in this path to ornithological excellence. Establishment of additional ornithological research institutes in Africa should further complement their efforts. The recent establishment of a master's degree course in ornithology conservation development at the Université Gaston Berger in Saint-Louis, Senegal (BirdLife International 2022), that will be taught in French is a promising start to a positive multilingual trajectory for ornithological research institutes in Africa.

References

A. P. Leventis Ornithological Research Institute. 2024. 23 years of capacity building for ornithology and conservation science in West Africa. Accessed December 5, 2024, from https://www.aplori.org/capacity-building.

Abu-Bakarr, I., M. I. Bakarr, N. Gelman, J. Johnny, P. J. Kamanda, D. Killian, A. Lebbie, M. Murphy, A. Ntongho, S. O'Connor, E. Sam-Mbomah, P. F. Y. Thulla, and R. Wadsworth. 2022. Capacity and leadership development for wildlife conservation in sub-Saharan Africa: Assessment of a programme linking training and mentorship. *Oryx* 56(5):744–752. https://doi.org/10.1017/S0030605321000855

African Bird Atlas Project. 2024. Coverage summary: Sierra Leone. Retrieved December 4, 2024, from https://www.birdmap.africa/coverage/country/Sierraleone.

Annorbah, N. N. D., N. J. Collar, and S. J. Marsden. 2016. Trade and habitat change virtually eliminate the Grey Parrot *Psittacus erithacus* from Ghana. *Ibis* 158(1):82–91. https://doi.org/10.1111/ibi.12332.

Awoyemi, A. G., and D. Bown. 2019. Bird conservation in Africa: The contributions of the Ibadan Bird Club. *Biodiversity Observations* 10:1–12. https://doi.org/10.15641/bo.v10i0.635

Awoyemi, A. G., and J. D. Ibáñez-Álamo. 2023. Status of urban ecology in Africa: A systematic review. *Landscape and Urban Planning* 233: Article 104707. https://doi.org/10.1016/j.landurbplan.2023.104707.

Beale, C. M. 2018. Trends and themes in African ornithology. *Ostrich* 89(2):99–108. https://doi.org/10.2989/00306525.2017.1407834.

Beaudry, C., J. Mouton, and H. Prozesky. 2018. The next generation of scientists in Africa. African Minds. https://scienceopen.com/hosted-document?doi=10.47622/978-1-928331-93-3

Beresford, A. E., F. J. Sanderson, P. F. Donald, I. J. Burfield, A. Butler, J. A. Vickery, and G. M. Buchanan. 2019. Phenology and climate change in Africa and the decline of Afro-Palearctic migratory bird populations. *Remote Sensing in Ecology and Conservation* 5(1): Article 1. https://doi.org/10.1002/rse2.89.

BirdLife International. 2018. *State of Africa's Birds 2017: Indicators for Our Changing Environment.* BirdLife International Africa Partnership, Nairobi, Kenya.

BirdLife International. 2022. The Africa report 2023. Retrieved November 20, 2024, from https://www.birdlife.org/wp-content/uploads/2023/07/BirdLife-Africa-Report-2022-ENG.pdf.

BirdLife International. 2023. The 2023 IUCN Red List update reveals hope for birds in crisis. Retrieved November 20, 2024, from https://www.birdlife.org/news/2023/12/12/the-2023-red-list-update-#reveals-hope-for-birds-in-crisis.

BirdLife International. 2024a. BirdLife Africa. Retrieved November 20, 2024, from https://www.birdlife.org/africa.

BirdLife International. 2024b. Globally threatened species for Africa. Retrieved August 1, 2025, from https://datazone.birdlife.org/species/dashboard.

BirdLife International. 2024c. South African Names for South African Birds (SANSAB). Retrieved November 4, 2024, from https://www.birdlife.org.za/what-we-do/empowering-people/what-we-do/#1654518246719-92be2b4b-2f54.

Bolick, J., A. Emmett, M. Greenberg, B. Rosenblum, and A. Peterson. 2017. How open access is crucial to the future of science. *Journal of Wildlife Management* 81:564–566. https://doi.org/10.1002/jwmg.21216.

Brito, D., and M. Oprea. 2009. Mismatch of research effort and threat in avian conservation biology. *Tropical Conservation Science* 2(3):353–362. https://doi.org/10.1177/194008290900200305.

Brooks, M., S. Rose, R. Altwegg, A. T. Lee, H. Nel, U. Ottosson, E. Retief, C. Reynolds, P. G. Ryan, S. Shema, T. Tende, L. G. Underhill, and R. L. Thomson. 2022. The African Bird Atlas Project: A description of the project and BirdMap data-collection protocol. *Ostrich* 93(4):223–232. https://doi.org/10.2989/00306525.2022.2125097.

Brooks, T., and H. S. Thompson. 2001. Current bird conservation issues in Africa. *The Auk* 118(3):575–582.

Buchanan, K. L., and S. Pruett-Jones. 2023. Ornithology and open access publishing. *Emu – Austral Ornithology* 123(4):265–267. https://doi.org/10.1080/01584197.2023.2270232.

Burt, A. J., J. Gane, I. Olivier, L. Calabrese, A. De Groene, T. Liebrick, D. Marx, and N. Shah. 2016. The history, status and trends of the Endangered Seychelles Magpie-Robin *Copsychus sechellarum. Bird Conservation International* 26(4):505–523. https://doi.org/10.1017/S0959270915000404.

Conservation Leadership Programme. 2024. Home. Retrieved December 11, 2024, from https://www.conservationleadershipprogramme.org.

Convention on Biological Diversity. 2020. Long-term strategic framework for capacity building. Retrieved December 11, 2024, from https://www.cbd.int/cb/strategic-framework.

Cresswell, W. 2018. The continuing lack of ornithological research capacity in almost all of West Africa. *Ostrich* 89(2):123–129. https://doi.org/10.2989/00306525.2017.1388301.

Crowe, T., and P. Ryan. 2024. History: Overview. FitzPatrick Institute of African Ornithology. https://science.uct.ac.za/fitzpatrick/about-us-history/history-overview.

De Vos, K., C. Janssens, L. Jacobs, B. Campforts, E. Boere, M. Kozicka, D. Leclère, P. Havlík, L.-M. Hemerijckx, A. Van Rompaey, M. Maertens, and G. Govers. 2024. African food system and biodiversity mainly affected by urbanization via dietary shifts. *Nature Sustainability* 7(7):869–878. https://doi.org/10.1038/s41893-024-01362-2.

Dowsett, R. J., and F. Dowsett-Lemaire. 1989. Liste préliminaire des oiseaux du Congo. *Tauraco Research Report* 2:29–51.

Fuller, R. A., J. R. Lee, and J. E. M. Watson. 2014. Achieving open access to conservation science. *Conservation Biology* 28(6):1550–1557. https://doi.org/10.1111/cobi.12346.

Gichuki, N. N., N. J. Adams, and R. H. Slotow. 1999. The contribution of indigenous knowledge to bird conservation. Pages 1345–1350 in N. J. Adams and R. H. Slotow, eds. *Proceedings of the 22 International Ornithological Congress, Durban.* BirdLife South Africa, Johannesburg.

Global Partnership for Effective Development Co-operation. 2016. Nairobi outcome document. https://www.effectivecooperation.org/content/nairobi-outcome-document.

Harnad, S., T. Brody, F. Vallières, L. Carr, S. Hitchcock, Y. Gingras, C. Oppenheim, C. Hajjem, and E. R. Hilf. 2008. The access/impact problem and the green and gold roads to open access: An update. *Serials Review* 34(1):36–40. https://doi.org/10.1016/j.serrev.2007.12.005.

Harrison, J. A., D. G. Allan, L. G. Underhill, C. J. Brown, M. Herremans, V. Parker, and A. J. Tree. 1997. *The Atlas of Southern African Birds* (Vol. 1). BirdLife South Africa, Johannesburg.

Harrison, J. A., L. G. Underhill, and P. Barnard. 2008. The seminal legacy of the Southern African Bird Atlas Project: Research in action. *South African Journal of Science* 104(3–4): 82–84.

iNaturalist. 2024. Home. Retrieved December 6, 2024, from https://www.inaturalist.org.

Kilpatrick, A. M., D. J. Salkeld, G. Titcomb, and M. B. Hahn. 2017. Conservation of biodiversity as a strategy for improving human health and well-being. *Philosophical Transactions of the Royal Society B: Biological Sciences* 372(1722): Article 20160131. https://doi.org/10.1098/rstb.2016.0131.

Kobori, H., J. L. Dickinson, I. Washitani, R. Sakurai, T. Amano, N. Komatsu, W. Kitamura, S. Takagawa, K. Koyama, T. Ogawara, and A. J. Miller-Rushing. 2016. Citizen science: A new approach to advance ecology, education, and conservation. *Ecological Research* 31(1):1–19. https://doi.org/10.1007/s11284-015-1314-y.

Kullenberg, C., and D. Kasperowski. 2016. What is citizen science? A scientometric meta-analysis. *PLoS One* 11(1): Article 1. https://doi.org/10.1371/journal.pone.0147152.

Kumdet, P. S., B. A. Danmallam, and S. T. Ivande. 2023. BirdPlus—User guide (Version 1.0) [Mobile application software]. Retrieved December 2, 2024, from https://www.birdplus.org.

Lee, A. T. K., M. Brooks, and L. G. Underhill. 2022. The SABAP2 legacy: A review of the history and use of data generated by a long-running citizen science project. *South African Journal of Science* 118(1–2). https://doi.org/10.17159/sajs.2022/12030.

Lee, A. T. K., and H. Nel. 2020. BirdLasser: The influence of a mobile app on a citizen science project. *African Zoology* 55(2):155–160. https://doi.org/10.1080/15627020.2020.1717376.

Li, W., T. Aste, F. Caccioli, and G. Livan. 2019. Early coauthorship with top scientists predicts success in academic careers. *Nature Communications* 10(1): Article 5170. https://doi.org/10.1038/s41467-019-13130-4.

Lord, J., and D. J. Munns. 1970. *Atlas of Breeding Birds of the West Midlands: Based on a Field Survey Covering the Counties of Warwickshire, Worcestershire and Staffordshire During the Years 1966–1968.* West Midland Bird Club, Stafford, UK.

Louette, M., and E. K. Urban. 2007. Overview of the eleven Pan-African Ornithological Congresses. *Ostrich* 78(2):xiii–xxiv. https://doi.org/10.2989/OSTRICH.2007.78.2.ii.162.

Maas, B., R. J. Pakeman, L. Godet, L. Smith, V. Devictor, and R. Primack. 2021. Women and Global South strikingly underrepresented among top-publishing ecologists. *Conservation Letters* 14(4): Article e12797. https://doi.org/10.1111/conl.12797.

Mammides, C., U. M. Goodale, R. T. Corlett, J. Chen, K. S. Bawa, H. Hariya, F. Jarrad, R. B. Primack, H. Ewing, X. Xia, and E. Goodale. 2016. Increasing geographic diversity in the international conservation literature: A stalled process? *Biological Conservation* 198:78–83. https://doi.org/10.1016/j.biocon.2016.03.030.

Miller, J., T. B. White, and A. P. Christie. 2023. Parachute conservation: Investigating trends in international research. *Conservation Letters* 16(3): Article e12947. https://doi.org/10.1111/conl.12947.

Mwampamba, T. H., B. N. Egoh, I. Borokini, and K. Njabo. 2022. Challenges encountered when doing research back home: Perspectives from African conservation scientists in the diaspora. *Conservation Science and Practice* 4(5): Article e564. https://doi.org/10.1111/csp2.564.

Negret, P. J., S. C. Atkinson, B. K. Woodworth, M. Corella Tor, J. R. Allan, R. A. Fuller, and T. Amano. 2022. Language barriers in global bird conservation. *PLoS One* 17(4): Article e0267151. https://doi.org/10.1371/journal.pone.0267151.

O'Connell, M. J., O. Nasirwa, M. Carter, K. H. Farmer, M. Appleton, J. Arinaitwe, P. Bhanderi, G. Chimwaza, J. Copsey, J. Dodoo, A. Duthie, M. Gachanja, N. Hunter, B. Karanja, H. M. Komu, V. Kosgei, A. Kuria, C. Magero, M. Manten, . . . E. Wilson. 2019. Capacity building for conservation: Problems and potential solutions for sub-Saharan Africa. *Oryx* 53(2):273–283. https://doi.org/10.1017/S0030605317000291.

Ogada, D. L., and R. Buij. 2011. Large declines of the Hooded Vulture *Necrosyrtes monachus* across its African range. *Ostrich* 82(2):101–113. https://doi.org/10.2989/00306525.2011.603464.

Perring, F. H., and S. M. Walters, eds. 1962. Atlas of the British *Flora*. Botanical Society of the British Isles, Durham, UK.

Pototsky, P. C., and W. Cresswell. 2021. Conservation research output in sub-Saharan Africa is increasing, but only in a few countries. *Oryx* 55(6):924–933. https://doi.org/10.1017/S0030605320000046.

Research4Life. 2024. Home. Retrieved December 8, 2024, from https://www.research4life.org.

Rodríguez, J. P., B. Sucre, K. Mileham, A. Sánchez-Mercado, N. De Andrade, S. B. Bezeng, C. Croukamp, J. Falcato, P. García-Borboroglu, S. González, P. González-Ciccia, J. F. González-Maya, L. Kemp, M. D. Kusrini, C. Lopez-Gallego, S. Luz, V. Menon, P. D. Moehlman, D. C. Raimondo, . . . Y. Xie. 2022. Addressing the biodiversity paradox: Mismatch between the co-occurrence of biological diversity and the human, financial and institutional resources to address its decline. *Diversity* 14(9): Article 708. https://doi.org/10.3390/d14090708.

Ryan, P. G., C. Bento, C. Cohen, J. Graham, V. Parker, and C. Spottiswoode. 1999. The avifauna and conservation status of the Namuli Massif, northern Mozambique. *Bird Conservation International* 9(4):315–331. https://doi.org/10.1017/S0959270900003518.

Shan, S. 2024. Help sought for *Taiwan Bird Atlas—Taipei Times*. Retrieved December 8, 2024, from https://www.taipeitimes.com/News/front/archives/2024/03/14/2003814894.

Sharrock, J. T. R. 1976. *The Atlas of Breeding Birds in Britain and Ireland*. T. and A. D. Poyser, London.

Shaw, P., D. Ogada, L. Dunn, R. Buij, A. Amar, R. Garbett, M. Herremans, M. Z. Virani, C. J. Kendall, B. M. Croes, M. Odino, S. Kapila, P. Wairasho, C. Rutz, A. Botha, U. Gallo-Orsi, C. Murn, G. Maude, and S. Thomsett. 2024. African savanna raptors show evidence of widespread population collapse and a growing dependence on protected areas. *Nature Ecology and Evolution* 8(1):45–56. https://doi.org/10.1038/s41559-023-02236-0.

Slater, C., and J. Mundava. 2024. Conference report: The 15th Pan African Ornithological Congress, 21–25 November 2022: A vulture perspective. *Vulture News* 84(1):58–87. https://doi.org/10.4314/vulnew.v84i1.6.

South African National Biodiversity Institute. 2024. Home. Retrieved December 7, 2024, from https://www.sanbi.org.

Sullivan, B. L., C. L. Wood, M. J. Iliff, R. E. Bonney, D. Fink, and S. Kelling. 2009. eBird: A citizen-based bird observation network in the biological sciences. *Biological Conservation* 142(10):2282–2292. https://doi.org/10.1016/j.biocon.2009.05.006.

Sunderland, T., J. Sunderland-Groves, P. Shanley, and B. Campbell. 2009. Bridging the gap: How can information access and exchange between conservation biologists and field practitioners be improved for better conservation outcomes? *Biotropica* 41(5):549–554. https://doi.org/10.1111/j.1744-7429.2009.00557.x.

Tatayah, V. 2022. Falco punctatus (Green Status assessment). The IUCN Red List of Threatened Species, 2022: E.T22696373A2269637320221.

Taylor, M., P. Perakakis, and V. Trachana. 2008. The siege of science. *Ethics in Science and Environmental Politics* 8:17–40. https://doi.org/10.3354/esep00086.

Tende, T., I. A. Iniunam, S. T. Ivande, A. G. Awoyemi, B. A. Danmallam, A. S. Ringim, L. A. Bako, F. J. Ramzy, N. W. Kazeh, A. I. Izang, P. S. Kumdet, J. I. Ibrahim, M. A. Haruna, K. Eyos, E. D. Iki, A. A. Chaskda, and U. Ottosson. 2024. Citizen science mitigates the lack of distributional data on Nigerian birds. *Ecology and Evolution* 14(4): Article e11280. https://doi.org/10.1002/ece3.11280.

The Tony Elumelu Foundation. 2024. The Tony Elumelu Foundation 2023 annual report. Retrieved December 8, 2024, from https://www.tonyelumelufoundation.org/wp-content/uploads/dlm_uploads/2024/04/TEF-2023-ANNUAL-REPORT-TEF-V.3-28-03-24-2.pdf.

Thiollay, J.-M. 2006. Severe decline of large birds in the Northern Sahel of West Africa: A long-term assessment. *Bird Conservation International* 16(4): 353–365. https://doi.org/10.1017/S0959270906000487.

Thiollay, J.-M. 2007. Raptor population decline in West Africa. *Ostrich* 78(2):405–413. https://doi.org/10.2989/OSTRICH.2007.78.2.46.126.

Thiollay, J.-M. 2017. Decline of African vultures: A need for studies and conservation. *Ostrich* 88(2):iii–iii. https://doi.org/10.2989/00306525.2017.1326004.

Thompson, H. S. 2001. Future directions for bird conservation in Africa. *Ostrich* 72(Suppl. 15):13–21.

Thompson, H. S., P. Bai-Sesay, A. Haffner, and M. Bai-Sesay. 2024. Common Birds of Sierra Leone: An Indigenous Language Guide. Sierra Leone Writers Series. Freetown Polytechnic Jui, Freetown, Sierra Leone, https://www.slwritersseries.org.

Tuyisenge, M. F., L. Kayitete, D. Tuyisingize, M. O'Malley, T. S. Stoinski, and Y. Van Der Hoek. 2023. Status of African authorship among conservation research output from sub-Saharan Africa: An African perspective. *Conservation Science and Practice* 5(10): Article e13013. https://doi.org/10.1111/csp2.13013.

United Nations. 2024a. Africa: Fastest growing continent. Retrieved December 8, 2024, from https://www.un.org/en/global-issues/population.

United Nations. 2024b. Young people's potential: The key to Africa's sustainable development. Retrieved December 8, 2024, from https://www.un.org/ohrlls/news/young-people%e2%80%99s-potential-key-africa%e2%80%99s-sustainable-development.

United Nations Department of Economic and Social Affairs. 2020. World Social Report 2020: Inequality in a rapidly changing world. https://doi.org/10.18356/7f5d0efc-en.

United Nations Development Programme. 2017. *Capacity Development: A UNDP Primer*. United Nations, New York.

Vickery, J., and P. Jones. 2002. A new ornithological institute in Nigeria. *Bulletin of the African Bird Club* 9(1):61–62.

Vickery, J., J. W. Mallord, W. M. Adams, A. E. Beresford, C. Both, W. Cresswell, N. Diop, S. R. Ewing, R. D. Gregory, C. A. Morrison, F. J. Sanderson, K. Thorup, R. E. Van Wijk, and C. M. Hewson. 2023. The conservation of Afro-Palaearctic migrants: What we are learning and what we need to know? *Ibis* 165(3): Article 3. https://doi.org/10.1111/ibi.13171.

Williams, M. M., U. Ottosson, T. Tende, and J. P. Deikumah. 2021. Traditional belief systems and trade in vulture parts are leading to the eradication of vultures in Nigeria: An ethno-ornithological study of north-central Nigeria. *Ostrich* 92(3):194–202. https://doi.org/10.2989/00306525.2021.1929534.

Zwarts, L., R. G. Bijlsma, and J. Van Der Kamp. 2018. Large decline of birds in Sahelian rangelands due to loss of woody cover and soil seed bank. *Journal of Arid Environments* 155:1–15. https://doi.org/10.1016/j.jaridenv.2018.01.013.

26

Does Climate Change Alter How We Should Do Bird Conservation?

Chris S. Elphick

By July 2022, the International Union for Conservation of Nature (IUCN) and BirdLife International listed one in eight bird species worldwide as being threatened with extinction (IUCN 2022). Of these species, 225 were considered Critically Endangered, 447 Endangered, and 773 Vulnerable. Another 159 species were judged to have already gone extinct, and more than 1,000 were ranked as Near Threatened. Despite these numbers, birds as a whole are faring relatively well compared to most groups for which detailed status assessments have been completed (e.g., Hoffmann et al. 2010, Barnosky et al. 2011). Nonetheless, there is little doubt that the widespread threats to biodiversity will alter the world's avifauna in substantial ways.

Of those threats, forestry and farming, and the land-cover change they cause, collectively affect the greatest number of species, with hunting and introduced species each also affecting hundreds of species (IUCN 2022). Climate change and the impacts of severe weather rank behind these threats, despite a sense that the shifting climate could fundamentally change how we view conservation. Given the potential mismatch between current and possible future threats, well-argued concerns have been raised that placing too much emphasis on the conservation implications of climate change could distract from more immediate conservation problems, particularly land-cover change and overexploitation (Caro et al. 2022). Supporting this perspective, few modern bird extinctions have been linked to climate change (nine, based on BirdLife International 2022), and there are few for which it is clearly the primary threat. It is, therefore, perhaps useful to ask whether, and how, bird conservation needs to be transformed in light of climate change.

Arguably, there are at least four, somewhat overlapping, reasons why climate change should garner our attention. First, although it is reasonable to wonder whether it should receive as much prominence over other threats as some have given it (Tingley et al. 2013, Caro et al. 2022), and there are few species for which it is the primary threat, there is no doubt that climate change could harm many species: more than 920 based on systematic Red List assessments (IUCN 2022) and 1,245 based on BirdLife's data portal (BirdLife International 2022)—approximately 1 in 10 bird species. Moreover, the nature of the threat posed by climate change is relatively new, and if it proves to be fundamentally different from other threats, then it might take some time to determine how best to act.

Second, although climate change is not the primary threat for most species, it is typical for rare and declining species to be affected by multiple factors

Chris S. Elphick, *Does Climate Change Alter How We Should Do Bird Conservation?*. In: *New Perspectives in Ornithology*. Edited by: Scott V. Edwards and J. Michael Reed, Oxford University Press. © Oxford University Press (2025).
DOI: 10.1093/oso/9780197787670.003.0026

(Wilcove et al. 1998, Yiming and Wilcove 2005) and for different threats to interact (Brook et al. 2008, Cahill et al. 2013). If climate change exacerbates, ameliorates, or simply complicates the effects of other things, then we might need to better understand how before we can identify appropriate responses.

Third, eventually climate change is likely to affect, in some way or another, almost all organisms on the planet: Some effects will be beneficial, many will be harmful, and a major reshuffling of the world's biota is almost certain. Of course, species are responding to changing conditions all the time, but the magnitude and rate of change in the face of climate disruptions are great, especially when combined with other aspects of global change. What this means for ecosystems, the species they contain, and the functions they provide clearly warrants attention.

Fourth, and perhaps most important, as humanity slowly recognizes how far-reaching the repercussions of climate change will be, societal responses are likely to permeate much that we do—from domestic policy to international relations and individual behavior. Even if these actions are less than ideal in their magnitude or timing, they will have implications for land use, energy production, water availability, and more. These societal changes themselves will affect our ability to conserve birds. This chapter explores these four ideas and considers what they mean for ornithologists seeking to influence future work on bird conservation (Table 26.1).

Is Climate Change Like Other Threats?

Climate clearly affects birds in many ways (Parmesan and Yohe 2003, Root et al. 2003, Crick 2004, Pearce-Higgins and Green 2014), and it would be surprising if changing conditions did not have ramifications across the world's avifauna. Nonetheless, the effects may simply mimic other things that have long been the subject of conservation attention. Land-cover change in particular is likely to be pervasive as terrestrial systems warm and rainfall patterns change. Climate change, then, may simply increase the frequency with which we need to think about existing threats.

There are, however, ways in which modern climate change is at least quantitatively, and perhaps qualitatively, different from other threats. The rate and global scale of climate change, for example, both point to more extreme shifts than have been typically addressed in conservation. Land conversion to agriculture, although pervasive and ongoing, has largely occurred as a gradual process over time frames of centuries to millennia. In contrast, some contemporary climate-driven changes have potential to occur rapidly, sometimes with change happening more or less simultaneously across large portions of a species range. Although these changes may not be harmful everywhere—and may be beneficial in some places—the potential for widespread disruption is likely greater than conservation practitioners have seen in the past. For example, coastal species threatened by offspring loss due to higher tides and more frequent storm surge are potentially affected throughout their breeding range because sea levels are rising in most coastal areas (Field et al. 2017, 2018).

The appearance of near-simultaneous range-wide threats to species is not unheard of, but it has previously been largely limited to species with small geographic ranges, such as those restricted to small islands that suffered rapid losses following species

Table 26.1 Recommendations for Bird Conservation Emanating Based on Documented and Likely Future Effects of Climate Change

Observation/Result	Implications for Bird Conservation
Simultaneous changes across a species range will likely become more common.	The need for coordinated range-wide action will increase, and this would be facilitated by the identification of successful conservation models that have been applied across large geographic scales.
Climate change will often cause changes that are effectively permanent.	Focusing on mitigation strategies, rather than attempts to reverse or maintain conditions, may become increasingly necessary.
Constraints on dispersal by plants that structure key vegetation types might limit the speed at which the habitats of some birds move.	Active habitat creation, and management that speeds the movement of dispersal-limited species, will become increasingly necessary to ensure that suitable conditions exist for birds to move into.
Movement of habitats with climate change may alter which sites are most important for species protection.	Land protection planning should proactively identify sites that are likely to become important in the future but are not currently. Degazettement of currently protected sites may be warranted if the conservation value of sites declines—especially when revenue can be generated for new conservation—but regulatory procedures may need to be developed to ensure this is done effectively.
Conservation actions will need to be applied at larger spatial scales and with more constraints.	Research on how to scale up local conservation solutions to large spatial scales is essential.
Disruption of ecological networks through local extirpations and colonization may have indirect consequences for many species.	Better information about exactly which species interact with which could allow better prediction of the network-wide consequences of climate change, especially if linked to species distribution modeling and conservation planning.
Actions to improve human well-being in the face of climate change might be harmful to bird conservation.	Preemptive efforts to predict, and mitigate, sources of conflict between bird conservation and human climate adaptation are likely warranted. Such work may be especially necessary in relation to energy production, agriculture and food security, and geoengineering. Actively seeking opportunities to integrate bird conservation with climate adaptation that benefits people will likely also produce benefits.
Uncertainty is considerable and may grow.	Conservation practitioners need to be willing to take risks and trial management actions that may fail, and to do so before problems become acute. Not everything that we attempt will work, and we need to be better at documenting our efforts no matter what the outcome so that we can learn from successes and avoid repeating failures.
The number of affected species and the challenges of protecting them will grow.	The need to take advantage of tools designed to improve the efficiency of conservation will become even more necessary than is currently the case, and coordination among those interested in conservation will become increasingly essential.

introductions (e.g., Stephen's Island Wren *Traversia lyalli*; Galbreath and Brown 2004). Examples from more widespread species are less common but include situations in which the introduction of chemicals designed to benefit humans resulted in rapid population declines—for instance, when the anti-inflammatory drug diclofenac became widely used to treat livestock, with devastating consequences for many vulture species (e.g., Green et al. 2004, Swan et al. 2006). With one-third of the planet's ice-free land farmed, and increased globalization, other agricultural innovations increasingly spread quickly across large areas—as seen with the production of oil palm, which has caused broadscale habitat conversion with widespread impacts (Vijay et al. 2016). The frequency of emerging range-wide threats is likely increasing at a rapid pace, but not just due to climate change.

Effects of climate change on basic physiological processes may also mark a way in which climate change is fundamentally different from other threats. Increasing temperature and changing humidity may have far-reaching physiological consequences for many species. Such effects might directly affect organisms by pushing them toward physiological limits (Albright et al. 2017), or they might have indirect effects by forcing behavioral changes that come with attendant costs (Conradie et al. 2019, Cunningham et al. 2021). The consequences of such changes potentially place hard constraints that underlie range limit shifts (Albright et al. 2017) and can explain ecosystem-wide changes in bird communities (Riddell et al. 2019). Climate-induced responses by species may also interact with other abiotic factors to affect species in unanticipated ways (Spence and Tingley 2020).

Conservation activities have often been successful when a threat can be removed, allowing a reversion to historical conditions. For instance, when DDT was banned in North America, declining raptor populations recovered, some quicker than expected (Grier 1982), and when introduced species have been eradicated from oceanic islands, many bird populations have rebounded (Jones et al. 2016). Threats that arise from permanent changes are much more difficult to solve and, unsurprisingly, have a poorer conservation track record. Many consequences of climate change are effectively irreversible and thus more akin to threats such as wholesale habitat change that precludes restoration (e.g., loss of old-growth forest and extensive soil erosion) or well-established species introductions over large geographic areas, such as the continent-wide introduction of foxes to Australia (Saunders et al. 2010). Solutions that focus on mitigating losses are, consequently, more likely to be effective than trying to reverse, or even maintain, current conditions. As these examples suggest, the existence of this type of conservation problem is not new, but it does point toward the growing need for the type of conservation actions that have historically been more difficult to identify and implement.

Complicating things further, climate-driven conservation needs will often exist at geographic scales that are larger than much historical conservation work, requiring considerable coordination and planning. Looking to past conservation efforts that share these characteristics might prove helpful. For instance, the North American Waterfowl Management Plan, which has effectively restored population numbers for multiple species through coordinated continent-wide planning involving multiple management agencies and conservation nongovernmental organizations, might be a good model (North American Waterfowl Management Plan Committee 2018).

How Does Climate Change Interact With Other Threats?

Land-cover change affects more threatened bird species than any other cause (IUCN 2022) and is projected to be the most important source of biodiversity loss in the near-term (Sala et al. 2000, Jetz et al. 2007). Land-cover change also continues to occur at extraordinarily high rates, with recent estimates suggesting that approximately one-third of the world's land surface has been altered since the 1960s (Winkler et al. 2021). Climate change will add to the rate of land-cover change, primarily via its effects on vegetation communities as temperature and rainfall patterns shift, but also by other means, such as altering flooding patterns in coastal areas, melting ice at high latitudes and elevations, and altering which crops are grown where. These changes will exacerbate the consequences of direct habitat destruction for many species, and they may be especially problematic for species that use habitats which are already rare (e.g., Correll et al. 2017, Lehikoinen et al. 2019) or that have narrow ranges (Jetz et al. 2007). Climate and land-cover change may also interact in ways that have demographic consequences. For instance, Srinivasan and Wilcove (2021) found evidence that understory insectivorous birds in the Himalayas have greater capacity to track climate change in primary montane forest than in disturbed forest subject to logging.

Some species may benefit by expanding their ranges as the changing climate increases the area of certain habitats or connects separated patches. For example, the movement of seabirds between the North Pacific Ocean and the North Atlantic Ocean appears to have increased in recent years (McKeon et al. 2016), as might be expected given the melting of Arctic sea ice. Such changes might open opportunities for some species to colonize new breeding or foraging sites, as has been seen by the northward range expansions of several herons and other wading birds in various parts of the Northern Hemisphere—spread that may result from some combination of habitat expansion, reduced hunting, and climate change (e.g., Ławicki 2014).

Expansion, however, requires both that the habitat moves and that species can move with it. Given the mobility that flight affords, for many birds the former is perhaps the larger problem—especially for species that use habitats formed by long-lived plant species or those that have limited dispersal. Forest birds are among those most likely to be adversely affected because the regeneration of forests can take decades, and the development of some forest types can take much longer (Rozendaal et al. 2019). Slow habitat movement can arise when key species have limited capacity to move their propagules to new places, but it could also be caused by the lack of suitable abiotic conditions for survival of those propagules—for instance, because of a lack of suitable soils or even a lack of land that species can move into for island, montane, and high-latitude species. Recent reviews suggest that many plant species are unlikely to attain range shifts that match the rate of climate change, except perhaps when current habitat is continuous and where altitudinal, rather than latitudinal, shifts are required (Corlett and Westcott 2013). For conservation practitioners, this means that ensuring birds have suitable places to move into will often require active management that speeds up the movement rate of key habitats. This in turn means that avian conservation will often involve better understanding the responses to climate of the plants that form the structure of habitats in which the birds of concern occur.

Climate change will also interact with other sources of land-cover change via land protection efforts. The importance of habitat loss to a large proportion of endangered species makes protection of habitat a high priority for many conservation programs. Because land protection is an efficient way to support conservation of many species simultaneously, while providing potential colonization sites for species undergoing range expansion (Hiley et al. 2013) and protecting a range of other ecosystem properties and services, it is also a cost-effective way to conduct conservation (Schwartz 1999). Climate-induced habitat change, however, threatens to interfere with land protection strategies by causing the places that are suitable for a given species to shift over time. Consequently, land protection requires not only that we protect areas that are currently important but also that we proactively build a land protection portfolio with climate-induced land-cover change in mind (Peters and Darling 1985, Heller and Zavaleta 2009, Klingbeil et al. 2021). In an ideal world, places that are important now will also be important in the future. One analysis of Important Bird Areas in Africa found considerable overlap, even with widespread species turnover (Hole et al. 2009), but it is not yet clear whether these projections can be generalized. Even if broadly true, it seems likely that land protection will become increasingly complicated as we try to protect areas that are important now but may not be in the future, while also protecting areas that are not yet important so that they remain suitable for future colonization.

Devising strategies for this level of planning will not only require sophisticated modeling of future conditions, so that we know where to protect preemptively and how to sequence land protection over time, but also require new approaches to how we think about reserve systems. For many people, protecting land means acting to ensure that it is conserved in perpetuity. Although the perpetual creep of development argues for this goal in principle, other approaches might prove better in practice. For instance, if protected areas become less suitable for species that are high conservation priorities as the planet warms, it might become prudent to sell the land to generate income that could be used to purchase properties of higher conservation value elsewhere. Such degazetting of protected land can be highly controversial—especially when land was initially protected via hard-won battles or when it provides other benefits, such as recreational opportunities. Modeling exercises have shown, however, that selling the least valuable conservation lands can allow considerable gains in what can be protected (Fuller et al. 2010), and the spatially shifting impacts of climate change seem especially well aligned with a system of rolling protection. Developing regulatory processes that avoid misuse of such systems, and the science to inform them, will require planning.

Climate change also interacts with a myriad of other conservation threats in ways that exacerbate extinction risk. Hawaiian forest birds, for example, have long suffered from habitat loss, overkill, and both introduced predators and disease (Pratt et al. 2009). As the climate warms, the influence of disease has grown because warmer conditions allow introduced mosquitos, which are vectors for avian malaria, to expand their range to higher elevations that had previously served as refugia (Atkinson et al. 2014). This combination of climate change, species introductions, and disease risk further interacts with past habitat losses to progressively shrink the area of suitable habitat in which species can persist, even if it is otherwise protected

(Benning et al. 2002). More generally, Garamszegi (2011) has shown that increased incidence of *Plasmodium* parasites in birds worldwide is associated with temperature increases, suggesting that climate change has potential to have widespread effects on patterns of avian malaria and illustrating the larger potential for disease and climate change to interact. This type of situation—in which the spread of other species increases risk for endangered birds and alters the geographic scope of such risks—will only become more common as both the rate at which species are introduced to new areas of the world and the effects of climate change accelerate.

What Are the Broader Ecosystem Consequences?

Ecological communities exist as networks of interacting parts, and changes in one part can be expected to have ramifications elsewhere in the network. Climate change, by acting on many species simultaneously, has potential to alter ecological networks substantially, not only by pushing some species toward extinction but also by benefiting other species and by redistributing where species occur. All species losses—local or global—have potential to influence the systems in which they occur, no matter what the cause. One possibility is that the loss of pairwise connections between species due to climate-driven extirpation will cause the remaining species to become more weakly connected within the network. If this process continues, species may eventually become disconnected entirely—with local extirpation a consequence when a species depends on those connections (Sandor et al. 2022).

For many birds, however, such effects are probably small because species are often buffered through diffuse systems of network interactions. Most species of birds eat many prey species (Billerman et al. 2022), for example, so the loss of individual prey species can likely be compensated for by increased use of other foods (cf. Mallord et al. 2017). Only when prey-switching is impossible because a species is dependent on a single prey type, or on a group of prey that are simultaneously affected by climate change, will such buffering fail. Unfortunately, our knowledge of the diet of many birds often lacks detail: Although we know generally what most bird species eat, we often lack specifics on which species and their relative importance (see Hurlbert et al. 2021). Better understanding of the network of dependencies among species, and the differential vulnerabilities of those other species, would improve predictions about when climate change will have effects via trophic interactions.

Climate-driven benefits to some species may also create conservation problems through network interactions. Just as species losses can cause network structures to change, so can the addition of new species or increases in their relative abundance. Network rewiring—where new connections are made or existing ones are strengthened—can alter the rest of the network (Kaiser-Bunbury et al. 2010, Bartley et al. 2019). For example, as sea ice has declined in the Arctic, polar bears *Ursus maritimus* spend more time on land during the summer, with concomitant increases in the degree to which they prey on nesting birds—rising to greater than 90% nest losses in some cases (Prop et al. 2015). More generally, the spread of generalists—likely to be the main beneficiaries of climate change—may increase connections within food webs, with possible benefits in terms of stability, but also homogenization of food webs across regions, which might increase their vulnerability to future changes (Bartley et al. 2019). Exactly what this means for bird conservation is unclear, but it further illustrates the need to better understand how ecological networks function and to link models of species redistribution and abundance with network models so that we can predict the consequences of species assemblage reorganization.

Specialist species are clearly more vulnerable than most to the loss of interacting species, and "chains of extinction," in which declines of one species can lead to the demise of another, have long been identified as one of the main causes of species vulnerability (Diamond 1989). Such strong dependencies between species, however, tend to be greatest for invertebrates (Koh et al. 2004), and it is more likely that losses of birds will affect dependent specialists than vice versa. Larger, system-wide changes that simultaneously affect many species—such as those caused by climate change—could, nonetheless, increase the extent to which bird populations are vulnerable. For instance, phenological mismatches—whereby the timing of key events in the life cycle of a species changes relative to other species on which it relies—have been widely documented (Renner and Zohner 2018) and are expected to become more common (Kharouba et al. 2018). Disruptions among species in different trophic levels have received much discussion (Kharouba and Wolkovich 2020) and, although avian studies often focus on a single bird species, the prey that they are associated with are often an amalgam of many species (e.g., Visser et al. 1998) that are presumably important to other bird species, suggesting potential for trophic level-wide disruptions.

The extent to which system-wide problems are occurring is unclear due to limitations in the ways that phenological mismatch has been studied (Kharouba and Wolkovich 2020), and there is a need to better integrate understanding of the effects of climate change across food webs. The possibility of such problems, however, is great. For instance, growing indications of widespread insect declines in many, although not all, ecosystems (Van Klink et al. 2020, Wagner 2020), combined with cross-system evidence that insect abundance directly affects songbird reproductive success (Grames et al. 2023), illustrate how such disruptions could affect bird populations. Better understanding how climate change is affecting insect populations (Wagner 2020, Forister et al. 2021) would improve our capacity to predict and mitigate consequences for birds. Similarly, system-wide changes are beginning to happen in marine settings in which ocean warming is altering the distributions and abundance of many species that seabirds prey upon. Birds are responding accordingly by adjusting where they forage and breed, and modeling suggests potential for larger shifts (Sydeman et al. 2015, Smith et al. 2019). Although some of these changes have benefitted species, ocean warming has often been associated with worsening vital rates (Sydeman et al. 2015), and the continuing risk of systemic changes in the world's oceans as waters warm and acidify is likely to play an increasingly important role in marine bird conservation.

Synthesis of the mechanisms underlying climate-driven effects on species suggests that biotic factors are more often documented as the direct cause than are abiotic factors (Ockendon et al. 2014). Although this study considered all taxa, birds comprised almost half of the data sets, and the general pattern was especially strong for species at higher trophic levels. There are, however, cases in which birds are affected directly by abiotic processes, some of which can have far-reaching effects on large swaths of the local avifauna. Sea-level rise and more frequent storm surge, for example, can increase nest failure due to flooding, with effects on a wide variety of species across coastal habitats (van de Pol et al. 2010, Bayard and Elphick 2011, M. Reynolds et al. 2015). Such physical changes will ultimately result in land-cover changes (see above), but sometimes the direct demographic effects on species—for example, via reduced reproductive success—may play out on shorter timescales than the rate at which visible change occurs. As noted above, direct effects of rising temperature and declining humidity also could increasingly have broadscale effects (Riddell et al. 2019, Cunningham et al. 2021).

Clearly, many other ecosystem-wide changes could arise from climate change. Phenological mismatch and other shifts in food webs could have substantial effects on carbon and nutrient cycling. For instance, earlier arrival of Brant *Branta bernicula* in the Arctic leads to increased grazing impacts on the vegetation, which in turn affect nitrogen and carbon cycling (Beard et al. 2019). When birds are exerting top-down effects such as this, they may not be directly affected, but the finding that the timing of goose migration can flip a system from being a carbon sink to a source (Kelsey et al. 2018) suggests that there could be feedbacks that have conservation implications, if only by exacerbating the rate of greenhouse gas emissions. Fully understanding the variety and complexity of these interaction networks remains a long way off, but it is clearly important if we are to fully comprehend how climate change will affect bird populations.

How Will Society's Response to Climate Change Affect Birds?

Perhaps the main way in which climate change may differ from other threats to biodiversity is in the nature of the human response. Most threats to biodiversity are met with either direct action to reduce harm or a decision—often made by default, rather than any explicit process—simply not to act. Actions to stem the magnitude of climate change or to respond adaptively to its effects, however, might themselves create conservation problems. Climate change is not unique in this respect—for instance, efforts to control introduced species may have unanticipated adverse effects, as seen when removal of an introduced predator reduces population control of another introduced species that it preyed upon (Bergstrom et al. 2009). With regard to climate change, however, societal responses will be dominated by efforts to improve outcomes for people, many of whom will suffer badly if no action is taken, and direct conflicts are likely to arise. A key challenge facing anyone interested in mitigating the effects of climate change on birds thus will be to identify management strategies that advance avian conservation goals but that are consistent with climate mitigation work designed to protect human health, well-being, and infrastructure.

The increasing use of renewable energy is one area in which conflicts have been apparent for decades (e.g., Crockford 1992) and are increasingly well-studied. Reducing carbon emissions requires a shift to other forms of energy production, but wind turbines frequently result in collisions (Loss et al. 2013, Smallwood 2013), and solar panels cause both collisions and burning (Walston et al. 2016). In addition to directly killing birds, renewable energy development often requires the destruction of existing habitat, can have other environment consequences such as altering local weather (Roy and Traiteur 2010), and can cause various types of behavioral shift that result in species avoidance of an installation (May 2015), all of which could impact bird populations. Developing strategies to mitigate these effects—via siting decisions, facility design, and operational details—while still moving forward with renewable energy production will continue to be a major, but necessary, challenge into the future. Explicitly framing debates on how to produce energy in ways that weigh the effects in terms of population consequences, rather than simply considering the number of birds killed or displaced, would help inform decision-making (Elphick 2008). If mortality is simply compensatory, for example, or if the effects have only a trivial effect on population trajectories, then concerns should be set aside because conservation work on other topics is likely to have greater impact. Assessing the consequences of energy production in a currency that rates harm relative to benefit—for example, by comparing options in terms of birds killed per megawatt produced (Walston et al. 2016) or relative to the reduction in carbon emitted—would also result in better-informed decisions.

Changes in agricultural production and attempts to ensure human food security are also likely to have widespread consequences for biodiversity (Springmann et al. 2018). As climate changes, the suitability of many areas for production will change, often becoming worse and triggering new conversion of natural habitat to farmland (Pugh et al. 2016). These effects are perhaps most likely in north-temperate areas as growing seasons lengthen in places that have historically been too cold for many forms of agriculture, but we also can expect shifts in which crops are grown within established agricultural areas. For many intensively farmed crops, crop shifts will have minimal effects on birds because field use is limited to relatively few, adaptable species. However, the consequences could be severe where species have become dependent on specific types of farmland as natural habitats have declined. For example, in many places, flooded rice fields have become important habitats for aquatic birds due to loss of natural wetlands (Elphick et al. 2010). In some cases, these fields are used by highly endangered species—for example, Sarus Cranes *Antigone antigone* (Sundar 2009) and Crested Ibis *Nipponia nippon* (Sun et al. 2014)—although the extent to which they require farmland may depend on the availability of other habitats. Because most crops are not grown under flooded conditions, conversion of rice to other forms of agriculture could dramatically change the conservation value of farm fields in many areas of the world.

Shifts in where crops are grown will be accompanied by technological changes in the ways that they are grown. Changes will involve genetic modifications to crops, new agricultural chemicals, and new approaches to field management—all of which are already routine but will likely be influenced increasingly by the changing climate. For example, methods of reducing water use, such as converting irrigation canals to buried pipes or growing rice in unflooded fields—both of which are likely to reduce the value of farmland to birds (Lane and Fujioka 1998, Longoni 2010)—may become commonplace in areas with declining or less predictable rainfall.

Many other technological changes are, of course, expected as the effects of climate change mount. Whether global geoengineering should or could be instituted remains an open question, but large-scale efforts to reduce how much solar radiation reaches the Earth's surface or to extract carbon dioxide from the atmosphere (Vaughan and Lenton 2011) will undoubtedly continue to receive attention in coming decades. Exactly how such changes will proceed remains highly uncertain (J. Reynolds 2019); however, any such action implemented at a scale large enough to have a real effect on climate should be expected to have considerable consequences for birds. For instance, proposals to increase the albedo of desert areas or to combine biofuel production with carbon capture would require massive-scale land-use change (Vaughan and Lenton 2011), with attendant consequences for current bird habitat. Identifying the likely consequences of alternative technological responses now, before widespread implementation happens, would help ensure that there are plans in place to minimize any negative impacts before it is too late for them to have any effect.

Another avenue in which technological innovation is likely is management of biodiversity itself. Many technological fixes for biodiversity loss have been proposed, ranging from gene editing as a means of facilitating adaptive responses (Phelps et al. 2020, Samuel et al. 2020) to the use of directed selection based on genomic information in captive breeding programs (Cassin-Sackett et al. 2019) and assisted colonization to help dispersal-limited species better track climate shifts (Lawler and Olden 2011). Some of these methods hold great promise, whereas others are much more controversial, but all likely have risks of some sort. Moreover, as the effects of climate change mount, it is likely that the frequency with which technology-based solutions are proposed will increase, and people may become increasingly willing to take risks.

Understanding the greatest uncertainties about how people will respond if climate conditions worsen will also require insights from the social sciences. For instance, the potential for widespread human migration due to rising seas, weather extremes, and the resulting food shortages and economic hardship is considerable, but much is unclear about whether and where it will transpire (Cattaneo et al. 2019). Even if people are not physically displaced, the risk of increased poverty, or for climate change to exacerbate the consequences of poverty, could have a tremendous human toll (Leichenko and Silva 2014). Such effects could cause great changes in both legislative and funding priorities, which in turn could have serious repercussions for conservation: As society acts to protect vulnerable human populations, other species could become less of a priority. On the other hand, growing recognition of the ecological services that nature provides to humans—whether through shade-producing green space in urban settings (Bowler et al. 2010) or carbon sequestration and land protection benefits of reforestation (Rudel et al. 2020)—could push society in the direction of seeking land-use solutions that have collateral benefits for birds. Achieving such benefits will not be simple, however, and will be more likely if ornithologists and conservation groups preemptively identify ways to achieve their goals that are consistent with other climate mitigation actions that are likely to be needed.

Moving Forward

Addressing the challenges of climate change is not going to be easy. Most difficult of all is the rapid pace at which change is happening relative to our ability to act, and the uncertainty over what outcomes are most likely. As the previous sections of this chapter suggest, much research on conservation for birds in light of climate change continues to focus on exactly what the effects will be. Although this work is often useful, and sometimes necessary, describing the problem is no longer the major challenge that we face. We know that many bird species are already in trouble, as the IUCN statistics show, and future projections suggest that many more will become so. We also know that climate change will make protecting species more difficult because it will be layered on top of other—often more immediate—problems. Although there are incremental benefits to refining our knowledge and narrowing the uncertainty, it seems unlikely that they will change our fundamental understanding of the situation. The effects of climate change, moreover, are largely extensions of existing problems that conservation biologists already have considerable experience addressing. The main task appears to be to determine how to do more, faster, and with more restrictive societal constraints.

Addressing the effects of climate change on birds, then, requires much more research attention to solutions—with a focus on what exactly they should be for a given species or system, on how to rapidly and effectively implement them at scales that are large enough to slow the declines of species which are affected over large regions, and on how to build in sufficient flexibility to allow managers to pivot when required by changing circumstances.

Recognizing the broad types of response available to affected species—acclimate, adapt, move, or die (Corlett and Westcott 2013)—is also useful for framing the nature of the problem we face in bird conservation. Because we lack clear evidence that adaptation can happen fast enough to benefit the bird species most at risk, conservation requires that we seek ways to maintain or manage habitats so that species can persist in them despite climate change and that we facilitate movement of species and their habitats when that is not possible. For instance, to conserve salt marsh–nesting birds threatened by rising seas (Bayard and Elphick 2011, Klingbeil et al. 2018), it is clear that we must either ensure that marshes gain elevations at a rate that matches sea-level rise or facilitate the ease with which new marsh is created to replace that being lost to the ocean. Various ideas have been proposed about how to achieve both of these goals, ranging from adding sediment to marshes to cutting down trees at the marsh edge in the hope that it will speed up new marsh development (Atlantic Coast Joint Venture 2019). Whether any of these methods will work on a timescale that will avert species losses, however, is highly uncertain. Even if the proposed management actions work, it is far from clear whether it is logistically or financially plausible to implement them at a large enough spatial scale to maintain even a small fraction of current populations. Such uncertainties over implementation are likely to be common in many conservation scenarios.

Facilitating acclimation or movement will be difficult without new land protection, either to make management possible when the goal is to help species adjust in place or to ensure that there are suitable areas to occupy if the goal is to enable range shifts. Better understanding of how to manage the land to maintain or create suitable conditions also is necessary. Land protection and management are, of course, central to most current conservation, so the basics of how to implement solutions exist. We must still, however, find ways to cope with the rapid growth in the needed response—both spatially and temporally. During the past two decades, considerable work has been done to make conservation planning more systematic and efficient (Margules and Pressey 2000, McIntosh et al. 2017). Given the scale of climate change, accelerating work in this area—both research on how to improve practice and greatly expanded implementation—may be the most important thing we can be doing in conservation.

Simultaneously, we need advances in the rate at which we identify methods for achieving management goals, and we must ensure that these methods become widely used. Systematic collation and synthesis of evidence about the effectiveness of conservation actions, and dissemination of that information, has improved considerably (Sutherland et al. 2019), but it has also revealed how often we lack clear evidence on what the best practices are for a given problem or how they differ among species or locations (Christie et al. 2021). Implementing solutions will take time, even when we know what to do, so waiting until a given climate-driven problem is upon us will likely mean that we have waited too long to be able to act as effectively as we might.

Preemptive horizon-scanning (Sutherland et al. 2011) to identify the suite of problems the world's avifauna might face would help, but this needs to be combined with an assessment of how widespread each problem will be (e.g., in terms of geography or proportion of species affected), how severe its effects will be, and how easy they are to reverse. Armed with this information, we can prioritize attempts to determine the ecological, engineering, and regulatory steps that need to be taken. Given the magnitude of climate change's likely effects, conservation biologists will not be able to avoid always having to play catchup, but strategic coordination across organizations, research groups, and countries will make the odds of success more likely. Planning ahead is especially important because funding for conservation—always limited—is likely to become more so as the effects of climate change on people mount and the need to protect human lives and livelihoods becomes ever more paramount. Having conservation plans ready and in place in advance will make it easier to act with diminishing resources as problems arise.

Acknowledgments

My thanks to all who have contributed to my thinking on this topic; to Michael Reed and Scott Edwards, both for providing the opportunity to try to organize those thoughts and for their infinite patience while I struggled to do it; and to Eliza Grames, Michael Reed, and an anonymous reviewer for comments on early drafts of the manuscript.

References

Albright, T. P., D. Mutiibwa, A. R. Gerson, E. K. Smith, W. A. Talbot, J. J. O'Neill, A. E. McKechnie, and B. O. Wolf. 2017. Mapping evaporative water loss in desert passerines reveals an expanding threat of lethal dehydration. *Proceedings of the National Academy of Sciences of the USA* 114:2283–2288.

Atkinson, C. T., R. B. Utzurrum, D. A. Lapointe, R. J. Camp, L. H. Crampton, J. T. Foster, and T. W. Giambelluca. 2014. Changing climate and the altitudinal range of avian malaria in the Hawaiian Islands—an ongoing conservation crisis on the island of Kaua'i. *Global Change Biology* 20:2426–2436.

Atlantic Coast Joint Venture. 2019. Salt marsh bird conservation plan. https://acjv.org/documents/salt_marsh_bird_plan_full_res.pdf.

Barnosky, A. D., N. Matzke, S. Tomiya, G. O. U. Wogan, B. Swartz, T. B. Quental, C. Marshall, J. L. McGuire, E. L. Lindsey, K. C. Maguire, B. Mersey, and E. A. Ferrer. 2011. Has the Earth's sixth mass extinction already arrived? *Nature* 471:51–57.

Bartley, T. J., K. S. McCann, C. Bieg, K. Cazelles, M. Granados, M. M. Guzzo, A. S. MacDougall, T. D. Tunney, and B. C. McMeans. 2019. Food web rewiring in a changing world. *Nature Ecology & Evolution* 3:345–354.

Bayard, T. S., and C. S. Elphick. 2011. Planning for sea-level rise: Quantifying patterns of Saltmarsh Sparrow (*Ammodramus caudacutus*) nest flooding under current sea-level conditions. *The Auk* 128:393–403.

Beard, K. H., K. C. Kelsey, A. J. Leffler, and J. M. Welker. 2019. The missing angle: Ecosystem consequences of phenological mismatch. *Trends in Ecology & Evolution* 34:885–888.

Benning, T. L., D. LaPointe, C. T. Atkinson, and P. M. Vitousek. 2002. Interactions of climate change with biological invasions and land use in the Hawaiian Islands: Modeling the fate of endemic birds using a geographic information system. *Proceedings of the National Academy of Sciences of the USA* 99:14246–14249.

Bergstrom, D. M., A. Lucieer, K. Kiefer, J. Wasley, L. Belbin, T. K. Pedersen, and S. L. Chown. 2009. Indirect effects of invasive species removal devastate World Heritage Island. *Journal of Applied Ecology* 46:73–81.

Billerman, S. M., B. K. Keeney, P. G. Rodewald, and T. S. Schulenberg. 2022. Birds of the World. https://birdsoftheworld.org/bow/home.

BirdLife International. 2022. DataZone. Retrieved July 5, 2022, from http://datazone.birdlife.org/home.

Bowler, D. E., L. Buyung-Ali, T. M. Knight, and A. S. Pullin. 2010. Urban greening to cool towns and cities: A systematic review of the empirical evidence. *Landscape and Urban Planning* 97:147–155.

Brook, B. W., N. S. Sodhi, and C. J. A. Bradshaw. 2008. Synergies among extinction drivers under global change. *Trends in Ecology and Evolution* 23:453–460.

Cahill, A. E., M. E. Aiello-Lammens, M. C. Fisher-Reid, X. Hua, C. J. Karanewsky, H. Y. Ryu, G. C. Sbeglia, F. Spagnolo, J. B. Waldron, O. Warsi, and J. J. Wiens. 2013. How does climate change cause extinction? *Proceedings of the Royal Society B: Biological Sciences* 280: Article 20121890.

Caro, T., Z. Rowe, J. Berger, P. Wholey, and A. Dobson. 2022. An inconvenient misconception: Climate change is not the principal driver of biodiversity loss. *Conservation Letters* 15: Article e12868.

Cassin-Sackett, L., A. J. Welch, M. X. Venkatraman, T. E. Callicrate, and R. C. Fleischer. 2019. The contribution of genomics to bird conservation. Pages 295–330 in R. H. S. Kraus, ed. *Avian Genomics in Ecology and Evolution: From the Lab into the Wild.* Springer, Cham, Switzerland.

Cattaneo, C., M. Beine, C. J. Fröhlich, D. Kniveton, I. Martinez-Zarzoso, M. Mastrorillo, K. Millock, E. Piguet, and B. Schraven. 2019. Human migration in the era of climate change. *Review of Environmental Economics and Policy* 13:189–206.

Christie, A. P., T. Amano, P. A. Martin, S. O. Petrovan, G. E. Shackelford, B. I. Simmons, R. K. Smith, D. R. Williams, C. F. R. Wordley, and W. J. Sutherland. 2021. The challenge of biased evidence in conservation. *Conservation Biology* 35:249–262.

Conradie, S. R., S. M. Woodborne, S. J. Cunningham, and A. E. McKechnie. 2019. Chronic, sublethal effects of high temperatures will cause severe declines in southern African arid-zone birds during the 21st century. *Proceedings of the National Academy of Sciences of the USA* 116:14065–14070.

Corlett, R. T., and D. A. Westcott. 2013. Will plant movements keep up with climate change? *Trends in Ecology & Evolution* 28:482–488.
Correll, M. D., W. A. Wiest, T. P. Hodgman, W. G. Shriver, C. S. Elphick, B. J. McGill, K. O'Brien, and B. J. Olsen. 2017. Predictors of specialist avifaunal decline in coastal marshes. *Conservation Biology* 31:172–182.
Crick, H. Q. P. 2004. The impact of climate change on birds. *Ibis* 146(Suppl. 1):48–56.
Crockford, N. J. 1992. A review of the possible impacts of wind farms on birds and other wildlife. Joint Nature Conservation Committee Technical Report JNCC-27.
Cunningham, S. J., J. L. Gardner, and R. O. Martin. 2021. Opportunity costs and the response of birds and mammals to climate warming. *Frontiers in Ecology and the Environment* 19:300–307.
Diamond, J. 1989. Overview of recent extinctions. Pages 37–41 in D. Western and M. Pearl, eds. *Conservation for the 21st Century*. Oxford University Press, New York.
Elphick, C. S. 2008. Editor's choice: New research on wind farms. *Journal of Applied Ecology* 45:1840.
Elphick, C. S., K. C. Parsons, M. Fasola, and L. Mugica. 2010. Ecology and conservation of birds in rice fields: A global review. *Waterbirds* 33(Special Publication 1).
Field, C. R., T. Bayard, C. Gjerdrum, J. M. Hill, S. Meiman, and C. S. Elphick. 2017. High-resolution tide projections reveal extinction threshold in response to sea-level rise. *Global Change Biology* 35:2058–2070.
Field, C. R., K. J. Ruskin, B. Benvenuti, A. Borowske, J. B. Cohen, L. Garey, T. P. Hodgman, R. A. Kern, E. King, A. R. Kocek, A. I. Kovach, K. M. O'Brien, B. J. Olsen, N. Pau, S. G. Roberts, E. Shelly, W. G. Shriver, J. Walsh, and C. S. Elphick. 2018. Quantifying the importance of geographic replication and representativeness when estimating demographic rates, using a coastal species as a case study. *Ecography* 41:971–981.
Forister, M. L., C. A. Halsch, C. C. Nice, J. A. Fordyce, T. E. Dilts, J. C. Oliver, K. L. Prudic, A. M. Shapiro, J. K. Wilson, and J. Glassberg. 2021. Fewer butterflies seen by community scientists across the warming and drying landscapes of the American West. *Science* 371:1042–1045.
Fuller, R. A., E. McDonald-Madden, K. A. Wilson, J. Carwardine, H. S. Grantham, J. E. M. Watson, C. J. Klein, D. C. Green, and H. P. Possingham. 2010. Replacing underperforming protected areas achieves better conservation outcomes. *Nature* 466:365–367.
Galbreath, R., and D. Brown. 2004. The tale of the lighthouse-keeper's cat: Discovery and extinction of the Stephens Island wren (*Traversia lyalli*). *Notornis* 51:193–200.
Garamszegi, L. 2011. Climate change increases the risk of malaria in birds. *Global Change Biology* 17:1751–1759.
Grames, E. M., G. A. Montgomery, C. Youngflesh, M. W. Tingley, and C. S. Elphick. 2023. The effect of insect food availability on songbird reproductive success review, chick body condition: Evidence from a systematic meta-analysis. *Ecology Letters* 26:658–673.
Green, R. E., I. Newton, S. Shultz, A. A. Cunningham, M. Gilbert, D. J. Pain, and V. Prakash. 2004. Diclofenac poisoning as a cause of vulture population declines across the Indian subcontinent. *Journal of Applied Ecology* 41:793–800.
Grier, J. W. 1982. Ban of DDT and subsequent recovery of reproduction in bald eagles. *Science* 218:1232–1235.
Heller, N. E., and E. S. Zavaleta. 2009. Biodiversity management in the face of climate change: A review of 22 years of recommendations. *Biological Conservation* 142:14–32.
Hiley, J. R., R. B. Bradbury, M. Holling, and C. D. Thomas. 2013. Protected areas act as establishment centres for species colonizing the UK. *Proceedings of the Royal Society B: Biological Sciences* 280: Article 20122310.
Hoffmann, M., C. Hilton-Taylor, A. Angulo, M. Böhm, T. M. Brooks, S. H. M. Butchart, K. E. Carpenter, J. Chanson, B. Collen, N. A. Cox, W. R. T. Darwall, N. K. Dulvy, L. R. Harrison, V. Katariya, C. M. Pollock, S. Quader, N. I. Richman, A. S. L. Rodrigues, M. F. Tognelli, J.-C. Vié, J. M. Aguiar, D. J. Allen, G. R. Allen, G. Amori, N. B. Ananjeva, F. Andreone, P. Andrew, A. L. A. Ortiz, J. E. M. Baillie, R. Baldi, B. D. Bell, S. D. Biju, J. P. Bird, P. Black-Decima, J. J. Blanc, F. Bolaños, W. Bolivar-G, I. J. Burfield, J. A. Burton, D. R. Capper, F. Castro, G. Catullo, R. D. Cavanagh, A. Channing, N. L. Chao, A. M. Chenery, F. Chiozza, V. Clausnitzer, N. J. Collar, L. C. Collett, B. B. Collette, C. F. C. Fernandez, M. T. Craig, M. J. Crosby, N. Cumberlidge, A. Cuttelod, A. E. Derocher, A. C. Diesmos, J. S. Donaldson, J. W. Duckworth, G. Dutson, S. K. Dutta, R. H. Emslie, A. Farjon, S.

Fowler, J. Freyhof, D. L. Garshelis, J. Gerlach, D. J. Gower, T. D. Grant, G. A. Hammerson, R. B. Harris, L. R. Heaney, S. B. Hedges, J.-M. Hero, B. Hughes, S. A. Hussain, J. Icochea M, R. F. Inger, N. Ishii, D. T. Iskandar, R. K. B. Jenkins, Y. Kaneko, M. Kottelat, K. M. Kovacs, S. L. Kuzmin, E. La Marca, J. F. Lamoreux, M. W. N. Lau, E. O. Lavilla, K. Leus, R. L. Lewison, G. Lichtenstein, S. R. Livingstone, V. Lukoschek, D. P. Mallon, P. J. K. McGowan, A. McIvor, P. D. Moehlman, S. Molur, A. M. Alonso, J. A. Musick, K. Nowell, R. A. Nussbaum, W. Olech, N. L. Orlov, T. J. Papenfuss, G. Parra-Olea, W. F. Perrin, B. A. Polidoro, M. Pourkazemi, P. A. Racey, J. S. Ragle, M. Ram, G. Rathbun, R. P. Reynolds, A. G. J. Rhodin, S. J. Richards, L. O. Rodríguez, S. R. Ron, C. Rondinini, A. B. Rylands, Y. Sadovy de Mitcheson, J. C. Sanciangco, K. L. Sanders, G. Santos-Barrera, J. Schipper, C. Self-Sullivan, Y. Shi, A. Shoemaker, F. T. Short, C. Sillero-Zubiri, D. L. Silvano, K. G. Smith, A. T. Smith, J. Snoeks, A. J. Stattersfield, A. J. Symes, A. B. Taber, B. K. Talukdar, H. J. Temple, R. Timmins, J. A. Tobias, K. Tsytsulina, D. Tweddle, C. Ubeda, S. V. Valenti, P. Paul van Dijk, L. M. Veiga, A. Veloso, D. C. Wege, M. Wilkinson, E. A. Williamson, F. Xie, B. E. Young, H. R. Akçakaya, L. Bennun, T. M. Blackburn, L. Boitani, H. T. Dublin, G. A. B. da Fonseca, C. Gascon, T. E. Lacher, G. M. Mace, S. A. Mainka, J. A. McNeely, R. A. Mittermeier, G. M. Reid, J. P. Rodriguez, A. A. Rosenberg, M. J. Samways, J. Smart, B. A. Stein, and S. N. Stuart. 2010. The impact of conservation on the status of the world's vertebrates. *Science* 330:1503–1509.

Hole, D. G., S. G. Willis, D. J. Pain, L. D. Fishpool, S. H. M. Butchart, Y. C. Collingham, C. Rahbek, and B. Huntley. 2009. Projected impacts of climate change on a continent-wide protected area network. *Ecology Letters* 12:420–431.

Hurlbert, A. H., A. M. Olsen, M. M. Sawyer, and P. M. Winner. 2021. The Avian Diet Database as a source of quantitative information on bird diets. *Scientific Data* 8: Article 260.

International Union for Conservation of Nature. 2022. RedList. Retrieved July 5, 2022, from https://www.iucnredlist.org/search/stats.

Jetz, W., D. S. Wilcove, and A. P. Dobson. 2007. Projected impacts of climate and land-use change on the global diversity of birds. *PLoS Biology* 5:1211–1219.

Jones, H. P., N. D. Holmes, S. H. M. Butchart, B. R. Tershy, P. J. Kappes, I. Corkery, A. Aguirre-Muñoz, D. P. Armstrong, E. Bonnaud, A. A. Burbidge, K. Campbell, F. Courchamp, P. E. Cowan, R. J. Cuthbert, S. Ebbert, P. Genovesi, G. R. Howald, B. S. Keitt, S. W. Kress, C. M. Miskelly, S. Oppel, S. Poncet, M. J. Rauzon, G. Rocamora, J. C. Russell, A. Samaniego-Herrera, P. J. Seddon, D. R. Spatz, D. R. Towns, and D. A. Croll. 2016. Invasive mammal eradication on islands results in substantial conservation gains. *Proceedings of the National Academy of Sciences of the USA* 113:4033–4038.

Kaiser-Bunbury, C. N., S. Muff, J. Memmott, C. B. Müller, and A. Caflisch. 2010. The robustness of pollination networks to the loss of species and interactions: A quantitative approach incorporating pollinator behaviour. *Ecology Letters* 13:442–452.

Kelsey, K. C., A. J. Leffler, K. H. Beard, R. T. Choi, J. A. Schmutz, and J. M. Welker. 2018. Phenological mismatch in coastal western Alaska may increase summer season greenhouse gas uptake. *Environmental Research Letters* 13: Article 044032.

Kharouba, H. M., J. Ehrlen, A. Gelman, K. Bolmgren, J. Allen, S. Travers, and E. M. Wolkovich. 2018. Global shifts in the phenological synchrony of species interactions over recent decades. *Proceedings of the National Academy of Sciences of the USA* 115:5211–5216.

Kharouba, H. M., and E. M. Wolkovich. 2020. Disconnects between ecological theory and data in phenological mismatch research. *Nature Climate Change* 10:406–415.

Klingbeil, B. T., J. B. Cohen, M. D. Correll, C. R. Field, T. P. Hodgman, A. I. Kovach, E. E. Lentz, B. J. Olsen, W. G. Shriver, W. A. Wiest, and C. S. Elphick. 2021. High uncertainty over the future of tidal marsh birds under current sea-level rise projections. *Biodiversity and Conservation* 30:431–443.

Klingbeil, B. T., J. B. Cohen, M. D. Correll, C. R. Field, T. P. Hodgman, A. I. Kovach, B. J. Olsen, W. G. Shriver, W. A. Wiest, and C. S. Elphick. 2018. A focal species for tidal marsh bird conservation in the northeastern United States. *Ornithological Applications* 120:874–884.

Koh, L. P., R. R. Dunn, N. S. Sodhi, R. K. Colwell, H. C. Proctor, and V. S. Smith. 2004. Species coextinctions and the biodiversity crisis. *Science* 305:1632–1634.

Lane, S. J., and M. Fujioka. 1998. The impact of changes in irrigation practices on the distribution of foraging egrets and herons (Ardeidae) in the rice fields of central Japan. *Biological Conservation* 83:221–230.

Ławicki, Ł. 2014. The Great White Egret in Europe: Population increase and range expansion since 1980. *British Birds* 107:8–25.
Lawler, J. J., and J. D. Olden. 2011. Reframing the debate over assisted colonization. *Frontiers in Ecology and the Environment* 9:569–574.
Lehikoinen, A., L. Brotons, J. Calladine, T. Campedelli, V. Escandell, J. Flousek, C. Grueneberg, F. Haas, S. J. Harris, S. Herrando, M. Husby, F. Jiguet, J. Atle, K. Åke, L. Romain, L. Blas, M. Clara, P. Gianpiero, C. Thomas, S. Hans, P. M. Sirkiä, N. Teufelbauer, and S. Trautmann. 2019. Declining population trends of European montane birds. *Global Change Biology* 25:577–588.
Leichenko, R., and J. A. Silva. 2014. Climate change and poverty: Vulnerability, impacts, and alleviation strategies. *WIREs Climate Change* 5:539–556.
Longoni, V. 2010. Rice fields and waterbirds in the Mediterranean region and the Middle East. *Waterbirds* 33(Special Publication 1):83–96.
Loss, S. R., T. Will, and P. P. Marra. 2013. Estimates of bird collision mortality at wind facilities in the contiguous United States. *Biological Conservation* 168:201–209.
Mallord, J. W., C. J. Orsman, A. Cristinacce, T. J. Stowe, E. C. Charman, and R. D. Gregory. 2017. Diet flexibility in a declining long-distance migrant may allow it to escape the consequences of phenological mismatch with its caterpillar food supply. *Ibis* 159:76–90.
Margules, C. R., and R. L. Pressey. 2000. Systematic conservation planning. *Nature* 405:243–253.
May, R. F. 2015. A unifying framework for the underlying mechanisms of avian avoidance of wind turbines. *Biological Conservation* 190:179–187.
McIntosh, E. J., R. L. Pressey, S. Lloyd, R. J. Smith, and R. Grenyer. 2017. The impact of systematic conservation planning. *Annual Review of Environment and Resources* 42:677–697.
McKeon, C. S., M. X. Weber, S. Alter, N. E. Seavy, E. D. Crandall, D. J. Barshis, E. D. Fechter-Leggett, and K. L. L. Oleson. 2016. Melting barriers to faunal exchange across ocean basins. *Global Change Biology* 22:465–473.
North American Waterfowl Management Plan Committee. 2018. North American Waterfowl Management Plan (NAWMP) update: Connecting people, waterfowl, and wetlands. U.S. Department of the Interior, Environment Canada, and Secretary of Environment and Natural Resources.
Ockendon, N., D. J. Baker, J. A. Carr, E. C. White, R. E. A. Almond, T. Amano, E. Bertram, R. B. Bradbury, C. Bradley, S. H. M. Butchart, N. Doswald, W. Foden, D. J. C. Gill, R. E. Green, W. J. Sutherland, E. V. J. Tanner, and J. W. Pearce-Higgins. 2014. Mechanisms underpinning climatic impacts on natural populations: Altered species interactions are more important than direct effects. *Global Change Biology* 20:2221–2229.
Parmesan, C., and G. Yohe. 2003. A globally coherent fingerprint of climate change impacts across natural systems. *Nature* 421:37–42.
Pearce-Higgins, J. W., and R. E. Green. 2014. *Birds and Climate Change: Impacts and Conservation Responses.* Cambridge University Press, Cambridge, UK.
Peters, R. L., and J. D. S. Darling. 1985. The greenhouse effect and nature reserves. *BioScience* 35:707–717.
Phelps, M. P., L. W. Seeb, and J. E. Seeb. 2020. Transforming ecology and conservation biology through genome editing. *Conservation Biology* 34:54–65.
Pratt, T. K., C. T. Atkinson, P. C. Banko, J. D. Jacobi, and B. L. Woodworth. 2009. *Conservation Biology of Hawaiian Forest Birds: Implications for Island Avifauna.* Yale University Press, New Haven, CT.
Prop, J., J. Aars, B. J. Bårdsen, S. A. Hanssen, C. Bech, S. Bourgeon, J. de Fouw, G. W. Gabrielsen, J. Lang, E. Noreen, T. Oudman, B. Sittler, L. Stempniewicz, I. Tombre, E. Wolters, and B. Moe. 2015. Climate change and the increasing impact of polar bears on bird populations. *Frontiers in Ecology and Evolution* 3: Article 33.
Pugh, T. A. M., C. Müller, J. Elliott, D. Deryng, C. Folberth, S. Olin, E. Schmid, and A. Arneth. 2016. Climate analogues suggest limited potential for intensification of production on current croplands under climate change. *Nature Communications* 7: Article 12608.
Renner, S. S., and C. M. Zohner. 2018. Climate change and phenological mismatch in trophic interactions among plants, insects, and vertebrates. *Annual Review of Ecology, Evolution and Systematics* 49:165–182.

Reynolds, J. L. 2019. Solar geoengineering to reduce climate change: A review of governance proposals. *Proceedings of the Royal Society A: Mathematical, Physical and Engineering Sciences* 475: Article 20190255.

Reynolds, M. H., K. N. Courtot, P. Berkowitz, C. D. Storlazzi, J. Moore, and E. Flint. 2015. Will the effects of sea-level rise create ecological traps for Pacific Island seabirds? *PLoS One* 10: Article e0136773.

Riddell, E. A., K. J. Iknayana, B. O. Wolf, B. Sinervo, and S. R. Beissinger. 2019. Cooling requirements fueled the collapse of a desert bird community from climate change. *Proceedings of the National Academy of Sciences of the USA* 116:21609–21615.

Root, T. L., J. T. Price, K. R. Hall, S. H. Schneider, C. Rozenzweig, and J. A. Pounds. 2003. Fingerprints of global warming on wild animals and plants. *Nature* 421:57–60.

Roy, S. B., and J. J. Traiteur. 2010. Impacts of wind farms on surface air temperatures. *Proceedings of the National Academy of Sciences of the USA* 107:17899–17904.

Rozendaal, D. M. A., F. Bongers, T. M. Aide, E. Alvarez-Dávila, N. Ascarrunz, P. Balvanera, J. M. Becknell, T. V. Bentos, P. H. S. Brancalion, G. A. L. Cabral, S. Calvo-Rodriguez, J. Chave, R. G. César, R. L. Chazdon, R. Condit, J. S. Dallinga, J. S. de Almeida-Cortez, B. de Jong, A. de Oliveirade Oliveira, J. S. Denslow, D. H. Dent, S. J. DeWalt, J. M. Dupuy, S. M. Durán, L. P. Dutrieux, M. M. Espírito-Santo, M. C. Fandino, G. W. Fernandes, B. Finegan, H. García, N. Gonzalez, V. G. Moser, J. S. Hall, J. L. Hernández-Stefanoni, S. Hubbell, C. C. Jakovac, A. J. Hernández, A. B. Junqueira, D. Kennard, D. Larpin, S. G. Letcher, J. C. Licona, E. Lebrija-Trejos, E. Marín-Spiotta, M. Martínez-Ramos, P. E. S. Massoca, J. A. Meave, R. C. G. Mesquita, F. Mora, S. C. Müller, R. Muñoz, S. N. de Oliveira Neto, N. Norden, Y. R. F. Nunes, S. Ochoa-Gaona, E. Ortiz-Malavassi, R. Ostertag, M. Peña-Claros, E. A. Pérez-García, D. Piotto, J. S. Powers, J. Aguilar-Cano, S. Rodriguez-Buritica, J. Rodríguez-Velázquez, M. A. Romero-Romero, J. Ruíz, A. Sanchez-Azofeifa, A. S. de Almeida, W. L. Silver, N. B. Schwartz, W. W. Thomas, M. Toledo, M. Uriarte, E. V. de Sá Sampaio, M. van Breugel, H. van der Wal, S. V. Martins, M. D. M. Veloso, H. F. M. Vester, A. Vicentini, I. C. G. Vieira, P. Villa, G. B. Williamson, K. J. Zanini, J. Zimmerman, and L. Poorter. 2019. Biodiversity recovery of Neotropical secondary forests. *Science Advances* 5: Article eaau3114.

Rudel, T. K., P. Meyfroidt, R. Chazdon, F. Bongers, S. Sloan, H. R. Grau, T. Van Holt, and L. Schneider. 2020. Whither the forest transition? Climate change, policy responses and redistributed forests in the twenty-first century. *Ambio* 49:1770–1774.

Sala, O. E., F. S. Chapin, III, J. J. Armesto, E. Berlow, J. Bloomfield, R. Dirzo, E. Huber-Sanwald, L. F. Huenneke, R. B. Jackson, A. Kinzig, R. Leemans, D. M. Lodge, H. A. Mooney, M. Oesterheld, N. L. Poff, M. T. Sykes, B. H. Walker, M. Walker, and D. H. Wall. 2000. Global biodiversity scenarios for the year 2100. *Science* 287:1770–1774.

Samuel, M. D., W. Liao, C. T. Atkinson, and D. A. LaPointe. 2020. Facilitated adaptation for conservation: Can gene editing save Hawaii's endangered birds from climate driven avian malaria? *Biological Conservation* 241: Article 108390.

Sandor, M., C. S. Elphick, and M. W. Tingley. 2022. Extinction of biotic interactions due to habitat loss could accelerate the current biodiversity crisis. *Ecological Applications* 32: Article e2608.

Saunders, G. R., M. N. Gentle, and C. R. Dickman. 2010. The impacts and management of foxes *Vulpes vulpes* in Australia. *Mammal Review* 40:181–211.

Schwartz, M. W. 1999. Choosing the appropriate scale of reserves for conservation. *Annual Review of Ecology and Systematics* 30:83–108.

Smallwood, K. S. 2013. Comparing bird and bat fatality-rate estimates among North American wind-energy projects. *Wildlife Society Bulletin* 37: 19–33.

Smith, M. A., B. K. Sullender, W. C. Koeppen, K. J. Kuletz, H. M. Renner, and A. J. Poe. 2019. An assessment of climate change vulnerability for Important Bird Areas in the Bering Sea and Aleutian Arc. *PLoS One* 14: Article e0214573.

Spence, A. R., and M. W. Tingley. 2020. The challenge of novel abiotic conditions for species undergoing climate-induced range shifts. *Ecography* 43:1571–1590.

Springmann, M., M. Clark, D. Mason-D'Croz, K. Wiebe, B. L. Bodirsky, L. Lassaletta, W. de Vries, S. J. Vermeulen, M. Herrero, K. M. Carlson, M. Jonell, M. Troell, F. DeClerck, L. J. Gordon, R. Zurayk, P. Scarborough, M. Rayner, B. Loken, J. Fanzo, H. C. J. Godfray, D. Tilman, J. Rockström, and W. Willett. 2018. Options for keeping the food system within environmental limits. *Nature* 562:519–525.

Srinivasan, U., and D. S. Wilcove. 2021. Interactive impacts of climate change and land-use change on the demography of montane birds. *Ecology* 102: Article e03223.

Sun, Y., A. K. Skidmore, T. Wang, H. van Gils, Q. Wang, B. Qing, and C. Ding. 2014. Reduced dependence of Crested Ibis on winter-flooded rice fields: Implications for their conservation. *PLoS One* 9: Article e98690.

Sundar, K. S. G. 2009. Are rice paddies suboptimal breeding habitat for Sarus Cranes in Uttar Pradesh, India? *Condor* 111:611–623.

Sutherland, W. J., E. Fleishman, M. B. Mascia, J. Pretty, and M. A. Rudd. 2011. Methods for collaboratively identifying research priorities and emerging issues in science and policy. *Methods in Ecology and Evolution* 2:238–247.

Sutherland, W. J., N. G. Taylor, D. MacFarlane, T. Amano, A. P. Christie, L. V. Dicks, A. J. Lemasson, N. A. Littlewood, P. A. Martin, N. Ockendon, S. O. Petrovan, R. J. Robertson, R. Ricardo Rocha, G. E. Shackelford, R. K. Smith, E. H. M. Tyler, and C. F. R. Wordley. 2019. Building a tool to overcome barriers in research-implementation spaces: The Conservation Evidence database. *Biological Conservation* 238: Article 108199.

Swan, G. E., R. Cuthbert, M. Quevedo, R. E. Green, D. J. Pain, P. Bartels, A. A. Cunningham, N. Duncan, A. A. Meharg, J. L. Oaks, J. Parry-Jones, S. Shultz, M. A. Taggart, G. Verdoorn, and K. Wolter. 2006. Toxicity of diclofenac to *Gyps* vultures. *Biology Letters* 2:279–282.

Sydeman, J. W., E. Poloczanska, T. E. Reed, and S. A. Thompson. 2015. Climate change and marine vertebrates. *Science* 350:772–777.

Tingley, M. W., L. D. Estes, and D. S. Wilcove. 2013. Comment: Climate change must not blow conservation off course. *Nature* 500:271–272.

van de Pol, M., B. J. Ens, D. Heg, L. Brouwer, J. Krol, M. Maier, K.-M. Exo, K. Oosterbeek, T. Lok, C. M. Eising, and K. Koffijberg. 2010. Do changes in the frequency, magnitude and timing of extreme climatic events threaten the population viability of coastal birds. *Journal of Applied Ecology* 47:720–730.

Van Klink, R., D. E. Bowler, K. B. Gongalsky, A. B. Swengel, A. Gentile, and J. M. Chase. 2020. Meta-analysis reveals declines in terrestrial but increases in freshwater insect abundances. *Science* 368:417–420.

Vaughan, N. E., and T. M. Lenton. 2011. A review of climate geoengineering proposals. *Climatic Change* 109:745–790.

Vijay, V., S. L. Pimm, C. N. Jenkins, and S. J. Smith. 2016. The impacts of oil palm on recent deforestation and biodiversity loss. *PLoS One* 11: Article e0159668.

Visser, M. E., A. J. van Noordwijk, J. M. Tinbergen, and C. M. Lessells. 1998. Warmer springs lead to mistimed reproduction in great tits (*Parus major*). *Proceedings of the Royal Society B: Biological Sciences* 265:1867–1870.

Wagner, D. L. 2020. Insect declines in the Anthropocene. *Annual Review of Entomology* 65:457–480.

Walston, L. J., Jr., K. E. Rollins, K. E. LaGory, K. P. Smith, and S. A. Meyers. 2016. A preliminary assessment of avian mortality at utility-scale solar energy facilities in the United States. *Renewable Energy* 92:405–414.

Wilcove, D. S., D. Rothstein, J. Dubow, A. Phillips, and E. Losos. 1998. Quantifying threats to imperiled species in the United States. *BioScience* 48:607–615.

Winkler, K., R. Fuchs, M. Rounsevell, and M. Herold. 2021. Global land use changes are four times greater than previously estimated. *Nature Communications* 12: Article 2501.

Yiming, L., and D. S. Wilcove. 2005. Threats to vertebrate species in China and the United States. *BioScience* 55:147–153.

Index

For the benefit of digital users, indexed terms that span two pages (e.g., 52–53) may, on occasion, appear on only one of those pages.